Resource Management Information Systems: Remote Sensing, GIS and Modelling

Second Edition

Resource Management Information Systems: Remote Sensing, GIS and Modelling

Second Edition

Keith R. McCloy

A CRC title, part of the Taylor & Francis imprint, a member of the Taylor & Francis Group, the academic division of T&F Informa plc.

Published in 2006 by
CRC Press
Taylor & Francis Group
6000 Broken Sound Parkway NW, Suite 300
Boca Raton, FL 33487-2742

Printed in the United States of America on acid-free paper
10 9 8 7 6 5 4 3 2 1

International Standard Book Number-10: 0-415-26340-9 (Hardcover)
International Standard Book Number-13: 978-0-415-26340-5 (Hardcover)
Library of Congress Card Number 2005053800

Library of Congress Cataloging-in-Publication Data

McCloy, Keith R.
Resource management information systems : remote sensing, GIS and modeling / by Keith R. McCloy.--2nd ed.
p. cm.
Includes bibliographical references and index.
ISBN 0-415-26340-9 (alk. paper)
1. Natural resources--Management. 2. Geographic information systems. 3. Resource allocation--Management. I. Title.

HC55.M315 2005
333.7'0285--dc22 2005053800

Visit the Taylor & Francis Web site at
http://www.taylorandfrancis.com

and the CRC Press Web site at
http://www.crcpress.com

To Anny.
And to
Sarah, Sophia and Simon,
With love and appreciation

Preface

The purpose of this book is to provide the reader with the knowledge and skills necessary to design, build, implement, operate and use spatial resource management information systems for the management of the physical resources of a region. Spatial resource management information systems are based on the technologies of remote sensing, geographic information systems and modelling. The integration of these technologies with appropriate field data provides the basis of these systems. Accordingly, this book provides the reader with the skills necessary to use these technologies in a spatial context, and provides the reader with the skills to integrate them into, and then to operate, such information systems.

This book is the second edition of a text by the same author, *Resource Management Information Systems: Process and Practice*. That title reflected the end goals of the material given in the book, but did not reflect the contents of the book, which are the material necessary to reach this end goal. The new title, *Resource Management Information Systems: Remote Sensing, GIS and Modelling in Resource Management*, better reflects the contents of the book, and hence the end goal. This edition offers extensive revisions of the material offered in most chapters of the book, reflecting the rapidly evolving nature of the technologies that underlay the management tools needed for the management of spatial resources. The authors believe that this edition reflects the current status of these technologies and their evolving role in the management of spatial resources.

The management of the physical resources of a region is increasingly becoming a balance between the competing needs to both optimise productivity and to meet community demands for the maintenance of resources. The competition for scarce resources is driving the need for better management information systems designed to provide decision support tools to resource managers. One such tool is spatial resource information systems. This type of decision support system can be used to support the management of a farm or park, through regional management to global resource management. They have the characteristic of providing up-to-date spatially extensive yet consistent information on aspects of the resources of interest to the manager. There are many issues where such spatial decision support systems become crucial including dealing with most forms of environmental degradation and pollution since the source of the damage or pollution is separated in time and space from the effects. There are many situations where productivity gains can also be made from the use of spatial information systems, such as the case where agricultural industries collaborate in the construction of harvesting, processing and packaging factories and where there are advantages to be gained from the careful scheduling of the harvest so as to minimise waiting time and storage costs after harvest and to minimise the deterioration of produce prior to processing and packaging. As there becomes increasing competition for the use of scarce land resources, the potential for conflict increases. Resource managers will face an increasing need to resolve such conflicts and spatial decision support tools are critical in providing resource managers with this type of support.

In providing a text that is meant for resource managers, or for those closely connected to the management of resources, it has been necessary to think very carefully about the level of depth and breadth that should be covered as well as the theoretical and practical treatment that should be given to the material in the book. Some readers will want to go further in the study of these topics; the treatment has also been designed so as to provide sufficient foundations for them to do so. Thus, some mathematical treatment is included for some of the more important methods, but the reader should be able to understand the use of the material and its implementation without getting into too much mathematical development if that is his or her wish. In addition, it was recognised that a book like this cannot do justice to its task without some real data to deal with. Accordingly, the text contains a CD that includes data, notes on applications and various Web addresses so that the interested reader can access software to analyse the data in various ways. The CD holds text material to guide the user through the application material and in that sense it stands on its own. But it is meant to be used as an adjunct to the learning that is done by reading the text. Many software systems also provide learning material, and this material is also worth using.

There are a number of different software systems that can be used in the analysis of spatial data. What is needed are software systems that seamlessly include image processing, GIS, modelling and statistical tools and decision support tools. Of the various software systems around only IDRISI has implemented a conscious plan to do this, and they should be complimented for having this vision. I look forward to seeing other systems going down a similar route in the future. All software systems involved in the spatial analysis of data are different and thus cater for different needs. No system is best for everyone. Accordingly, no one system is recommended in this text. By the time that you get to the end of this text, you should know what to look for in selecting a software system to meet your needs.

The importance of information in the management of resources, and the importance of having sets of information of equivalent resolution, timeliness and accuracy for the different managers so as to minimise conflicts arising out of differences in perception of the nature of the problem, cannot be overestimated. These spatial resource management types of information are an important component in meeting these needs. Thus, this text is an important resource for all of those concerned with the management of physical resources, including those involved in agriculture, forestry, marine and aquatic resources, environmental science, landuse planning, valuation, engineering and geography.

A number of people have assisted in the preparation of this book and their assistance is greatly appreciated. Professors Steven De Jong, David Atkinson and Henning Skriver as well as Drs. Jurgen Boehner, Thomas Selige, Niels Broge and James Toledano have all very carefully reviewed various chapters and provided greatly appreciated comments and corrections. Dr. Susanne Kickner has gone much further and contributed valuable sections to the Chapter 6 on GIS. The support of the publishers has been greatly appreciated. Very special thanks must go to Anny for her support throughout the revision of this text.

Keith McCloy
Tørring
Denmark

Authors

Keith R. McCloy is a specialist in remote sensing, GIS and modelling, with over 30 years of experience in conducting research into, teaching, developing applications and using remote sensing, GIS and modelling for the management of renewable resources. He has a degree in surveying and a Ph.D. in geography, working 18 years in academia, 13 years in a research environment and 4 years in an operational environment, all focussing on the use of remote sensing, GIS and modelling in resource management. He is currently a senior scientist (remote sensing) with the Danish Institute for Agricultural Sciences.

During his career, he has implemented a number of operational applications in Australia, some of which continue to this day, developed postgraduates course programs in The Philippines and Australia, and developed a number of image analysis tools and techniques in classification, estimation and time series analysis of image data. This textbook reflects his belief in the need for the types of information provided by these spatial technologies, and the importance of integrating them for the best management of renewable resources.

Susanne Kickner graduated from the University of Salzburg in geography and gained her Ph.D. in nature science from the Technical University Karlsruhe in 1998. Her thesis was titled "Cognition, attitude and behaviour — a study of individual traffic behaviour in Karlsruhe." Since then she has focused on the use of geographic information systems for regional and site analyses for a variety of resource management purposes. Until 2004 she was scientific assistant at the Department of Economic Geography of the University of Göttingen, and is currently self-employed as an adviser to businesses as a GIS expert on spatial-economic questions.

List of Tables

List of Figures

List of Photographs

Contents

Chapter 4

Chapter 7

1 Introduction

1.1 THE GOALS OF THIS BOOK

There is clear, readily accepted evidence of resource degradation in many parts of the globe, and there are many other examples of claimed resource degradation that have yet to receive general acceptance. What are the cause or causes of this degradation? The underlying causes are to do with the shear level of pressure that is being placed on the resources of the globe by the combination of the number of people, their expectations in terms of goods and services and our technical capacity to utilise those resources to meet this demand. The level of degradation is also to do with our levels of consumption and wastage associated with production, distribution and sale of products. Some of this degradation results from inappropriate land use practices both in the use of the land for appropriate purposes and the use of inappropriate practices on the land. Thus, lands with steep slopes continue to be used for cropping even though extensive erosion results, and even moderately sloped land is cultivated with furrows down the slope by some farmers, again exacerbating the soil erosion problem. Our land use allocations and the practices adopted on those lands continue to be the major factors in exacerbating land degradation. It follows that improved land allocation and improved land using practices will significantly reduce land degradation. Changing land allocations and land using practices are management activities and so the way that we manage our rural lands is a major factor in the fight to reduce land degradation and to strive to achieve the goal of sustainable land using practices.

The title of this book was chosen to convey its dominant theme: *to understand the role of, as well as to develop and use, spatial information systems for the proper management of physical resources*. There are a number of critical components to such Resource Management Systems including remote sensing, geographic information systems (GIS), modelling resource management and decision support. This text covers the principles and practices associated with these components as well as their integration into a system so as to emphasise the holistic way in which the management of resources is going to evolve, if we are to achieve the twin goals of maximising productivity and maintaining the resource base.

In some way these physical resources may be managed adequately using current tools and techniques; however, this is rarely the case if the second criterion of resource maintenance is adopted. How much does resource degradation have to do with resource management? Does our method of management, the quality of that management, the nature of the resource management tools and how they are used to influence the level of resource degradation? This chapter explores some of these issues:

- Exploring the status of physical resources, specifically, "Is there a reason to be concerned?"
- Moving on to the reasons why this situation exists, or "What is the nature of this concern?"
- Considering which types of these concerns can be addressed by the use of better management tools or techniques, or "How can better tools and techniques help deal with these issues?"
- Addressing the question of how can resource management use these tools and techniques, or "What are the characteristics of resource management that may be part of the problem and how can these tools and techniques address some of these imitations?"

The chapter then moves on to consider the nature of the information that is required for good resource management, and how geographic information fits into this role. It finishes with a description of how the book is structured to assist the reader in understanding the need and role of Resource Management

Information Systems (RMIS) and then in developing the skills that they need to meet the goal of developing and implementing these systems.

1.2 THE CURRENT STATUS OF RESOURCES

There is widespread recognition that the finite resources that constitute the Earth — its soils and land covers, its oceans and its atmosphere — are being consumed at such alarming rates that they have already become or will become seriously depleted in the foreseeable future. While at the same time, we are also producing wastes faster than they can be absorbed and broken down by the natural ecosystems upon which we depend, and that supply many of the resources we use. There are many examples of this situation and current responses to them. Three of them will be considered here.

1.2.1 THE OZONE HOLE

One of the most notable examples of degradation and response at the global level concerns the identification and response to the existence of the ozone hole. The British Antarctic Survey first noticed the dramatic losses of ozone over the Antarctic in the 1970s. When the first instruments were set up to take accurate readings of the ozone levels over the Antarctic in 1985, the readings were so low that the scientists thought that there must be something wrong with their instruments. They sent them back to be replaced, and it was only when the replacement instruments gave similar readings that the results were accepted.

Ozone forms a layer in the stratosphere that is naturally thinnest at the equator and thickest over the poles. It is formed when ultraviolet light from the sun strikes the atmosphere. Some of the ultraviolet light strikes oxygen molecules in the atmosphere, splitting them into two oxygen atoms.

$$O_2 \rhd O + O$$

The oxygen atoms are unstable in the atmosphere and will quickly combine with other chemicals that they come in contact with. They are most likely to either combine with another oxygen atom, thereby just reversing the equation, or to combine with an oxygen molecule to form ozone.

$$O + O_2 \rightarrow O_3$$

Ozone absorbs ultraviolet light and so the ozone in the atmosphere filters out most of the ultraviolet light before it reaches the surface of the Earth. All plants and animals are vulnerable to high levels of ultraviolet light, and so the absence of ozone will create very dangerous conditions for all plants and animals. The dramatic reductions in ozone that were occurring were soon recognised as being a serious threat to life on Earth. As a consequence, global agreement on a response was achieved at the Montreal Accords in 1987 that planned to halve the production of the offending chemicals the were causing the loss of ozone by 2000. In fact this commitment has been exceeded with most countries stopping production of the chemicals, except for minor medical needs, by 1995. Currently, it is expected that the ozone hole will have disappeared by about 2045.

Man-made chemicals that contain chlorine or bromine can, when released into the atmosphere and under certain conditions, decompose releasing the chlorine or bromine that reacts with the ozone to form oxygen. These atmospheric conditions are much more likely to occur over the Antarctic continent than other parts of the globe and hence the appearance of the ozone hole over Antarctica.

It could be said that the ozone issue has been a successful environmental fight. However, it was a very simple fight; there was an obvious culprit, the solution was clearly identifiable and those dependent on this industry were sufficiently small in number so that stopping production did not have significant economic implications for national budgets.

Most environmental issues are not this clear cut. There are often a number of causes. Some of those causes are man induced and some may be due to natural processes. It is often difficult to identify all the actions by man and the natural processes which are contributing to the degradation, particularly where the degradation is due, at least in part, to complex loop and interconnection mechanisms. As a consequence it is often difficult to identify all the culprits and allocate responsibility. There are often a number of potential solutions, many of which are partially or even totally in conflict, either amongst themselves or with solutions to other issues. Finally, it is often the case that some of the interested parties have a large investment, of one form or another, that may be at risk of being lost or degraded depending on the solutions that are adopted. When this occurs, then the economic implications need to be taken into account in developing a solution.

1.2.2 Water-Borne Soil Erosion

The basic principles of water-borne soil erosion are conceptually well understood by most land managers. Rainfall, hitting the Earth's surface either directly or as drops from leaves, tends to dislodge small amounts of soil from the surface and mobilise them into the surface flow. The rain accumulates into runoff, and this water tends to both carry the water-borne sediment down slope, as well as collecting more sediment from the soil surface. The size, density and volume of soil particles that are dislodged from the soil surface and then carried away in the surface water depend on the energy in the water droplets in the rain and in the surface flow. By far, the greatest impact is due to the droplets hitting the surface of the Earth, hence the greatest single factor in dislodging and mobilising soil particles is the impact of the rain on the surface. The effect of this energy is reduced by vegetation and any other matter that absorbs most of the impact energy of the rain that it intercepts, so that vegetative cover and mulch are important factors in reducing soil erosion. The energy in the surface water affects the capacity of the flow to carry the particles away from the site of displacement. As the volume and the velocity of the water increase, the capacity of the water to hold the sediment increases. The net result is an increasing load of sediment being carried by the water as its volume and/or velocity increases. Once the water starts to loose energy, most typically when it starts to slow down as the slope decreases or it is resisted by denser vegetation, it will loose the capacity to carry sediment. The heavier sediments, such as gravel and sand, are deposited first, followed by the finer sediments such as clay and silt.

Despite this well-established understanding of the process, water-borne soil erosion still continues at a very high rate. The International Soil Reference and Information Centre (http://www.isric.nl/) in Holland conducted a global mapping of soil degradation in 1991 and concluded that water-borne soil erosion is the single most important factor affecting the soils of the globe. They estimated that 56% of the global soil degradation was due to soil erosion. It has been estimated that 200 years of farming in the United States has meant that most of the lands in the United States (http://www.seafriends.org.nz/enviro/soil/erosion.htm) have had 25–75% of the topsoil lost. Some have lost so much as to be unproductive. They have further estimated that soil formation takes place at about 25 mm every 1000 years or about 0.6 tons/ha/year. This compares to the 10–40 tons/ha/year that is lost from the so-called sustainable agricultural production systems that do not use soil tillage at all, the amount depending on the soil slope. Clearly agriculture, in all the forms currently practised, causes levels of soil erosion that are not sustainable in the long term.

One of the problems with soil erosion is that it appears to be happening very slowly. The effects may be difficult to see over a period of 30 years or so. This slowness is exacerbated by the way that the application of fertiliser tends to mask the impacts of soil erosion on productivity. Thus, farmers, faced with a competitive market situation, unable to see the significant impact of erosion and having this impact masked by the use of fertiliser, are often reluctant to take appropriate remedial and preventative actions. Soil erosion has always occurred. It is part of the natural weathering processes that wear down landforms as they are created. What is critical is the significantly higher rates of soil erosion that have been introduced by agriculture and other land using activities.

It should be pointed out that this amount of soil may be lost, but some of it is recovered as deposits on flood plains and some of it ends up as nutrients in the sea. The deposits on flood plains may well be more productive than the areas that lost the soil, due to our capacity to irrigate flood plains at lower costs than most other areas, the suitability of flood plains for agriculture and the concentration of fertile soils in the flood plains. But only a small percentage of that which is caught in this way can be used for agriculture since most of the alluvium is too deep to be of use. The sediment that goes into the sea provides nutrients to the coastal zone, but again, too much of it can destroy many forms of marine life, as well as being buried under subsequent layers of silt.

Even though the processes of soil erosion have been well understood for more than 50 years, soil erosion is clearly a continuing source of soil degradation. Why is this the case? Are the problems technical, economic, social or political? In the sense that agricultural research has not found an alternative to cultivation that is sustainable and is acceptable to most farmers around the globe, the problem is technical. That the process is slow so that the farmers are only partially aware of the scale of the problem, and how its impact varies considerably with management practices, primarily because of the differences in landcover with the different practices, the problem is technical. There are alternatives to cultivation including no-tillage cultivation and stepping away from the monoculture ethic that dominates agriculture. But neither alternative has been widely adopted because they are, for many situations more expensive than cultivation. So, in this sense the problem is economic and political.

Sustainable means that the process to which it refers can be continued in the same way and at the same level into the foreseeable future. Sustainable agriculture has been defined (Ikerd, 1990), as resource conserving, socially supportive, commercially competitive and environmentally sound. Sustainable agriculture thus has three main foci:

- Ecological sustainability
- Economically sound
- Socially supportive

Ecological Sustainability. This requires that it be resource conserving, in harmony with the environmental processes that operate in the area and maintains biodiversity. If it is resource conserving then it will be at lower economic and environmental cost, it will not result in wastes that cannot be used and consumed within the ecological cycles that are in operation and it will not remove resources permanently from the environment. If it is in harmony then the processes involved in production are harmonised with the ecological processes in the environment and the wastes become part of the food chain in a way that contributes to those ecological processes. If it supports biodiversity then it supports the maintenance of a rich complexity of life.

Economic viability. Capitalism has been shown to be a good, but incomplete or inadequate mechanism for the conduct of normal trading business in our society. Its strengths include its resource conserving nature, its independence of political power and its traceability through society. It tends to be resource conserving because all resources have a cost and one of the main goals of capitalist processes is to minimise the costs of production. Independence of political power means that the market is freer of political manipulation and corruption than most alternative systems that have been tried. Its traceability means that efficient and equitable taxation systems can be established that ensure that all members of society can carry their correct proportion of the taxation burden and it provides the basis of accurate quantifiable information on cash flows throughout society that can form the basis of more sound financial management than is possible without this information.

However, it is inadequate because it does not adequately value resources that are not owned by individuals or by organisations in the commercial system (the problem of the commons), it does not take into account future implications of proposed or actual actions and often it does not properly address the costs and benefits of an action to parties outside of the trading, particularly where the relationship between these parties and the trading has non-commercial aspects.

One of the fundamental characteristics of capitalism is its need to grow. The bottom line for every business is to show that it has grown over the last reporting period. Yet we are striving towards sustainable land using practices and systems. The need to find a balance between the aggressive growth goals of capitalism on the one hand and the goals of balance and continuity implicit in sustainability on the other is one of the big challenges that our society is facing.

Socially supportive. Sustainable systems should treat all livings things with reasonable respect and care. Animals should not be kept or slaughtered under conditions that are cruel in any way. Sustainable practices should not be ethically or morally reprehensible. Thus, many practices in agriculture that cause pain and anguish to the animals do not have a place in sustainable agricultural systems.

1.2.3 Loss of Biodiversity

The third aspect of modern resource management to which I wish to refer is biodiversity. Diversity is defined as the variety of conditions that can exist within the object of attention. High diversity means great variety. The biological scientific community currently recognises three types of biodiversity: habitat, genetic and species (http://www.defenders.org/).

Habitat diversity refers to the variety of places where life exists — coral reefs, old-growth forests in the Pacific Northwest, tall grass prairie, coastal wetlands and many others. Each broad type of habitat is the home for numerous species, many of which are utterly dependent on that habitat. When a type of habitat is reduced in extent or disappears, a vast number of species are placed under pressure and may disappear.

Genetic diversity is the variety in the genetic pool of the species, and this is related to the number of populations that exist in the species. Species live in population clusters. Generally, the individuals in one population will breed with another individual in the same population, simply because of the distance between populations, although they can, and will mate with individuals in other populations. Populations only diverge into different species when individuals will no longer mate across the population boundaries, because of genetic differences between the individuals. The genetic diversity within a species is primarily the variety of populations that comprises it. Species reduced to a single population (like the California condor) generally contain less genetic diversity than those consisting of many populations. Song sparrows, found over much of North America, occur in numerous populations and thus maintain considerable genetic diversity within the species.

Species diversity refers to the diversity of species that exist. Thus, one can consider the species diversity within a habitat, a continent or the globe. There are about one and a half million identified and classified species on Earth, but we know that many unnamed species exist. It is estimated that the total number of species is probably between 5 and 15 million. Most of the evidence for numerous unnamed species comes from studies of insects in tropical forests: when the canopy of a tropical tree is fumigated and all the dead insects are collected, large numbers of hitherto unknown insects are frequently collected.

Species are categorised as *variety: species: genus: family: order: class: phylum: kingdom*. Thus the categories for humans are

Species	Homo sapiens
Genus	Homo
Family	Hominidae (Man and apes)
Order	Primates
Class	Mammalia
Phylum	Chordata (vertebrates)
Kingdom	Animal

The relatively small percentage of the species that are actually known, estimated to be between 10 and 30% of the total number of species, means that it is impossible to accurately estimate the decimation that is actually occurring. Biologists estimate that about 20% of the birds species have become extinct over the last 2000 years and that about 11% of the remaining known 9040 species are threatened with extinction. But these estimates cannot start to take into account the vast majority of species that we do not know about. Since these species are unknown, they are likely to be animals of very small size, to be either in very small populations or to exist in small numbers, or to exist in unique habitats such as the deep ocean. Some of these conditions mean that many of these unknown species could be very vulnerable, particularly those that exist in small numbers and live in unique habitats if these habitats are limited in scope, as many of them will be. As a consequence, the extinction rate among the unknown species may well be higher than amongst the known species.

In 1992 the World Resources Institute (http://www.wri.org/index.html) proposed that there are seven causes for the loss of biodiversity:

1. Population growth
2. Increased resource consumption
3. Ignorance of species and ecosystems
4. Poorly conceived policies and poor resource management
5. Global trading systems
6. Inequities in resource distribution
7. Low valuation of biodiversity in economic systems

What are the implications of loss of biodiversity? Whilst it is far from clear what all of the implications are, and the significance of these impacts, some are understood:

1. *Higher levels of Ecological instability*. The lower the biodiversity in an area, the more vulnerable the area is to predation or attack by invasive plants or animals, and the less likely that the area can adapt to these attacks and to changes in environmental conditions. Areas of low biodiversity are agricultural cropping areas and cities, where, in both cases, the environment is dominated by one species. The presence of large numbers of a species in one location is attractive to other species that can use that species for food, as a host, or for some other reason. This is clearly shown in the responses of specific insect populations to the existence of monocultures in their vicinity that are beneficial to them in some way. Nature, of itself, promotes or encourages biodiversity, as can be seen by the growth of biodiversity after mass species destructions in the past. But man, with his significant capacity to influence the biosphere, is often inhibiting this process.
2. *The loss of species of benefit to man*, including the possibility of new drugs and medicines, and better food sources.

It is generally accepted that high levels of biodiversity usually indicate complex ecological systems. Complex ecological systems contain species that overlap in the domains of influence, so that the destruction of one species leaves others that can step in, albeit at not quite the same level of efficiency and effectiveness. Thus, complex ecological systems with high levels of biodiversity represent systems that are capable of high levels of adaptation to and absorption of forces of change. Such systems are inherently stable over a wide range of conditions of the forcing functions that are affecting that ecological system. However, even such stable systems will degrade and eventually collapse if the level of force applied to the ecological system exceeds that capacity for absorption of the ecological system for long enough. Any debate on biodiversity adequacy or inadequacy must therefore hinge on the sustainable level of adaptation of the system. The sustainable level of adaptation does not imply no change, and so it allows for changes in the species composition of an ecological system, but it does imply that these changes are occurring through sustainable adaptation and not due to unsustainable destruction of species.

Monocultures are ecological systems with very low biodiversity. Stability is only maintained in man-made monocultures by the application of significant levels of resources in the form of insecticides, herbicides and physical management. Agriculture and forestry are thus major factors in reducing biodiversity, but they are not the only factors. Many other activities of man affect the biodiversity of an area. Hunting and the introduction of new species also affect biodiversity, certainly in the short term, but whether they do in the long term depends very much on the nature of the environments being dealt with and the other pressures being exerted on those environments. The Australian Aboriginals are thought to have introduced both fire and the dog into Australia. They used both in their hunting of animals for food, clothing, utensils and shelter. Both had a significant impact on the ecology on the Australian environment, however, the rate of impact was not too great for the communities to adapt, although it may have been too great for specific species to adjust and they have subsequently become extinct. This is the fundamental difference between the impact of the Aboriginals and white man on the Australian environment. In the former case the impact was not too large for the environment to adapt and change, even though some species could not adapt and became extinct in the process. In the latter case the level of pressure is far too high for the environment to adapt in the time available.

Thus, evaluation of the significance for the loss of biodiversity can be very complex. There are usually a variety of causes, where these causes can vary in significance over time and space, and there are usually a variety of solutions, where the more obvious solutions that may include reestablishment of the status quo prior to the impact, may not be the best solution, nor even viable. Often the solutions will be in conflict, so consideration and choice between solutions involve taking into account the impacts on others who are not directly involved in the processes being considered.

There are many forms of resource degradation, with only three being briefly considered here. All of them are, in some way, created or exacerbated by man. Often there are multiple causes where the influence of a cause on the degradation can vary spatially and temporally. It is often difficult to properly assess all the causes of degradation and their relative significance. The impact of degradation usually occurs at different locations and times to the initiation of the degradation. Degradation thus often affects different communities to those causing the degradation. It can thus be difficult to evaluate the effects of degradation on an individual or a community. Finally, the proposed solutions can have impacts on both these groups, as well as other groups in society, where these impacts may not be exactly in proportion to their involvement with the degradation. Thus, the development of solutions that are satisfactory within society can also be very difficult and complex.

The range, level of severity and ongoing nature of many forms of degradation indicate that there is a significant cause for concern. With many, but not all forms of degradation, the basic principles underlying the degradation are well understood. However, the interdependencies between the degradation and the broader environment are often not well understood, just as the detailed understanding that is often necessary for good management is often not yet available.

1.3 THE IMPACT OF RESOURCE DEGRADATION

As resources decline, more resources, albeit of different kinds are used to maintain production. There are many examples of this process. In agriculture and forestry, if production declines, then more fertiliser, herbicide or pesticides are used, depending on the cause of the loss in productivity. In mining, when the ore becomes very deep, or of low quality, then production can only be maintained by the use of bigger machines, longer lines of production or more miners, or combinations of all three. As these inputs increase, so do the costs of production. However, if the resulting production is the same, then this means that productivity decreases, where productivity is defined as the amount of output for one unit of input.

Increasing inputs often masks the decline in production, so that the decline in productivity may not be obvious outside of the units of production. Even within the units of production, the decline may be gradual over time and may be masked by changes in management practices. As a consequence the implications of declining productivity are often very subtle, and may be well advanced before these implications are well understood. However, the units of production that are suffering degradation continue to suffer declining productivity relative to units of production that are not suffering a degrading resource base. This means that the units of production with a degrading resource base are becoming poorer compared to those not affected. They are becoming poorer in two ways:

- *Reduced wealth or loss of assets.* Low or declining productivity is reflected in the budgets of the affected units of production. As a result these units of production cannot sell their resource base at the same value as those units with higher productivity. This means that these units of production have a low or declining capital asset base, reducing the wealth of the owner, and reducing their capacity to borrow money, using the asset base as security.
- *Reduced income.* Lower productivity reduces the profit margins and hence the income. A lower income reduces the investment capacity of the business, further exacerbating the problem.

The same principles apply to the nation as to the individual unit of production within the nation, but the effects have historically been quite different. When a unit of production declines, then the owners become progressively poorer until such time as the unit is sold. Historically, with nations, when one became weak (or poor), then it becomes vulnerable to forced takeover or invasion. The degradation will have reduced national productivity. The ruling class are usually in the best position to resist this loss in productivity, shifting the burden for carrying it onto the poorer members of society. The result is a growing disparity between the rich and the poor, with the percentage of those in poverty increasing, and the level of poverty declining. This leads to frustration, anger and desperation when conditions become very bad. This weakening of the fabric of society, on top of the weakening economic conditions, makes the country vulnerable to internal or external takeover. This is one of the various causes for revolution or invasion.

History has shown that degradation of the resource base of a society can be a cause of the decline of that society. The Mayan society was a flourishing society in the central Yucatan Peninsular of Mexico from about 400 A.D. until about 900 A.D. At about this time the society suffered a serious and permanent decline. There is no evidence of massive destruction by invasion, as is the case with the Aztec civilisation in the valley of Mexico City. It is known that the Mayan civilisation was very dependent on irrigated agriculture. The Yucatan Peninsular is a limestone formation, so that the limestone would have allowed rainfall to permeate through the stone and away from the topsoil. Areas with these characteristics will have limited agricultural capacity without irrigation if rainfall is insufficient to maintain soil moisture in the root zone. However, the water that has drained down to the water table is accessible across most of the Yucatan Peninsula through bores since it is not too deep to remove in this way. The water is thus stored, ready for use for irrigation and for human and animal consumption. The same characteristic that makes the area unsuitable for non-irrigated agriculture makes it very suitable for irrigated agriculture. However, such limestone systems can have another disadvantage. If the surface water is polluted for some reason, then this pollution is readily carried down into the water table. If the groundwater becomes polluted then it cannot be used for any activity where the pollution in the water has a deleterious effect.

Whilst the actual reasons for the decline are not known, historians have ruled out the conventional ones of warfare and invasion. It seems reasonable to suggest that the Mayan society declined and then collapsed because of some form of environmental and institutional degradation, where these may well have been related. The irrigated agricultural system may have collapsed because of the inability of the Mayan society to maintain the monocultures on which their agriculture was based. Alternatively, the groundwater may have become polluted in ways that were harmful to the crops, to the people of society or to both. Either of these situations could have brought on institutional decline, which may have resulted in a decline in maintenance of the irrigation system, contributing to further declines in the production base of society.

The ancient societies of the Mesopotamian valleys were also dependent on irrigated agriculture. Extensive areas in these valleys are now salinised in the root zone or at the surface. If irrigation caused excessive salinisation in the root zone or on the surface, then the resulting decline may have been a factor in the decline and fall of these societies. When these lands started to become salinised, their productivity may have started to decline.

Most soils contain some sodic salts, and all plants can accept some sodic salts in their water. Many animals, including man, need some salt in their diet. Soils that have been beneath the sea can contain more sodic salt from this source. Another source of sodic salt is the atmosphere. When water is evaporated from the sea, some salt is carried aloft in this process. This salt is then carried on by the wind, and will eventually fall back to Earth. If it falls on the land, then it is deposited on the surface. Sodic salts are highly soluble, and so rainfall will dissolve the salt and carry it down into the soil. Some of it will be deposited within the soil layers, and some will end up in the groundwater. Eventually, a balance is achieved between the salt being deposited, the amount that is held in the soil profile and the amount that is being lost from transport within the groundwater or removal from the overland water flow. The higher the rainfall and the lower the evaporation in the area, the lower the amount of salt that is retained in the soil profiles in this balance and vice versa. As the level of salt increases, it favours species that are salt tolerant; as it increases further even these species cannot tolerate the levels of salt, and all vegetation dies.

Irrigation usually deposits water into the groundwater, because more water is applied than is used by the plants. This particularly applies to flood irrigation because of the difficulty fine tuning water allocations by means of flood irrigation, and to permanently flooded irrigation as is used with rice. In these situations the groundwater level rises, mobilising the salt in the soil horizons that have been saturated, making the groundwater more saline. If the groundwater approaches or breaks the surface, then this salt is carried into the root zone and to the surface. If there is sufficient salt in the groundwater, then the result is toxic levels of salt in the root zone and on the surface. This process may well have happened in Mesopotamia.

Currently, there are no realistic methods available for the removal of salt from the soil horizon of salinised soils other than for the relatively slow process of the natural flushing of the salt out of the soil horizons by rainwater. Clearly, this process is much faster in areas of high rainfall and low evapo-transpiration than it is in areas of low rainfall and high evapo-transpiration. It is no surprise that most of the salinisation problems are in those parts of the globe that experience low rainfall and high evapo-transpiration, such as Australia and Mesopotamia.

There are, of course, major differences between conditions as they now apply and those that were applied in earlier times. These differences take three forms. The first is the shear numbers of people right around the globe, and the pressure that this applies to the natural resources of the globe. Whereas with fewer people, there is the possibility of leaving some of the land to lie fallow for long periods, as the numbers increase, so the opportunities to leave the fallow land deceases. Thus, land has less and less opportunity to be left unused and for it to recover during this period. Population densities would have been as high, or higher, in some areas in the past than what they are now. However in the past, if these areas could not sustain their population, then the opportunity existed to migrate to other, less populous areas. This would have been the basis of some of the mass migrations in the past. However, this safety valve mechanism no longer exists since there are few areas that are significantly under-populated. This could be one of the reasons for resistance to immigration that exists in some countries around the globe.

The second major dimension is the advent of technologies that enable us to place much greater demands on the land. In both agriculture and forestry the advent of modern machinery has allowed fewer people to plant, manage and harvest larger areas in less time than in the past. Other advances have increased the yield per plant or per area, but with the costs of this being the withdrawal of more resources from the soil. This cost is partially offset by the use of other technologies that allow the application of nutrients to the soil, where these nutrients have been taken from other places around the globe. Still other technologies have improved our mobility, consuming resources in the process and requiring packaging in the case of foodstuffs, where that packaging then ends up as a waste product from our society.

The third major dimension is the advent of modern forms of information dissemination and transmission. Conditions on the land are not now exclusively the knowledge of those directly affected and a small elite who may have been aware of the problem. These conditions are now known throughout the society. One effect of this dissemination of information on the extent and degree of resource degradation is that society as a whole becomes involved in the issue, and society often takes on the responsibility for the maintenance of the resource base of the society. Thus, modern media is an active participant in the process of societies rising concern for and action in relation to the environment. Not only does this have implications for the protection of resources, but also for the way resources are viewed and managed in our society.

1.4 THE NATURE OF RESOURCE DEGRADATION

Resource degradation can be considered at least at two levels — at the process or regional and the local levels.

1. *At the regional or process level.* The region is defined as the area covered by the cause and the effects of the degradation. Most forms of degradation have a cause at one spatial and temporal set of coordinates and the effects are imposed at other spatial and temporal coordinates. Clearly, the size of the region is a function of the spatial extent of the cause and effect coordinates, and the temporal duration is a function of the time that it takes for the processes to occur. Thus, in the first example discussed above, the causes of the ozone hole come from the release of the chemical at all latitudes lower than about 75°, on the basis that there are negligible populations at higher latitudes. The effects are global, but are focused on latitudes higher than 75°. The temporal dimension is continuous due to the nature of the release. Remove the release of gases containing chlorine and bromine into the atmosphere and the hole will gradually disappear.

For water-borne soil erosion, the normal source of the erosion is cultivation within a watershed, and the deposition is within that watershed or the sea. So, watersheds would be reasonable regions to consider for water-borne soil erosion. It should be noted in passing that all other forms of soil degradation including wind-borne soil erosion and chemical pollution of the soil and groundwater cannot usually be dealt within watersheds, since the areas of the process for these other forms of degradation are not the watershed.

The complexity of the issue of biodiversity loss has been discussed above. If individual issues of biodiversity loss are addressed, then the process regions may be a habitat, or a cluster of habitats, where these can vary significantly in size depending on the characteristics of the species of interest. The issue of region of impact depends very much on the migratory patterns of the species of interest and their dependencies.

2. *At the unit of management or the local level.* While most forms of degradation have to be considered at the process or regional level, implementation will have to be at the unit of management or local level. If the process or regional level is smaller than the unit of management level, then the source and the effects of the pollution or degradation are contained within units of management and the individual manager holds the power to both introduce remedies and to reap the benefits. Unfortunately, this is not usually the case. Usually, the cause of the degradation are removed from the jurisdiction that has to deal with the effects, so that the manager who introduces the degradation

may not suffer all, or even some of the consequences of his or her actions. Sometimes the region or process area is larger than an administrative area within a country, or may even cross national boundaries. The difficulties of dealing with this situation has meant that many forms of degradation have taken a long time to be accepted by all parties, and then a solution found that is acceptable to all parties.

The community concern for natural resources and consequent pressure to implement sustainable practices requires responses at a number of levels:

- *The adoption of more sustainable tools, techniques and practices* at the field or the farm. Clearly significant savings in degradation can occur with the adoption of more resource conserving practices. However these, on their own, are likely to be insufficient, particularly if they are costly relative to the alternatives. It has, for example, been estimated (http://www.seafriends.org.nz/enviro/soil/erosion.htm) that no tillage agriculture still incurs losses of 10 to 40 tons/ha/year of soil compared to the 0.6 tons/ha/year of soil creation in the United States. This represents the technological component of the solution.
- *Placing more realistic values on many resources* and recognising both the fragility of the environment and its transitory nature. Many of our attitudes to natural resources arise from assumptions about their permanence and pervasiveness, neither of which is correct under the levels of resource use and demand that is occurring with current population and technology pressures. This represents the attitudinal component of the solution.
- *Improving the way we allocate and use resources* in a region so as to provide the most sustainable use of resources as possible within the region. Thus, some areas are suitable for some activities, and less suitable for other activities. A simple case is that tillage should not be practised on steep slopes, and not even on slight slopes with downhill furrows. So, practices that are acceptable under certain conditions are not acceptable under other conditions. In addition, the degraded resources from one unit of production may well be a valued input for another unit, and so these units need to be managed in a symbiotic manner. This is the resource management component of the solution.

A system designed to sustainably manage the physical resources of an area has to function at the regional or process level, yet be capable of providing information at the local level so as to contribute to local resource management decisions. Such a system needs to cover the spatial extent of the region and derive and supply information at a resolution that can be used at the local level. The nature of the process region has been discussed. Clearly, their extent varies with the issue being addressed. Such areas will usually be different to the areas of local government, even if local government is structured so as to have boundaries that are compatible with the more common forms of resource degradation. Fortunately, these differences do not cause significant problems if the resource management system is based on GIS as these systems can easily join together sets of information and then repartition that information into new areas. Thus, geographic information can be structured so that it is stored in a way that is readily manageable and understood by all users, and it can then be readily merged and partitioned into new areas of interest from a point of view of resource management. It is likely that the most effective way to store digital geographic data is in the form of map sheet layers, that can then be readily merged to cover the extent of the area of interest, and that area masked out from the map data outside the area, so that all the analysis deals just with the information within the area of interest.

Since resource management typically occurs at the field level, those dealing with the process regions need to supply information at the field level, and of sufficient accuracy for it to be of use at the local management level. For example, information on landuse, may supply one landuse class for a field, but the bounds of the field, and the area of the field, will need to be supplied at much higher resolution and accuracy.

1.5 THE NATURE OF RESOURCE MANAGEMENT

Once upon a time, a long time ago, or so it seems these days, resource management was very simple. It meant simply the management of resources so as to make the best return on your investment. To put it more simply, it meant making the largest possible profit from the resources that you had at your disposal. This was how it would have been stated in those days. This is not to say that the average landowner did not care about the resources that he was using; quite the contrary most landowners cared quite a lot. But the stated aims were to maximise profit. The unstated aims usually included maintaining the farm or the resource for future generations of their family. But something has happened over the last 50 years or so to change that perspective, in fact to turn that perspective on its head.

I can still remember, when, in Australia, a farmer would view his farm as his castle. When he got home he was away from, safe from, protected from all those pernicious things that the average farmer typically considered plagued those who live in cities — to many people, no privacy, controlled by too many rules, told what to do by too many people, in a few words, no freedom. When the farmer got home he could forget all those details and he was free to just get on with his struggle against nature and the markets to "make a quid."

But no longer do farmers; at least in Australia and Denmark, think that way. No longer does a farmer view the farm as his castle to which he could flee from the complexities of modern living. These days farmers think much more like most other people making a living in a society — they have to deal with intrusions into their life, with more rules on what they can and cannot do, with the need to supply more information to diverse bureaucracies, all of which have some form of life or death hold over them. So, now many of them do not view their farm as their own, but rather as a business in which they have a major stake, even though their children may not do so.

These attitudes have changed in a revolutionary way since the middle of the 20th century. This change in attitudes has largely come about due to an increasingly wider societal recognition that we are consuming increasingly large quantities on non-renewable resources, and creating significant pollution in the process. This recognition has been brought about primarily by the revolution in information and communication that has occurred since the middle of the 20th century. This revolution has provided stronger evidence of the effects of societies on the environment on the one hand, and enabled pressure groups to more rapidly and extensively mobilise support for action against significant environmental damage and impact.

Increasingly, resource managers cannot ignore the environmental and sustainability implications of their actions, and this trend will increase as the information tools available to resource managers, and society, provide more up to date and accurate information on the environmental and sustainability implications of management actions. Thus, society is now becoming much more involved in the way all resources are being managed, and this involvement will increase.

Given this trend, there is a need to understand the nature of resource management, so as to ensure that the tools available to the resource manager, and to the society, are the best that can be built, in terms of their usability and economy. To design and develop such tools requires that we first analyse the nature of resource management, the information that resource managers require for the management of resources, how the decisions get implemented and finally how to best support such a process.

All resource management can be seen to fit into one of three broad categories, which I will call strategic, process or regional and local management.

1.5.1 STRATEGIC MANAGEMENT

> Strategic planning is defined as planning the future directions for the corporation through establishing the corporations long-term objectives and mobilising its resources so as to achieve these objectives.

Developing corporate objectives or goals will normally do this, establishing the structure appropriate to the achievement of these goals, and then setting the goals of the units within the corporation. This type of activity is the primary concern of the board of directors and senior managers.

Strategic planning activities are concerned with the environments (physical, economic, social and political) within which the corporation operates. It involves assessing how these may change in the future, the potential impact of these changes on the corporation, and how the corporation needs to respond and operate under evolving conditions. *Strategic planning requires extensive information about environments that are external to the corporation.* Strategic planning, in considering these external environments, will need to consider many factors over which the corporation has no control, and many of which are quite unpredictable since the corporation will usually have incomplete information on the factors and their environments. *Strategic planning requires the use of models that can accommodate the impact of unexpected and unpredictable factors. It generally requires the use of unstructured models.*

Generally, the information that is used in strategic planning will be of low spatial resolution, generalised and qualitative because of the way that the information is to be used, and the extensive range of information that has to be considered.

Typical information used in strategic planning includes:

1. Statistical information as collected by central Statistical Bureaus providing national and regional level information on different aspects of the society and the economy. Remote sensing and GIS can improve the collection and accuracy of this information by providing sampling strata appropriate to the information being collected, and in some cases by the actual collection of the sampled information itself.
2. Broad-scale monitoring information as collected by satellite at the global and regional level. Remote sensing with field data is the basis of most of these monitoring programmes, such as with weather prediction.
3. Institutional information as collected from documents and publications including economic, social and political information about the institution and its relationships with other institutions, including governments.

1.5.2 Process or Regional Management

Process or regional management is concerned with translating strategic plans and programmes into action plans whilst ensuring the maintenance or improvement of the productivity of the units of production in the region.

Strategic policies and programmes will often include implementation of environmental and social policies. Such policies may have to do with the sustainability of production systems, maintenance of biodiversity and social equity. As has been discussed earlier, the sources of degradation are often removed spatially and temporally from the effects. A major component of process management is to ameliorate the effects of degradation. Often the best way to achieve this, and ensure the long-term viability of the resource base is to reduce degradation at the source. One of the main foci of resource management at the process level will thus be to reduce the degradation of resources.

Regional management is also concerned with optimising productivity within the corporation or region. It will achieve these goals by influencing all the units of production. Regional management is not concerned with the over arching focus of strategic planning, but with a more direct focus on the corporation as a whole, and its interactions with its immediate environment.

Regional management is concerned with more localised external and internal information on all aspects of the physical, economic and sociological environments within which the corporation operates.

Regional management is concerned with predicting future effects of proposed actions so as to assess decision options for the region or corporation. This prediction must come from the use of quantitative models of processes, requiring quantitative information to drive these models. The resolution of the information must be compatible with assessing the impacts of decisions on the individual units of production. The resolution required in this information is a function of the size of the local management units.

> Regional management requires that the quantitative and qualitative information that it uses has a resolution sufficient to provide information about, or within, the smallest units of management within the corporation.

At present only Australia has attempted to seriously develop this level of resource management. Landcare Australia (http://www.landcareaustralia.com.au/) is the umbrella organisation formed for the many landcare groups that have been established across the country. This concept has taken off in Australia with wide political and financial support. The general modus operandi of a landcare group is that the group itself is community based, with representatives from the major economic and resource using groups in the community. They identify resource degradation issues that are of concern in the community. They access scientific and technical assistance from government, academic and scientific institutions to formulate action plans designed to address the individual degradation issues. They then implement the plan of action. The general method of implementation is to identify the main causes of the problem and then help those managers to find solutions that are less destructive of resources. The community pressure is the main form of influence that such groups have on the actions of individual land managers.

The landcare groups themselves are voluntary organisations, and the members are not paid for their participation. Whilst this approach has gained a very significant level of community support in Australia, it also raises the issue of how will it deal with declining membership should that occur in the future, and it does not properly reflect the costs of this level of management. It may thus work for some considerable time in Australia where the dominant forms of resource degradation of the land are very obvious on the surface. However, the degradation in resources in other areas can be subtler, resulting in declining quality or quantity of the groundwater and slow reductions in the fertility of the soil. Will there be the political and community will, in these situations, to implement this type of system?

An alternate approach could be developed along the lines of a taxation and incentives system. Regional management assesses the costs of production, including the costs of the use of all resources and the costs of the degradation, less the benefits of using degradation coming from other sites and sets a tax on these costs. The goal of such a system would be to provide a financial incentive to local managers to adopt sustainable practices. Such a system will require similar information systems as those properly required in the first approach, but they will be used in a different way. In this approach the information systems would be used to assess the taxation owed by the individual local manager. A significant disadvantage of the taxation approach is the general attitude of the community towards taxation and those responsible for implementing it. The information systems that are developed for such a purpose could, and should, be used to provide advice to individual land managers on the alternatives open to them.

The European Union (EU) is slowly moving towards an incentive form of this approach. The EU decides on the criteria that it will set for environmentally friendly resource use and management, and then provides financial incentives to resource managers to implement methods that abide by these criteria. The rationale behind this approach being that society as a whole requires that the natural resources be used in a sustainable manner, so the society as a whole should pay for the costs of implementation. Such an approach requires that adequate monitoring systems be established so as to verify that the system is not being abused. Information system used to verify claims for incentive payment should be constructed so that they can also be used to support resource management activities and thus have their costs distributed across both the community, in terms of resource maintenance, and resource management in terms of production.

Two broad approaches to the implementation of regional management have been described. Other approaches may evolve or adapt these ideas. However, the information systems required by both approaches are similar, even if they are used in quite different ways. These information systems will need to have the following characteristics:

- *Resolution* in the information content down to the unit of management (the field in agriculture) or better, and to each period or season of management.
- *Accuracy* to be compatible with the information used at the field level by the local manager.
- *Spatially extensive*, so that it covers the area of both cause and effect in relation to the resources being managed.
- *Predictive*, so that it can provide information on the likely effects of proposed actions.
- *Neutral* in its display of information and predictions, so that it cannot be interpreted as being critical of, or negative about, individual resource managers, except in terms of the sustainability of the proposed management.
- *Provides alternatives*, that is to show the costs and benefits of various options that are available to the resource manager. Many agricultural areas use rotation practices, where the high costs of one rotation are offset, to some extent, by mitigation in other rotations. The advice provided needs to cover a period long enough to consider this strategy as part of the advice.

The information used in such systems will need to be of two main forms:

- *Stable information*, or information that does not change during the process being investigated. Such information may include digital elevation models, some soil chemical characteristics, administrative data on field and farm boundaries, and navigation data.
- *Dynamic information*, such as weather data, landuse and land condition data, soil moisture and humus content, and so forth.

1.5.3 Operational Management

> To control and direct the current operations of individual units of production so as to achieve maximum productivity within the constraints and guidelines set by regional action plans and programmes.

Operational control involves the direction of staff in activities directly involved in managing the resources of the unit of production, whether it be a farm, a factory or a shop. Operational control is concerned with day-to-day decisions. It requires quick decisions, often on site, using primarily on site or internal information. It requires models that can be implemented by the manager on site, and that can respond to events as they occur. Information requirements at this level are for high resolution, yet simple, information since the decisions are usually made quickly using empirical models.

1.5.4 Relationship between these Levels of Management

The characteristics of these three levels of resource management and their relationship are discussed in more detail in Chapter 8, and shown in Figure 1.1. The figure is meant to emphasise the differences between the management decisions that are being taken and the information needs of each level of management, yet show that there are very fuzzy boundaries between the levels of management, and that there is an interdependence between these levels of management.

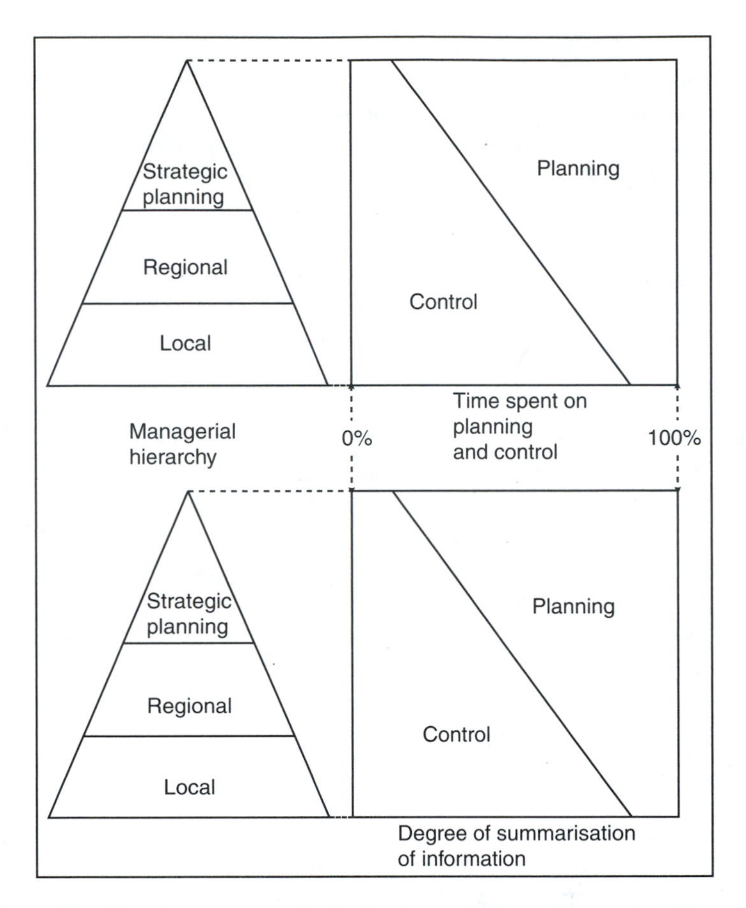

FIGURE 1.1 The relationship of management levels, the types of information used and the activities conducted by management.

Thus, even though the information systems needed at each level of management are quite different, there none the less needs to be an extensive communication among these different levels and there needs to be a consistency in the information provided to and used by each level of management. It is important that the local manager uses, in field management, information that is consistent with the information on the status of that field that is being used by the regional manager, just as it is also important that the information being used by the regional managers is consistent with that used by the policy-makers in establishing the policies that the regional managers will need to implement.

Thus, regional resource management systems need to provide information down to the unit of production, and this information needs to be accurate enough to be consistent with that used by the local resource manager. It also needs to be extensive enough to provide evidence that it is compatible with the information used by policy-makers for the construction and monitoring of the effectiveness of those policies.

1.6 THE NATURE OF REGIONAL RESOURCE MANAGEMENT INFORMATION SYSTEMS

Section 1.5 identified and discussed the three broad levels of resource management that can be found in all countries, albeit with some significant differences in their relative significance and impact. The section showed that the information needs of each form of management are quite different.

It is clear from this discussion that the key level of management in terms of resource sustainability is the regional level. If a form of degradation is to be brought under control, then controlling the degradation has to be done at its source. However, the local resource managers at the source may be unaware of the impact of their actions on other managers, and they may have little incentive to control the degradation if its impacts are exported elsewhere. This is where they need to see themselves as part of a community, where they may have responsibilities towards others in the community, but where the community also has responsibilities towards them. If regional management is to be able to have this impact on the individual resource managers, then they require information systems that provide information down to the individual unit of production, typically the field. If this information is to have credibility with the local managers, then it needs to be accurate enough to be credible to the local manager, it needs to be up to date and it needs to be detailed enough to tell the regional managers about the general conditions in the field. It does not have to be detailed enough to show the local manager how to better manage his field, but detailed enough to show whether his management is having the desired effect.

However, another critical characteristic of regional resource management information systems (RMISs) is that they must contain a predictive capacity. This is not meant to suggest the creation of long-term projections that are the focus of the strategic manager, but rather short-term predictions of the effects of proposed actions. What are the impacts of different forms of regional resource management on the resources over a season? Such predictive information also needs to be of adequate accuracy to earn the respect of the local managers, if they are to take them into account in the development of their resource management plans. The only way to achieve these goals is by the establishment of rigorous, quantitative information systems that contain embedded models and spatial information systems when dealing with spatially distributed resources. An RMIS must also be capable of providing options to the regional managers, that they can suggest for implementation by the local resource managers. Thus, faced with problems arising from current land using practices, the regional and local resource managers can work together to consider alternative management options that are available to the local resource manager.

An important component of these activities have to do with the resolution of conflict. In attempting to resolve conflicts among community desires, as reflected in policies, and local managers, or among local managers in terms of land uses, the first response should be to consider those alternative options that are available to the local managers. It may, for example, be possible to adopt other practices that remove or significantly reduce the causes of the conflict of interest.

The functions of an RMIS are thus to:

1. Be able to *display information on the status of resources* across an area as the basis for information, discussion, decisions and further analysis.
2. Show or *predict the effects of degradation* under different scenarios of variations in weather conditions and management across the area of interest.
3. Show or *predict the effects of different proposed actions* or management decisions, under different scenarios.
4. *Provide option advice* on the options available to the manager, their expected costs and benefits.
5. *Optimise specific criteria*, given constraints on physical conditions, weather, management, legal and economic constraints or conditions.

To be able to implement these functions, an RMIS needs to have the characteristics of:

1. Providing spatial and temporal context across the region of interest for the information being analysed in the system. The RMIS thus needs to be able to analyse and display temporal geographical information.
2. It needs to be able to provide information down to the level of resource management so as to link to the local managers. It thus has to have spatial information that has a resolution

down to the unit of management, or smaller. The unit of management is usually the field in agriculture or forestry, but may be some other form of area for other land using activities.

3. It needs to have a temporal resolution that is also compatible with the local management, so that typically it would need to provide information within the season.
4. It needs to be able to record decisions as a layer that is retained in the system so that it can be subsequently used to influence the analysis tools implemented within the RMIS.
5. The RMIS will require a suite of physical, economical and sociological models, all of which can be implemented individually or together, responding to the data in the system in the construction of predictions based on that data and other parameters as are necessary to drive the models.
6. Provide estimates of the reliability of the derived information.

RMIS are defined as systems that use temporal geographic data and information, with associated ancillary data and information, to derive information required by resource managers in a cost-effective manner.

An RMIS is intended to provide information to regional resource managers, presenting that information in a way that allows the manager to understand the information and its implications so as to construct a tactical knowledge model before making a management decision. An RMIS is therefore a decision support system for the regional resource manager and does not in any way take the actual decision making process away from that manager, nor his responsibilities in the management of resources. Since the RMIS is providing a comprehensive set of relevant information to the manager, it should facilitate the implementation of more timely decisions, and should be accompanied by an improvement in the quality of the decisions made by the manager. However, an RMIS is not a panacea that will relieve the manager of his responsibilities, nor the challenges, in making management decisions.

Since an RMIS is a management decision support system, it is essential that the system interface to the manager be easy for the manager to learn, use and understand. Clearly, the more that this language mimics the language of the manager the more convenient it will be, and so it is important to phase questions in the RMIS in terms used by the resource manager. The current trend in computer interfaces, using the concepts of windows and pointers or a mouse, improves the convenience of the interface, requiring minimum training by those who are not familiar with computers. The RMIS must also communicate its information to the manager in a manner that reflects their concerns and priorities if it is to be of most economic use to those managers. An RMIS is often based on sophisticated technology; it often requires sophisticated technological skills to implement, in consequence the workshop component of an RMIS may contain staff with highly specialised technical skills. Whilst this workshop component must be accessible to the manager, it will not be through the workshop that the manager will communicate with the RMIS. It is more likely that he will communicate directly with the RMIS. In consequence the interface in the RMIS to the manager must be suitable for use by the manager.

An RMIS will consist of at least four parts:

- Input
- Analysis
- Estimation, prediction, option evaluation and optimisation
- Decision support

Input includes acceptance by the system of all of the data and information used by the system to derive management information. These inputs can include aerial photographs, field data, maps and remotely sensed images. The maps can include topographic, property or cadastral (land ownership), geologic, soil, landuse and other relevant maps in either analog or digital forms. If the system is to use analog data then the input is likely to include the facilities necessary to convert analog data into digital data, as well as the facilities necessary to store both the analog and digital data.

Input must also consider error checking and validation of the data that is being entered into the RMIS. The adage, "garbage in, garbage out," is quite correct. Users may be more nervous about digital data than analog data because they cannot "see" the digital data and hence they cannot assess the quality of the information for themselves, as they can with an analog document. In consequence, if they come across errors in the information given to them by an RMIS then they are likely to be less forgiving of the system; they will place less reliance on the system when making management decisions. The ongoing viability of an RMIS depends on its credibility with its users, and this in turn depends on users understanding the accuracies that they can both expect and are actually getting, in the information provided by the system.

Analysis involves the derivation of raw information from the input data and information, as well as the derivation of management information from the raw information. The derivation of raw information is usually done from the analysis of remotely sensed images, derived from maps or from field data. The derivation of this raw information is likely to be done by image analysis specialists, sometimes using advanced image processing and analysis techniques. Whilst this raw information can be used on its own, it is more frequently used in conjunction with other information. The processing of sets of raw information to derive management information is usually done within a GIS environment. The resource manager or his assistants may derive the management information, so that this is where a user-friendly interface is essential.

One goal of an RMIS would be to create raw information that can be used by a range of different resource managers. Once the raw information is supplied to the individual managers, they integrate it with their other raw information layers to derive the management information that they require. Such an approach minimises the costs of data acquisition and analysis, and spreads this cost across a number of users. It also means that the different managers have consistent information sets in the management of the resources in a region. This is critical when resolving conflicts in resource use or allocation.

Typical types of management information that might be derived using a GIS could be to answer questions of the form, "What fields in County Alpha have been cropped in each of the last four years and what is the average yield of these fields relative to the average for the county," or "Show me all areas that have been cropped at least twice over the last five years and are within x Km of railhead Beta. What is the area involved in cropping each year and the average production."

The raw information for both of these questions is the cropping history, yield information and land administration information. The cropping history information can be readily taken from remotely sensed images. The yield information may come from remotely sensed data or from field data in the form of annual returns. The administrative information will have come from digitisation of maps that may be maintained as a digital database. Each layer of raw information described for the above example can be used on its own, but they become capable of being used to provide much more powerful and useful information when they are combined with the other layers of raw information. These same layers can be used to address many other management questions, either on their own or in conjunction with other layers of raw information such as soil or slope information. The manager is the person who is in the best position to define clearly the questions that he wants addressed, he will also want the system to be responsive to his needs; to provide the requested information when required. This component of an RMIS must therefore be very accessible to, and usable by, the resource manager himself.

Estimation and prediction are defined as the extrapolation from data or information on specific resource attributes to derive estimates of those or other attributes by making assumptions about the behaviour of parameters that affect the estimation or prediction. Estimation is the extrapolation from a specific set of attributes to another set at the same time. It thus involves models that estimate the second set of parameters from values given to the model for the parameters in the first set. Prediction is the extrapolation over time and thus normally includes assumptions about climatic events over the period of time of extrapolation. Both estimation and prediction are modelling tasks, and as such similar considerations are involved in the design and development of both types of models. But the method of use of both types of models can be quite different within an RMIS. For example, predictive models are usually implemented to address "what-if" type questions as a precursor to making a management decision. For example, "what would happen to the condition of my pasture if my property receives x mm of rainfall over the next y days?" or "how long will I have adequate fodder without rain?" Estimation models are most often used to derive estimates of physical parameters of the environment such as above ground vegetative mass or biomass, Leaf Area Index (LAI) or other parameters of interest to the resource manager.

Decision support is the interface between the regional resource manager faced with the need to resolve conflicts arising out of the use of unsustainable land using practices, conflicts among land using practices and societal policies and conflicts among the desires of different local managers. The tools that the regional resource manager needs to deal with these conflicts include:

- *Estimates of the magnitude of the conflicts* and their potential ecological, economical and sociological implications. The information required for this role is derived in the Analysis and Estimation components of the RMIS.
- *Considerations of the alternatives* that are available to the local managers and to the community should that be necessary, available within the Prediction and Options components of the RMIS.
- *Prediction of the potential impacts* on proposed actions on the environment, the economies of the local managers and the region and on other parties that may not be participants in the conflict up until this stage.

Bringing these tools together, using them in the appropriate way and in the appropriate sequence and presenting the required information to the regional resource manager is the decision support component of the RMIS.

All information in an RMIS contains errors. The way that the manager uses the information in the RMIS must both acknowledge the existence of these errors and protect the resources, the community and the manager from errors of judgement that arise due to errors in the information base. This can be done in a variety of ways as will be discussed further in subsequent chapters in the text.

1.7 GEOGRAPHIC INFORMATION IN RESOURCE MANAGEMENT

Data are defined as the recorded values of an attribute or parameter at a time in the past, even if it is the immediate past. They are the raw measurements that are often made by instruments or collected by field staff, and form the basis on which information is created. They generally do not constitute information of themselves. Most data are an objective historical record of specific attributes of the surface being observed under the conditions that existed at the time the data were acquired. For example, the data acquired by aerial photographs are related to the energy emanating from the surface in the wavebands being sensed by the film in the camera at an instant in the past. The data used in the RMISs are often of values that are not of much direct interest to managers in their form of reflectance or radiance data or individual point measurements. They have to be transformed or converted into something that is more meaningful and useful in terms of the information needs of the resource manager and which is less constrained to the historical perspective or the conditions that pertained to the acquisition of the source data. This transformation or conversion creates "raw" information.

Information is defined as specific factual material that describes the nature or character of the surface or aspects of that surface, which are important to the resource manager. For a variety of reasons, information is often processed to an intermediate stage, called "raw" or base information from the data. This "raw" information can be used directly by the resource manager, but more usually the manager requires information that is an integration of different sets of "raw" information each of which are often derived from different sources. This integrated information, uniquely required by a particular manager to address a specific management issue, is called "management" information. For example, an aerial photograph may be used to identify crop areas. Various managers can use such raw information, but they frequently require more complex information for making management decisions. For example, the manager may want to know the area of crop on a particular soil type or property. Integration of the raw cropping information with either property or soil information provides the required management information.

Knowledge is defined as an understanding of the characteristics and processes that occur in an object or surface. Knowledge has two dimensions: a conceptual or generalised dimension, and a detailed or parameterised dimension. Generally, resource managers build their conceptual or generalised models through education, learning and discussion; indeed many people in society have conceptual models of many environmental processes. Thus, many people have a conceptual understanding of the processes involved in soil erosion. Conceptual or generalised models of conditions or processes are essential in the making of decisions, but they are not sufficient. The resource manager also needs a detailed or parameterised model that provides him with details on the status of the object or surface of interest, or how the actual object or surface is acting or responding to the surrounding conditions. Thus, whilst most farmers can tell you how soil erosion occurs, many continue to have erosion occurring on their property, partly because they do not have adequate parameterised models of the process itself and its expression on their property. The parameterised models that are required to make decisions are constructed by using information to define the parameters for quantified versions of the conceptual or generalised models. Once this parameterised model is constructed then the manager should be in a position to make decisions in relation to that specific issue. Clearly, the information that is needed by the resource manager is that information that will allow him to construct an adequate detailed knowledge base on which to make the relevant decisions; other decisions by other managers will require different parameterised models and hence other information.

Geographic data or information are defined as those forms of data or information that incorporate spatial, temporal and attribute information about the features or resources of interest. Some types of geographic data and information have the same temporal information value for each attribute value in the data set, as occurs with an aerial photograph or a remotely sensed image, since the image is acquired at the one time. Other types of geographic data have one spatial coordinate but many temporal coordinates such as data recorded on a logging device. Other types of information have variable spatial and temporal coordinates. This particularly applies to information that is related to dynamic processes, and how they change over time.

A map is the classic form of geographic information. Each symbol on the map represents one type of feature, indicating that the feature represented by the symbol occurs at that location, at the time that the map was constructed at some specified level of accuracy. The distribution of symbols on a map indicates the relative locations of the different features in the area covered by the map, to another level of accuracy. The absolute accuracy of location of a feature is usually different to the relative accuracy between features, giving two versions of accuracy. Different resource managers are interested in different types of information, for example, one may be interested in the area of crop, whilst a second may be interested in the location of the cropping fields. Both are interested in the accuracy of identifying fields, but one is interested in the accuracy of mapping the contents of the field to estimate the field area, whilst the second is interested in the location of the boundary. There are thus different perspectives on the issue of accuracy. Different types of accuracy are measured in different ways and thus accuracy assessment may need to be conducted in a number of ways to meet the needs of the resource managers who will be using the derived information. The accuracy of a map must be considered in spatial, informational and temporal terms. Most maps are derived from remotely sensed images and thus have accuracy characteristics typical of these data sources as discussed in the text. Maps that are constructed from other data sources, such as point observations, will have other accuracy characteristics. Again, maps depict information that is correct at a temporal coordinate value, or time, prior to the date of publication of the map. This temporal information can also be important to users and so it is usually printed on the map.

Accuracy can be defined as the closeness of the estimated values of an attribute to the true values in that estimate, where modern physics shows us that it is not possible to ever measure the "true" value of any physical quantity. Thus, measures of accuracy are estimates of closeness of the estimated value to the best estimate that can be made of that value. The spatial accuracy of a map depends on many factors. The scale of the map is the major constraint on accuracy since that affects the policy for the depiction of details on the map, and the ability to properly convey the size and shape of those details. Thus, the symbols used affect the accuracy of the map as well as the accuracy of plotting the map detail. The level of interpolation of features can affect map accuracy, and finally the stability of the medium on which the map is drawn affects the accuracy of the map. Informational accuracy also refers to the likelihood of features being correctly depicted on the map: that they are in the locations indicated, that all of the features are shown and that no extraneous features are wrongly depicted. Temporal accuracy is affected by the changes that occurred between acquisition of the imagery and the printing of the final map. The actual temporal accuracy is also affected by the changes that have occurred between the printing of the map and its use by the reader. These aspects of accuracy will vary from feature to feature because of the differences in accuracy in identifying the different features and because of the frequency of changes in them over time.

A map is a record of the spatial relationships between selected types of objects at an instant in time across a spatial area or extent. It is thus a snapshot of specific resource conditions at a moment in time. It becomes out of date. Whilst this is adequate for some type of resources that do not change significantly within normal time periods, it is not adequate for those attributes of the environment that change significantly within significant time periods. Thus, the mapping of the geology and landforms in an area may be a satisfactory way to geographically describe these resources, particularly for geologically stable areas. However, for many resource management purposes, the issues being addressed involve resources that change over time in response to a range of natural and man-induced driving forces. When this situation arises, then the temporal dimension becomes important, and thus requires monitoring of those resources over time.

Monitoring is defined as maintaining a watch over or keeping track of the values and spatial distribution of specific resource attributes within a geographic area. Monitoring is concerned with changes in the resources over time, and so it includes an additional dimension over that of mapping. There are other important differences between mapping and monitoring. A mapping program is conducted on resources that exist at the time the mapping program is started. The program can thus be designed with a full knowledge of the conditions that will be met during the mapping work, and so data sets can be selected and used to make the best use of them in the mapping task. With monitoring, by contrast, the nature of the resources of interest are not known at the time that the monitoring program is established, since by its nature it will be recording information on future conditions. Of course, the best estimate will be made of the data sets that are required for the monitoring program, but these data sets cannot be optimised in the same way as they can be with mapping. Thus, monitoring systems need to be robust; they need to continue to provide useful information over a range of environmental and data-acquired conditions that may not be expected to occur. Thus, monitoring systems will need to operate over a range of weather conditions that would not be met in a mapping program, and the system will be expected to operate using data sets that would not be considered acceptable for mapping tasks. This places significant constraints on the design and operation of monitoring systems that do not exist with mapping tasks.

If monitoring systems are also required to provide timely information, then it is possible that they will be expected to provide this information before full accuracy assessment has been conducted.

Monitoring systems therefore not only need to be robust in their capacity to provide information over a range of data and environmental conditions, but they must also provide information at an acceptable accuracy. They must therefore also be robust in the accuracy of the information provided. These additional demands that are placed on monitoring systems mean that the design of monitoring systems has to be conducted with much more care than is the case with mapping systems. It also means that such systems will need to be conservative in their design, and they can be expected to need much more field data than would be the case with a mapping system. Integration of extensive information from local managers into the RMIS has the advantage that the local and regional resource management systems will tend to converge and by the supply on information in both directions, both forms of management should be strengthened and the costs of these systems can be minimised.

Monitoring systems are also characterised by continuity in contrast to the once off nature of mapping tasks. There is thus a much greater need to ensure that the costs of the monitoring are minimised. This is another reason for looking for means to create closer ties between local and regional managers so as to optimise the use of information in both systems.

The main drivers for the dynamic state of the resources of a location come from the basic forces that operate on all resources:

- *Incident radiation*. Whilst the incident radiation changes throughout the day, and from season to season, it does not change spatially, except for the effects of clouds and aerosols. This force thus has primarily a temporal dimension.
- *Rainfall and temperature*. Both have high temporal variability and moderate spatial variability, due to the way rainfall can be distributed, and due to the effects of topography on temperature and rainfall.
- *Soils* can vary dramatically in a spatial sense, but change slowly in a temporal sense.
- *Management* can vary dramatically in both the spatial and temporal domains.

These drivers create responses in the environment that then create secondary drivers and feedback loops influencing the main drivers. As one becomes involved with these secondary and tertiary effects, one realises that the spatial and temporal complexities become compounded. Thus with most, if not all environmental processes, there are strong spatial and temporal components to the processes. Both, understanding these processes, and subsequent management of them requires an appreciation of and allowance for the spatial and temporal dynamics that are implicit in most, if not all, of these processes. The resources can only be adequately mimicked in models, and managed in practice if the spatial and temporal complexity of the resources is taken into account. It is for this reason that regional resource management systems need to be based on the use of geographic data.

So, now we can see that resource management systems must have the characteristics of:

- *Using spatially extensive, temporally rich yet cheap sources of data*. This type of data can only be supplied by the use of remotely sensed data.
- *Having cheap yet robust methods for the extraction of information from this data*. This can only be achieved by the integration of extensive local management level information into the information extraction process.
- *Incorporating information on the spatial and temporal changes that occur in the resources and integrating this information* with many other sets of information on critical aspects of the environment being managed. This can only be achieved by the use of geographic information systems.
- *Providing resource managers with option and prediction information* that allow for the spatial and temporal complexity yet provide advice in a way that allows the proper management of those resources. This requires models of the processes that are operating in the area and it requires that these technologies be integrated into a system that supports the management of resources whilst supporting the drive for the sustainable use of resources.

1.8 THE STRUCTURE OF THIS BOOK

The dominant theme of this book was given right at the beginning of this chapter as "Understanding the role of, as well as to develop and use, spatial information systems for the proper management of physical resources." This leads naturally into the major goal of this book, which is to "impart the knowledge, skills and attitudes necessary to design, develop, evaluate and implement operational Regional RMIS or components of such systems." To achieve this objective it is necessary to impart knowledge and skills on the tools and techniques available to an RMIS, where these include remote sensing, GIS, modelling and decision support. It is also important to impart an understanding of the need to appreciate the human and natural environments in which the RMIS will work, as these environments will affect the design and use of the implemented RMIS. This text cannot provide the information necessary on all of these human and physical environments as this task is well addressed in other disciplines, but will address the need for this information. This text will cover the principles of and methods used in the design and implementation of RMIS. Finally, it is important for the reader to appreciate how this information can be used in the actual management of resources.

Whilst this text cannot address the many other disciplines concerned with the physical environment, it is assumed that students using this text have an adequate appreciation of the elements of these environments that might influence their implementation or use of an RMIS. All students using this text must expect to apply their knowledge of these other disciplines during the course of study using this text. The text will, however, deal with the three facets of RMIS in detail:

- Tools and techniques
- Principles of design and use
- Applications

These scientific fields of endeavour are quite new. Many of the basic paradigms or rules that control the use or application of the tools and techniques are still being developed and evaluated. As the process of development continues, existing tools and techniques will evolve and others will be developed. In consequence all students of this science must accept that they cannot just learn about RMIS and leave it at that. They will need to continue to learn and explore as these systems evolve.

The newness of this science also means that all students of this science must treat all working hypotheses and conventional wisdom with considerable scepticism as many of them will be either extensively modified or replaced as the science develops. Whilst similar statements can be made about all sciences, it is important to emphasise them in this case because the immaturity of the science leaves it particularly vulnerable to errors of judgement. It is also important to recognise that this science, by bridging across from the physical to the mathematical and biological sciences in quite integrated ways, creates opportunities for both growth and errors of judgement. Both can be seen in the development of the science.

The science is also developing at the time that the web is developing and evolving. This text plans to both exploit and complement the web in ways that are to the best advantage of you, the user of this book. This written text thus contains no practical application material, and it contains little information on operational systems. Application material is provided in a CD, so that the reader can both understand the application and conduct some application tasks themselves. The contents of the CD are shown in the directories of the CD, each of which contains a word document that describes the contents of that directory.

The author is well aware of the pitfalls and risks associated with the conduct of research and development of this science, having participated in remote sensing and GIS research, development and implementation since 1973. Everything given in this text is given in good faith and the belief that it is the best information and advice that is currently available. Nevertheless, some of the material in this text will prove to be wrong in the fullness of time as our knowledge and experience in using this science evolves. It is therefore important that all readers of this text do so with care, caution and

I would ask with constructive criticism in mind. I welcome constructive comments and discussion with readers on the contents and presentation of this text.

Having said all of this, I wish all readers a rewarding study of the scientific disciplines of remote sensing, GIS and modelling and hope that this text contributes in improving your understanding, skills and competence in developing and using RMIS.

REFERENCE

Ikerd, J., 1990. "Sustainability's promise," *Journal of Soil and Water Conservation*, 45, 4.

2 The Physical Principles of Remote Sensing

2.1 INTRODUCTION

Remote sensing is defined as the *acquisition* of data using a remotely located sensing device, and the *extraction* of information from that data. In the simplest context the eye and the brain could be considered as a remote sensing system. Light enters the eye through the lens, which focuses the light on the retina at the back of the eye. At the retina, the light is converted into an electrical signal by the light sensitive rods and cones that is then transmitted to the brain. On receipt of these signals the brain conducts a process of analysis, or signal processing, to extract information from the data, in this case about the world around us. In practice the term remote sensing is restricted to data acquired by man-made sensing devices, such as a camera and to electromagnetic radiation, such as light. The acquisition of photography using a remotely located camera, and then the interpretation of the data to extract information by viewing the photographs, is the simplest form of remote sensing.

The definition of remote sensing contains two main components; those of *data acquisition* and *information extraction*. The theoretical basis for data acquisition lies primarily within the domain of the physical and mathematical sciences whereas information extraction is based primarily on the geographic, mathematical, statistical, biological, physical and psychological sciences. However, the nature of the data used in that analysis depends upon the scientific basis and technological characteristics of the sensor system and the acquisition conditions.

A user, requiring the acquisition of specific data, is most likely to arrange for a specialist group to acquire the data because of the specialised, highly technical and capital-intensive nature of the work. These groups have the expertise to provide advice on the acquisition of data. However, the characteristics of the imagery used in the analysis must be thoroughly understood if it is to provide accurate, reliable and economical information. The user must therefore have a good understanding of the characteristics of remotely sensed imagery, but not a detailed understanding of the technical aspects of data acquisition itself. There are some exceptions to this position, specifically in relation to those aspects of data acquisition that the user has some control over, including films, filters and flight planning for airborne image data. These aspects will be dealt with in more detail than the remainder of data acquisition.

This partitioning of remote sensing into two broad components is illustrated in Figure 2.1, which depicts the progression from energy source through data acquisition, information extraction to management decision. All of the sub-components that are depicted in Figure 2.1 need to be discussed and understood if the user is to be successful in the design and implementation of remote sensing projects. All of these sub-components will be covered in the next several chapters in this textbook.

The aim of any operational remote sensing program is to provide information as an input to resource management information systems (RMIS). The nature of the required information will specify the constraints in establishing the remote sensing program. It follows that proper design of a program is dependent upon accurately specifying the information that is to be provided by the program. The information required by the management can often be acquired by other means, but it should be acquired by remote sensing when this method provides either better, more cost effective, or new but necessary information. The user is interested in determining whether remote sensing is the best available method of getting the information that he requires, and he can do this by asking

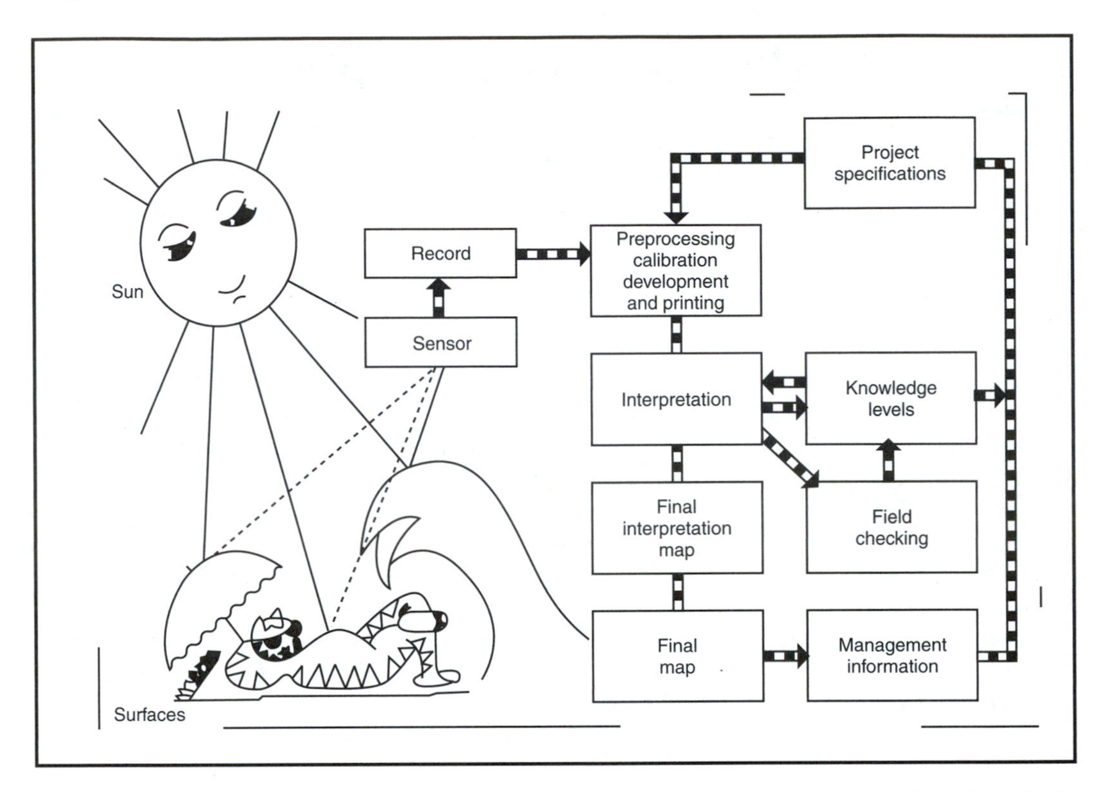

FIGURE 2.1 The components of a remote sensing project based on image data acquired using reflected solar radiation. Some sensors use other forms of radiation as will be discussed in this chapter, but the conceptual structure shown in the figure remain essentially the same.

four questions:

1. *Does the information that is required about the physical environment change with spatial position in that environment?* If the answer is yes and particularly if this spatial location is important then remote sensing might be the best way of getting the required information. There are other methods of acquiring this sort of information, but they all depend upon point observations by human observers, with interpolation between these point observations to create a continuous map of the information type. They suffer the tyranny when high accuracy requires a dense network of points, achieved at great expense; but the greater difficulty is the time required to create the map under differing weather and other conditions where the information is actually changing whilst the map is being made. Remotely sensed data provides a permanent record of the whole area of interest at a point in time, so that the extraction of information does not involve interpolation between points (but does involve other types of interpolation and extrapolation that will be discussed later). Information derived from remotely sensed data is also relatively consistent across the image area at one instant in time. It is consistent within the variations imposed on the image data by variations in the atmospheric conditions that have not been corrected for, and relative errors in the sensor. These sources of errors are discussed later in this chapter.

2. *Is the required information dependent on time?* Remotely sensed data has the advantage that the derived information is relatively consistent spatially and temporally, particularly if using similar sensors. Many types of important information in the management of renewable resources do vary with time, and therefore require this consistency, but just as importantly the information may need to be collected at a regular interval to see how it changes with time. Remotely sensed data allows the comparative analysis of changes in condition over time much better than most other methods.

With point observations by human observers it is unlikely that the point observations will be at the same locations, and there is a degree of error associated with the interpolation between the points that precludes accurate comparison. These criticisms of alternatives to remote sensing on technical grounds are quite separate to the very great cost that would be associated with repetitive mapping of an area using observers, a cost that has prohibited the collection of this sort of information on renewable resources up until now.

3. *Does the collection of the required information either affect the resource being monitored, cause undue time delays, or create unacceptable administrative or cost structures to be created for its collection?* Many methods of data collection interfere with, or cause responses by, the resources being measured or observed. Remote sensing does not interact with the observed resources in this way. This can be a significant advantage, particularly where the interaction introduces changes that either reduce the accuracy of the observations or affect the value of the resources. Unacceptable time delays can arise because physically based methods of data collection take time to collect the data. If the information is required within a set period of time, then the time of data collection may preclude the use of that method. Thus, methods of manually mapping vegetative status throughout a season, the advance of floodwaters or changes in terrain height as a result of Earth crustal deformations, may all take far too long for the information to be of use to resource managers. In all of these cases, remote sensing can offer a much more rapid method of mapping.

4. *Can the required information be reliably extracted from image data?* Clouds, for example, totally block solar radiation that is the source of energy for most optical imaging systems. If an area has frequent cloud cover, then optical imagery may not be acquired at a suitable rate so as to derive the required information. If the information required is of a night-time process, then optical energy may not be suitable if imagery cannot be acquired at night and which contains information on the process. One case where this is not the case, is where the imagery is acquired of night-time lighting and can be acquired of night temperatures, in both cases using other sources of energy rather than the sun. All imagery acquires data in selected wavebands; if the required information does not affect these wavebands, then the information sought may not be capable of being derived from the data, either directly, or through its relationship with other physical characteristics in the environment.

Remote sensing is thus a tool in the armoury of the manager of renewable resources, to be used when its particular characteristics make it suitable to the task at hand. It has its own set of advantages and disadvantages relative to other methods of collecting spatial information. However, it is the first time that we have had a tool with the characteristics particular to remote sensing; it will take us a while to properly learn to understand and exploit these characteristics.

The historical development of the science of remote sensing started with the development of photography and photo-interpretation. Photo-interpretation remained the term of choice whilst most images were acquired using cameras with film. The advent of digital sensors that could sense wavebands outside the visible part of the electromagnetic spectrum and record this radiation using digital devices that converted the incident energy into an electrical signal as well as digital image analysis techniques made the term photo-interpretation inappropriate. At this stage the term remote sensing was coined so as to reflect the much larger range of the activities involved in the acquisition of image data and the extraction of information from that data.

Daguerre and Niepce took the first photographs in about 1822. The photographs taken by Daguerre and Niepce were made using the Daguerreotype process that required bulky and awkward to use cameras and facilities. The emulsion film had to be laid on glass plates just before taking the exposure, and then developed soon after. Despite these disadvantages the French demonstrated the potential of the tool in topographic mapping. Laussedat took extensive ground-based or terrestrial photographs for mapping purposes. He showed that two photographs of the same area but from slightly different positions allowed the analyst to construct a three-dimensional model of the surface which could then be used to map the elevations in the surface and the surface detail could be mapped onto a plane surface. This application of photography, called stereoscopic analysis, has significantly increased the value of the photographs for mapping. Tournachon demonstrated that photographs could be taken from a balloon in 1852.

The next important step was the development of a gelatine base supporting and securing light sensitive silver halide salts as the means of recording an image. This process meant that development did not have to occur immediately after the photograph was taken, and was much simpler and easier to handle than the Daguerreotype process cameras. Many photographs were taken from balloons and kites once the technique was developed. However, the acquisition of aerial photography across extensive areas had to wait for the development of the aircraft to provide a suitable platform. Wilbur Wright took the first aerial photographs in 1909.

Hauron developed the first colour separation film in 1895. From this, and the work of others, evolved Kodachrome in 1935. Colour film was first used from an aircraft in 1935, but it took a number of years to develop reliable methods of acquiring colour aerial photography, due to haze, vibration and the need for good navigation facilities.

The interpretation of aerial photographs was first appreciated by the military in World War I. The skills then acquired were applied and extended for topographic and geologic mapping between the two world wars, with the art of photo-interpretation reaching full maturity in World War II. Photo-interpretation has continued to be used extensively since then, as well as being applied to many other applications.

The idea of using thermal sensing to detect men and machines was first proposed during World War II. Devices to detect thermal radiation gradually led to the development of scanners sensing in the visible and infrared wavebands. These instruments had evolved to a reasonably satisfactory level by the 1950s, and received considerable impetus after this time because of the use of digital sensors or scanners in satellites, particularly the TIROS and NIMBUS meteorological satellites. Unlike photography, the data acquired by a scanner can be readily converted into a signal suitable for radio transmission, much more convenient for getting data from a satellite than canisters containing film. Satellite-borne scanners were first used for Earth resource related purposes with the launch of Earth Resources Technology Satellite (ERTS) 1 in 1972. ERTS-1, renamed Landsat 1, was the first of a series of experimental Earth monitoring satellites launched by the United States. Up until 2003, seven Landsat satellites had been launched, with some variations in sensors on the different satellites. Landsat 1 contained a successfully used Multispectral Scanner (MSS) with 80 m resolution data and four video cameras that proved to be unsuccessful. The video cameras were not included in the payloads of Landsats 2 and 3. Landsats 4 and 5 included a Thematic Mapper (TM) scanner with 30 m resolution as well as the MSS at 80 m resolution. Landsats 6 and 7 used an enhanced TM sensor as the primary sensing device. In 1986 France launched SPOT 1, a new scanner design that contained fewer moving parts than the Landsat scanners. SPOT 1 could also be pointed off track, so that it represented a significant development on the Landsat technology. In 1999 the first truly commercial remote sensing satellite was launched. The Ikonos satellite acquired images with a panchromatic resolution of 1 m, and a multispectral resolution of 4 m, with the capacity of acquiring images every 3 to 4 days. It showed that the commercial sector was going to focus on the acquisition of high resolution, short response time imagery, suitable for many commercial and disaster coverage purposes. The Ikonos launch was soon followed by the launch of Quickbird in 2001, with a multispectral resolution of 2.4 m and a panchromatic resolution of 0.6 m.

The launch of commercial satellites has not slowed the launch of research-oriented satellites, but they exhibit quite different characteristics. Research satellites usually contain sensors designed to address specific issues, or to extend the capabilities of sensors into new data dimensions. The first hyperspectral sensor was launched in 2001 on board the NASA Earth Observing-1 Platform. The Hyperion sensor sensed the Earth at 30 m resolution in 220 bands from the visible through to the short wave infrared part of the spectrum. The first Earth Observing System (EOS-AM1) containing a suite of four Earth resources monitoring sensors was launched in 2000. These sensors are designed primarily to extend our capacity in monitoring and modelling Earth systems sciences. In 2002 the EOS-PM1 was launched with a complimentary orbit to that of EOS-AM1.

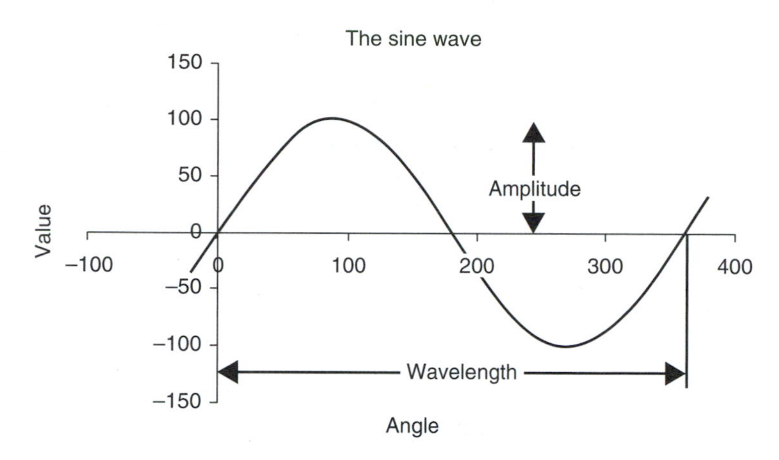

FIGURE 2.2 The sine wave and its parameters of wavelength, λ and amplitude, A.

The first sensors to be developed that depended on their own energy source to illuminate the target surface was RADAR (RAdio Detection And Ranging) developed in the United Kingdom during the World War II. The first imaging radars were developed in the USA in the 1950s. These imaging radars were side looking airborne radar using real aperture antenna (SLAR). A major difficulty of these radars was the long antenna required to get a reasonable resolution at the wavelengths being sensed. This problem was solved with the development of synthetic aperture radar (SAR) in the 1960s in the United States. The first satellite-borne radar accessible to the civilian community was Seasat A in 1976, followed by the Canadian Radarsat and the European ERS-1 sensors.

2.2 ELECTROMAGNETIC RADIATION

2.2.1 The Nature of Electromagnetic Radiation

Electromagnetic radiation is transmitted through space in the form of an oscillating sine wave as shown in Figure 2.2. The parameters that define the shape of the sine wave are the wavelength, λ, and amplitude, A, where the wavelength is the distance between identical points on adjacent symmetrical wave segments, and the amplitude is the magnitude of the oscillation from the mean value.

The frequency, f_c of electromagnetic radiation is the number of oscillations or wave segments that pass a point in a given time. Electromagnetic radiation always travels at the speed of light in the medium. This speed, however, may change with the physical properties of the medium compared to the speed of light in a vacuum. The speed of light at sea level in the atmosphere is about 299,792,000 m/sec depending on, for example, the moisture content in the atmosphere. The frequency of radiation is inversely proportional to the wavelength in the equation:

$$f_c \times \lambda = c = 299{,}792{,}000 \text{ m/sec} \tag{2.1}$$

Electromagnetic radiation varies from very short to very long wavelengths. Only a subset of these are of use in remote sensing because absorption or scattering makes the atmosphere opaque to the others as will be discussed in Section 2.2.4. The metric units of measurement used to define wavelength are given in Table 2.1.

All physical bodies, at temperatures above absolute zero (0K or −273°C) radiate electromagnetic radiation. A blackbody is a perfect absorber and emitter of radiation and will look absolutely black

TABLE 2.1
Metric Units of Distance

Unit Name	Abbreviation	Relationship to the metre
Angstrom unit	Å	1 Å = 10^{-10} m
Nanometre	nm	1 nm = 10^{-9} m
Micrometre	μm	1 μm = 10^{-6} m
Millimetre	mm	1 mm = 10^{-3} m
Centimetre	cm	1 cm = 0.01 m
Kilometre	Km	1 Km = 1000 m

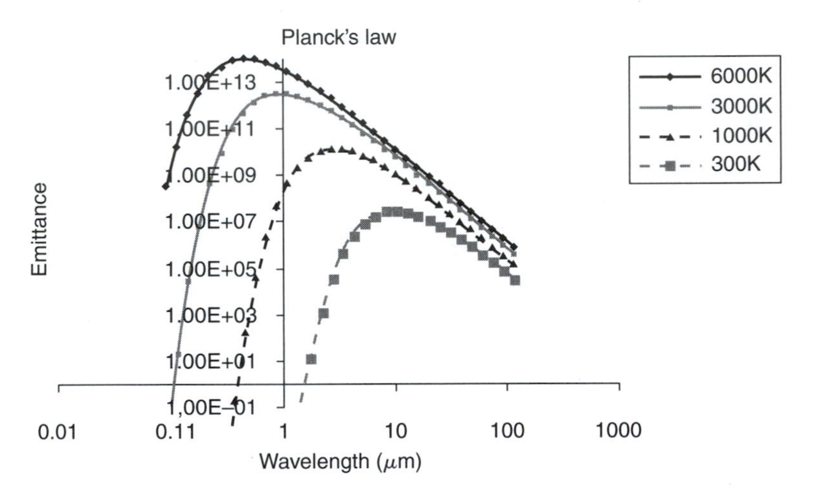

FIGURE 2.3 Energy emitted by blackbody with parameters of wavelength, λ and amplitude, *A*.

to the human eye. The energy that it emits obeys Planck's Law.

$$E_\lambda = \frac{2\pi hc^2}{\lambda^5(e^{hc/\lambda kT} - 1)} \mathrm{W/m^2/\AA} \tag{2.2}$$

where h is the 6.626069×10^{-34} J/sec (Planck's constant), c is the 299,792,458 m/sec (speed of light in a vacuum), k is the $1.3806503 \times 10^{-23}$ J/deg (Boltzmann constant), λ is the wavelength in metres and T is the temperature in degrees Kelvin.

Typical energy radiation curves, for blackbodies of various temperatures, are shown in Figure 2.3, showing the variation in radiated energy with wavelength. The curves depicted in Figure 2.3 illustrate the characteristics of the distribution of energy radiated from a body:

1. As the temperature of the source increases, the wavelength of the peak level of radiated energy gets shorter. Thus, bodies at a temperature of 300K will have peak radiation levels at longer wavelengths than bodies at temperatures of 6000K. To the human eye, as the temperature of a source increases, the colour changes from dark to bright red, yellow, white then blue, where these colours match changes to shorter wavelengths with higher energy levels.
2. As the temperature of a source increases, the amount of radiated energy increases. The total amount of energy radiated from a body is given by integrating the area under the

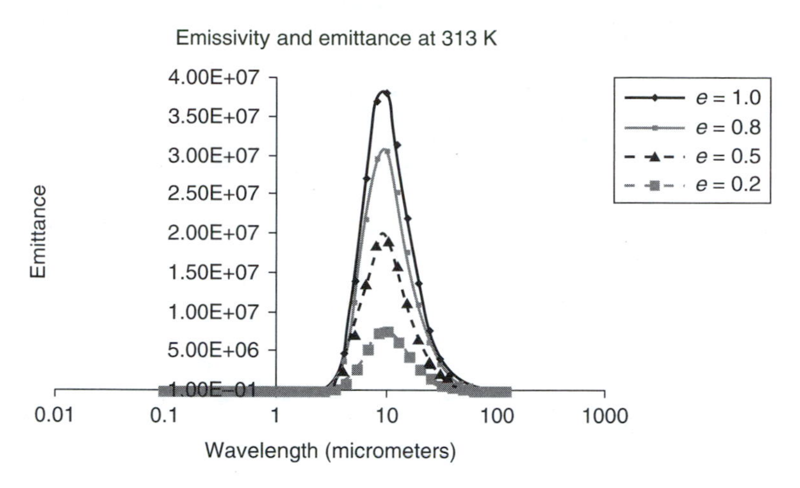

FIGURE 2.4 The effect of emissivity on the energy radiated by a surface at a temperature of 313K (40°C), showing the energy radiated for surfaces with emissivities of 1.0, 0.8, 0.5 and 0.2.

energy curve for that body. This area is larger for the body radiating at 6000 K than it is for the body at temperature 300K.

Wien's Displacement Law gives the wavelength of the peak energy radiated by a body from the equation:

$$\lambda_p = a/T \tag{2.3}$$

where λ_p is the peak wavelength in millimetres and a is the constant, 2.898 mm/K.

Equation (2.2) and Equation (2.3), depicted in Figure 2.3, mean that the energy emitted by a body rises rapidly to a peak, and then tails off much more slowly as the wavelength increases. Most bodies have surfaces that are not perfect emitters or radiators of energy, that is, they are not perfect blackbodies. The efficiency of a surface as an emitter or radiator is termed as the emissivity of the body:

$$E_\lambda = e_\lambda \times E_{b,\lambda} \tag{2.4}$$

where e_λ is the emissivity at wavelength λ, E_λ is the energy emitted by surface at wavelength λ and $E_{b,\lambda}$ is the energy emitted by blackbody at wavelength λ.

All surfaces that have an emissivity less than 1.0 are called greybodies. The effect of variations in the emissivity of a surface on the emitted radiation is shown in Figure 2.4, in which the emissivity is assumed to be constant at all wavelengths for each curve. In this discussion, it is assumed that the object, or body has the temperature that creates the radiation, but it is the surface of the body that has the capacity to radiate or absorb energy, and thus it is the surface that has the emissive characteristics. In this text, surface is defined as the infinitely thin external border of a body or object. Electromagnetic radiation that is incident on a body penetrates some way into the body as a function of wavelength and the characteristics of the body, as will be discussed in more depth in succeeding sections in this chapter. This layer is called the boundary layer. A body consists of its boundary layer and the material beneath this layer.

Figure 2.4 shows that reducing the emissivity of a surface does not change the wavelength of the peak radiation from that surface, or the wavelength distribution of energy from the surface but it reduces the amount of energy radiated at each wavelength. If the energy being radiated from a surface is only recorded in the one waveband, then both the temperature and emissivity of the surface

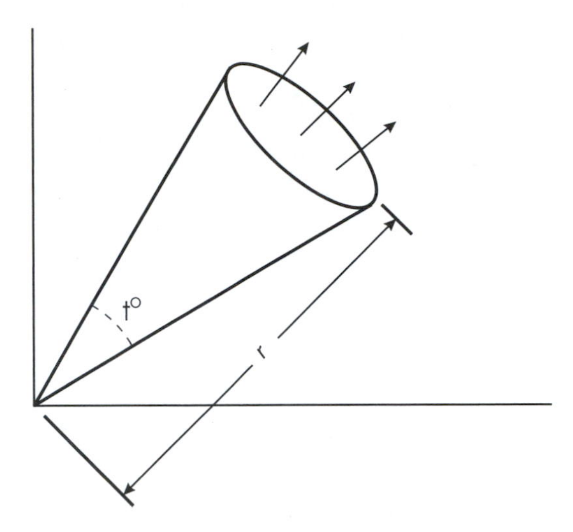

FIGURE 2.5 Radiant intensity from a point source.

will affect the actual energy levels received at the sensor. As the wavelength changes, the emissivity of a surface changes. As a consequence, the energy emitted over a broader waveband is the sum of the energies emitted at each waveband within this broader waveband.

2.2.2 Radiometric Terms and Definitions

Radiometric energy, Q, is a measure of the capacity of radiation to do work, for example, to heat a body, move objects or cause a change in the state of the object. The unit of measurement of radiometric energy is the joule or kilowatt-hour.

Radiant flux, ϕ, is the rate of radiometric energy per unit time (joules per second or watts). It can be considered to be like the flow of energy past a point per unit time or conceptually similar to the flow of water in an ocean current past a point in unit time.

Radiant flux density, E or M, is the radiant flux that flows through unit area of a surface. In the previous analogy of the ocean, it is the flow of water per unit cross-sectional area. Just as the flow of water will change from place to place, so can the radiant flux density change from surface element to surface element. Radiant flux density incident on a surface is called the irradiance, E, and the flux density that is emitted by a surface is called the exitance, M. Both irradiance and exitance are measured in units of watt per square metre.

Radiant intensity, I, is the radiant flux per unit solid angle leaving a point source, as shown in Figure 2.5. The radiant energy from a point source, being radiated in all directions, has a radiant flux density that decreases, as the surface gets further away from the source. The intensity, however, does not decrease with increasing distance from the source, as it is a function of the cone angle. Measured in units of watts per steradian (W/sr).

Usually, the point source is being observed by a sensing device with a fixed aperture area, A. As the distance between the point source and the aperture A increases, that is r increases (Figure 2.5), the solid angle decreases and the flux passing through the aperture is reduced. The decrease is proportional to the square of the distance from the source. This is because the flux has to be distributed over the hemispherical surface at distance r from the source. Since this area, $4\pi r^2$ is proportional to the square of the distance from the source, so is the flux density.

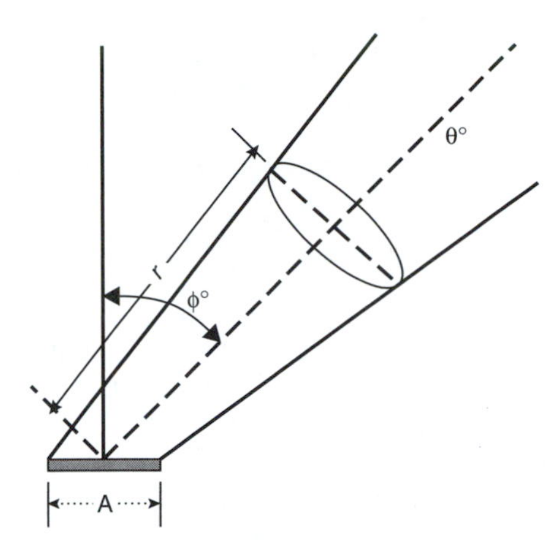

FIGURE 2.6 Radiance into an extended source area, *A*.

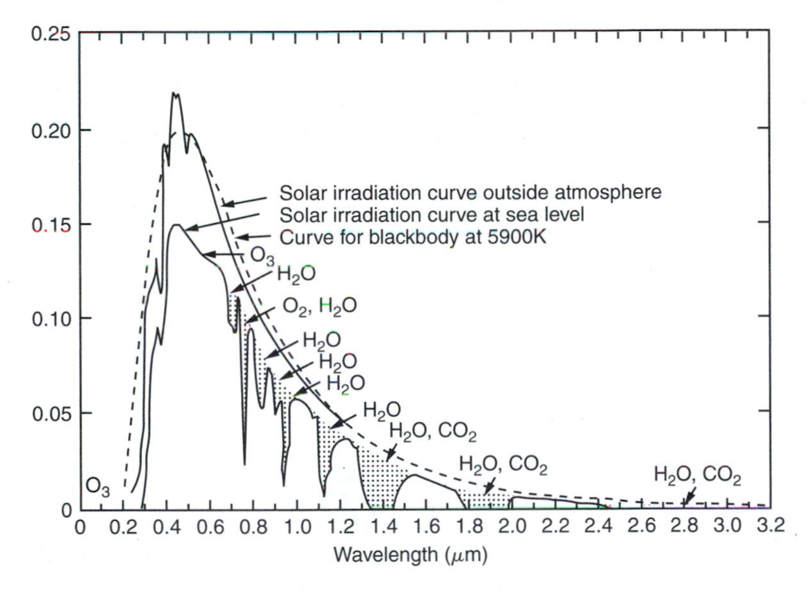

FIGURE 2.7 Solar irradiance before and after transmission through Earth's atmosphere, with atmospheric absorption, for the sun at the zenith (from Lintz, and Simonett, D.S. (Eds), 1976. *Remote Sensing of Environment*, Addison-Wesley, Reading, M.A. With permission).

Radiance L (Figure 2.6), is the radiant flux per unit solid angle leaving an extended source in a given direction per unit projected source area in that direction, measured in watts per steradian per square centimetre (W/sr/cm^2).

All of the terms defined above are wavelength dependant, and this can be indicated by the use of the term "spectral" in front of the term, the term then applying to a unit of wavelength such as a nanometre or micrometer. For example:

$$\text{Spectral radiance} = L\,\text{W/sr/cm}^2/\text{nm}^{-1}$$

2.2.3 ENERGY RADIATED BY THE SUN AND THE EARTH

The energy distributed by the sun has a spectral distribution similar to that of a blackbody at a temperature of 6000K (Figure 2.7). The temperature of the Earth's surface varies, but is often at about 30°C or about 300K. The peak radiance from the sun is at about 520 nm, coinciding with the part of the spectrum visible to the human eye as green light. The human eye can detect radiation from about 390 nm, or blue in colour, through the green part of the spectrum at about 520 nm, to the red part of the spectrum up to about 700 nm. Thus the human eye is designed to exploit the peak radiation wavelengths that are radiated by the sun.

2.2.4 Effects of the Atmosphere

The atmosphere is transparent to electromagnetic radiation at some wavelengths, opaque at others and partially transparent at most. This is due to scattering of the radiation by atmospheric molecules or particulate matter (aerosols), absorption of energy by the constituents of the atmosphere, often as molecular resonance absorption, and emission of radiation by matter in the atmosphere. The contributions of each of these depend on atmospheric conditions and the wavelength being considered.

The effect of this scattering and absorption is to reduce the amount of energy that penetrates the Earth's atmosphere to the surface, and then is reflected back to a sensor. The wavelength regions for which the atmosphere is relatively transparent are called windows, the main windows of interest in remote sensing being shown in Figure 2.8. Scattering and absorption of electromagnetic radiation occurs predominantly at the shorter wavelengths due to molecular or Rayleigh scattering, and at longer wavelengths due to absorption and scattering by particulate matter in the atmosphere. Rayleigh scattering causes the characteristic blueness of the sky.

The effects of scattering and absorption are to attenuate, or reduce, the direct energy that will be incident on a surface by the transmissivity of the atmosphere, t_λ, and to add a constant radiance level,

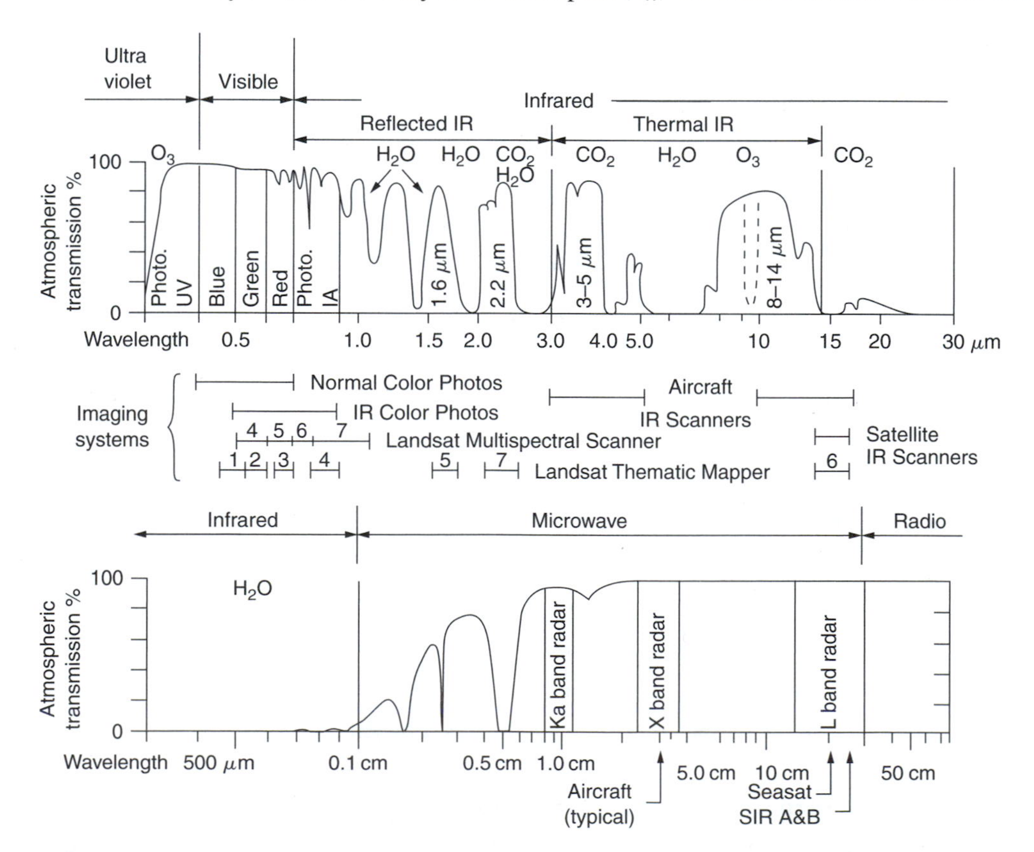

FIGURE 2.8 Atmospheric transmissivity, common absorption bands, and the spectral ranges of selected remote sensing systems (from Sabins, F.F., 1987. With permission).

$E_{A,\lambda}$ due to scattered radiation from the atmosphere being incident on the surface. In the normal practice of acquiring imagery under cloud free conditions the contribution from the atmosphere is small compared to the attenuated contribution from the sun, for the windows that are normally used. The radiation incident at the surface is thus of the form:

$$E_{i,\lambda} = t_{\lambda} \times E_{s,\lambda} \times \cos(\theta_s) + E_{A,\lambda} \tag{2.5}$$

Where t_{λ} is the average atmospheric spectral transmittance for the ray path through the atmosphere, $E_{i,\lambda}$ is the solar radiance incident on the surface, $E_{s,\lambda}$ is the solar radiance at the top of the atmosphere, θ_s is the solar zenith angle, $E_{A,\lambda}$ is the atmospheric spectral radiance incident on the surface and λ is the wavelength (nm).

Atmospheric attenuation and scattering have both stable and highly variable components, and these components can have quite different magnitudes. The stable components are due to the chemical constituents of the atmosphere that change only very slowly, if at all. The effects of these constituents can be modelled and corrected either without the need for actual data or with easily acquired data. The dominant highly variable constituents are water vapour, aerosols and ozone. Ozone is more stable than either the atmospheric water vapour or aerosols, and its concentrations are routinely measured, so that it can be readily corrected. The highly variable components can only be accurately corrected by the use of detailed and appropriate data taken close to or at the time of data acquisition. Water vapour primarily causes absorption in the known water absorption bands. The density of water vapour in the atmosphere changes with weather, including temperature, and surface conditions. It can thus change significantly with both time and space. If the water vapour density in the air column between the sensor and the ground is known, then the effects of water vapour can be readily computed and corrected. Aerosols come from many sources, and they all have different absorption and reflectance characteristics. The amount of suspended aerosols in the atmosphere depends on the weight of the aerosol components and the turbulence in the atmosphere. Thus, smoke and some forms of air pollution are light and can stay in the atmosphere until washed out with rain or blown away by the wind. On the other hand particles of dust are relatively heavy, and often will only stay in the atmosphere whilst conditions are windy. To further complicate the issue, aerosols tend to attract moisture over time, the moisture changing the reflectance characteristics of the aerosols and causing them to tend to coagulate into larger particles with quite different shape characteristics, further affecting their reflectance characteristics. Since the reflectance of aerosols can change significantly with both time and location, and their reflectance properties are difficult to quantify, the effects of aerosols on imagery is the hardest to estimate and hence to correct in an accurate way.

The known absorption spectra of water can be used to form water vapour corrections. If energy is detected in water absorption bands and on the shoulders of these bands, then the relative difference between the reflectance in the water absorption band relative to that on the shoulders can be used to estimate the water vapour density in the atmospheric column. Once the water vapour density in the column has been estimated, then this can be used to estimate the effect of the water vapour on other wavelengths. This strategy cannot be used with dust and other aerosols since the reflectance and transmittance of these varies with their type and possibly their source. There are a number of alternative ways that have been developed to correct image data for the effects of atmospheric aerosols. Some of these techniques will be discussed in the next section.

The transmittance of an absorbing medium is defined as the ratio of flux emitted to that entering the medium over unit distance travelled in the medium. Lambert developed the law governing this when he noted in experiments that equal distances of travel in an absorbing medium caused equal proportions in the loss of flux due to absorption by the medium, that is

$$\frac{\Delta L}{L} = -\beta_{\lambda} \Delta z \tag{2.6}$$

where L is the radiance, β is the volume extinction coefficient and ΔL, Δz are small changes in the radiance and the distance travelled. If we take the small changes to approach infinitely small then we can replace them with $\mathrm{d}L$ and $\mathrm{d}z$, and then integrate the equation over the distance travelled, z, from a starting radiance value of L_s, and ending with a value of L_e:

$$\int_{L_s}^{L_e} \frac{\mathrm{d}L}{L} = -\beta_\lambda \int_0^z \mathrm{d}z \tag{2.7}$$

$$\therefore \ln(L_e) - \ln(L_s) = -\beta_\lambda z$$

And from this, by re-arrangement we get

$$L_e = L_s \, \mathrm{e}^{-\beta_\lambda z} = L_s t_\lambda \tag{2.8}$$

So that it can be seen that the transmissivity, t_λ is

$$t_\lambda = \frac{L_e}{L_s} = \mathrm{e}^{-\beta_\lambda z} \tag{2.9}$$

It is normal practice to take the z values as vertical distances, and so when dealing with radiation from a zenith angle, θ, we need to modify this equation to:

$$t_\lambda = \mathrm{e}^{-\beta_\lambda z/\mu} = \mathrm{e}^{-\tau/\mu} \tag{2.10}$$

where $\mu = \cos(\theta)$ and τ_λ is the optical thickness, so that $\tau_\lambda = -\mu \ln(t_\lambda)$.

The optical thickness is a measure of distance. As the atmosphere becomes more opaque, that is the rate of extinction per unit distance increases, the optical thickness increases and the transmissivity is reduced. If radiation is passing through a medium with transmissivity t_1 and then a second with transmissivity t_2, then the total transmissivity through both is the product of the two, that is $t_{\mathrm{total}} = t_1 \times t_2$. The optical distances due to the various components can be added together, thus:

$$\tau_{\mathrm{total},\lambda} = \tau_{\mathrm{Rayleigh},\lambda} + \tau_{\mathrm{ozone},\lambda} + \tau_{\mathrm{water\ vapour},\lambda} + \tau_{\mathrm{aerosols},\lambda} \tag{2.11}$$

If the spectral reflectance, ρ_λ, at the surface is defined as the ratio of the reflected to incident energy, then the spectral radiance, L_λ, reflected from the surface due to the incident irradiance, $E_{\mathrm{i},\lambda}$ is given by the equation:

$$L_\lambda = \rho_\lambda E_{\mathrm{i},\lambda}/\pi \tag{2.12}$$

The radiance from the surface is further attenuated during transmission to the sensor due to atmospheric transmissivity, and gains an additive component due to atmospheric scattering:

$$L_{\mathrm{O},\lambda} = t'_\lambda L_\lambda + E_{\mathrm{a},\lambda}/\pi \tag{2.13}$$

where t'_λ is the average atmospheric spectral transmittance for ray path from the surface to the sensor, $E_{a,\lambda}$ is the atmospheric radiance incident on the sensor and $L_{\mathrm{O},\lambda}$ is the radiance incident on the sensor.

The energy, $E_{\mathrm{O},\lambda}$ that enters the sensor optics will be modified by the efficiency of the sensor optics, e_λ, in transmitting the energy through the optics to the detector units. The energy that activates the detectors in the sensor is given by the equation:

$$L_{\mathrm{d},\lambda} = e_\lambda L_{\mathrm{O},\lambda} \tag{2.14}$$

Substituting Equation (2.5), Equation (2.12) and Equation (2.13) into Equation (2.14) gives:

$$\begin{aligned} L_{d,\lambda} &= e_\lambda\{t'_\lambda[\rho_\lambda(t_\lambda E_{S,\lambda}\cos(\theta_S) + E_{A,\lambda})/\pi] + E_{a,\lambda}\}/\pi \\ &= e_\lambda\{\rho_\lambda t_\lambda t'_\lambda E_{S,\lambda}\cos(\theta_S) + \rho_\lambda t'_\lambda E_{A,\lambda} + E_{a,\lambda}\}/\pi \end{aligned} \quad (2.15)$$

Equation (2.15) is a simplification of reality since it ignores multiple reflections including reflection from the surface, to the atmosphere and then to the sensor, or from the atmosphere to the surface and then to the sensor. More accurate formulations need to take these effects into account, although their impact is small.

Atmospheric conditions change from place to place as well as from day to day. These changes can be due to point sources, such as a fire or the emission of steam into the atmosphere, in which case the upwind changes in atmospheric conditions can be large over small distances and with the downwind changes being localised, or remaining in the atmosphere for some time depending on atmospheric conditions. Alternatively, the changes can be due to extended sources, such as evaporation of moisture from the ocean, or the injection of dust from large desert areas into the atmosphere. In consequence, atmospheric conditions will change from image to image, and indeed will change across the area covered by an image. Changes in atmospheric conditions can cause changes in atmospheric transmissivity and atmospheric path radiance or skylight. Increasing levels of atmospheric path radiance are normally accompanied by a reduction in transmissivity, but molecular absorption means that reduced transmissivity need not be accompanied by increased atmospheric path radiance.

Lowering transmissivity and raising the atmospheric path radiance cause an increase in the noise level in the data compared to the signal value. This makes it more difficult to extract information from the data. In consequence, it is important to use wavebands that are sufficiently transparent to ensure that the signal to noise ratio is high enough to extract the required information from the acquired data.

In Equation (2.15) there are eight unknowns, of which one is due to the sensor (e_λ), four are due to either atmospheric transmissivity (t'_λ, t_λ) or radiance ($E_{A,\lambda}$, $E_{a,\lambda}$), two are due to the incident solar radiation ($E_{s,\lambda}$, θ_o) and only one is due to the surface, ρ_λ. In practice the equation is often simplified by:

1. Replacing radiance at the top of the atmosphere with incident irradiance at the surface as measured in the field. This is done by substituting Equation (2.8) and Equation (2.9) into Equation (2.10) to give

$$L_{d,\lambda} = e_\lambda\{t'_\lambda \rho_\lambda E_{i,\lambda} + E_{a,\lambda}\}/\pi \quad (2.16)$$

2. Assuming perfectly efficient sensor optics, that is make $e_\lambda = 1.0$

The resulting equation is

$$L_{d,\lambda} = \{t'_\lambda \rho_\lambda E_{i,\lambda} + E_{a,\lambda}\}/\pi \quad (2.17)$$

So that

$$\rho_\lambda = \{\pi L_{d,\lambda} - E_{a,\lambda}\}/\{t'_\lambda E_{i,\lambda}\} \quad (2.18)$$

With field instruments the transmissivity, t'_λ can be taken as 1.0, and the atmospheric component, $E_{a,\lambda}$ as zero because of the short distance between the field instrument and the target. Equation (2.18)

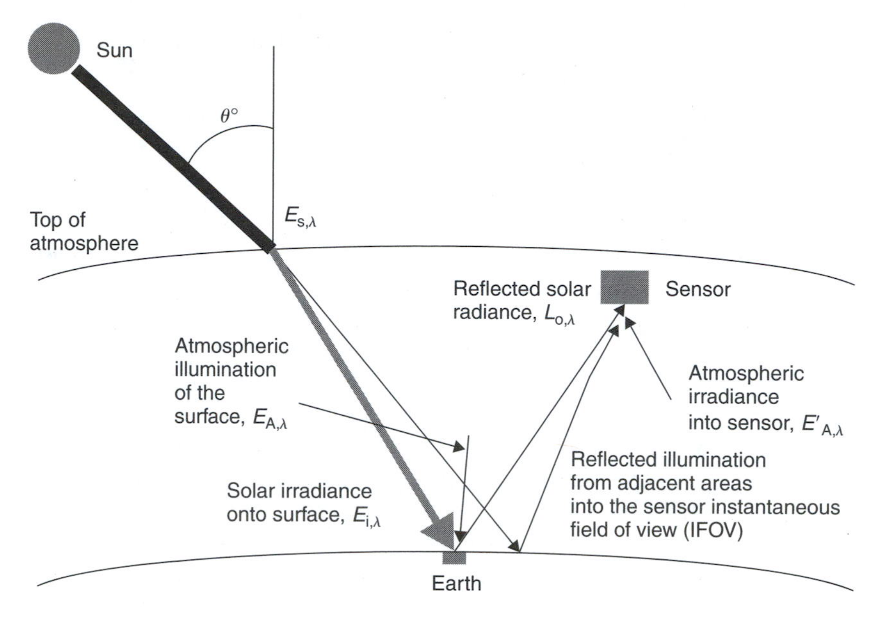

FIGURE 2.9 Transmission of solar irradiance through the atmosphere to the sensor.

can be re-arranged into the form:

$$\rho_\lambda = \pi L_{d,\lambda}/E_{i,\lambda} \tag{2.19}$$

in accordance with the definition of reflectance (Figure 2.9). With field instruments the reflectance from the surface can thus be determined by measuring both the incident irradiance, the reflected radiance and computing the ratio of the two, as long as the field instruments are calibrated so as to allow for e_λ. For airborne and spaceborne sensors the sensor efficiency is normally taken as unity. Equation (2.17) is re-arranged as:

$$\begin{aligned}\rho_\lambda &= \{\pi L_{d,\lambda}/t'_\lambda E_{i,\lambda}\} - E_{a,\lambda}/t'_\lambda E_{i,\lambda} \\ &= c_1 L_{d,\lambda} + c_2 \end{aligned} \tag{2.20}$$

In which the incident radiance, atmospheric path radiance and transmissivity are assumed to be constant during data acquisition and where c_1 and c_2 are constant values. This shows that a linear model is a simplification of the relationship between reflectance, incident irradiance and radiance incident on a sensor that assumes constant path radiance incident on the sensor, constant atmospheric transmission and reflectance between the surface and the sensor and ignores the effects of multiple reflections, across the area to which the linear correction is applied. Such assumptions are often reasonable within a small area, but may not be valid across an image.

2.2.5 Correction of Remotely Sensed Data for Attenuation through the Atmosphere

Equation (2.15) shows that the atmosphere has a very significant impact on the signal incident in a sensor by:

1. Attenuating the incoming radiation
2. Contributing radiance incident on the surface due to the scattered radiation in the sky

3. Attenuating the reflected beam prior to its being incident on the sensor optics, and
4. Contributing direct radiance from scattered energy in the sky, as well as
5. Multiple reflection effects between the sky and the surface, prior to entry into the sensor

$$L_{d,\lambda} = e_\lambda \times \{\rho_\lambda \times t_\lambda \times t'_\lambda \times E_{S,\lambda} \times \cos(\theta_S) + \rho_\lambda \times t'_\lambda \times E_{A,\lambda} + E_{a,\lambda}\}/\pi \qquad (2.15)$$

All four of these processes are affected in some degree by the three sources of attenuation and scattering:

1. *Scattering and absorption by the chemical constituents of the atmosphere, particularly nitrogen, oxygen and ozone*. The amounts of nitrogen and oxygen in the atmosphere tend to be very stable spatially, temporally and vertically, so that their impact on images of the same area will tend to be very similar. The density of these gases decreases exponentially with altitude. Ozone concentrations vary with elevation, the season and latitude. It is concentrated in the 10–30 km elevation range, so that correction of airborne data usually does not need to take this into account. Ozone has significant absorption features between 550–650 nm, and limits Earth observation at wavelengths shorter than 350 nm. Oxygen has a strong absorption feature at about 700 nm. The transmissivity of these gases at sea level is shown in Figure 2.7. The effects of atmospheric chemicals can thus be dealt with by the use of numerical correction models, as long as the density of ozone is known or assumed.

2. *Water vapour*. The amount of water vapour in the atmosphere is very dependent on climatic and meteorological conditions. Climatic conditions define the range of water vapour conditions that can be expected. Thus, the air column above cold dry environments may contain less than 1 g/m^3 of water, whilst moist tropical conditions usually have about 4 g/m^3. The meteorological conditions indicate the specific moisture conditions that can be expected at a site, or in a locality. Water vapour is vertically stratified, so that the greatest impact is at low altitudes. Water vapour conditions can show local spatial and temporal variations and water vapour absorption mainly affects wavelengths longer than 700 nm. Corrections for water vapour can be made using numerical correction models if information on the water vapour in the atmospheric column is known. Such information can come from meteorological observations, by the use of standard atmospheric conditions, or by satellite measurement.

Techniques have been developed to estimate atmospheric water vapour from satellite observations by the use of three wavebands in the near infrared part of the spectrum. Solar radiation between 0.86 and 1.24 mm on the sun–surface–sensor path is subjected to atmospheric water vapour absorption, atmospheric aerosol scattering and surface reflection. Rayleigh scattering and ozone absorption are negligible in this part of the electromagnetic spectrum. The effects of aerosols is assumed to be a linear trend across this spectral region, just as the reflectance of many land surfaces can be approximated by a linear trend with respect to wavelength in this spectral region. In this case, if an absorption band, (λ_o) and two shoulder wavebands, (λ_1) and (λ_2), adjacent to and either side of the absorption band are used, then it can be shown that the transmission at the absorption band can be expressed as Equation (2.21):

$$t_{abs}(\lambda_o) = L(\lambda_o)/(C_1 L(\lambda_1) + C_2 L(\lambda_2)) \qquad (2.21)$$

where:

$$C_1 = (L(\lambda_2) - L(\lambda_o))/(L(\lambda_2) - L(\lambda_1)) \quad \text{and} \quad C_2 = (L(\lambda_o) - L(\lambda_1))/(L(\lambda_2) - L(\lambda_1))$$

Once the double path water transmissivity has been obtained using Equation (2.21), then the columnar water vapour density must be determined. The columnar water vapour density is estimated by inverting a numerical atmospheric transmissivity model that is normally used to estimate atmospheric transmissivity from columnar water vapour data. Once this has been done then the model can be used in the forward way to estimate the transmissivity at each wavelength of interest from the computed water vapour data. Such models are implemented in software in many image-processing systems.

3. *Particulates*. Particulates exhibit a great range of size and reflectance values as well as a great range of densities in the atmosphere. They are lifted into the atmosphere at a location, transported in the atmosphere for periods from minutes to weeks, and then deposited again, often with rain. The period of transportation depends on the density of the particulates and the velocity of the wind. Since the particulates lifted into the atmosphere will usually exhibit a range of densities, the density of aerosols in the atmosphere, and sourced from the one site, will vary with time after lifting. As a consequence, the density of aerosols in the atmosphere can be highly variable. The more common particulates are salt (from the ocean), dust (from deserts and agricultural areas) and smoke. All of these can tend to clump in the atmosphere by the use of atmospheric moisture as a cementing agent, changing their density, shape, size and reflectance. Thus, not only is the density of aerosols highly variable, but also are their reflectance and scattering characteristics.

Whilst a number of methods of correction for aerosols have been developed, two are mostly used at the present time. The first method (*The Dark Background Method*) assumes that the surface has negligible reflectance, and so it is used routinely over the ocean, using the near infrared wavebands to estimate atmospheric attenuation. From Equation (2.15) it can be seen that setting $\rho_\lambda = 0$ yields:

$$L_{\mathrm{d},\lambda} = \frac{e_\lambda E_{\mathrm{a},\lambda}}{\pi} \tag{2.22}$$

from which single path atmospheric path radiance can be estimated assuming an optical efficiency for the sensor optics. Using standard atmospheres, the aerosol load that would give this atmospheric reflectance is estimated, and from this and by making assumptions about the aerosol reflectance, path transmissivity and path radiance can be estimated at other wavelengths. The second method uses simultaneous *multiple path observations* of a target. Observations are taken from a satellite, forward, at nadir and behind the satellite of the same target. The images are corrected for Rayleigh scattering by the atmospheric constituents, including ozone, and for attenuation by water vapour. Since the target may reflect differently in different directions, the images are corrected for these variations in reflectance. Once this has been done, then the only sources of variation between the different image values for the same ground location are variations in the atmospheric path transmissivity and variations in the path length inducing different levels of absorption and scattering for the different paths. The method assumes that the transmissivity is the same, leaving the effects of the different path lengths as the sole source of variations between the signals. Since the path lengths are known, the observations can be used to estimate atmospheric transmissivity at the wavebands at which measurements were taken. Once the atmospheric transmissivity has been computed, then atmospheric irradiance is estimated using standard atmospheric models, and thus the imagery can be corrected for the effects of aerosols.

There are two broad approaches that can be used for correcting image data for atmospheric effects. These approaches are based on using field data, and the use of atmospheric data with numerical models.

2.2.5.1 Atmospheric Correction Using Field Data

Equation (2.20) shows that a linear model is a first approximation of the relationship among reflectance, incident radiation and radiance incident on the sensor, at a point. The assumptions made in the derivation of this model show that it does not account for multiple reflections between the surface and the atmosphere; it may not be accurate over the area of an image due to variations in atmospheric conditions, nor for images with significant differences in elevation. However, it is a simple way to achieve an approximate correction of image data for atmospheric effects, particularly for airborne data where some of these assumptions are more likely to be correct. To derive a linear model, at least two surfaces are required in close proximity to each other; one to be highly reflective and the other to have low reflectance. Measurement of the reflectances of each on the ground, and measurement of their response values in the image data are then used to determine the parameters of the linear model. Since the assumptions of this model are weak over significant distances, it is a good idea to repeat the process at a number of locations in the image. If significant differences are found between the linear parameters, then either the values should be interpolated across the area or an alternative approach should be used.

The ground reflectances used to calibrate the image data may come from a number of sources. A number of surfaces may be selected and their reflectance measured using a field spectrometer at the same time that the image is acquired. If this approach is adopted, then the area sampled with the spectrometer has to be large enough to ensure that at least one pixel is located within the sample area. To ensure this, the area sampled needs to be of size:

$$\text{Dimension of sample area} = (p + (k \times \sigma)) \tag{2.23}$$

Where p is the pixel size in metres and σ is the standard deviation of the image rectification in metres. The constant value of k in Equation (2.23) is to ensure, with some confidence, that the sample area contains at least one pixel. If an image can be perfectly rectified, so that $\sigma = 0$, then the sample area can be the size of a pixel. However, this can never happen in practice where the rectified position of a pixel is only known to some level of accuracy, which is usually stated in terms of the standard deviation of the fitting or σ. The more confidence that one wants to have that the sample area contains a pixel, the larger the area needs to be. If these residual errors in the rectification are assumed to be normally distributed, as is usually accepted, then it can be shown that $\pm\sigma$, or $k = 2$ in Equation (2.23) gives an area such that the analyst can have a 68% confidence that the sample contains at least one pixel. It is usual to set $k = 4$ so that the analyst will have a 95% confidence that a pixel will fall within the sample. Clearly, the more uniform the surface, and the more stable it is in reflectance, the better. Another source of reflectance data could be an image that has previously been calibrated. If this approach is adopted, then care needs to be taken to ensure that the surfaces used for calibration will not have changed significantly in reflectance between the dates of acquisition of the calibrated and the second images. Most surfaces also exhibit some variations in reflectance with changes in the sunsurface — sensor geometry, so that care also needs to be taken to ensure that differences in look direction between the images does not introduce differences in reflectance. These assumptions cannot usually be met with vegetated surfaces; with other surfaces it can be difficult to verify. Thus, the surfaces may have been moist at one date but not at the second date, affecting the reflectances in all wavebands, but by different amounts. With airborne data, and high-resolution satellite image data, standardised surfaces may be constructed of known reflectance characteristics and taken out into the field.

The approach is relatively simple and easy to understand but it is not that easy to implement in practice, particularly for satellite image data when large field sites need to be used. It is for this reason that most work focuses on using numeric atmospheric models.

2.2.5.2 Atmospheric Correction Using Numerical Atmospheric Models

A number of models have been developed to correct image data for atmospheric effects. These models are implemented in a two-stage process. In the first stage, data are used to derive the atmospheric conditions that applied at the time of acquisition so as to provide input parameters to the models. Once this has been achieved, the model is inverted and applied to the image data to correct that data to reflectance at the surface.

2.2.5.2.1 Stage 1 — Derivation of the Model Parameters
Most models use a version of Equation (2.17)

$$\rho_\lambda = (\pi \times L_{d,\lambda}/t'_\lambda \times E_{i,\lambda}) - (E_{a,\lambda}/t'_\lambda \times E_{i,\lambda}) \tag{2.17}$$

In this equation, the image data response values are converted to values of $L_{d,\lambda}$ by the use of sensor calibration tables. These calibration tables are usually supplied by the sensor owner, and are also usually updated throughout the life of the sensor. There are four ways that can be used to find the other three unknowns, t'_λ, $E_{i,\lambda}$ and $E_{a,\lambda}$. It should not be forgotten that the optical depths due to the various sources are summed to get the final atmospheric path transmissivity, as shown in Equation (2.11). In the same way the atmospheric irradiance due to the individual sources can also be summed to give the total atmospheric irradiance.

1. *By the use of standard atmosphere tables.* Most models supply a set of standard atmosphere tables, and the table most appropriate to the area is chosen.
2. *The use of field data.* Field observations of the reflectance of a number of fields can be taken and this data can be used with the satellite data for the same fields to derive the model parameters as discussed earlier.
3. *By incorporation of atmospheric data*, as may be collected by radiosonde. This data is then used to parameterise the model, rather than using a standard atmospheric set of conditions. Atmospheric data can also be used to provide better estimates of the effects of particulates.
4. *Using satellite data.* Whilst the first three techniques are usually satisfactory for the correction of the atmospheric attenuation by the atmospheric gases, they are often not satisfactory for correction of the variable causes of attenuation, notably water vapour and aerosols. Satellite data can be used to estimate ozone concentrations, columnar water vapour and aerosol load for use in the model.

2.2.5.2.2 Stage 2 — Correction of the Image Data
Once the model parameters are known, then the model can be inverted and used to correct the image data, each pixel in turn.

Most models, in implementing this approach, also discriminate between satellite and airborne data, and between flat and mountainous terrain. Satellite sensors generally have a small field of view, so that the sun–surface–sensor geometry can be assumed to be constant across the area of the image. This is not the case with some satellite sensors, and neither is it the case with airborne data. In these cases the changes in the geometry have to be taken into account in the implementation of the model. Variations in altitude have a significant impact on the path transmissivity and path radiance since many of the factors that affect the transmission of radiation through the atmosphere have a greater impact near the Earth's surface. If the image exhibits significant differences in elevation, then this needs to be taken into account in the parameterisation of the model and the correction of the data.

2.2.6 Measurement of Radiance and Irradiance

Radiance and irradiance are measured using spectrometers. Versions of these instruments are also called spectroradiometers or spectrophotometers; the latter usually being restricted either to that part of the spectrum that is visible to the human eye or to reflected solar radiation. Some of these instruments contain an internal calibration source. A schematic figure of a spectrometer is shown in Figure 2.10.

A spectrometer consists of four main components:

1. The collecting optics and aperture
2. Filter or spectrometer unit

3. Detector or detectors
4. The recording device or display

2.2.6.1 Collecting Optics

The collecting optics defines a cone shaped field of view (FOV). For all spectrometers the solid angle of this cone is constant at the instrument, so that the area sensed on the target depends on the distance between, and the orientation of, the sensor relative to the target surface (Figure 2.10).

The collecting optics collects the energy within the instrument's FOV, directs it through the aperture that controls the amount of energy entering the spectrometer per unit time and onto the filter or spectrometer unit. The energy emerging from the filter unit is then directed at the detectors in the sensor. The energy incident on the detectors is recorded or sampled at specific times, the area actually sensed at that instant is called the instrument's ground coverage or footprint.

Imaging spectrometers need to focus the energy entering the instrument through its FOV to create individual picture element values or pixel values. This is done by means of the scanner unit. Non-imaging spectrometers do not need a scanner unit. The aperture controls the amount of energy entering the spectrometer unit per unit time, much as they do in a camera. Imaging devices will be discussed in Section 2.4, whilst in this section we will concentrate on field instrumentation.

2.2.6.2 Filter Unit

This unit filters the required wavebands for projection onto the detector units. It can do this in a number of ways:

1. Split the incoming beam into separate wavebands using a prism.
2. Split the incoming beam into a number of sub-beams that are individually filtered using either gelatine or metal oxide filters.
3. Filter the required wavebands from the incident beam using a diffraction grating.

Prism. The prism glass refracts light entering a wedge prism, where the degree of refraction changes with the type of glass used and the wavelength of the radiation (Figure 2.11). In this way the incoming beam into the prism is split into sub-beams. The sub-beams are then detected to provide data at finer spectral resolution. Prisms cannot, however, provide very fine spectral resolution in the derived data since the physical size of the detectors limits the resolution that can be detected. Thus, for very fine resolution field spectrometers, diffraction gratings are used.

Gelatine filters. A gelatine filter allows a specific set of wavebands to be transmitted through the filter, with different types of filters allowing different wavebands to be transmitted. Gelatine filters are relatively cheap, but they cannot provide narrow transmission spectra and hence cannot be used

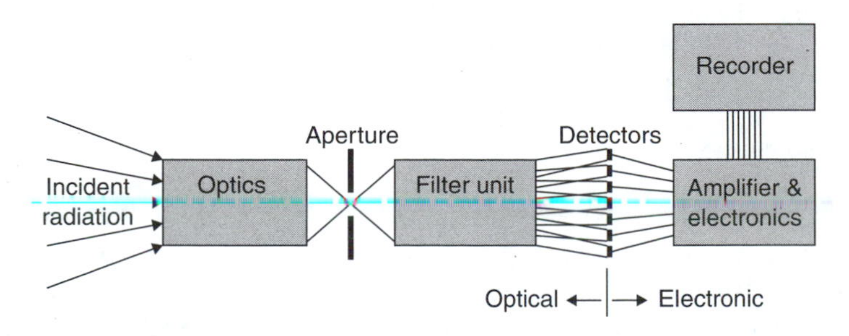

FIGURE 2.10 Schematic diagram of the construction of a spectrometer, consisting of an optical system, aperture, filter unit, detectors, amplifier and recorder.

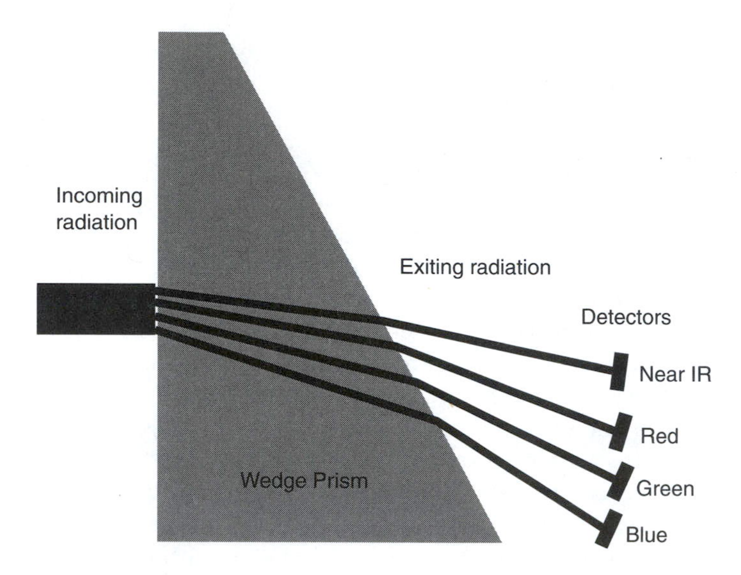

FIGURE 2.11 The wedge prism filter unit.

when the user wants to detect narrow wavebands. It is also often difficult to achieve a sharp cut-off in transmission on the edges of the sensed band, as shown in Figure 2.12.

Metal oxide filters. These filters can be used to provide either much narrower or sharper transmission curves than gelatine filters. They operate by combining the absorption characteristics of various metallic oxides to create the desired transmission spectra. The selected metallic oxides are sprayed as a fine coating onto a glass base and welded to the base using heat. Metal oxide filters are very stable.

Gelatine or metal oxide filters have the advantage that they are relatively cheap. However, since the incident beam has to be split before filtering, there is less energy available to create each waveband. The amount of energy that is available is often a significant issue in sensor design.

Diffraction grating. A diffraction grating is a fine graticule of parallel lines etched onto glass. The glass is chosen so that it will not significantly interfere with the radiation being sensed. The graticule allows light of a specific wavelength to be transmitted through the grating, where this wavelength depends on the grating interval normal to the incident radiation. Rotation of the grating changes the grating interval in this normal plane and hence the wavelength that can be transmitted through the graticule. The graticule interval and range of rotations controls the range of wavelengths that can be detected by means of this type of spectrometer. A grating can be used in field instruments where there is time to change the orientation of the grating between measurements, but they are not usually suitable for sensors on moving platforms such as an aircraft or satellite.

2.2.6.3 Detectors

The detectors contain light-sensitive chemicals that respond to incident electromagnetic radiation by converting that radiation into an electrical charge. The detector electrical charge is taken, or sampled, by electronic circuitry at a regular interval, and reset to zero after sampling. The charge accumulates due to the incident radiation between each sampling of the detector. Each sample charge becomes a detector reading that is then amplified and transmitted to the storage device. The magnitude of this charge is a function of the amount of incident radiation accumulated between samples, the sensitivity

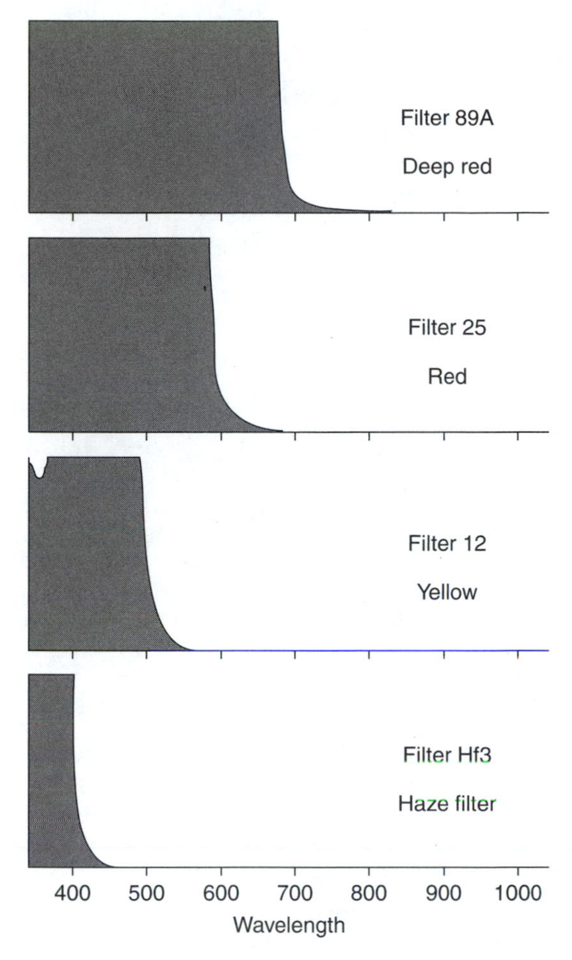

FIGURE 2.12 The transmission spectrum of selected gelatine filters, Haze filter 3,12 — Yellow, 25 — Red and 89A — Deep red (from Kodak. With permission).

of the detector and the detector response speed. The latter of these three factors causes a lag between change in incident radiation and detector response. This lag will influence the values that are recorded from the detector.

The response of a detector is not linearly related to the incident radiation, so that it is usually necessary to calibrate the electrical signal before use. In addition, as detectors age, or change their temperature, their response might change. It is therefore essential to calibrate instruments at regular intervals to ensure that the results being obtained are correct. Figure 2.13 shows the typical response of a charged couple device (CCD) detector and how it varies with wavelength. Different types of detectors have different response characteristics to radiation and as a consequence are used to acquire data in different parts of the electromagnetic spectrum.

2.2.6.4 Output Device

The simplest output device is a meter attached to the output of the detectors, to be read by the operator. In many systems this output is sampled electronically, and recorded automatically onto a suitable recording device, such as a cassette tape, a disc unit or optical disc. This approach has significant benefits as it allows the more efficient collection of field data. Creation of a digital record also means that the raw data can be easily calibrated, allowing faster and more convenient analysis.

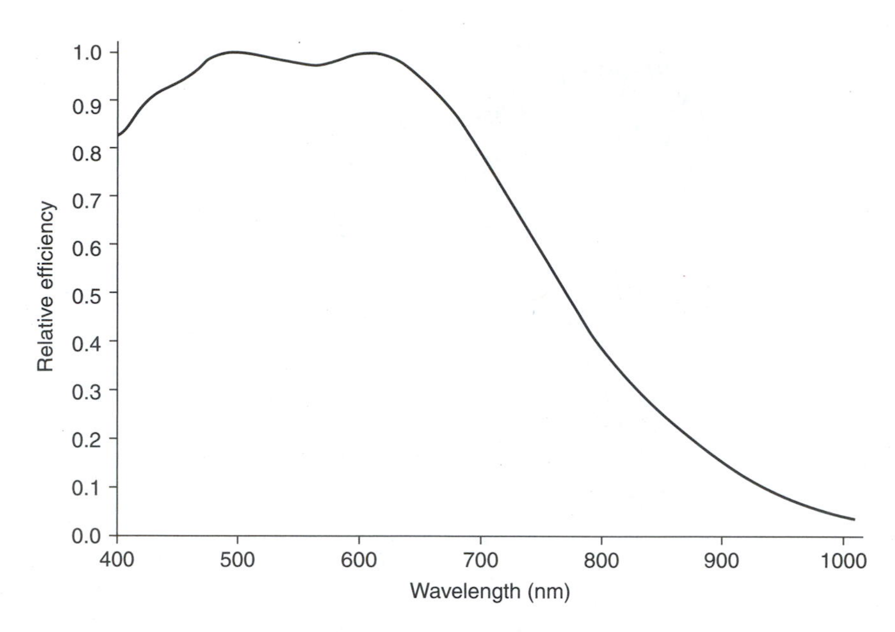

FIGURE 2.13 The relative quantum efficiency of modern (CCD) as used in spectrometers and cameras.

Spectrometers can be grouped into three main categories:

1. *Simple fixed waveband spectrometers.* These record the radiance or irradiance in set wavebands that usually match the wavebands of a particular scanning instrument using filter or prism units as the basis of separating the incident radiation into the required wavebands. This type of instrument is usually cheaper, they can be more robust and are usually simpler to use than the other types of spectrometers. The disadvantage of these instruments is that the results obtained are only used for those specific wavebands that have been sensed.

2. *High spectral resolution field instruments.* These spectrometers provides high-resolution data over an extensive range of wavebands, using very precise and delicate optical and filter systems that are usually based on diffraction gratings. Whilst they are usually more expensive and may be more difficult to use, they have the advantage that the acquired data can be integrated over any broader bands of interest to the analyst. They can thus provide data that can match existing or proposed wavebands in either current or potential sensing systems. They provide a more powerful data set than that collected using the first type of spectrometer.

3. *Very high-resolution laboratory instruments.* These spectrometers are very precise laboratory instruments that measure the reflectance and transmittance of elements of the surface of interest, often for areas of 1 cm^2 or smaller. These instruments provide the analyst with information on the reflectance, absorptance and transmittance of the surface type. The biggest disadvantages of these instruments is that they provide information on the components of the surface, and not on the reflectance of the surface itself and they interfere with the surfaces being measured.

The use of spectrometers for measuring the reflectance of surfaces in the field is discussed in Chapter 6 under the collection of field data.

2.3 INTERACTION OF RADIATION WITH MATTER

2.3.1 THE NATURE OF REFLECTANCE

Solar energy is attenuated and scattered by the atmosphere and its aerosols prior to being intercepted by the ground or by the vegetative canopy. When energy contacts the surface of an object, it is absorbed, transmitted or reflected in accordance with the Law of the Conservation of Energy.

$$E_{i,\lambda} = E_{a,\lambda} + E_{r,\lambda} + E_{t,\lambda}$$
$$= E_{i,\lambda} \times (a_\lambda + \rho_\lambda + t_\lambda) \quad (2.24)$$

therefore

$$a_\lambda + \rho_\lambda + t_\lambda = 1 \quad (2.25)$$

where E represents the incident, absorbed, reflected and transmitted spectral energy and a_λ, ρ_λ and t_λ are the spectral absorptance, reflectance and transmittance of the surface.

The transmitted energy is then available for further absorption, reflection or transmission within the body. Some of this energy is eventually reflected out of the body to contribute to the reflected component from the surface of the body. The total amount of energy that is reflected from a body is an integration of the myriad interactions that occur at the surface of and within the body and within the sensor FOV. This energy is further attenuated and scattered by the atmosphere before entering the sensor optics.

Of these various factors that influence the final calibrated radiance values in the sensor, the impact of the body on radiance is of most interest to users, because it is the component that contains information on the features of the body within the depth of the body that is penetrated by the electromagnetic radiation. This depth depends on the wavelength of the radiation and the nature of the body. In the visible part of the spectrum, metals essentially reflect energy just from the surface of the body, whereas organic materials reflect energy from depths of up to millimetres, whilst with water, penetration can be up to metres, depending on the wavelength, as will be discussed further in this section. With radar energy, greater penetration can be achieved, depending primarily on the wavelength and moisture content of the medium, as will be discussed later in this chapter. The depth penetrated by the radiation is called the boundary layer of the body.

The other components are of concern to users because they interfere with our capacity to extract information from the data. Thus from (2.15):

$$L_{d,\lambda} = [e_\lambda t'_\lambda t_\lambda \rho_\lambda E_{s,\lambda} \cos(\theta_S) + e_\lambda t'_\lambda \rho_\lambda E_{A,\lambda} + e_\lambda E_{a,\lambda}]/\pi \quad (2.16)$$

The parameter of most interest is ρ_λ, the surface reflectance.

2.3.1.1 Reflectance within the Boundary Layer

Radiation incident on a body is initially reflected, absorbed or transmitted at the surface. The reflected energy is either an approximation of specular or Lambertian reflectance. Specular reflectance is defined as reflectance where all the energy that is incident on the surface is reflected at the same angle of exitance from the surface as incidence, that is $I = R$ in Figure 2.14. Such surfaces obey Snell's Law. Lambertian reflectance is defined as reflectance that scatters or reflects the energy such that the radiance is equal in all directions, independent of the angle of incidence of the incident energy (Figure 2.14). The nature of the reflection at the surface depends upon:

1. *The angle of incidence* of the energy.
2. *Surface roughness* as a function of wavelength. The smoother the surface the more specular or mirror like the reflectance is likely to be. The reflectance of smooth surfaces obeys Snell's Law. The rougher the surface the more scattered in direction the reflection is likely to be. This reflectance is more Lambertian in form.

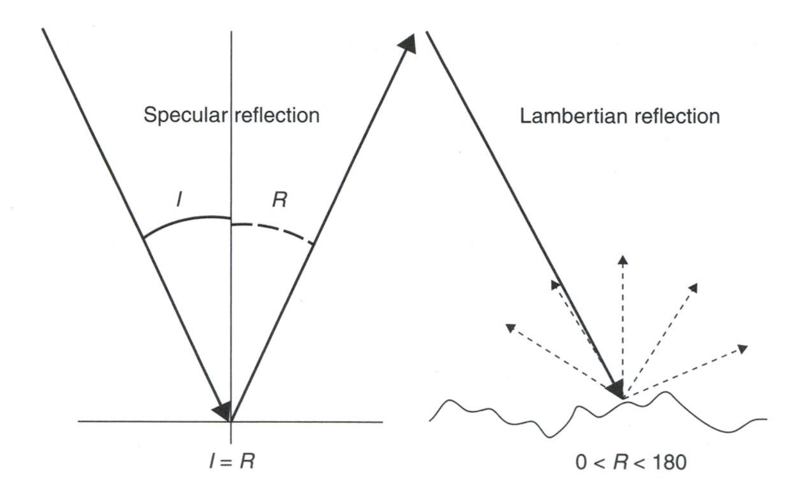

FIGURE 2.14 Specular reflection in accordance with Snell's Law and Lambertian reflectance.

3. *The materials on either side of the surface*, as they will affect the angle of refraction and the percentages that are reflected, absorbed and transmitted at the interface.
4. *The wavelength of the radiation.*

With specular surfaces, the closer the incident ray is to the normal from the surface, or the smaller the angle of incidence, the lower the level of reflectance that is likely to occur. As the angle of incidence increases, the reflectance is likely to increase, particularly when the ray becomes a grazing ray, or the angle of incidence approaches 90°. It is for this reason that water is more reflectant at large angles of incidence. The reflectance of specular surfaces is highly dependent on the geometric relationship among the source, usually the sun, the surface elements and the sensor. As surfaces become more Lambertian their reflectance becomes less dependent on this geometric relationship. Since, few surfaces are perfectly Lambertian reflectors, few have reflectance characteristics that are completely independent of this geometric relationship. It will be seen that the closer the surfaces are to the Lambertian ideal, the better they are from the remote sensing perspective. Whatever the nature of the surfaces, their reflectance characteristics are an important consideration in remote sensing.

Energy transmitted into the boundary layer is then available for further absorption, reflection or transmission. During transmission through the boundary layer the energy is selectively absorbed by the chemical constituents of the layer, and is reflected or scattered by the surfaces of elements within that layer, such as the surfaces of soil grains and the cell structure in leaves. Thus, for example, energy transmitted into a green leaf will undergo selective absorption by the pigments in the leaf, such as the chlorophyll pigments and be reflected from the cell walls within the leaf.

Selective absorption means that the energy eventually emerging from the surface will be deficient in those wavelengths that have been selectively absorbed. The reflection from the surfaces of elements within the boundary layer will tend to scatter the energy in all directions and so the reflected energy emerging from the surface will tend to be Lambertian in form. The closeness of this reflectance to the Lambertian ideal depends on the nature of the reflections within the boundary layer. If these reflections are from elements that are oriented in all directions in an apparently random manner, then the resulting reflectance will be a close approximation of a Lambertian surface, as is the case with reflectance from many soil surfaces. If the internal elements have strong correlation with certain orientations then the resulting reflectance may not be a good fit to the Lambertian ideal. Reflection that primarily occurs within the boundary layer is called body reflection, in contrast to surface reflection that occurs at the surface of objects.

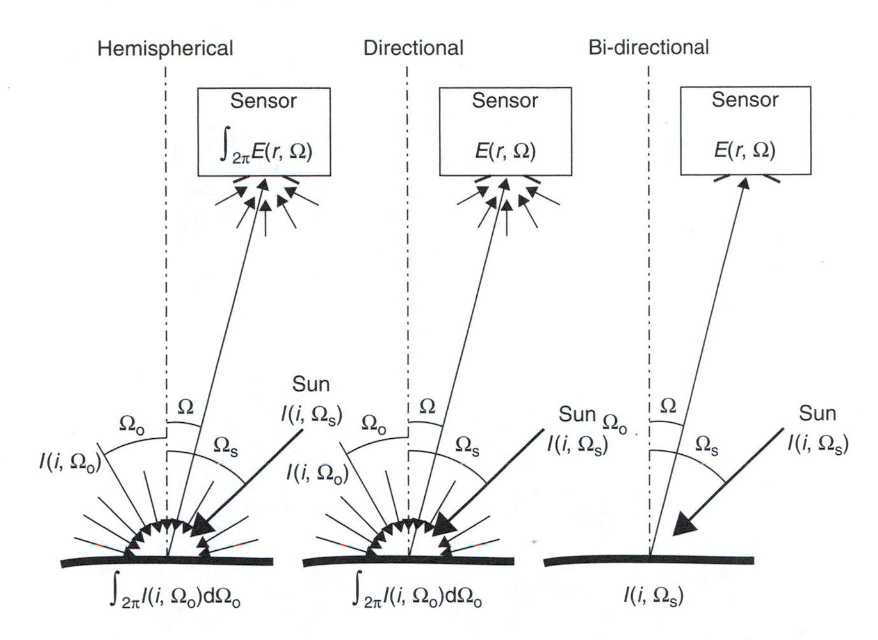

FIGURE 2.15 Hemispherical, directional and bi-directional reflectance.

The characteristics of near-Lambertian reflectance, and selective absorption, are critical to the detection of objects in remote sensing. Remote sensing assumes the existence of these characteristics in the collection and analysis of remotely sensed data. Consider how remote sensing would operate if the majority of objects exhibited specular reflectance. Images of these surfaces would be predominantly black, when reflection is away from the sensor or white when reflection is towards the sensor, exhibiting very little gradation between these limits. The elements of the surface that were reflecting towards the sensor would change dramatically as the sensor moved, so that each image would appear quite different, and indeed sequential images, including sequential aerial photographs would be quite different to each other. If there were no body reflectance then reflection would have more contrast than is the case, leading to images of maximum or minimum contrast for all elements within the image area. The result would be images that could not be interpreted very easily, if at all, so that specular reflectance would mean that remote sensing would probably be of little use in resource management.

Both the reflected and incident radiation at a surface can be measured in three different ways (Figure 2.15):

1. *Bi-directional reflectance.* If an observer is looking at a surface that is illuminated by a torchlight then the source of energy comes from one direction, and the surface is being observed at another direction. The ratio of these two energies is called the bi-directional reflectance factor of the surface, being dependent on the two directions involved. Illumination of surfaces by the sun, involving predominantly illumination from the one direction, and observation at a sensor is a close approximation to bi-directional reflectance. It is an approximation because the definition of bi-directional reflectance assumes a negligible FOV for both the incident and reflected radiation, and because skylight contributes to illumination of the target. For non-Lambertian surfaces, the bi-directional reflectance may change with changes in the sun–surface–sensor geometry. In consequence the reflectance of some surfaces may change across an image and from image to image. Most forms of active sensors, or sensors that have their own energy source to illuminate the target, use a point source and sense in one direction so that they represent perfect bi-directional reflectance. The dependence of the

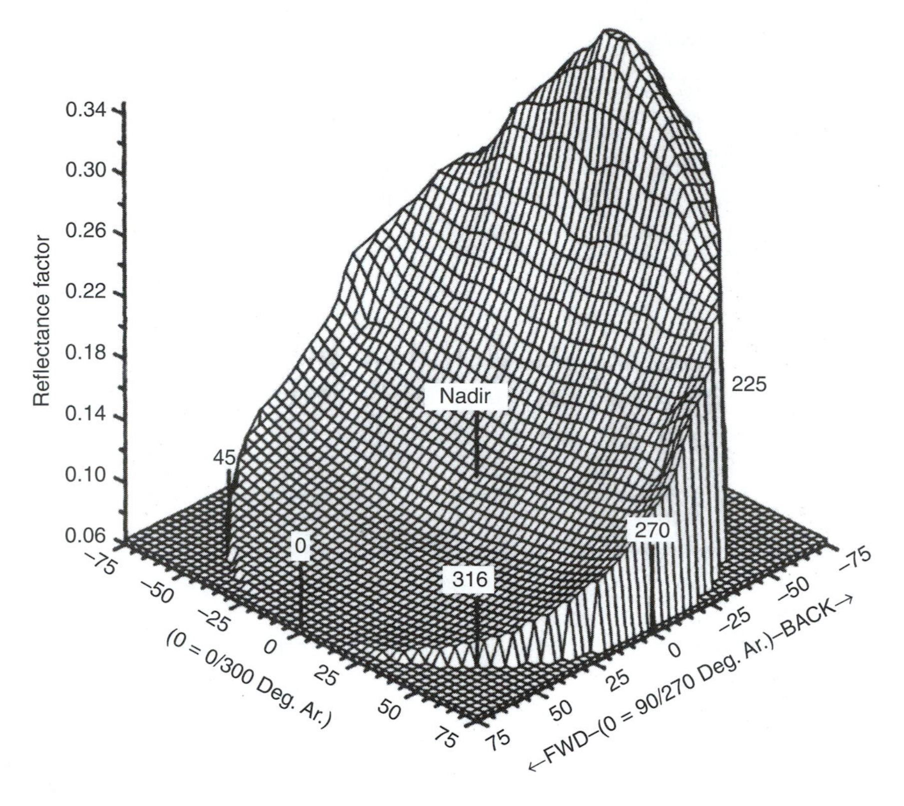

FIGURE 2.16 The bi-directional reflectance function (BRDF) of a recently plughed bare field at 25° solar elevation for 662 nm wavelength (from Asrar, G. (Ed.), 1989. *Theory and Applications of Optical Remote Sensing*, John Wiley & Sons, New York. With permission).

bi-directional reflectance on the incident and reflected angles is shown in the formula:

$$\text{Bi-directional reflectance factor} = \frac{L(\Omega)}{\pi \times I(\Omega_0)}$$

where L is the radiance of the reflected energy in direction Ω and I is the irradiance of the incident energy from direction Ω_0. The bi-directional reflectance factor will only be the same for all combinations of azimuth and elevation for both the incident and reflected radiation, if the surface is a perfectly Lambertian reflector. This has given rise to the concept of bi-directional reflectance distribution function (BRDF). The BRDF is a map of the bi-directional reflectance factor for all combinations of altitude and azimuth of both the incident and reflected radiation. Such a map is difficult to portray graphically. As a consequence, only the portion of the BRDF is often portrayed, usually the bi-directional reflectance factor with variations in the reflected or sensed angle, at constant incidence angle and at one waveband. The BRDFs of three surfaces are shown in Figures 2.16 to 2.18.

The BRDF shows how a surface varies from the Lambertian ideal, how the surface varies from other surfaces, and indeed how the surface can change over time. With vegetation, the BRDF can also vary during the day and over longer time periods, since the canopy structure does change over time, and can change during the day. Analysis of variations in BRDF over time, both within a cover type, and between cover types, can indicate strategies for the acquisition of data so as to maximise the differences in reflectance between selected surface types.

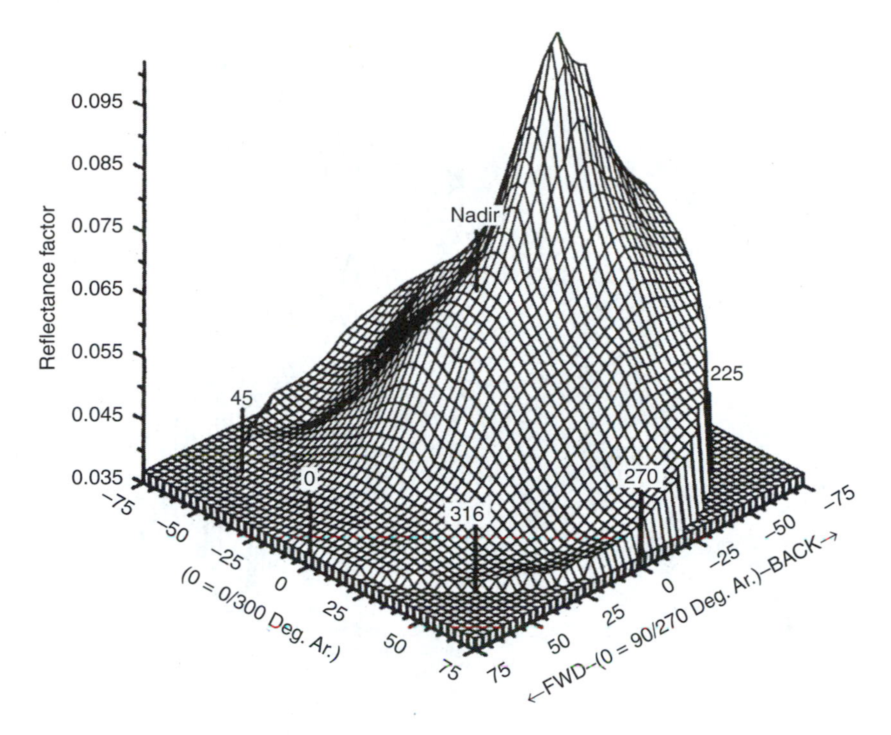

FIGURE 2.17 The bi-directional reflectance function of a sand shinnery oak rangelands community at 31° solar zenith angle at = 662 nm (from Asrar, G. (Ed.), 1989. *Theory and Applications of Optical Remote Sensing*, John Wiley & Sons, New York. With permission).

2. *Directional reflectance*. If the observer views the surface under overcast conditions when the main source of illumination is scattered skylight, then the illumination is approximately equal from all directions, but the sensor is in one direction relative to the surface. Reflectance of this form is called directional reflectance. The directional reflectance is the ratio of the single directional reflected energy to the total irradiance incident on the surface giving the hemispherical directional reflectance factor as:

$$\text{Hemispherical directional reflectance factor} = \frac{L(\Omega)}{\pi \times \int_{2\pi} I(\Omega_o)\,\mathrm{d}\Omega_o}$$

where L and I are as defined above and the irradiance is integrated across the hemisphere.

3. *Hemispherical reflectance*. Hemispherical reflectance occurs when both the incident and reflected energy is measured over the whole hemisphere. Hemispherical reflectance is of little concern in relation to sensors, but is of great interest in the modelling of light interactions within the surface and in assessing the impact of reflectance on aerial photography and other imagery that is taken with sensors that have a wide FOV. Hemispherical reflectance should be identical to bi-directional reflectance for Lambertian surfaces, and provides an average reflectance value for specular surfaces. Comparison of hemispherical and bi-directional reflectance values can thus indicate how close surfaces are to the Lambertian model. For canopy modelling and inversion, hemispherical reflectance is defined as the ratio of the total (hemispherical) reflected energy from a location to the total (hemispherical) incident energy. This bi-hemispherical reflectance factor is given by:

$$\text{Bi-hemispherical reflectance factor} = \frac{\int_{2\pi} L(r,\Omega)\,\mathrm{d}\Omega}{\int_{2\pi} I(i,\Omega_o)\,\mathrm{d}\Omega_o}$$

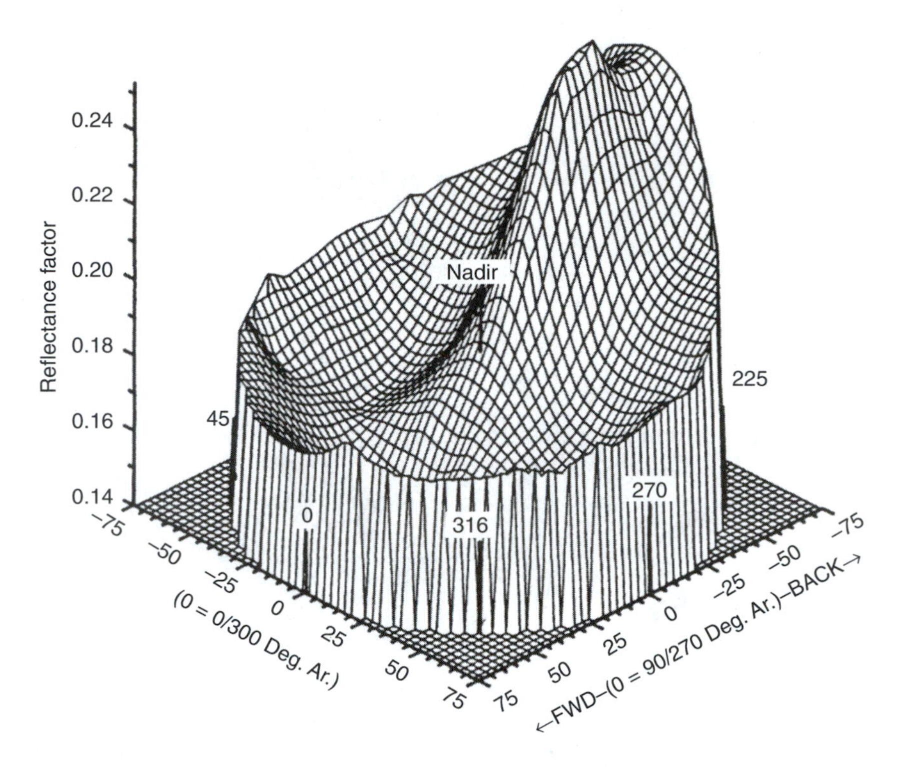

FIGURE 2.18 The bi-directional reflectance function of a sand shinnery oak rangelands community at 31° solar zenith angle at = 826 nm (from Asrar, G. (Ed.), 1989. *Theory and Applications of Optical Remote Sensing*, John Wiley & Sons, New York. With permission).

Field spectrometers usually measure an approximation to bi-directional reflectance, or directional reflectance under cloudy conditions. Most laboratory spectrometers, usually large and very precise measuring instruments, require the samples to be set in a small sample container such that they are uniformly distributed in the container and then they measure the hemispherical reflectance, and transmittance of the sample.

2.3.2 The Reflectance of Water Surfaces

Energy incident on a water surface is primarily reflected from or transmitted into the surface, where the proportions of each depend upon the angle of incidence between the incident energy and the surface. The surface reflected energy is specular in character, usually as grazing rays from wave elements. Most of the energy is transmitted into the water body where it is absorbed by the chemical constituents of the water body, scattered by particulate matter in the water or reflected from the floor of the water body. The absorptance of water increases with increasing wavelength from a minimum absorptance at about 420 nm (Figure 2.19) until it is greater than 90% in the reflected infrared portion of the spectrum. Most scattering and reflection from water is therefore in the blue–green part of the spectrum, particularly for relatively pure water. Both chlorophyll and suspended soil sediments significantly affect the reflectance of water, with both of them absorbing energy in accordance with their individual absorptance characteristics. The effect of chlorophyll on absorptance is shown in Figure 2.20. The effects of sediments depend on the reflectance characteristics of the sediment particles (see Section 2.3.3), the density of the particles in the water and the depth of the particulate layer. The effects of some types of sediment are shown in Figure 2.20. Bottom reflectance affects the reflectance of the water body when sufficient energy is transmitted through the body to be reflected

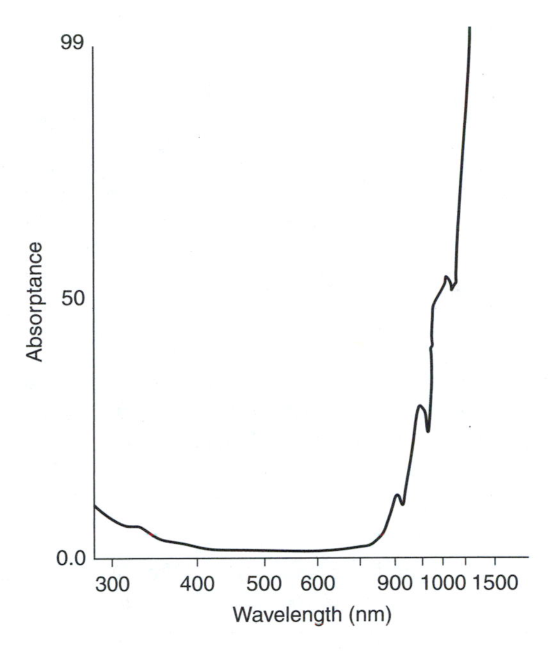

FIGURE 2.19 The absorption characteristics of pure water.

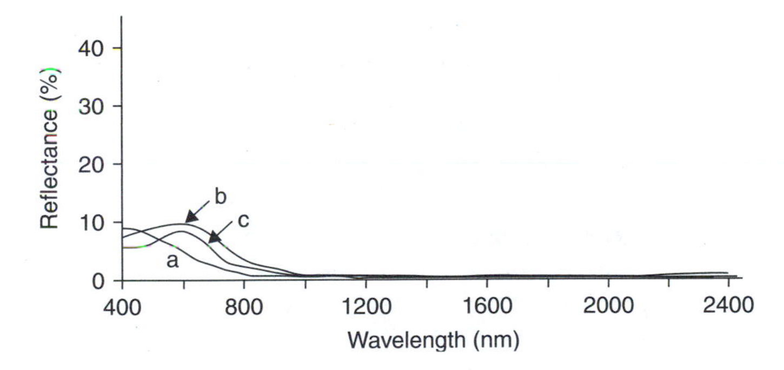

FIGURE 2.20 The reflectance of pure water (a), and the effects of sediments (b) and chlorophyll (c) on the reflectance of water.

from the floor and returned to the surface. The bottom reflectance affects the amount of energy reflected and hence available for transmission through the water body.

When waves break they create foam which will also affect the surface reflectance of the water, with the nature of this effect depending on the resolution of the image data relative to the wave size and area.

Patterns in water bodies are thus primarily due to variations in suspended sediment, chlorophyll concentrations and variations in depth to the floor and bottom reflectance, whilst texture may be introduced due to surface wave shape and foaming characteristics. Because most of the factors that affect the body reflectance can have similar effects on reflectance, it is often necessary to use auxiliary data, such as field visits and bathymetric data, to identify the actual cause in each situation.

2.3.3 THE REFLECTANCE CHARACTERISTICS OF SOILS

Soil reflectance occurs in the surface layers of the soil, with a significant proportion occurring as surface reflectance from the soil grains themselves. The factors that affect the reflectance of

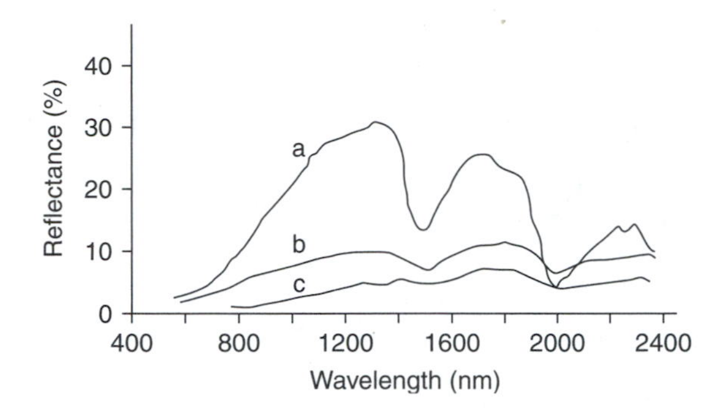

FIGURE 2.21 Representative spectra for the reflectance of soils with (a) minimally altered (fibric), (b) partially altered (ferric) and (c) fully decomposed (sapric) organic matter (from Stoner, E.R. and Baumgardner, M.F., 1981. *Soil Science of America Journal*, 45, 1161–1165. With permission).

soils are:

1. *Chemical composition*. The chemical composition of the soil affects soil colour due to selective absorption. Different chemicals have different absorption bands and in consequence the existence of different chemicals in the soil will affect the selective absorption of energy during transmission through the soil. The most obvious effect of this type is the high red reflectance of soils rich in ferric iron.

2. *Humus content*. Humus, the organic products derived from the breakdown or decomposition of vegetables and animal matters, usually exhibits absorption at all wavelengths in the visible region, but with slightly higher absorption in the blue–green part of the spectrum due to the antho-cyanin compounds created by the breakdown of chlorophyll in leaf elements. The level of humus in soils varies over time so that the reflectance of soils can also vary over time.

The decompositional state of plant litter significantly affects the reflectance curves of organic soils. Fully decomposed vegetative material gives a concave curve as illustrated in curve (c) of Figure 2.21. Soil with a significant component of preserved fibres has a curve of type (b) in the figure, or an organically affected reflectance curve. Soils with minimally decomposed litter exhibit a concave curve up to 750 nm and then quite high reflectance beyond 900 nm as shown for type (a) soils in Figure 2.21.

3. *Surface soil moisture*. Water selectively absorbs at all wavelengths, but with increasing absorptance at longer wavelengths as previously discussed. Since, most of the reflectance of soils occurs in the top layers of soil particles, only the water in these top layers will affect the reflectance of the soil surface. If these layers are moist then the water will have a strong influence on reflectance. However, this layer dries out quite rapidly, hence normally the influence of surface water on soil reflectance is quite minimal. The influence of three levels of moisture content on the reflectance of soil type is shown in Figure 2.22.

4. *Surface roughness*. The combination of surface roughness and solar elevation affects the amount of shadowing that occurs in the soil and in consequence the soil reflectance. Surface roughness can be due to the soil grain size for soils and sands that do not form clods, or the soil texture due to the clod size for other soils. For example, claypans or flat surfaces with a very finely textured surface are more highly reflective than rougher sandy surfaces even though sand itself is often more reflective than the particles of clay in the claypan.

Surface roughness can also be due to cultivation or other human activities on the soil surface that break up the surface and either increase or decrease its texture. Cultivation increases the surface clod size immediately after cultivation, thereby increasing the texture and reducing the reflectance due to increased shadowing by the clods. The marked reduction in the reflectance of cultivated soils is primarily due to this cause. As the cultivation ages so the clods are smoothed, the texture decreases

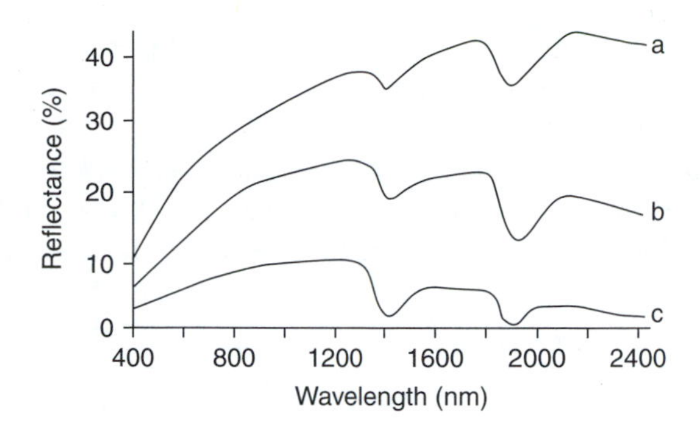

FIGURE 2.22 Changes in reflectance of a typical soil with changes in moisture content.

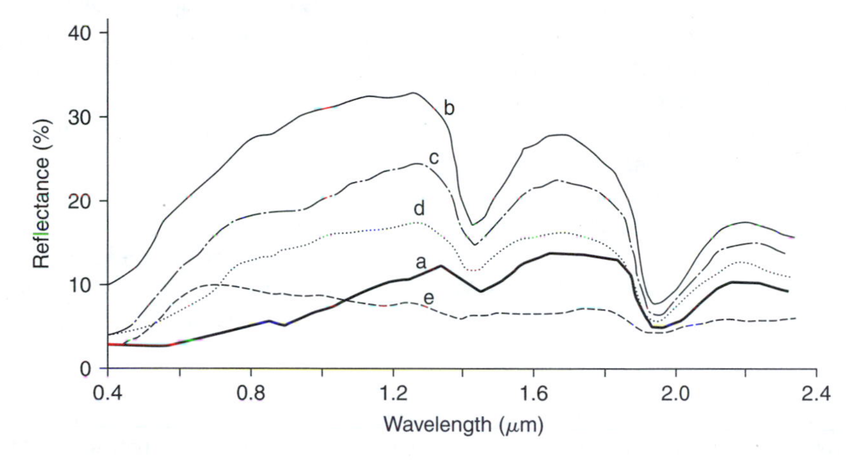

FIGURE 2.23 Reflectance spectra of surface samples of five mineral soils; (a) organic dominated (high organic content, moderately fine texture), (b) minimally altered (low organic and iron content), (c) iron altered (low organic and medium iron content), (d) organic affected (high organic content and moderately coarse texture) and (e) iron dominated (high iron content with fine texture) (from Stoner, E.R. and Baumgardner, M.F., 1981. *Soil Science of America Journal*, 45, 1161–1165. With permission).

and the reflectance increases. Images of areas under cultivation will thus exhibit variations in soil colour that are a function of the period since cultivation.

Work by Condit (1970) and Stoner and Baumgardner (1981) indicates that the reflectance curves for soils of the United States can be grouped into five distinctive shapes as shown in Figure 2.23. These classes have been termed:

Type 1 Organic dominated (a). Has a low overall reflectance with characteristic concave shape from about 500 to 1300 nm. Water absorption bands are noticeable at 1450 and 1950 nm, the broadness of the bands indicating that the absorption is due to both chemically bound and free water.

Type 2 Minimally altered (b). High reflectance at all wavelengths with characteristic convex shape in the range 500 to 1300 nm. The water absorption bands are usually noticeable at 1450 and 1950 nm unless there is negligible bound water as occurs with sandy soils.

Type 3 Iron affected (c). Type 3 soils have lower reflectance than type 2 soils, with slight absorption evident at the ferric iron absorption bands of 450 and 900 nm. The water absorption bands are usually present.

Type 4 Organic affected (d). Higher overall reflectance than Type 1 soils, maintains the concave shape in the 500 to 700 nm region and with a convex shape at longer wavelengths. Usually exhibits the water absorption bands at 1450 and 1950 nm.

Type 5 Iron dominated (e). Has a unique spectral curve shape with reflectance peaking in the red at about 720 nm, and then decreasing due to iron absorption. The iron absorption dominates at the longer wavelengths so that the water absorption bands are often obliterated.

2.3.4 The Reflectance of Vegetation

A major aspect of remote sensing is to better understand the relationship between reflectance from a plant canopy and the physical attributes of that canopy so as to extract more accurate and reliable information on the physical status of the canopy. The leaves of the plant are the main components used in photosynthesis. Although, the structure of leaves varies with species and environmental conditions, most leaves contain mesophyl cells enclosed by their upper and lower epidermis. The mesophyl is composed of cells that contain the chloroplasts for capturing the carbon dioxide (CO_2) absorbed from the atmosphere and, airspaces that can contain moisture. The epidermis protects the leaves and contains the stomata or openings that allow moisture, CO_2 and oxygen (O_2) to enter the leaf. Leaves are categorised into two major classes:

- *Dicot leaves* contain the main central vein and many branching veins. The mesophyl is usually composed of long cylindrical upper palisade cells and below these are the spongy mesophyl cells. The spongy mesophyl also usually contains much intercellular airspace.
- *Monocot leaves* generally have parallel veins and the mesophyl usually contains only the spongy mesophyl cells.

There are three methods by which most plants convert the sunlight, moisture, atmospheric gas and nutrients from the soil, into plant tissue, known as the C3, C4 and CAM pathways. Most higher order plants use the C3 pathway. In this process both the photosynthesis that converts the gas and energy into intermediate chemicals and the subsequent "dark reactions" that create the plant materials occur in the same chloroplast cells. The C3 plants represent most temperate species such as spinach, wheat, potato, tobacco, sugar beet, soya and sunflower. They contain one type of chloroplast with photosynthesis saturating at about 20% of full sunlight. They are not water use efficient. The C4 process is conducted in two stages, in different parts of the leaf. The initial photosynthesis traps the CO_2 in the form of a 4-carbon atom acid, such as oxaloacetic or malic acid, which is then transported to the other site for subsequent conversion into plant material in those plants "dark reactions." The C4 plants are represented by most tropical and sub-tropical species, such as sugar cane, maize and sorghum. They are highly productive and are characterised by high water use efficiencies. They contain two types of chloroplasts, and do not readily photo-saturate at high radiation levels. The CAM plants are typically found in desert to semi-desert conditions and include the cacti, orchids and agaves. They typically absorb radiation and carbon dioxide during the day, but transform them into complex carbohydrates during the night. Leaf cross-sections of C3 and C4 plants are shown in Figure 2.24. The figure shows that C4 plants typically have a more compact leaf structure than C3 plants, with fewer inter cell airspaces (Figure 2.25 and Figure 2.26).

Light entering the leaf is subjected to multiple reflections at the cell walls within the leaf and absorption by the pigments within the chloroplasts in the palisade cells, by the leaf water and by the chemical constituents of the leaf. The light is scattered in many directions due to the multiple reflections and scattering. The resulting reflectance will be approximately Lambertian, if all angles are represented in equal proportions by the cell walls in the leaf. However, if some angles dominate the orientations of the cell walls, then the reflected light is not likely to be Lambertian. Further, smooth or waxy leaves can have a strong specular component to their reflectance, particularly if the incident radiation is a grazing ray to the leaf surface. Finally, the absorption that occurs in a

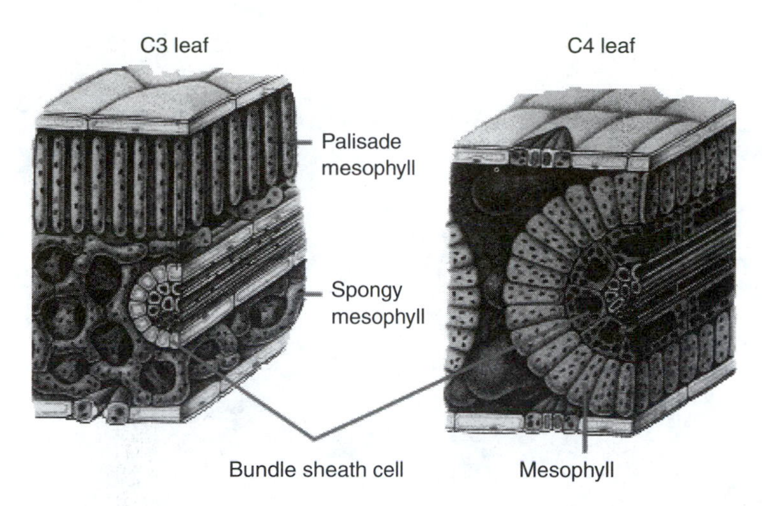

FIGURE 2.24 Cross-sections through typical C3 and C4 leaves, showing the arrangement of the upper and lower epidermis, containing the stomata, and the mesophyll consisting of palisade and spongy mesophyll cells, airspaces, and the veins.

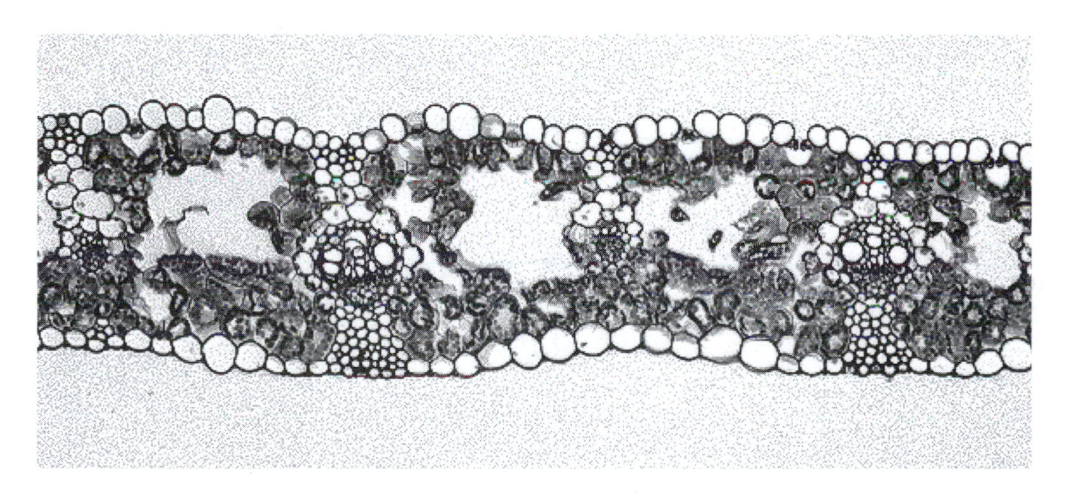

FIGURE 2.25 Cross-section of a C3 grass leaf.

leaf means that the light reflected out of the leaf is deficient in those wavelengths that have been selectively absorbed in the leaf.

Most of the absorption in a leaf occurs in the chloroplasts of the palisade cells. The chloroplasts contain chlorophyll and carotenoid pigments. The two types of chlorophyll, a and b, are distributed in a relative ratio of 3:1 in most plants. Their relatively low absorption in the mid-visible or green part of the spectrum results in the green colour of vegetation. The carotenoids are accessory photosynthetic pigments that support photosynthesis with yellow or orange pigments that are highly absorbent in the blue part of the spectrum. In autumn the yellow and red carotenoid pigments survive longer than the chlorophylls, giving the trees the characteristic yellow to red in autumn.

The absorption spectra of the principal plant pigments and water are shown in Figure 2.27. In this figure the pigments are shown to have high, but different levels of absorption in the blue part of the spectrum (375–495 nm), and the chlorophyll A and B pigments also absorb in the red part of the spectrum (600–700 nm). Leaf water exhibits increasing absorption with increasing wavelength, with peaks coinciding with the moisture absorption bands at about 990, 1200, 1550, 2000 nm, 2.8 and 3.5 μm.

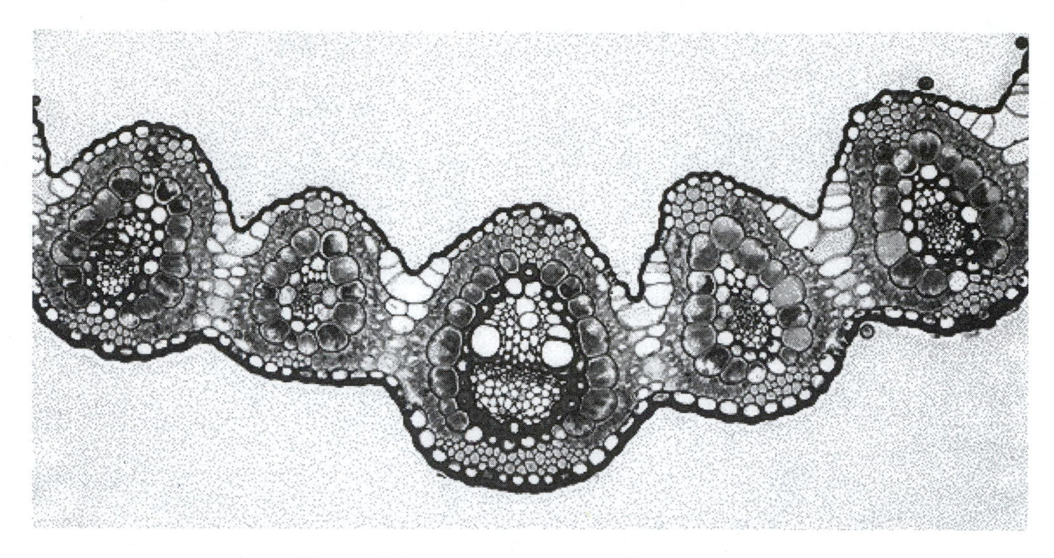

FIGURE 2.26 Cross-section of a C4 grass leaf.

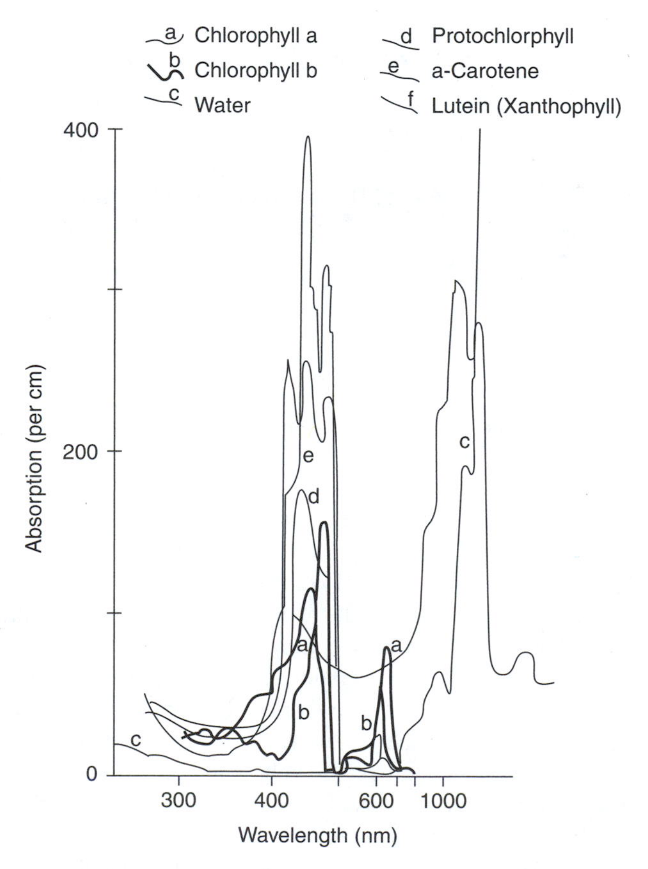

FIGURE 2.27 Specific absorption versus frequency and wavelength of the principal plant pigments and liquid water (from Gates, D.M., 1965. *Applied Optics*, 4, 11–20. With permission).

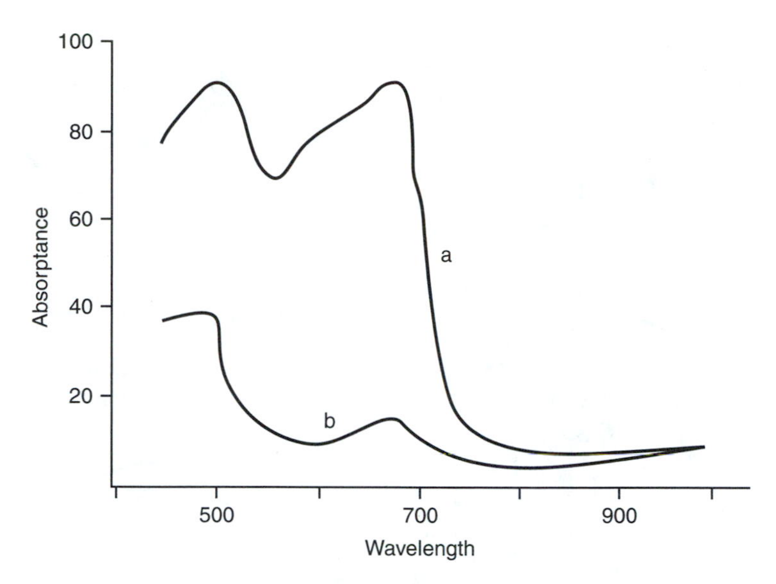

FIGURE 2.28 The spectral absorption of green (a) and albina (b) leaves of a *Hedera helix* plant (from Gates, D.M., 1965. *Applied Optics*, 4, 11–20. With permission).

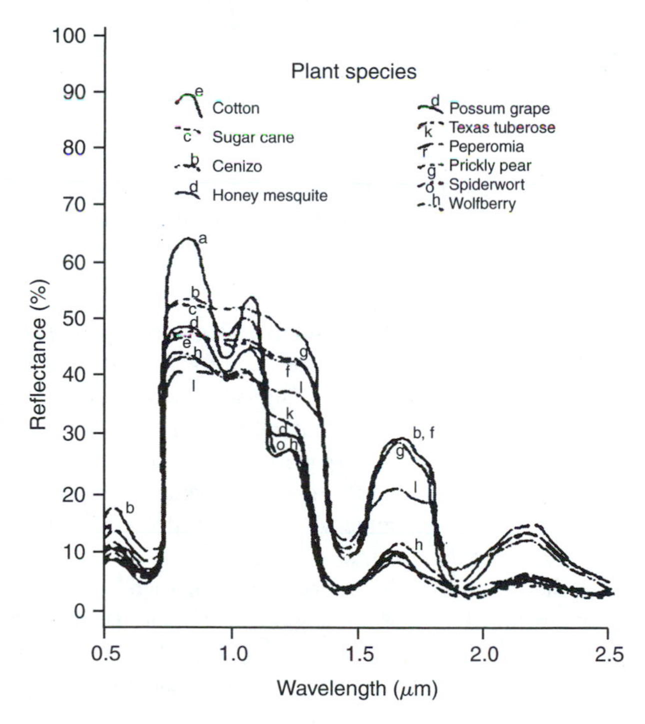

FIGURE 2.29 Laboratory spectrophotometric reflectances measures over the waveband 500 nm 2500nm for the leaves of dorsiventral — oleander and six succulent and four nonsucculent species (Gausman et al. 1978a).

The importance of selective absorption in green leaves is indicated in Figure 2.28 and Figure 2.29. In Figure 2.28 the spectral absorption of green and albino *Hedera helix* leaves from the same plant show significantly greater absorption for the green leaf due to the absorption characteristics of the leaf pigments that are absent from the albino leaves.

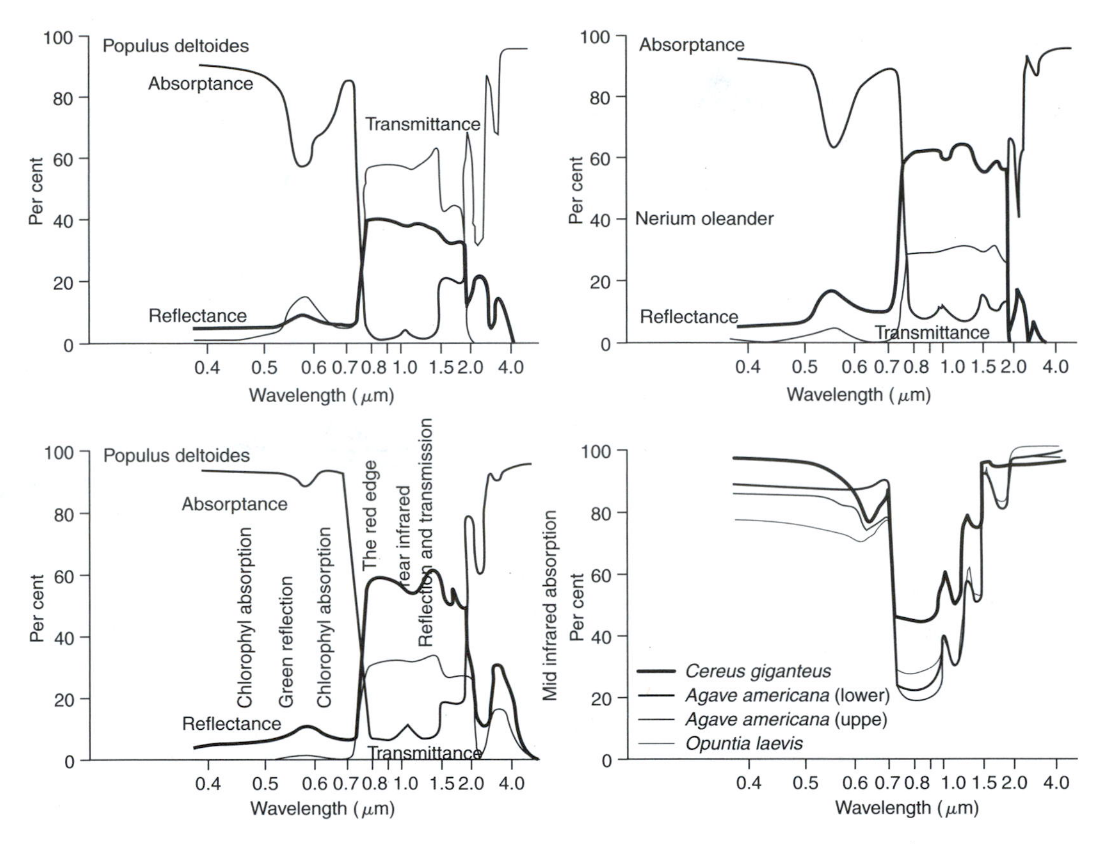

FIGURE 2.30 The reflectance, transmittance and absorptnce of *Populus deltoides* leaves, being moderately thin and light coloured, *Nerium oleander* a thick green leaf, *Raphilolepis ovata* a thick dark green lead and four selected desert succulent plants (from Gates, D.M., 1965. *Applied Optics*, 4, 11–20. With permission).

Figure 2.29 shows the reflectance, transmittance and absorptance curves in the range 400–2500 nm for three different types of green leaves, and four desert succulent plants. The figure shows that green vegetation is highly absorbent in the visible part of the spectrum due to the plant pigments, but with varying amounts of absorption of green light (500–580 nm). In all cases there is negligible transmission in the visible part of the spectrum. At about 700 nm, the "red edge," changes the curve in a most dramatic way. At this wavelength the absorption due to the plant pigments ceases and absorption by the plant becomes negligible. All species reflect and transmit highly in the near infrared (NIR) (760–1300 nm), but the extreme thickness of succulent plant tissue means that the transmitted component of the energy is either reflected or absorbed by these plants. Beyond about 1200 nm the absorption characteristics of water begin to dominate. Beyond about 2000 nm leaves become efficient absorbers and emitters, a characteristic that allows them to also act as efficient radiators of energy at these wavelengths so as to expel accumulated thermal energy.

2.3.5 The Reflectance Characteristics of Green Leaves

The reflectance spectra of the leaves of a variety of vegetation species and conditions are shown in Figure 2.30 and Figure 2.31. Figure 2.30 shows the reflectance, absorptance and transmittance of three species. All three species, and in fact, all vegetation species, exhibit the dominant features of absorption in the visible part of the spectrum, the red edge, negligible absorption in the NIR and good absorption and reflection in the thermal infrared. They absorb energy efficiently in the chlorophyll absorption bands where they use the energy for photosynthesis. They absorb poorly in

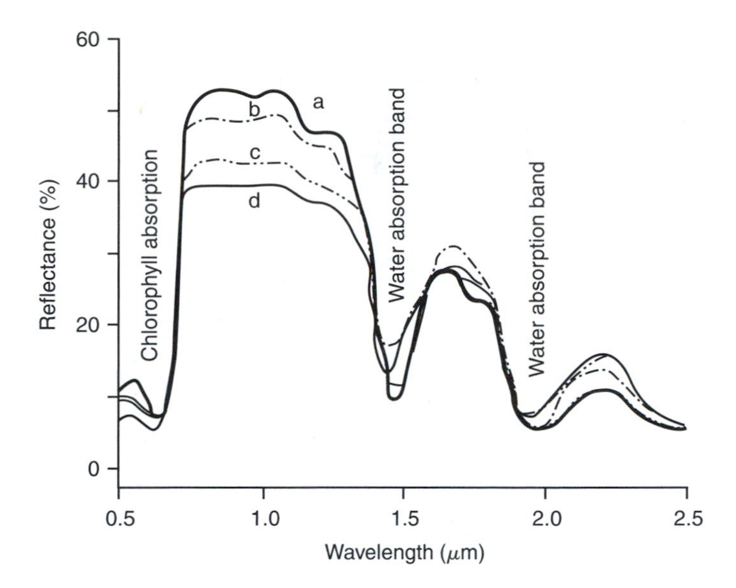

FIGURE 2.31 Reflectance spectra of four plant genera differing in plant leaf structural or mesophyl arrangement, being dorsiventra — (a) oleander and (b) hyacinth and isolateral — (c) eucalyptus and (d) compact (from Gaussman et al, 1970. *Technical Monograph* no. 4, Texas A&M University, College Station, Texas. With permission).

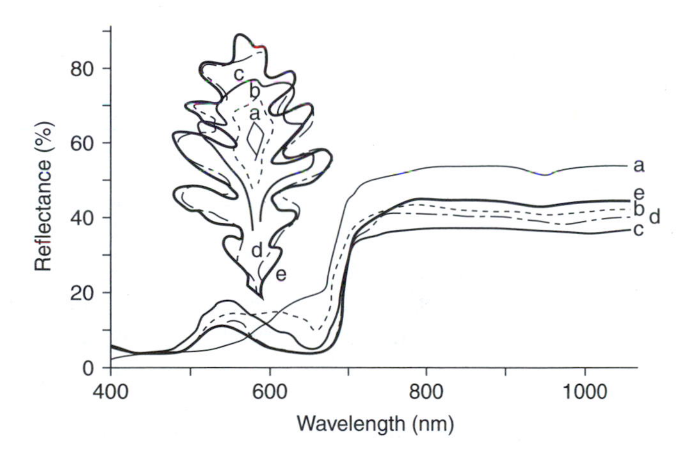

FIGURE 2.32 Evolution of the spectral reflectance of a Quercus alba leaf during the growing season for the dates shown in the figure (from Gates, D.M., 1965. *Applied Optics*, 4, 11–20. With permission).

the NIR where the energy is of no use to their photosynthetic activity and would only accumulate as heat and they absorb and radiate efficiently in the mid and far infrared so as to efficiently expel excess heat. The figure shows that there are subtle differences between the species, due to variations in the mix of chlorophylls, the plant leaf structure and the water content in the leaves. These differences are often larger in the NIR than in the visible part of the spectrum as shown in Figure 2.30, because the high absorption at visible wavelengths tends to mask the differences due to the variations in the mix of chlorophylls, and because the high level of reflectance and transmittance in the NIR leads to multiple reflections in the leaf, emphasising the differences between the species.

The evolution of the spectral reflectance of plant leaves during the life of a leaf is illustrated for *Quercus alba* in Figure 2.32. Initially, the young leaves exhibited the absorption characteristics of proto-chlorophyll; as the leaves develop chlorophyll a and b there is an increasing absorption

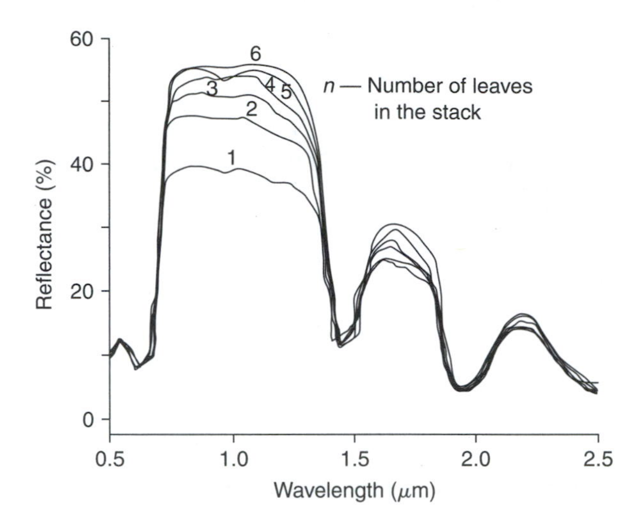

FIGURE 2.33 The reflectance of corn leaves for each increase in stack thickness by one leaf, up to a thickness of six leaves.

in the red part of the spectrum. The increasing concentration of chlorophyll causes a deepening of the red absorption band and a narrowing of the green reflectance band. In early maturity the leaf is lighter with higher green reflectance. The leaf darkens with age, this ageing coinciding with increasing intercellular air spaces and a corresponding increase in NIR reflectance due to lower water absorption. The leaf reflectance changes with leaf maturity as a function of chlorophyll density in the leaves and changes in leaf structure, with the mean "red edge" value shifting from about 700 to 725 nm.

The transmissive and reflective characteristics of green leaves in the visible and NIR part of the spectrum are reinforced when considering the reflectance from a stack of leaves as illustrated in Figure 2.33. Increasing the stack thickness by adding one leaf at a time to the stack up to a thickness of about six leaves, causes the reflectance to increase in the NIR but shows negligible change in the visible part of the spectrum. This phenomenon arises because of the high transmittance of green leaves in the NIR (Figure 2.30), leading to multiple reflections that contribute to the overall reflectance out of and further transmission into the leaf stack. This situation is in contrast to the visible region where the high absorptance means that there are negligible transmission through the leaf and hence negligible multiple reflections within the canopy. Changes in crop canopy reflectance as crop leaf area and soil cover increase, are shown in Figure 2.34. The extent of these changes are greater in the NIR than they are in the visible part of the spectrum, as would be expected from the low transmission in the visible component compared to that in the NIR.

The reflectance characteristics of the upper and lower surfaces of a leaf can vary. Figure 2.35 shows the reflectance characteristics of the upper and lower surfaces of leaves. It shows that for this, and many species, the differences in reflectance between the upper and lower surfaces are largest in the visible part of the spectrum and becomes negligible in the NIR part of the spectrum.

It is not surprising that the reflectance of leaves varies both over time and between species, as well as being non-Lambertian. It has been found empirically that in general the angular distribution of reflectance and transmittance from a leaf can be approximated by the relationships:

$$\begin{aligned} \rho &= a_0\theta_s^2 + b_0\theta_s + c_0 \\ \tau &= a_1\theta_s^2 + b_1\theta_s + c_1 \end{aligned} \tag{2.26}$$

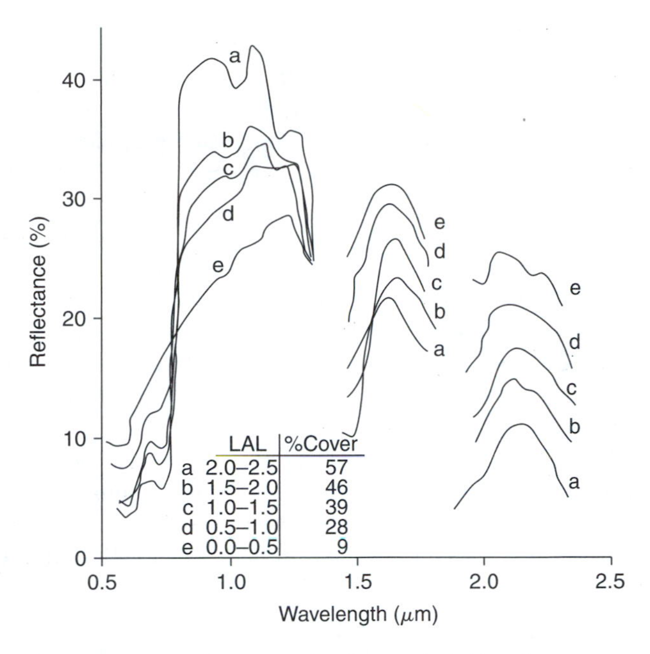

FIGURE 2.34 Change in crop canopy reflectance with changes in LAI and percent cover (from Bauer, M.E., Daughtry, C.S.T. and Vanderbelt, V.C., 1981. *AgRISTARS Technical Report SR-P7-04187*, Purdue University, West Lafayette, IN. With permission).

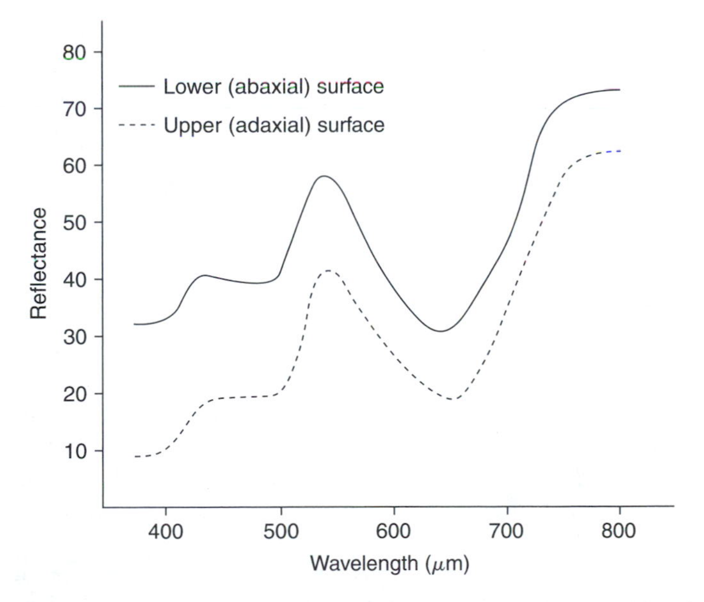

FIGURE 2.35 The reflectance characteristics of the upper and lower surfaces of wheat leaves (from Schutt et al, 1984. *International Journal of Remote Sensing*, 5, 95–102. With permission).

In which the constants $a_0, b_0, c_0, a_1, b_1, c_1$ are dependent on species and wavelength, and θ_s is the zenith angle in radians of the incident radiation.

The reflectance of green leaves can thus be summarized:

1. *The reflectance, absorptance and transmittance of green leaves is wavelength dependent.* It exhibits relatively broad high-absorption bands in the blue and red part of the spectrum,

with a narrow area of somewhat lower absorption in the green part of the spectrum. There is then a very abrupt change at the red edge to low absorption and high reflectance and transmittance in the near IR, subject to moisture absorption bands at specific parts of the spectrum. The absorption of radiation by chlorophylls in a green leaf is the dominant factor defining leaf reflectance, absorptance and transmittance, however, the level of dominance depends on the influence of these other factors.

2. *The reflectance, absorptance and transmittance of green leaves are age and species dependent.* This arises because of differences in leaf construction and chemical composition between species, and because of changes in these with the age of the leaf.
3. *The reflectance and transmittance of leaves are not Lambertian.* This arises primarily because of differences in the surface coating of leaves of different species and the orientation of the cell walls within the leaf are not random but have a somewhat regular pattern, that also varies with species.
4. *The reflectance of the lower surfaces of leaves can be different to that of the upper surface.* This effect is usually wavelength dependent, being more obvious in the visible part of the spectrum when the reflectance from the lower surfaces is usually higher than from the upper surfaces. The differences are usually negligible in the NIR part of the spectrum.

The reflectance of the green leaves will also change with moisture stress. As leaves initially dehydrate, they close their stomata so as to reduce moisture loss to the atmosphere, causing a temperature imbalance to occur in the leaves, leading to a rise in temperature in the leaves. If the moisture deficit continues, then the leaves loose turgidity and start to wilt, affecting the canopy geometry. This change in canopy geometry changes the canopy reflectance. Since, the leaves will droop more than in their healthy status, this usually causes a reduction in reflectance in the NIR due to less leaf reflectance and more shadowing or more soil being exposed to the incident radiation. It may cause an increase in the visible part of the spectrum, depending on the changes in shadowing and soil exposure that occur. Individual leaves initially suffer a loss of moisture, and this will affect leaf reflectance, primarily by reducing absorption by moisture in the NIR. Later the plant pigments start to decompose, at which time the reflectance of the leaves changes dramatically from that of a green leaf towards that of a dead leaf.

2.3.6 The Reflectance Characteristics of Dead Leaves

There is negligible transmission through dead leaves and vegetation at all wavelengths as shown in Figure 2.36, in contrast to the situation with green leaves as discussed in the previous section. The reflectance spectra of dead leaves are similar for all species, with significant absorption in the blue–green waveband, and higher reflectance than green vegetation at longer wavelengths. The high reflectance of the brown leaf elements is most likely due to both the hardening of the cell walls, making them more reflective, and the reduction in cell water, reducing absorption in the moisture absorption bands.

As brown vegetation decomposes the absorption levels increase, particularly at shorter wavelengths, primarily due to the absorption by antho-cyanin compounds that are formed as the breakdown chemicals from decomposition of the chlorophyll plant pigments. In consequence dead vegetation tends to be very reflective soon after it has browned off, and to become darker with age.

2.3.7 Vegetative Canopy Reflectance

A vegetative canopy consists of the canopy elements, their arrangement in space and the soil or water background. The canopy elements will usually include green and brown leaf elements and may include stem, bark, flower and seedpod elements. All of these canopy elements, and the background, reflect, absorb and transmit in accordance with the discussion in the previous sections.

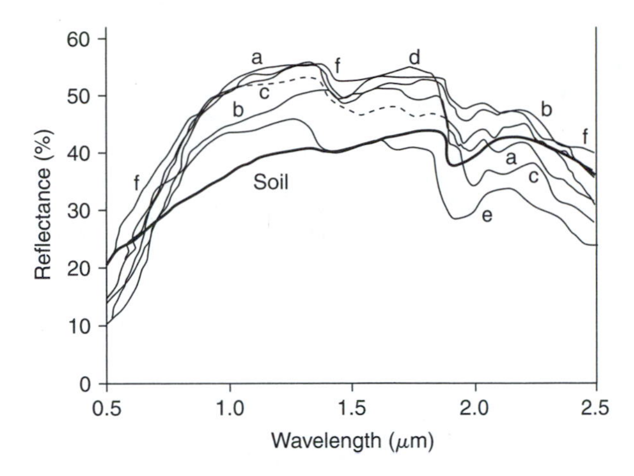

FIGURE 2.36 Laboratory spectrophotometrically measured reflecatnces of dead leaves and bare soils for six crop species, (a) avocado, (b) sugarcane, (c) citrus, (d) corn, (e) cotton and (f) sorghum (from Gausman et al., 1976. *Journal of Rio Grande Valley Horticultural Society* 30, 103–107. With permission).

The energy incident on the canopy typically comes as direct solar energy, as well as the diffuse skylight component. These interact with the canopy elements, and the background. Some of this energy is reflected, out from the canopy, but some is reflected further into the canopy. The energy that is reflected into the canopy is then available for further interactions with the canopy and background elements. Some of the energy incident on the canopy elements is transmitted through the vegetative components. It is then available for further interceptions with elements of the canopy or the background. The discussion in Section 2.3.5 showed that changes in the reflected energy continue to occur up to about six such interceptions in the NIR part of the spectrum, but for at most one such interception in the visible part of the spectrum.

Clearly, the interactions that can occur in the canopy are complex, although they can be categorised as being of the three forms:

1. Reflectance from canopy elements themselves
2. Reflectance from the background
3. Reflectance from a combination of the canopy and the background

The proportions of each that dominate depend on the canopy density and structure. Thus, sparse canopies, typical of young crops and some land covers, will have a high component of background reflectance. As the leaf area index (LAI) increases, the proportion of the energy that intercepts vegetation for the first, and multiple times increases. Once the canopy becomes denser than about an LAI of 2, the vegetative elements intercept all the incident energy at least once; with multiple interceptions continuing to increase up until the LAI is about 6. However, the mix of these interactions also depends on the source–surface–sensor geometry and on the wavelength. Consider the two simple canopies shown in Figure 2.37. This figure shows a planophile canopy (B) in which the canopy elements are approximately horizontal, and an erectophile canopy where the canopy elements are approximately vertical (A). Both canopies have an LAI of about 1. Consider just the one situation of incident radiation at 30° zenith angle (60° elevation), and the sensor directly overhead.

In the planophile canopy (B), the ground is covered by vegetation, yet there is negligible shadowing obvious to the sensor. The leaves absorb most of the radiation in the red part of the spectrum, with a very small percentage of about 5% being reflected and transmitted. There is only one layer in the canopy, so the transmitted red energy is then incident on the ground, from which some will be reflected back into the canopy, where most of it will be absorbed. For NIR radiation, about 45%

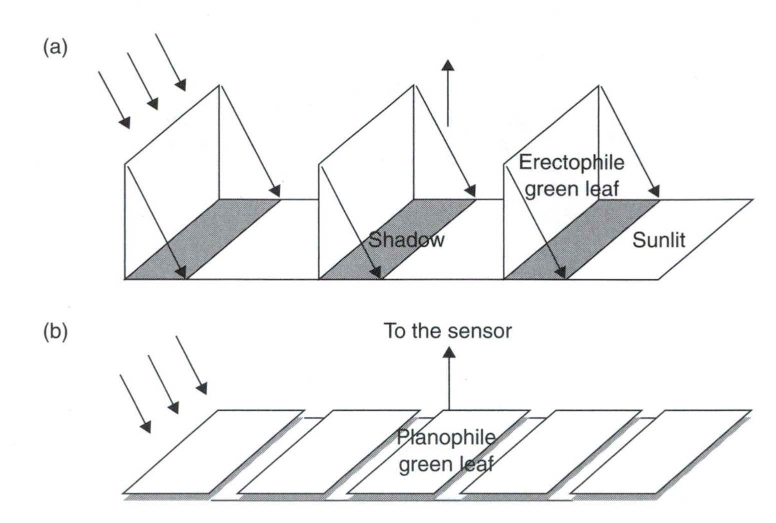

FIGURE 2.37 The geometry of energy interactions in vertical and horizontal canopies.

of the radiation will be reflected back into the atmosphere, whilst another 45% will be transmitted through the leaf to the soil background. At the soil there will be some reflectance, typically between 20 and 40% depending on the soil reflectance, and of this reflected energy, about 45% will be transmitted through the leaf to the atmosphere, contributing about 4.1 to 8.1% of the total source energy to the reflected energy going out of the canopy. The reflectance of this planophile canopy will thus be about 5% in the red part of the spectrum and between 49.1 and 53.1% depending on the soil reflectance in the NIR part of the spectrum.

For the erectophile canopy, the sensor detects areas of leaf, shadow and soil, where the proportions of these depend on the geometry of the canopy and on the sun–surface–sensor geometry. The reflectance can thus be expected to vary significantly within the sensor FOV. At 30° nadir angle, about 58% of the ground surface will be in shadow, and thus 42% will be sunlit. There is negligible direct reflectance from the canopy. The sunlit soils will give about half the reflectance in the red part of the spectrum than in the NIR, and so they will contribute (10–20% by 0.42) or about 4.2–8.4% of the source energy in the red and about 8.4–16.8% in the NIR. The 58% that are in shadow will have a single transmission through the vegetation and a single reflection from the soil. In the red part of the spectrum about 5% of the energy is transmitted through the leaf yielding (5% × 58% × 10–20%) or 0.1 to 0.2% in the red and 5.2 to 10.4% in the near infrared. The total reflected energy from the erectophile canopy would thus be about 4.3–8.6% in the red part of the spectrum and 13.6–27.2% in the near infrared, ignoring secondary reflections, which in fact will occur. The erectophile canopy will thus give higher red reflectances and lower NIR reflectances than the planophile canopy. It can be seen that changing the solar elevation and the sensor zenith angle will cause only very small changes in the reflectance of the planophile canopy if the leaf elements are assumed to have near-Lambertian reflectance, whereas significant changes will occur in the reflectance of the erectophile canopy. The changes that occur will be larger in the NIR than in the visible region, because of the higher transmissions and multiple reflections in the NIR. It can also be shown that the reflectances are lower when there is a shadow component in the canopy, giving higher reflectances when there is negligible shadow, as occurs at that point in the image where the source of illumination is behind the sensor. This is one factor in the "hotspot" as can occur in aerial images when the shadow of the aircraft would fall on the image. Instead, what is often seen is an area of higher reflectance. The other factors that contribute to the hotspot are the increased reflectance through dense canopies as can occur when looking diagonally through the canopy, and refraction of the light around the aircraft, reducing the shadow of the aircraft.

Of course, this is an over simplification of what happens in a canopy, but it shows clearly that the reflectance of a canopy depends on the canopy shape and LAI, the sun–surface–sensor geometry as well as the reflectance characteristics of the leaf and soil elements. This situation is made more complex due to the fact that the leaf elements are not perfect Lambertian reflectors, and with some species, the leaf elements can orient themselves, to some degree, to follow the sun, all of which will change the reflectance characteristics of the canopy throughout the day and from day to day.

Since the scene geometry, specifically the sun–surface–sensor geometry changes across an image, the reflectance can be expected to change across the same. This change is measured by the bi-directional reflectance function (BDRF) of the canopy as will be discussed in more detail in Section 2.3.8.

In considering these two simple canopies quite a few assumptions have been made concerning canopy geometry and the reflectance characteristics of leaf elements. In practice, the canopy is much more complex than this, however, the modelling shows:

1. The reflectance from a canopy dominated by erectophile components is much more likely to show changes in reflectance with changes in solar elevation and sensor zenith angle than will a canopy with predominantly planophile components, as long as the leaf elements are approximately Lambertian.
2. The variations in canopy reflectance are much more obvious in the NIR than in the visible region.
3. Changes in the proportions of vertical and horizontal leaf components can affect canopy reflectance.
4. Erectophile canopies will often exhibit higher red reflectance and lower NIR reflectance than will planophile canopies.
5. Vegetative canopies are rarely Lambertian, even if the leaf elements are approximately Lambertian, and the closeness of the canopy to the Lambertian ideal depends on the wavelength, the mix of vertical and horizontal leaf elements in the canopy, and the canopy geometry.

Consider how this might assist us in explaining the reflectance that is characteristic of different canopies. The canopies of *Eucalyptus* sp. have much lower reflectance values than the canopies of many deciduous and tropical trees, even though the reflectance of the individual leaf elements are similar as shown in Figure 2.38. *Eucalyptus* sp. trees have leaf elements that hang vertically or much more so than the predominantly horizontal orientation of the leaves of most deciduous and tropical trees. In consequence, a Eucalyptus canopy is quite open when viewed (sensed) directly from above. When the sun is at low angles of elevation the leaves create considerable shadow, so that vertical imagery will contain significant shadow and soil reflectance. When the sun is at higher elevations the shadowing is reduced, but the contribution of soil reflectance increases. As a consequence, eucalyptus species are likely to diverge significantly from the Lambertian ideal, as will other canopies that exhibit open structures with dominantly vertical leaf elements.

All plants use varying proportions of the same pigments for photosynthesis and the same chemical materials for plant building. For those plants that occupy similar environmental regimes, they all have to handle similar energy budgets. To compete successfully with other species a plant will need to develop unique characteristics that will allow it to occupy an ecological niche or niches. Plants do this in many ways; changing the mix of pigments used in photosynthesis, modifying their canopy geometry so as to optimise their ability to use energy or to minimise environmental impact on them, adapting their growth patterns and processes and modifying their leaf structure. All of these techniques affect the energy balance of the plant, and all of them affect the reflectance from the canopy in some way during their phenological life cycle. Thus analysis of diurnal and seasonal changes in canopy reflectance is an important strategy in attempting to discriminate between different plant communities.

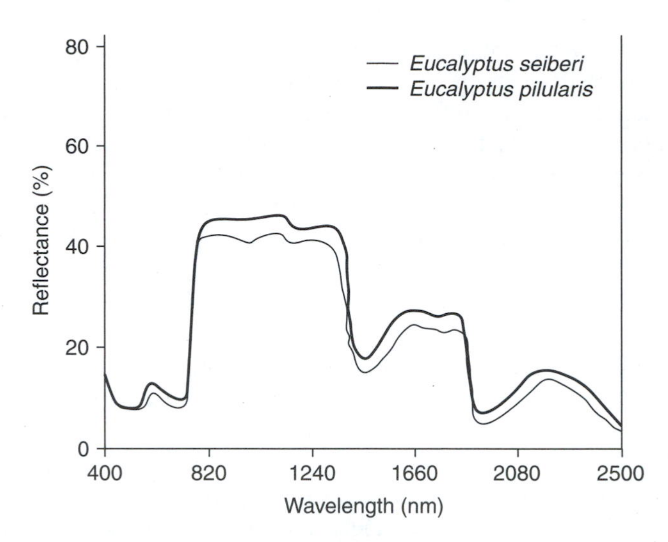

FIGURE 2.38 The relectance of selected *Eucalyptus* sp. plants (from O'Niel, A., 1990. *Personal communication*. With permission).

Consider winter and summer growing herbage or grasses that occupy different ecological niches by growing at different seasons. The two communities will look quite different on the one image if the imagery is chosen to ensure that the phenological differences are maximised. Agricultural crops have a different phenological cycle to most of the other land covers due to the physical interference by man. This interference is most obvious at cultivation and at harvest when the crop status changes from a mature crop to litter and straw. Both of these activities are quite unique to crops and can be used to improve the discrimination of crops from other land covers. The changes at harvest are often more useful because they are usually more dramatic in their effect in the imagery, the period of harvest covers a shorter period than cultivation and sowing and there is a greater probability of cloud-free conditions suited to the acquisition of image data.

The above discussion has shown that the reflectance of a canopy is due to:

1. *The reflectance, absorptance, and transmittance characteristics of the canopy elements*, how they change across the canopy elements, how close they are to being Lambertian and how they change with the age of the canopy element.
2. *The density, distribution and orientation of the canopy elements* in the canopy.
3. *The sun–surface–sensor geometry* and the relative proportion of the incident radiation that comes from the one point source (the sun).

In theory one can model all the components and thus derive exact estimates of the canopy reflectance. In reality this is impossible to do precisely, because it is impossible to accurately model the occurrence of the environmental actions that influence the plants being imaged. For example, a canopy is subject to breezes, and these change the distribution and orientation of the canopy elements, thus changing the canopy reflectance. The canopy elements are also subject to deposition of moisture and dust, and these will affect the reflectance of the canopy elements, and hence the reflectance of the canopy. Unfortunately, the magnitude of the variation introduced by these unpredictable factors is significant, and will sometimes be of the same order, or larger, than the differences that are being used to estimate canopy conditions. There is a practical limit to the accuracy with which canopy reflectance can be estimated from canopy parameters, no matter how good the model being used. Of course, this means that the inverse problem, of estimating canopy variables from reflectance data, will always contain a level of error and thus will limit our capacity to estimate canopy variables from remotely sensed

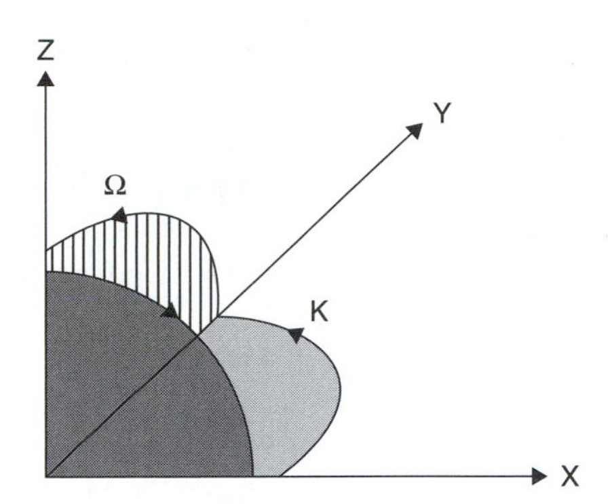

Figure 2.39 Polar co-ordinate system used in the observation of bi-directional reflectance functions.

data. The absolute problem is always larger than the relative problem; better information will always be derived when the image-based data is integrated with field-based data.

2.3.8 Bi-Directional Reflectance Distribution Function of Surfaces

The definition of BRDF was given in Section 2.3.1. It can be seen from this definition that field instruments do not give perfect estimates of BRDF, since they observe the surface through a finite FOV, even if the solar energy source was considered to be a point source. In practice skylight contributes to the reflectance, further removing actual observations from the ideal definition of BRDF. The BRDF varies as a function of the orientation and elevation of both the source and the sensor. Construction of a full BRDF would thus require observations of reflectance at a range of orientations and elevations, for each observed orientation and elevation of the source and the sensor. In practice most BRDF functions are constructed assuming that the sensor is at the nadir position. Even with this assumption, the construction of a full BRDF is a time-consuming task.

In general BRDFs are measured using field instruments with narrow FOVs, over a range of angles of orientation and elevation, under conditions of bright sunlight. If the sensor orientations are made relative to the orientation of the source, then polar co-ordinates can be used as shown in Figure 2.39 and following the right-handed rule. With the right-handed rule, a right hand is closed as a fist, and then the thumb, main and second fingers are extended, representing the x, y and z-axes, respectively. From the origin at the intersection of these axes, a clockwise rotation will bring one axis into the position of the second axis. Thus, a clockwise rotation of the z-axis about the y-axis will bring it into the position of the x-axis. This represents a positive rotation. The same logic applies to the other axes.

Typical BRDFs are shown in Figure 2.16 to Figure 2.18. Figure 2.16 is of a recently ploughed bare field, with the sun at an elevation of 25° at wavelength 662 nm. The BRDF shows a variation of about 28% in the reflectance factor between the minimum and maximum values that occur in the plane of solar illumination (bearing of 0°). The lowest reflectance values are towards the sun, and the highest values are when looking away from the sun. This suggests that the soil particles are highly opaque, so that, when looking towards the source, the non-illuminated parts of the particles are contributing significantly to the reflectance, and since they are not illuminated, their contribution is much less than the illuminated parts. The reverse applies when looking away from the source.

Figure 2.17 shows a rangelands brush species at a solar elevation of 59° at two wavelengths, 662 and 826 nm. The red wavelength BRDF shows a strong ridge in the plane of the source, with maximum reflectance when looking away from the source, with a variation of about 6% between minimum and maximum reflectance. In the NIR, the same ridge feature exists; with about the same difference between minimum and maximum values, but the BRDF function has a different shape away from this ridge.

Figure 2.18 shows the NIR BRDF for the same rangelands community as depicted in Figure 2.17. The variation in the NIR is more marked than are the variations in the red waveband.

The BRDF shows how reflectance is dependent on orientation and elevation, and these examples have shown that the BRDF can vary significantly with variations in these factors. If the sensor has a narrow FOV, and comparative analyses between images use images at similar look and solar angles, then the BRDF may not have a strong effect on variations in reflectance in the one cover type. If, however, sensors are used with a wide FOV, the sensor can be pointed at different look angles or if significant differences in solar direction of azimuth and elevation exist between the images, then the BRDF may be one of the causes of variation in reflectance within each cover type.

2.4 PASSIVE SENSING SYSTEMS

Sensing systems collect energy from the surface, and convert that energy into an analog or digital record. The important distinguishing characteristics of image data are:

1. *The relative spatial locations of all data elements are known to a high degree of accuracy*. This is essential if the data are to be used to create an image that accurately depicts the imaged surface. Theoretically, the relative spatial locations are known because the geometric model of the sensor is known. For example, the theoretical camera model states that the lens is an infinitely small pinhole and there is a one-to-one correspondence between a point in the object and its recording in the image, or that there is a unique line joining this object point to its image point, where this line passes through the lens point (Figure 2.40). However, a sensor is not a perfect fit to such a model, but contains various types of errors. Again, to consider the camera, the lens is not infinitely small but has a physical size, and the lens itself contains distortions so that there is not a perfectly straight line between the object and its image point, where this error can be different in the direction that is radial from the lens axis to the direction that is tangential to this line. Such errors introduce errors in the relationship between an object and its recording in the image. The closer the sensor fits the geometric model the more spatially accurate the image will be. With analog sensors, such as the camera, the data are preserved in their correct relative position in the recording medium or film, within the accuracy standards of the camera. With digital sensors the data are stored without preservation of spatial organisational information attached to the individual image element, but the data are preserved in such a way that allows the reconstruction of the data into an image.

2. *The map co-ordinates can be determined*. The map co-ordinates of image data depend on the sensor geometry as discussed in (1) above, as well as the location and orientations of the sensor relative to the object co-ordinate system. Control points with known map and image co-ordinates are used to determine the orientation, displacement and scaling that needs to be introduced into the image co-ordinates so as to give map co-ordinates to the image points. This process is called absolute orientation. Since, absolute orientation depends upon both the internal and external geometry, the accuracy of establishing map co-ordinates for image data is usually inferior to the internal accuracy of the data. Techniques to rectify image data, or create a new image that is a close fit to an absolute co-ordinate system, will be discussed in Chapter 4 under rectification

3. *Image depicts conditions at an instant of time*. Images are acquired in seconds to minutes, depending on the type of platform and sensor being used. If this is fast relative to the speed of change on the surface then the image is effectively acquired at an instant in time. A set of temporal images record conditions at different times. The use of temporal data sets requires comparison of the images and hence requires good geometrical accuracy between the images.

4. *Image data is a record of specific energy radiation* from the surface, not of the surface features themselves. Unlike field data, that usually measures the surface features of interest, remotely sensed data records the effect of the surface on electromagnetic radiation under specific illumination and

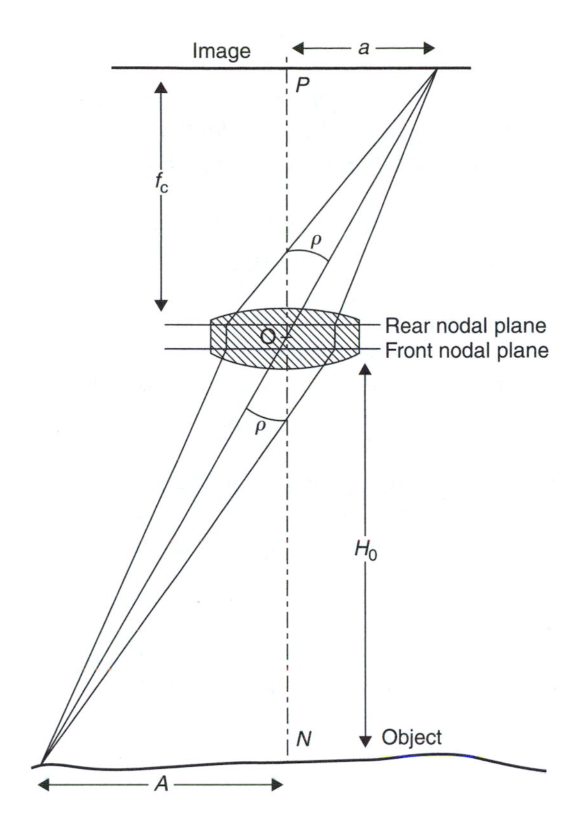

FIGURE 2.40 Relationship between an object and its image in a perfect camera that exactly obeys the principles of the Central Perspective Projection.

sensing conditions. The sensor records the incident radiance in response units, where the sensor calibration provides the relationship between the recorded response values and the incident radiance values. The incident radiance values can be readily converted into reflectance at the satellite by using the known or measured irradiance incident on the Earth's upper atmosphere from the sun. However, neither the incident radiance nor the reflectance at the top of the atmosphere are the data of interest, but rather the reflectance at the surface. To convert the incident radiance to surface reflectance requires that the effects of the atmosphere be taken into account, both for the incident radiation on the surface and for the radiation reflected from the surface. Some of these atmospheric effects are stable and do not change significantly from place to place or over time. However, some can change quite dramatically, both spatially and temporally. Atmospheric correction is essential when the imagery is to be used for estimation and for the production of temporal image sets. Techniques for conducting atmospheric correction, estimation and the analysis of temporal image data sets are discussed in Chapter 4.

Remote sensing devices are either passive, using existing sources of energy to illuminate the object, or active systems that generate their own illumination. Passive systems are usually simpler and cheaper, but they are dependent on the characteristics of the natural source of energy that they use. Active systems are independent of the normally used naturally occurring energy sources such as the sun and the Earth itself.

The most commonly used passive sensor systems in remote sensing are:

1. The camera
2. The scanner

2.4.1 The Camera

Cameras can be either photographic or digital in form. A photographic camera forms an image by focusing the incoming electromagnetic radiation from the scene onto a film that is sensitive to electromagnetic radiation in the sensed wavebands. A digital camera operates in exactly the same way, except that the film is replaced with a light sensitive surface. This surface is usually either a phosphor coating that is activated by electromagnetic radiation or a two-dimensional array of very small light sensitive detectors, the most common being charged couple devices (CCDs) from their method of activation. The phosphor coating or CCD may be charged electrically so as to amplify the energy levels prior to converting them into an electrical potential. The surface is scanned and the potential on the surface used to modulate an electrical signal that is then amplified before being stored as either an analog signal on a videotape recorder or as a digital image. As the surface is scanned, the potential is reset to zero, and will subsequently accumulate potential as a function of the levels of light energy that are incident on the surface.

All cameras are designed to obey the principles of the Central Perspective Projection. They achieve this with some errors that result in geometric and radiometric errors in the data. For precision cameras, the effects of these errors are minimized by fitting the images to ground control in a process called calibration and rectification. These will be discussed in subsequent sections of this chapter and in Chapter 4. Cameras meet the conditions of this model by obeying the rules that:

1. There is a one to one correspondence between the object, the perspective centre (point O in Figure 2.40) and the image point. These three points all lie on the same line, so that objects at different spatial locations must be imaged at different positions in the image. This definition, of course, ignores the effects of the atmosphere on the radiation, including the bending of the rays due to refraction.
2. Straight lines in the object space must be imaged as straight lines in the image space. If they are recorded as curved lines in the image space due to distortions in the camera lens then they need to be calibrated to remove the effects of these distortions before they can be used for analytical work. The magnitude of the corrections are normally conducted in camera calibration and then applied in rectification.
3. The light entering the camera must be controlled in both duration and intensity to ensure proper exposure of the film or on the electronic cell array.

A camera consists of the following components.

2.4.1.1 The Lens Cone

Light reflected off the object is focused onto the film by the lens cone. There are some important definitions:

Focal length (f_c) The length of the normal from the perspective centre or rear nodal plane of the lens to the image plane when the camera is focused on infinity so that the incoming radiation from a point is effectively parallel across the lens aperture.

Principal point (P) The foot of this normal on the image plane. It is defined by the intersection of the lines joining the fiducial marks on an aerial photograph.

The *fiducial marks* are V-shaped indentations in the four sides of the frame of each photograph. They are used to define the principal point by joining them across the photograph.

The lens of a camera must strive to maintain close fidelity to the geometric model that is the basis of the camera's construction, so as to maintain geometric accuracy, and provide as fast an exposure as possible so as to minimise image blur. These requirements place competing demands on the design and construction of the camera. Smaller lens apertures come closer to the geometric model of a camera in which the perspective centre is defined as a point. As the aperture size is increased, more light enters the camera in a given time, allowing shorter exposures. These competing demands cannot be solved perfectly. As a consequence there are distortions present in all cameras. The various forms of camera distortion are:

1. *Chromatic aberration*. If light of different wavelengths is refracted by different amounts in the lens then it will be focused at different distances from the lens. This means that some wavelengths may not be in focus on the film plane, since the plane is at a constant distance from the lens.
2. *Spherical aberration and coma*. Refraction at the edges of the lens causes more bending of the rays than at the centre of the lens. There are effectively different focal distances at different distances out from the central axis of the lens causing image blur. Spherical aberration is this effect on radii from the centre of the lens, whilst coma is the same effect but on chords that do not pass through the centre of the lens.
3. *Astigmatism*. The curvature of the lens is different in planes through the lens axis to the curvature on chords that do not pass through the axis. This difference in curvature creates differences in the focusing distance on the plane through the centre of the lens to the plane through the chord. The result is that a point may be in focus in one direction, but be out of focus in the direction at right angles or 90°.

The lens of most cameras contains the diaphragm and shutter. The diaphragm controls the amount of light that can enter the camera in a unit period of time. As the size of the aperture increases, the camera geometry diverges more and more from the perfect camera, and the larger the geometric distortions that are created. The intensity of the light in the camera decreases as the square of the distance travelled from the lens to the image plane (because the same amount of energy must cover a larger area). The F-number or ratio of area of aperture to focal length indicates equivalent light intensities on the image plane of cameras of any focal length. The F-number or stop is therefore used as the standard measure of the light intensity that will fall on the image plane for all cameras, and hence is used as the measure of the aperture. The larger the F-number the smaller the aperture relative to the focal length and hence the less light that can enter the camera in a given time and vice versa.

The range of F-numbers used in most cameras are shown in Table 2.2.

The camera shutter is designed to control the length of time that light is allowed to enter the camera, and hence control the total exposure. Shutters do not operate with perfect precision, taking a period of time to open, and a period to close, during which time the shutter does not allow the full amount of light into the camera. Figure 2.41 illustrates the typical efficiency that is achieved with many modern camera shutters.

2.4.1.2 Magazine or Digital Back

The magazine holds the film such that it can be supported in the image plane during exposure, and wound on in preparation for the next exposure. The film is held so that it is sealed from light other than that entering to expose the film by the shutter and aperture. The magazine also has counters to record the exposures taken or still available, and facilities to record critical information about the film on each exposure such as the aircraft altitude, time of exposure, date, flight or sortie number and the exposure number. Digital cameras do not contain film, but rather contain a light sensitive cell array. The most common type of light sensitive cell array is a CCD. A CCD array has the disadvantage that exposure to excessive levels of radiation causes the adjacent cells to become energised, causing flaring in the image. The cell array records the energy incident on it and this record is then sampled and stored in a digital storage device such as a computer hard disk. If multiple wavebands are

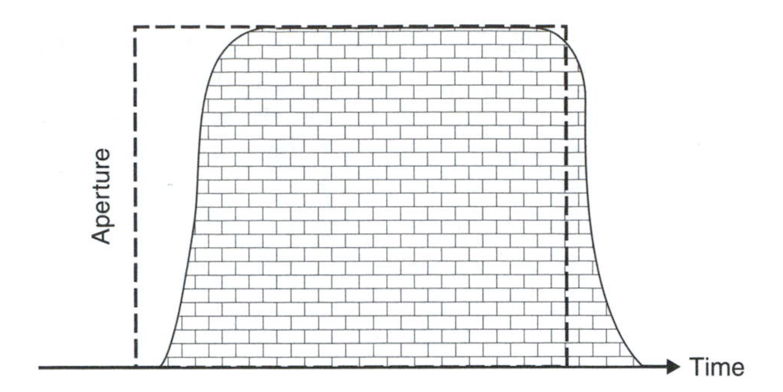

FIGURE 2.41 Shutter efficiency. The hatched rectangular figure represents ideal shutter action, open in zero time, stay completely open, close in zero time. The shaded figure shows the actural shutter aperture.

TABLE 2.2
***F*-number, Ratio of Aperture Area to Focal Length and Relative Size**

F-number	Area of Aperture/ Focal Length	Relative Aperture Size
32	1/32	Very small
16	1/16	Medium
2.8	1/2.8	Large

recorded, then adjacent cells in the array can have different filters imposed over them so that the different cells record energy in the different wavebands. This has the disadvantage that resolution is degraded, but the different bands are all recorded simultaneously. The exposure in the different wavebands will also be different for different targets. The camera solves this problem by measuring the energy levels on the cell array, and finding an exposure that provides the best possible trade-off between the wavebands.

2.4.1.3 Camera Body

The camera body holds the lens cone in the correct position relative to the image plane and the magazine so as to minimise distortions in the images taken by the camera. It is a light tight compartment so as to seal the film from stray light. The camera body includes the register glass and a flat ground glass plane that is just behind the image plane and against which the film is pressed during exposure (Photograph 2.1).

2.4.1.4 Suspension Mount

The suspension mount secures the camera to the platform. The mount usually contains mechanisms to damp vibrations from the platform being transmitted through to the camera, as well as levelling and orientation adjustments. The level and orientation of the camera will vary with aircraft payload, the wind direction and velocity relative to aircraft velocity and direction.

In a camera, unlike other sensors, the sensing and recording functions are undertaken by the same elements of the sensor, the emulsion surface on the film. This characteristic of cameras simplifies their construction and hence reduces their cost relative to other sensors, as well as eliminating a number of sources of geometric error. Because of the fixed exposure and the development steps that occur in creating a photograph, it is difficult to relate densities on a photograph with radiance that

PHOTOGRAPH 2.1 A modern wild RC30 film mapping camera to acquire large format (23×23 cm^2) images that are suitable for topographic mapping, geologic and many other interpretation task set on its aircraft mount and its associated navigation sight.

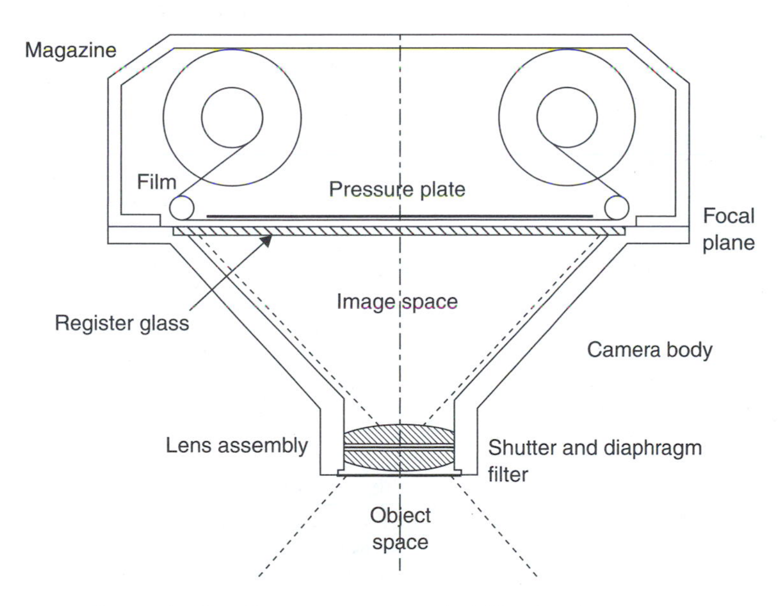

FIGURE 2.42 Construction of a typical film framing camera.

is incident on the camera. Exposure panels, of known or constant reflectance characteristics can be used to help control and calibrate the density values in the film.

The most common type of camera is the framing camera as shown in Figure 2.42, or a camera that simultaneously exposes the whole of the image area for each fixed exposure. Typical framing cameras are all 35 mm cameras, various 70 mm cameras, and 23 cm format mapping cameras. Mapping cameras are designed to minimise distortions so as to produce high precision negatives that are 23×23 cm^2 in size. They are flown in aircraft properly set up for the purpose, with properly installed mounts, intervalometers to ensure the proper interval between exposures and navigation sights. Suitable 70 mm cameras can be used for taking aerial photography from a light aircraft, but with better resolution and quality than can be achieved with most 35 mm format cameras. The smaller format cameras can be set up in banks to take simultaneous photography, for example, simultaneous colour and colour infrared photography. They can also be set up to take black and white images in different wavebands so as to create multispectral images. The major problem with this approach is the need for high co-registration accuracies if the resulting images are going to be suitable for analysis. Such accuracies are difficult to achieve in the aircraft mount, and so additional registration costs are usually associated with the subsequent processing of these images, prior to co-registration. Consider a digital camera with 10 mm focal length supporting a 10×10 mm^2 CCD array with 2000 by 2000 elements. Each cell on the array is 0.005 mm in size. Consider Figure 2.40 where a distance, a in the image plane represents a distance A in the ground plane. By similar triangles it can be shown that:

$$\frac{a}{f_c} = \frac{A}{H_0} \tag{2.27}$$

If this camera acquires images at an elevation of 1000 m then it will have a pixel size of 500 mm, since

$$A = \frac{H_0}{f_c} \times a = \frac{1000 \cdot 1000}{10} \times 0.005$$

With aerial cameras the focus is set at infinity so that the distance from the rear nodal plane of the lens to the image plane is set at the focal length of the lens. It is essential to use high-quality framing cameras for accurate measurement work as in the production of topographic maps. Other types of framing cameras of lower geometric quality can often be used where accurate geometric information is not being sought from the photographic data.

Often photo-interpretation does not require accurate geometric measurements and can therefore use other, or cheaper cameras in relation to geometric quality. However, any work involving detailed analysis of images requires photographs that are of the highest possible image quality, in terms of resolution, focus, contrast and colour rendition. For this reason photo-interpretation requires cameras that are of high image making quality, but can be somewhat inferior to mapping cameras in geometric quality.

The image is focused by the camera lens onto light sensitive film, housed in the magazine, and kept flat during exposure by being pressed against the register glass. There are four main types of film used in remote sensing, identified by their sensitivity to light of different wavelengths and how they record this radiation.

The light sensitive layer of all films depends on the reaction of silver halide salts to electromagnetic radiation. Silver halide grains are naturally sensitive to light in the blue and ultraviolet parts of the electromagnetic spectrum. Their sensitivity is extended into the remainder of the visible part of the spectrum, and into the NIR, by the use of dyes called colour sensitisers. The grains are still sensitive to the shorter wavelengths after they are sensitised, and so this radiation needs to be filtered out to stop it entering the camera and exposing the film. Without filters this shorter wavelength radiation would contribute to, and often dominate the exposure, masking the contribution due to longer wavelength radiation.

The four common types of film emulsion (Figure 2.43 and Figure 2.44) are:

1. *Panchromatic film.* A one-layer film that creates black and white photographs, recording the radiation in the visible part of the spectrum, from about 350 to 700 nm, in accordance with the spectral sensitivity of the film.
2. *Black and white infrared film.* A one-layer film that creates black and white photographs recording the radiation from the blue part of the spectrum through into the NIR, at about

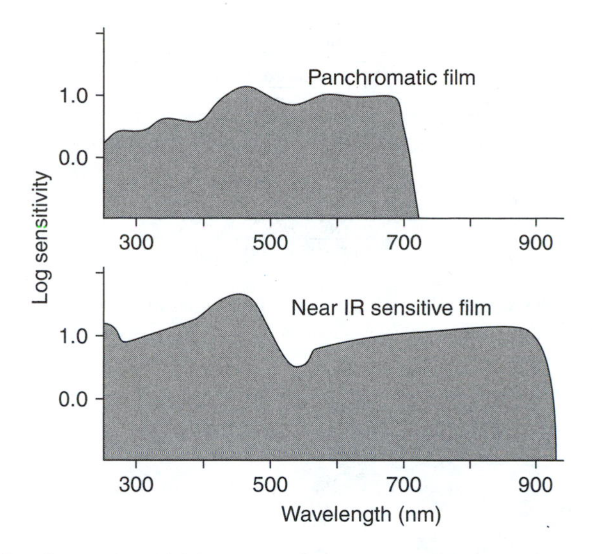

FIGURE 2.43 Spectral sensitivity curves for two common one layer emulsion films.

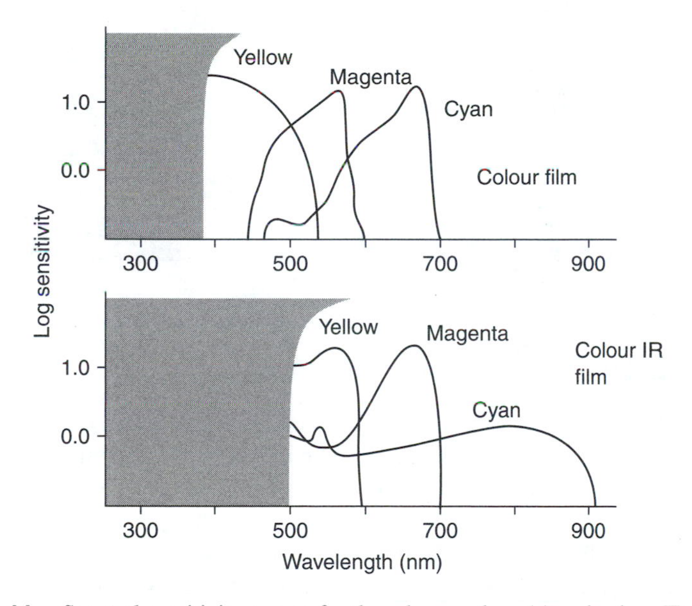

FIGURE 2.44 Spectral sensitivity curves for three layer colour (a) and colour IR (b) films.

900 nm in accordance with spectral sensitivity of the film. These films are usually used to record the incident energy in the NIR, and so a filter must be used to stop energy below about 700 nm being transmitted to the film.

3. *Colour film.* A three-layer emulsion film constructed similarly to that shown in Figure 2.45. The layers are sensitised to record energy up to the blue, green and red wavebands, respectively. An yellow filter is inserted between the blue and green sensitive layers to filter out blue light from activating the green and red sensitive layers (Figure 2.44). Each layer contains a dye, usually of yellow, magenta and cyan, respectively, so that exposure and development of that layer creates an image in that colour.
4. *Colour infrared film.* A three-layer emulsion film with the layers sensitised to the green, red and NIR wavebands. It is of similar construction to that of colour film, but without the

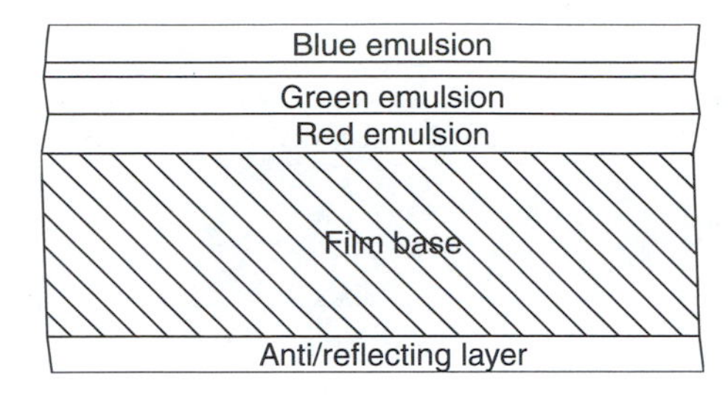

FIGURE 2.45 Cross section through a colour film emulsion showing the emulsion layers.

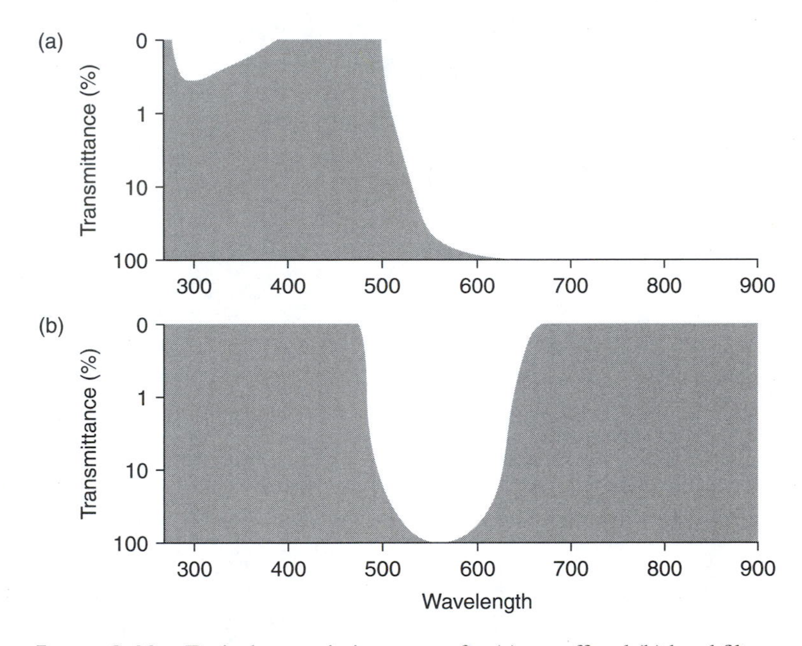

FIGURE 2.46 Typical transmission curves for (a) cut-off and (b) band filters.

yellow filter. All the layers of this film retain their sensitivity to blue light, and so a filter must be used to stop the blue light from being transmitted onto the film.

Yellow or neutral density filters are used to filter out some of the blue light and hence correct for the effect of haze. A yellow filter with cut-off at about 500 nm is normally used with colour infrared film, and a deep red (Wratten 25) filter is normally used with black and white IR film to record incident NIR radiation on the film. Filters are in two forms (Figure 2.46):

1. *Cut-off filters.* These are filters that have either a lower or upper cut-off, only transmitting radiation above, or below the cut-off waveband. Yellow filters are typical cut-off filters.
2. *Band filters.* Band filters transmit energy within a selected waveband.

Filters only transmit part of the radiation to which the film is sensitive. Sometimes the manufacturer expects the user to utilise the filter, and will publish the film speed as if the film will be used with the filter. However, sometimes the user may wish to record radiation in only part of the normally used waveband. Thus, manufacturers of black and white IR film expect users to use a deep red filter with the film and so the film speed is set accordingly. Panchromatic film on the other hand is normally used to record radiation in the whole of the visible part of the spectrum. If the user wishes to record only part of that spectrum, such as green light, then the green filter will reduce the amount of radiation

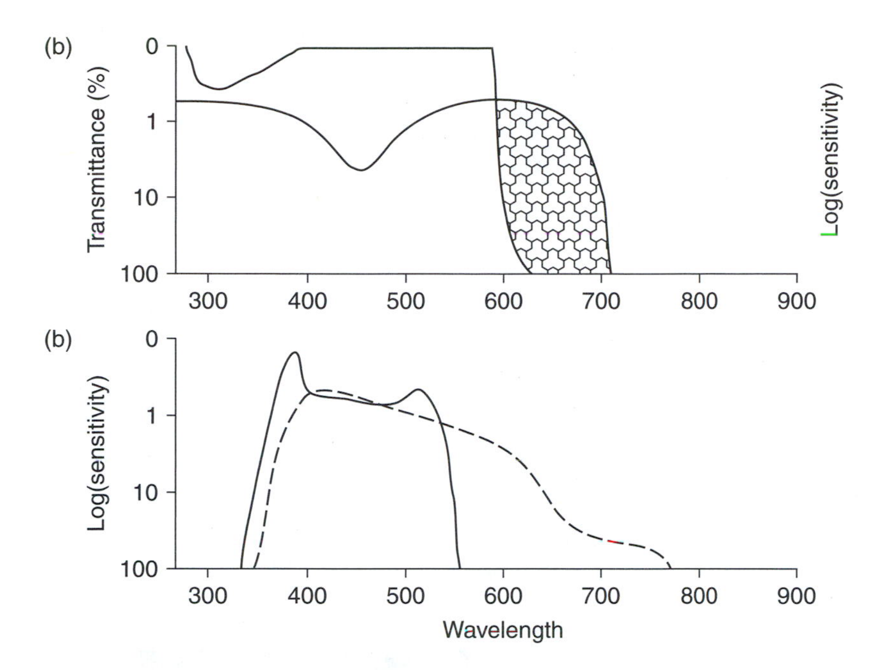

FIGURE 2.47 Determination of the FF as the ratio of the area under the film curve, as normally exposed, over the area under the curve with the filter (a) and the responsivity versus wavelength for several typical photon detectors (b).

available to activate the film, relative to the amount anticipated by the manufacturer in determining the speed of the film. This restriction effectively makes the film slower, requiring a longer exposure. This change in exposure due to the use of a filter is known as the filter factor (FF) for that film and filter combination. An yellow filter, with a cut-off at 500 nm, restricts the transmission of blue light. It would normally have an FF of 1.5 or 2 for use with panchromatic film; but an FF of 1 with colour IR film as the manufacturer assumes the use of this filter with this film.

The FF can be calculated by plotting the filter transmission curve on the film sensitivity curve as in Figure 2.47(a). The FF is the ratio of the area bounded by the film sensitivity curve in normal usage over the area bounded by the film sensitivity and filter transmission curves when the filter restricts transmission. The FF is normally a multiplicative factor greater than one. The filter factor can also be determined by photographing a target without the filter, then with the filter at a series of exposures. Compare the film densities to select the filtered exposure that gives a density closest to the unfiltered density. The ratio of the two exposures is an estimate of the FF.

2.4.1.5 Light Sensitive Cell Arrays

The most common type of light sensitive cell array is the CCD. Silicon-based CCD arrays are sensitive in the visible and NIR as shown in Figure 2.48. They are thus suitable for this part of the spectrum, but not for longer wavelengths. Other types of CCD array need to be used if they are to be sensitive to other wavelengths.

2.4.1.6 Measurement of Resolution in Image Data

Resolution of detail in an image is of critical importance in visual interpretation. Resolution has historically been measured as the number of lines per mm that can be resolved in a photograph of a bar chart like that shown in Photograph 2.2. The use of bar targets to determine resolution is easily understood, but it does not show the loss of resolution as the detail becomes finer, nor does

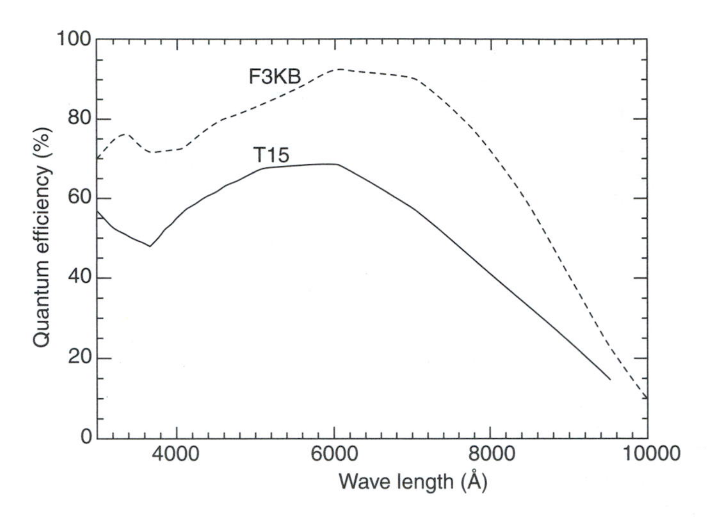

Figure 2.48 The quantum efficiency of tow digital camera CCD arrays.

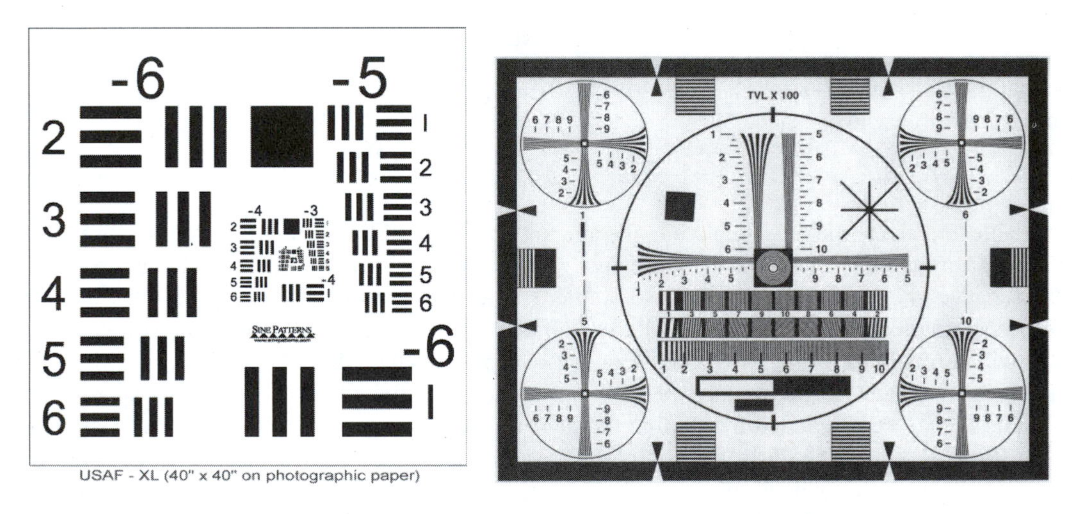

Photograph 2.2 (a) The USAAF 1951 bar target designed to determine the resolution characteristics aerial photographs and (b) the IEEE Resolution Chart designed to test camera performance.

it indicate the differences in resolution that occur when contrast is reduced. These difficulties are overcome by the use of the optical transfer function (OTF). If a bar chart with intensities that varied in a sinusoidal fashion is photographed then the resulting image densities would vary in a sinusoidal way (Figure 2.49). As the frequency of the bars increases, or the wavelength of the sine curve is reduced, then the photograph would exhibit a reduction in the amplitude of the sine wave recorded in the image. At long wavelengths the reduction in contrast is negligible but as the wavelength decreases, the reduction in contrast increases until eventually the sine wave cannot be discriminated. The ratio of the recorded to the incident signal at each wavelength is recorded, the OTF value at that wavelength, and being plotted as shown in Figure 2.49. This shows how the sensitivity of the camera decreases as the sine wave becomes finer, equivalent to higher resolution detail.

The OTF indicates the percentage of the information that will be recorded for different sized objects, at maximum contrast. The level of contrast between objects and their surroundings is determined as a proportion of this maximum contrast. The product of this proportion and the percentage

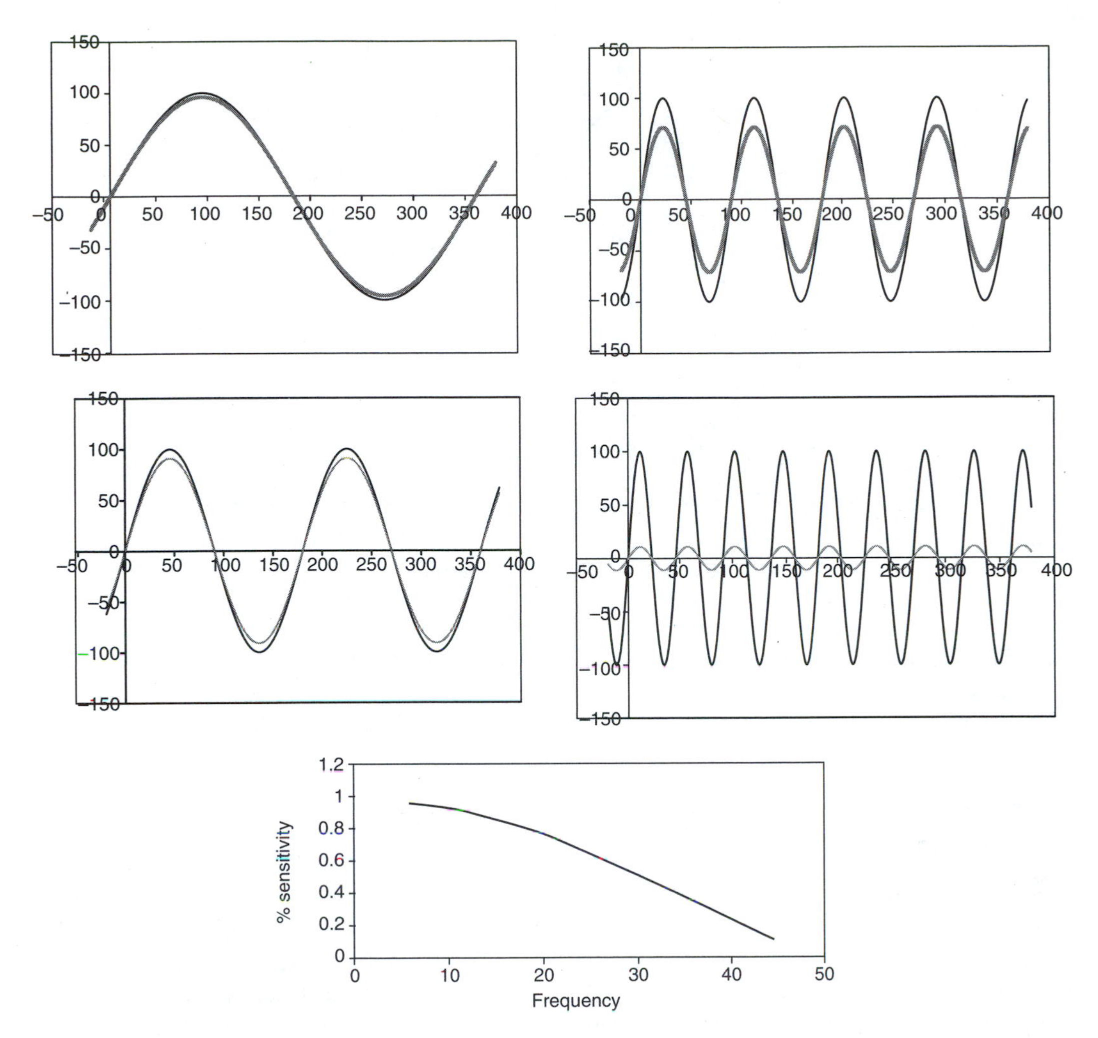

FIGURE 2.49 The OTF for a panchromatic film derived from the use of four sources sine wave signals and their response signals.

recorded in the OTF gives the percentage that will be recorded at the actual contrast. Clearly, the higher the contrast in the original scene, the better the resolution in the information that can be extracted.

Modern digital cameras use a square or rectangular lattice of light sensitive elements to record the incident radiation instead of film. In most digital cameras these arrays are constructed of silicon chips on which are superimposed the light sensitive material. The size of these light sensitive elements controls the resolution that can be achieved by the camera. A filter can be used with the camera, as with panchromatic film, to record the energy in the specific waveband transmitted by the filter. With colour digital cameras, filters are inserted into the array, usually covering four elements in the array with three filters. Thus, colour cameras will usually have only half the resolution of the equivalent black and white camera. The spectral sensitivity of typical CCD arrays used in digital camera is shown in Figure 2.48. It shows that these cameras are not equally sensitive at all wavelengths. Since this material is much less sensitive at longer wavelengths, in colour cameras, two cells are often used with the red filter, compared with one cell in the green and the blue. This variation in sensitivity also means that the exposure will vary with wavelength. A significant advantage of digital cameras for aerial imagery is that the exposure can be made to be a function of the energy incident on the digital

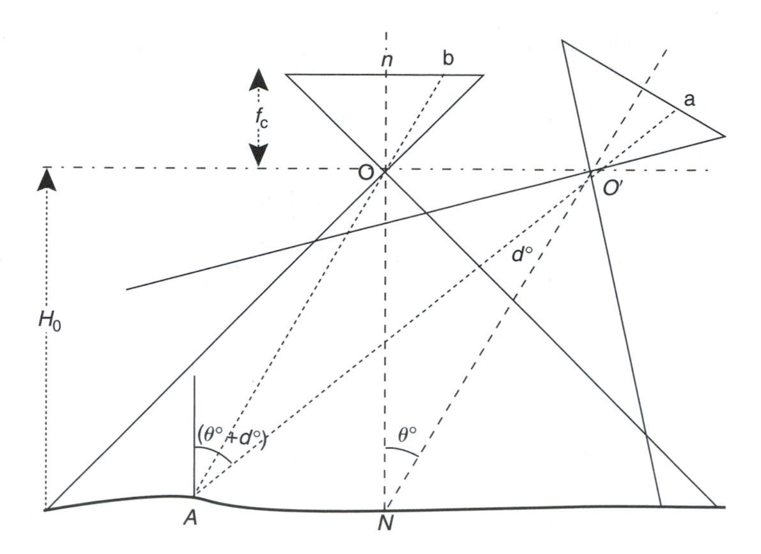

Figure 2.50 Tilted (oblique) and vertical photographs.

back. This means that automated exposure of a digital camera should ensure good exposure, as long as there is sufficient energy to activate the CCD array. By contrast, the exposure of a film camera is estimated by the use of an exposure metre that is independent of the film in the camera.

2.4.2 Acquisition of Aerial Photography with a Framing Camera

Aerial photography can be acquired either as vertical aerial photographs or as oblique photographs (Figure 2.50). With vertical aerial photographs, the camera back is approximately horizontal and the camera axis is vertical. With vertical photographs the scale in the photograph is approximately constant across the photograph, but varies as a function of elevation distances in the terrain. With oblique photographs the camera axis is tilted to the vertical axis. With oblique photographs the scale varies across the photograph as a function of the tilt, position in the photograph and the height differences as is discussed in more detail below.

Aerial photography is usually required of a larger area than can be covered with the one exposure of the camera. This problem is usually solved by covering the area with a number of photographs, arranged in sets or strips along flightlines of the aircraft as shown in Figure 2.51. Within each strip the photographs are taken at a regular interval designed to ensure that there is 10% or more overlap between the adjacent photographs. Usually, the overlap is either 60 or 80%, to provide sets of photographs in which all points are photographed on at least two adjacent photographs. The reason for this is to provide stereoscopic coverage of the area of interest, as will be discussed in Chapter 3. About 15% of the coverage of one strip is also covered on the adjacent strip, to ensure no gaps between adjacent strips of photography.

An aerial photograph contains not only the image, but also essential information about the image, including:

1. The agency that took the photograph
2. Flight or sortie number
3. Photograph number, within the flight or sortie
4. Focal length of the camera
5. Altimeter reading at the time of exposure of the photograph

6. Time and date of the photograph
7. Levelling bubble to indicate the photograph tilt at the time of exposure

In the centre of each side of an aerial photograph are the fiducial marks. The intersection of the lines joining the fiducial marks on opposite sides of the photograph indicates the position of the principal point of the photograph.

The scale of an aerial photograph, as in a map, is defined as the ratio of one unit of distance on the photograph, or map, to the number of units of distance to cover the equivalent distance on the ground. It is specified by the ratio 1:x indicating that 1 unit of distance on the map or photograph represents x units on the ground. The parameter x is known as the scale factor. Thus, if an aerial photograph is at a scale of 1:48,600 then 1 mm on the aerial photograph represents 48,600 mm or 48.6 m on the ground.

The scale of a vertical aerial photograph can be determined in either of two ways. The first method is to measure the distance between two features on the aerial photograph and measure the equivalent ground distance. The equivalent ground distance can be calculated as the product of the map distance and the map scale factor. The scale of a vertical aerial photograph can also be calculated from the flying height above ground level and the focal length. In Figure 2.52 the scale at point b is given by:

$$\mathrm{Ob} : \mathrm{OA} \quad \text{or} \quad 1 : (\mathrm{OA}/\mathrm{Ob})$$

by similar triangles:

$$\frac{\mathrm{OA}}{H_0} = \frac{\mathrm{Ob}}{f_\mathrm{c}}, \quad \text{so that} \quad \frac{\mathrm{OA}}{\mathrm{Ob}} = \frac{H_0}{f_\mathrm{c}} \tag{2.28}$$

Where H_0 is the flying height above ground level so that the scale of a vertical aerial photograph is:

$$1 : H_0/f_\mathrm{c} \tag{2.29}$$

For flat, horizontal ground surfaces a vertical aerial photograph has constant scale at all points in the photograph. Height differences introduce changes in H_0 and hence introduce changes in the scale of the photograph. Tilts in a photograph introduce changes in scale across the photograph. Under normal conditions where there are variations in topographic height and small tilts in the photograph,

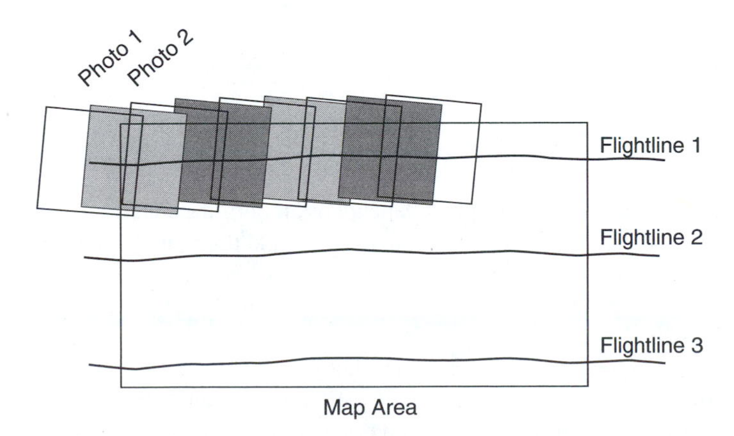

FIGURE 2.51 Typical layout of an aerial sortie, with flight lines of photography by 60–80% and with sidelap between the photographs in the different flight lines of 10–20%.

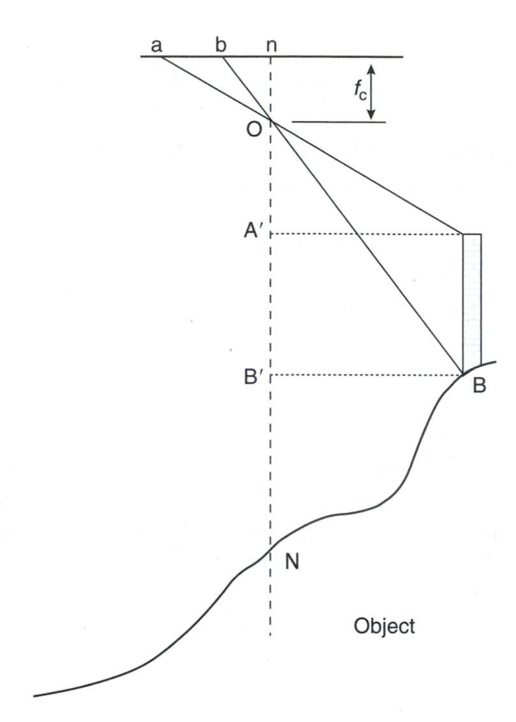

FIGURE 2.52 The effects of height differences on position in an aerial photograph.

calculation of the scale using the flying height and focal length will give only an approximate value for the scale.

Photographs that are taken with large tilts are known as oblique photographs. High oblique photographs contain the horizon whilst low oblique photographs do not. Consider the tilted photograph in Figure 2.50. Scale at A is $1 : \text{O}'\text{A}/\text{O}'\text{a}$.

Now

$$\text{O}'\text{A} = H_0 \times \sec(\theta + d)$$

Now

$$\text{O}'\text{a} = f_c / \cos(d)$$

Scale at A is

$$1 : (H_0 \times \cos(d))/(f_c \times \cos(\theta + d)) \tag{2.30}$$

The scale of an oblique aerial photograph is dependent on both the magnitude of the tilt, θ and the position of the point in the photograph, relative to the principal point, in the direction of tilt. The scale of an oblique aerial photograph is continuously changing in the direction of tilt, but is constant at right angles to the direction of tilt. If there are variations in altitude in the terrain then these variations will also affect the scale.

An advantage of oblique aerial photographs is that very large areas can be recorded in an oblique photograph. The disadvantage is that it is very difficult to take quantitative measurements from such photographs. In consequence, oblique photographs can be valuable when the user wishes to conduct some interpretation without measurement, or where a permanent record is required and the change in scale is not a serious disadvantage to that record.

2.4.2.1 The Effects of Height Differences on an Aerial Photograph

A height difference AB (Figure 2.52) introduces a displacement ab in the aerial photograph, as well as introducing a change of scale between a and b. The displacement ab is radial from the nadir point in the aerial photograph. The nadir point is defined as the point on the ground surface vertically below the front nodal plane (principal point) of the lens cone, and its corresponding point on the image plane. In a vertical aerial photograph the nadir and principal points coincide so that height displacements are radial from the principal point. The greater the height difference, and the further out the point is from the nadir, the greater the height displacement. Consider the height displacement ab in Figure 2.52.

$$\frac{\mathrm{AA}'}{\mathrm{OA}'} = \frac{\mathrm{na}}{f_\mathrm{c}}$$

then

$$\mathrm{OA}' = \mathrm{AA}' \times f_\mathrm{c}/na$$

also

$$\frac{\mathrm{BB}'}{\mathrm{OB}'} = \frac{\mathrm{nb}}{f_\mathrm{c}}$$

then

$$\mathrm{OB}' = \mathrm{BB}' \times f_\mathrm{c}/\mathrm{nb}$$

The height difference

$$\begin{aligned}\mathrm{AB} = \Delta H_{\mathrm{AB}} &= \mathrm{OB}' - \mathrm{OA}' \\ &= f_\mathrm{c} \times (\mathrm{BB}'/\mathrm{nb} - \mathrm{AA}'/\mathrm{na})\end{aligned}$$

Now $\mathrm{BB}' = \mathrm{AA}'$ by definition so that:

$$\Delta H_{\mathrm{AB}} = \mathrm{BB}' \times f_\mathrm{c} \times (\mathrm{na} - \mathrm{nb})/(\mathrm{na} \times \mathrm{nb})$$

Let $\mathrm{OB}' = H_0$ the flying height, then:

$$\Delta H_{\mathrm{AB}} = H_0 \times \mathrm{ab}/\mathrm{na} = H_0 \times \mathrm{ab}/(\mathrm{nb} + \mathrm{ab}) \tag{2.31}$$

When a point is recorded in two photographs then the height displacements in the two photographs will be of different magnitude, and will be in opposite directions as is shown in Figure 2.53. This characteristic can be used to estimate the difference in height between two points in an image. It is also the basis for stereoscopic vision, or the ability to simultaneously view an overlapping pair of photographs, with each eye viewing different images on the pair, and to see a three dimensional model of the overlap. Most aerial photographs are taken with 60% overlap so as to be able to view the area covered by each photograph in two stereo pairs. Stereoscopy will be discussed in Chapter 3.

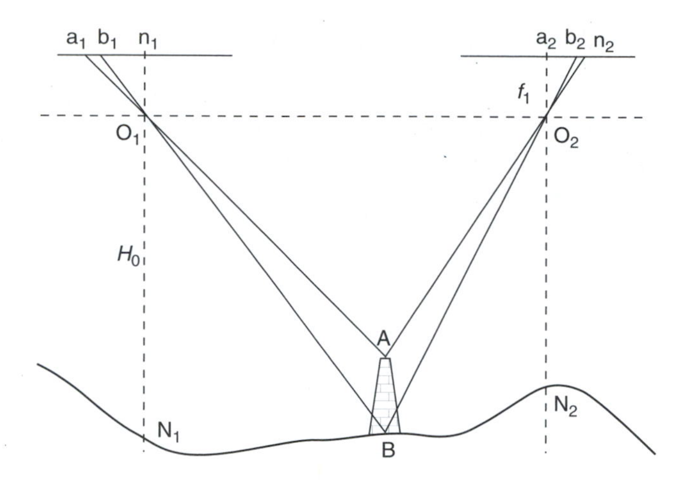

FIGURE 2.53 Height displacements for a point in an overlapping pair of aerial photographs.

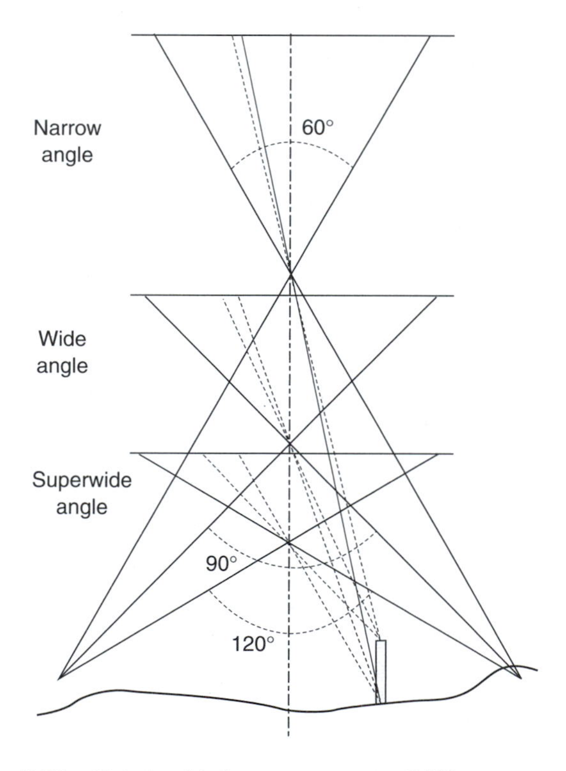

FIGURE 2.54 Relationship between cameras of different cone angles.

2.4.2.2 Types of Lens Cones

Aerial cameras are normally designated as being *Normal*, *Wide Angle* or *Superwide Angle* depending on the FOV of the camera, in accordance with Table 2.3. For a given image size and scale, as the FOV increases, the focal length and flying height decrease, as shown in Figure 2.54. Constant flying height would increase the coverage and reduce the scale as the focal length is reduced.

The figure also shows that, for a given scale photography, as the focal length decreases, the height displacement increases for a given height difference because the distance between the camera and the object decreases. In consequence, superwide angle photography will contain the largest

TABLE 2.3
Relationship Among Lens Cone Type, FOV and Focal Length for Aerial Cameras

Class	Field of View (Cone Angle)	Focal Length (23 x 23 cm² format)
Normal	Less than 60°	150 mm or longer
Wide angle	90°	115 mm
Superwide angle	120°	90 mm

displacements due to height differences and will thus be best for measurement of height differences. This same characteristic makes superwide angle photography the least suitable when displacements due to height differences interfere with the task being conducted, such as the construction of a mosaic or measurement of planimetric distances.

Variations of the standard camera are stereoscopic and multispectral cameras. Stereoscopic cameras contain two camera lenses side by side in the camera, so that simultaneous exposure of both cameras gives a stereoscopic pair of images on the camera film.

In a multispectral camera a series of cameras are aligned within a frame. Normally, at least four cameras are involved. Each camera uses a selective filter to record incident radiation in that waveband. The images from each camera are acquired simultaneously, providing a synchronized set of images in the wavebands being recorded. The images can be integrated together to give false colour images. One of the main problems with multispectral cameras is the need to accurately register all the cameras since slight variations in pointing direction will give large differences in coverage.

2.4.3 THE SCANNER

In scanning sensors the reflected radiance from the surface is optically scanned and electronically recorded. The reflected radiance is focused onto light sensitive detecting devices that convert the electromagnetic energy into an electrical signal. The electrical signal is then converted into either a photographic image or more usually into a digital signal for use in computer analysis. There are two commonly used forms of scanners:

1. The moving mirror scanner
2. Pushbroom scanners

2.4.4 THE MOVING MIRROR SCANNER

A moving mirror scanner uses a rotating or oscillating mirror to scan the surface, and direct a narrow beam of energy onto a small set of detectors, after passing through a filter unit. The moving mirror scanner consists of five main components:

1. *The scanning element*. A mechanically driven mirror scans across a strip approximately at right angles to the direction of movement of the platform (Figure 2.55). The *instantaneous field of view* (IFOV) of the scanner is the FOV of the mirror at the instant that the energy is sensed on the detector. The IFOV controls the area on the surface that is being sensed at any instant by the mirror of the scanner, where this area depends upon the angular field of view and the distance between the sensor and the surface. The area of the mirror controls the total amount of energy that can come into the scanner, and thus larger mirrors are required

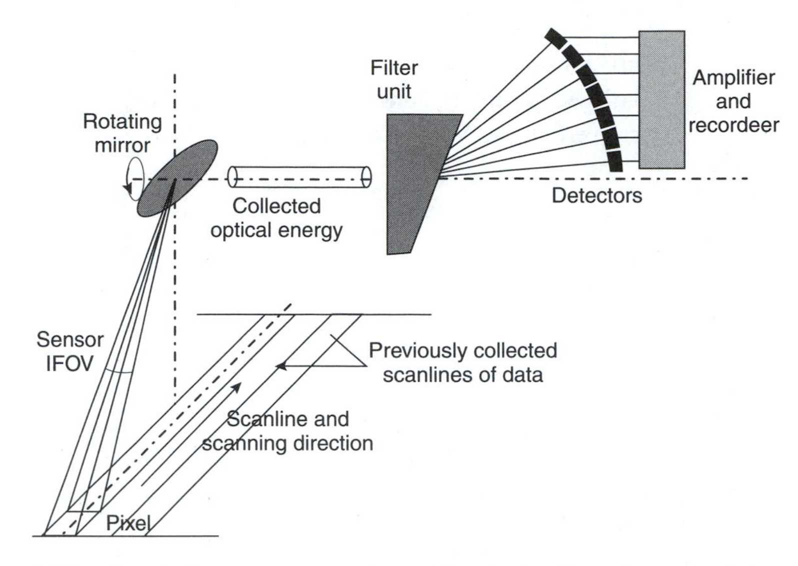

FIGURE 2.55 A typical scanner construction and its relationship to the surface being sensed.

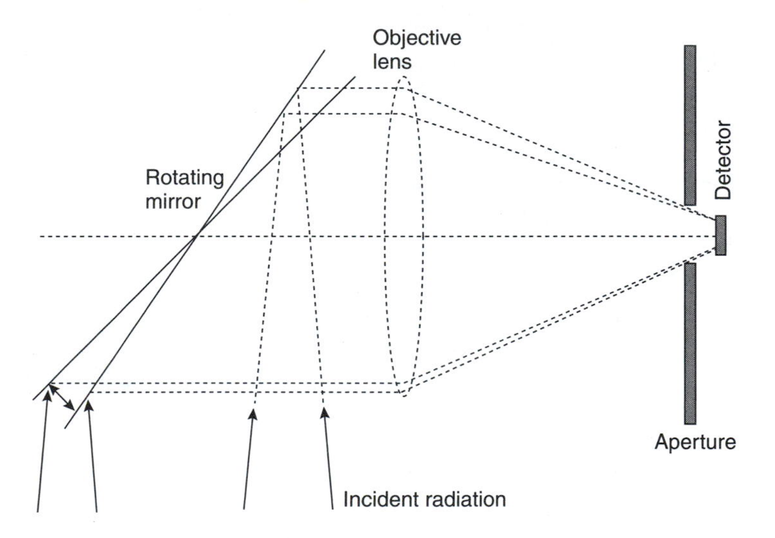

FIGURE 2.56 The object space scanning unit.

if either narrow wavebands or low light levels are expected. The scanning unit may also collect energy from a reference source to calibrate the detector and electronics.

The scanning unit can be of two types:

(a) *Object space scanners.* The majority of mechanical scanners are of this type because they always generate an image on the axis of the objective lens, and the optical aberrations that affect the signal are minimised. For these scanners the scanning mirror is either an oscillating (Figure 2.56) or conical mirror (Figure 2.57). The conical mirror is rarely used because mirror imbalance is a difficult problem to eliminate. The oscillating mirror generates a linear scan, unlike the conical rotating mirror; a most desirable characteristic. However, it is extremely difficult to keep the velocity of the mirror constant across the whole of the sweep, so that positional errors are common.

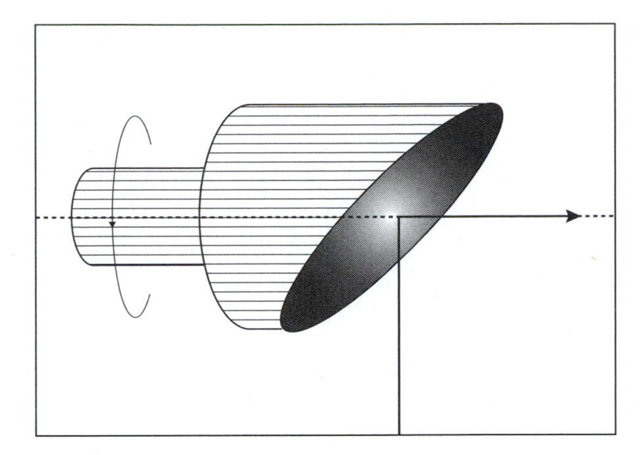

FIGURE 2.57 The rotating conical mirror scanning unit.

   (b) *Image space scanning units.* In the object space scanners the scanning unit comes before the objective lens of the scanner, whereas in image space scanners the objective lens precedes the scanning unit.
2. *Objective.* The objective lens focuses the incoming radiation onto the spectrometer or filter unit. The objective must be optically corrected to minimise radiometric and geometric distortions.
3. *Spectrometer or filter unit.* The purpose of the spectrometer or filter unit is to split the beam into a number of specified wavebands. This can be done in a number of ways as discussed in Section 2.2.6.
4. *Detectors.* Convert the incident electromagnetic radiation into an electrical signal.
5. *The Recorder.* The electrical signal is normally amplified before it is used to create a digital signal for subsequent analysis by computer. The digital signal is created by sampling the electrical signal at a regular interval by means of an analog to digital (A to D) converter to create digital values for the IFOV at a regular interval across the scanline.

2.4.4.1 The Resolution of Scanner Data

The scanner records incident radiation as a series of scanlines at about right angles to the flight line of the platform. Within each scanline there is a set of recorded values that are called picture elements or pixels, with each pixel recording the radiance entering the scanner optics from an IFOV, with the pixel representing the radiance originating from the ground area beneath that IFOV. The pixel is thus one measure of the spatial resolution limit of scanner data. The digital values normally range between 0, representing negligible radiance into the sensor, and $(2^n - 1)$ at the full scale or saturation response of the detector where each data value is stored in n bits. The values normally have a range of 2^n so as to get the maximum number of values or shades for a given storage capacity. The radiometric resolution of the data is defined as the difference in radiance between adjacent digital count values in the data. It can be computed by dividing the (radiance at the maximum count-radiance at the minimum count) by the number of digital values in the data. All scanners contain some internal current or dark current that has nothing to do with the detector response to incident radiation. The value of this dark current is the radiance recorded by the sensor when there is negligible radiance coming into the sensor.

The number of wavebands being sensed, and their spectral range, specifies the spectral sensitivity or spectral resolution of the scanner. The frequency of acquisition, particularly of satellite systems, defines the temporal resolution of the data.

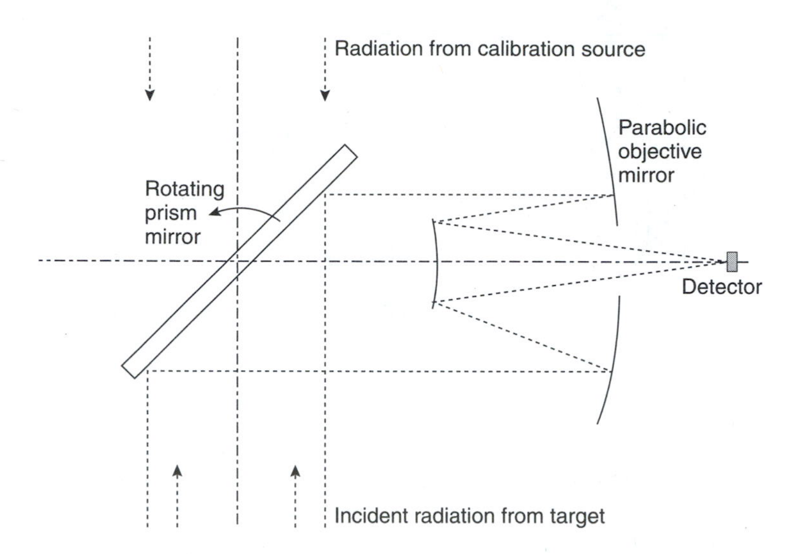

FIGURE 2.58 Diagram illustrating of the construction of a typical thermal scanner unit.

The actual level of discriminality of the information that can be derived from image data depends on a complex mix of these four resolutions; spatial, spectral, radiometric and temporal, as well as the discriminability of that information from other types of information in the data.

2.4.4.2 Thermal Scanner Data

Thermal scanners use the energy radiated by the object as their source, rather than radiation radiated by the sun or some other object. Thus, thermal imagery of the Earth uses the energy radiated by the Earth as its source. As was noted at the start of this chapter, all surfaces at temperatures above absolute zero radiate energy, where the level of energy radiated is a function of the temperature and the emissivity of the surface (Section 2.2.1). Because of this, and the warming of both the Earth and the atmosphere by the sun during the day, much thermal imagery of the Earth is taken at night.

The thermal characteristics of the surfaces being sensed need to be considered in designing thermal data acquisition and analysis. Not only does the temperature and the emissivity of the surface need to be considered, but also the capacity of the surface to absorb heat as a function of its temperature. Thus, water, with a high latent heat, or a high capacity to absorb heat before a change in temperature occurs, stays colder during the day and is warmer at night than the surroundings.

Scanners used to acquire thermal data have to address some unique problems. All units in the scanner, including the detector, radiate energy as a function of their temperature in accordance with Planck's Law (Section 2.2.1). The energy radiated by the sensor will activate the detector, thereby creating a large noise or dark current signal from the scanner that will usually swamp the signal coming from the target. The solution is to cool the detector, and all other elements of the scanner that may introduce a significant signal, sufficiently to reduce this noise to an acceptable level. This cooling is normally done by immersing the detector in liquid nitrogen with a boiling point of $-120°C$.

The other difficulty is in fabricating optics that will focus the radiation on the detector. The common solution is to use a reflecting mirror as the objective optics as shown in Figure 2.58.

2.4.4.3 Sources of Error in Oscillating Mirror Scanner Imagery

As a scanner sweeps across the object, it builds up the image as a series of scanlines, recording a set of pixel values across each scanline. Because of the serial nature of data acquisition, changes in the altitude or orientation of the platform during or between scans will affect the location of pixels

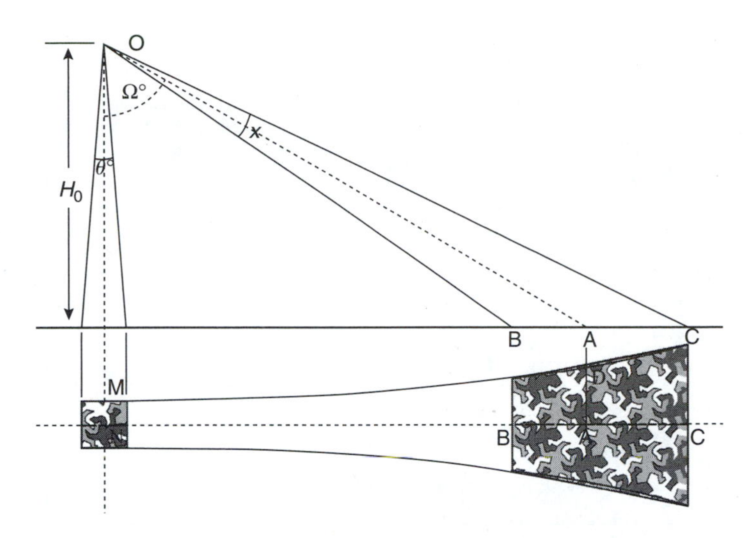

FIGURE 2.59 Geometric relations in a scanline.

and scanlines on the surface of the object. The more common sources of geometric error in scanner data are:

1. *Changes in elevation or orientation of the platform.* The IFOV and angular interval between pixels remains constant, independent of the altitude of the scanner. Changes in elevation will change both the size of, and the ground interval between, pixels. None of the current methods of monitoring variations in altitude or orientation in the aircraft or satellite are to a sufficient accuracy to calculate the ground co-ordinates of image data to the precision required for most purposes. Image rectification is required for this purpose.

2. *Changes in pixel size with sweep angle.* As the mirror rotates out from the nadir, so the distance between the object and the sensor increases (Figure 2.59). From Figure 2.59 the geometric relations are

$$\begin{aligned}\text{Pixel size at nadir} &= 2\text{MN}\\ &= 2H_0 \tan(\theta/2) \qquad (2.32)\end{aligned}$$

$$\begin{aligned}\text{Pixel size at inclination}&\\ \text{in scanline direction} &= \text{NC} - \text{NB}\\ &= H_0 \times [\tan(\Omega + \theta/2) - \tan(\Omega - \theta/2)]\\ &= \frac{H_0 \times [2\tan(\theta/2) \times (1 + \tan^2 \Omega)]}{[1 - \tan^2 \Omega \times \tan^2(\theta/2)]} \qquad (2.33)\end{aligned}$$

$$\begin{aligned}\text{At right angles} &= 2\text{AD}\\ &= 2H_0 \sec \Omega \times \tan(\theta/2) \qquad (2.34)\end{aligned}$$

Overlap between adjacent scanlines increases as a function of sec (Ω). Since the sampling interval or angle along the scanline is constant at $\theta/2$, the pixel size and interval will both increase with increasing Ω. In consequence there will be negligible overlap between pixels along the scanline, but there will be increasing overlap at right angles to the scanline.

3. *Finite scan time.* A finite scan time will introduce a shift or displacement across a scanline in the same direction as the direction of travel of the platform. The size of this displacement, D, depends on

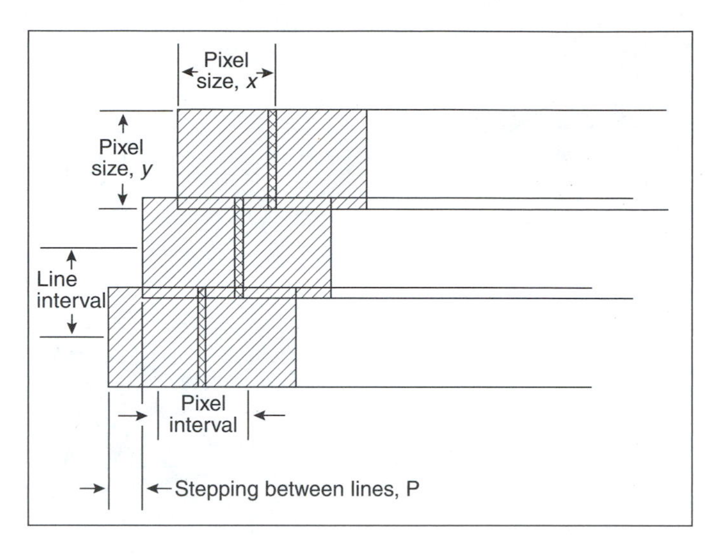

FIGURE 2.60 Stepping of satellite scanner data due to earth rotation.

the time from the start of the scan to the pixel i, t_i sec and the platform forward velocity, V according to the equation $D = Vt_i$ at pixel i.

4. *Variations in mirror velocity.* Rotating mirror scanners can be kept to a uniform velocity across the scanline. It is very difficult to maintain constant mirror velocity in oscillating mirror scanners, due to the acceleration at the start of the scan and deceleration at the end. This change in velocity creates errors in position of the pixels across the scanline if the pixels are assumed to be at a constant separation. The scanner mirror velocity can be calibrated and corrections applied for this source of error.

5. *Earth rotation.* Introduces displacements due to the relative velocities of the Earth and the platform, ignoring the forward velocity due to the platform. With aircraft there is little relative velocity and so there is little error from this source. However, there is considerable relative velocity between a satellite and the Earth and in consequence large displacements are incurred. If the Earth radius is given as $R = 6{,}370{,}000$ m then in time t sec a point on the Earth's surface at the equator moves an angular distance of d sec of arc. The Earth completes one rotation of $360 \cdot 60 \cdot 60$ seconds of arc in 24.60.60 sec of time. In time t

$$\begin{aligned} d &= 360 \cdot 60 \cdot 60t/24 \cdot 60 \cdot 60 \\ &= 13.84t \text{ seconds of arc} \\ &= 1 \cdot 098 \cdot 10^{-4}\text{t rad} \end{aligned}$$

At latitude θ the effective radius $R' = R \times \cos(\theta)$ so that the distance covered is:

$$\begin{aligned} D &= d \times R' \\ &= 1.098 \times t \times R \times \cos(\theta) \times 10^{-4} \end{aligned} \quad (2.35)$$

where D is in the same units as the radius. This creates a step between adjacent scans as shown in Figure 2.60.

6. *Earth curvature and height differences.* Curvature of the Earth introduces displacements in the position sampled by a scanner as shown in Figure 2.61. With both Earth curvature and heights, the errors increase as the angle of the mirror increases out from the nadir.

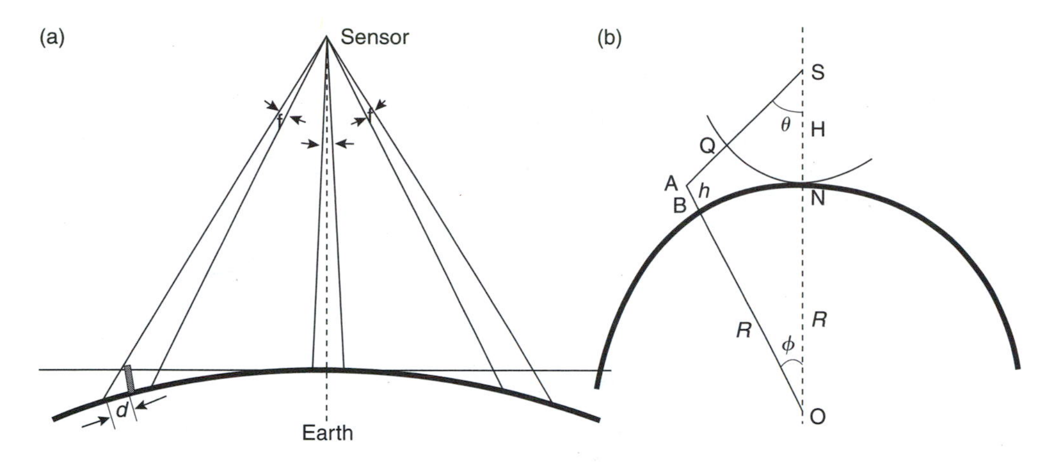

FIGURE 2.61 Displacements due to earth curvature and height displacements and the geometric relationships used to solve this problem.

Consider Figure 2.61(b), in which an image scanline is acquired across the page, sweeping out angle θ in the process. The scanline is constructed as a series of pixels, of a specific size and at a specific interval, so that they define a distance on the arc QN. A height difference AB $= h$ at a distance arc QN out from the centre of the scanline introduces errors into the data. The point A will be imaged further out than B, but it should be imaged at point B, at distance arc BN from the centre of the scanline.

We know the Earth radius, R, the satellite height, H, and we want to find out the error introduced by a difference in elevation, h at distance arc QN in the scanline. The error in the distance, $\varepsilon =$ arc BN − arc QN.

Arc QN = pixel count from the centre of the scanline × pixel interval $= \theta \times H$, so that:

$$\theta = (\text{pixel count} \times \text{pixel interval})/H \tag{2.36}$$

In triangle SAO:

$$\frac{R+h}{\sin(\theta)} = \frac{R+H}{\sin(180-\theta-\phi)} = \frac{R+H}{\sin(\theta+\phi)} \tag{2.37}$$

So that:

$$\frac{\sin(\theta+\phi)}{\sin(\theta)} = \frac{R+H}{R+h} = k$$

Expand the left-hand side to give:

$$\cos(\phi) + \cot(\theta)\sin(\phi) = k \tag{2.38}$$

Replace $\cos(\phi)$ with $(1-\sin^2(\phi))^{1/2}$, so that:

$$(1-\sin^2(\phi))^{1/2} = k - \cot(\theta)\sin(\phi)$$

Expand and simplify to get an equation in $\sin(\phi)$, so as to solve for ϕ:

$$\sin^2(\phi) - 2k\cos\theta\sin\theta\sin\phi + (k^2-1)\sin^2\theta = 0,$$

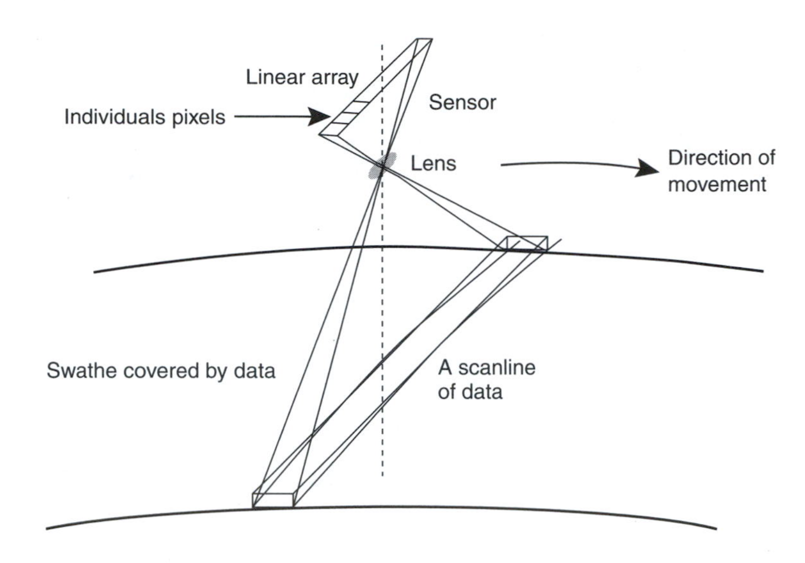

FIGURE 2.62 Schematic layout of a pushbroom scanner.

from which:

$$\sin\phi = k\sin\theta\cos\theta \pm \sin\theta\sqrt{1 - k^2\sin^2\theta} = A \pm B \tag{2.39}$$

In practice we are looking for the solution $\sin\phi = A - B$, and arc BN $= R\phi$, so that we can now find the error introduced due to h at a distance from the centreline. In 2.33, if $h = 0$, then the equation devolves to the error created by Earth curvature.

Radiometric errors are another major source of error in scanner data. The sensitivity of the detectors in a scanner changes over a period of time. There is electronic noise in the detector circuitry that can affect the signal attributed to the detector. Whilst these errors do not affect the geometric quality of the data, they do affect the validity of the data received, and hence the accuracy of any derived information. The major sources of radiometric errors are:

1. *Changes in detector response* with time or environmental conditions.
2. *Detector lag*. There is a finite lag between the time when energy falls on the detector and the detector responds. Similarly, once a detector is responding then there is a lag in the response after the incident radiation ceases. If the incident radiation is intense and saturates the detector then the lag time can be increased.
3. *Sensor electronic noise*. All detectors and circuitry generate electronic noise or dark current. The dark current can change with age of the circuitry in the scanner.

2.4.5 Pushbroom Scanners

This sensor uses a wide-angle optical system that focuses a strip across the whole of the scene onto a linear array of detectors. The signal from each detector is sampled to create a record for the pixels in that line. The detector array is sampled again when the platform has moved forward by the size of the pixels on the Earth's surface. In this way the sensor can be visualised as a broom that sweeps out a swath of data (Figure 2.62).

If only one band is to be imaged then a single array of detectors is adequate. When multiple wavebands are to be recorded then a linear array of detectors is required for each band, with appropriate filters set in front of each array. The arrays are arranged so that they image adjacent scanlines.

The main disadvantages of the pushbroom scanner are the number of detectors and their calibration. The major advantage is that the scanner does not contain a moveable mirror and so a major source of malfunction or wear, particularly in a satellite, is removed. The other advantage is that there is a long dwell time between sampling the detectors compared to the time available in mechanical scanners. This dwell time can be used to accumulate a larger signal and hence used to either sense in narrower wavebands or develop a better signal to noise ratio.

2.5 ACTIVE SENSING SYSTEMS

2.5.1 INTRODUCTION

Active sensing systems illuminate the target with their own source and sense the reflected radiation. This is in contrast to passive systems as previously discussed that depend upon independent, usually naturally occurring, sources of illumination, such as the sun or the Earth itself. This makes active systems much more independent of natural events than passive systems. The most common type of active sensing system is radar. The name comes from the early use to which radar was put in detecting and estimating the distance to aircraft. This type of radar depicts the reflecting objects as bright spots on a black screen. In contrast remote sensing uses radars that create an image of the surface. The image is depicted as variations in density or greyscale on a photograph, or variations in digital values across the area imaged.

British physicist James Clerk Maxwell developed equations governing the behaviour of electromagnetic waves in 1864. Inherent in Maxwell's equations are the laws of radio wave reflection. The German physicist Heinrich Hertz first demonstrated these principles in experiments in 1886. Some years later a German engineer Chistian Huelsmeyer proposed the use of radio echoes in a detecting device designed to avoid collisions in marine navigation.

The first successful radio range-finding experiment was carried out in 1924, when the British physicist Sir Edward Victor Appleton used radio echoes to determine the height of the ionosphere, an ionised layer of the upper atmosphere that reflects longer radio waves. The British physicist Sir Robert Watson-Watt produced the first practical radar system in 1935, and by 1939 England had established a chain of radar stations along its south and east coasts to detect aircraft and ships. In the same year two British scientists were responsible for the most important advance made in the technology of radar during World War II. The physicist Henry Boot and biophysicist John T. Randall invented an electron tube called the resonant-cavity magnetron. This type of tube is capable of generating high-frequency radio pulses with large amounts of power, thus permitting the development of microwave radar, which operates in the very short wavelength band of less than 1 cm, using lasers. Microwave radar, also called LIDAR (light detection and ranging) is used in the present day for communications and to measure atmospheric pollution.

In the 1950s side looking radars (SLAR) were developed in the USA, that created images as a function of the strength of the return signal, corrected for the different paths lengths of the slant range signal. The problem with these radars was the long antenna required to get reasonable resolution in the resulting image data, leading to the development of synthetic aperture, side looking radars (SAR) as will be discussed in the next sections.

The theoretical resolution limit of radars are coarser than that of imagery in the visible region because the wavelengths of the radiation used is much longer in the microwave portion of the spectrum than in the visible and NIR portion (Table 2.4). Another constraint is the energy required to generate the illumination signal. The power requirements of radars are many times greater than the energy requirements of passive systems. Generally this does not inhibit the acquisition of airborne radar imagery, but it may limit the capability of satellite-borne sensors. However, the independence of radar from natural illumination means that radar images can be acquired during the day and night. The atmosphere is usually transparent to the radars used in remote sensing, and so radar is much less affected by weather conditions than are optical sensors.

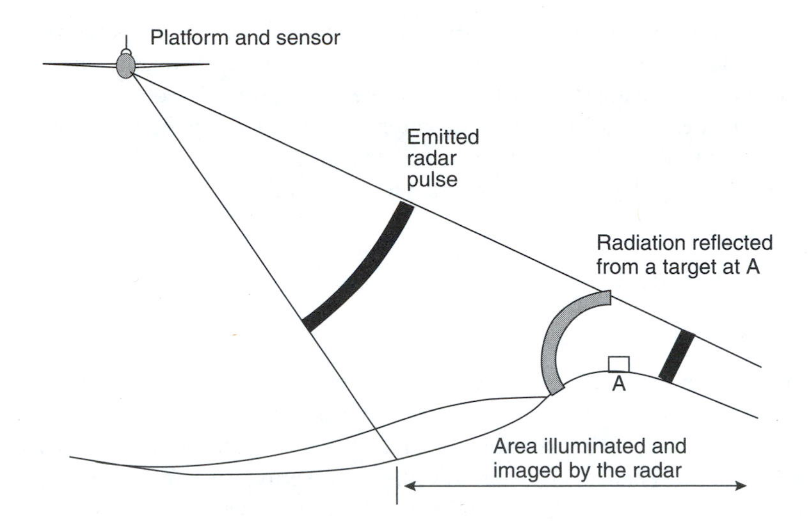

FIGURE 2.63 Geometry of a side looking imaging radar.

TABLE 2.4
The Common Radar Wavebands, Their Wavelength Ranges and Frequencies

Radar Band Name	Wavelength (cm)	Frequency (GHz)
P	30.0–100.0	1.0–0.3
L	15.0–30.0	2.0–1.0
S	7.5–15.0	4.0–2.0
C	3.8–7.5	8.0–4.0
X	2.4–3.8	12.5–8.0
Ku	1.67–2.40	18.0–12.5
K	1.19–1.67	26.5–18.0
Ka	0.75–1.18	40.0–26.5
Q	0.6–0.75	46.0–40.0

2.5.2 THE GEOMETRY OF RADAR SYSTEMS

Radars used in remote sensing are Side Looking Airborne Radars (SLAR), as shown in Figure 2.63. The radar antenna transmits a pulse of energy on a slant angle from the side of the aircraft down to the ground surface. Some of the signal is reflected from the surface back towards the radar antenna. The radar antenna receives the returned energy, converts the total travel time out and back, into a slant distance and then into a ground distance, and compares the transmitted with the return signal strength. The radar creates an image that is proportional to the strength of the return relative to the transmitted signal and corrected for the path length.

Radars are categorised by the wavelength of the microwave waves transmitted by the instrument. The wavelength of the radar affects the way that the radar images the surface. It is important to select the most appropriate radars for a task, and to be aware of the radar wavelength in the interpretation of radar imagery. The radars are grouped into categories as a function of wavelength as shown in Table 2.4. The relationship between wavelength and frequency is the same as for optical

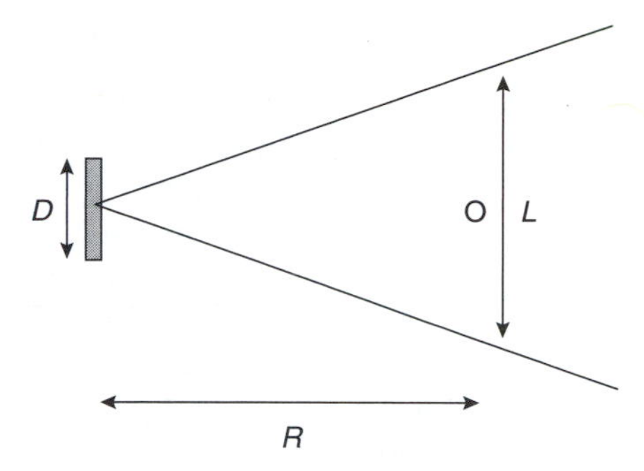

FIGURE 2.64 The approximated antenna pattern for side looking radar.

electromagnetic energy (Section 2.2), that is:

$$F \times \lambda = C \tag{2.40}$$

where C is the speed of light.

The radar antenna for side looking radars is normally a rectangular planar antenna, with the largest dimension in the along track direction. In Figure 2.64 is shown a top-view of the approximated antenna pattern, where the width of the pattern, L, depends on the length of the antenna, D (larger D means smaller L), the radar frequency (larger frequency means smaller L), and the distance from the radar to the target, R (larger R means larger L). For an SLAR the resolution in the along track direction is directly related to the width of the beam.

2.5.2.1 Resolution of Radar Data

A pulse of the real aperture radar beam illuminates a strip [area A on Figure 2.65(a)] at a particular instant. The duration or bandwidth of the pulse defines the resolution of the data in the cross-track or range direction. This resolution is constant at all distances from the sensor, since the bandwidth is constant. The situation shown in Figure 2.65(a) is typical of real aperture radars.

The pulse beam width is shown as area B, so that the width of B defines the resolution in the along track direction. The along track resolution decreases with distance from the aircraft due to the increasing beam width. The angular width of the beam is inversely proportional to the length of the antenna so that a longer antenna will reduce the angular beam width. The intersection of areas A and B indicates the resolution of the data. The strength of the return signal is dependant upon the path length, the antenna pattern and the reflectance or backscatter from the surface. The return signal slant distance is converted into a horizontal distance to create the image data. This type of radar is known as real aperture radar (RAR). Its coarse along track resolution will always be constrained by the largest antenna that can be carried on the platform. Since, even moderate resolution image data will require antenna of 10s of metres in length, in practice no current space borne radars and only a few airborne radars are real aperture radars.

Synthetic Aperture Radar (SAR) was developed to provide the resolution of very long antenna radar, but by the use of short antenna radar. An understanding of SAR can be considered from a number of perspectives, and two will be considered here. The measurements done along the flight path by the physical radar antenna can be combined to form a long synthetic antenna (also called aperture) (Figure 2.66). The individual measurements must be phase corrected corresponding to the difference in distance between the radar and the target at the different positions along the flight path. This correction will ensure that the long synthetic antenna is focused in the range distance corresponding to the target.

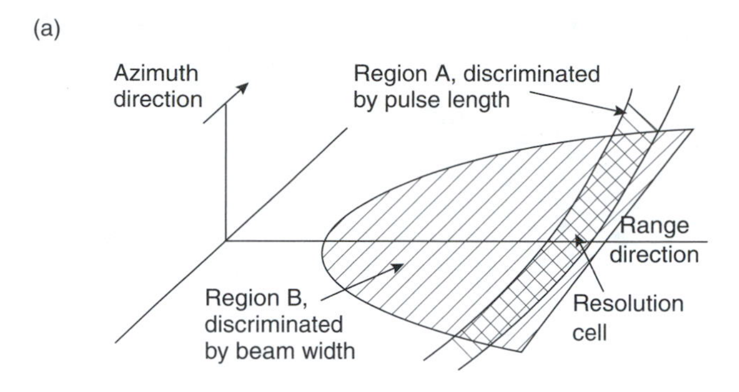

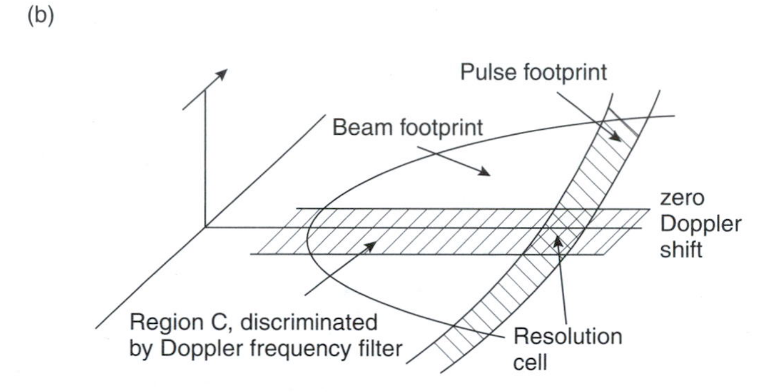

FIGURE 2.65 Resolution constraints on radar image data (a) real aperture radar, and (b) synthetic aperture radar.

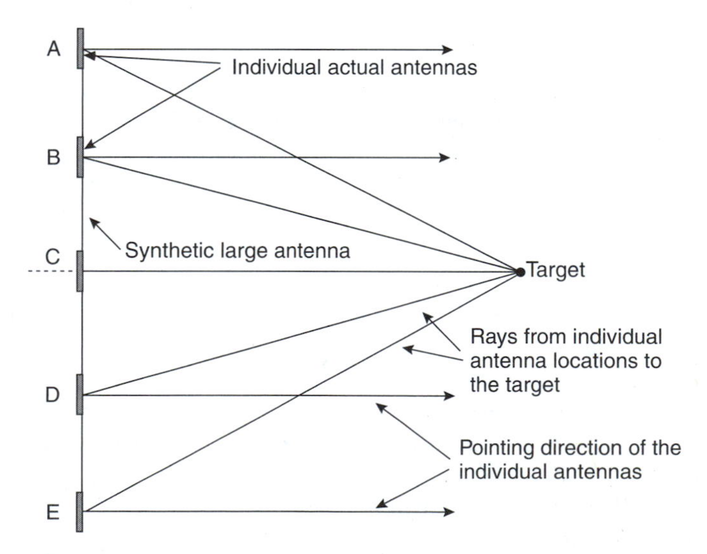

FIGURE 2.66 Concept of synthetic large antenna represented by the observation of individual rays from that antenna by individual actual antennas.

Another way to consider SAR is in terms of Doppler shift. Everyone is familiar with the Doppler effect in relation to the whistle of an approaching and receding train. The whistle has a higher frequency as it approaches, and lower as it recedes. The whistle emitted by the train has a particular frequency and wavelength. If the listener was moving at the same speed in the same direction as

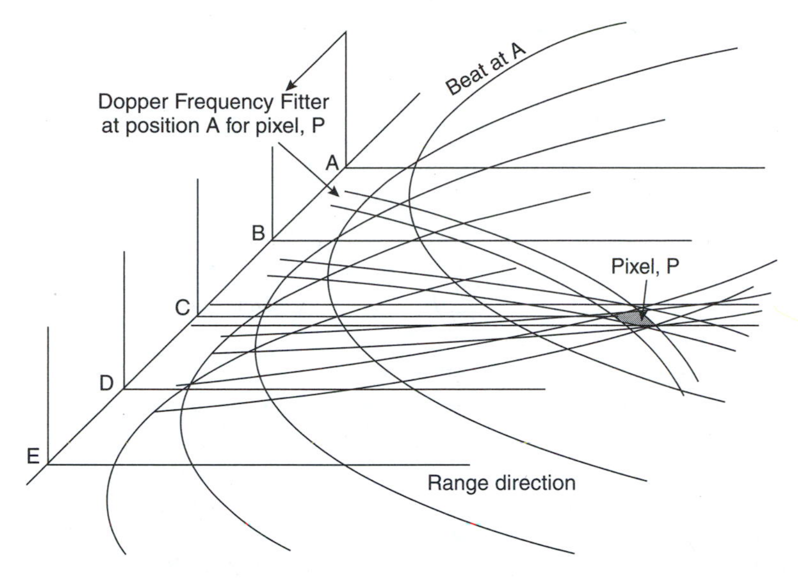

FIGURE 2.67 Incorporation of Doppler shift into the construction of SAR image data.

the train, the two would have zero relative motion, and the listener would hear the signal at the same frequency as it was emitted. However, when the listener is stationary, he is effectively moving through the signal wave as he is hearing it, giving the effect of changing the apparent frequency of the signal. If the train is approaching, he moves through the signal faster than if there was no relative motion, and so achieves a higher frequency. Conversely, when the train is receding, he is moving more slowly thorough the signal and a lower frequency is the result. The same concept applies in radar. Because of the Doppler effect, the frequency of the signal coming back to the radar will change depending on the relative velocity of the two. At right angles to the aircraft there is no relative motion, and so there is no Doppler shift. In front of the aircraft, the aircraft is approaching and the result is higher frequencies. Behind the aircraft, the shift is to lower frequencies. The magnitude of the Doppler shift is a function of the relative radial velocity of the radar and the object. It would be possible to introduce a Doppler filter, to only accept signals within a specific range of frequencies. However, this has the disadvantage that most of the beam power is wasted. Since the Doppler shift is different at each of the positions A, B, C, D and E, these can all be used, with their Doppler shift, to improve the resolution of the data. Figure 2.67 illustrates the concept.

To construct the image either using the synthetic antenna principle or the Doppler shift principle, both the amplitude and the phase of the return signal at each point in the image must be stored for each look of the radar. These amplitude and phase images are processed in a digital SAR processor to derive the images that we use from the source images.

2.5.2.2 Effect of Height Displacements

The slope of the radar signal can cause radar shadows and radar layover (Figure 2.68). Radar shadows occur when the slope of the terrain is at steeper gradients away from the sensor than the radar signal. As a consequence, there is a long interval in the signal between adjacent sections of the image, and this gap is shown as black in the image, or as a shadow. Radar layover occurs when the top of a hill is closer to the sensor than the bottom, and so is imaged first. Since it is imaged first, the surface below and in front of it is effectively imaged beneath and behind it, creating confusion in the image between the response from this area, and the image from the top of the hill.

Radar images contain displacements due to height differences, much like aerial photographs. These displacements can be used in pairs of radar images to construct stereoscopic images. Suitable

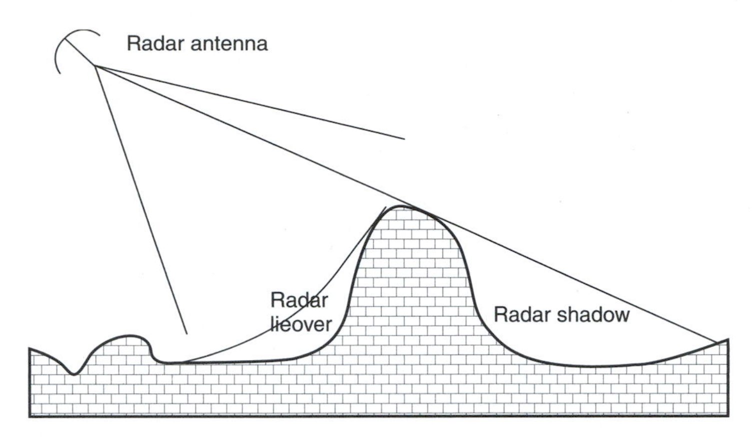

FIGURE 2.68 The effects of topographic heights on radar imagery.

sets of radar imagery can therefore be used stereoscopically to visually analyse landforms. An alternate approach to the determination of elevation models from radar data is by the use of radar interferometry, discussed in Section 2.5.5.

2.5.3 The Attenuation and Scattering of Radar in the Atmosphere

The atmosphere is transparent to radar wavelengths, making these wavelengths a good window for the acquisition of image data. Particulate matter in the atmosphere affects radar waves, in the same way as are optical wavelengths, but the longer wavelengths of radar data mean that Rayleigh scattering from the atmospheric gases is insignificant and most particulate matter is too small to cause significant Mie scattering. Clearly the shorter the wavelength, the larger is the effect of atmospheric particulate matter. However Mie scattering does occur when there are larger particulates in the atmosphere, such as rain and hail and so shorter wavelength radars are used to detect rain, hail and ice in clouds.

This transmissivity of the atmosphere under most atmospheric conditions gives radar one of its important advantages. Imagery can usually be acquired when it is required, and this can be important in the detection and accurate mapping of some environmental conditions, which change dynamically in ways that have an impact on the interpretation or analysis.

2.5.4 The Information Content of Radar Imagery

The radar equation is the fundamental equation showing the amount of energy received by a radar system from a target:

$$W_r = \frac{[W_t \times G_t]1}{4R^2} \times \sigma \times \frac{[1] \times (A)}{4R^2} \tag{2.41}$$

where W_r is the received power, W_t is the transmitted power, G_t is the gain of the transmitting antenna, R is the slant range to the target, σ is the effective radar cross-section and A is the effective aperture of the receiving antenna.

The first term in the equation indicates the power per unit area at the target. The second term, σ, is the effective backscatter cross-section. The product of these two terms is the reflected energy radiated from the target. The third term introduces the losses that occur on re-transmission back towards the sensor. This occurs because the signal is transmitted as a spherical wave front and the energy is equally distributed across this wave front. As the spherical surface increases in size, so the energy density must reduce, in proportion to the spherical surface area. The product of the first three terms

TABLE 2.5
Factors that Affect the Effective Backscatter Cross-Section Area per Unit Area

Properties of the source
Wavelength
Polarisation
Incidence angle
Azimuth look angle
Resolution
Properties of the target
Roughness
Slope
Inhomogeneity
Dielectric properties
Resonant-sized objects

is thus the energy received at the sensor. The product of this energy and the effective aperture of the sensor gives the actual power measured by the sensor. Equation (2.41) can be written in the form:

$$W_r = \frac{W_i G_i^2}{(4\pi)^2 R^4} \sigma \times \lambda^2 \tag{2.42}$$

Since

$$G_i = 4\pi A/\lambda^2 \tag{2.43}$$

In Equation (2.43) the effective aperture has been expressed in terms of gain, G_t and wavelength of the radar, λ. All the factors in Equation (2.41) and Equation (2.42) can be controlled to some degree by the designer, except for σ. With imaging radars the area being sensed will not usually have homogeneous targets within each resolution cell in the image data, but will consist of a number of different objects, each with their own backscattering and absorption characteristics. Because of this Equation (2.41) and Equation (2.42) are often modified into the form:

$$W_r = \int \frac{W_t G_t^2}{4\sigma^0 R^4} \sigma^0 \lambda^2 \, dA \tag{2.44}$$

σ^0 is called the differential scattering cross-section or cross-section per unit area. In this way the effects of area on the backscattering cross-section have been addressed, so that σ^0 deals only with the characteristics of the surface. There are a number of properties of the radar and of the target that influence σ^0. These properties are listed in Table 2.5 and will be discussed in more detail in the remainder of this section.

2.5.4.1 Surface Roughness and Slope

Most surfaces are neither perfect Lambertian nor specular reflectors to radar signals, but act somewhere between these limits. One way to consider reflectance by radar is to consider that a surface consists of many facets. If the facets are small relative to the radar wavelength then the reflection due to normally incident radiation would be in all directions in response to the orientation of the

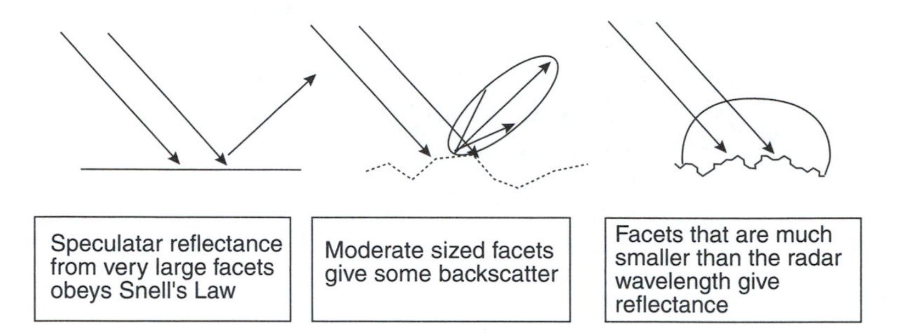

FIGURE 2.69 Backscattering cross-section angular distribution for incident radiation at angle of incidence, θ, as a function of facet size, L.

facets, and the resulting reflectance will be approximately Lambertian (Figure 2.69). As the facet size increases, the reflection becomes more specular until very larger facets, relative to wavelength, give specular reflectance.

A surface is considered to be smooth in radar if the height of the features in the surface, δh are given by the equation, $\delta h < (1/8) \times \lambda / \cos(\theta)$. Surface roughness is thus a function of both the wavelength of the radar, λ and its angle of incidence, θ. If the surface consists of large facets relative to wavelength then most of the energy is reflected in accordance with Snell's Law. This generally means that most of the energy will be reflected away from the antenna and the surface appears dark in the image. If the facets are at right angles to the radar signal, then most of the energy will be reflected back towards the radar, giving a very large response. Surfaces that have the characteristics of large facets include smooth water, airport runways, claypans, etc., and these surfaces are typically very dark to black in a radar image. As the surface becomes rougher, or the facets become smaller, the backscatter towards the receiving antenna increases, and the signal increases. Ploughed or cultivated fields, areas of gravel or stone appear rough to shorter wavelength radars and will thus reflect strongly back towards the sensor.

Figure 2.69 shows that the slope of the surface relative to the direction to the sensor, has an effect that is a function of the size of the facets. With very large facets the effect of slope is negligible. However, as the facet size decreases, the effect of slope increases, until at the extreme of very small facets the slope will control the direction of reflectance of all the energy. The effect of slope can dominate the effect of facet size, so that smooth fields on slopes facing the radar can have higher returns than ploughed fields that are sloping away from the sensor.

A particular aspect of surface slope is the corner angle effect, shown for two and three dimensions in Figure 2.70. The corner angle effect occurs when two facets that are large relative to the wavelength reflect the incoming radiation back towards the sensor. This will occur through a large range of angles of incidence on the facets as long as the facets are at right angles. The same effects can be seen in three dimensions, if three large facets are used, such that they are all at right angles to each other. Such situations often occur in man-made structures, so that the corner reflector effect is often seen from buildings, engineering structures such as bridges and power line frames.

The corner reflector effect can be used to create control points in the radar image data to rectify the image data. Corner reflectors are made of a lightweight metallic surface (Photograph 2.4) and oriented in approximately the direction of the incident radiation from the sensor. Since such targets are less than 1 m in size, their position can be accurately established by global positioning system (GPS) for the accurate and automated rectification of radar image data. Whilst the corner will reflect back to the sensor through a wide range of angles, the effective aperture of the corner reflector changes with orientation, with the largest aperture when the radiation is at 45° to the reflecting surfaces. As the radiation moves away from this the size of the effective aperture decreases, reducing

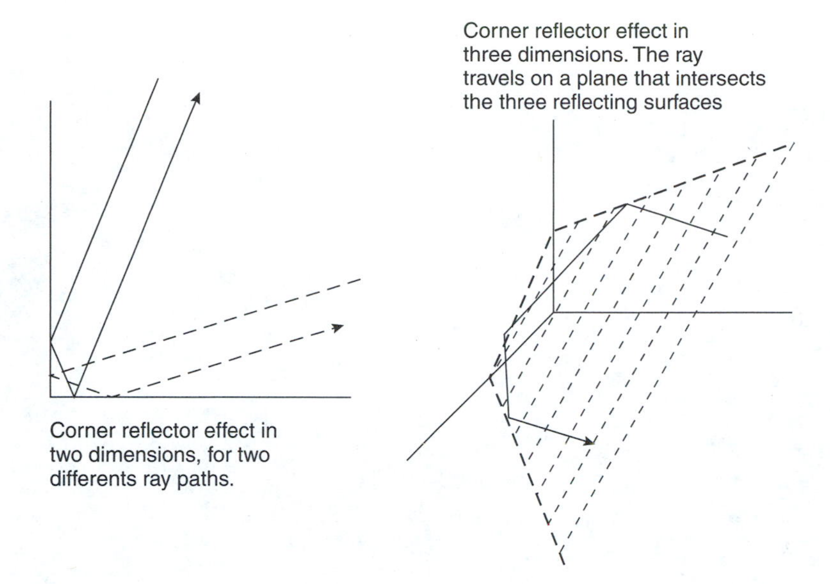

FIGURE 2.70 The corner reflector effect in two and three dimensions.

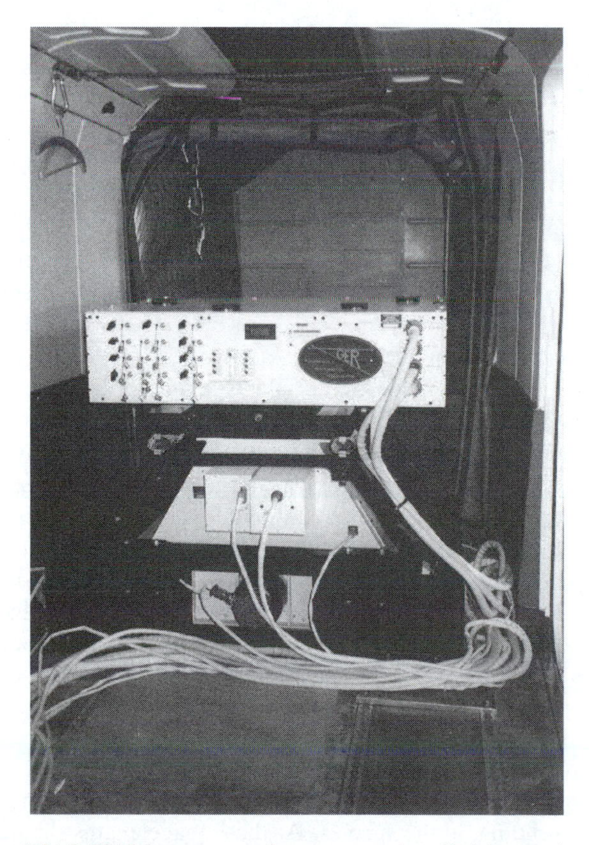

PHOTOGRAPH 2.3 The DIAS 7915 hyperspectral scanner as installed in an aircraft. The DIAS scanner, constructed at the German Aerospace Center, contains 79 wavebands of data for each pixel.

Photograph 2.4 The corner reflectors used by the Danish Technical University (photograph supplied by Dr. Henning Skriver. With permission).

the reflecting area and hence reducing the power return to the sensor. Corner reflectors are also used to calibrate the radar signal, to ensure an accurate estimate of the backscattering coefficient.

2.5.4.2 Inhomogeneity

The effect of the inhomogeneity of targets can be considered in terms of horizontal and vertical inhomogeneity. Horizontally inhomogeneous targets are those that are not homogeneous across the spatial extent of the pixel. This form of inhomogeneity can be due to variations between cover types, and due to inhomogeneity within cover types. Variations between cover types are dealt with in terms of different σ^0 values for the different cover types. Inhomogeneity within a cover type can be in the form of ploughing in an agricultural field, rows of plants in an orchard or vineyard, rock banding in geological formations and other forms of inhomogeneity. These forms of inhomogeneity introduce variations in signal strength as a function of the direction of flight relative to the orientation of the inhomogeneity. Thus, ploughed fields can give a large return when the ploughing is parallel to the flight line (at right angles to the range direction) and low responses at right angles to the line.

The effect of vertical inhomogeneity depends on the capacity of the radar to penetrate into the volume of the surface. Radar signals penetrate up to about 1 m into dry soils for low radar frequencies, but much less into wet soils for reasons that will be discussed later, so that inhomogeneities in soils are rarely detected by, or affect, radar image data. However, radars can penetrate into vegetative canopies; inhomogeneities that exist in the canopy can affect the radar signal. The radar signal penetrates into the canopy, and is then reflected from components in that canopy. The proportions of each depend very much on the density of the canopy, the wavelength of the radar and the size of the facets relative to the wavelength. Shorter wavelength radars, such as C band radars, generally reflect primarily from the top of the canopy since the leaf facets can vary from being smaller than, to larger than the wavelength, depending on the species. At these wavelengths the penetration is also less, contributing to the relatively higher return from the top of the canopy relative to the reflectance from within the canopy. With longer wavelength radars the penetration into the canopy is greater. This is in part because the facets are small compared to the wavelength of the radars and in part because

TABLE 2.6
The Dielectric Properties of Some Materials, with Most at 3 x 10^9 Hz

Material	ε'_r	$\varepsilon''_r/\varepsilon'_r$
Dry, sandy soil	2.55	0.006
Dry, loamy soil	2.44	0.001
Fresh snow	1.20	0.0003
Distilled water	77	0.157
Saturated sand	30	0.08

there is greater penetration by the radar wave. If the trunk and branch elements are moderate to large relative to wavelength, then strong specular reflections can occur, giving high response from within the canopy. If, however, the trunk and branch elements are small relative to wavelength, then the returns will not usually be so strong.

Radars usually show a complex mix of surface and volume scattering from vegetative canopies. Since these reflectances are strongly affected by the wavelength relative to the element sizes in the canopy, radar can often provide a lot of information about canopy structure.

2.5.4.3 Dielectric Properties

Each type of medium has its own unique characteristics in terms of its ability to absorb, store and transmit electrical energy. This characteristic is determined by the roughness of the surface of the medium and the permittivity (or complex dielectric constant) of the medium, defined by ε_c:

$$\varepsilon_c = \varepsilon'_c - j \times \varepsilon''_c \tag{2.45}$$

In which $j^2 = -1$.

To provide a standard reference, this is divided by the permittivity of free air, ε_0, to give the relative dielectric constant:

$$\varepsilon_r = \varepsilon'_r - j \times \varepsilon''_r \tag{2.46}$$

The relative dialectrical properties of various materials are listed in Table 2.6 for one wavelength [most are taken from Lintz and Simonett (1976), for 3 MHz radars] since the dielectric constant of surface materials changes with the frequency of the radar.

All the permittivity (ε'_r) values listed in Table 2.6 are quite small except for distilled water. The presence of water in the medium, whether it is vegetation or soil, will dramatically affect the permittivity of the medium, as can be seen by comparing the values for the dry sandy soil and the saturated sand in Table 2.6.

The real part of the dielectric constant controls the size of the reflection at and the transmission through a surface between two media. The fraction of the reflected signal that is backscattered towards the radar is determined by the surface roughness, as described above. The soil moisture, for instance, has a large influence on the real part of the dielectric constant, and thereby on the backscatter from the soil.

The imaginary part of the dielectric constant, ε''_r, determines the ability of the medium to absorb the wave. The ratio $\varepsilon''_r/\varepsilon'_r$ also indicates the depth of the material that might be penetrated by the radar as a function of wavelength, and still return a signal to the antenna. If $\varepsilon''_r/\varepsilon'_r$ is less than 0.16 then the distance at which the penetrating wave is reduced to 37% of its original value exceeds a wavelength in the material, assuming 100% reflection in the material. Because the signal has to return the same distance, a value of 0.08 should be used. Thus, for dry soils and dry snow, radar should penetrate several wavelengths into the surface.

2.5.4.4 Resonance-Sized Objects

Resonance-sized objects are those objects that are of a composition, size and construction that give a very high backscatter coefficient. As a consequence, their radar cross-section can appear to be much larger than that for typical objects in the image, that are actually much larger in size. The change in backscatter coefficient is caused by resonance of the objects when they are about the same size as the wavelength of the radiation. The effect of the objects resonating is to reinforce the radiation when the frequency of resonation matches that of the incident radiation, and to reduce the radiation when they are out of synchronisation. Resonant-sized objects often act as inhomogeneous surfaces.

2.5.4.5 Wavelength

The wavelength of the radar affects the signal in a number of ways. It affects whether the surface is seen to be rough or smooth, and it affects the penetration into the surface by the signal, as have been discussed in previous sections. The dielectric properties of the surface are also a function of the wavelength of the radar.

2.5.4.6 Polarisation

An electromagnetic plane wave consists of an electric field component and a magnetic field component. These two components are orthogonal, and they are both orthogonal to the direction of propagation. The polarisation of such a wave is determined by the direction of the electric field component as a function of time. Radars can be polarised so that the energy transmitted is oriented about the horizontal or vertical axes, and the returned signal can be measured in the same two directions, so that four combinations are possible, HH, VV, HV and VH. The first two are like polarised and the second two are cross-polarised. Due to symmetry in backscattering, HV = VH, so that the use of dual polarised radars gives three sets of information about the area being imaged. It is conventional to polarise radar waves horizontally and vertically. Measurement of the horizontal and vertical polarised signal and the cross-polarised signal, allows the computation of the signal strength at any other orientation. In this sense, the signals could be polarised at any pair of orientations that are at right angles, and the signal strengths at any other orientation can be computed. However, some orientations will be better at depicting the responses from some types of surfaces, hence some orientations are better than the others. For vertically and horizontally orientated objects, such as man-made objects, then horizontal and vertical polarisation are the best planes of polarisation to use. However, for objects with different orientations, then different orientations of the polarisation would be better to use, as can occur, for example, on hill slopes at gradients across the signal from the radar.

Horizontally polarised radars are more sensitive to the horizontal dimensions of the target, whilst vertically polarised radars are more sensitive to the vertical dimensions. The reflection of the radar signal in a vegetative canopy, such as a forest is complex, as shown in Figure 2.71. There is surface reflectance from the canopy. This occurs more strongly at shorter wavelengths as these are predominantly reflected from the top of the canopy, and do not penetrate far into the canopy. Then there is reflectance from the vegetative components within the crowns of the canopy. Whilst there is some of this at shorter wavelengths, this tends to be stronger at mid to longer wavelengths. The reflectance depends very much on the size of the branches and other components in the canopy. The diverse

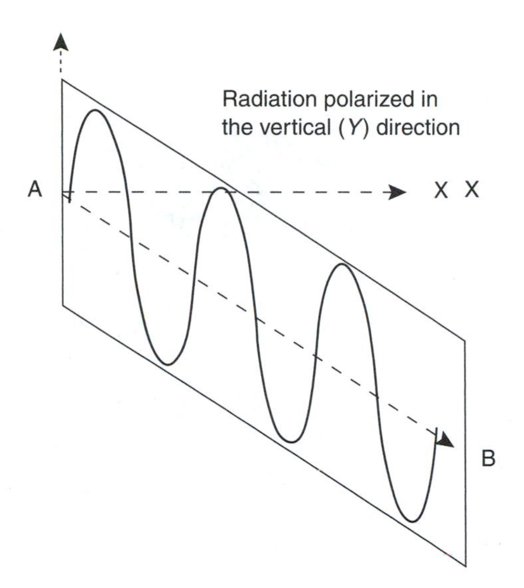

FIGURE 2.71 Polarisation of a sine wave. Normal radiation in the direction AB will have its sine wave at all orientations about the line AB. Polarised light has its radiation at the one orientation. In this figure the radiation is polarised in the vertical or *Y* direction.

orientations on the canopy elements means that cross-polarisation often occur in the scattered radiation. The soil or water background then contributes the third level of reflectance. Water is a strong specular reflector, and so most of the energy will be reflected away. However, individual tree trunks will operate as corner reflectors, giving a strong response. Vegetation in the water tends to dampen the reflectance characteristics of the water. The surface texture of the soil affects the level of backscatter from the soil. As the surface becomes rougher, the backscatter increases and the less radiation that is available for corner reflector effects. As the surface becomes rougher, the cross-polarisation component of the signal also increases.

Figure 2.72 and Figure 2.73 illustrate some of the discussion above in relation to forest covers. Figure 2.73 shows typical digital counts for HH, VV and HV polarised C, L and P band radar from a softwood forest area in Canada. The C band shows slightly higher response at higher tree densities, primarily due to lower canopy gap effects. Most of the reflectance is coming from the leaf layer in the canopy, and so the HV component is less than the HH and VV components. In the L and P bands, the volume scattering from within the whole of the canopy is much more dominant and so there is a larger HV response than either HH or VV. In both the VV components are reduced relative to the HH components. In these wavebands the HV signal component increases markedly with forest density relative to the HH and VV signals.

2.5.5 RADAR INTERFEROMETRY

A very innovative development in the application of radar is the use of radar images acquired over a short time period, of an area that has been subjected to some disturbance that causes a differential shift in the surface of the area imaged, between the two acquisitions. Such shifts can be due to volcanic action, earthquakes or the movement of ice or a glacier. Other changes, such as the movement of a car, cause very large changes between the acquisitions, and these changes introduce noise into the resulting interferometric image.

The concept can be likened to optical interferometry. If a film of oil is placed on a surface, the oil refracts light as it both enters and leaves the oil. If the oil is the same thickness, then all of the light will be refracted by the same amount at all locations. However, if the oil is slightly different

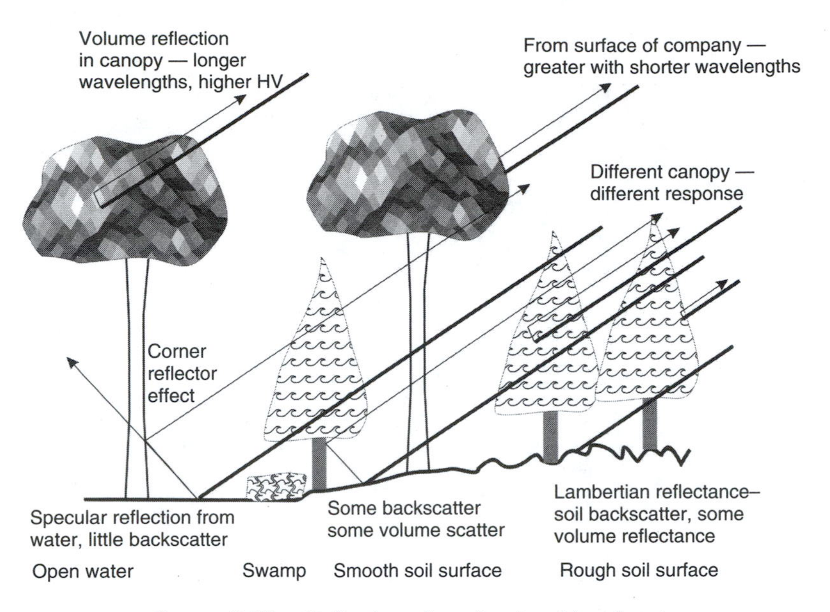

FIGURE 2.72 Reflection of a radar signal in a forest.

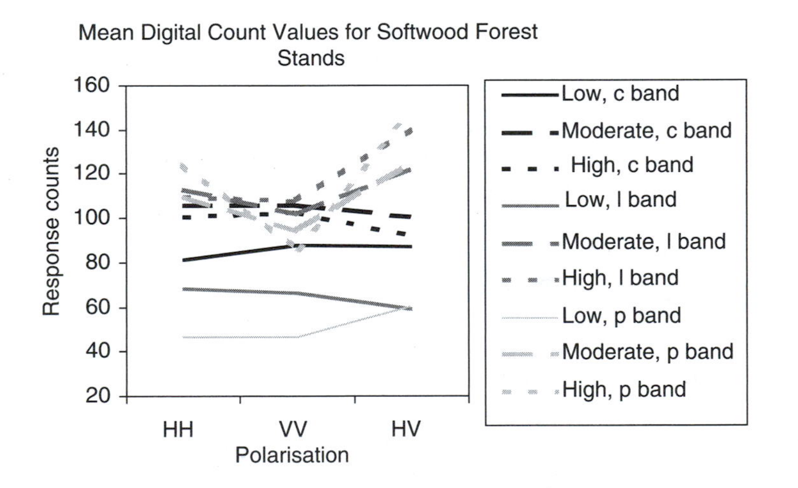

FIGURE 2.73 Typical radar responses from a softwood forest with low, moderate and high density tree cover.

in thickness, then slight differences in refraction occur, and the result is that the light is refracted slightly differently at different locations. This results in interference between the signals that are refracted by different amounts, giving interference rings of different colour in the oil.

The same concept can be used with the two radar images. Over most of the area imaged, there is no displacement between the images, and so no interference can be constructed. However, when there is some displacement between the images, this displacement can be converted into interference fringes between the images, giving interference contours joining points of equal displacement, and the gradient across these shows the gradient of greatest change in displacement between the images. In this way the areas of greatest displacement can be found, and the areas of greatest deformation are those areas where the contours are closest together.

Interferometry is being used extensively in the analysis of the deformations that occur as a result of earthquakes, volcanoes and of ice flows in glaciers.

2.5.6 Summary

In this section the principles of radar data and how it interacts with the environment have been introduced. The reader will appreciate the complexity of radar data from the material presented here. From the design of the radar, through the collection and processing of the data, radar theory is soundly based on complex physical and mathematical laws. This theoretical development is beyond the scope of this book, and is not necessary for practitioners of remote sensing, although it is obviously important for those who wish to get immersed in the development of radar theory and practice. For those who wish to understand this theory in more detail I recommend that they chose from the range of books that are available on radar theory and practice. Suggested additional reading is given at the end of the chapter.

Many things are influencing the radar return, and there are usually only a few channels of radar information to use in the analysis and interpretation of that data. It is therefore difficult, and sometimes impossible, to discriminate between the potential causes of variations in response in a radar image. However, the situation is gradually improving as radar systems provide more information, through the use of multiple bands and polarisation, integration of radar with optical data and the development of better analysis techniques. At present radar has the advantage that it can be used at any time during the day or night and it is relatively free of atmospheric effects, so that good radar imagery can be acquired under cloudy, and even under rainy conditions. This is a significant advantage for the monitoring of dynamic processes, including flooding and vegetation dynamics since imagery can be acquired at the most important phenological or environmental stages. In addition to these advantages, the radar signal is very sensitive to the water content and structure of vegetation. It thus also provides complementary information to that derived from optical and infrared data. It is also usually possible to extract first order or the dominating cause of the response in the data. However, the data will retain a tantalising amount of information that as yet cannot be reliably extracted. The reader should thus approach the analysis of radar data with caution.

This situation is changing. A number of workers have shown that good quality radar image data can provide land cover and land use information of similar accuracies to that derived from optical image data. Radar has also shown that it can be used very successfully for specific applications, such as the detection of sea ice, oil spills, the mapping of deformations that arise from earthquakes and in the vicinity of volcanoes, mapping the velocity of glaciers and in topographic mapping. Other potential applications are showing promise, such as the estimation of timber volumes in forests and the mapping of near surface soil moisture.

2.6 HYPERSPECTRAL IMAGE DATA

2.6.1 Definition

Imaging spectrometers produce hyperspectral image data. The data occupies many contiguous spectral bands over the spectral region of interest. The ultimate goal is to produce laboratory like reflectance spectra for each pixel in the image. Most imaging spectrometers operate in the visible to NIR spectral region, with some sensing into the longer wavelength mid IR wavebands. Within this region the instruments record the radiation incident on the sensor in many wavebands, typically between 100 and 300 wavebands. Ideally, these wavebands will all be of the same spectral width, but in practice this does not happen because of the differences in sensitivity of the detector or detectors at different wavelengths, and the differences in energy levels at different wavelengths. Some current airborne imaging spectrometers are listed in Table 2.7.

The different entries in Table 2.7 for each sensor represent different detectors used to sense radiation in the different wavebands. These different detectors have different sensitivity characteristics both to each other and across their spectral range. Imaging spectrometers meet the same conditions

TABLE 2.7
Characteristics of Selected Aircraft-Based Imaging Spectrometers

Sensor	Spectral Range	Number of Bands	Spectral Resolution (nm)	IFOV (mrad)	Spatial Resolution	Swathe Width
AVIRIS	400–720 nm	31	9.7	1	20 m (at 20 km)	11 km
	690–1300 nm	63	9.6	1	20 m (at 20 km)	11 km
	1250–1870 nm	63	8.8	1	20 m (at 20 km)	11 km
	1840–2450	63	11.6	1	20 m (at 20 km)	11 km
DAIS	500–1050 nm	32	15–30	3.3		51.2 km
	1500–1800 nm	8	45	3.3		51.2 km
	1900–2500 nm	32	35	3.3		51.2 km
	3000–5000 nm	1	2000	3.3		51.2 km
	8.7–12.5 μm	6	900	3.3		51.2 km
HYMAP	450–890 nm	31	15	2		60°
	890–1350 nm	31	15	2		60°
	1400–1800 nm	31	13	2		60°
	1950–2480 nm	31	17	2		60°

as other scanners. The atmospheric effects affect them in the same way, their detectors age and decay in the same way, and they are affected by the same geometric considerations.

2.6.2 APPLICATIONS OF HYPERSPECTRAL IMAGE DATA

Currently, there are three main avenues of analysis of hyperspectral data:

1. *Comparison of spectra with known spectra.* There are a number of established libraries of spectra for different types of materials. Most of these spectra are for rocks and soils since some of the constituents of these cover types have absorption spectra within the spectral regions that are sensed by these instruments. The spectra from the spectrometer can be compared with either these spectra or with other spectra from the image data and tested for similarity using suitable statistical tests. The conduct of such tests will show areas that are similar in the image data to the spectra being used as a basis for comparison. If image data is to be compared with library spectra, then the image data needs to be calibrated for atmospheric effects before the comparison can be made.
2. *Correction for atmospheric absorption.* If the spectrometer records data in the water absorption bands, and on the shoulders of these bands, then this data can be used to estimate the impact of atmospheric moisture on the image data. This application of image data is discussed under atmospheric correction in Chapter 4.
3. *Construction of vegetation indices.* Hyperspectral data provides the opportunity to exploit information on vegetation condition that is contained in the red edge and its associated red and NIR parts of the spectrum. The use of data to construct such indices is discussed in Chapter 4.
4. *End member and mixing analysis.* The hyperspectral data provides a rich data set for mixture analyses. This type of analyses will be discussed under Mixture Analyses in Chapter 4.

2.7 HYPERTEMPORAL IMAGE DATA

2.7.1 INTRODUCTION

Hypertemporal image data is defined as the data that is acquired at a much higher frequency than that required by Shannon's sampling theorem (Chapter 4) to describe the cyclic nature of the process, and covering at least one cycle of the process. A set of hypertemporal data can be viewed as an image or video clip. A video clip will give some appreciation of the rapidity and range of variation or change in the data, but it does not allow the observer to understand the dynamic processes that are expressed in the image data. To gain this appreciation, other forms of display and analysis are required as will be discussed in Chapter 4.

Hypertemporal image analyses is primarily concerned with understanding the dynamic processes that are affecting the image data. As such, a hypertemporal image analysis builds on the analyses conducted on single or small sets of image data, in the form of image classification and estimation. Using hypertemporal image analysis in this way means that this form of analysis is complimentary to rather than competive with these other forms of analysis. The use of these assumptions also assists in reducing the dimensionality of the hypertemporal image analyses problem.

Hypertemporal image data is four-dimensional data:

1. Spatial (x, y) information (two dimensions)
2. Spectral (R) information
3. Temporal (t) information

Hypertemporal image data involves the acquisition of many images of the same area over a period of time. For such images to be useful, the image data needs to be consistent from image to image, in the same way as internal image consistency is often assumed in the analysis of single images. This consistency can be violated in a number of ways:

1. *Lack of geometric consistency*. Geometric consistency can be established by means of rectification of the images. At present most image processing systems rectify image data, however, few of them have automated rectification processes as are essential for rich temporal data sets. Geometric inconsistency can be due to variations in the normal orientation elements of the sensor. They can also be due to variations in the satellite orbit characteristics.

2. *Sensor calibration*. Sensor sensitivity changes over time, so that variations in sensor sensitivity can affect the temporal record much more than it is likely to affect the analysis of one or a few images. Calibration of the image data to correct for variations in sensor response is thus very important in the analysis of temporal imagery.

3. *Correction for atmospheric effects*. Variations in atmospheric conditions impact on each image in a temporal data set. For some purposes, the effects of the atmosphere can be ignored when analysing single images, simply because the analysis techniques provide a means of correcting for these variations. This often occurs in image classification, for example. However, these effects cannot be ignored in the analysis of temporal image sets, and so atmospheric correction always needs to be implemented prior to the analysis of temporal image sets.

2.8 PLATFORMS

Platforms are the means of supporting and moving a sensor so as to acquire data or imagery as required. Different platforms have characteristics that make them suitable for some tasks and not for others. An appreciation of these characteristics allows a user to select the platform most appropriate for the task.

2.8.1 Terrestrial Platforms

Terrestrial platforms are those rigidly secured to the Earth's surface, including the tripod, boom, tower and cherry picker. They are suitable for the acquisition of large-scale photography, for temporal data sets and for observations with spectrometers. The advantage of terrestrial platforms is that the precise position and pointing of the sensor can be known relative to the surface. The other advantage of

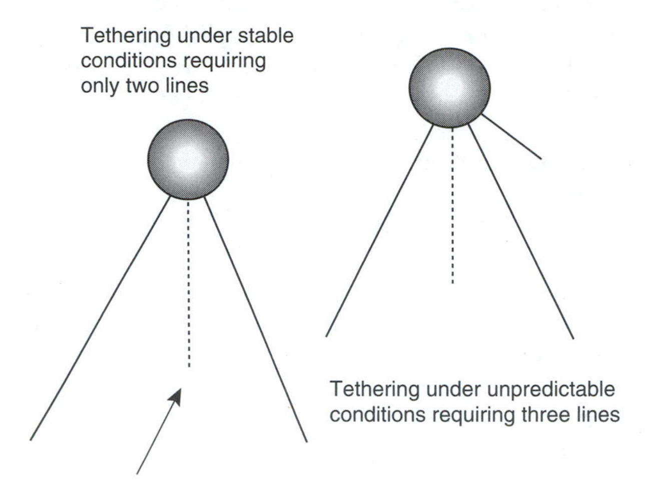

FIGURE 2.74 Tethering of a ballon under steady and fluctuating conditions.

terrestrial platforms is the ability to precisely re-establish the position of the sensor relative to the surface. This means that sets of images or spectral observations can be taken over a period of time for the same area. To ensure precise re-establishment it is necessary to establish permanent marks that can be used to position and point the sensor.

2.8.2 The Balloon

Balloons can be used to lift a payload of sensors above an area of study. Balloons would normally be tethered using either two or three lightweight but high tensile lines (Figure 2.74). A radio-controlled device can be used to activate the sensors on the balloon. Balloons move about in the air and so the sensor should be located in a gimbal mount to ensure that it is pointing vertically down.

The advantage of the balloon is the ability to establish a sensor in a fixed position for relatively long periods so as to monitor specific events. The disadvantages include the relatively high cost of launching the balloon, and the difficulty of replacing magazines and correcting faults.

2.8.3 The Helicopter or Boat

Both are highly manoeuvrable platforms that can also be held relatively stationary over the object being sensed and can be stopped so as to collect field data. However, they sacrifice stability, the helicopter in particular have high vibration levels that can seriously interfere with the acquisition of data.

2.8.4 Manned and Unmanned Aircraft

The aircraft provides a platform that is fairly stable and moves at a relatively constant high velocity in a specified direction. It is used to acquire data covering large areas as strips or runs of photographs along each flightline, with a number of parallel flight lines covering the whole area (Figure 2.51).

Within each flightline, photographs are usually taken so as to have 60% overlap between adjacent photographs in the run, and at least 10% sidelap between adjacent strips. The exposures along the flightline are triggered by means of an intervalometer. The time between exposures can be calculated in either of two ways. The simplest is to use the aircraft ground speed and the distance between

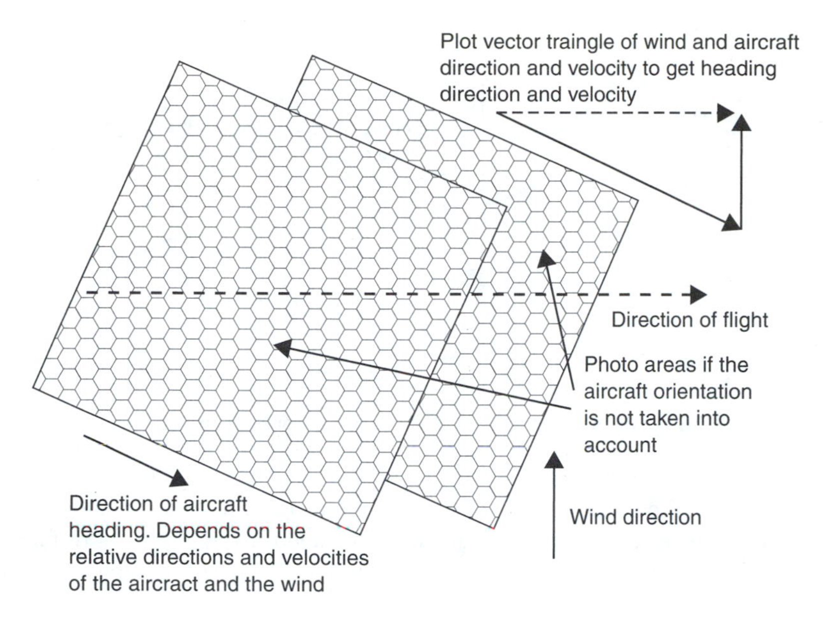

FIGURE 2.75 The effect of differences between aircraft heading and direction of flight on photography.

adjacent photograph centres to compute the time between exposures. However, this technique suffers from the difficulty of estimating the aircraft ground speed. A better method is by the use of a viewing telescope. This telescope displays the ground under the aircraft, and includes a moving line projected onto the FOV. The projected line is adjusted so as to move across the screen at such a rate that it stays stationary relative to the ground features. When it does this it provides a very good estimate of the aircraft ground velocity. A navigation site is used to maintain the aircraft heading (Figure 2.75).

The sensor may be placed in a gyroscopic mount to minimise the effects of aircraft movement during flight. These movements may not create serious distortions in photography because of the relatively short exposure time. Significant movements of the platform will introduce errors in scanner data, which is continuously acquiring data, and the distortions are hard to correct. The motions that are of interest are:

1. *Forward motion* can cause image blur since the object moves relative to the camera during exposure. With aircraft velocity, V (m/sec), and exposure E (sec), the distance moved on an image of scale factor S will be:

$$\text{Image movement, } I(\text{mm}) = V \times E \times 1000/S \tag{2.47}$$

If the resolution of the film, R is set in lines/mm then the maximum exposure that will maintain this resolution can be calculated from:

$$R = 1/I \leq S/(V \times E \times 1000)$$

so that

$$E \leq S/(V \times R \times 1000) \tag{2.48}$$

The maximum exposure time can be increased by (1) reducing the aircraft speed or (2) reducing the acceptable resolution in the photography.

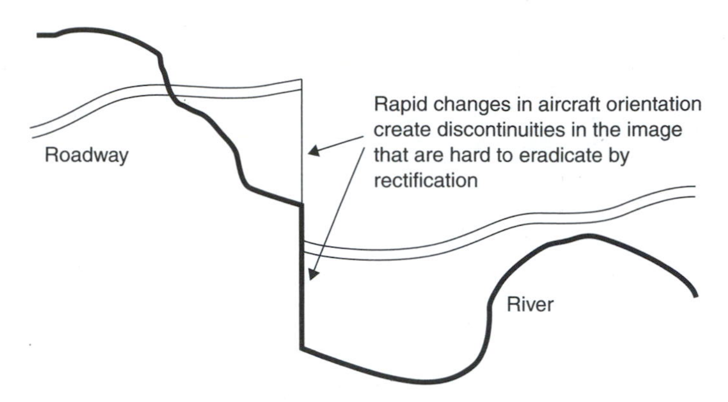

FIGURE 2.76 The effect of sudden changes in roll rotation on scanner imagery.

2. *Sideways motion*. Side wind causes the direction of heading to be different to the direction of flight. Modern camera and scanner mounts can be rotated and tilted to allow for differences in both heading and nose-to-tail tilt. If the mount does not allow for it then each photograph will be rotated relative to its neighbours as shown in Figure 2.76.

3. *Rotations of pitch and roll*. With good aircraft platforms these rotations are kept in the range 2 to 5°.

With cameras these rotations introduce constant tilts across each photograph, which can be corrected using appropriate rectification techniques. With a scanner, where the imagery is acquired continuously during the flight, these rotations will introduce significant displacements into the imagery, as shown in Figure 2.76 that can be difficult to eradicate. They can be partially corrected using information on the orientation of the sensor held within the sensor itself. However, this information is usually not as accurate as would be required for full correction for these errors. With aerial photography this problem is overcome by using a distribution of control points across the image in rectification, discussed in Chapter 4. However, such a spatial distribution of control points is only possible in the one direction in a scan line of data, and the orientations of one scanline may be quite different to that in each other scanline. As a consequence, the accurate geometric rectification of scanner data is very difficult to achieve.

2.8.4.1 Hot Spots

Hotspots are areas on a photograph that appear, as a brightening on the photograph in a way that is not related to the detail on the photograph (Figure 2.77). Hotspots are caused by either:

1. *Lack of shadow*. If the sun is directly behind the aircraft, then at this location in the image, no shadows will occur. This results in higher reflectance. As the orientation between the sun and the sensor changes, more and more shadowing will be contributing to the reflectance, and the total reflectance will tend to decrease.

2. *Refraction around aircraft*. This hotspot effect may be further exaggerated since the aircraft shadow at this location, is generally lost at aircraft altitudes, due to refraction of light around the aircraft.

2.8.5 Planning an Aerial Sortie

Prior to planning an aerial sortie the following information must be known:

1. Scale and resolution of the required imagery.
2. Film and filters required or the wavebands required for scanner imagery.

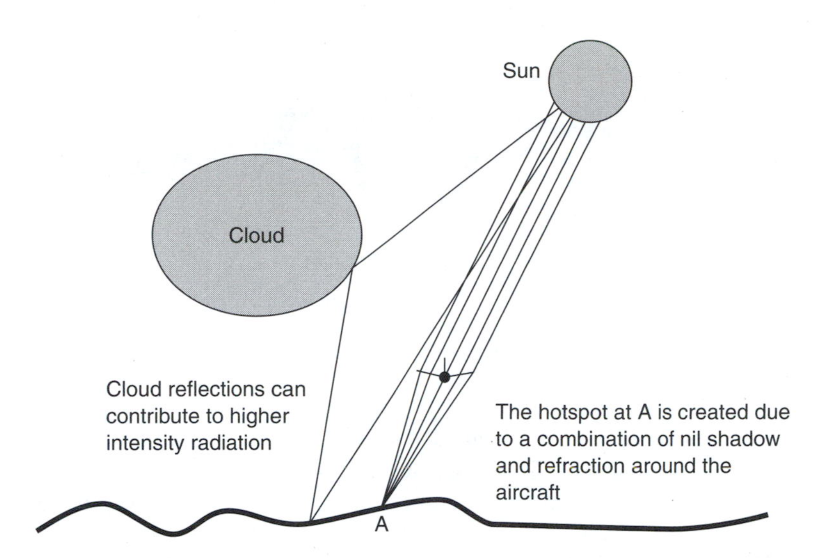

FIGURE 2.77 Creation of hotspots due primarily to the lack of shadowing in the surface at this direction and partly to refraction around an aircraft.

3. Area of coverage and overlap requirements; 60% for stereoscopy. Does the information sought from the imagery influence the direction of the flight lines? For example, bi-directional reflectance effects will be minimised with scanner image data if that data is acquired in scanlines at right angles to the direction of the sun.
4. The essential or preferred time of day or season for acquisition of the photography.
5. Any specific geometric requirements of the imagery, for example the use of normal angle camera versus super wide angle or maximum scan angle so as to control the effects of variations in BRDF.

Given this information the flight planner will have to ascertain:

1. Whether the available aircraft, camera exposure range with the required film type will provide the required resolution, or that the available scanners are suitable.
2. Are hotspots likely to occur, and if so, how to deal with them.
3. That the available equipment is suitable and that no undue limitations exist on flying over the area.
4. The best flight lines for the sortie to achieve user needs at lowest cost, the number of exposures per flightline and total number of exposures so as to prepare adequate film for the sortie.

As well as using standard aircraft and light aircraft, ultra-light, motorised wings and pilotless aircraft can be used. The principles are the same for all of them, although the solutions may be somewhat different.

2.8.6 The Satellite Platform

Orbiting satellites are under the influence of three forces, gravity (g) or attraction to the Earth mass, centripetal (c) or the outward force due to the angular change in momentum of the satellite, and drag (d) or the resistance of matter in the orbit (Figure 2.78).

These forces must be in balance if the satellite is to stay in the orbit for any length of time. A satellite orbit is normally elliptical, with the satellite moving much faster when it is closer to the

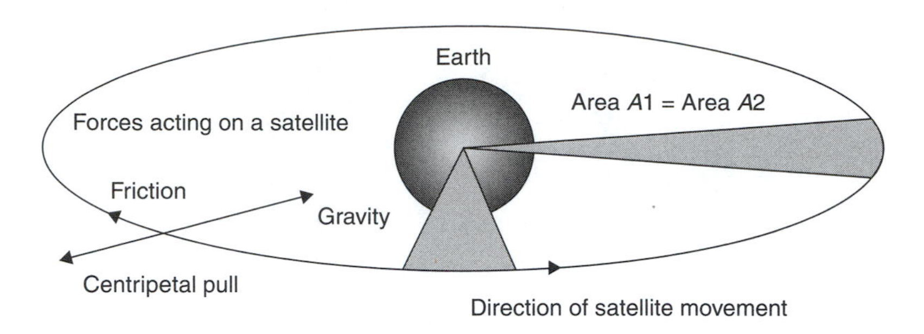

FIGURE 2.78 The forces acting on a satelite.

Earth (higher gravity pull needs to be counteracted by higher centripetal throw), and slower as it moves away from the Earth. An elliptical orbit causes great difficulty with sensor systems because of the variations in altitude and velocity. In consequence, satellites that carry remote sensors are launched into orbits that are as close as possible to being parallel to the Earth's surface.

The period, T, of a satellite in a circular orbit is given by the expression:

$$T = 1.4 \times [1.0 + (E/R)]^{1.5} \text{ h} \tag{2.49}$$

where R is the radius of the Earth, 6370 km, E is the elevation of the satellite above the Earth's surface (km).

The lowest orbit capable of supporting a satellite for a significant period of time is about 400 km above the Earth due to atmospheric drag. The highest altitude that is probably of interest for polar orbiting satellites is about 1,000 km, whilst geostationary satellites, at orbits of about 36,000 km are used to provide high frequency repetitive coverage at coarse resolutions.

Polar orbiting satellites have their orbit in a plane that goes close to the poles. A polar orbit can be designed to precess about the globe, thereby achieving full coverage of the globe. If this plane is set to precess so that the plane retains the same angle to the sun, then imagery will always be acquired at the same solar time. This provides consistent illumination conditions, except for seasonal variations. The disadvantage is that the sensor cannot acquire imagery at other times during the day.

Consider the requirements of a satellite system that must image the full globe once every day. The Earth completes one rotation about the sun once every 365 days, and so each day the plane of the satellite orbit needs to precess or rotate by 1/365th of 24 h. Since this annual rotation is in addition to the 365 rotations, in effect the Earth completes 366 rotations of the sun in 365 days. Thus, the satellite needs to pass over the same location on the Earth's surface every (24–(24/365)) h, or 23.934 h. Let the satellite complete 14 orbits in this time, of 1.70958 h per orbit. From (2.29) above a period of 1.70958 h would mean a satellite altitude of 909 km. The coverage per orbit is 360/14°, or 25° 42′ 51″. To ensure full coverage, the swathe for an orbit could be set at about 27° or 3000 km. It can be seen that this is a very large area to cover, with considerable distortion out from the edges of the imagery due to Earth curvature. Alternatively, the satellite can be in lower Earth orbit, have more orbits and acquire less coverage per orbit.

Another option is to acquire full coverage every n days, completing partial coverage every day in N orbits such that the Nth orbit, 23.934 h after the first orbit, is stepped by the swathe of the scanner relative to the first orbit. These steps are designed so that the swathes acquired each day progressively completes coverage of the Earth's surface between the swathes acquired on the first day, in n days. Once coverage is complete then the cycle is repeated.

The advantage of this approach is that the swathe width per orbit can be significantly reduced. Consider the previous case where the satellite, acquiring 14 orbits every 23.934 h has to sweep out 25° 42′ 51″ per orbit. If the sensor is to have a swathe width of 4° 30′ (27°/6) then the satellite will need to cover 364° 30′ in the 14 orbits. Now the plane of the satellite precesses 360° in 23.934 h, so that the duration of the 14 orbits will need to be $23.934 \times 364.5/360$ h if the 14th orbit is to be at the same solar time. This requires a satellite orbit period of 1.7309 h.

Whilst a satellite provides a very stable platform it is still subject to rotations about its three axes. Satellites use a reference to determine their orientation and to decide when corrections need to be made. The reference is usually by fix on a star; as the satellite oscillates so the direction to the star changes and retrorockets are used to re-establish orientation.

Because of the cost of loading and recovering film from a satellite, most sensors on satellites will be electronic sensors such as video, scanners or radar. The data acquired by these systems is converted into digital data to minimise the effects of atmospheric attenuation on the data, before transmission to the Earth.

The forward velocity of the satellite is used to provide the advance during acquisition of data. Both the rotation velocity of the Earth relative to the satellite and the forward velocity of the satellite introduce significant errors during sampling of the data. Other difficulties that can arise are geometric distortions due to Earth curvature and radiometric variations in the returned signal due to variations in path length.

Despite these limitations satellite data are very suitable for the ongoing monitoring of an area because of the repeatability of the data, the small errors due to rotations in the data and the small effects of height displacements on the data.

2.9 SATELLITE SENSOR SYSTEMS

Current satellite systems have been placed in orbits by a variety of countries and country-based consortia, as well as by commercial organisations. These satellites and their payloads are usually designed for:

1. Military purposes, in which case access to the data is strictly restricted.
2. Scientific purposes, in which case the data is restricted to the conduct of scientific and technological development activities, but for which the data are usually relatively cheap.
3. Commercial purposes, in which case the data are available on a commercial basis.

The range of available data for the analysts is very large, particularly if the purpose of the analysis is a form research and development purposes. Most owners of satellite systems provide information on the products that are available from their systems on the web, and seeing that these are changing and evolving at a rapid rate, it is not realistic to list some of them in a textbook of this type. A partial listing is provided as part of the CD that goes with this book.

ADDITIONAL READING

The Manual of Remote Sensing provides a comprehensive treatment of most aspects of remote sensing. The Manual is currently divided into four volumes, dealing with:

Volume 1 — *Theory, Instruments and Techniques*,

Volume 2 — *Principles and Applications of Imaging Radar*,

Volume 3 — *Remote Sensing for the Earth Sciences*,

Volume 4 — *Remote Sensing for Natural Resource Management and Environmental Monitoring*.

The first two volumes are relevant to the material given in this chapter. As can be seen, Volume 1 is quite old. Even though much of the basic physical theory does not change, there are none the less significant changes that have occurred since this date in many of the instruments and techniques in remote sensing.

More focused books on aspects of remote sensing include that of Charles Elachi (1987) on the physical principles of remote sensing, Ghassem Asrar on the interaction of radiation with vegetative canopies and the detailed treatment of radar can be found in Ulaby et al. (1981).

There are many web addresses that focus on the supply of satellite image data. Some of these are given in the CD provided with this text.

REFERENCES

Asrar, G. (Ed.), 1989. *Theory and Applications of Optical Remote Sensing*, John Wiley & Sons, New York.

Bauer, M.E., Daughtry, C.S.T. and Vanderbelt, V.C., 1981. "Spectral-agronomic relationships of corn, soybean and wheat canopies," *AgRISTARS Technical Report SR-P7-04187*, Purdue University, West Lafeyette, IN.

Bodechtel, J. and Sommer, S., 1992. "The European Imaging Spectroscopy Airborne Campaign — EISAC," *Imaging Spectrometry: Fundamentals and Prospective Applications*, Toselli, F. and Bodechtel, J. (Eds.), Kluwer Academic Publishers, Dordrecht.

Cocks, T., Jenssen, R., Stewart, A., Wilson, I. and Shields, T., 1998. "The HyMap airborne hyperspectral sensor: the system, calibration and performance," *First EARSEL Workshop on Imaging Spectroscopy*, Schaepman, M. (Ed.), University of Zurich, Switzerland.

Colwell, R.N. (Ed.), 1984. "Manual of remote sensing: volume 1 — theory, instruments and techniques," American Society of Photogrammetry and Remote Sensing, Falls Church, VA.

Condit, H.R., 1970. "The spectral reflectance of American soils," *Photogrammetric Engineering*, 36, 955–966.

Elachi, C., 1987. *Introduction to the Physics and Techniques of Remote Sensing*, John Wiley & Sons, New York.

Gates, D.M., 1965. "Spectral properties of plants," *Applied Optics*, 4, 11–20.

Gaussman, H.W., 1973. "The leaf mesophylls of twenty crops, their light spectra and optical and geometric parameters," *Technical Bulletin 1465*, USDA, USA.

Gausman, H.W., Allen, W.A., Scxhupp, M., Weigand, C.L., Escobar, D.E. and Rodriguez, R.R., 1970. "Reflectance, transmittance and absorptance of light of leaves for eleven plant genera with different leaf mesophyll arrangements," *Technical Monograph No. 7*, Texas A&M University, College Station, Texas.

Gausman, H.W., Gerberman, A.H., Weigand, C.L., Leamer, R.W., Rodriguez, R.R. and Noriega, J.R., 1975. "Reflectance differences between crop residues and bare soils," *Proceedings of the Soil Science Society of America*, 39, 752–755.

Gausman, H.W., Rodriguez, R.R. and Weigand, C.L., 1976. "Spectrophotometric reflectance differences between dead leaves and base soils," *Journal of the Rio Grande Valley Horticultural Society*, 30, 103–107.

Gausman, H.W., Escobar, D.E., Everitt, J.H., Richardson, A.J. and Rodriguez, R.R., 1978a "Distinguishing succulent plants from crop and woody plants," *Photogrammetric Engineering and Remote Sensing*, 44, 487–491.

Gausman, H.W., Escobar, D.E. and Rodriguez, R.R., 1978b. "Effects of stress and pubescence on plant leaf and canopy reflectances," *Proceedings of the International Symposium on Remote Sensing for Observation and Inventory of Earth Resources and the Endangered Environment*, International Archives of Photogrammetry, XXII-7.

Henderson, F.M. and Lewis, A. J. (Eds), 1998. *Manual of Remote Sensing: Volume 2 — Principles and Aplications of Imaging Radar*, John Wiley & Sons, New York.

Horler, D.N.H., Dockray, M. and Barber, J., 1983. "The red edge of plant leaf reflectance," *International Journal of Remote Sensing*, 4, 273–288.

Lintz, J. and Simonett, D.S., (Eds), 1976. *Remote Sensing of Environment*, Addison-Wesley, Reading, MA.

Muller, A., Oertal, D., Richter, R., Strobel, P., Beran, D., Fries, J., Boehl, R., Obermeier, P., Hausold, A. and Reinhaeckel, G., 1998. "The DAIS 7915 — three years operating airborne imaging spectrometer," *First EARSEL Workshop on Imaging Spectroscopy*, Schaepman, M. (Ed.), University of Zurich, Switzerland.

O'Niell, A., 1990. Personal communication.

Rencz, A.N. and Ryerson, R.A. (Eds), 1999. *Manual of Remote Sensing: Volume 3 — Remote Sensing for Earth Sciences*, John Wiley & Sons, New York.

Sabins, F.F., 1987. "Remote Sensing: Principles and Interpretation", W.H. Freeman and Company, New York, USA.

Schutt, J.B., Rowland, R.R. and Heartly, W.H., 1984. "A laboratory investigation of a physical mechanism for the extended infrared absorption ('red shift') in wheat," *International Journal of Remote Sensing*, 5, 95–102.

Stoner, E.R. and Baumgardner, M.F., 1981. "Characteristic variations in the reflectance of surface soils," *Soil Science of America Journal*, 45, 1161–1165.

Ulaby F.T., Moore, R.K. and Fung, A.K., 1981. *Microwave Remote Sensing: Active and Passive*, *Artech House*, Norwood, MA.

Ustin, S. (Ed.), 2004. *Manual of Remote Sensing: Volume 4 — Remote Sensing for Natural Resource Management and Environmental Monitoring*, John Wiley & Sons, New York.

Vermote, E.F., Tanr'e, D., Deuze, J.L., Herman, M., and Morcrette, J.J. 1997. Second simulation of the satellite signal in the solar spectrum, 6S: an overview. *IEEE Trans Geosci. Remote Sens.* 35:675–686.

Wiegand, C.L., 1972. "Physiological factors and optical parameters as bases of vegetation discrimination and stress analysis," *Proceedings of the Seminar on Operational Remote Sensing, American Society of Photogrammetry*, Falls Church, VA.

3 Visual Interpretation and Map Reading

3.1 OVERVIEW

3.1.1 REMOTELY SENSED DATA AND VISUAL INTERPRETATION

Remotely sensed data can either be in the form of visual or digital images. Different types of visual images include photographs and other forms of hard copy images. Digital images are stored on computer and optical discs, tapes or cassettes. The digital analysis of the digital image data will be the subject of subsequent chapters. Part of this focus will be in the display of digital images for visual analysis and interpretation. The visual analysis of digital images is thus part of the focus of this chapter, but the student need not be concerned at this stage with how digital images get to be displayed; that will come in Chapter 4.

Visual interpretation of an image or photograph is one of the two important ways of extracting information from remotely sensed data. Visual interpretation uses skills that are much more readily available to the human interpreter than through the processing of digital image data in a computer. The most important of these skills is the interpreter's ability to conduct analytical reasoning and identify and explain complex patterns. A major reason for the continuing need for visual interpretation is because of the linkage between our understanding of a problem and our ability to formulate hypotheses as to the causes of the problem. Understanding is essential to this formulation. With some problems, this understanding can only come from understanding spatial and temporal interrelationships. When this situation exists then visual interpretation is an essential component in the formulation of hypotheses.

Digital image processing is the second method of extracting information from data. It will be discussed in Chapter 4. Digital processing and extraction of information is assuming increasing importance, particularly when used in conjunction with visual interpretation. Currently, it is used to conduct a lot of the more routine tasks, like the pre-processing of images, including rectification to ground control, sensor calibration, correction for the effects of the atmosphere on the image data, filtering and enhancing images for visual interpretation and the classification of land covers and land uses. Increasingly, digital image processing will be used to derive estimates of specific physical attributes, once methods of routinely correcting image data for atmospheric effects become standardised. Sometimes such information, when integrated with other information in a Geographic Information System (GIS), discussed in Chapter 6 is sufficient to the user. Sometimes further analysis needs to be conducted. It has been seen that visual interpretation may be an essential step in formulating hypotheses about a situation or a process. However, validation of these hypotheses will frequently use digital image processing so as to provide a quantitative assessment of the validity of the hypotheses.

The first section of this chapter studies how height displacements occur and how they can be used to estimate differences in elevation. It is followed with an introduction to techniques for manually mapping planimetric detail. The central sections of the chapter deal with the techniques of photo-interpretation and how to conduct an interpretation task. Photo-interpretation certainly benefits from innate skills in observation and analysis, but just as important is the attitude of the analyst to the task. Interest, a desire to learn, confidence and a preparedness to work are prerequisites for successful interpretation. Competence in interpretation also comes from practice, so it is important to consolidate study of this chapter with practical experience in the techniques and methods discussed.

The middle sections of the chapter are completed by dealing with the interpretation of specific types of image data, such as thermal and radar images. The final section deals with maps and map reading.

3.1.2 Effects of Height Differences on Remotely Sensed Images

In analysing photographs it is important to recognise that all images are affected to some degree by the differences in height of the terrain. Many aerial photographs are taken to deliberately exploit this characteristic for the estimation of height differences. This is the basis of photogrammetric mapping where a major objective is to map elevations in an area by means of contours. These height differences can be estimated:

1. With great precision using ground survey techniques, photogrammetric instruments with aerial photographs and airborne laser ranging. Differential height differences can also be estimated with great precision using differential radar images.
2. With less precision when replacing the aerial photographs with satellite images acquired for this purpose.
3. Using simple (approximate) techniques.

The existence of displacements due to height differences means that single images will not be spatially accurate. The introduction of these displacements can be ignored when they are negligible, either because the terrain is flat or the geometry of the sensor introduces negligible displacements at that position. These displacements will be interpreted as differences in spatial location if only a single photograph is used unless:

1. The image has been corrected for their effects as with orthophotographs.
2. The process of transferring the information from the image to a map makes corrections for the displacements.

Photogrammetric processes can map spatial position from images with great precision. There are also simple techniques that can map spatial position, but within limited areas due to height displacements.

3.2 STEREOSCOPY

3.2.1 Introduction

Stereoscopy is concerned with the visual perception of the three-dimensional space around us. Stereoscopy covers a wide range of phenomena including the physiological and psychological basis of visual perception, as well as the principles involved in the construction of equipment used to facilitate stereoscopic vision and perception.

Binocular vision is the ability to observe objects simultaneously with two eyes set at different positions. All animals that have two functioning eyes have the ability to observe with binocular vision within the field of view (FOV) that is common to both eyes. The differences in position of the two eyes mean that the images they acquire are different. These differences are translated into differences in depth, or depth perception, by the observer. Most people with binocular vision are familiar with the perception of depth, and in fact take it for granted. Just as the perception of depth is important in day-to-day living, so is it important in many aspects of remote sensing.

The study of visual perception within psychology has not resolved all aspects of the visual perceptive process and so some of these processes remain imperfectly understood. Visual perception provides the theoretical and practical basis for stereoscopic perception. As our understanding of the

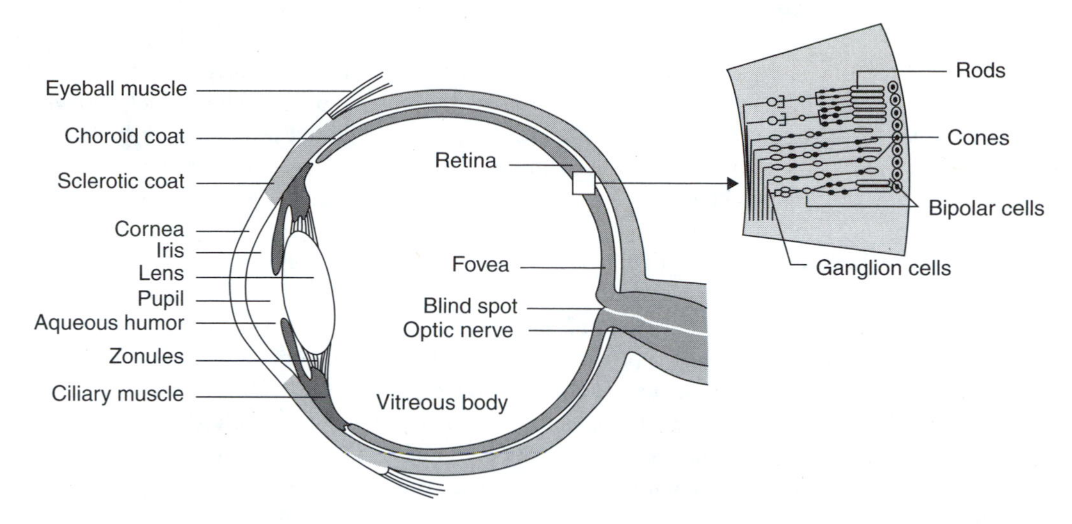

FIGURE 3.1 Cross-section of the human eye showing the major components that affect vision.

processes of visual perception increase, our understanding of binocular vision will also increase and can be used to improve stereoscopic perception.

3.2.2 MONOCULAR VISION

The major components of the eye are shown in Figure 3.1. Light propagated from the object falls on the lens of the eye, which then focuses that light onto the retina. This occurs in a similar way to that which occurs in a camera where the lens focuses the light entering the camera onto the focal plane. The retina contains light sensitive elements that respond to the incident radiation by sending signals to the brain, in contrast to a camera where the film in the focal plane is exposed by this energy. The lens of a camera changes position relative to the film to focus objects at different distances from the camera onto the film. The lens of the eye cannot do this, but alternatively it changes its shape and hence focal length when focusing objects at different distances onto the retina. This ability to change shape is called accommodation. Thus, when you look at this page the lens will adopt one shape to focus objects at short distances onto the retina. When you then look out of the window the lens of the eye changes shape so as to focus objects at far distances onto the retina. As a person ages the lens of the eye becomes more crystalline and tends to loose its ability to change shape. For this reason older people often have difficulty focusing on close objects.

The retina consists of a matrix of light sensitive filaments of two types, called rods and cones that are connected to nerve endings of the optic nerve system. The rods are sensitive to light of all wavelengths in the visible region and so can be considered to be like panchromatic sensors. There are three types of cones, sensitive to narrower wavebands that approximate the blue, green and red parts of the spectrum. The cones provide our perception of colour. The rods predominate around the periphery of the retina and the cones towards the centre. At the most sensitive part of the retina, the fovea centralis, there are no rods and each cone is connected to a single nerve ending of the optic nerve. Out from the fovea centralis groups of rods and cones are connected to single nerve endings. The greatest resolution is therefore achieved at the fovea centralis, but good illumination is required to fully use this part of the retina. The eye is usually oriented so that the object of interest is focused on the fovea centralis.

At lower light levels the cones more rapidly loose their effectiveness, being sensitive to narrower wavebands. With increasing distance away from the fovea centralis more rods and cones are connected to each nerve ending, with only rods being connected at the periphery of the retina. The effect of connecting more rods and cones to a nerve ending is to increase the response of the nerve ending to low light levels. In consequence, the periphery of the retina is much more sensitive to low light levels than is the centre of the retina. Objects tend to loose colour and become more black and white at night because of the domination of the rods. Indeed at very low light levels the fovea centralis is effectively blind. The rods and cones are connected to the optic nerve through simple pattern recognition processing centres called synapses (Figure 3.2). The optic nerve leaves the eye at the blind spot where there are no rods or cones.

The iris controls the amount of light that enters the eye in a given time by changing the diameter of its aperture or pupil, in a similar way to that which occurs in a camera. The pupil is usually about 3 to 4 mm in diameter, but it can vary between 2 and 8 mm. Animals that are nocturnal in habit, such as cats, can exhibit a greater range of pupil diameters, thereby having more control over the amount of light that enters the eye. A clear layer and aqueous solution protect the components of the eye.

The image that is focused onto the retinal surface of a single eye does not contain information on the distance to the object from the observer, but it does contain clues to that distance that we have learnt to interpret as distances, with individual levels of reliability. The clues to depth in monocular vision are:

1. *The relative sizes of objects*. We usually interpret identically shaped objects of different sizes to be identical objects but at different distances from the observer. Similarly the size of familiar objects is usually used to estimate the distance to that object. This characteristic can be used to trick an observer, for example by showing a miniature version of the object, such as a toy car, next to a full scale model in which case the observer may decide that he is viewing two full sized cars but at different distances.
2. *Shape or position.* Parallel lines appear to come together the further away they are from the observer.
3. *Interference.* If an object is masking a second object. then the first is assumed to be closer to the observer than the second.

3.2.3 Binocular Vision

Binocular vision occurs when we view a scene simultaneously with two eyes at different locations. The images formed in either eye are flat images on the retina that contain displacements due to differences in depth to different objects within the FOV, in a similar way as it occurs in an aerial photograph.

As the two eyes are at different locations when viewing the object, the displacements in either eye due to differences in distance or depth will be different. These displacements will be parallel to the eye base in accordance with Figure 3.3. The brain, on receiving these two different images, integrates the two by interpreting the differences as depths or distances from the observer, and so uses the two images to create the one three-dimensional image of the scene observed by both eyes.

Conventionally, the eyebase is taken as the x-axis, and so these displacements are called x parallaxes. The size or magnitude of the x parallaxes is a function of the eyebase, the distance from the observer to the object and the depth of the object. Focusing on a distant object, such as a wall, and placing a finger in the line of sight can demonstrate the existence of x parallaxes. Two images of the finger will be seen. The displacement between these two images is a function of the x parallax difference due to the difference in distance between the observer and the fingers, and the observer and the wall. The displacement between the two fingers should be parallel to the eyebase, as can be demonstrated by tilting the head whilst viewing the wall.

When an object is viewed simultaneously with both eyes, the eyeballs are rotated so that the line of sight to the object passes through the fovea centralis of both eyes. This phenomenon is known as convergence of the optical axes. Focusing on a second object will cause the same phenomenon to occur, changing the convergence of the eye axes. The closer the objects are to the observer, the

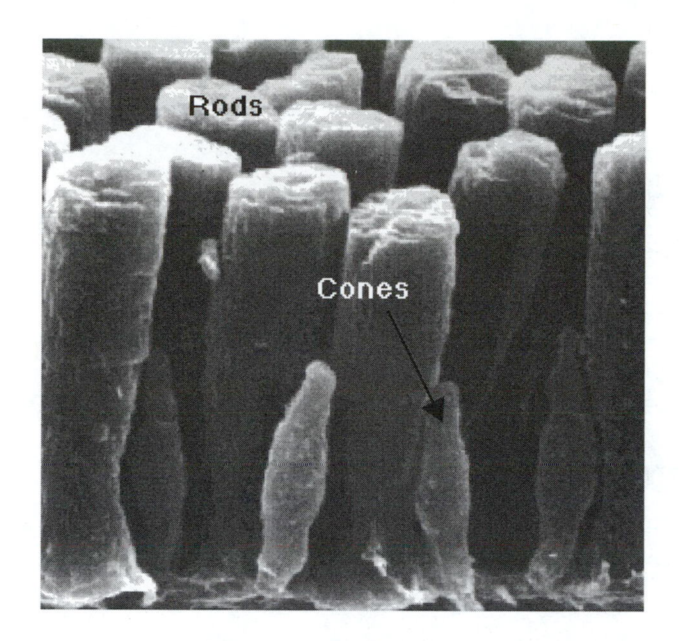

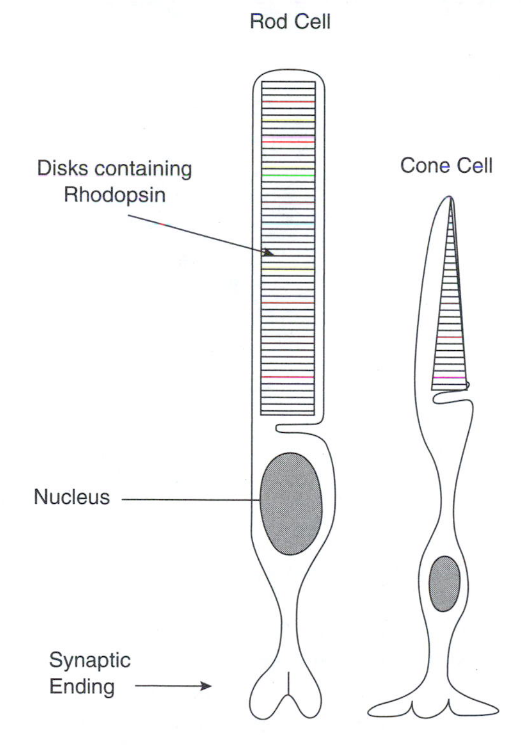

FIGURE 3.2 The rods and cones in the human eye.

greater is the angle of convergence. Viewing a target means that both eye axes should bisect the target, so that the two eyes and the target define a plane known as the epipolar plane. There are clearly an infinite number of epipolar planes, with all of them having the eyebase as a common line, as shown in Figure 3.4.

If the eye axes do not intersect in the target then the observer will not be observing in epipolar planes. In this case the shortest distance between the axes will be at right angles to the eye axes in the

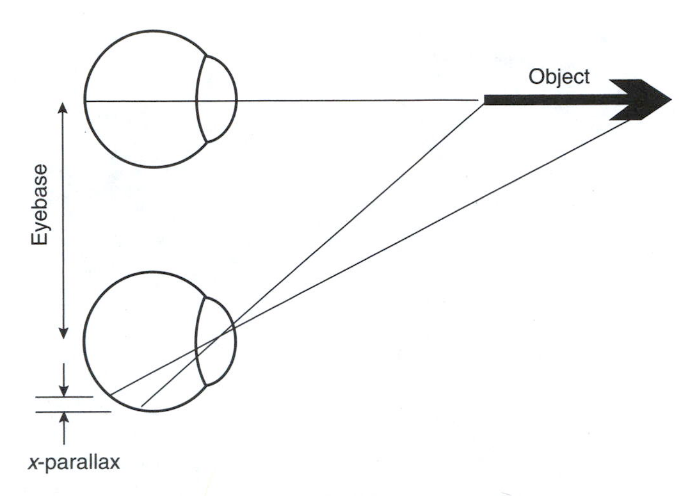

FIGURE 3.3 Binocular vision and displacements in those images due to the eyebase during observation.

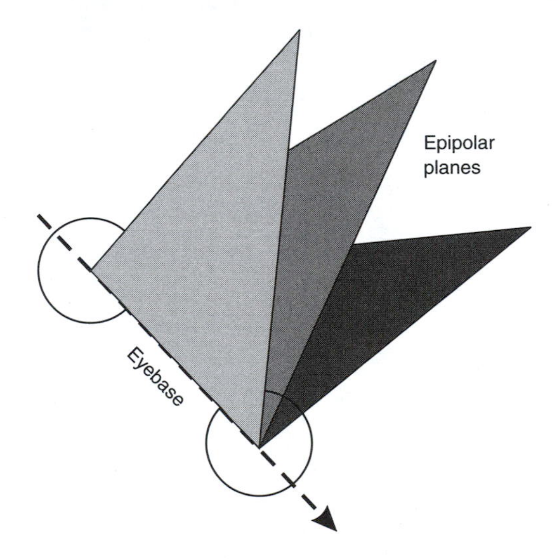

FIGURE 3.4 Viewing in epipolar planes.

direction vertical to the eyebase. These vertical displacements are called *y*-parallaxes, in accordance with the co-ordinate system shown in Figure 3.5. The brain can accommodate small *y*-parallaxes between features in an image. For larger *y*-parallaxes the brain ceases to be able to match the detail recorded by either eye and hence cannot create a perceptual three-dimensional model of the object or surface. When the brain cannot rationalise the two images into a three-dimensional image, then the two images compete for domination. This competition is called retinal rivalry. It means that either one of the images dominates, or the two images alternatively dominate perception of the scene. In both cases three dimensional or stereoscopic perception is destroyed.

3.2.4 The Binocular Perception of Colour

The visual perception of colour is discussed in detail in Section 3.5. Here we are interested in the influence of differing colours, as seen by each eye, on the perceptual process.

If the images received by the left and right eyes contain small differences in colour or tone then the brain will allocate a colour or tone to the object that is some mixture of the two colours or tones.

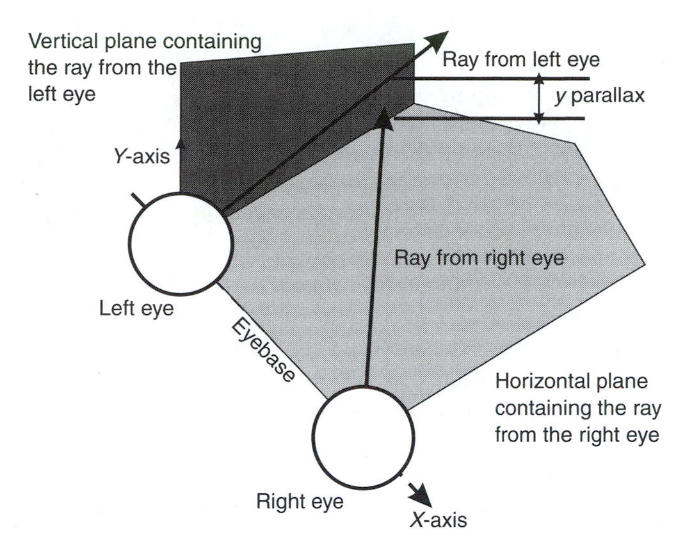

FIGURE 3.5 The co-ordinate system used for viewing and associated x- and y-parallaxes.

The perceived image may thus contain colours or tones that do not occur in either of the original images. It is important to ensure that this is a negligible source of error in visual perception, by ensuring good rendition of colours in the viewed images, and ensuring that the analyst views images under good quality illumination.

The brain cannot accommodate large variations in tone or colour between the two images, setting up retinal rivalry when this occurs. A relatively common case of retinal rivalry occurs when water is recorded as being highly reflecting on one image and very dark on the other. In this case, observers will note that either one of the tones will dominate or the surface will appear to glitter. In either case, the retinal rivalry is likely to cause stress to the observer, resulting in headaches and lethargy. Such a situation often occurs with water in optical wavelength images, but it can occur with other types of surfaces in radar image data, particularly when the corner reflector effect gives a strong response into one of the radar images and a weak response in the other image.

3.2.5 General Principles of Stereoscopic Vision

Binocular vision allows us to perceive a three-dimensional model from two flat images acquired at different positions so that they contain different displacements due to variations in the distance between the observer and the object or surface. We can also create three-dimensional models of an object using a pair of photographs of that object or surface, as long as both the photographs and the observer obey certain rules. These rules are:

1. The photographs of the same object detail must be taken, but from different positions where these positions are the ends of the camera base. The camera base is analogous to the eyebase of the observer.
2. Viewing of the photographs must be oriented so as to reconstruct the geometry that is applied during acquisition of the photographs. It is also important to ensure that both photographs are of similar quality and that both are well illuminated.
3. The two photographs must be viewed simultaneously, with the observers eyebase parallel to the photo base, where the photobase is defined by a line passing through the principal points of the two photographs and their conjugate principal points in the other photograph (Section 3.2.8). This requirement ensures that the photographs are being viewed in epipolar planes.

3.2.6 Methods of Stereoscopic Viewing

There are a number of ways of satisfying the general principles of stereoscopic vision listed above. All methods require the separation of the two images, so that each image is only seen by one eye.

It is possible to simultaneously view a pair of photographs with the unaided eyes and perceive a three-dimensional model. Viewing the photographs either with parallel or with crossed eye axes can achieve this goal. With parallel eye axes the left eye views the left photograph and the right eye views the right photograph. With crossed eye axes, the left eye views the right eye and vice versa. The point of focus with both these techniques is different to that which the brain would expect from the convergence of the eye axes. Parallel eye axes are usually accompanied by focus at infinity but in this case the focus is at a shorter distance. Crossed eye axes means that the eye expects to focus at shorter distances than the actual distance to the photographs. For this reason the procedure takes some practice. It is, however, handy to use in the field and does not appear to incur damage to the eyes in its execution.

Devices have been designed to eliminate this inconvenience in viewing stereoscopically. These devices achieve separation of the images either physically or optically.

3.2.7 Physical Methods of Separation using Stereoscopes

A stereoscope is a device that achieves physical separation of the images by directing the left eye to receive either reflected or transmitted light from the left image and the right eye from the right image. Stereoscopes are convenient to use because their optical systems are designed to allow the observer to focus at the most comfortable viewing distance, about 300 mm for most observers, as well as providing magnification of the viewed photographs.

The two most common forms of stereoscope are the lens (Figure 3.6) and mirror (Figure 3.7) stereoscopes. The lens stereoscope is light and compact so that it is convenient for field use. However, the eyebase limits the separation of the photographs, so that often one photograph will lie above the second photograph. This limitation can be overcome by bending the masking photograph so as to reveal the detail beneath.

The mirror stereoscope overcomes this limitation by reflecting the light from the photographs off a mirror and a prism before it enters the binocular eyepiece. This design achieves wider separation of the photographs (Figure 3.7). The use of a binocular eyepiece allows a range of magnifications to be incorporated into the eyepiece in contrast to the lens stereoscope that has limited magnifications. The magnifications in most binocular eyepieces mean that only part of the stereo-pair can be viewed

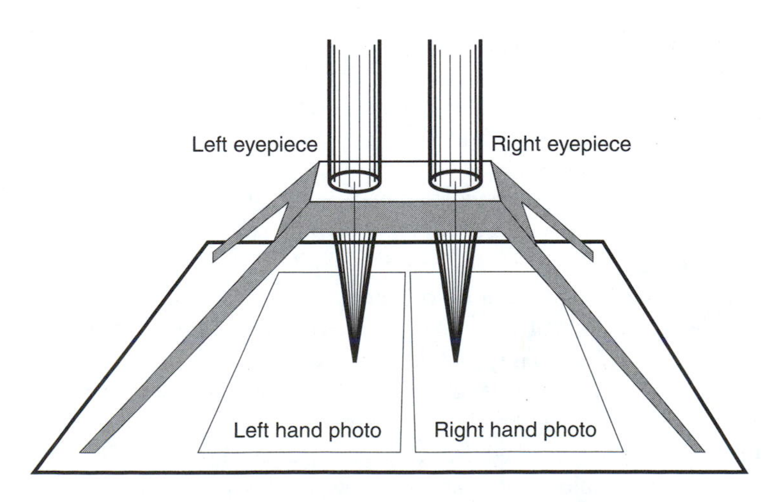

Figure 3.6 The lens or pocket stereoscope.

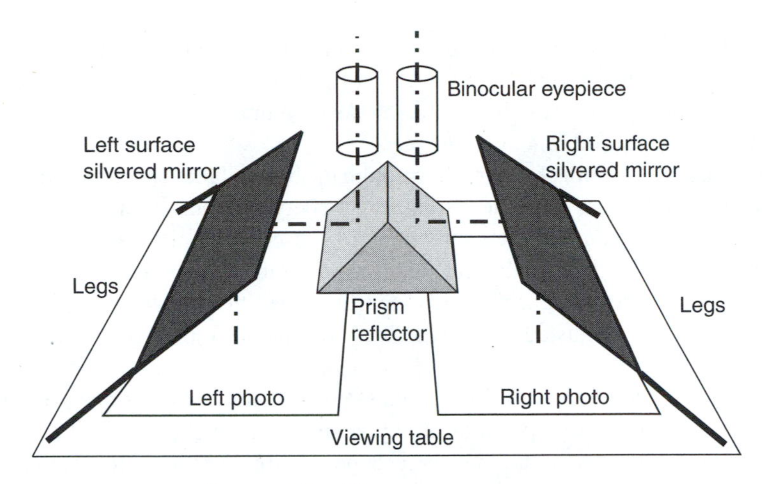

FIGURE 3.7 The mirror stereoscope.

at the one time. Most mirror stereoscopes overcome this disadvantage by allowing the observer to remove the binocular eyepiece so as to view the whole of the stereo-pair. The greater separation of the photographs means that they can be secured to a base during viewing and the stereoscope moved across the pair of photographs to view different parts of the stereo-pair in detail. Being able to fix the photographs to a base also allows the observer to use a parallax bar to measure x-parallaxes and then calculate height differences as discussed in the next section. The mirrors of the mirror stereoscope are surface silvered to eliminate double reflection. These mirrors are susceptible to etching by the acids of the skin; it is very important not to touch or otherwise handle the mirrors of the stereoscope with fingers. If the mirrors are inadvertently touched then they must be very carefully cleaned with alcohol and a soft cloth.

3.2.8 Viewing with a Stereoscope

Proper stereoscopic viewing of an area common to a pair of photographs requires the observer to simultaneously view the pair of photographs in epipolar planes. These conditions are achieved as follows:

1. Mark the principal point (Section 2.4) of each photograph. Using the details in the photographs, visually locate the principal points in the adjacent photograph. These two points are called the conjugate principal points.
2. Orient the photographs on the desktop or hard surface so that the parts of the photograph with common detail are closest together, and shadows are towards the observer. In this arrangement the conjugate principal points will be closer together than will the principal points. Secure one of the photographs to the baseboard.
3. Orient the second photograph to the fixed photograph so that the four principal and conjugate principal points all lie on the same line. Place a ruler or straightedge along this line. Position the stereoscope above the photographs and orient so that the eyebase of the stereoscope is parallel to the photo-base as represented by the ruler or straightedge. The eyebase of the stereoscope is defined as a line through the axes of the binocular eyepiece. This arrangement sets the conditions necessary to view the photographs in epipolar planes.
4. Ensure that comfortable focus is achieved for each eye by viewing the respective images through the corresponding eyepiece of the stereoscope and adjusting the focus for that eyepiece as necessary. It is now necessary to set the convergence so that the viewer will see the same detail in both photographs. Select an obvious feature in the fixed photograph

and move the stereoscope so that this feature is in the centre of the FOV of the binocular eyepiece, ensuring that the photo-base remains parallel to the eyebase. Move the second photograph in or out along the line of the photo-base until the same detail is in the centre of the FOV of the second binocular eyepiece.

5. Simultaneously, view both photographs through the binocular eyepieces of the stereoscope. If the images do not fuse then the displacements between them will be in the x and the y directions. Displacements in the y direction are caused when the observer is not viewing in epipolar planes. In this case repeat step 3 to re-establish viewing in epipolar planes. If the displacements are in the x direction then the separation between the images as done in step 4 needs to be adjusted by moving the loose photograph either in or out along the photo-base.
6. Once fusion has been achieved find the most comfortable viewing position by carefully moving the second photograph in or out by small distances, to get the best stereoscopic perception, and carefully rotate the stereoscope by small amounts to eliminate displacements in the y direction. Check whether a comfortable viewing position has been achieved by leaving the stereoscope for 5 min and then coming back to it. If a comfortable viewing position has been achieved then fusion will be created immediately on viewing through the stereoscope. If this does not occur, but the images gradually fuse then further adjustment is required. Finally, move the stereoscope across all the pair of photographs to check on the stereoscopy across the whole area of overlap.
7. The observer creates a perceptual model from fusion of the two separate images. The observer should become aware of differences in height in the model due to x-parallax differences created during acquisition of the photographs.

Most people can view a pair of photographs in this way and achieve stereoscopic vision in a reasonably short time. Some people, however, have a dominant and a lazy eye, the dominant eye doing most of the work. To achieve stereoscopy it is essential that both eyes do equal work; if this is not happening then the observer will see only one image, that which the dominant eye is observing. Initially, the person with a dominant eye may only be able to maintain stereoscopic vision for a short period or have difficulty establishing stereoscopic vision in the first place. When this occurs the student needs to persevere until the lazy eye becomes used to operating with the other eye and stereoscopy can be easily maintained for long periods.

3.2.9 Optical Methods of Separation

The two images are projected optically onto the same screen using an appropriate filter system. They are separated using the same filter system before entering the respective eyes of the observer. This optical separation can be achieved using either:

1. *Coloured filters in two complimentary colours* usually called the anaglyph method.
2. *Polarising filters* with the two filters set to accept light at polarisations that are at right angles to each other.

3.2.9.1 Coloured Anaglyph

Two black and white images are projected through complimentary filters of blue–green and yellow–red onto the same screen. The observer, wearing glasses containing similar filters, then views the images. The filters used in the glasses will separate the viewed images so that the image projected in blue–green will be received by the eye behind the blue–green filtered glass and the yellow–red image will be received by the other eye.

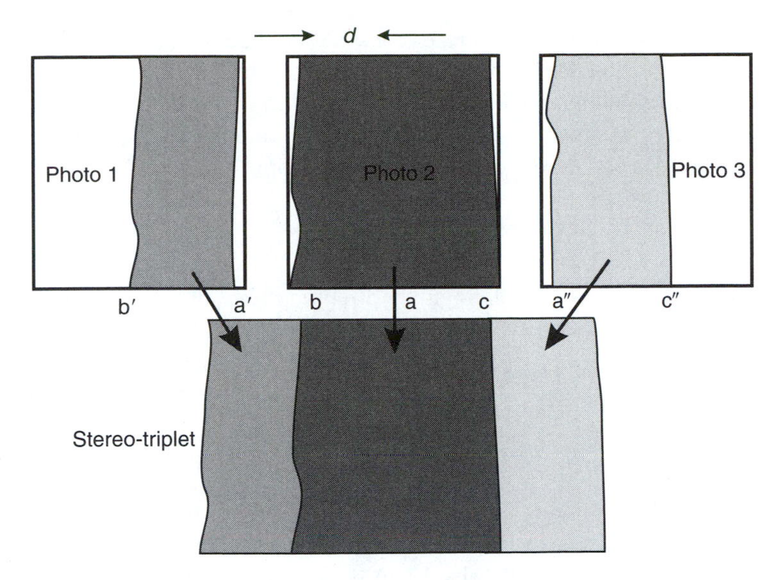

FIGURE 3.8 Construction of a stereo-triplet.

The disadvantages of this method are that only black and white images can be used since colour is used to achieve separation; the levels of illumination can be quite low and the complimentary colours often create retinal rivalry.

3.2.9.2 Polarising Filters

The images are projected through polarising filters set at right angles to each other onto a screen that reflects the images back towards the observer without a significant loss of polarisation. The observer views the images through spectacles containing polarising filters matching those used in projection.

Polarisation can be used with colour or black and white photographs. The main disadvantage of this method is that the screens on which the images are projected often significantly degrade the polarisation causing mixing of the images and degradation of the stereoscopic effect. Metallic paint or lenticular screens are suitable for reflecting polarised light without a serious loss of polarisation.

3.2.10 Construction of a Stereo-Triplet

When an interpreter wishes to take photography out into the field, he may wish to view those photographs stereoscopically. To do this he will normally choose a lens stereoscope because of its compactness and lightness. A stereo-triplet is a way of creating two adjacent stereo-pairs using the three adjacent photographs, fixed so that they are suitable for viewing with a lens or pocket stereoscope. A stereo-triplet may be constructed as follows:

1. Select three overlapping photographs of the area of interest. On the center photograph draw two vertical lines, b and c at distance *d* on either side of the principal point (Figure 3.8) and passing through line a. The distance *d* is slightly less (about 5 mm) than the eyebase of the observer, and is usually set to about 60 mm.
2. Visually transfer lines b and c to the left and right adjacent photographs at b′ and c′ respectively. Line a can be transferred to both of the other photographs at a′ and a″ respectively. Cut away the portions of the center photograph outside lines b and c and the outer photographs outside of lines b′, a′ · a′ and c″ respectively.

3. Secure the center photograph to a piece of card along line a, the line joining the fiducial marks. Visually position the right-hand photograph, in the correct orientation, on the card with portion of the photograph left of line a, beneath the central photograph.
4. Using the stereoscope adjust the position of the right-hand photograph on the card so that the pair can be viewed comfortably during stereoscopic analysis. Secure the right-hand photograph to the card.
5. Repeat with the left-hand photograph.

This gives a stereo-triplet. If only two photographs are used the result is a stereo-doublet.

3.3 MEASURING HEIGHT DIFFERENCES IN A STEREOSCOPIC PAIR OF PHOTOGRAPHS

3.3.1 Principle of the Floating Mark

The height difference AB in Figure 3.9 introduces displacements a_1b_1 and a_2b_2 into the left and right-hand photographs, respectively. If small marks are placed immediately over the photographs at points a_1 and a_2 then the brain is deceived into believing that these marks are in the left and right-hand photographs, respectively. As such the brain interprets the two marks as being images of the same mark, so that the observer perceives one mark at A in the fused model. If the marks are moved, say to b_1 and b_2 then the observer will perceive the fused mark moving from A to B in the model. Because of this characteristic the two points are called a floating mark.

The floating mark can be made to move up and out of, or down and into, the perceived three-dimensional model. At some stage the x-parallax differences between the mark and the model detail will become too great for the brain to accommodate, and the floating mark will break into two individual marks.

The principal of the floating mark allows us to place a point in the perceptive model as a fusion of two points, one imposed into each image. The co-ordinates of each point can be measured relative to the principal point of the respective image, and these distances can then be used to estimate height differences in the model. Because of this the floating mark is used as the basis for extracting quantitative information on height differences and positions of points in the pair of aerial photographs. One simple way to measure height differences is by the use of a parallax bar.

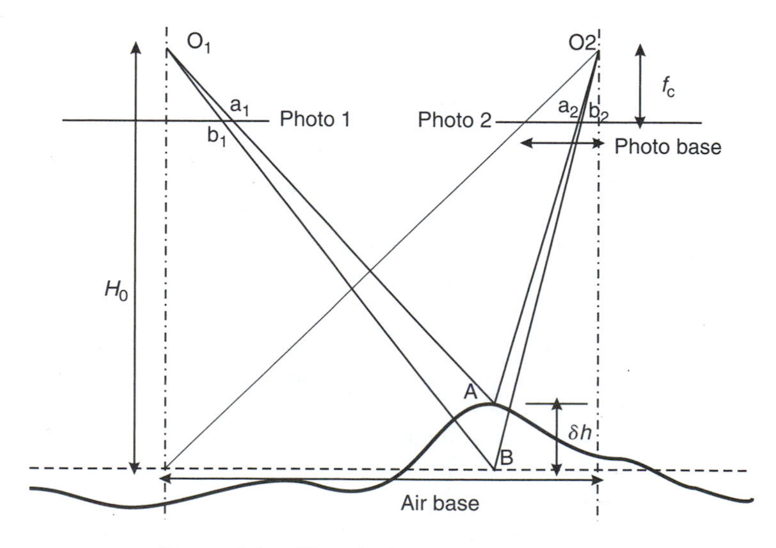

FIGURE 3.9 The principle of the floating mark.

3.3.2 Parallax Bar

A parallax bar consists of a rod supporting a mark etched onto a glass tablet at one end and with a micrometer, supporting a second glass tablet with etched mark at the other end. The micrometer allows the distance between the floating marks to be read to the nearest 1 mm from the main scale and to the nearest 0.01 mm from the vernier scale of the micrometer.

The main scale of the micrometer is usually graduated with values that increase inward for reasons that will be discussed later. In operation the etched marks on the glass tablets are placed over corresponding detail in the two photographs. When the two marks exactly coincide with the separation of the photographic detail then the two marks will fuse into a floating mark at the same elevation as the perceived model. Slight changes in the x-parallax or separation of the floating marks, by rotating the micrometer screw, will cause the floating mark to either move up and out of, or down and into the model. The separation between the marks can be measured by means of the micrometer. Since differences in x-parallax between two points are a function of the difference in height between the two points, observing the two distances by means of the micrometer can be used to estimate the height difference as will be discussed in Section 3.3.4.

3.3.3 Vertical Exaggeration

When looking at a stereoscopic model, the vertical distances, or height differences, are exaggerated relative to corresponding horizontal distances. This difference between the vertical and horizontal scales is called vertical exaggeration, V. Vertical exaggeration occurs when photographs are taken at one base length (B) and are viewed at a shorter base length (b). The longer base length gives larger x-parallaxes. These larger x-parallaxes are interpreted by the brain as greater height differences than is actually the case. The existence of this vertical exaggeration enables the observer to readily appreciate vertical shapes in the terrain. However, it also means that it is very difficult to estimate height differences when viewing a stereo model without an aid such as a parallax bar.

3.3.4 Displacements due to Height Differences in an Aerial Photograph

Height differences introduce displacements in an aerial photograph that are radial from the nadir point as shown in Figure 3.10. The magnitude of these displacements, $\delta p = (p_a - p_b)$ is a function of the distance of the point from the nadir point and the ratio of the height difference to the flying height of the aircraft above ground level.

In near vertical aerial photography the nadir is usually assumed to coincide with the principal point.

At the nadir, NA = NB = 0 so that (Figure 3.10):

$$p_a = p_b = 0.0$$

Hence,

$$\delta p = 0.0$$

At distance NA = NB:

$$\frac{p_a}{f_c} = \frac{\text{NA}}{(H_0 - \delta h_{AB})}$$

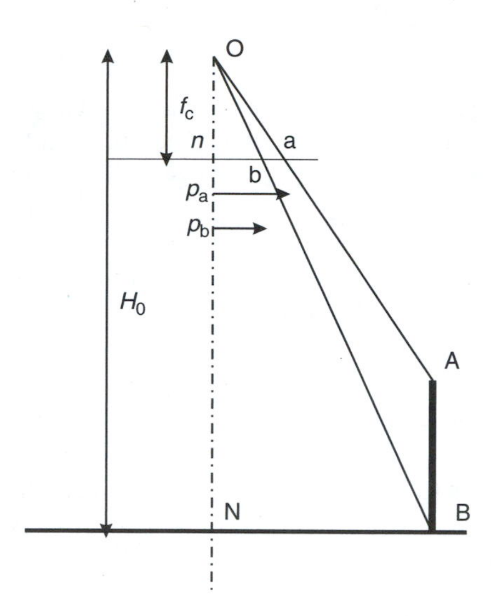

FIGURE 3.10 Magnitude of displacements due to height differences.

and

$$\frac{p_b}{f_c} = \frac{NB}{H_0}$$

By combining these equations we get:

$$\delta p = (p_a - p_b) = \frac{f_c\{NA \times H_0 - NB(H_0 - \delta h_{AB})\}}{H_0(H_0 - \delta h_{AB})}$$

Re-arrange to get:

$$\delta p = \frac{f_c NB \delta h_{AB}}{H_0(H_0 - \delta h_{AB})} \tag{3.1}$$

Since NA = NB. Now $p_b = NB \times f_c/H_0$ so that after substituting this into Equatin (3.1) we get:

$$\delta p = \frac{p_b \delta h_{AB}}{H_0 - \delta h_{AB}}$$

Which can be rearranged to give:

$$\delta h_{AB} = \frac{H_0 \delta p}{p_a} \tag{3.2}$$

3.3.5 Derivation of the Parallax Bar Formulae

The displacements introduced by a difference in height in two adjacent aerial photographs will be in opposite directions when the height difference is located between the nadir points of the two photographs, as is normally the case (Figure 3.11). Using the principal of the floating mark the distance between points of corresponding detail can be measured. The difference in these measurements, known as the parallax difference, is a function of the difference in height between the two points,

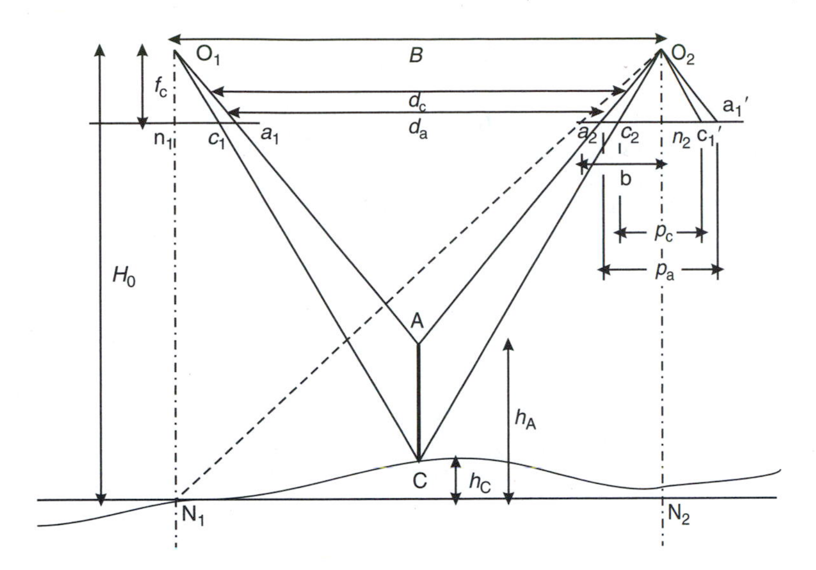

FIGURE 3.11 Derivation of the parallax bar formula.

as shown in Figure 3.11. The relationship between the parallax difference and the height difference will now be developed making the assumptions that:

1. Photographs are taken with vertical camera axes that are parallel.
2. Both photographs are taken at the same altitude so that the air base, *B* in Figure 3.11 is at right angles to the camera axes.

O_2a_1' is drawn parallel to O_1A and O_2c_1' is drawn parallel to O_1C. The distances d_a, d_c are measured by means of the parallax bar. The parallax of A is given by:-

$$p_a = n_1a_1 + n_2a_2 = a_2a_1'$$

For C

$$p_c = n_1c_1 + n_2c_2 = c_2c_1'$$

The parallax difference, δp_x is defined as:

$$\delta p_x = p_a - p_c = (B - d_a) - (B - d_c) = d_c - d_a$$

The height difference AC is given by:

$$h_{AC} = h_A - h_C$$

By similar triangles O_1O_2A and $O_2a_2a_1'$:

$$\frac{B}{H_0 - h_A} = \frac{a_2a_1'}{f_c} = \frac{p_a}{f_c} \tag{3.3}$$

So that:

$$(H_0 - h_A) = \frac{Bf_c}{p_a} \tag{3.4}$$

By similar triangles O_1O_2C and $O_2c_2c'_1$ we can also derive:

$$(H_0 - h_C) = \frac{Bf_c}{p_c} \tag{3.5}$$

From which we can get:

$$\Delta h_{AC} = (H_0 - h_C) - (H_0 - h_A) = B \times f_c \times \left(\frac{1}{p_c} - \frac{1}{p_a}\right) \tag{3.6}$$

Re-arrange to give:

$$\Delta h_{AC} = B \times f_c \times \left(\frac{\delta p_x}{p_a p_c}\right) \tag{3.7}$$

Substitute Equation (3.5) into Equation (3.7) to give:

$$\Delta h_{AC} = \frac{(H_0 - h_C) \times \delta p_x}{(p_c + \delta p_x)} \tag{3.8}$$

In Equation (3.8) we either know, or can measure H_0, δp_x, but not h_C or p_c. If C is placed on the datum surface then $h_C = 0$ and $p_c = b$ as can be proved using similar triangles, so that:

$$\Delta h = \frac{H_0 \delta p_x}{b + \delta p_x} \tag{3.9}$$

known as the *parallax bar formula*, where Δh is the difference in height of the point relative to the height of the model at the left nadir point.

3.3.6 Characteristics of the Parallax Bar Equation

To have δp_x positive for positive height differences, most parallax bars read values that are increasing inward. The parallax difference, δp_x, is given by:

$$\delta p_x = p_a - p_c = d_a - d_c \tag{3.10}$$

If the parallax bar reads increasing values inward then reading $d_c < d_a$ will given Δh is positive when δp_x is positive. If the parallax bar does not read increasing inward then the sign of δp_x will need to be changed in the parallax bar formulae. The important data are the parallax differences, not the actual parallax bar readings themselves.

By convention the datum is taken as the height of the left-hand principal point, that is the photo-base, b, is measured in the right-hand photograph as the distance between the principal point and the conjugate principal point in that photograph. If the principal points are at different elevations then the bases measured in the two photographs will be different. The calculated height differences are relative to the elevation of the left-hand principal point.

When δp_x is small the equation can be approximated by:

$$\Delta h = H_0 \times \delta p_x / b \quad (3.11)$$

In this situation H_0/b is constant for a photo-pair so that all height differences are proportional to δp_x.

When using the approximate equation, $\Delta h = (H_0/b_0) \times \delta p_x$ then height differences can be calculated between two points without reference to the datum. In this case the parallax bar readings are taken at the two points of interest, and the parallax difference between the two is used to calculate the height difference.

Tilts in the photographs introduce warping into the perceived stereoscopic model created whilst observing a pair of stereoscopic photographs. Aerial photographs contain small tilts and slight differences in altitude from photograph to photograph. These tilts create significant height displacements in the perceptual model because the photographs, when set up lying flat beneath the stereoscope, are not replicating the conditions that existed during acquisition. They thus introduce errors into the Δh values derived using the above formulae.

3.4 PLANIMETRIC MEASUREMENTS ON AERIAL PHOTOGRAPHS

3.4.1 INTRODUCTION

It is often necessary to make planimetric measurements from aerial photographs, either during the interpretation process or during the transfer of interpreted information from an aerial photograph to a map or grid of known scale. The planimetric measurements that may need to be done include:

1. Determination of the scale of a photograph or a map.
2. Measurement of distances in a photograph or a map.
3. Measurement of areas in the photograph or a map.
4. Transfer of detail from a photograph to either update a map, add specific information to a map or to construct a simple map if suitable ones do not exist.

3.4.2 DETERMINATION OF SCALE

All the techniques that are discussed in this section deal with single photographs. They cannot measure, and therefore cannot take into account, displacements in the photograph due to height differences. In consequence, displacements due to height differences are normally converted into errors when using these techniques. Because of this source of error, these techniques are most accurate either when there are only small height differences in the area being mapped or if the area is mapped in segments to meet this condition. If these restrictions are too harsh for a particular situation then sophisticated photogrammetric mapping techniques, routinely used in the production of topographic maps, must be used.

Scale is defined as the ratio of 1 unit of distance on the map or photograph to the equivalent x units of distance on the ground. The scale is thus depicted in the form 1:x where 1 unit of distance on the photograph or map represents x units of the same distance on the ground. The larger the scale factor, x, the smaller is the map scale and the larger the ground area covered within a given map area as shown in Table 3.1. The smaller the scale factor, the larger the map scale and the smaller the ground area covered by a map. Maps and photographs are often categorised as being small, medium or large scale.

TABLE 3.1
Typical Map Scales and Some of their Geometric Characteristics

Scale	Description of this Scale	Ground Distance Covered by 1 mm on the Map (m)	Map Distance Covered by 1 m on the Ground	Typical Area of a Map Coverage (km^2)
1:250,000	Small scale	250	0.004 mm	14,400
1:50,000	Medium scale	50	0.02 mm	600
1:25,000	Medium scale	25	0.04 mm	200
1:10,000	Large scale	10	0.1 mm	50

3.4.3 Measurement of Distances

Ground distances can be calculated from a photograph or map as the product of the map or photograph distance and the scale factor. Distances on a map or photograph can be calculated by dividing the ground distance by the scale factor of the map or photograph. Distances can be measured on a map or photograph by:

1. Graduated ruler or scale
2. Paper strip
3. A length of string
4. An odometer

3.4.3.1 Graduated Rule or Scale

A graduated rule or scale can be used to measure straight-line distances on a map or photograph. The measured distance is either compared with the scale bar on the map to give the equivalent ground distance or the measured distance is multiplied by the scale factor to calculate the ground distance.

3.4.3.2 Paper Strip

The paper strip is oriented along chords of the curved line, marking both the positions on the strip and on the map at the start and end of each chord. In this way the strip is gradually tracked around the total distance to be measured. The total length of the segments is measured and used to calculate the ground distance.

3.4.3.3 Length of String

The string is laid over the curved line, with one end of the string placed at the start of the line. The length of string used to cover the curved distance is compared with the scale bar to estimate the ground distance. Alternatively, the length of the string can be measured and the distance calculated knowing the scale of the photograph or map.

3.4.3.4 Odometer

The odometer is a device to measure distances on a map or photograph. It consists of a small wheel linked to a rotating dial against a graduated scale. When the odometer is tracked around a line, the wheel rotates, rotating the dial against the scale. The difference in scale readings at the start and end of the line is a function of the distance travelled.

3.4.4 Measurement of Areas

There are three different types of areas to be measured on an aerial photograph: regular boundaries, irregular boundaries and ill-defined boundaries. Typical regular boundaries include paddock and fence lines, roads and administrative boundaries. Creeks and watercourses have irregular boundaries. Scattered vegetation and soil associations have ill-defined boundaries. There are two manual ways of determining areas, all of which are suitable for measuring areas with regular and irregular boundaries. Only the dot grid method is suitable for measuring areas with ill-defined boundaries. Areas can also be measured by the use of remote sensing and GIS techniques, as will be discussed in later chapters. The methods of manually measuring areas are

1. The dot grid
2. By the use of a digitiser

3.4.4.1 Dot Grid

A dot grid consists of a grid of dots at some set interval on a clear film base. The product of the distances in the x and y directions separating the dots is the area represented by one dot. Counting the number of dots within the boundaries of the area and multiplying this by the area represented by one dot, determines the area. To reduce the workload a line grid may be drawn around set blocks of dots, say 10×10 or 100 dot blocks and counting the blocks as representing the number of dots contained in a block, when all of the dots within a block are within the area of interest.

The accuracy of the technique can be increased by increasing the density of dots in the grid, but at the cost of increasing the workload. Dot grids are suitable for measuring areas with both well and ill-defined boundaries. It could, for example, be used to measure the area of scattered trees across the image.

3.4.4.2 Digitiser

Digitising is a method of recording boundary detail that is associated with digital data. The digitised boundary detail can then be used in a GIS, in addition to using it to estimate the area within a boundary. Digitising is described in Chapter 6.

3.4.5 Transfer of Planimetric Detail by the Use of the Anharmonic Ratio

There are accurate stereo-photogrammetric techniques, based on the use of stereoscopic photo pairs, as well as digital techniques for the transfer of planimetric detail from images to maps. The photogrammetric methods are the basis of most topographic maps and some digital elevation models. They are not dealt with in this text, but the reader is referred to the photogrammetric literature, should they wish to know more about this type of activity. Some of the digital techniques include laser ranging from aircraft and image interferometry. The reader is referred to the literature on these topics should they wish to know more.

In this chapter we are interested in learning about simple, approximate methods that a photo-interpreter may use to transfer detail from a photograph or image to a map. These techniques will use single images in the task; since each image contains displacements due to height differences, then the techniques will contain errors. We thus need to consider how to deal with these errors when we learn about the techniques themselves, as we progress through the chapter.

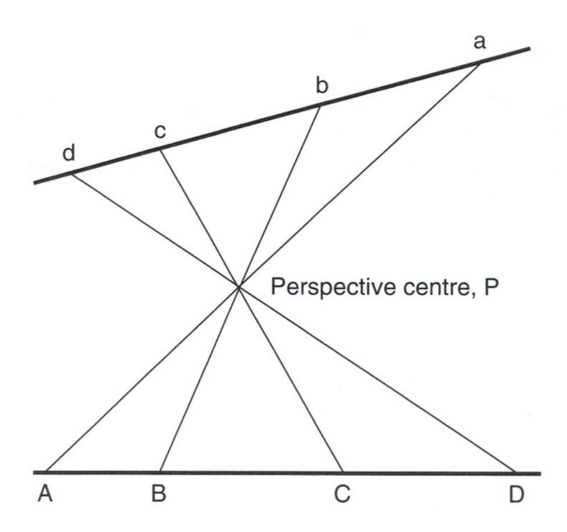

FIGURE 3.12 The theoretical basis of the anharmonic ratio.

The techniques that we will consider are based on the anharmonic ratio (Figure 3.12), or the use of proportional dividers.

The anharmonic ratio states that

$$\frac{\mathrm{AC}}{\mathrm{BC}} \times \frac{\mathrm{BD}}{\mathrm{AD}} = \frac{\mathrm{ac}}{\mathrm{bc}} \times \frac{\mathrm{bd}}{\mathrm{ad}} = \text{constant}, k$$

The method is suitable for an image derived from a camera or a device based on the central perspective projection but not for scanner or radar data. There are two methods of transferring details based on the anharmonic ratio. These methods are

1. The paper strip method
2. Projective nets

3.4.5.1 Paper Strip Method

A set of four points that can be identified in the photograph (a,b,c,d) and on the map (A,B,C,D) are selected so that they bound the area of interest (Figure 3.13). The aim is to transfer other points from the photograph to the map. Select the next point to be transferred (x). Place a paper strip on the photograph to cross the three lines radiating from one of the control points (point a in this case) to the other control points (b, c and d) and the line to x. Mark where the four lines pass beneath the paper strip. Transfer the paper strip to an equivalent position on the map. Orient the strip so that the marks to the three control points coincide with the radial lines on the map to the other three control points. Mark on the map where the fourth line passes to X. Draw a line from the control point, A through this mark. Repeat the process from another control point to give an intersection at X. Repeat this process for all new points being transferred.

The process of using a paper strip is time consuming and untidy if more than a few points are involved. The best results will be achieved if the points used give intersections at about right angles and if the four control points and the area enclosed are a close approximation to a planar surface that may be tilted to the horizontal.

3.4.5.2 Projective Nets

Identify four corresponding points on the map and the aerial photograph, ensuring that the points confine the detail to be transferred. Join the points and create a grid by linearly interpolating a set

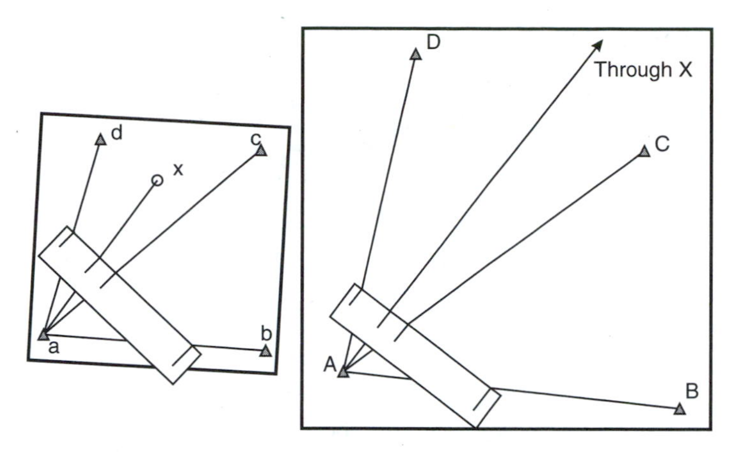

FIGURE 3.13 The paper strip method of transferring detail.

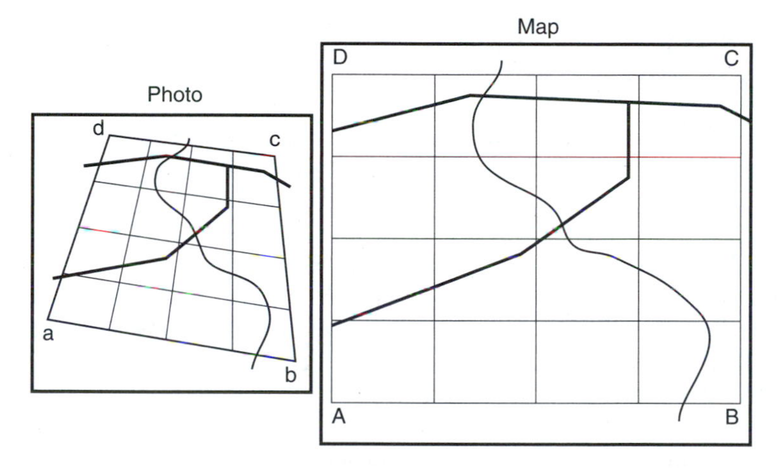

FIGURE 3.14 Construction of projective nets between four corner points on an aerial photograph and a map.

number of intermediate points between the corner points, as shown in Figure 3.14 and joining the appropriate intermediate points. The density of the grid should be sufficient to transfer the detail as will be explained below. The grid is created by linearly interpolating along the sides of the grid, and then joining the equivalent points on opposite sides. The same procedure is used to create the grid for both the map and the photograph. Once the grids are drawn then detail is transferred by the eye from individual grid squares on the photograph to the equivalent squares on the map. As for the paper strip method, the area covered by the projective net should be a close approximation to a planar surface, albeit tilted to the horizontal.

3.4.6 Proportional Dividers

Detail can also be transferred by the use of proportional dividers. Proportional dividers have an adjustable pivot point so that the distance from the pivot point to the pointers at either end can be changed. The ratio of the two distances gives the change in scale between the pairs of pointers. Once the scale of an aerial photograph and a map are known, the relative distances to the center pivot on the proportional dividers can be calculated and set. Measuring a distance on the photograph or map will then allow the same distance to be marked on the other image (Figure 3.15).

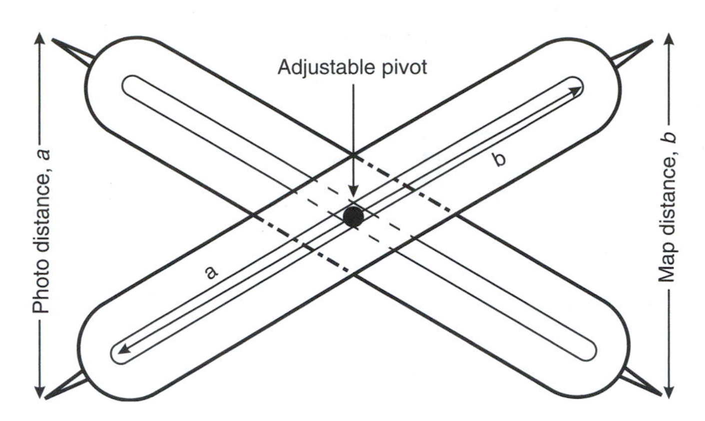

FIGURE 3.15 Proportional dividers.

3.5 PERCEPTION OF COLOUR

The generally accepted explanation for the perception of colour in the human eye was given in Section 3.2 as being due to the sensitivity of three types of cones to blue, green and red lights, respectively. If blue, green and red lights are combined then the resulting ray contains light of all three wavelengths. This process is called additive colour mixing, because it entails the addition of wavebands. The three colours, blue (470 nm), green (520 nm) and red (650 nm) are called the three additive primary colours.

Additive mixing of the three primary colours will produce a range of spectral colours. These colours are all depicted within the chromaticity figure (Figure 3.16 and in colour on the CD, Miscellaneous directory). In this figure the three primary colours are depicted on the outer edge of the colour domain at 4700 (blue), 520 (green), and 650 (red) where the numbers refer to the mean radiation wavelength that is perceived as being that colour in the human eye. The centre of the figure is white, created by mixing full intensities of all three primary colours. The colours in the chromaticity triangle can be expressed in other co-ordinate systems than just the intensities of blue, green and red. The most common is in terms of hue, saturation and intensity where:

Hue is the dominant wavelength in a colour, indicated as the number on the outside of the chromaticity triangle when drawing a line from the white point, C, through the location of the colour in the triangle, to the edge of the triangle.

Saturation is the degree of spectral purity of the colour or the proportion of the colour due to the dominant wavelength. The closer the colour is to the outer edge of the triangle, the more pure it is, and the more saturated the colour. Saturation is measured as the ratio of the distances from the neutral point to the colour point and the neutral point to the edge of the triangle on the same line. The maximum value of saturation is 1.0, and the minimum is zero at the neutral point. Pastel colours, containing a lot of white are desaturated, whereas intense, vibrant or strong colours are highly saturated.

Intensity is the total amount of illumination from the scene or the scene brightness. Low levels of illumination affect the effectiveness of the cones and hence the perceived colours. It is important to have good illumination when viewing imagery.

Black and white images acquired in any three wavelengths can be projected through red, green and blue filters to create a coloured image. The relative intensities of the three additive primaries at each point in the coloured image control the resultant hue, saturation and intensity. If two of the primaries are full intensity and the third is zero, then the resulting colour will have a hue value at the edge of the chromaticity triangle and at the point of bisection of the line joining the two contributing primaries. It is called additive colour because the colours add together to give the resulting colour. Thus, the additive colour resulting from the combination of green and red contains both wavelengths, and will be viewed as being yellow by the human eye. A combination of full intensities of green and red will

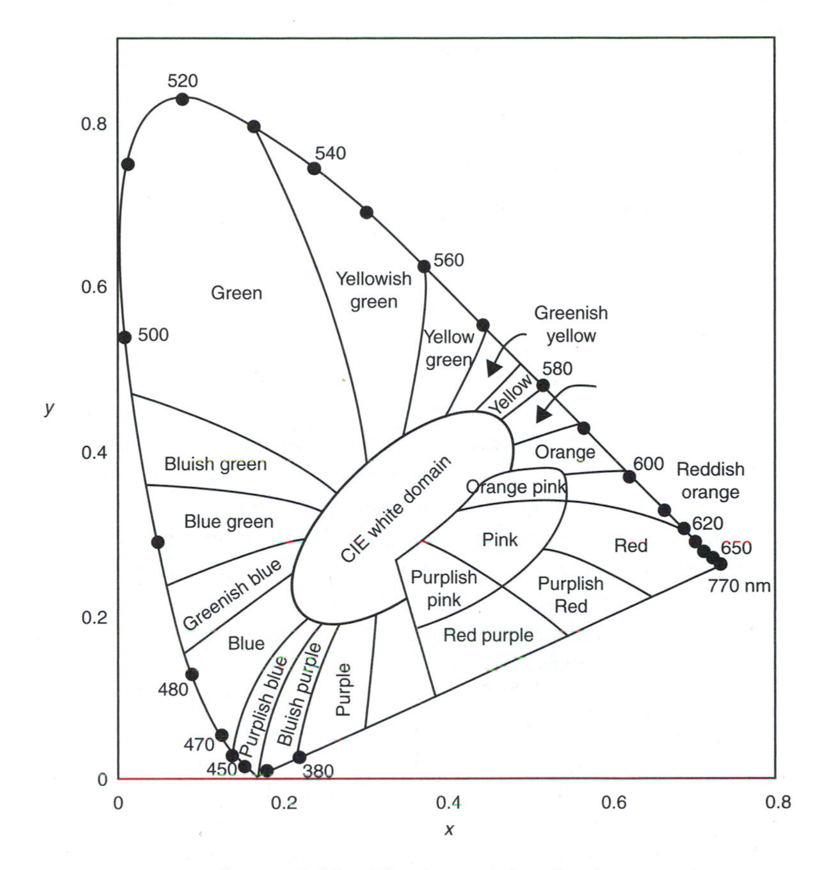

FIGURE 3.16 The chromaticity triangle.

lie at the bisection of the 520 and 650 nm points. The hues that result from the additive combination of the three primaries can be plotted on the chromaticity figure (Figure 3.16) at 580nm. The blue, green and red primary colours lie on the edge of the triangle. As the intensities of the other primary colours are increased, the combined colour moves from the edge of the triangle towards the white point. Equal intensities of the three primaries will yield white, or grey, depending on the intensity.

This form of colour is in contrast to subtractive colour as occurs in a painting. Red paint is deficient in blue and green colours because they are selectively absorbed by the pigments in the paint. Similarly, cyan paint is deficient in red colour. If these two paints are combined, then the resulting paint absorbs blue, green and red. If the levels of absorption are the same then black paint will result. Subtractive colour processes occur in the pigments of film.

Figure 3.16 shows that the additive combinations of pairs of saturated primary colours will yield the colours:

Red + Green = Yellow
Red + Blue = Magenta
Green + Blue = Cyan
Blue + Green + Red = White

Consider selecting panchromatic (black and white) images that record the scene radiation in the blue, green and red wavebands. Combine these additively using blue, green and red colours for the blue, green and red waveband images. The result will be a true colour image, or an image depicting the scene in the same colours as would be seen by the human eye. Panchromatic images acquired in sets of any three wavebands can be additively combined to produce coloured images. They will only be in true colour under the above conditions. Under all other conditions they will produce what are called as false colour images.

3.6 PRINCIPLES OF PHOTOGRAPHIC INTERPRETATION

3.6.1 INTRODUCTION

Image interpretation is defined as the extraction of qualitative information on the nature, origins and role or purpose of the objects or surfaces viewed in the imagery. This qualitative information transcends the estimation of simple quantitative information of size, position, number and distribution of objects. Indeed, extraction of this quantitative information may be an essential prerequisite to proper image interpretation. However, there are some types of quantitative information, such as estimates of leaf area index (LAI) or biomass that cannot be estimated by visual interpretation, but can be estimated by digital image processing techniques discussed in Chapter 4.

The importance of quantitative information in image interpretation can be assessed by example. An adequate description of the geology of an area may depend on measures of the dip and strike of bedding planes. Assessment of the importance of the different agricultural practices in an area may require measurement of the proportions of the different land uses. The interpreter, in conducting visual interpretation, must assess what quantitative information is required to develop an understanding of the phenomena influencing those objects of interest in the interpretation task.

But interpretation is not only dependent upon the imagery. The knowledge level of the interpreter influences the perceptions gained from the imagery and hence will affect both the quality and quantity of information gained during any interpretation process. As an interpreter learns about an area, perceptions change, causing modification of the interpretation and increasing the depth and range of information that is being extracted. Knowledge of the sciences that influence the objects or surfaces of interest and hence better understanding relationships depicted in the imagery will also have a considerable impact on the quality of interpretation. Such knowledge can be supplemented by information from other sources, including maps and discussion with colleagues.

The interpretation process can be seen as a cyclical process of learning and deducing. The learning involves collecting information so as to formulate a hypotheses, (inductive reasoning), and then testing that hypotheses (deductive reasoning). Once the hypotheses has been accepted then it is used to extract information about the surface or object depicted in the image, before the process is repeated for another aspect of the image.

There is an analogy between the activity of interpretation and hunting or fishing. In both hunting and fishing it is necessary to:

1. Have appropriate tools
2. Know how to properly use those tools
3. Understand how your quarry acts and responds
4. Understand the environment in which the quarry lives

The same four components are essential in good photo-interpretation. The purpose of this book is to help you with the first two of these requirements. The third requirements comes from the specific disciplines associated with the types of information being sought, typically geology and soil science, agriculture, engineering, urban studies and others as are appropriate to the task. The last requirement is addressed through an understanding of the environment in which the information resides, where this environment may be physical, sociological or economical, involving an understanding of geography, history, sociology and economics. For these reasons I often view interpretation as the hunting or mining of information from image data.

3.6.2 Levels of Interpretation

There are three broad levels of analysis in the interpretation of imagery. These levels differ in the degree of complexity in the interpretation process, and the depth of information extracted from the images. These levels are:

1. Image reading
2. Image analysis
3. Image interpretation

3.6.2.1 Image Reading

A relatively superficial visual examination of an image is undertaken to quickly identify the main features on the image and determine their distribution and likely relationship. It is the level of analysis employed when, for example, attempting to locate the position and orientation of an image in an area on a map. It is the usual level of analysis of photographs in a magazine or book.

Image reading is an important preliminary level of analysis. It allows the analyst to select appropriate imagery and to assess the relative merits of the different images available for more detailed analysis.

3.6.2.2 Image Analysis

Image analysis is image reading with longer and more detailed visual inspection of the imagery to understand the more general and obvious features in the imagery, as well as the measurement of those quantitative features deemed necessary. In geology it is likely to involve the marking of bedding planes, determination of dip and strike, delineation of drainage patterns and identification of discontinuities. In agriculture it is likely to involve identification and measurement of the area of the different land uses, marking of drainage patterns, mapping slope classes, eroded areas and access roads.

Image analysis provides a lot of very important information about the area, but it does not contain a lot of synthesis of that information. For some purposes image analysis is quite sufficient, but it may not be adequate if an understanding of causes of events is required.

3.6.2.3 Image Interpretation

Image interpretation is the inductive and deductive analysis of imagery, in conjunction with other data, to gain as full an understanding as possible of the features in the imagery, their functions and their inter-relationships. Image interpretation involves the processes of inductive and deductive reasoning in an iterative loop until such time as the interpreter is satisfied that the formulated hypotheses have been adequately tested, and then used to provide information on the surfaces or processes that are of interest to the analyst.

For example, an accumulation of facts about a building on a farm may allow the analyst to hypothesise that the building is a house rather than a shed or barn. This is the inductive reasoning process. However, a house would exhibit certain characteristics, if a majority of these exist then the hypothesis is likely to be accepted. Some of these typical characteristics are clothes lines, power and telephone lines, play area for children, barbecue areas, car parking areas, stove with chimney, washing and toilet facilities with waste outlet area, etc.

The context in which the building occurs can also be important. A property in an area could be expected to have a number of houses, where the number might depend on the size of the property and the landuses within the property. The number of houses that are occupied might also vary and may be important in the analysis. Thus, information on all the buildings may need to be collected before the analyst can finally test and either accept or reject the hypothesis.

Testing of the hypothesis that the building is a house is the deductive interpretation process. If this testing shows that the hypothesis is unsatisfactory, then it needs to be amended or replaced. The deductive process results in the collection of additional information that can then be used in a second inductive phase, to create a new hypothesis. The inductive phase thus uses all of the accumulated information to formulate a new hypothesis.

Image interpretation is thus a continuous learning process, consisting of formulating and testing hypotheses, before extracting information. It requires high levels of concentration by the analyst, and as such must be done under conditions that are conducive to concentration and learning. It is also an extremely exhilarating and satisfying process of discovery of information about an area.

3.6.3 Principles of Object Recognition

We identify objects on visual imagery primarily by use of the characteristics of those objects as recorded in the imagery. These image characteristics concerning the object include:

1. Size
2. Shape
3. Shadow associated with the object
4. Colour and tone
5. Pattern and texture

These features are modified by the geometric, radiometric, spectral and temporal characteristics of the sensor system being used and by the processing done prior to interpretation. It is very important that the image interpreter be fully aware of, and understand, the characteristics of the imagery being used in the interpretation. If the analyst is not aware of these characteristics then the wrong conclusions may be drawn as to why certain phenomena appear to exist in the imagery.

3.6.3.1 Size

The size of objects is a function of the scale of the imagery, the perspective and geometric characteristics of the imagery, as well as the size of the actual object itself. Size can refer to the objects three-dimensional size. An interpreter should always measure the size of objects using suitable techniques and not depend on his subjective judgement.

3.6.3.2 Shape

The shape of objects in visual imagery is a function of the perspective of that imagery. The shape of objects as seen from the ground is quite different to the shape as seen from the air, particularly when viewing vertically downwards. Oblique perspective images provide a transition between the horizontal and the vertical perspectives. Cultural features are designed with the horizontal perspective in mind. In consequence, the depiction of cultural features in the vertical perspective will often be quite different to their depiction in the horizontal perspective. There may, therefore, be a considerable dichotomy between these two perspectives, and the interpreter will often need to look for quite unfamiliar clues or characteristics when interpreting cultural features. Most other features are not designed with a particular perspective in mind, and indeed may favour the vertical perspective as in the case of agricultural features. In consequence, there is usually a closer correspondence between the depiction of these other types of surfaces in both perspectives.

3.6.3.3 Shadow

Shadows can indicate the nature of an object by displaying its profile, such as of trees and buildings. Imagery taken either early or late in the day can be particularly valuable for forestry and geology because the shadow effects can emphasise features of interest in the interpretation, such as the branching profiles of trees.

Shadows mask details that are contained within the shadow. Shadows in imagery are a function of the solar elevation and height of the object; the interpreter has to consider their effects in selecting the best time of day to acquire imagery.

3.6.3.4 Colour or Tone

Tone is the greyness in black and white photographs. The radiance at the sensor equation in Chapter 2 shows that the absolute values of the tones of an image are influenced by many factors other than the surface itself. Most of these external factors tend to influence the whole of the image in a similar way; they usually have little influence on relative tonal changes from location to location within the one image. Relative tonal changes are primarily due to changes in surface conditions and consequently are most important in the visual analysis of photographs. One situation that is different to this is concerned with the shadow created by clouds. The separation of the cloud and its shadow is a function of the solar elevation, the height of the cloud above the terrain and the view angle to the sensor. The detail that can be detected in these shadow areas is much less than can be seen in sunlit areas.

In photography the relationship between tone and radiance is difficult to establish because:

1. The shutter speeds of cameras are set at fixed values, so that subtle variations in total radiance incident on the camera may not cause a change in exposure setting, but may still cause subtle changes in the resulting density on the film.
2. The light meter controlling exposure may not have the same response characteristics as the filtered film for the given conditions.
3. The temperature and duration of development and processing as well as the condition of the chemicals used in these processes all affect the final tones in the photographs.

With electronic sensors the incident radiation levels are quantised in an analog to digital converter to discrete values so that the sensor optical efficiency and detector error are the major factors affecting the relationship between incident radiance and response. Enhancement of the digital data can be done in a number of ways (Chapter 4), many of which create complex relationships between radiance and the final tone. These tones are further modified in producing hard copy images. In consequence, the hue, saturation and intensity relationship within hard copy images of satellite data can be quite complex and difficult to appreciate.

Changes in surface conditions, such as changes in water turbidity, the amount of green vegetation in the canopy, cultivation of the soil, surface soil moisture and the existence of fire scars, affect the response in the images. The analyst is concerned with information that is contained in these changes, but these changes can make it difficult to calibrate one image to another, and in consequence, can make it difficult to quantify the changes that have occurred.

Tone and colour are important clues because of their relationship to the energy received in specific wavebands. Where relative tone or colour in an image can be used, they can be very reliable indicators of variations in surface conditions as long as local variations in atmospheric conditions, for example, due to smoke from fires, are either non-existent, or are taken into account. When interpretation has to be done between images then they may not be so reliable, unless the images have been radiometrically calibrated.

Since tones and colours are a function of the spectral and radiometric characteristics of the surface being imaged, it is important to be fully aware of the spectral bands that are depicted in the imagery, and how the different surface types will be recorded in that imagery. The discussion on the reflectance characteristics of surfaces in Chapter 2 is important in providing the basic reflectance information necessary to make the link between reflectance and surface condition.

3.6.3.5 Pattern and Texture

Texture is defined as the degree of roughness or smoothness of an object or surface in the image. Pattern is the regular repetition of a tonal arrangement across an image. Drainage lines form a pattern across the image since the arrangement is replicated from stream to stream, with different patterns indicating different important characteristics of the climate history, landform, soil and geology of the area. The paddocks of an agricultural area also form a pattern. Changes in this pattern may indicate changes in land use activities and thus may indicate changes in physical, sociological or economic factors or processes that are at work. Differences in soil, climate, landform or costs of production are amongst the more usual of these factors that may be changing. Within a paddock the surface texture can be smooth (e.g. for good crops), rougher (for cultivated fields) or very rough for fields with variations in cover conditions. Sometimes, the variations become sufficiently obvious as features to form patterns in their own right. Texture and pattern are related through scale. As the scale is reduced, patterns will degrade into textures. Scattered trees form a distinct pattern on moderate to large-scale photography; as the scale is reduced the trees tend to merge and loose their pattern creating a rougher texture on the image.

Texture and pattern are very important identifiers on visual imagery. It is essential that imagery selected for a task be at a scale that ensures that the features sensed by the image are depicted as appropriate textures and patterns.

3.6.4 Interpretation Strategies

The principles of object recognition provide us with the techniques necessary to identify individual objects or features on imagery, and understand many of their characteristics and functions. These principles focus on the object of concern, and its immediate environs and not the broader perspective in the analysis. This broader perspective is often essential to really understand the linkages between objects and functions within an area or to resolve ambiguities in the interpretation. These broader perspectives can be considered by the use of what are called interpretation strategies.

3.6.4.1 Location and Association

The interdependence of objects or features frequently controls their relative locations. Analysis of the nature and function of the features in the vicinity of the object in question will often provide important clues as to the nature and function of the object. This appreciation of the interdependence of features starts at the regional scale and can proceed down to the microscale if required. It is the reason why interpretation should start with a broad overview of the environs of the area of interest. In this way the interpreter is assessing features within the context of its environment. The overview is then followed by a more detailed interpretation that can be interrupted by a revision of the overview if the interpreter believes that his subsequent analysis is raising issues that are in conflict with the currently held overview.

The location and association of buildings relative to transport corridors, such as roads or railway lines, can indicate the role or function of those buildings, just as the location and association of a set of buildings can be important in assessing the roles or functions of the different buildings. More subtle relationships can be even more important. The location of a stand of trees, and their association with particular slope values and soil classes, can be indicative of soil conditions; the soils may be either too thin or poor to justify clearing of the trees. Other information can strengthen or weaken such hypotheses, for example, the mix of species within the stand of trees and the condition of pastures around the area, or on other areas that have similar physical attributes, can be valuable information in this type of assessment.

3.6.4.2 Temporal Change

Imagery of an area can be quite different at different times of the day, and from season to season. These differences are due to changes in surface conditions, in the sun–surface–sensor geometry, in shadowing and in the effect of the atmosphere. Many activities or phenomena have diurnal and seasonal patterns that can be used to improve the interpretation of these phenomena. The interpreter can often select, and must be aware of, the time of day and the date of the imagery so as to use these characteristics to improve his interpretation. Vegetation is also dependent on its meteorological history so that an interpreter must be aware of this history before conducting detailed interpretation of the condition of vegetation.

3.6.4.3 Convergence of Evidence

The interpreter collects a set of circumstantial evidence about the features of interest in the imagery. When sufficient circumstantial evidence has been collected then it is used to construct a hypothesis. For example, the circumstantial evidence that the analyst may collect about a feature could include:

1. The tone of the feature is dark in the image
2. The feature is linear in shape and creates a branching pattern
3. The feature is in the bottom of valleys

Clearly, this evidence will suggest a watercourse as the nature of the feature rather than a hedgerow, railway corridor or road. Similar processes occur with each feature in the image even though the process may occur quite automatically for many features.

There will always be cases where the interpreter does not believe that the evidence is strong enough to construct a hypothesis. In this case the interpreter has to gather further evidence from other sources. The most common sources of additional information are discussion with colleagues, archival information in the office and the collection of evidence in the field, or field data. Discussion with colleagues is a valued source of additional information because they, by approaching the problem from a different perspective, may identify evidence in the imagery that was missed by the first analyst. If this source of information, and other information available in the office, do not resolve the difficulty then a field visit may be necessary.

Field visits are best conducted prior to the start of the interpretation as part of the overview process. Subsequent field visits may be required as part of the interpretation process, depending primarily on the complexity of the analysis and the experience of the analyst with this type of situation and task. The analyst will also need to visit the field for other purposes such as accuracy assessment. If ambiguities arise during the interpretation that the analyst considers are not of major importance, then their resolution can wait until the accuracy assessment stage. If the ambiguities are major then they may influence the whole interpretation process. In this case, it is usually best to conduct the fieldwork as soon as possible.

3.6.5 Interpretation Procedure

In approaching an interpretation task the analyst should proceed in accordance with a well-established method in the conduct of the interpretation. Whilst there is scope for adapting this method to the satisfaction of the individual interpreter, the following provides the logical steps that can be followed:

1. Understand fully the information that is required; its type, levels of discrimination, spatial resolution and accuracy.

KNOW THY TARGET

2. Review the characteristics of the information sought and how it might be depicted on the imagery. Consider other classes and how they might also be depicted.

HOW DOES THY TARGET OPERATE

3. Evaluate the suitability of the imagery, facilities and staff skills to extract the information from the data.

ARE THY TOOLS SUFFICIENT TO THE TASK

4. Conduct an overview interpretation of the area using maps, photographs and other data or discussions. A preliminary visit to the area may be necessary.

WHAT ARE THE CHARACTERISTICS OF THE TARGETS ENVIRONMENT

5. Conduct the interpretation.

HUNT THY TARGET

6. Conduct accuracy assessment on completion of the analysis.

EVALUATE THY PERFORMANCE

3.7 VISUAL INTERPRETATION OF IMAGES

Visual interpretation should proceed from the general to the specific, from the overview or strategic, to the detail or tactical. Have a general appreciation of the physical, cultural and social characteristics of the area within which the interpretation is to be conducted. Reinforce this understanding with a broad overview analysis of maps and images of the whole area, supplemented with field visits. Proceed from this general level to the more detailed. The analyst's task is made easier by first interpreting the clearly identifiable or relatively unambiguous objects or features first, such as the drainage and transport corridors. This interpretation is also an aid to subsequent analysis since drainage features tell a lot about the geomorphology in the area, and the transport corridors can tell a lot about land use intensity. The analyst progressively addresses more complex features with the benefit of knowing more about the area from the prior interpretation.

During interpretation the analyst needs to record the results of the analysis either on the photograph or on a clear film overlay to that image. The way to approach this issue depends on whether the interpretation is being done on the computer or on hard copy images. Both have advantages and disadvantages, and so both may be used at different stages in the analysis. Use of the computer provides great flexibility in the enhancement and integration of images for the interpretation. It also allows the interpreter to create an overlay of the interpretation that is superimposed on the image, but which can be readily removed if the interpreter wants to look at detail below prior interpretations. The use of images for interpretation in a computer exploits the capacity of image processing and GIS software to readily enhance and overlay different sets of images and maps to assist the analyst. The capacity to conduct this type of work will be introduced in subsequent chapters in this text. At this stage you just need to be aware that the computer provides a very powerful tool for the interpretation of image data.

Hard copy images and maps provide the capacity to take the material into the field to compare interpretations with conditions that are met in the field. They can also provide a broader coverage, facilitating contextual interpretation of the area of interest in its surroundings since the area that can be displayed on a computer screen can be limited either by the limitations of the screen, or of scale and resolution. If a hard copy image is to be used for the interpretation then it is better to avoid marking the interpretation results onto a map, as the act of changing focus from the image to the map tends to break concentration. Marking directly onto the photograph has the disadvantages that the annotation can mask details in the images and it can be difficult to erase without damaging the photographs. Marking onto a clear film is easier to erase, using water-soluble pens, and it can be lifted off the photograph as required. However, the clear film can cause reflection that may interfere with visual analysis, scratching can occur on the surface of the film, and it is more difficult to take into the field.

In practice a combination of the two may prove to be the most suitable. Conduct the preliminary interpretation using a hard copy image or images as this provides a better means of getting an overview of the area, and provides a reference whilst conducting the subsequent detailed interpretation. Next, conduct the detailed interpretation using computer facilities, since this gives the analyst the maximum capacity to integrate different sources of data in the interpretation, and to enhance the images to get the maximum benefit during the interpretation. Create an overlay of the interpretation in the computer. Finally, take the most suitable image or images into the field as hard copy images, complete with overlay printed into them. Support this data with the digital image data in a portable computer, so that the detailed sets of image data can be referred to in resolving ambiguities that arise.

Scanner data may have lower spatial resolution than aerial photographs, but usually have better radiometric and spectral resolution. Spatial resolution affects the shape of features, the texture and patterns that will be depicted in the imagery. Shape, texture and pattern are important interpretation clues in the visual analysis of photographs. Interpretation tasks that depend very heavily on these clues can often use black and white photography as successfully as colour imagery. The improved spatial resolution that is occurring in scanner image data will lead to these clues becoming more important in the interpretation of high-resolution satellite and scanner image data.

Radiometric and spectral resolutions affect the colours and tones in image data. Scanner data usually has higher radiometric and spectral resolution than film data. Scanner data, for example, can contain hyperspectral data with many bands of recorded information. As you will see in the next chapter, these bands can be combined in different ways to create new images that do not correspond exactly to any one band, but contain information that is distilled from many bands. Such images can contain a lot more information than is contained in single bands of data. Scanner data can also be calibrated to reflectance more easily than can film data. As a consequence, the colours in scanner data can often be more precisely linked to conditions on the ground than can the colours in film images. These characteristics of scanner image data mean that colour is an important clue in the interpretation of these images.

The enhancement of images is discussed in detail in Chapter 4. You will then appreciate the capacity that the computer provides to enhance images, so as to display the information that you seek in the image data. In general, every image that you view in a computer is enhanced. Some of these enhancements are linearly related to the radiance incident on the scanner, and some are not. The enhancements affect the brightness and colours that are displayed on the computer screen. This means that the analyst can be mislead as to the relative impact of features or characteristics on the image data since some features may be downplayed by the enhancement and others may be emphasised. It is essential that the analyst be aware that the image being viewed is enhanced, that this enhancement will change the emphasis placed on different features in the image, and allow for this in the analysis.

The loss of fine resolution detail that occurs in some scanner images can be very disconcerting to an interpreter who is used to seeing fine detail both through normal viewing and through analysis of aerial photographs. A change in approach is required in the analysis of these images. The analyst can no longer depend on the fine detail to assist him in drawing conclusions as to the nature of the features imaged in the scene, but rather uses shapes, texture and pattern to interpret broader, more macro level features in the photograph. The image, however, contains a lot more spectral and radiometric information, and so the analyst must learn to depend much more on the colours and tones in the image to identify smaller and more detailed features in the image. Proper exploitation of the radiometric and spectral information in the image requires the interpreter to learn a lot more about the interaction of radiation and matter to properly exploit this information source than is required with the visual interpretation of aerial photographs.

It is recommended that an interpretation task be conducted in a set sequence:

1. Understand the goals of the interpretation task.
2. Describe the characteristics of the process to be interpreted and of the area to be interpreted. Use existing knowledge to gather this generalised information about the area and about the features of interest in the area. Thus, if one is interested in the hydrology of the area, one is interested in the geologic and climatic history of the area as well as current soils and land uses.
3. Validate this generalised understanding of the area and the problem. Use existing images, maps and field visits as necessary to validate this generalised model of the area and the problem.
4. Use this knowledge to plan the information sources that you will need to address the problem, and to develop a plan of action to extract the information that is sought from the analysis.
5. Acquire the image data and the facilities required for the interpretation task. Thus, to study the hydrology in an area, the analyst may be satisfied with the existing geologic and topographic maps, but not the soil maps, and so he may decide that the project needs to derive accurate geomorphic, soil, drainage and land use maps.
6. Conduct the interpretation. This should be done by first conducting the simpler, or more obvious, or the more fundamental analyses first. The advantage of the simpler or more obvious analyses is that the analyst becomes familiar with the area in conducting this analysis. The advantage of the more fundamental analyses is that they influence subsequent analysis tasks. Thus, in the case of the hydrologic study, the analyst may chose to delineate the drainage pattern first, as these will be obvious in good quality stereoscopic images, and they help to define the geomorphology and soils of an area. The second analyses is likely to be of the landforms, then the soils and subsequently the land uses.
7. Conduct field inspections to resolve conflicts or confusions that have arisen during the interpretation. Once this has been done, revise the analyses.
8. Conduct accuracy assessment, so as to provide an estimate of the accuracy of each interpretation result.

3.7.1 Visual Interpretation of Thermal Image Data

Thermal imagery can be acquired in the two thermal windows that exist between strong water absorption bands. These windows cover the range 3.4–4.2 m and 8.0–13.3 m. The shorter wavelength window is near the transition from solar radiation domination to earth radiation-dominated energy. During the day it is generally dominated by solar radiation. The longer wavelength window typically covers the peak radiation from the earth and the atmosphere, including aerosols.

Changes in tone in a thermal image are due to variations in the temperature, emissivity and roughness of the surface. The first two affect the level of radiation from each point on the object

TABLE 3.2
The Emissivities of Common Materials in the 8 to 13.3 m Waveband

Material	Emissivity
Granite	0.898
Sand, quartz	0.914
Basalt	0.934
Asphalt paving	0.959
Concrete paving	0.966
Water, pure	0.993
Water, with thin oil film	0.972
Polished metal	0.06

as discussed in Chapter 2. The roughness generally lowers the radiation from the surface due to shadowing; in much the same way as occurs in imagery in visible wavebands. As the emissivity of the surface decreases, the amount of energy radiated decreases, with the sensing instrument interpreting this decrease as a drop in temperature. Within an image, variations within the one cover type, such as within a water surface, can usually be assumed to be due to variations in the temperature of the surface, as long as that surface is consistent in its emissivity and texture. The temperature of a surface depends on the incident radiation, the thermal capacity and emissivity of the surface. The level of incident radiation affects the heating of the surface. The thermal capacity of a surface is the ability of the surface to store heat. It takes into account the specific heat and density of the different materials in the surface. Specific heat is a measure of the amount of energy that needs to be absorbed to raise the temperature of the surface by 1°. Surfaces, such as water, that need to absorb a lot of energy to raise their temperature by 1° have a high specific heat. The temperature of a material is also dependent on its ability to transmit heat to its neighbours. Good conductors can transmit heat in this way, whilst poor conductors (or good insulators such as air) cannot.

In Chapter 2 the nature of radiation from a source, and emissivity, were discussed. It was shown that the curve of radiation of a blackbody source is a function of the temperature of the source, rising to a peak defined by Wein's Displacement Law and then dropping again. It was also shown that the emissivity of a body, reduces the energy emanating from the body in proportion to the emissivity of the surface. If the emissivity is similar across the thermal region, then the effects of differences in temperature are different to the effects of differences in emissivity. Multispectral thermal imagery should thus be able to discriminate between the effects of temperature and emissivity, leading to the possibility of estimating both from the data (Table 3.2).

Since thermal imagery is affected by incident radiation, the time of day of the imagery and the orientation of the surfaces is very important. Water has a high specific heat and so it tends to stay cooler during the day than other surfaces, but it does not change its temperature at night as much as the surroundings. Surfaces that contain a lot of moisture thus generally appear cool during the day and warm at night. Surfaces that face the sun will warm up more quickly than surfaces that face away from the sun (Figure 3.17).

Thermal inertia is defined as the time it takes to raise a surface by 1° when that surface is illuminated at a set level of radiation. The thermal inertia of a surface depends upon the ability of the surface to absorb and radiate energy as well as the specific heat of the material in the surface. If a thermal image is taken just after midday, and a second image is taken just prior to sunrise, then the difference between the two will normally be due to differences in these factors. If the ability of the surface to absorb and radiate energy is measured or estimated, then the thermal inertia of the surface

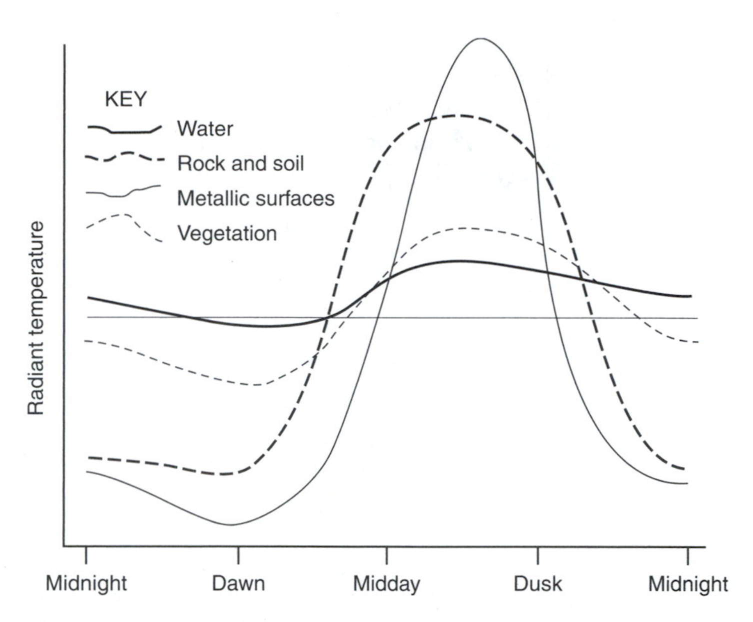

FIGURE 3.17 Typical diurnal variations in temperature for different types of surface.

can be calculated. The ability of the surface to absorb and radiate energy is normally estimated by acquiring visible imagery at the same time as the daytime thermal imagery. The reflectance of the surface in the visible region, ρ_λ, is used to estimate the surface absorptance, a_λ, in the visible region, where $a_\lambda = 1.0 - \rho_\lambda$ since the transmissivity $= 0.0$ and hence the emissivity in the thermal region. Alternatively, the emissivity can be estimated using standard values, if the nature of the surfaces is known. Surfaces that contain a significant amount of water will have high thermal inertia values.

3.7.2 Visual Interpretation of Radar Image Data

Radar image data is a very powerful imaging system that is marred by complex interactions between the surface and the sensor that gain expression in the image data in ways that are not always consistent, and which can give similar effects in the image data from quite different causes. The use of polarised, multi-wavelength radars will make this complexity more amenable to analysis, but it will not eliminate some of the inconsistencies that arise in the data. It may ultimately prove to be necessary to use multi-look angle image data to do this.

The tones in radar images are affected by variations in the surface and near surface as well as by variations in the radar geometric characteristics in relation to surface characteristics as were discussed in Chapter 2. Some of the more dominant or common features of radar image data include:

1. Corner reflector effects due to metallic and building surfaces creating bright spots in the image data.
2. Variations in response due to facet orientation relative to the sensor. This effect gives higher response for surfaces facing the sensor, and lower response for those facing away.
3. Radar layover of ridges and other geomorphic features.
4. Variations in response due to variations in surface roughness and volume scattering. These effects are modified by the wavelength of the radar, the look angle of the radar signal at that location in the image, particularly for smoother surfaces and the dielectric constant. The smoother the surface, the more reflection that will occur closer to the platform than at the other side of the image. In contrast, volume scattering from vegetative canopies may

give stronger signals from the opposite side of the image due to increased reflectance from the stems and trunks.

5. Variations in response with variations in moisture content, due to the variations in the dielectric constant.

These multiple factors can give similar responses in radar image data. Thus, high response levels in the data can be due to surface texture, the existence of corner reflectors or the orientation of large objects in the image. Since all of these can be oriented at quite different angles to the sensor, their effect on the image is very dependent on this orientation. Their effects are thus not consistent, and surface texture due to cultivation, for example, will give a strong response over a range of orientations, and negligible reflectance over another range of orientations. The same applies to corner reflectors and the existence of larger objects in the image.

The speckle in the data also affects these characteristics of radar images. Filtering can reduce speckle, but care needs to be taken that the filtering does not also reduce or eliminate fine details. The complex nature of radar image data, with changes of tone being a function of a number of interacting sensor and surface characteristics, makes the interpretation of the image data, for anything other than first-order effects, difficult and time consuming. The most significant advantage of radar data is its ability to penetrate most cloud and rain conditions, providing image coverage when this cannot be done using visible imagery. One strategy is thus to use the radar imagery with optical data. The radar images, being acquired when the optical data cannot be acquired, can be used to identify changes in surface conditions.

In addition to these characteristics, the geometry of radar image data, with its capacity for radar lie-over and radar shadows means that radar images will be distorted if there is significant height differences in the terrain in the image area. This may make it most important to rectify the radar image data prior to interpretation, but it complicates this process, particularly if there is not a good digital elevation model (DEM) on which to base the rectification.

The effects of heights on radar image data means that the data contains parallaxes, much like stereoscopic optical data, that can be used to view the terrain in three-dimension, and to create a DEM of the area. A significant development along these lines is the capacity to acquire radar images from similar orientations over a short time period and compare these for deformations of the earth's surface, such as earthquakes and volcanic effects. An image is acquired before and a second immediately after the event. The subtle differences in elevation that occur during the event, introduce displacements into the second radar image relative to the first. When the two images are simultaneously projected onto a screen using collimated (single wavelength parallel) light, the light emanating from the images either combines or subtracts, depending on the phase of the two at each location. The result is an image with phase fringes showing the addition and subtraction of the two and these phase fringes are related to the differences in displacement between the two images. The technique thus can be used to map the extent and severity of the displacements that occur due to the earthquake or volcanic eruption.

The images need to be taken as close together in time as possible as differences in surface conditions interfere with our ability to develop the phase fringes between the images, and make analysis and interpretation more difficult.

3.8 MAPS AND MAP READING

3.8.1 MAP PROJECTIONS

A map is a depiction of a part of the earth's surface onto a flat sheet. There are two major components to the creation of this relationship between the features on the earth's surface and the features on the map:

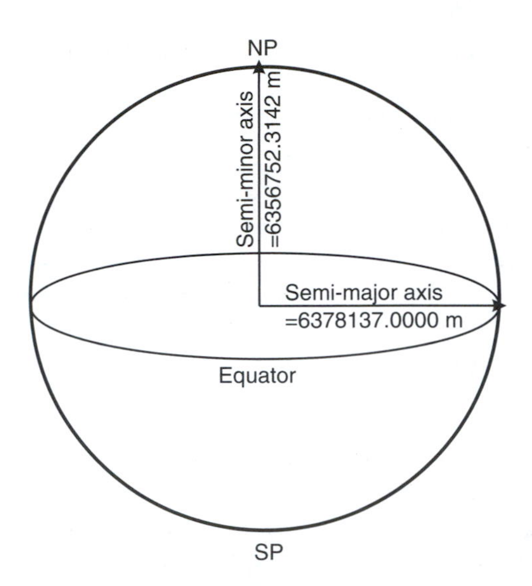

FIGURE 3.18 The parameters of the World Geodetic System (WGS), 1984 spheroid, showing its semi-major and semi-minor axes values. The flattening of the spheroid, $f = (a - b)/a = 1/298.257223563$.

1. Definition of the mathematical shape of the portion of the earth
2. Specification of how the curved surface of the earth is to be unfolded or projected onto a flat sheet

3.8.1.1 Definition of the Mathematical Shape of the Portion of the Earth

The earth's shape is defined by a surface of equal gravitational force called the geoid. The geoid is usually set at about sea level. The shape of this figure is not a sphere, but is more like a sphere that has been squeezed in at the poles and thus bulges out at the equator. A spheroid or an ellipsoid that is rotated about the north–south axis of the globe, is usually used to represent the geoid in the development of all map projections (Figure 3.18). The geoid is not a perfect mathematical figure, and so different countries use different spheroidal figures as providing the best fit of the numerical model to actual data for the mapping in their country. What is more, as surveying accuracy improves, many countries refine the spheroid they use to define the shape of the earth's surface within the area that they conduct mapping, leading to the adoption of different spheroid names and parameters at different times.

All spheroids are referred to one location on the earth's surface, called the origin for that spheroid. At the origin the spheroidal surface is fitted exactly to the geoid. Obviously, the mathematical shape of the spheroid does not exactly match the shape of the geoid, and so small errors will occur away from the origin. In practice, these errors are very small compared to the resolution that is normally used in a map, or within a GIS, although they may not be negligible for some surveying and engineering tasks.

Most spheroids are defined in terms of their semi-major and semi-minor axes and their origin relative to the earth's centre of gravity.

3.8.1.2 Specify How the Curved Surface of the Earth is to be Unfolded onto a Flat Sheet

There are many ways of transferring the curved information on the surface of the earth onto a flat sheet. All of them incur errors in this transformation process. The nature and magnitude of the errors depends on the projection used and the size of the area to be mapped. The main types of error that occur are:

- Errors of scale in both directions
- Errors of orientation or direction
- Errors in area

Cartographers may decide to minimise or eliminate one or more of these errors from a map, but that will be at the cost of making other errors larger. For some purposes it is more important to have some types of these errors smaller than others, and so this leads to the four categories of maps:

- *Equal area* — Areas are preserved so that the size of objects is consistent across the map.
- *Conformal* — Shape is preserved; such maps are often used for navigation and weather maps.
- *Equidistant* — Preserves some distances, but cannot preserve all distances.
- *Minimal errors* — Minimise all of the errors.

In addition to these categories of maps, there are three broad methods of projection of the spheroidal surface of the globe onto a flat sheet, or types of projections:

- *Conic projections* — The surface is projected onto a cone, such as the Albers Equal Area and the Lambert Conformal projections (Figure 3.19). Conic projections are often suitable for mapping land areas that are wider in the east–west direction than in the north–south direction.
- *Cylindrical projections* — The surface is projected onto a cylinder, such as the Mercator and the Transverse Mercator projections. The Transverse Mercator projection has been adopted as the standard projection for topographic mapping in most countries. It is well suited for mapping areas that are larger in the north–south direction than in the east–west direction (Figure 3.20).
- *Azimuthal projections* — project part of the globe onto a flat sheet. Can be good for part global maps.

Analyses of the azimuthal projections are an easy way to understand the nature of errors in a map, and so we will use this form of projection to analyse how the errors can change with map projection. Similar types of analyses can be conducted using the other types of projections.

There are three alternate common ways to create azimuthal projections as are shown in Figure 3.21:

1. Projected from the centre of the earth (gnomic projection)
2. Projected from the circumference of the earth (stereographic projection)
3. Projected from infinity (orthographic projection). This is the projection used if one assumes that the earth is flat in making a local map

The correct distance on the spheroid is the arc distance, TP, where this distance is:

$$\text{Arc}_{\text{TP}} = R \times 2\theta \tag{3.12}$$

To determine the accuracy of each projection, it is necessary to first determine how the projected distances change with θ in comparison with this arc distance. As shown on Figure 3.21 the three projected distances are:

$$\text{Gnomic projection} \qquad TS_o = R\tan(2\theta) \tag{3.13}$$

$$\text{Stereographic projection} \qquad TS_c = 2R\tan(\theta) \tag{3.14}$$

$$\text{Orthographic projection} \qquad TS_i = R\sin(2\theta) \tag{3.15}$$

These three projections are depicted in Figures 3.22 to 3.24 inclusive. The three distances represented by Equation (3.14) to Equation (3.16) are divided by the arc distance in Equation (3.13), and plotted

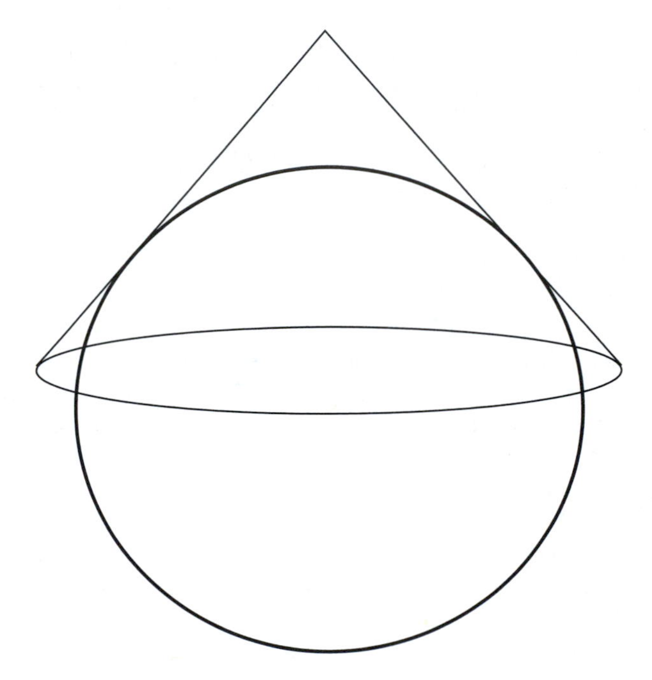

FIGURE 3.19 The conic projection.

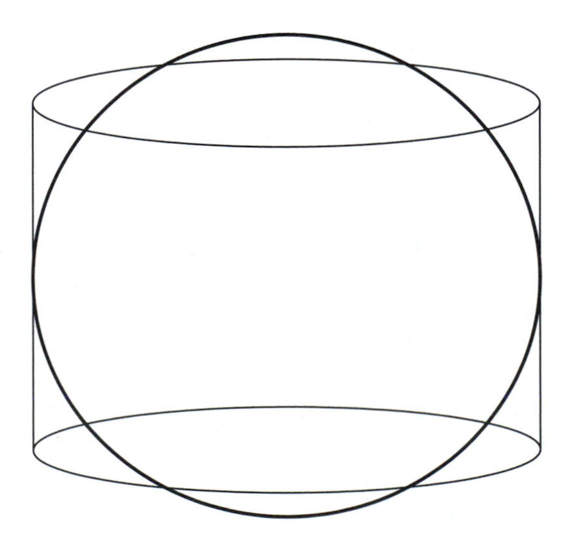

FIGURE 3.20 The cylindrical projection.

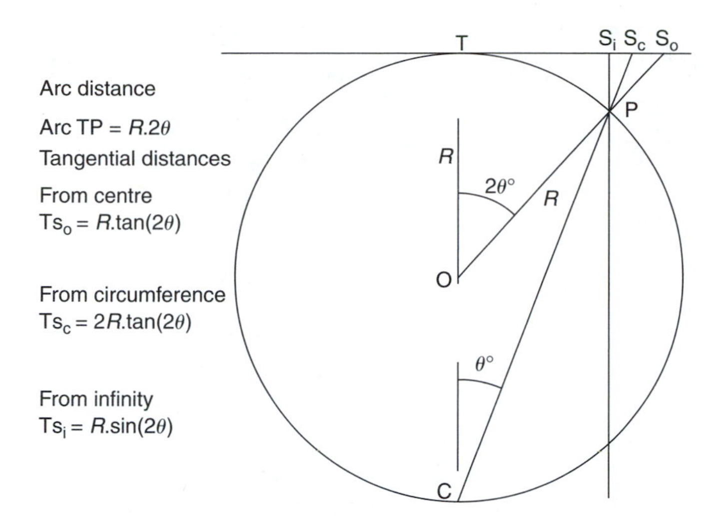

FIGURE 3.21 Three common alternative forms of azimuthal map projection.

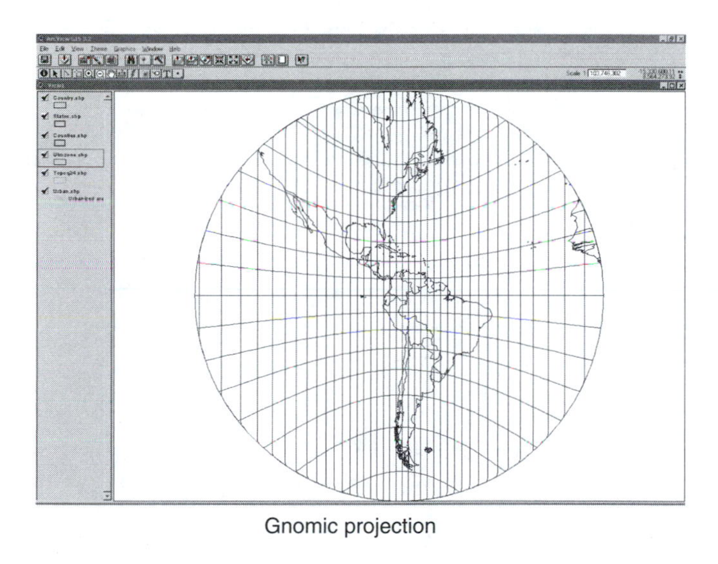

FIGURE 3.22 The gnomic projection with axis in the equator and nadir point in South America.

against θ in Figure 3.25. It can be seen that directions and scale will be correct at T. Scale changes as angle θ increases as shown in the formulae (3.14 to 3.16), and as plotted in Figure 3.25. Both of the projections from the centre and the circumference of the earth cause the projected distance TS to increase faster than the arc distance, TP, whilst the projection from infinity causes TS_i to increase at a slower rate than the arc distance as can be seen in the figure. The projection from the centre increases most quickly and then accelerates dramatically. It can be seen that projection from the circumference gives the most accurately scaled distances out to an angle of about 40°. The scale in the direction at right angles to TP is different to that for TP, as can be established by deriving similar formulae for these projections. This means that distortions occur in direction and area away from T.

One of the more common projections used in mapping is the Transverse Mercator projection (Figure 3.26). In the Mercator projection the flat surface commences as a cylinder with axes through the poles. This is one of the typical projections used in atlases. In the Transverse Mercator projection the axis of the cylinder passes through the equator. The cylinder is at a slightly smaller

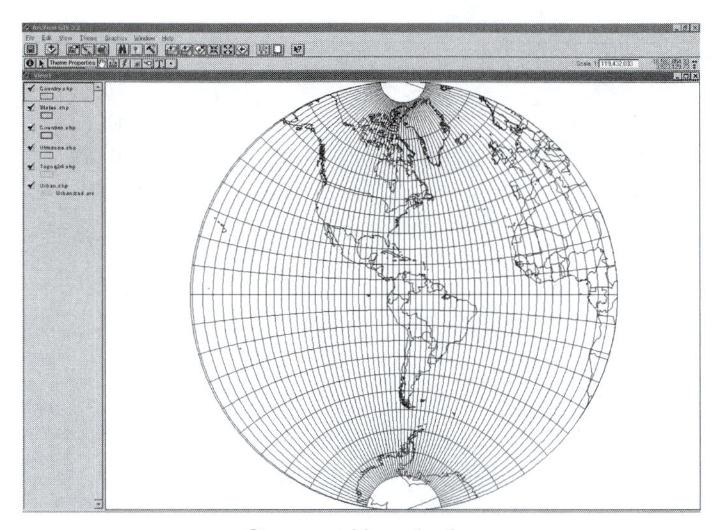

FIGURE 3.23 The stereographic projection using the same axis and nadir point as for the gnomic projection.

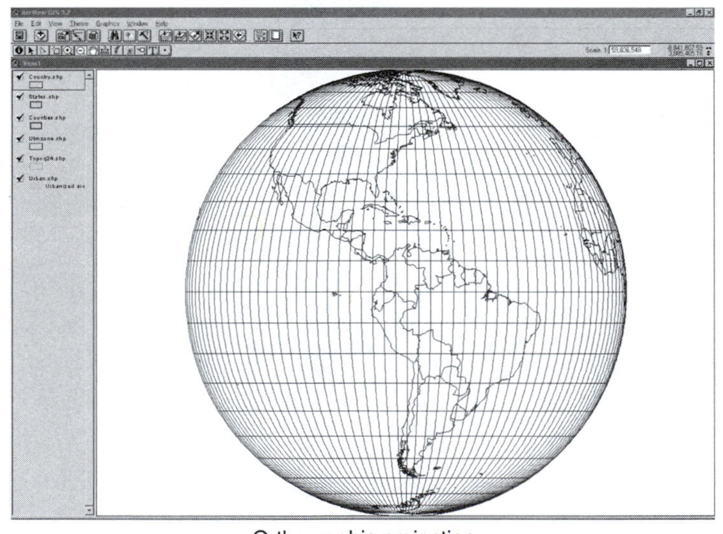

FIGURE 3.24 The orthographic projection using the same axis and nadir point as for the other two projections.

diameter than the earth, so that it cuts through the earth on two circles, creating two central meridians on which the scale is correct. Between the two central meridians the earth detail is squeezed to fit onto the cylinder, and outside them the detail is stretched. The map scale thus contains errors away from the Central Meridians, with the scale being reduced (smaller scale) between them and being magnified (larger scale) outside of them. The diameter of the cylinder is set so that the largest errors in scale are equal but opposite at the centre and the edges of the zone. The cylinder is used to project a zone of the earth from the opposite circumference of the earth, much like the stereographic projection. Once the cylinder is opened out it gives the flat map of that part of the globe. The cylinder is rotated about the line through the poles by the zone width less an overlap so as to produce the next

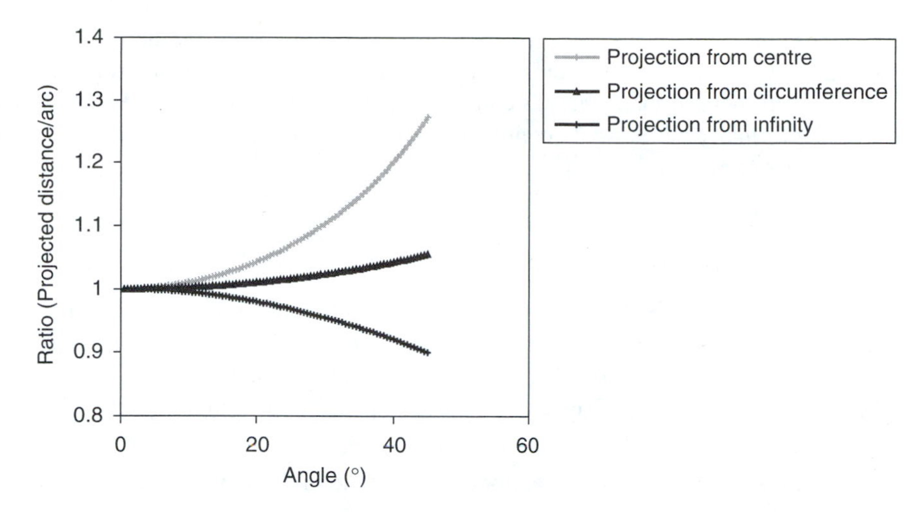

FIGURE 3.25 Ratio of the three standard alternative ways of deriving projected distances to the arc distance when projecting from a sphere onto a plane.

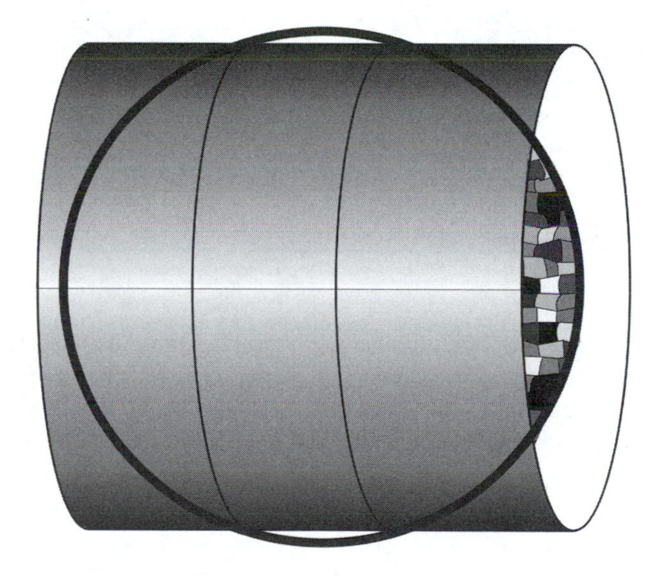

FIGURE 3.26 Construction of the transverse mercator projection.

zonal coverage of the globe. In this way full coverage of the globe is achieved without incurring very large errors or distortions.

Most standard topographic mapping is produced using the Transverse Mercator projection, with zones of 6° at an interval of 5°, giving some overlap between the zones. Lines of longitude and parallels of latitude bound each map, with the area covered depending on the scale of the map. On such maps the normal co-ordinate system used is a Cartesian co-ordinate system (CCS), with (x, y) co-ordinates set so that x increases to the east and y increases to the north and the co-ordinate axes are at right angles to each other. Such co-ordinate systems are easy to use in making calculations such as the computation of bearings, distances and areas. This is in contrast to geographic co-ordinates of latitude and longitude, where the lines of longitude are great circles that converge to the poles and the parallels of latitude are small circles. On such a system it is hard to make the same sorts of calculations.

When transforming digital map data from one projection to another, the normal process is to transform the projection Cartesian co-ordinates into geographic co-ordinates of latitude and longitude, and then convert these to the Cartesian co-ordinates in the new projection. Most GIS contain the transformation formulae necessary to convert from most projections into latitude and longitude, and then into another projection.

3.8.2 Mapping Systems and Map Types

Most countries establish a standard system for the mapping of topographic and other features. These systems contain a hierarchy of map scales, with the coverage at each map scale being standardised. Typical map scales and coverages are given in Table 3.3. As can be seen lines of longitude and parallels of latitude bound the maps. The contents of a map usually include (Figure 3.27):

1. The name of the map. This is conventionally the name of a dominant feature or town in the area covered by the map.
2. The map scale. Determination of scale for a map is identical to that discussed for images.
3. The series number. Most countries establish a series number at the smallest scale and then sub-divide this for the progressively larger scale maps.
4. The actual map contents, in the central part of the map sheet.
5. Scale bar, usually set immediately below and centred on the map.
6. North point, often on one side of the map. It will include information on conversion of grid to magnetic bearings.
7. Details on the method of construction and accuracy of the map.
8. The spheroid and projection of the map.
9. Legend describing the symbols used in the map.
10. Location diagram.

Topographic maps usually display information on the landform, the major cultural and natural features in the landscape. In addition to the topographic maps, many countries have established other suites of maps, including geologic, soils, cadastral or landownership and land use maps. With all these maps the legends will change, and indeed the legends may change from one scale to another, within the one suite of maps.

Table 3.3
Standard Mapping Scales and Coverages

Map Representative Fraction or Map Scale	Coverage (Longitude)	Coverage (Latitude)	Description
1:10,000	3′45″	3′45″	Large scale
1:25,000	7′30″	7′30″	Medium scale
1:50,000	15′	15′	Medium scale
1:100,000	30′	30′	Medium scale
1:250,000	1°30′	1°	Small scale
1:1,000,000	6°	4°	Small scale

3.8.3 Map Co-ordinates and Bearings

Co-ordinate systems evolved from the need to locate features relative to each other so as to produce maps and to navigate from one location to another. Historically, two types of co-ordinate systems evolved: Cartesian co-ordinates for local maps and Geographic co-ordinates for navigation over long distances. Geographic co-ordinates are defined as the position of a feature on the earth's surface in terms of its latitude and longitude. The latitude of a feature is its angle relative to the earth's axis at the centre of the earth and is specified as an angle north or south of the equator. It is an absolute measurement of location since the angle is defined relative to an absolute characteristic of the earth. The longitude, in contrast, is the angle between the feature and the line of zero longitude, which passes through Greenwich in England. It is relative to the longitude of Greenwich (Figure 3.28).

Historically, the latitude of a place was found by measuring the maximum altitude of the sun during the day, and then knowing the time of the year so as to determine the tilt on the Earth's axis, the latitude was determined. Longitude is more difficult. Since it is a relative measurement, it can only be measured using another form of angle measurement that is simultaneously known at Greenwich. If the time of transit of the sun (i.e., the time of the maximum elevation of the sun) is measured in Greenwich time units, then the time difference between the local time and the Greenwich time gives the longitude of the place of observation.

Latitude and longitude are very good for measuring positions on the globe, but they are difficult to use when bearings and distances between points have to be calculated, as is typically the case for local maps. Thus, the CCS evolved for use in the construction of local maps.

The CCS, contains two axes at right angles. Distances are measured along these axes from the origin of the co-ordinate system. Bearings are taken relative to one of these axes (Figure 3.29). If the normal convention is adopted, then the zero bearing is set on the vertical axes so that:

$$\text{The bearing of P} = \alpha = \tan^{-1}\left(\frac{X_{\mathrm{p}}}{Y_{\mathrm{p}}}\right)$$

However, we are usually interested in the bearing and distance between two points, thus: The bearing

$$\mathrm{QP} = \alpha = \tan^{-1}\left(\frac{(X_{\mathrm{p}} - X_{\mathrm{q}})}{(Y_{\mathrm{p}} - Y_{\mathrm{q}})}\right)$$

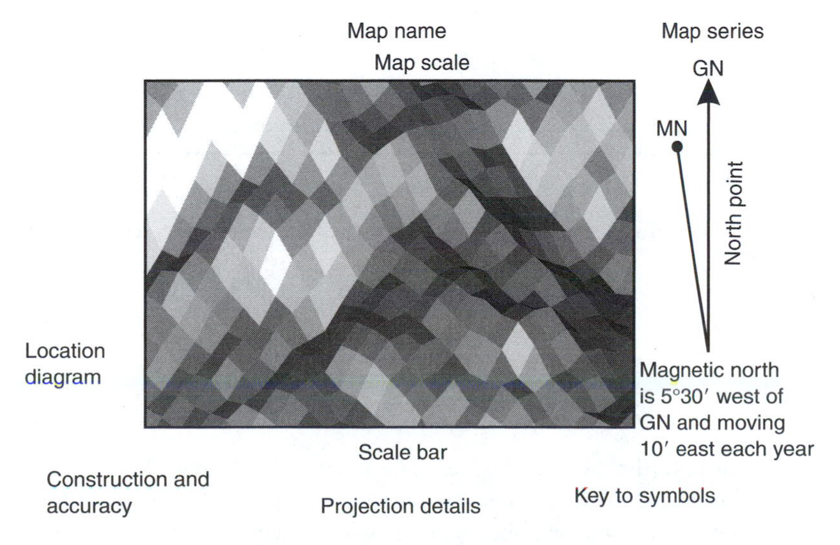

Figure 3.27 The contents of a map sheet.

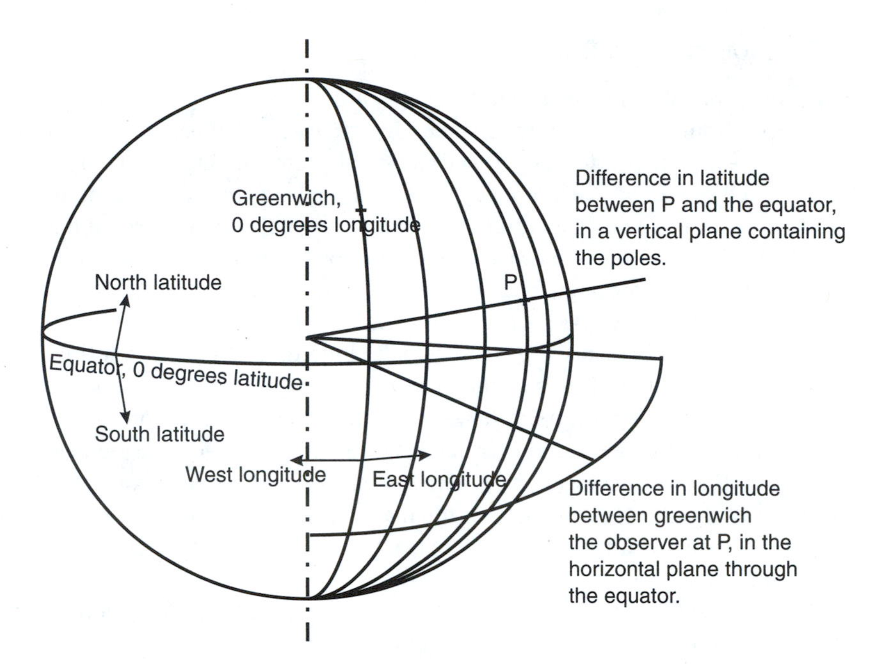

FIGURE 3.28 Geographic co-ordinates of latitude and longitude around the globe.

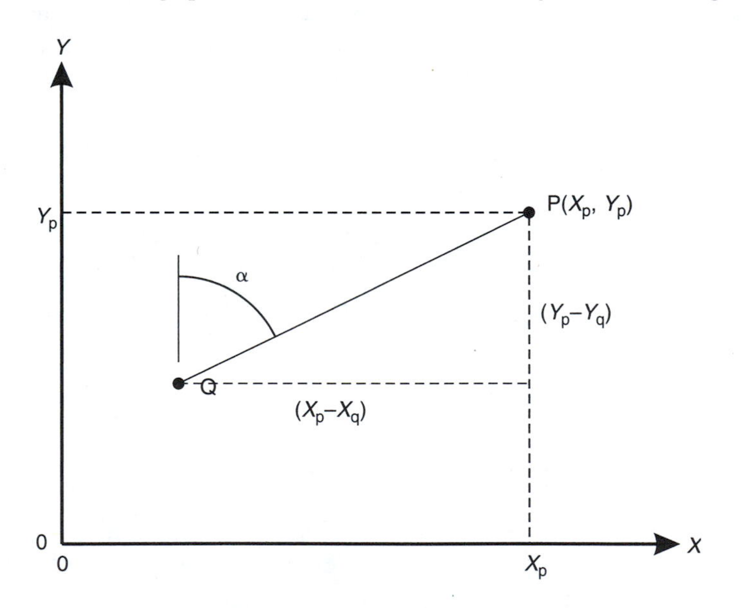

FIGURE 3.29 The CCS of two axes at right angles, distances measured along these axes, and bearings measured relative to one of the axes.

and the distance

$$QP = (X_p - X_q)\sin\alpha = (Y_p - Y_q)\cos\alpha$$

Historically, Cartesian co-ordinates were only used for maps of small areas. However, as our areas of interest expanded, the need to provide maps with the ease of use of Cartesian co-ordinates drove the development of national and global mapping systems. With these mapping systems, the conventional way to depict and read co-ordinates is as given in Figure 3.30.

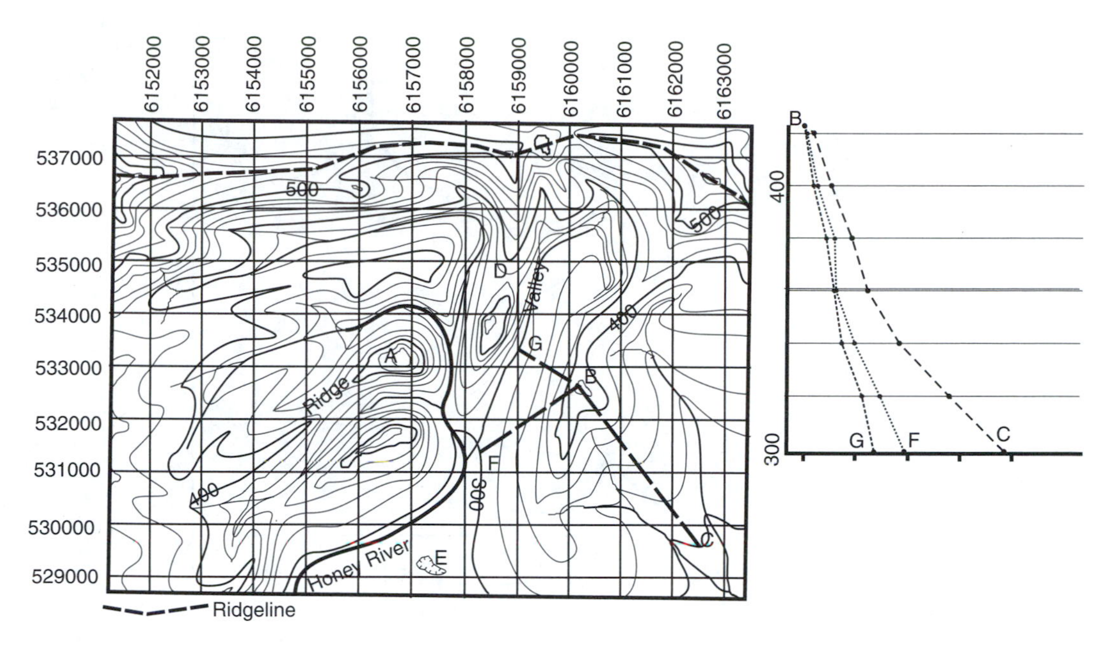

FIGURE 3.30 Part of a topographic map and three profiles from the map.

In Figure 3.30 it can be seen that the eastings values increase to the east and the northings values to the north. The origin for the bearings is the northern axis, increasing clockwise to 360° at the north point again. The grid lines displayed on the map are at 1000 m interval, hence the four figure grid reference for B is the co-ordinates of B to the nearest 1000 m, using just the two figures that change most rapidly at that interval, with the eastings given first and the northing second (6032). A six figure grid reference takes the co-ordinates to 1/10th of a grid cell. For B the six figure grid reference is (602326). In the same way an eight figure grid reference estimates the position to 1/100th part of a grid cell. The bearing from one point to another is calculated in the same manner as discussed earlier. Thus, if the six figure grid reference for C is (625296) then the bearing BC is given by:

$$\tan(\alpha) = \frac{(625 - 602)}{(296 - 325)} = -0.7931034 \text{ so that } \alpha = 128^0 25'$$

This is the grid bearing, which differs from the magnetic bearing by the grid to magnetic angle. The location of the magnetic north pole is not fixed, but moves with time. Topographic maps always display the grid to magnetic angle that applied at the date given on the map, usually the date of production of the map, and the direction and rate of change of the angle. Since the angle and the rate of change will both change over time, it is better to use as up to date map as possible when converting between grid and magnetic bearings. The grid bearing is also different to the geographic bearing, or the direction relative to the direction to the north pole, except at the centre of each zone.

3.8.4 Establishing One's Location on a Map

The map reader can locate himself accurately on the map by:-

1. Orienting the map using map detail and map reading
2. By the use of global positioning system (GPS),(discussed in Chapter 6)
3. By the use of a compass in compass resection

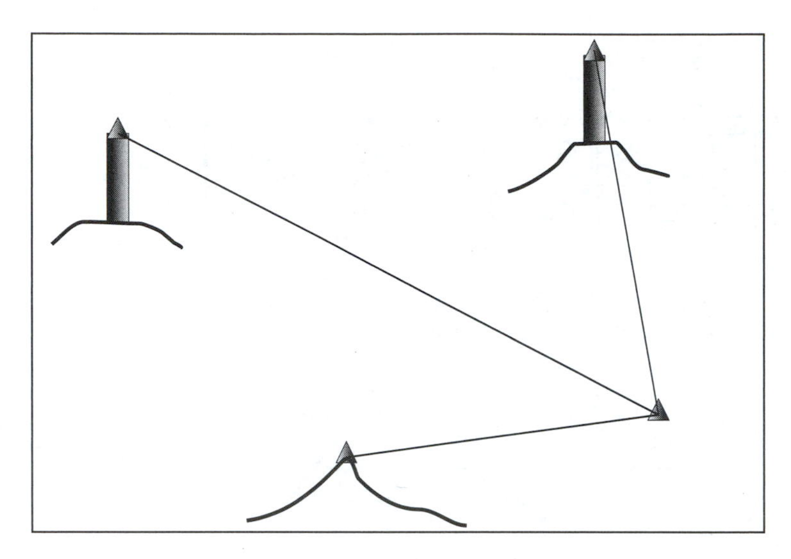

FIGURE 3.31 The compass resection.

Orienting the map using map detail and map reading. The map reader uses features that he can identify in the field and on the map to locate his position. He can then orient the map by sighting a feature on the map and rotating the map until the line to the feature on the map coincides with the direction to the feature in the field. Alternatively, he can mark the direction of magnetic north at his location and orient the map until the compass needle points along this line in the map.

By the use of a compass in compass resection. The map reader identifies three features from his location in the field, that he can identify on the ground and in the map. He then measures compass or magnetic bearings to these three locations, converts these bearings to grid bearings by using the conversion information on the map, and computes the back-bearing, or the bearing from the feature to the observer. The back-bearing is 180° different to the forward bearing. He then draws these lines on the map at the correct bearing from the correct features as marked on the map using a protractor, and the intersection of the three lines is the observers location. Once the observer knows his position, he can orient the map, either by identifying features, and orienting the map until those features are in the same direction as lines to them on the map, or by using the compass to orient the map (Figure 3.31).

3.8.5 Map Reading on a Topographic Map

Heights or elevations are normally depicted on a map by means of contours and spot heights. A contour is a line joining points of the same elevation. A contour will eventually close on itself, and it can never split into two contours although contours can overlap at vertical cliffs.

Contour lines (Figure 3.32) have the characteristics that:

1. Contour lines are concave in the uphill direction in valleys and are concave downhill along ridges.
2. When contours are close together the slope of the ground is steep, and when they are far apart the ground surface has little slope.
3. Concave surfaces have wider apart contours at their base, and close together contours near their top, and the reverse applies to convex surfaces.
4. Contours can only join at a cliff, and at some stage they must separate again, although they may not be within the area covered by the map.

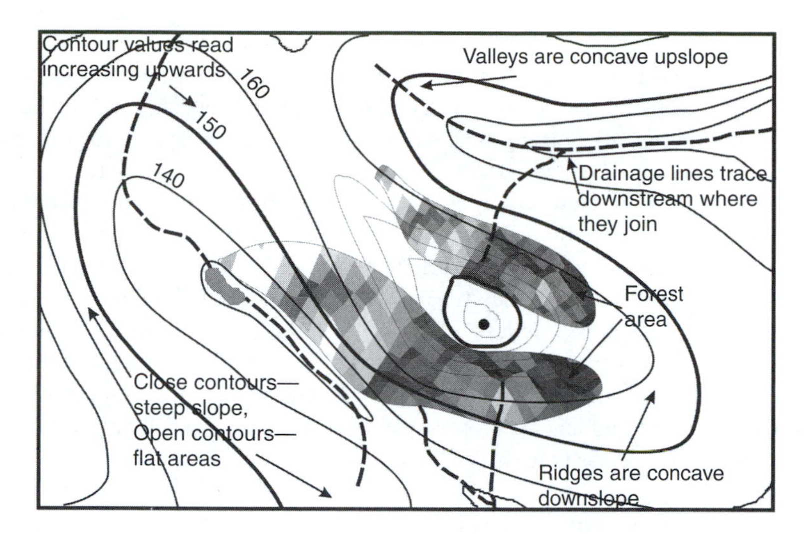

FIGURE 3.32 The characteristics of contours.

Contours allow the map reader to identify five major topographic features on a map, and three minor features:

1. *The hill.* A hill is defined as a high feature in the landscape that has down slopes in all directions away from the feature. A hill is located at A in Figure 3.30. A solitary hill will have contours that close on themselves and the contours at higher elevation will cover a smaller area than those at lower elevations.
2. *The ridge* is defined as a long high feature that varies in height along its length, sometimes rising, sometimes falling, and which always has down slopes at right angles to the direction of the ridge. A ridge is shown in Figure 3.30. A ridge will have contours running approximately parallel to each other along the sides of the ridge and the end of the ridge will have contours that are concave when facing the end of the ridge, or facing downhill.
3. *The saddle* is a location of lower elevation along a ridge, so that it is located between parts of the ridge that have higher elevations. A saddle is located at D in Figure 3.30.
4. *The valley* is the reverse of a ridge, so that the floor of the valley is a linear feature of low elevation and the ground slopes up at right angles to the direction of the valley. Valleys contain streams and so the Honey River (Figure 3.30) is flowing in a valley. Unlike ridges, valleys usually start at a higher elevation and continue at gradually lowering elevations to the end of the valley. The contours again, run approximately parallel to the sides of a valley and they are concave when viewing the uphill end of the valley, that is streams in the valley run down towards the convex contours.
5. *Depressions* are areas of lower elevation than their surroundings, so that they are the opposite to a hill. Depressions can also have contours that close on themselves within the map, but they can be discriminated from hills by the dashes along the contours in the depression, as at E in Figure 3.30.
6. *A draw* is a small valley that may be too small to be depicted in the contours of the map, or the start of a valley.
7. *A spur* is a small ridge that again may be too small to be depicted on the map. A spur ridge is depicted at B in Figure 3.30.
8. *A cliff* is an area where the contours merge.
9. *A false crest* is where the observer, walking up or down a hill or a ridge, sees the edge of the hill or ridge and may believe that it is the top of the ridge. However, if the surface is convex

in shape, then the false crest is a minor feature at some point between the observer and the actual crest of the ridge. False crests can be often detected on a map because the contours will be closer together, and then be further apart higher up the ridge or hill. Consider the profile BC in Figure 3.30. Starting at C, the contours are further apart and getting closer as one rises up the slope. The land surface is concave and the walker will have a good view of the land in front of him. However, once the walker reaches contour 360, then the steepest slope is immediately in front of them, and the contours start to increase in separation on the map so that the slope of the land starts to decrease. The walkers will see a false crest at about contour 380, but once they reach this elevation, they should see the goal of the top of the ridge. False crests are less common when the slopes on the sides of ridges and hills are concave, as is the case for many of the slopes in Figure 3.30. All three profiles show concave slopes, but the most concave of the three is that between B and C. Convex slopes will have contours that are closer together at the lower elevations and further apart at higher elevations, opposite to that of concave slopes.

Planimetric detail in a topographic map includes transport corridors, houses and buildings, major vegetation associations, lakes and streams. The centre of the linear features is meant to be the centreline of the actual feature. However, for buildings all are often not shown, but only some are shown so as to represent the buildings in the area. Thus, the centre of a building symbol cannot be taken as being the centre of a specific building.

3.8.6 Terrain Classification

Terrain classification is based on three main premises. The first premise is that landforms evolve as a result of the geology of the area being acted upon by the climate during the history of the area. Different types of geology will create different landforms under the same climatic conditions, and different climatic conditions will produce different landforms for the same geology. The second premise is that the geology of an area can and often will show sharp discontinuities, but the climate does not do so, except in the extreme cases of high mountain ranges interfering with the climatic regimes creating moist conditions on one side of the mountains and dry on the other side. The third premise is that stereoscopic aerial photographs are ideal for detecting and classifying landform shapes.

Given these premises, the task of terrain classification is to map the landform shapes at the micro-scale, and then to use these shapes with the base geology of the area to categorise the shapes into shape classes. This gives the basic classification of the landforms of the area. Field visits are then used to allocate characteristics to each of these landform classes and to derive from this the map of the features of interest to the analyst. Thus, if the analyst is interested in soil mapping, he could proceed by conducting the terrain classification and then use field visits to sample each landform category and assign soil characteristics to each landform class. Consider the area depicted in Figure 3.30. In this area the analyst can readily detect:

1. Pinnacle hills
2. Flat topped hills
3. Steep, concave sloping surfaces
4. Steep, planar sloping surfaces
5. Moderately steep, concave sloping surfaces
6. Moderately steep, planar sloping surfaces
7. Gentle planar sloping surfaces
8. Flat planar surfaces
9. Rounded ridge tops
10. Peaked ridge tops

11. Rounded valley floors
12. Sharply troughed valley floors

However, whilst these can be detected in the map, inspection of stereoscopic aerial photographs will detect subtle variations that may introduce other landform classes but will improve the mapping of the classes through better definition of the class boundaries. Once the landform features have been mapped, then each feature is given a classification code that includes digits for:

1. The base geology of the area
2. Landform shape
3. Slope class
4. Elevation class

Terrain classification exploits the capacity of stereoscopic imagery in particular to provide insights into the shape characteristics of the geomorphic units of the landscape, and provides a means of stratifying the area into units that are consistent in their base geology and climatic history. Such strata will normally also be consistent in the soil characteristics and native fauna and flora. Terrain classification does not provide this information, but it provides a basis for the most economical yet accurate basis for the collection of field data to complete the mapping of the physical characteristics of the area.

3.9 FURTHER READING

There is a wealth of literature on the topics dealt with in this chapter.

The human visual system continues to be an area of research and development. Ever since the eye's rods and cones were discovered, scientists have been trying to observe them in action. But the retinal photoreceptors, which change light into electrical pulses that the brain can process, are so tiny and their flashes of activity so brief that they have eluded researchers. Finally, a team led by David R. Williams of the University of Rochester managed to peek at and photograph human cones (Holloway, 1995 and Hubbel, 1988). This work builds on our understanding of how light energy is changed into neural signals at the individual photoreceptor cells of the eye which register the absorption of a single photon, or quantum of light (Schnapf, et al. 1987) and how dozens of kinds of cells have specialised roles in encoding the visual world (Masland, 1986).

The ultimate manual on photo-interpretation has to be the "Manual of Photo-Interpretation," even though it is now very out of date concerning the visual interpretation of non-film acquisition images, nor does it incorporate our latest understanding of the human visual system and how it works.

For those who wish to gain more insight into maps and map projections there are a number of books that are worth reading. Starting with that of Maling (1973) who describes co-ordinates systems and map projections, and Keates (1982) who covers map reading in some detail before then moving onto the more complex or detailed texts of Canters or Richardus (1972).

For those who wish to learn more about how to classify the landscape in terms of geomorphic units, for use in soil, vegetation and habitat analyses, I suggest that they look at Mitchell (1991).

REFERENCES

American Society of Photogrammetry, 1960. *Manual of Photo-Interpretation*, American Society of Photogrammetry, Washington, DC.

Canters, F. and Decleir, H., 1989. *The World in Perspective: a Directory of Map Projections*, John Wiley & Sons, London.

Holloway, M., 1995. "Seeing the cells that see," *Scientific American*, 272, 27.

Hubbel, D.H., 1988. "Eye, brain, and vision," *Scientific American Library*, Washington.

Keates, J.S., 1982. *Understanding Maps*, Longman, London.

Maling, D.H., 1973. *Coordinate Systems and Map Projections*, George Phillip, London.

Masland, R.H., 1986, "The functional architecture of the retina," *Scientific American*, 102.

Mitchell, C.W., 1991. *Terrain Evaluation*, Longmans, London.

Richardus, P., 1972. *Map Projections for Geodesists, Cartographers and Geographers*, North Holland, Amsterdam.

Schnapf, J.L. and Baylor, D.A., 1987. "How photoreceptor cells respond to light," *Scientific American*, 272, 40.

4 Image Processing

4.1 OVERVIEW

The purpose of both image processing and visual interpretation is to derive information from remotely sensed data. Image processing is defined as the computer processing of digital image data to either extract, or facilitate the extraction of, information from that data. Visual interpretation is the extraction of information by the visual analysis of image data. In extracting information it is often advantageous to utilise both methods of extracting information because of their relative strengths and limitations, as listed in Table 4.1. This information will usually then be entered into a geographic information system (GIS) for integration with other information so as to create management information.

Image processing is the better of the two for the conduct of numerically intensive and more routine tasks; visual interpretation is better where the intuitive and creative skills of the analyst are required. It is important, therefore, to see visual interpretation and image processing as complementary techniques in the extraction of information. The designer of resource management systems needs to continuously look for opportunities to optimise the extraction of information from remotely sensed data. Often these opportunities involve integration of both techniques into the information extraction process.

TABLE 4.1
The Advantages and Disadvantages of Visual Interpretation and Digital Image Processing

Visual Interpretation	Image Processing
Relative strengths	
Ability to utilize a range of both qualitative and quantitative information both structured and unstructured in form	Fast and cheap in processing large amounts of complimentary quantitative data
Exploits interpreter's creative aptitude in analysing images	Can conduct routine tasks in a fast and consistent way
Ability to use and build on intuitive skills in analysis	Can produce quantifiable information
Provides good broad-scale information	Can compare results across different classifications
Can easily integrate other sources of information, including subjective sources	Easy to integrate different quantitative information sets in the analysis
Facilitates development of an understanding of the situation; supports the formation of hypotheses	Supports the rigorous quantitative evaluation of hypotheses
Relative weaknesses	
Perishable skills and knowledge base	Cannot resolve unexpected situations; "uninspired"
Unpredictability due to variability in interpretation	Cannot learn intuitively, only through formalisms that can be represented in computer code
Labour intensive	Does not support the formation of hypotheses, except when used with subsequent visual interpretation
Expensive to derive quantitative information	Cannot easily handle either unstructured data sets or qualitative data

There are some facets of image processing that are not, as yet, particularly useful in the extraction of information from remotely sensed images about physical resources including many artificial intelligence techniques developed for topics like robotic vision and identification of writing. These facets of image processing are not covered in this text. However there are many facets of image processing that are important in the derivation of resource information and these are covered in this text. Those facets that will be included are introduced in this section, and are then developed fully in subsequent sections of this chapter. These facets include:

1. The pre-processing of image data
2. Enhancement of image data
3. Classification of image and other associated data
4. Estimation of physical parameter values from image data
5. Spatial and temporal analysis of imagery.

Pre-processing has the role of transforming the data into a more standard form of data that can then be more readily used or integrated with other types of data. Enhancement has the role of improving the contrast and range of tones or colours in the image so as to support the more productive visual analysis of that data. Classification is designed to address the question, "What is it?" that we see in the image. Estimation addresses the question, "What is its condition?" Temporal analysis addresses the question, "How does it respond to dynamic processes?" Spatial analysis is designed to support the derivation of information from image data and as such may play a role in both classification and estimation. The categories of classification, estimation and temporal analysis represent a progressive deepening of the level of environmental and resource information that is derived from image data.

4.1.1 Pre-Processing

Remotely sensed data is either uncalibrated or only partially calibrated in the spatial and radiometric dimensions when it is transmitted from the satellite to a ground receiving station. The calibration that may have been conducted includes the use of reference standards in the satellite, and the use of the stars for orientation of the satellite and sensor. This means that the image data will usually contain significant sources of error, and often the image data has to be corrected for some or all of these errors prior to it being used for image analysis. It will need to be calibrated if the errors in the data may introduce errors in the information that is derived from the image data. The most important calibrations that may need to be taken into account are to correct the data for radiometric and spatial errors. Small errors can also arise due to errors in the spectral characteristics of the sensor, or due to errors in the time of acquisition of the data.

Radiometric errors affect the digital values in the data for each picture element or pixel. They are caused by errors in the optics or the electronics of the sensor as well as by atmospheric scattering and absorption. The effect of these errors is the introduction of a signal into the data that comes from sources other than reflectance at the surface, and the reduction of the strength of the signal from the surface. They thus reduce the signal-to-noise ratio, but in quite different ways depending on the nature of the error. Radiometric errors that arise in the sensor are corrected by calibration. Errors introduced due to atmospheric effects are corrected by the application of atmospheric corrections.

Spatial errors arise due to spatial distortions in the sensor, due to the orientation of the sensor relative to the object and due to variations in the objects that are not adequately allowed for in correcting the data for these spatial errors. The existence of spatial errors affects the ability of the analyst to relate field data to the imagery, to relate the imagery to other data and to provide spatially accurate information or advice. Spatial errors are corrected by rectification of the data.

Spectral errors arise when the actual spectral transmission characteristics of the sensor filters are different to those expected, causing differences in the response values. The spectral transmission characteristics of the filters are known at instrument construction, and they rarely change or deteriorate. This is rarely a source of error in image processing.

Temporal errors arise due to differences between actual and expected acquisition times. These differences are usually small relative to the rate of change of events in the environment so that temporal errors are usually insignificant. One source of temporal errors concerns current methods of interpolating temporal sequences of images so as to remove the effects of clouds in that image sequence. The nature of this technique and its associated errors are discussed later in this chapter under the Temporal Analysis of image data.

Spatial, radiometric and temporal errors involve systems or processes outside of the sensor. Spatial errors are affected by the altitude and orientation of the platform as well as by the effects of topographic relief on the image data. Radiometric errors are affected by the state of atmospheric transmission during acquisition. Neither of these sources of error can be fully corrected without external sources of information. The most likely source of temporal errors will arise in the creation of the temporal image sets in the post-acquisition phase.

> Pre-processing is defined as the calibration of the data so as to create ratio data, that is, the processed data is in commonly used, standard, physical units of measurement, both spatial and radiometric.

4.1.2 Enhancement

Enhancement techniques are processes that enhance subsets of the data so as to improve the accuracy, reliability, comfort and speed of visual interpretation. Most digital image data have a much greater dynamic range and sensitivity than the human eye. Consequently, there are levels of information in the data that cannot be seen visually when all of the data are being depicted. For example, the human eye can detect about 20 grey tone levels in black and white images, but there are normally many more than these in scanner data when the full dynamic range of the data is considered. Most modern sensors such as TM (Thematic Mapper) and SPOT have more than 256 radiometric levels in the data and whilst the actual data may not cover the full 0 to 255 range, it will be contained in a significant portion of this range. To see all of the information embedded in the data requires enhancement of subset ranges of the image data before display either on a screen or in a photograph.

> Enhancement is defined as the emphasis or suppression of specific sub-ranges of the data so as to facilitate visual interpretation. The emphasis or suppression of a sub-range of the data can be at the expense of the remaining sub-ranges of the data.

The main focus of most enhancements is concerned with enhancement of the radiometric data as discussed above. However, the analyst may wish to enhance other aspects of the image data, including the spatial, spectral and temporal context of each pixel. Thus the analyst may wish to interpret narrow features in the image data, in which case the spatial context is incorporated in a way that enhances these narrow features. An inverse enhancement occurs when the analyst wishes to suppress noise in the image data, and again the spatial context can be used to implement an enhancement that reduces the noise level in the image data. This is a commonly used approach in reducing the effects of speckle on radar data.

4.1.3 Classification

If remotely sensed images are considered to be an m-dimensional data set, where there are m wavebands in the image data, then classification is the process of partitioning the m-dimensional response domain into a discrete number of class sub-domains. Each of these class sub-domains will ideally have boundary surfaces set so that the range of response values in the class sub-domains match those that come from one landcover or landuse class. Each of the other class sub-domains would ideally each represent the range of response values for one landcover or landuse class. In practice there is usually some overlap of the response values from different landcover or landuse classes, and so classification is greatly concerned with techniques, methods and strategies to use so as to set the boundaries or decision surfaces between classes in order to minimise errors in classification.

Classification can provide either a final information layer, say of landcovers, or it can provide an indicator or pointer to specific conditions to assist visual interpretation. Use of classification as a "pointer" allows the interpreter to quickly identify areas of similar radiometric conditions. This technique exploits the advantages of image processing, the fast processing of large sets of remotely sensed data, and the advantages of visual interpretation in using all clues in the data to make decisions.

In a strict sense, a landcover applies at a moment in time; at a later moment some of the landcovers may have changed. In this sense, landcover classification should be conducted on a single image, and the landcover is then correct (within the accuracy of the classification) at the moment of acquisition of the image used for the classification. In practice, more than one image may be used in landcover classification, but this means further complication and possibly higher error levels as the changes that occur will create difficulties in assignment of the pixel to a class.

Landuse, by contrast, typically remains constant for a season, although the landcovers that constitute that landuse may change during the season. The classical example of this is observed in agriculture where the season may last from 4 months to 3 years, with twelve months being typical. During this season the landcovers will change from bare soil, through the full cultivation phase, to the growing crop, the mature crop to the stubble and then back to bare soil again. For this reason landuse mapping should use imagery within the seasonal cycle so as to capture information on the landcovers in the cycle in order to improve discrimination between the different landuses. Strictly, therefore, it should not be conducted using a single image, but a set of images covering key phases within the seasonal cycle.

> Classification is defined as the partitioning of the digital data domain into a discrete number of class sub-domains, which have meaning in terms of the physical covers within the image area at the time or period of acquisition.

In collecting information on renewable resources, the most important first matter to be resolved is, "What is the land cover or land use at the location being considered?" Addressing this question is the province of classification. Once this question has been resolved, the analyst can address the second question of, "What is its condition?" Estimation is concerned with this second question.

4.1.4 Estimation

Estimation is the transformation of the data, sometimes with other data, into an estimate of one or more physical parameters using numerical modelling techniques. Most applications of estimation use the radiometric values in the spectral bands, or some combination of these, as the independent variable or variables in the estimation of the dependent variable or variables. These models often vary from one landcover to another, for example, models estimating leaf area index (LAI) from remotely sensed data vary with the species being sensed. These models are also dependent on the correct calibration and correction of the radiometric values in the data since these radiometric values are transformed into the estimate of the physical parameter. It follows that errors in the radiometric values in the data will introduce errors in the estimates of the derived parameters. Radiometric errors and atmospheric attenuation thus introduce errors into estimation models, some models being more affected by this source of error than others. An alternative way to conduct estimation is to use interpolation between field observations of the variable; this avoids some of the difficulties that occur with this type of numerical model, but at the cost of more field data.

Estimation focuses on condition at a moment in time, and as a consequence, estimation is conducted on individual images, and the derived estimates are valid at the date of acquisition of the image data.

A few other applications of estimation use spatial information as the basis of the estimation, for example, in the estimation of the texture in the image. Spatial analysis has been constrained by the relatively coarse resolution of image data relative to the size of the objects of interest. This situation is now changing so that much greater use of spatial information can be expected in the future.

Estimation is defined as the process of transforming the digital data into estimates of one or more physical condition parameters.

Enhancement, classification and estimation are the classical applications of digital image processing. A new dimension to the digital analysis of image data is to address the question, "How does the sensed surface respond to specific forcing processes?" or the domain of temporal image analysis.

4.1.5 Temporal Analysis

An image captures information on the nature and condition of the surface at a moment in time. We have seen that classification and estimation have the goals of extracting this information from the data. A series of images acquired over time contains information on the nature and condition of the surface at each acquisition date. Some of the changes that occur in the nature and condition of the surface are due to the way the surface responds to dynamic temporal forcing mechanisms, including the seasonal cycles of temperature and rainfall. A temporal sequence of images thus contains information on the way that the surface responds to these forcing functions. However, images contain other types of information that are not related to the dynamic forcing functions, such as the activities of man and the effects of some natural phenomena such as storms. The information on the response to the dynamic forcing functions is thus embedded in a range of other types of information. Temporal analysis focuses on extracting information on how the surface responds to dynamic forcing functions. Since the image data contains information relevant to this goal that is embedded in other information and noise, a major focus in the temporal analysis of image data is to filter out the relevant information from the other information in the images.

Temporal analysis is defined as the activity of extracting information on the way the surface responds to dynamic temporal forcing functions from the other information in the image data.

The characteristics of these dynamic forcing functions, and how these characteristics will be expressed in the image data are important considerations in the analysis of temporal image data sets. Because of this, the temporal analysis conducted in relation to some forcing function analysis may be quite different to that conducted in relation to other forcing functions. An important goal of temporal image data analysis is to facilitate improvement in the analyst's understanding of the physical environment as a dynamic system. Since the complexity of the data tends to make this difficult to achieve, an important component of temporal analysis is the simplification of the image data, facilitating the analysis of the derived temporal images by the analyst.

Remotely sensed data is a sampled record of part of the Earth's surface at a moment in time. The data is sampled spectrally, temporally, radiometrically and spatially. The spectral and temporal sampling is easy to see and understand. The radiometric sampling occurs in two ways. The first arises because the signal is accumulated over time on the light sensitive surface, between samplings, giving a temporal dimension to the radiometric sampling. The send arises in sampling the analog signal to create the digital signal in the analog to digital (A/D) converter. The spatial sampling occurs because of the finite size and location of the image pixels across the object.

All current digital image systems acquire images in the form of a rectangular grid of cells covering the area of interest, whether the data is acquired by a remote sensor, or by sampling of an aerial photograph. The location of the cells in an image, relative to the object, are quite unique; no other image will have exactly the same pixel locations. An image is also taken at a specific moment in time, so that each image is a unique sampling of the surface. In addition, each image contains unique errors, and so the pixel locations and response values will obey a statistical distribution that reflects the characteristics of the sensor, modified by the effects of the atmosphere on the radiation at the time of acquisition. There are many implications of the sampled nature of remotely sensed data, but the first and most important is that all methods of processing or analysing digital image data must be statistically valid. This means that all data processing methods must abide by appropriate statistical rules as apply to the analysis of statistical data, and the assumptions on which statistical procedures are based must be met within statistically acceptable tolerances. If a procedure assumes that the data being analysed is normally distributed then it is important to ensure that the data actually used has a distribution that is a statistically valid approximation of the normal distribution; otherwise erroneous results may be derived.

Considering the spatial characteristics of an image, the size of each cell, the separation between cells and finally the orientation and scale of the image can contain errors. These errors come from three major sources:

1. *Internal errors.* Internal errors are errors introduced within the sensor due to lens distortions and internal displacements of the sensor elements from their nominal positions. These errors can be quite significant in size, but since they can be subjected to calibration prior to launch of the satellite, they can usually be corrected with a high degree of accuracy.
2. *External errors.* External errors arise due to errors in the orientation elements of the sensor relative to the external world at the time of image acquisition, errors in the position of the sensor and distortions that arise due to the effects of the atmosphere on the path followed by the incident radiation. These sources of error are totally dependent on the sensor–platform orientation and position during acquisition, as well as the characteristics of the atmosphere and so they cannot be removed by calibration prior to the launch of the sensor.
3. *Quasi-random errors.* Quasi-random positional errors arise primarily during the process of transforming the data to fit ground control in the rectification process. These errors can arise due to misidentification of control points in the image, or due to the introduction of errors in the control point co-ordinates and due to errors in the transformation model. Errors in the position of a cell relative to the adjacent cells can also arise owing to displacements introduced due to height differences.

In consequence, the actual position of a cell during acquisition has systematic and random error components associated with it, where both can change from one image acquisition to the next. Thus, a second image would have cells in different positions even if all of the known and measured parameters were identical for both images.

Consider now the response values at each pixel in the image. The response within a grid cell or pixel is dependent upon the reflectance within the field of view and its immediate environs due to the point spread function at the time of sampling the pixel in that waveband. The point spread function states that light emanating in a direction from a source is scattered in accordance with the $\sin(x)/(x)$ function as depicted in Figure 4.17. This curve shows that energy from a source is emanated in such a way that it can affect the response of adjacent pixels. In addition to this characteristic of optical systems and electromagnetic radiation, there are the effects of the atmosphere on electromagnetic energy and on the derived image data.

4.2 STATISTICAL CONSIDERATIONS

Remotely sensed data is acquired in accordance with the specific geometric model of the sensor, in specific wavebands, at specific times and the response is quantified into a finite number of levels. In consequence, remotely sensed data is sampled in four ways:

1. *Spatially*
2. *Radiometrically*
3. *Spectrally*
4. *Temporally*

Slight differences in canopy geometry or surface conditions introduce differences in reflectance which cause differences in the data values recorded in remotely sensed data. If these differences in canopy reflectance are large enough then they cause a change in the pattern of response in the image data that can be explained in terms of changes in the physical conditions of the surface. However, subtle changes in canopy geometry or surface conditions introduce an apparent or quasi-random scattering in the data.

Atmospheric effects also cause scattering of the data. The net result is that the data for a surface, which may be considered as being homogeneous or uniform by an observer, will usually have data values that are scattered in some way.

A histogram records the number of occurrences of each data value, or the count of each data value, in a set of data as a graph with the horizontal axis being the range of values that the data can have, and the vertical axis being the count for each value. A typical histogram is shown in Figure 4.1. A cumulative histogram has similar axes, but the count recorded at a value is the sum of all of the counts up to and including the count for that value. A cumulative histogram must therefore have

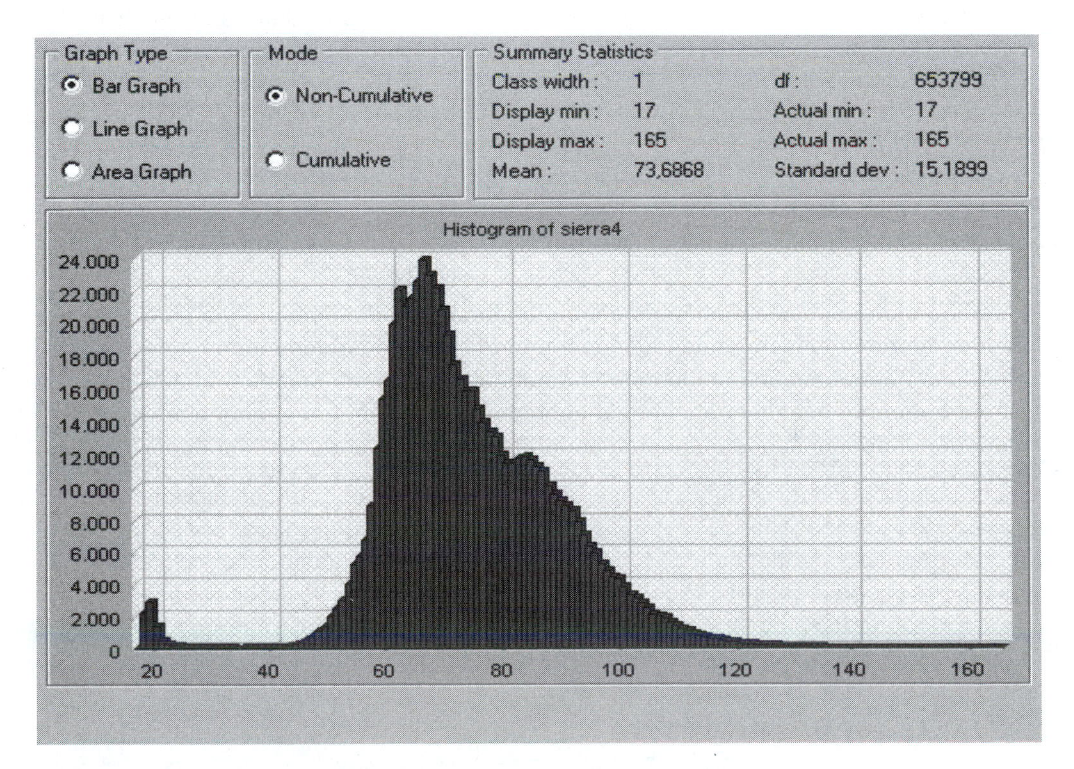

FIGURE 4.1 Histogram of an image as displayed in Idrisi. The display shows the range of values in the image, the mean value and the standard deviation.

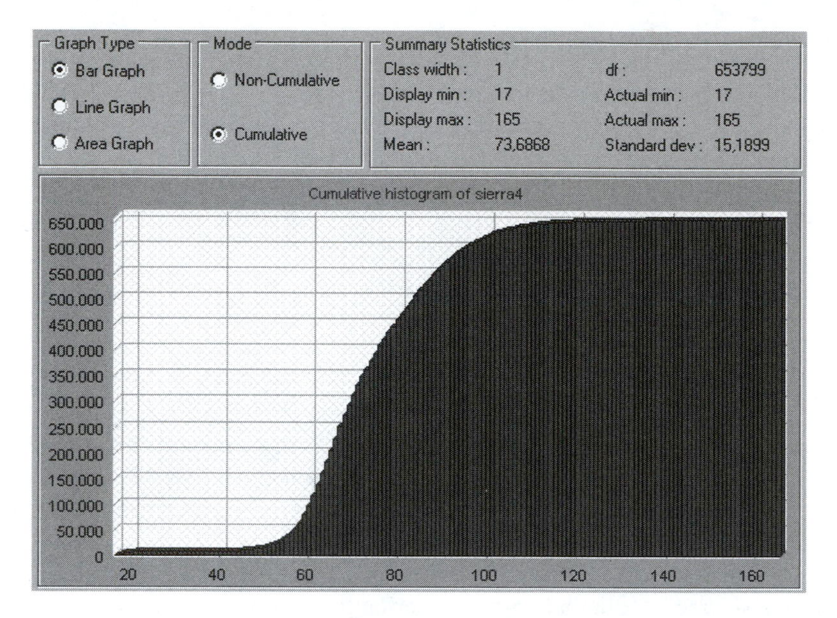

FIGURE 4.2 The cumulative histogram for the histogram shown in Figure 4.1, as displayed in Idrisi.

values that either increase or remain constant across its full extent. The cumulative histogram for the histogram in Figure 4.1 is shown in Figure 4.2. The histogram in Figure 4.1 contains two dominant peaks and several smaller peaks. The peaks show response values that occur more frequently in the image. The significance of the peaks depends on the waveband that is being considered, but they show frequently occurring values in that waveband. The cumulative histogram shows steep gradient values at the locations of the peaks in the histogram, and flat areas where the histogram has low values.

A two-dimensional histogram records the number of occurrences of each combination of data values in two separate parameters. Two-dimensional histograms are created by accumulating the number of occurrences of each pair of values in two dimensions of data, such as two wavebands. They can be displayed by printing the count values at each location in the two-dimensional space. They can also be displayed as scattergrams as shown in Figure 4.3. The scattergram in Figure 4.3 has the red band along the x-axis and the near-infrared (NIR) band along the y-axis. The sharp edge of the scattergram at about 45° to both bands is typical, representing the edge of the "Line of Soils." The values that are high in the NIR and low in the red represent vigorous green vegetation. The cluster with very low NIR values and low red values represents water.

All statistical methods assume that data has a distribution or spread function of some form, called a probability density function. The probability density function can be described using the values in the histogram. Parametric methods assume that the shape of the distribution fits a model that can be described using a few parameters. To use parametric methods it is necessary to derive estimates of these parameters and ensure that the actual data fits this theoretical distribution within acceptable tolerances. Non-parametric methods of statistical analysis do not assume knowledge of the shape of the distribution. They are thus more flexible in that they are not constrained to closely match one of the models used with parametric methods. However, they usually require a lot more data to define the nature of the distribution during processing or to establish an alternative model to an assumed distribution. As a result they can be a lot more complex and costly to use.

4.2.1 PROBABILITY DENSITY FUNCTIONS

A histogram graphically depicts the probability distribution of the data in the sample. In remotely sensed data the data observations for a class are usually distributed over a range of values as shown

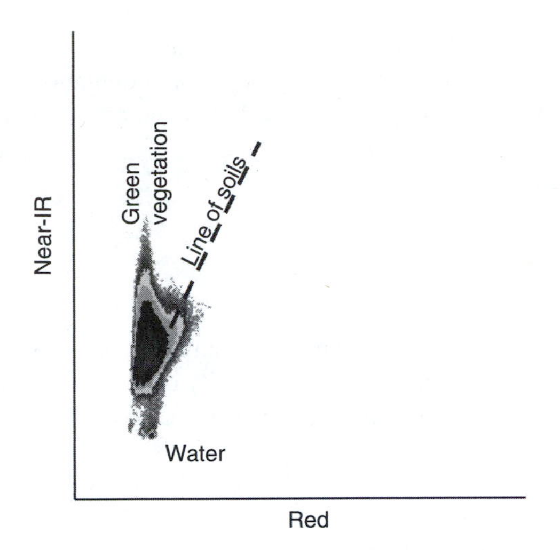

FIGURE 4.3 A Scattergram of Red (x-axis) versus Near-IR (y-axis) for the same image for which the Near-IR histogram is shown in Figure 4.1, as displayed in Idrisi image processing software.

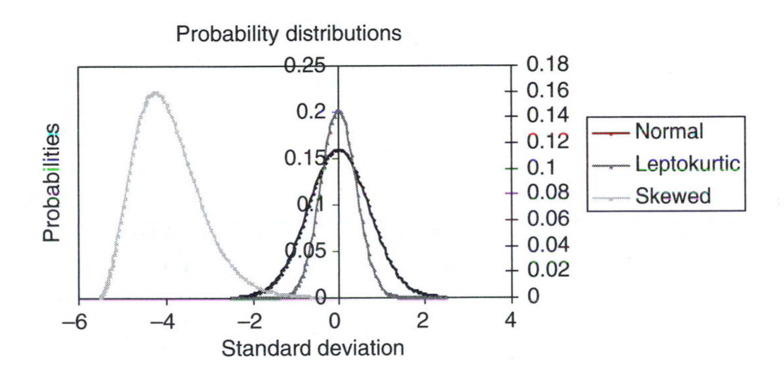

FIGURE 4.4 Probability distributions for three different sets of observational data, (a) skewed unimodal, (b) normal unimodal and (c) leptokurtic unimodal.

in Figure 4.4. The probability that a particular value will occur is the count of that value from the histogram divided by the total count in the histogram. The probability distribution is thus the histogram normalized for the sample population used to create the histogram.

Probability density functions (PDFs) that have been taken from observed data can have many shapes. If the data are of a homogeneous surface type then their PDF would usually be unimodal and have data values that are distributed equally about the mean value as shown in Figure 4.4. Data sets that contain the data collected for many different types of surface will often be multi-modal as in Figure 4.1.

All PDFs can be defined by the probability at each point in the distribution. But this method of specification is untidy and difficult to use. If there is some symmetry in the PDF, then a mathematical function might adequately describe the PDF by means of a few parameters. This approach is much more convenient and can lead to powerful solutions that can be used for a range of conditions, and not just those met in one area of one image. The use of a standard model in processing usually results in lower processing costs and increases the flexibility of the processing system. If the sample data is a good fit to the model and the sample is representative of the whole population, then the model can be assumed to be a good fit to the data for the whole population.

The use of models to create PDFs for a population requires training data to estimate the model parameters with sufficient reliability to represent the whole population and to test that the sample population is a good fit to the data. Some models of PDFs that are commonly used include:

1. Binomial distribution
2. Normal distribution

4.2.1.1 Binomial Distribution

Consider the situation where there is a probability p of an event occurring, and a probability $(1 - p)$ that the event will not occur. To determine the probability of x occurrences from n events, we need to analyse the binomial distribution, which is given as the equation

$$p(x) = p^x \times (1 - p)^{(n-x)} \times n!/[x! \times (n - x)!] \tag{4.1}$$

where x is the number of successes in the n trials with probability p of success in any one trial and $n!$ is factorial n, that is, $n! = n \times (n - 1) \times (n - 2) \times \cdots \times 2 \times 1$. A binomial distribution concerned with the probabilities associated with simultaneous occurrences of two or more events is called the multinomial distribution. A use of the multinomial distribution is in the analysis of the accuracy of visual interpretation or classification as will be discussed in Chapter 5. In the comparison of a map of landuse with field data on landuse, each class in the map has probabilities of being wrongly classified into the various classes. These probabilities should be a good fit to the multinomial distribution.

4.2.1.2 Normal Distribution

The normal distribution is unimodal and symmetrical about the central value as shown in Figure 4.5. The value of the central peak is the mean or average value. The spread is measured by the standard deviation, which is the distance from the mean value out to the point of inflection on the curve, where the point of inflection is the point at which the slope changes from increasing downwards, to decreasing downwards or increasing upwards. The importance of these parameters is indicated by the equation for the normal distribution:

$$p(x) = \frac{1}{\sigma\sqrt{2\pi}} e^{(-0.5(x-\mu)^2)/\sigma} \tag{4.2}$$

where μ is the mean and σ is the standard deviation. These parameters are estimated using a sample of n observations $\{x_1, x_2, x_3, \ldots, x_n\}$:

$$\text{Sample mean} = x' = \Sigma x_i/n \quad \text{for } (i = 1, \ldots, n) \tag{4.3}$$

$$\text{Variance} = s^2 = \Sigma(x_i - x')^2/(n - 1) \quad \text{for } (i = 1, \ldots, n) \tag{4.4}$$

$$= \{(x_1 - x')^2 + (x_2 - x')^2 + \cdots + (x_n - x')^2\}/(n - 1)$$

$$= \{\Sigma x_i^2 + nx'^2 - 2 \times (\Sigma x_i) \times x'\}/(n - 1)$$

$$= \{\Sigma x_i^2 - nx'^2\}/(n - 1) \tag{4.5}$$

Since the values for the mean and the variance that are derived from the sample data will not be the same as the actual values for the whole population, they are strictly given different symbols. In practice the two sets of symbols are often used interchangeably, but the reader should always be aware that the values estimated from a sample of data are strictly sample mean and variance values, not population mean and variance values. Either form of the variance equation [Equation (4.4 or 4.5)]

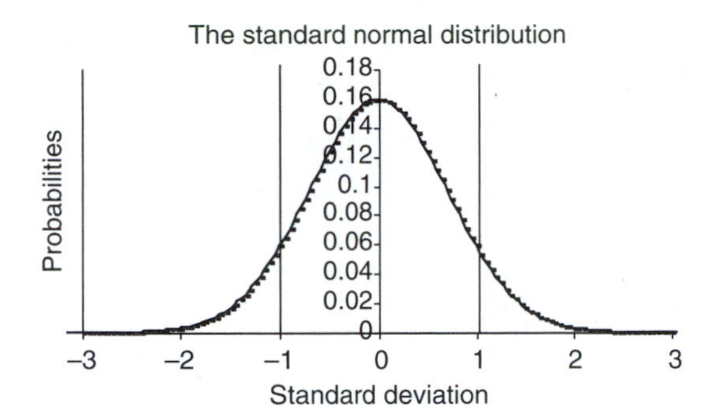

FIGURE 4.5 The Standard Normal Distribution, $N(0, 1)$.

can be used to calculate the variance. The form given in Equation (4.5) is usually more convenient since the accumulation of the two quantities can be used to compute both the mean and the variance from one pass through the data.

The normal curve is usually expressed in the form:

$$P_r(x) = N(x', s^2) \tag{4.6}$$

The total area under the normal curve and above the x-axis represents the probability, for that distribution. The area under the curve and above the x-axis between the values a and b represents the probability that x lies within the range a to b, written in the form, $P_r\{a < x < b\}$. This could be put another way, that if any random observation or data value is taken from the population, then $P_r\{a < x < b\}$ is the probability that the observation lies within the range a to b.

If the variable x is expressed in units of $y = (x - x')/s$, then the normal distribution associated with y is the standard normal distribution with a mean value of 0.0 and a variance of 1.0, that is $P_r(y) = N(0.0, 1.0)$, as shown in Figure 4.5. Values in any normal distribution can be converted into values in the standard normal distribution, and so probability density tables need only be kept for the standard normal distribution.

In the standard normal curve a value $y_a = \pm 1.0$ is one standard deviation either side of the mean. $P_r(-1.0 < y < +1.0) = 0.6826$ when y is $N(0.0, 1.0)$. This means that there is a 68.26% probability that an observation will fall within one standard deviation of the mean, or 68.26% of the population falls within one standard deviation (σ) of the mean. This distance out from the mean also indicates the point of inflection of the curve, or that the rate of change of the slope of the curve changes sign at this point. In the same way it can be shown that 95.44% percent of the observations fall within 2σ of the mean, and 99.72% within 3σ.

4.2.2 CORRELATION

Correlation is an indication of the predictability of the value of one parameter relative to the values of one or more other parameters. If field data are collected on two or more parameters (reflectance in wavebands, LAI or biomass), then correlated parameters will usually see increases (or decreases) in one parameter accompanied by proportional increases (or decreases) in the other parameter(s). With uncorrelated data an increase or decrease in one parameter has equal probability of being matched by increases or decreases in the other parameters, with the magnitudes of the changes in the parameters being quite random.

When there is a physical linkage between two parameters, then observations of these parameters would be correlated. The reverse, however, cannot be assumed. Thus correlated data may not be due to physical linkages between the parameters.

The correlation between observations of different parameters is determined by first calculating the variance and covariance. For two parameters (x, y), pairs of observations $(x_1, y_1), (x_2, y_2), \ldots, (x_n, y_n)$ are used to compute the mean and covariance arrays:

$$\text{Mean array} = U = \begin{vmatrix} x' \\ y' \end{vmatrix}$$
$$\text{Covariance array} = \Sigma = \begin{vmatrix} S_{xx} & S_{xy} \\ S_{xy} & S_{yy} \end{vmatrix} \tag{4.7}$$

where

$$S_{xx} = \Sigma(x_i - x') \times (x_i - x')/(n-1) = \text{Variance of } x,$$
$$S_{xy} = \Sigma(x_i - x') \times (y_i - y')/(n-1) = \text{Covariance between } x \text{ and } y, \text{ and}$$
$$S_{yy} = \Sigma(y_i - y') \times (y_i - y')/(n-1) = \text{Variance of } y.$$

If the values in each column and each row in turn are divided by the standard deviation in the column and row, then the covariance matrix becomes a correlation matrix:

$$\text{Correlation matrix} = C = \begin{vmatrix} [S_{xx}/(S_x \times S_x)] & [S_{xy}/(S_x \times S_y)] \\ [S_{xy}/(S_x \times S_y)] & [S_{yy}/(S_y \times S_y)] \end{vmatrix}$$
$$= \begin{vmatrix} 1.0 & S_{xy}/(S_x \times S_y) \\ & 1.0 \end{vmatrix} \tag{4.8}$$

The variance of a set of observations can never be negative since the variance is the sum of a series of second-order power terms. Similarly, the diagonal elements in the covariance array can never be negative. However, the elements off the diagonal can be negative. The values that are mirror imaged about the diagonal are identical in value. As a consequence, there is no need to save these extra terms, and the array is often shown in the diagonal form as shown in Equation (4.8).

The correlation value will vary between 1.0 and -1.0 where 0.0 indicates zero correlation. Positive values mean that the semi-major axis is oriented so that positive differences $(x_i - x')$ in one parameter are accompanied by positive differences $(y_i - y')$ in the second parameter. Negative values occur when positive differences in one parameter have negative differences in the other parameter. The data distributions shown in Figure 4.6 are depicted as plots at one standard deviation from the mean as indicated by the covariance arrays. The plot is circular in Figure 4.6(a) because the standard deviations are equal and there is negligible correlation. If the standard deviations are different and there is negligible correlation then the distribution will be elliptical, with the semi-major and semi-minor axes being parallel to the co-ordinate axes (Figure 4.6[b]). In Figures 4.6(c) and (d), respectively, the variances are different and the covariances are positive and negative, respectively.

Covariance and correlation matrices can be extended to n-dimensional space. They are a valuable means of analysing the correlation between sets of parameters as occurs with multispectral remotely sensed data.

The relationship between data values in old dimensions (x, y) and new dimensions (u, v) of rotation θ between them such that the new axes coincide with the semi-major and semi-minor axes of the ellipse are

$$\begin{aligned} u_i &= x_i \cos(\theta) + y_i \sin(\theta) - [\bar{x}\cos(\theta) + \bar{y}\sin(\theta)] \\ v_i &= y_i \cos(\theta) - x_i \sin(\theta) - [\bar{y}\cos(\theta) - \bar{x}\sin(\theta)] \end{aligned} \tag{4.9}$$

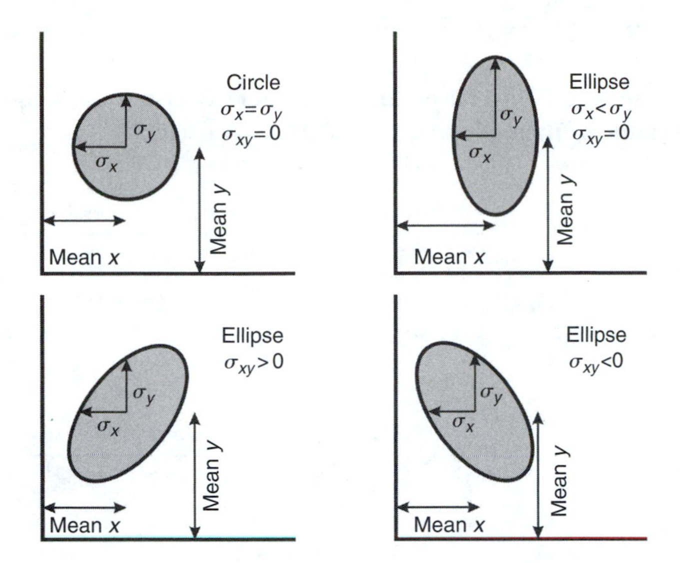

FIGURE 4.6 Diagrammatic illustration of the shapes of the normal distribution at one standard deviation for populations in two variables, with different correlations between the parameters. Figure 4.3 illustrates the distribution of actual data in two wavebands for a specific sensor at a specific site on a specific day. Such data, representing many cover types, rarely exhibits the circular or elliptical shapes discussed here.

By the law of propagation of error:

$$\begin{aligned} \mathrm{Var}\{u_i\} &= \cos^2(\theta)\mathrm{Var}\{x_i\} + \sin^2(\theta)\mathrm{Var}\{y_i\} + 2\sin(\theta)\cos(\theta)\mathrm{Cov}\{x_i, y_i\} \\ \mathrm{Var}\{v_i\} &= \sin^2(\theta)\mathrm{Var}\{x_i\} + \cos^2(\theta)\mathrm{Var}\{y_i\} - 2\sin(\theta)\cos(\theta)\mathrm{Cov}\{x_i, y_i\} \\ \mathrm{Cov}\{u_i, v_i\} &= \sin(\theta)\cos(\theta) \times [-\mathrm{Var}\{x_i\} + \mathrm{Var}\{y_i\}] + [\cos^2(\theta) - \sin^2(\theta)] \times \mathrm{Cov}\{x_i, y_i\} \end{aligned} \tag{4.10}$$

Now θ is selected such that $\mathrm{Cov}\{u_i, v_i\} = 0.0$, so that:

$$\tan(2\theta) = 2\mathrm{Cov}\{x_i, y_i\}/[\mathrm{Var}\{x_i\} - \mathrm{Var}\{y_i\}] \tag{4.11}$$

since $\tan(2\theta) = 2\sin(\theta)\cos(\theta)/[\cos^2(\theta) - \sin^2(\theta)]$ by definition. Equation (4.11) can be used to calculate θ and hence solve for $\mathrm{Var}\{u_i\}$ and $\mathrm{Var}\{v_i\}$. It follows that $Cov\{u_i, v_i\} = 0.0$ at the value of θ found by application of Equation (4.11).

4.2.3 Statistical Characteristics of Satellite Scanner Data

The radiance incident on a sensor is converted into an electrical signal by the light sensitive detectors, as discussed in Chapter 2. This signal is then amplified in the sensor electronics and sampled by an analog-to-digital converter to derive the response values that are recorded from the sensor. Because of differences in the sensor optics, the light sensitive surface and electronics, as well as the amplification and sampling, the same radiance values that are incident on the sensor will be recorded as different values by different sensors. Sensor calibration, discussed later in this chapter, is designed to convert these sensor specific response values back to radiance values. The data shown in Table 4.2 has not been calibrated in this way, and thus the data will give different values from the different sensors as listed in the table. This table holds statistical data derived from training areas for four different cover types that have been identified on images acquired by three different sensors. The image data have been analysed to derive mean and covariance arrays as shown in Table 4.2. The values given in

TABLE 4.2
Mean, Covariance and Correlation Arrays for Four Cover Types as Acquired Using the Three Scanners, Thematic Mapper, SPOT XS and MOS-1

Water

			Covariance Array				Correlation Array			
Sensor	**Channel**	**Mean**	**Ch1**	**Ch2**	**Ch3**	**Ch4**	**Ch1**	**Ch2**	**Ch3**	**Ch4**
MOS-1	1	20.19	0.27	0.01	0.03	0.00	1.000	0.068	0.118	0.000
	2	14.08		0.08	0.00	0.00		1.000	0.000	0.000
	3	8.58			0.24	0.00			1.000	0.000
	4	7.00				0.00				1.000

			Covariance Array/Correlation Array						
			Ch1	**Ch2**	**Ch3**	**Ch4**	**Ch5**	**Ch6**	**Ch7**
TM	1	64.42	3.44	0.37	0.01	0.11	0.03	0.12	0.02
	2	19.39		1.13	0.11	−0.02	0.04	−0.04	−0.02
	3	13.85			0.68	0.05	0.02	−0.04	0.00
	4	7.51				0.34	0.00	0.02	0.03
	5	4.40					0.57	−0.01	0.05
	6	134.62						0.24	−0.01
	7	3.14							1.25
	1		1.00	0.188	0.07	0.102	0.021	0.132	0.010
	2			1.00	0.125	−0.032	0.050	−0.077	−0.017
	3				1.00	0.104	0.032	−0.099	0.00
	4					1.00	0.000	0.070	0.046
	5						1.00	−0.027	0.059
	6							1.00	−0.018
	7								1.00

			Covariance array			Correlation array		
			Ch1	**Ch2**	**Ch3**	**Ch1**	**Ch2**	**Ch3**
SPOT XS	1	32.71	0.58	0.13	0.06	1.00	0.267	0.158
	2	20.69		0.41	0.14		1.00	0.437
	3	13.71			0.25			1.00

Urban

			Covariance array				Correlation array			
Sensor	**Channel**	**Mean**	**Ch1**	**Ch2**	**Ch3**	**Ch4**	**Ch1**	**Ch2**	**Ch3**	**Ch4**
MOS-1	1	31.57	5.02	3.88	1.36	1.02	1.00	0.817	0.532	0.529
	2	32.31		4.49	1.48	1.06		1.00	0.613	0.582
	3	21.21			1.30	0.79			1.00	0.805
	4	15.60				0.74				1.00

			Covariance array/correlation array						
			Ch1	**Ch2**	**Ch3**	**Ch4**	**Ch5**	**Ch6**	**Ch7**
TM	1	95.10	74.13	31.57	46.37	21.04	33.66	1.44	27.47
	2	37.23		16.96	22.85	13.43	19.52	0.32	14.36
	3	46.85			40.21	20.13	34.10	1.46	26.03
	4	43.49				40.76	33.37	−2.90	15.26
	5	61.02					68.01	0.07	40.65
	6	156.93						1.97	1.37
	7	36.38							31.37

(Continued)

Table 4.2 (Continued)

Urban

Sensor	Channel	Mean	Covariance array				Correlation array			
			Ch1	Ch2	Ch3	Ch4	Ch1	Ch2	Ch3	Ch4

Sensor	Channel	Mean	Covariance array/correlation array						
			Ch1	Ch2	Ch3	Ch4	Ch5	Ch6	Ch7
	1		1.00	0.890	0.849	0.383	0.474	0.119	0.570
	2			1.00	0.875	0.511	0.575	0.055	0.623
	3				1.00	0.497	0.652	0.164	0.733
	4					1.00	0.632	−0.324	0.427
	5						1.00	0.006	0.886
	6							1.00	0.174
	7								1.00

Sensor	Channel	Mean	Covariance array			Correlation array		
			Ch1	Ch2	Ch3	Ch1	Ch2	Ch3
SPOT XS	1	48.16	38.76	32.94	16.24	1.00	0.926	0.543
	2	42.28		32.63	15.09		1.00	0.550
	3	40.60			23.06			1.00

Forest

Sensor	Channel	Mean	Covariance array				Correlation array			
			Ch1	Ch2	Ch3	Ch4	Ch1	Ch2	Ch3	Ch4
MOS-1	1	20.64	0.71	0.07	0.36	0.39	1.00	0.163	0.314	0.283
	2	15.45		0.26	0.11	0.11		1.00	0.159	0.132
	3	24.39			1.85	1.91			1.00	0.858
	4	21.76				2.68				1.00

Sensor	Channel	Mean	Covariance array/correlation array						
			Ch1	Ch2	Ch3	Ch4	Ch5	Ch6	Ch7
TM	1	64.70	3.10	0.87	1.20	6.78	3.57	0.24	1.27
	2	23.11		1.07	0.89	6.34	3.61	0.15	1.21
	3	19.92			1.96	6.47	4.00	0.04	1.54
	4	73.45				86.49	38.34	1.70	10.98
	5	44.21					33.08	0.85	9.01
	6	137.20						0.49	0.19
	7	13.48							4.16
	1		1.00	0.477	0.487	0.414	0.353	0.195	0.354
	2			1.00	0.615	0.659	0.607	0.207	0.574
	3				1.00	0.497	0.497	0.041	0.539
	4					1.00	0.726	0.261	0.579
	5						1.00	0.211	0.768
	6							1.00	0.133
	7								1.00

Sensor	Channel	Mean	Covariance array			Correlation array		
			Ch1	Ch2	Ch3	Ch1	Ch2	Ch3
SPOT XS	1	30.17	0.98	0.46	4.08	1.00	0.615	0.703
	2	19.61		0.57	2.03		1.00	0.458
	3	61.77			34.41			1.00

TABLE 4.2 (Continued)

Rural

Sensor	Channel	Mean	Covariance array				Correlation array			
			Ch1	Ch2	Ch3	Ch4	Ch1	Ch2	Ch3	Ch4
MOS-1	1	28.77	4.05	5.37	0.23	−0.16	1.00	0.911	0.093	−0.078
	2	26.69		8.58	0.42	−0.32		1.00	0.117	−0.107
	3	28.29			1.50	1.00			1.00	0.797
	4	24.30				1.05				1.00

Sensor	Channel	Mean	Covariance array/correlation array						
			Ch1	Ch2	Ch3	Ch4	Ch5	Ch6	Ch7
TM	1	84.63	26.23	19.55	41.81	34.75	50.62	5.08	28.47
	2	37.71		18.22	35.89	40.88	50.01	2.86	22.65
	3	52.14			40.21	20.13	34.10	1.46	26.03
	4	68.63				40.76	33.37	−2.90	15.26
	5	91.88					68.01	0.07	40.65
	6	149.74						1.97	1.37
	7	41.92							31.37
	1		1.00	0.890	0.849	0.383	0.474	0.119	0.570
	2			1.00	0.875	0.511	0.575	0.055	0.623
	3				1.00	0.497	0.652	0.164	0.733
	4					1.00	0.632	−0.324	0.427
	5						1.00	0.006	0.886
	6							1.00	0.174
	7								1.00

Sensor	Channel	Mean	Covariance array			Correlation array		
			Ch1	Ch2	Ch3	Ch1	Ch2	Ch3
SPOT XS	1	48.16	38.76	32.94	16.24	1.00	0.926	0.543
	2	42.28		32.63	15.09		1.00	0.550
	3	40.60			23.06			1.00

Table 4.2 are not exactly comparable since the images were acquired in 1987 (XS) and 1991 (TM and MOS-1), so that differences in cover, atmospheric and solar zenith conditions will introduce further variations in the data values. However, the data do provide an insight into a number of characteristics of the data as acquired by these three sensors. Table 4.2 shows:

1. The response data acquired by a satellite follows the general shape of the spectral reflectance curves for the different surfaces. This is particularly noticeable in the case of the water and forest cover types for the TM and SPOT XS data.
2. The variances of the TM data are larger, in almost all cases, than the variances associated with either the XS or MOS-1 data. Busy surfaces, such as urban cover could be expected to have very large variances in the SPOT XS data because the smaller pixels have less opportunity to average across the range of surface conditions. This occurs for SPOT XS band 1, which has about twice the variance of TM Channel 2, but does not occur for the other bands. The variance in this TM data is about 1.5 to 2.5 times the variance in this SPOT

XS data, suggesting that this TM data is more sensitive than this SPOT or MOS-1. The XS data are next most sensitive, with variances that are usually more than twice that of the MOS-1 data, although the averaging effect of the larger MOS-1 pixels will be contributing to the difference in variance.

3. The correlations between bands for one sensor are generally matched by similar correlations in the other sensors. Thus high correlations occur between the visible bands for all three sensors for the urban and rural cover types. This suggests consistency between the sensors.

4.2.4 Measures of Distance

There are many ways to measure distances, some of which are relevant to the study of remotely sensed and geographic information system (GIS) data. There may be the need to consider the unit of measurement that is being used, and there is the need to consider the units of distance that are employed.

If an irregular feature is measured using a unit of measurement with a finite size, then features that are smaller than the unit of measurement will not be recorded in their entirety, but some approximation of the feature may be retained. How much is retained depends on the size of the unit of measurement relative to the feature, the position of the unit of measurement during the measuring process and the relative response values in the object relative to other features being measured.

There are two major units of distance that are of concern in image processing: Euclidean units and variance-based units. Euclidean distances are based on a constant physical unit of distance, and all objects, being measured in the same units, can thus be compared in those units, whether they be millimetres, grams or seconds. However, sometimes it is more important to be aware of the distance between two objects in other terms than those of the geometric distance separating the objects. In image processing it is often important to be aware of the separation between objects in terms of their variance and covariance. Including Euclidean distance, there are four measures of variance based distances:

1. *Euclidean distance*. The Euclidean distance between two points with array values x_1 and x_2 where these arrays contain the waveband response values for the particular sensor used to acquire the image is defined as

$$D_{\mathrm{e}} = \sqrt{\sum_{i=1}^{n} (x_1(i) - x_2(i))^2} \tag{4.12}$$

where n is the number of wavebands in the data.

2. *Mahalonobis distance*. The Mahalonobis distance between a point $(x(i), i = 1, n)$ and the mean value of a class $(\bar{y}(i), i = 1, n)$ with covariance array, Σ, is defined as

$$D_{\mathrm{m}} = \{(x - \bar{y})^{\mathrm{T}} \Sigma^{-1} (x - \bar{y})\}^{0.5} \tag{4.13}$$

where the $()^{\mathrm{T}}$ is the transpose of the array. The Mahalonobis distance is the Euclidean distance between the point and the class mean divided by the class standard deviation in the direction of the point from the class mean. It is thus the distance in units of the standard deviation in that direction.

3. *The divergence.* The divergence between two classes with means $\bar{x}_1$ and $\bar{x}_2$ and covariance arrays Σ_1 and Σ_2, respectively, is defined as

$$D_\mathrm{d} = 0.5\mathrm{Tr}\left\{(\Sigma_1 - \Sigma_2)\left(\Sigma_1^{-1} - \Sigma_2^{-1}\right)\right\} + 0.5\mathrm{Tr}\left\{\left(\Sigma_1^{-1} + \Sigma_2^{-1}\right)(\bar{x}_1 - \bar{x}_2)(\bar{x}_1 - \bar{x}_2)^\mathrm{T}\right\} \quad (4.14)$$

where Tr is the trace of the related matrix. The first term in the divergence is a function of the covariance arrays, whilst the second involves the square of the distance between the class mean values normalised by the covariance in the direction joining the mean values. The divergence is similar to the Euclidean distance between two class mean values, divided by the average of the standard deviations of the two classes along the line joining the means.

4. *Jeffries-Matusita distance.* The Mahalonobis distance is suitable for use in determining the distance of a point from a class, but is unsuitable for the determination of the separation of two classes, both with their own mean and covariance arrays, $\bar{x}_1$ and $\bar{x}_2$ and Σ_1 and Σ_2, respectively. The Jeffries–Matusita (JM) distance is suitable for this task, where it is defined as

$$D_\mathrm{JM} = 2(1 - \mathrm{e}^{-B}) \quad (4.15)$$

where

$$B = 0.125\,(\bar{x}_1 - \bar{x}_2)^\mathrm{T}\left\{(\Sigma_1 + \Sigma_2)/2\right\}^{-1}(\bar{x}_1 - \bar{x}_2) + 0.5\ln\left\{\frac{2\,|(\Sigma_1 + \Sigma_2)|}{\sqrt{|\Sigma_1|}\sqrt{|\Sigma_2|}}\right\}$$

The first term in B is the square of the Euclidean distance between the class means normalised by the average standard deviation in this direction. The use of the exponential factor in Equation (4.4) causes the JM distances to saturate at a value of 2; that is, no matter how large the physical separation of the classes, the JM distance will not exceed 2. The JM distance computed for normally distributed classes is also called the Bhattacharyya distance.

4.2.5 Shannon's Sampling Theorem

A series of photographs taken as a person walks across a field, timed such that they are taken at intervals of about 100 m, may give a good indication of the general direction of travel and average speed of the walker. But they will not tell the observer anything about the fluctuations in direction or velocity that occurred between the photographs. In the same way, annual images may indicate general landuse and land quality, but will tell the analyst very little about what actually happened in between the images.

In renewable resources there are certain frequencies about the events being monitored: imagery must be acquired at a frequency that allows this cyclical pattern of physical events to be adequately recorded. Shannon's sampling theorem is concerned with the development of a sampling frequency sufficient to adequately represent a population that exhibits a cyclic fluctuating pattern.

Shannon's sampling theorem states that a band-limited function at frequency W can be reconstructed exactly from samples taken at a frequency interval of $W/2$ or higher using a sinc function, sinc $= \sin(x)/x$, as the shape of the sample function. An image is band limited if the variability in the data does not contain variations that have frequencies higher than the band limit, that is, higher than W. A strict definition is that a picture function, $g(i,j)$, is band limited if its Fourier transform, $G(i,j)$ (Section 4.2.6), is zero whenever either $|f_i|$, or $|f_j|$ is greater than W, where f_i and f_j are frequencies in the data.

In a spatial sense, the importance of Shannon's sampling theorem lies in its ability to assist us in selecting suitable pixel sizes to sample the surface, or to assess the likelihood that data of a known pixel size can adequately record the radiance from the surface. In a temporal sense the theorem can be used to assess whether the sampled data is at a sufficient frequency to represent cyclic data, either to reconstruct that data or to derive information from the data, at a suitable level of accuracy.

If seasonal events are to be monitored then at least two images per season are required, assuming that higher frequency cycles, for example due to weather events, do not introduce significant variations from the seasonal cycle. If in-season variations are significant then more frequent images are required.

4.2.6 Autocorrelation and Variograms

The discussion of correlation has focused on the correlation between different attributes of a feature, such as the correlation between bands of image data. However, there is also spatial or temporal correlation between pixels. For some purposes it is important to measure this spatial and temporal correlation. For this text both spatial correlation between pixels and over time will be called autocorrelation. If it becomes necessary to discriminate between spatial autocorrelation and temporal autocorrelation, then these terms will be used.

The sample autocovariance function for a linear sequence of data $(z_1, z_2, \ldots, z_n)$ is given by the equation

$$\alpha(k) = \left(\frac{1}{2(n-k)}\right) \sum_{i=1}^{n} (z_i - z')(z_{i+k} - z') \tag{4.16}$$

where z' = mean value of the series, n = number of observations in the set of data and k = step value.

For spatial data, consider a line of pixels with values as shown in Figure 4.7.

For this set of data, the mean $z' = 37.31$

$$\begin{aligned}\alpha(1) &= (30 - 37.31) \times (32 - 37.31) + (32 - 37.31) \times (28 - 37.31) + \ldots + (41 - 37.31) \\ &\quad \times (37 - 37.31)/12 \\ &= 19.87\end{aligned}$$

If the autocovariance values are divided by the autocovariance value at a step of 0, then this gives the sample autocorrelation function. Note that the autocovariance value at a step of 0 is the same as the variance for the set of data points, 25.73 in this case. The autocorrelation function values for the above data set are 1.00 at step 0, 0.7721 at step 1, etc. The absolute values in the autocorrelation function are plotted in Figure 4.8.

The variogram function is related to the autocorrelation function, but is preferred by many scientists in the physical resource sciences. The variogram in one dimension is computed using

$$\delta(k) = \frac{1}{2(n-k)} \sum_{i=1}^{n-k} (z_i - z_{i+k})^2 \tag{4.17}$$

30	32	28	32	38	43	41	41	42	41	39	41	37

Figure 4.7 A line of 13 pixels and their response values.

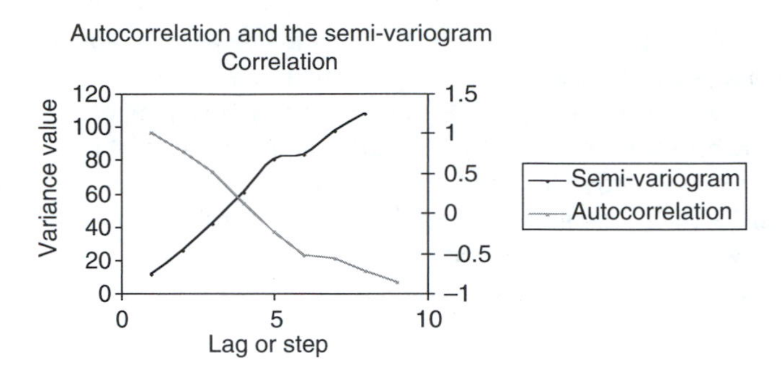

FIGURE 4.8 Autocorrelation and variogram functions for the data in Figure 4.7.

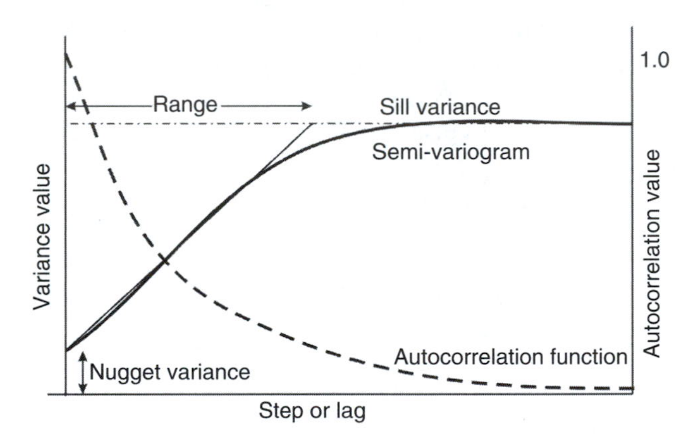

FIGURE 4.9 Typical shapes that can be expected for autocorrelation and variogram functions derived from remotely sensed data.

where k is the lag or separation between data values. For the above data set the variogram value at a lag of 1 is

$$\begin{aligned}\delta(1) &= \{(30-32)^2 + (32-28)^2 + (28-32)^2 + \cdots + (41-37)^2\}/12 \\ &= 11.55\end{aligned}$$

The variogram can never contain negative values. Typical autocorrelation and variogram functions are shown in Figure 4.9. The autocorrelation function starts at 1.0 for a step of zero. The value 1.0 indicates perfect correlation. The variogram cannot start at step or lag zero, but at step or lag one.

Adjacent pixels normally are of similar cover conditions, and so will normally be highly correlated. As the pixels become separated in distance or time, the level of correlation can be expected to decrease. This is shown as lower values in the autocorrelation function, and as a plateau in the variogram. The variogram shows the existence of correlation until the curve reaches the plateau, and after that there is assumed to be negligible correlation. The distance out to the sill indicates the range or distances over which correlation was found to exist. Theoretically, the variogram should have a value of zero at a lag of zero since the computation is the difference between exactly the same values. However extending the variogram to a lag of zero will usually show a positive value where the variogram cuts the y-axis. This variance value is called the nugget variance. The nugget variance is usually interpreted as the random noise in the data.

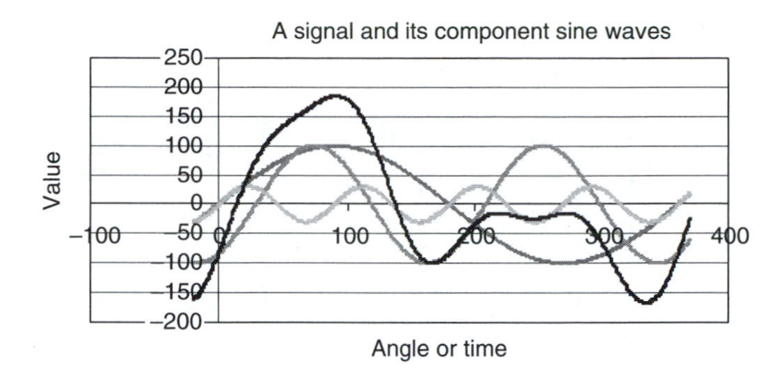

FIGURE 4.10 A set of data values and its component sinusoidal values.

With both Equation (4.16) and Equation (4.17), small samples will introduce instability. The functions are thus usually not computed when the sample sizes are less than 30, that is, when $(N - k)$ is less than 30.

Both autocorrelation functions and variograms can be computed for spatial and temporal data. For spatial data the variograms can also be computed for both the x- and the y-dimensions to produce two-dimensional variograms.

4.2.7 FREQUENCY DOMAIN

Most of the discussion in this section so far has focused on the statistical properties of data in the normal spatial and temporal domains. Data can also be defined in the frequency domain. Any set of data values, such as a time series or a line of pixel values, can be transformed into a power spectrum in the frequency domain. Conceptually, a data set samples a continuous curve. This curve can be represented as the summation of a set of sinusoidal curves, of different wavelengths and amplitudes (Figure 4.10). For a sine wave, the rate of oscillation is its frequency, f, and the period of oscillation is its wavelength, λ. The frequency and period or wavelength of a sine wave are related in the equation

$$f \times \lambda = \text{constant} \tag{4.18}$$

The black curve shown in Figure 4.10 can be represented by just three sinusoidal waves (grey curves). In the frequency domain the power spectrum curve can be represented as having contributions at three frequencies, the weighting of those contributions being a function of their amplitudes. The three sine waves in Figure 4.10 have amplitudes 100, 100 and 30, respectively, so that the power spectrum of the data in Figure 4.10 would appear as shown in Figure 4.11. In this figure, the frequencies are shown symmetrically set around the central vertical axis. Each contributing sine wave contributes a column that is mirrored to the left and the right of the central axis. The largest contributions are due to the two sine waves with the longest wavelengths, as can be seen from their amplitudes. As these are equal, the sizes of their columns are equal in the power spectrum. The three columns are equally spaced as their wavelengths are a doubling of each other. The longer the wavelength, or the lower the frequency, the closer the contributing signal is to the central axis. In the data shown here, the three sine waves have zero mean values. If the waves contain a non-zero mean value, then this would appear as a central spike of a sine wave with infinitely long wavelength.

Now the data shown in Figure 4.10 consists of three sine waves that all start at the same position in their waveform at the start of the curve. They all have zero phase shift. What happens if one or more of the individual sine waves are shifted along the x-axis, that is, they contain a phase shift? It so happens that any sine with a phase shift can be represented by a combination of a sine and a cosine wave of zero phase shift but different amplitudes. Consider two sine waves of the same wavelength

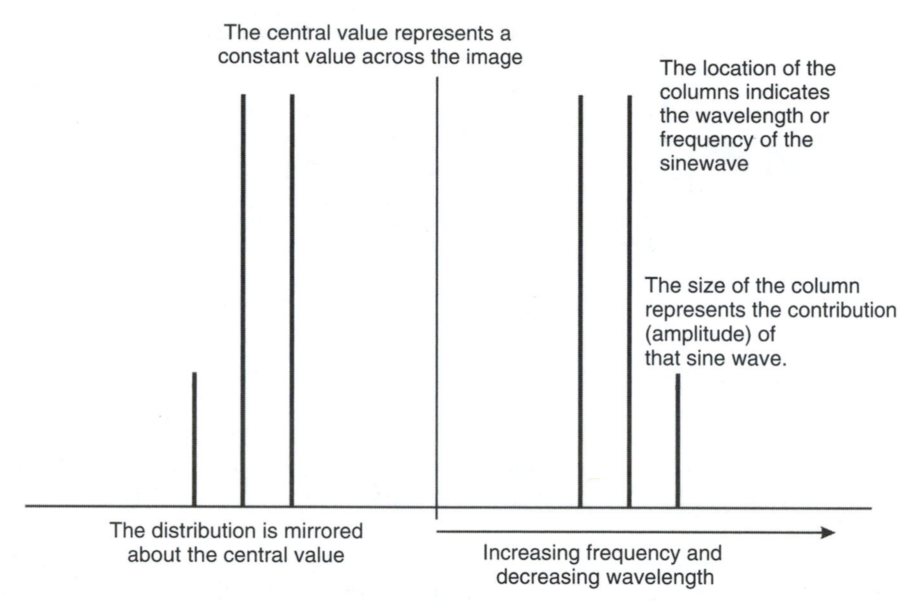

FIGURE 4.11 The frequency domain showing the power spectrum.

but different phase shifts and different amplitudes:

$$y = a_0 \sin(b_0 x) \quad \text{and} \quad z = a_1 \sin(b_0 x + c_0)$$

If we expand the second sine wave, we get

$$z = a_1 \sin(b_0 x + c_0) = a_1 \sin(b_0 x) \cos(c_0) + a_1 \cos(b_0 x) \sin(c_0)$$

Now that the phase shift is constant, we can make $a_1 \cos(c_0) = k_0$ and $a_1 \sin(c_0) = k_1$ to give

$$z = k_0 \sin(b_0 x) + k_1 \cos(b_0 x) \tag{4.19}$$

Now Equation (4.19) shows that a sine wave of wavelength $b_0 t$ at any amplitude and phase can be represented by the sum of the products of constants by sine and a cosine waves at the same wavelength but zero phase shift, where the two constants tell us about the amplitude and phase of the original wave. This is the basis of the Fourier transform. Since we are interested in all of the wavelengths that constitute the original wave, we can now form the general form of the Fourier transform. If the source signal is of the form $f(x)$, then

$$f(x) = \int_k a_k \sin(2\pi sx) + b_k \cos(2\pi sx) = \int_k F(s) \sin(2\pi sx) + F(s) \cos(2\pi sx) \tag{4.20}$$

where $F(s)$ is the Fourier transform values at the kth harmonic, with s being a scalar. Equation (4.20) can be put in the form:

$$f(x) = \int_k F(s) e^{i2\pi sx} \tag{4.21}$$

since $e^{i2\pi sx} = \cos(2\pi sx) + i \sin(2\pi sx)$ in which $i^2 = -1$ so that the cosine part of the equation gives the real part of the Fourier transform and the sine part gives the imaginary part of the transform.

Without proof, the inverse will now be stated as being

$$F(s) = \frac{1}{N}\int_k f(x)\mathrm{e}^{-i2\pi sx} \tag{4.22}$$

in which N is a scalar to maintain consistency between the two equations, Equation (4.21) and Equation (4.22).

However, all digital image data are discrete data rather than continuous data, and so the discrete version of the Fourier transform is more appropriate:

$$F(r) = \frac{1}{N}\sum_{k=0}^{N-1} x_k \mathrm{e}^{-12\pi rk/N} \tag{4.23}$$

and

$$x_k = \sum_{r=0}^{N-1} F(r)\mathrm{e}^{i2\pi rk/N} \tag{4.24}$$

If the source signal consisted of just one sine wave, then $F(r)$ contains values at its zero location ($r = 0$) equal to the mean value in the signal, and at its location corresponding to the frequency of the sine wave it will have a value corresponding to the amplitude of the wave, that is, $F(0) = \mu$ and $F(\lambda) = \text{Amplitude}$. Consider the situation where there are nine possible values in the Fourier transform, the first location having a value of 10 and the fourth location having a value of 8:

$$F(0) = 10, F(1) = F(2) = F(3) = 0, F(4) = 8, F(5) = F(6) = F(7) = F(8) = 0, \text{then :}$$

$$\begin{aligned} x_k &= \sum_{r=0}^{8} F(r)(\cos(2\pi rk/9) + i \times \sin(2\pi rk/9)) \\ &= F(0) + F(4)(\cos(8\pi k/9) + i \times \sin(8\pi k/9)) \end{aligned} \tag{4.25}$$

In Equation (4.25) if for each value of k in the range 0 to 8 there is a zero phase shift, that is, the imaginary component is zero:

$$\begin{aligned} X_0 &= F(0) = 10 \\ X_1 &= F(0) + 8\cos(8\pi/9) = 2.48 \\ X_2 &= F(0) + 8\cos(16\pi/9) = 16.12 \\ X_3 &= F(0) + 8\cos(24\pi/9) = 6.00 \\ X_4 &= F(0) + 8\cos(32\pi/9) = 11.39 \\ X_5 &= F(0) + 8\cos(40\pi/9) = 11.39 \\ X_6 &= F(0) + 8\cos(48\pi/9) = 6.00 \\ X_7 &= F(0) + 8\cos(56\pi/9) = 16.12 \\ X_8 &= F(0) + 8\cos(64\pi/9) = 2.48 \end{aligned}$$

It is simpler to demonstrate the conversion of the Fourier transform back to the sine wave than the reverse, simply because there are a lot more calculations to do in the reverse direction. However, the concepts apply equally in both directions.

The discrete Fourier transform is used to convert a set of data in the spatial or time domain into the frequency domain. This transformation can also be used to convert data in the frequency domain into the spatial or time domains. Thus, data can be converted into the frequency domain, analysed in some way in that domain, and then converted back into the spatial domain. This capacity to convert data readily to and from the frequency domain allows the data to be readily filtered to remove cyclic noise from the data. If image data contains a feature with a strong cyclic signal, then this will show very clearly in the power spectrum. A sine wave component in the power spectrum can be filtered out in the frequency domain, and the data then transformed back into the spatial domain. The transformed and filtered data will be similar to the source data, but without the strong sine wave signal.

Data that has been converted in this way into the frequency domain using the Fourier transformation have the following characteristics:

1. The data are almost uncorrelated.
2. The data have variances equal to the power spectrum values.
3. The data are approximately normally distributed.

There are a number of important characteristics of the Fourier transform. These will be given here without proof. A reader seeking a more comprehensive treatment of the Fourier transform is referred to the excellent texts that exist on the subject.

4.2.7.1 Scaling

If $F(s)$ is the Fourier transform of a function $f(x)$ that is scaled by a real, non-zero constant, a, then the Fourier transform of the scaled function, $f(ax)$ is equal to $(1/a)F(s/a)$. If the width of a function is decreased while its height is kept constant, then its Fourier transform becomes wider and lower. Conversely, if its width is increased, then the Fourier transform becomes narrower and taller.

4.2.7.2 Shifting

If $F(s)$ is the Fourier transform of a function $f(x)$ and x_0 is a real constant, then the Fourier transform of the function shifted by x_0 is $F(s)e^{(i2\pi x0s)}$, that is, the Fourier transform of a shifted function is the same as the Fourier transform, multiplied by an exponent term having a linear phase.

4.2.7.3 Convolution

Suppose that $g(x) = f(x) \times h(x)$, then the Fourier transform of $g(x) = G(s)$ is the product of the Fourier transforms of $f(x)$ and $h(x)$, that is, $G(s) = F(s) \times H(s)$. Convolution is the process of passing one function through a second function. An example in image data is the creation of images of a scene by a sensor where the scene function is $f(x)$ and the sensor point spread function is $h(x)$. Passing the scene values through the point spread function gives the image function, $g(x)$. Convolution is thus a method of filtering image data, and it can be used for many other purposes.

4.2.8 Least Squares Method of Fitting

Least squares is a method of finding a solution to a set of equations when there are more equations than there are unknowns. In this situation there is no one unique solution. Least squares finds a solution that meets the criterion that the sums of squares of the residuals (or the variance of the residuals) is a minimum. The least squares method starts from the assumption that there exists a linear relationship between the dependent variable and the independent variables. The independent variables drive the relationship and the dependent variable carries the result. Least squares can also be used for non-linear data as will be discussed at the end of this section.

A linear relationship is one of the form

$$y = a_0 + a_1x_1 + a_2x_2 + \cdots + a_{n-1}x_{n-1} \tag{4.26}$$

Such a relationship usually contains n unknowns, with one unknown constant and $(n-1)$ unknowns associated with the $(n-1)$ independent variables. If we have m observations of the independent and dependent variables then we can solve the resulting set of equations to find the constants, $\{a_i, i = 0, (n-1)\}$ if $m >= n$. If $m = n$ then there is one unique solution. If $m > n$ then there are a number of different solutions of subsets of n of the resulting equations, and different values of the constants will be found for each solution. Least squares addresses this problem by using all m equations to find one, optimised solution, which however will not usually solve each equation perfectly, but will give a residual for each observation. Least squares optimises by minimising the sums of squares, or variance, of the residuals.

As a consequence, Equation (4.26) can be written in the form

$$\begin{aligned}
y_1 &= a_0 + a_1x_{1,1} + a_2x_{2,1} + \cdots + a_{n-1}x_{n-1,1} + \varepsilon_1 \\
y_2 &= a_0 + a_1x_{1,2} + a_2x_{2,2} + \cdots + a_{n-1}x_{n-1,2} + \varepsilon_2 \\
&\vdots \\
y_m &= a_0 + a_1x_{1,m} + a_2x_{2,m} + \cdots + a_{n-1}x_{n-1,m} + \varepsilon_m
\end{aligned} \tag{4.27}$$

in which (x_i, y_i) is a set of observations of the independent variables, x_i, and the dependent variable, y_i. Such equations illustrate the common ways of forming the observation equations, called this because each observation of the set of independent and dependent variables creates one equation. There are other ways of forming the equivalent of observation equations that may be more appropriate when dealing with other adjustment issues. However, formation of observation equations is the typical way to use least squares in remote sensing, and so it will be the only approach dealt with in this book. The interested reader is directed to texts on adjustment for more detailed discussion on the alternative ways to implement least squares. The observation equations can be re-arranged:

$$\begin{aligned}
a_0 + a_1x_{1,1} + a_2x_{2,1} + \cdots + a_{n-1}x_{n-1,1} &= y_1 - \varepsilon_1 \\
a_0 + a_1x_{1,2} + a_2x_{2,2} + \cdots + a_{n-1}x_{n-1,2} &= y_2 - \varepsilon_2 \\
&\vdots \\
a_0 + a_1x_{1,m} + a_2x_{2,m} + \cdots + a_{n-1}x_{n-1,m} &= y_m - \varepsilon_m
\end{aligned} \tag{4.28}$$

In these equations the ε_i are the residuals left after the constants a_i have been found by means of least squares. It can be seen that the residuals are in terms of the dependent variable. These equations can be arranged in matrix format:

$$\begin{bmatrix} 1 \cdots x_{1,1} \cdots x_{2,1} \cdots x_{n-1,1} \\ 1 \cdots x_{1,2} \cdots x_{2,2} \cdots x_{n-1,2} \\ \vdots \\ 1 \cdots x_{1,m} \cdots x_{2,m} \cdots x_{n-1,m} \end{bmatrix} \times \begin{bmatrix} a_0 \\ a_1 \\ \vdots \\ a_n \end{bmatrix} = \begin{bmatrix} y_1 - \varepsilon_1 \\ y_2 - \varepsilon_2 \\ \vdots \\ y_m - \varepsilon_m \end{bmatrix} \tag{4.29}$$

or in matrix notation:

$$O \times U = D \tag{4.30}$$

in which $O[m,n]$, $U[n]$ and $D[m]$ are arrays of the size indicated in the square brackets. If both sides are then multiplied by the transpose of O or O^{T}, then n normal equations are formed:

$$\begin{bmatrix} 1 \cdots 1 \cdots 1 \cdots\cdots\cdots\cdots\cdots 1 \\ x_{1,1} \cdots x_{1,2} \cdots\cdots\cdots x_{1,m} \\ \vdots \\ x_{n-1,1} \cdots x_{n-1,2} \cdots x_{n-1,m} \end{bmatrix} \times \begin{bmatrix} 1 \cdots x_{1,1} \cdots x_{2,1} \cdots\cdots x_{n-1,1} \\ 1 \cdots x_{1,2} \cdots x_{2,2} \cdots\cdots x_{n-1,2} \\ \vdots \\ 1 \cdots x_{1,m} \cdots x_{2,m} \cdots\cdots x_{n-1,m} \end{bmatrix} \times \begin{bmatrix} a_0 \\ a_1 \\ \vdots \\ a_n \end{bmatrix}$$

$$= \begin{bmatrix} 1 \cdots 1 \cdots 1 \cdots\cdots\cdots\cdots 1 \\ x_{1,1} \cdots x_{1,2} \cdots\cdots x_{1,m} \\ \vdots \\ x_{n-1,1} \cdots x_{n-1,2} \cdots {}_{n-1,m} \end{bmatrix} \times \begin{bmatrix} y_1 - \varepsilon_1 \\ y_2 - \varepsilon_2 \\ \vdots \\ y_m - \varepsilon_m \end{bmatrix}$$

or in matrix notation:

$$O^{\mathrm{T}}[n,m] \times O[m,n] \times U[n] = O^{\mathrm{T}}[n,m] \times R[m] \tag{4.31}$$

giving

$$N[n,n] \times U[n] = S[n] \tag{4.32}$$

after the appropriate matrix multiplications, $O^{\mathrm{T}} \times O$ and $O^{\mathrm{T}} \times R$. An interesting thing about Equation (4.32) is that this equation cannot be solved if the determinant of N, $\det(N)$ or $|N|$ is zero. The determinant of a square matrix is a singular value that can be derived as given below.

If a square matrix, M[2,2] is of the $\begin{vmatrix} a & b \\ c & d \end{vmatrix}$ then the determinant is $(ad - bc)$. The determinant for larger square matrices can be found as follows for a matrix, $M[3,3]$:

$$\det(M) = \det \begin{vmatrix} a & b & c \\ d & e & f \\ g & h & j \end{vmatrix} = a \times \det \begin{vmatrix} e & f \\ h & j \end{vmatrix} - b \times \det \begin{vmatrix} f & d \\ j & g \end{vmatrix} + c \times \det \begin{vmatrix} d & e \\ g & h \end{vmatrix}$$

and so forth for larger matrices. The determinant of a matrix will be zero when two or more of the rows in the matrix are functionally the same, that is, they are either identical or they differ by a constant multiplier.

If we expand Equation (4.32), then we can see:

$$\begin{bmatrix} n \cdots \sum x_{1,i} \cdots \sum x_{2,i} \cdots \sum x_{n-1,i} \\ \sum x_{1,i} \cdots \sum x_{1,i} \times x_{1,i} \cdots \sum x_{1,i} \times x_{2,i} \cdots \sum x_{1,i} \times x_{(n-1),i} \\ \vdots \\ \sum x_{(n-1),i} \cdots \sum x_{(n-1),i} \times x_{1,i} \cdots \sum x_{(n-1),i} \times x_{2,i} \cdots \sum x_{(n-1),i} \times x_{(n-1),i} \end{bmatrix} \times \begin{bmatrix} a_0 \\ a_1 \\ \vdots \\ a_n \end{bmatrix}$$

$$= \begin{bmatrix} \sum (y_i - \varepsilon_i) \\ \sum (y_i - \varepsilon_i) \times x_{1,i} \\ \vdots \\ \sum (y_i - \varepsilon_i) \times x_{(n-1),i} \end{bmatrix}$$

Now in this equation, $\sum(y_i - \varepsilon_i) = \sum y_i - \sum \varepsilon_i$ and $\sum(y_i - \varepsilon_i)x_i = \sum y_i x_i - \sum \varepsilon_i x_i$. If we make $\sum \varepsilon_i = 0$ then we can solve this set of equations in accordance with the least squares optimisation criteria. To prove this, consider the case of the mean of a set of observations as a unique example of least squares with one unknown the constant value or mean.

If we have a set of observations, $(x_1, x_2, \ldots, x_n)$ then $\bar{x} = \left(\sum_{i=1}^{n} x_i\right) \Big/ n$ and so:

$$\begin{aligned} \bar{x} &= x_1 + \varepsilon_1 \\ &= x_2 + \varepsilon_2 \\ &\vdots \\ &= x_n + \varepsilon_n \end{aligned} \tag{4.33}$$

so that

$$\begin{aligned} \varepsilon_1 &= \bar{x} - x_1 \\ \varepsilon_2 &= \bar{x} - x_2 \\ &\vdots \\ \varepsilon_n &= \bar{x} - x_n \end{aligned} \tag{4.34}$$

Now the sums of squares, S, is

$$S = \sum_{i=1}^{n} (x_i - \bar{x})^2 = \sum_{i=1}^{n} \varepsilon_i^2 \tag{4.35}$$

This is minimised when its gradient is made zero since it forms a parabolic surface. Differentiation gives

$$\frac{dS}{d\bar{x}} = 2(x_1 - \bar{x}) + 2(x_2 - \bar{x}) + \cdots + 2(x_n - \bar{x}) = 0 \tag{4.36}$$

so that $\bar{x} = \left[\sum_{i=1}^{n} x_i\right] \Big/ n$ is the value of the mean for which the sums of squares is minimised. Other mean values will have larger sums of squares.

If both sides of Equation (4.32) are then multiplied by N^{-1}, the inverse of matrix N, then we get

$$N^{-1} \times N \times U = N^{-1} \times S, \quad \text{or} \quad U[n] = N^{-1}[n,n] \times S[n] \tag{4.37}$$

giving values for the unknowns in terms of the observational values and setting the sum of the residuals to be zero and the sums of squares of the residuals to be a minimum.

In practice, the relationship between the independent variables and the dependent variable may be non-linear. When this is the case, then the relationship has to be linearised before it can be solved by means of least squares. Developing a linear approximation to the relationship usually does this. Consider a simple case involving one independent variable, x, of the form:

$$y = a_0 + a_1 \sin(a_2 x) \tag{4.38}$$

This is linearised by

$$y + \delta y = (a_0 + \delta a_0) + (a_1 + \delta a_1) \times \sin((a_2 + \delta a_2) \times x)$$

in which the δ represents a small change. Expansion of this equation gives

$$y + \delta y = a_0 + \delta a_0 + (a_1 + \delta a_1)\,[\sin a_2 x \cos \delta a_2 x + \cos a_2 x \sin \delta a_2 x]$$

In this equation, $\delta a_2 x$ is small and so $\cos \delta a_2 x \approx 1.0$ and $\sin \delta a_2 x = \delta a_2 x$ so that

$$\begin{aligned} y + \delta y &= a_0 + \delta a_0 + (a_1 + \delta a_1) \times (\sin a_2 x + \delta a_2 x \times \cos a_2 x) \\ &= a_0 + \delta a_0 + a_1 \sin a_2 x + \delta a_1 \sin a_2 x + \delta a_2 x \times a_1 \cos a_2 x + \delta a_1 \times \delta a_2 x \times \cos a_2 x \end{aligned} \tag{4.39}$$

In Equation (4.39) the product of the two δ variables is very small and so is ignored giving us a linearised version that is an approximation:

$$y + \delta y \approx a_0 + \delta a_0 + a_1 \sin a_2 x + \delta a_1 \sin a_2 x + \delta a_2 x \times a_1 \cos a_2 x \tag{4.40}$$

which can be re-arranged into the form of an observation equation:

$$\delta y - \delta a_0 - \delta a_1 \sin a_2 x - \delta a_2 x a_1 \cos a_2 x = a_0 + a_1 \sin a_2 x - y \tag{4.41}$$

In such a situation, initial values have to be chosen for the variables and the least squares adjustment seeks to find small changes to these that should give a better fit to the data. Since the linearisation is an approximation, this may need to be conducted a number of times, each time modifying the values of the variables by the changes found in the previous solution. This processes is called iteration.

The coefficient of determination (R^2) gives an indication of the quality of the regression if the sample used is of sufficient size. The total sums of squares (SST) is the sums of squares computed about the mean and the sums of squares due to error (SSE) is the residual sums of squares after conducting the regression:

$$\mathrm{SST} = \sum_{i=1}^{N} (y_i - \bar{y})^2 = \sum_{i=1}^{N} y_i^2 - \left(\sum_{i=1}^{N} y_i \right)^2 \Big/ N \tag{4.42}$$

and

$$\mathrm{SSE} = \sum_{i=1}^{N} (y_x - \hat{y})^2 \tag{4.43}$$

where $\hat{y} = a_0 + a_1 x_i$, the estimated value of y_i from the regression equation, shown for the simple linear case here. The coefficient of determination $(R^2) = (\mathrm{SST} - \mathrm{SSE})/SST$. If $\mathrm{SSE} = 0$, as would only occur if all of the data points fall exactly on the regression line, then $R^2 = 1.0$. Conversely, if $\mathrm{SSE} = \mathrm{SST}$, then $R^2 = 0.0$ when there is no correlation between the parameters. Thus, it can be seen that the coefficient of determination is an indicator of the quality of the fit, with higher values indicating a better fit, as long as an adequate sample was used in the derivation of the model.

Whilst the coefficient of determination is an indication of the quality of the fit of the model to the data, it is not adequate as it does not take into account the sample size used in deriving the fit. As a consequence, it is important to test the significance of the derived parameters. It is usually tested by use of the hypotheses:

$$\begin{aligned} H_0 &: a_1 = 0 \\ H_1 &: a_1 \neq 0 \end{aligned}$$

The H_0 hypothesis is formulated as the proposal that $a_1 = 0$; that is, that the derived value is not statistically significant. If this hypothesis is accepted, then the regression equation is not statistically significant. For a linear regression it is tested by use of the F statistic where:

$$F = \frac{\text{MSR}}{\text{MSE}} = \frac{\text{SSR/Number of independent variables}}{\text{SSE/(Number of observations} - \text{Number of unknowns)}} = \frac{\text{SSR/NDoF}}{\text{SSE/DDoF}} \tag{4.44}$$

where NDoF are the numerator degrees of freedom and DdoF are the denominator degrees of freedom.

The MSE is an estimate of the variance of the residuals, as can be seen from its formulation. If H_0 is true, then MSR is a second, independent estimate of the variance. Since these should have similar values, when they are very different we can assume that MSR is not an estimate of the variance, thus the H_0 hypothesis is not valid and the regression parameter a_1 can be accepted. In general, the size of MSR will increase as more of the variability is explained by the regression curve, and MSE will decrease. Thus large values of F will cast doubt on the hypothesis and lead to its rejection and acceptance of the alternate hypothesis, H_1. If H_0 is true, then F follows the F distribution. If the value of F is less than the F-distribution value at a set probability for given numerator and denominator degrees of freedom, then the null or H_0 hypothesis can be accepted. Larger values are not acceptable.

4.3 PRE-PROCESSING OF IMAGE DATA

4.3.1 INTRODUCTION

Remotely sensed image data is acquired using a sensor that is an approximation of some geometric model. The collected data is correlated to some degree to the radiance or energy incident on the sensor. There are thus at least two sources of error in the recorded data: geometric and radiometric errors. There are usually errors related to the spectral bands recorded, and there may be errors related to the time of acquisition. However, spectral and temporal errors are usually very small for most applications in the management of physical resources.

The geometric errors are of two types:

1. The approximation of the sensor to the geometric model for the sensor or the internal geometric errors
2. The errors in the location and orientation of the sensor during exposure or external sources of error

Rectification is the process of correcting for both of these types of geometric error. It will be dealt with in this section of the book.

Radiometric errors have three main causes:

1. Variations in the atmospheric path reflectance, transmittance and absorptance
2. Optical and electronic sources of error in the sensor
3. Detector sensitivity in the sensor

It will be recalled that the atmosphere affects image data through:

1. Absorption and scattering from the molecules of the atmosphere
2. Absorption and scattering by moisture in the atmosphere
3. Ozone absorption
4. Absorption and scattering by aerosols in the atmosphere

Since the proportion of the gases in the atmosphere, other than ozone and moisture, are very stable, the effects of these gases can be modelled. The density of ozone in the atmosphere can be measured by satellite, and so this information can be used to correct for ozone absorption at the time of data acquisition. Atmospheric moisture levels can vary dramatically from day to day depending on weather conditions. The effects of atmospheric moisture can be estimated, and so adequate corrections can be made, if suitable data exists for this purpose. Aerosols in the atmosphere come from various sources including sea salt, soil and sand dust, organic dust, smoke and industrial sources of aerosols. These particles can also change in the atmosphere as a result of agglomeration, often caused by atmospheric moisture if the humidity exceeds about 35%. Aerosols coming from organic matter are often spheroidal in shape, but this is not the case with aerosols due to soil and sand. The density, size distribution reflectance and height of the aerosols all influence the effect of the aerosols on incident radiation. The reflectance can also be changed when the aerosol particles are moistened when the humidity exceeds about 35%. Aerosols can stay in the atmosphere for periods from a day or so to weeks. They are thus complex phenomena to attempt to model. At present there are no satisfactory routine techniques for the correction of the effects of atmospheric aerosols on the radiance values of image data.

The sources of error in the radiometry of the sensor contain both systematic and random components. Systematic components include:

1. *The response curve for the detector.* No detectors have perfectly linear response with energy at all wavelengths, and signal strengths. Detectors are calibrated prior to launch to determine this sensitivity. However, the sensitivity of a detector can change with changes in environmental conditions, for example with temperature, and with age.
2. *The efficiency of the sensor optics.* The optical components of a sensor are an approximation of a theoretical optical model. As the energy is transmitted through the optics, some of the energy is scattered at the surfaces of the optical components, some is refracted by different amounts as a function of the wavelength of the radiation, and some is absorbed by the optical components. All of these introduce errors in the incoming signal. The resulting distortions are larger for sensors with optics that need to simultaneously cover the whole of the sensor field of view, because of the more severe design constraints imposed on such optical components. These sensors include cameras and pushbroom scanners. The distortions in moving mirror scanners are smaller since the optics only need to cover the instantaneous field of view of the scanning element.
3. *Dark current.* All electronic systems create internal currents that affect the electronic functioning of the instrument. Such dark currents can affect the recorded signal from the detector. The dark current can change with age of the sensor and also with changes in environmental conditions.
4. *Sampling lag.* The signal coming onto the detector through the sensor optics is a continuous signal. It will be changing continuously in response to variations in surface reflectance and atmospheric conditions. As this signal changes, it creates changes in the detector response. There is, however, a time lag between the occurrence of a changed signal, and its expression as a change in the detector output. This lag can be affected by the magnitude of the incoming signal. The detector is sampled to create discrete digital data, and there may thus be lags in this sampling.

All sensors are calibrated before they are launched on a satellite, and airborne sensors should be calibrated at regular intervals. However, changes in the platform's environment between those at the time of calibration and those applying when the data is acquired, can introduce errors into the radiometry of the sensor. In addition, sensor components change with age, so that the calibration will also change with age. For these reasons, most satellite sensors incorporate means of internal calibration whilst in flight.

In this section, we will deal with the issues involved in the pre-processing of image data, specifically rectification, sensor calibration and atmospheric correction.

4.3.2 Rectification

4.3.2.1 Theoretical Basis for Rectification

Rectification is the process of transforming image data so as to create an image oriented to map co-ordinates in a specific map projection. The original image pixels will contain distortions due to the geometry of the scanner, its orientation relative to the surface being imaged at the moment of acquisition and due to relief in the terrain. Because of these errors scanner image data will not be a good fit to map data, and indeed one image may not be a good fit to another image. However it is essential to be able to integrate data sets, either by registering one image to another image or to a specific map projection. It is most usual to adopt the projection used for the topographic maps of the area.

Rectification involves three processes:

1. *Correction for systematic errors.* Establish the correct internal geometry in the scanner data.
2. *Absolute orientation.* Transform the data to a specified map projection by fitting the image data to ground control and then resampling the original scanner data to create rectified image data.
3. *Windowing and Mosaicing.* Sub-setting the rectified data to create images that correspond to map areas. The resulting data is called geocoded data.

Whilst it may be possible to correct for the systematic errors prior to absolute orientation, in practice this is not done for two main reasons. The first is that the shape of the systematic error correction surface may be known, but the actual values may not be known. Thus a correction may be a second order polynomial, but the parameters in that second order polynomial may not be known. The second reason is that conducting the correction involves resampling of the data values to create new pixel values. This process would thus be done twice. The resampling is not satisfactory as both resampling processes introduce some errors into the data.

Therefore, correction for systematic errors and absolute orientation are normally done together. The conduct of these two processes normally involves five steps:

1. *Step One — The choice of the projection to be adopted for the rectification of the image.* If the rectification is to transform the image to another (reference) image, then the projection to be adopted is that of the reference image. If the rectification is to be to map co-ordinates, then the projection has to be selected, and the control co-ordinates will need to be in the co-ordinates for this projection. For this form of rectification, the projection chosen will usually be that which suits most of the GIS data that will be used with the image data. Since this will often include navigational and topographical data, the map projection used for the standard topographic maps of the area is often chosen as the most suitable projection to use in the rectification.
2. *Step Two — The derivation of the formulae for the systematic errors and hence the corrections that need to be applied.*
3. *Step Three — The development of the formulae defining the mathematical relationship between the pixels to be created in the selected projection and the pixels in the image.* This relationship will take into account the systematic errors as well as the corrections that need to be applied to correct for the orientations of the sensor relative to ground control. Such

a relationship could be a second order polynomial of the form

$$X_{old} = k_0 + k_1 \times X_{new} + k_2 \times Y_{new} + k_3 \times (X_{new})^2 + k_4 \times (Y_{new})^2 + k_5 \times X_{new} \times Y_{new}$$
$$Y_{old} = k_6 + k_7 \times X_{new} + k_8 \times Y_{new} + k_9 \times (X_{new})^2 + k_{10} \times (Y_{new})^2 + k_{11} \times X_{new} \times Y_{new} \quad (4.45)$$

where X_{old}, Y_{old} are functions of the pixel co-ordinates (P, L) that may have been converted into standard distance units and for which corrections can be made prior to adjustment with corrections for the systematic errors, and the k_i values are constants that need to be found. The equation is of this form since the actual resampling of the image data will start with specified new pixel co-ordinates. It will need to determine the location of this point in the actual, or old image co-ordinates, to determine which pixel values will be used to compute the value at the new pixel location.

4. *Step Four — The use of ground control to solve for the k_i* values in this mathematical transformation (the fitting process).
5. *Step Five — Estimation of the values in the transformed or new pixels (the resampling process).* The final step is to use this relationship to determine the pixels in the actual image data to use in a filter for computation of the new pixel response values in a process called *resampling* of the image data.

   These components will now be dealt with in turn in the next three sections.

4.3.2.2 Correction for Systematic Errors

Whilst the *internal systematic errors* vary by type and magnitude between the various sensors, they will often include:

1. *Mirror velocity errors* due to variations in the mirror velocity as it sweeps to acquire the scanlines of data if the scanner contains a moving mirror.
2. *Platform velocity error* causing displacements along the scanline as it is acquired due to the velocity of the satellite relative to the surface during acquisition for each scanline.
3. *Distortions in the objective lens* as these will incur small systematic errors in the data.
4. *Variations in the resampling rate* caused by variations in the electronics or in the spacing of the detectors in the sensor. These errors are usually very small.
5. *Detector lag* means that the response signal from the detector lags behind the incident radiation signal. Whilst the effect of this is to introduce radiometric errors in the data, it can affect the perceived positions of features in the image data thereby affecting the positions chosen for those features in the image data.

External systematic sources of error are those that are independent of the internal characteristics of the sensor. The external sources of error include:

1. *The orientation of the sensor* relative to the Earth's surface.
2. *Position of the sensor* relative to features on the Earth's surface.
3. *Rates of change* in the orientation and position of the sensor relative to the surface.
4. *Height displacements* created in the imagery due to either relief on, or Earth curvature of, the surface. With some sensors height differences can introduce significant errors whilst with others Earth curvature can be significant.
5. *Earth rotation effects* during acquisition of the data.

The mathematical shape of the internal systematic errors and the effects of Earth rotation can usually be constructed as a correction formula or formulae with known or computable parameter values.

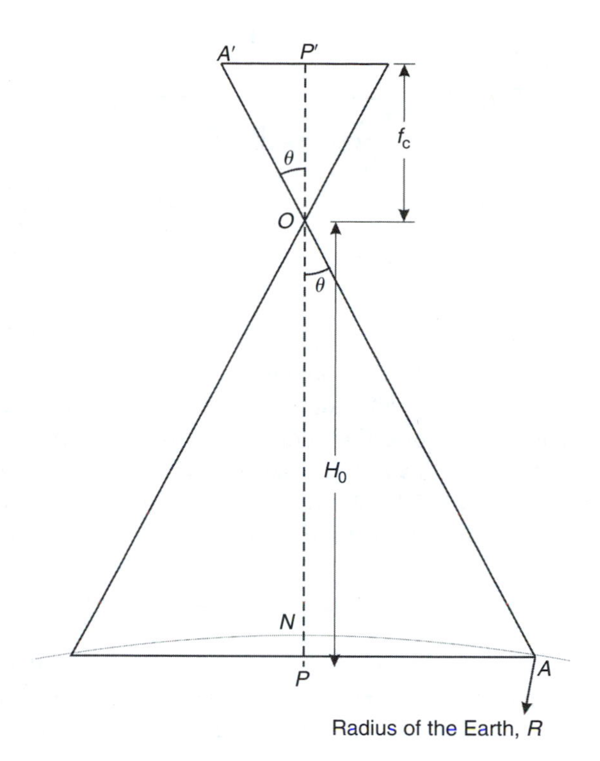

FIGURE 4.12 A simple example of the relationships between an image and the ground, typical of a camera in both directions, and of scanners in one direction, but not the other direction.

They should then be applied as a correction to the pixel co-ordinates to derive X_{old} and Y_{old}. The shape of the external systematic errors are usually known, but the parameters are not known, simply because the orientations and location of the sensor at the time of acquisition are not known accurately. When this occurs, then the shape is included in the mathematical model relating old and new image co-ordinates. Consider the case shown in Figure 4.12 where the basic geometry has to take into account Earth curvature.

Point A on the Earth surface is recorded at A' in the image. The sensor records the distance AP, whereas the correct distance is around the curve NA. An error of (NA − AP) is thus created due to Earth curvature:

$$\text{Error} = \text{NA} - \text{AP} = R \times \phi - R \times \sin(\phi) = R \times \phi - R \times \{\phi - (\phi^3/6) + (\phi^5/120) - \cdots) \tag{4.46}$$

where ϕ is the angle subtended at the centre of the Earth. When ϕ is small then only the first two terms in the expansion of the sine function need to be considered. Thus:

$$\text{Error} = R \times \phi^3/6 \tag{4.47}$$

The radius of the Earth, R, is known and θ can be calculated from the number of pixels out from the centreline of the scanline of data. The error can thus be corrected so as to give:

$$X_{\text{old}} = \text{Pixel} - R \times \phi^3/6 \tag{4.48}$$

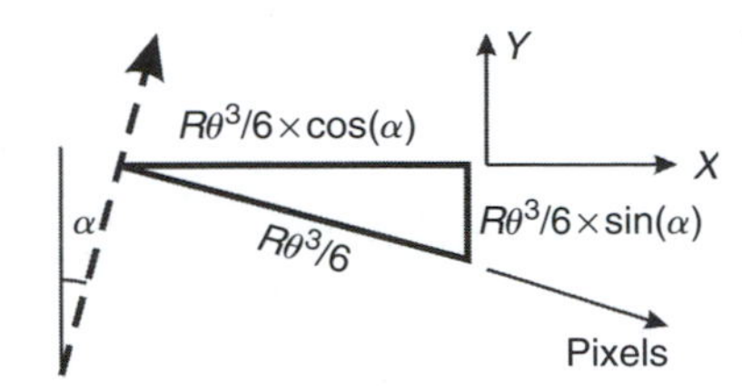

FIGURE 4.13 Errors in the x and y directions when the sensor orientation is at a bearing to these axes.

if the pixel orientation coincides with the x-axis direction. Normally this does not, and so the correction has to be applied to the x and the y co-ordinates, as shown in Figure 4.13. In this figure the platform orbit is at a bearing of α to the north direction, and so the error in the x-direction is equal to $R\theta^3 \cos(\alpha)/6$ and the error in the y-direction is $R\theta^3 \sin(\alpha)/6$. These errors will be taken from the pixel and line co-ordinates to give X_{old} and Y_{old} in accordance with Equation (4.42). However they will not be exact, and so any residual will be of the same form, and can be corrected in the least squares adjustment if a third order polynomial is included in the adjustment, as shown in Equation (4.49).

$$X_{\text{old}} = X_{\text{new}} - k \times (X_{\text{new}})^3 \tag{4.49}$$

This is a very simple example that uses one of the various sources of error that may exist in the data being used. Even in this situation the scanner will not be perfectly vertical, so that an orientation about the vertical axes has also to be taken into account in the adjustment. There will also be differences in elevation that need to be taken into account.

Theoretically, once the data are corrected for the internal and external errors, the image data should be a perfect fit to ground detail. In practice this will not be the case because:

1. The actual nature of the systematic errors is not known precisely, particularly as some may change with time and with environmental conditions.
2. The final effect of many small errors will not be known where these errors can accumulate or cancel, with possibly both occurring in different parts of the image.

The effect of unknown errors cannot be known. Generally these will be very small and so can be treated as for two above.

In most image processing systems the different geometric models, representing the different types of sensors, are available to the analyst. The analyst chooses the appropriate model, and the systematic errors relevant to the model are then automatically incorporated into the rectification process. Such geometric models will include the camera, the scanner and radar. However, there are important differences between some of the scanner geometric models, requiring a number of models for the different scanner systems.

4.3.2.3 Fitting Image Data to Ground Control

Once models of the internal systematic errors have been incorporated into the mathematical transformation, it is necessary to correct for the external sources of error by fitting the image data to ground control. Fitting image data to ground control is usually done by the use of the least squares technique. Least squares uses all of the equations to find solutions for the unknowns in accordance with the criterion that the sums of squares are minimised. It will be recalled that the variance of a set of observations has a minimum value when the mean is used in its calculation. If another value is used, the derived variance will be larger. Computation of the mean and variance of a set of observations of a quantity is thus compatible with least squares adjustment.

It has been noted that the least squares technique assumes that the observed data are normally distributed. The residuals will also be normally distributed. As this is not usually the case with systematic errors, it is a good reason why models of the systematic errors need to be included as corrections to the image co-ordinates prior to the actual least squares adjustment. Once the systematic corrections have been identified, then they need to be incorporated into the selected model. This transformation is usually a polynomial equation of the form:

First order

$$P_x = a_0 + b_0E + c_0N$$
$$P_y = a_1 + b_1E + c_1N \tag{4.50}$$

where P_x, P_y are the pixel co-ordinates and (E, N) are the map co-ordinates in Eastings and Northings. Map co-ordinates are dealt with in Chapter 6.

A first order polynomial contains six unknowns. A minimum of three points needs to be identified in both the image and on the ground to solve the equations. In practice a lot more points should be used and the unknowns determined by use of least squares.

Second order

$$P_x = a_0 + b_0E + c_0N + d_0EN + e_0(E)^2 + f_0(N)^2$$
$$P_y = a_1 + b_1E + c_1N + d_1EN + e_1(E)^2 + f_1(N)^2 \tag{4.51}$$

The second order polynomial requires a minimum of six control points to solve the 12 unknowns. A second order polynomial is the most common form used in rectification. A third or higher order polynomial can be used, having more unknowns and requiring more control points for their solution. Generally, the higher the order of the polynomial the smaller the residuals as the curve can provide a better fit to the data. However, the trade-off is that the polynomial can also fit distortions in the data caused by errors in the control point identification or co-ordinates.

Many of the known systematic errors can be described by a second order polynomial function. In consequence, a second order polynomial will usually provide a good fit between image and ground control once the other systematic errors, that is, those that have a shape that is not a second order polynomial function, have been corrected. There is only one major geometric error in Landsat data that is not a reasonable fit to a second order polynomial: error introduced due to variable mirror velocity. The only major errors in the SPOT data that are not a reasonable fit to a second order polynomial are those due to height differences in the terrain being imaged. A second order polynomial is the recommended correction surface for most types of image data.

Using ground control points to solve the polynomial equations by means of least squares derives the parameters in Equation (4.50) or (4.51). Control points are identified in the image and on the ground or on a map, with the points being distributed evenly across the area of the image or that part that is to be used. If the transformation contains N unknowns then a minimum of $N/2$ points need to be identified. However, for least squares a minimum of $5 \times N/2$ control points should be used. For a second order polynomial of 12 unknowns, at least 30 points should be identified and used to calculate the parameters in the transformation.

Control points that are clearly identifiable on both the image and the map must be used. One difficulty in control identification is the influence of the pixel structure on the images of ground features. This pixel structure often makes it difficult to accurately identify point features on the image. The best solution is to use intersections of linear features if they exist, or points on a linear feature where they do not. Thus road intersections are good control points, particularly if the analyst can interpolate along the linear features and use the intersections of these interpolations as the control points. The intersections will not usually coincide with the centres of pixels in the image data, so that the co-ordinates of the control points need to be determined to the sub-pixel level, usually to one decimal place, and to the equivalent resolution in the map data.

The least squares adjustment will provide estimates of the transformation parameters that minimise the variances of the residuals. These residuals can be printed as shown in Table 4.3 and plotted as shown in Figure 4.14. When plotting the residuals, plot the co-ordinates of the control points at one scale, and then plot the residuals at a much larger scale, so that they can be seen on the plot. The analyst should inspect these residuals to identify large individual residual values, as these will usually indicate errors in identification of a control point or calculation of the co-ordinates in either the map or the image. A plot of this type quickly indicates patterns in the residuals that may be due to misidentification of control or the existence of systematic errors that have not been adequately taken into account. Once the analyst is satisfied that the magnitude of both the individual residuals and their sums of squares are sufficiently small, then resampling can be implemented.

TABLE 4.3
Map, Adjusted Co-ordinates and Residuals in *X* and *Y* (Metres) after Rectification of an Image

Point	Map *X*	Map *Y*	Adjusted *X*	Adjusted *Y*	Residual in *X*	Residual in *Y*
1	486375.02	1274149.64	486376.00	1274148.62	+0.98	−1.02
2	496429.36	1273092.21	496429.65	1273093.44	+0.29	+1.23
3	507492.29	1269877.48	507491.78	1269878.35	−0.51	+0.87
4	483621.05	1264419.27	483619.58	1264420.50	−1.47	+1.23
5	493782.51	1261522.37	493783.10	1261522.12	+0.59	−0.25
6	504989.96	1259567.09	504990.49	1259565.98	+0.53	−1.11
7	481523.35	1254622.86	481524.41	1254623.06	+1.06	+0.20
8	491627.71	1249633.00	491627.24	1249632.28	−0.47	−0.72
9	502828.60	1249705.62	502827.58	1249705.19	−1.02	−0.43

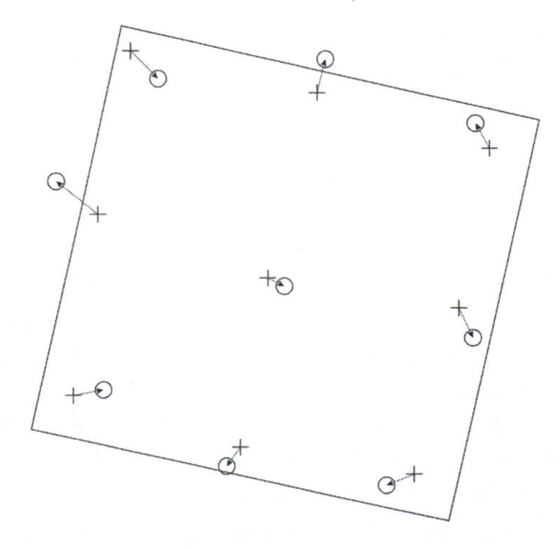

FIGURE 4.14 Plot of residuals after least squares adjustment and given in Table 4.3. The plot scale for the residuals is much larger than the plot scale for the control points. A pattern in the magnitude and direction of these residuals would be of concern to the analyst.

4.3.2.4 Resampling the Image Data

The pixels in the original image will be in a different position, orientation and of a different size to the pixels to be derived for the rectified image as shown in Figure 4.15. Resampling is the process of calculating the response values for the pixels in the rectified image by use of the response values in the original image data. The new pixels can overlap a number of the pixels in the original data, the number depending on the size and position of the new pixels relative to the original pixels. Usually the rectified pixels will be of sizes similar to those given in Table 4.4. Each size is generally an integer multiple of each smaller size, making the merging of data sets with different sized pixels much easier and faster for data analysis.

Resampling involves selection of the pixels in the original image that will influence the resampling process and then calculation of the new pixel response using the response values of these influencing pixels and the location of the new pixel relative to the location of the old pixels. The image-to-control transformation is used to calculate the co-ordinates of each new pixel in turn in the original data, deriving fractional image co-ordinates for each pixel. Knowing the original co-ordinates for the new pixel allows the resampling process to identify those pixels whose response values will be used to calculate the response of the new pixel. Once the response has been calculated it is entered as the response value for the new pixel and processing proceeds to the next new pixel. There are three commonly used methods of calculating the response of the new pixel from the original image

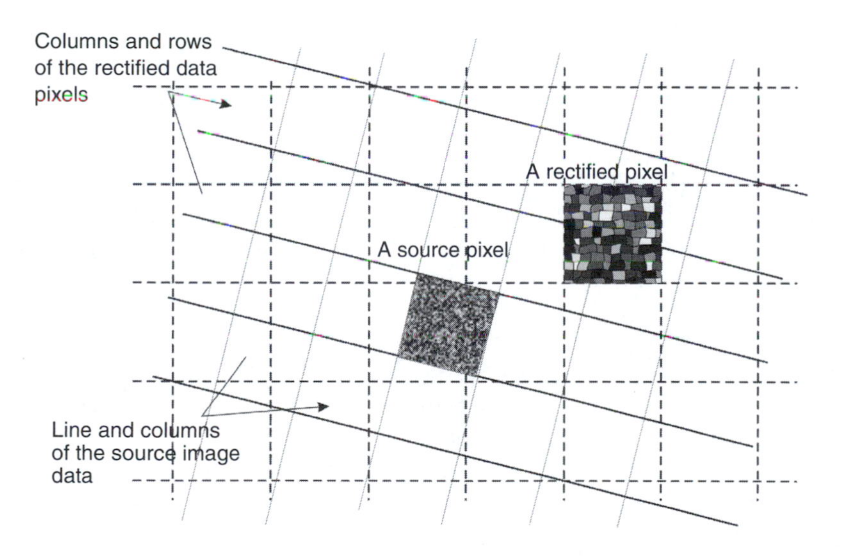

FIGURE 4.15 Relationship between the source pixels in the image data and the rectified pixels.

TABLE 4.4
Typical Rectified Pixel Sizes for Different Sensors

Sensor	Original Pixel Size	Resampled Pixel Size Typically Used in Practice
AVHRR	1.1 km by 1.1 km	1000 m by 1000 m
TM	30 m by 30 m	25 m by 25 m
Spot-XS	20 m by 20 m	25 m by 25 m or 20 m by 20 m
Ikonos MS	4.6 m by 4.6 m	5 m by 5 m
Quickbird Pan	0.6 m by 0.6 m	0.5 m by 0.5 m

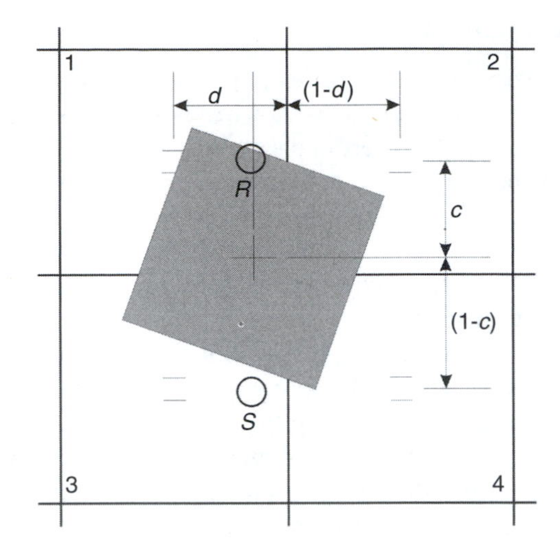

FIGURE 4.16 Relationship between the source pixels and the rectified pixels in the bi-linear interpolation.

data values:

1. *Nearest neighbour*
2. *Bi-linear interpolation*
3. *Cubic convolution*

4.3.2.4.1 Nearest Neighbour

The value of the nearest original pixel to the required pixel location is adopted as the response value for the new pixel. It is fast and cheap, appropriate for theme or classified image data, but is generally not considered to be suitable for image data.

4.3.2.4.2 Bi-Linear Interpolation

A linear interpolation uses the two pairs of pixels of (1 and 2), (3 and 4) in Figure 4.16 and the proportional distances of the new pixel centre between these original pixel co-ordinates to calculate the intermediate response values from the equations:

$$R_\mathrm{r} = R_3 \times (1 - d) + R_4 \times d$$
$$R_\mathrm{s} = R_1 \times (1 - d) + R_2 \times d \tag{4.52}$$

A second linear interpolation computes the response between these intermediate response values R_r and R_s using the proportional distances, c and $(1 - c)$, to the new pixel location.

$$R_{\mathrm{New}} = R_\mathrm{r} \times c + R_\mathrm{s} \times (1 - c) \tag{4.53}$$

These equations can be combined to give:

$$R_N = R_1 + cd \times (R_1 + R_4 - R_2 - R_3) + c \times (R_3 - R_1) + d \times (R_2 - R_1) \tag{4.54}$$

4.3.2.4.3 Cubic Convolution

Sampling theory states that a continuous signal can be sampled at discrete intervals and the sinc function $(\sin(x)/x)$ can be applied to the sampled data to completely reconstruct the continuous

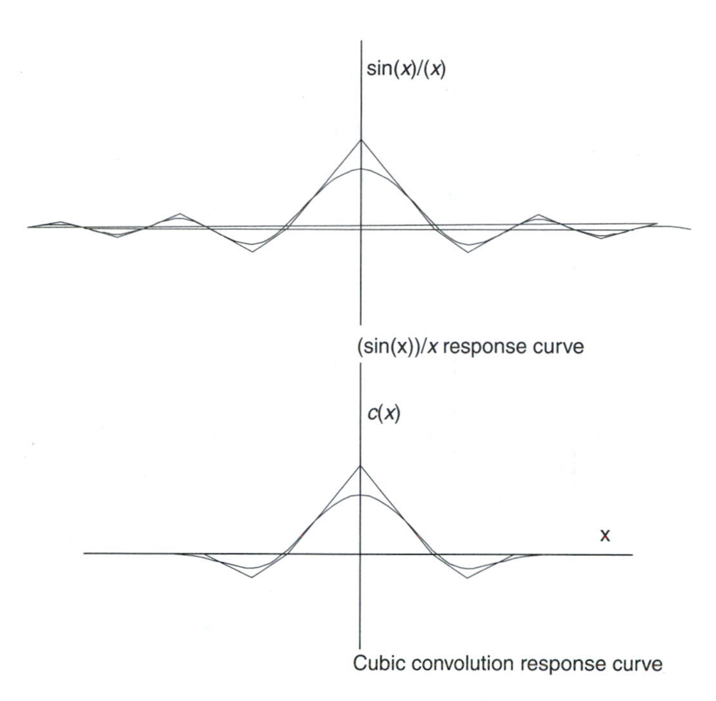

FIGURE 4.17 The $\sin(\pi \times x)/\pi \times x$ and cubic convolution filters.

signal, as long as the sampling frequency obeys Shannon's sampling theorem (Section 4.2). The cubic convolution filter is an approximation of the sinc function in which the main positive and the first negative lobes are retained as shown in Figure 4.17.

Cubic convolution is accomplished using a 4×4 box of pixels surrounding the location of the new pixel in the original data. A vertical axis through the new pixel location intersects lines along the four scanlines of data as shown in Figure 4.18. Interpolation values are computed for these four intersections using the data along each scanline in turn in the cubic convolution algorithm. Finally, the response for the new pixel is calculated using the cubic convolution algorithm with the four interpolated values as a transect at right angles to the scanlines. The interpolation formula for each of the four lines and for the final vertical line is

$$\begin{aligned} R_i = {} & R_1 \times \{4 - 8 \times (1+d) + 5 \times (1+d)^2 - (1+d)^3\} \\ & + R_2 \times \{1 - 2 \times d^2 + d^3\} + R_3 \times \{1 - 2 \times (1-d)^2 + (1-d)^3\} \\ & + R_4 \times \{4 - 8 \times (2-d) + 5 \times (2-d)^2 - (2-d)^3\} \end{aligned} \tag{4.55}$$

where R_1, R_2, ..., etc are numbered in accordance with Figure 4.19, and d is a measure of distance between pixel centres as shown in Figure 4.19. In practice the assumptions on which the cubic convolution is based are rarely met because the radiance from the surface frequently contains higher frequency components than those recorded in the image data. This means that the requirements of Shannon's sampling theorem are usually not met and so the resampling cannot properly reconstruct the data. Despite this limitation and the higher processing cost associated with use of the cubic convolution, it is the most commonly used method of resampling image data.

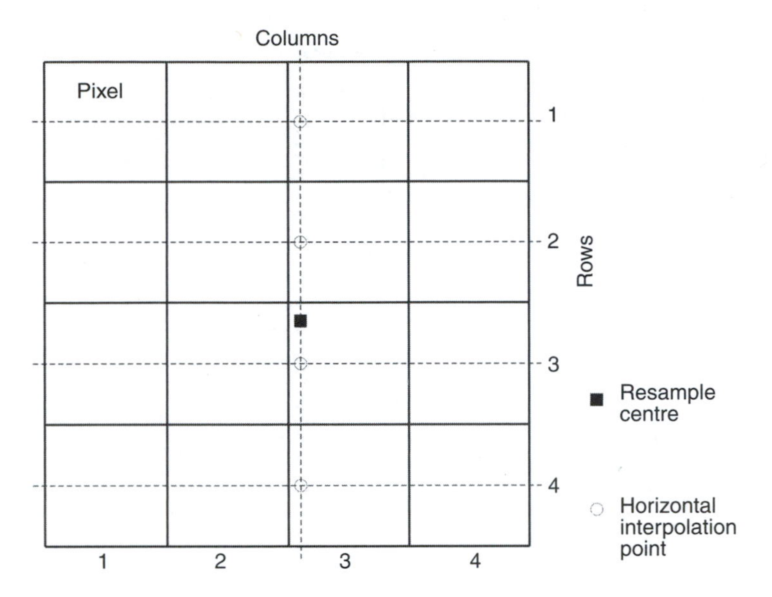

FIGURE 4.18 The interpolation axes for cubic convolution resampling.

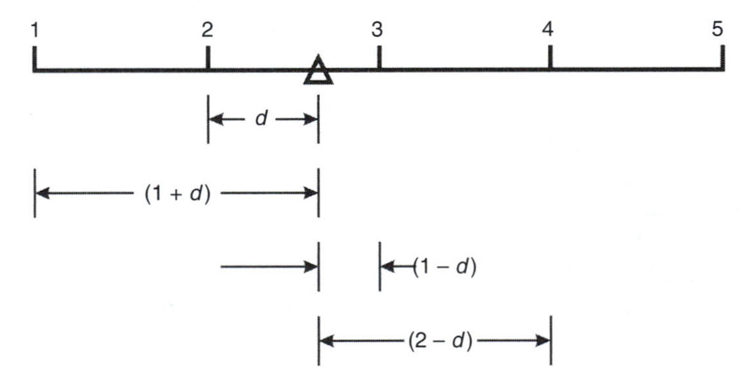

FIGURE 4.19 Relationships in using the cubic convolution interpolation equation.

4.3.2.5 Windowing and Mosaicing

The resampled image data will have pixels of a specified size and interval oriented along the co-ordinate axes of the map co-ordinate system, that is, the lines of data will be parallel to the x co-ordinate axis and the paths will be parallel to the y-axis. The rectangular area coverage created during resampling will be partially or wholly covered by image data; some areas within the rectangle may not be covered by image data and will be assigned the default value used for this condition. In char (byte) data such pixels are normally assigned 0 or 255, that is, black or white, in the rectified image data. The rectangle may not, however, coincide with the coverage of the topographic maps of the area. For many operational purposes it is better to create image areas that coincide with map areas. This can be done by mosaicing together image files to cover more than the map area and then defining a window subset of the data that coincides with the map area as is discussed in Chapter 6 on GIS.

Map areas have boundaries that are constant lines of longitude and parallels of latitude. These boundaries do not provide a straight edge parallel to the map co-ordinates. The solution is to create a mask of the map boundaries by digitising them, and using this mask in a GIS to create an image file of the data within the mask. As the mask will often cover two or more images, the procedure is to:

1. *Mosaic the relevant files* to cover the whole area of interest.
2. *Window this mosaic file* using a mask so as to create image files that correspond to map coverages.

4.3.2.6 Rectification in Practice

Rectification can be done in one of two ways: image to map and image to image. In image-to-map rectification, the user identifies control points in the image and in map co-ordinates. The map co-ordinates can be points that are either identified on a map or determined using global positioning systems (GPSs). In image-to-image rectification, one image is first rectified by image-to-map rectification. This rectified image then provides the control for the subsequent rectification of the other images. In general, image-to-map rectification should provide better absolute accuracy, since the ground control used should be much more accurate than image co-ordinates. However, it is sometimes difficult to identify features on the image that can be either identified on a map or that are accessible for GPS measurements. Image-to-image rectification is often easier to accomplish because it is usually easier to identify sufficient points on both images to use as control points. However, the co-ordinates of the control points are not more accurate than the image co-ordinates, so that absolute rectification accuracy will usually be lower.

Most image processing software systems incorporate models for the more common sensor geometric model systems. The user thus needs to specify the sensor used to acquire the data. The user then needs to select ground control points that can be identified on the image. The co-ordinates of these control points are determined either from a map, by use of GPS, or from the rectified image. The co-ordinates should be distributed in a systematic way across the image. If a second order polynomial surface is to be used, then a minimum of six control points are required to solve for the unknowns. In practice many more points should be used. For a second order polynomial, it is recommended that 25 or more points be used where these points could be distributed in a similar manner to that shown in Figure 4.20.

The ground co-ordinates would normally be in the same co-ordinate system as that used for the topographic maps of the area. They should use the same spheroid and the same map projection. Projections were introduced and discussed in Chapter 3. Features usually are not recorded with sharp boundaries in image data due to pixels overlapping the boundaries as shown in Figure 4.21 depicting a river or road junction.

The issue is: where should you locate the ground control point? If the software being used allows map detail to be superimposed on the image, then superimposition as shown here will allow the analyst to establish a best fit. However, this is no help when GPS is being used. In this situation the analyst needs to identify features in the image first, then get the co-ordinates of those features. Linear features are usually suitable since the analyst can visually fit a line to the edge or the centre of the feature, and in this way find the intersection of either two edges or two features. In the above case the analyst may have estimated the centrelines as shown, giving a point of intersection of the two as the control point. A difficulty that arises in this case is to find this point on the ground so as to take the GPS observations.

Sharply defined features that provide good corners include buildings, fields, airport runways, water tanks and dams. Features that can provide good intersections include road junctions, creek or river junctions and bridges over rivers.

Once the control points have been identified in the image and ground co-ordinate systems, the two are used to fit the model using least squares. This results in a least squares best fit of the two, with residuals for each control point. The residuals will be of the form shown in Table 4.3. The residuals should be randomly distributed in both size and direction, and the sums of squares should be within

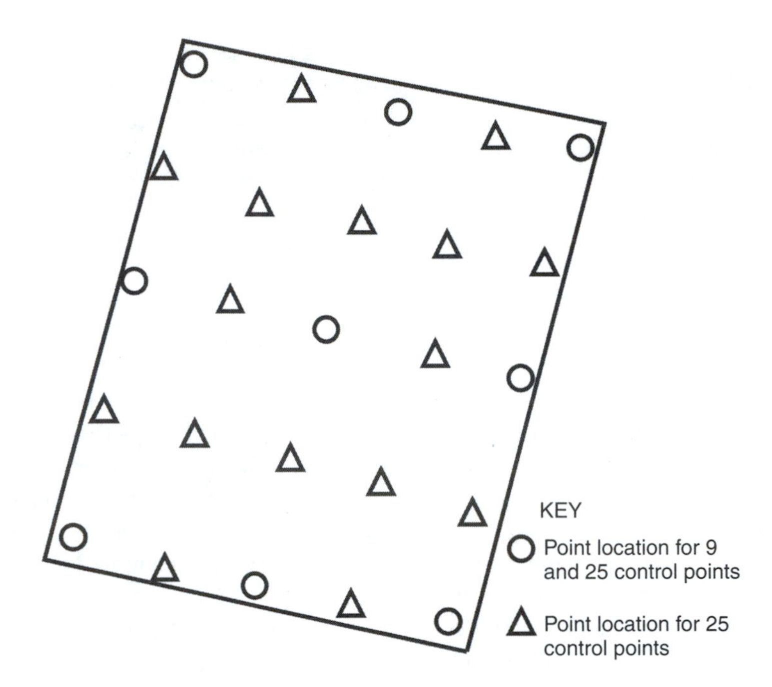

FIGURE 4.20 Distribution of 9 and 25 ground control points for image rectification.

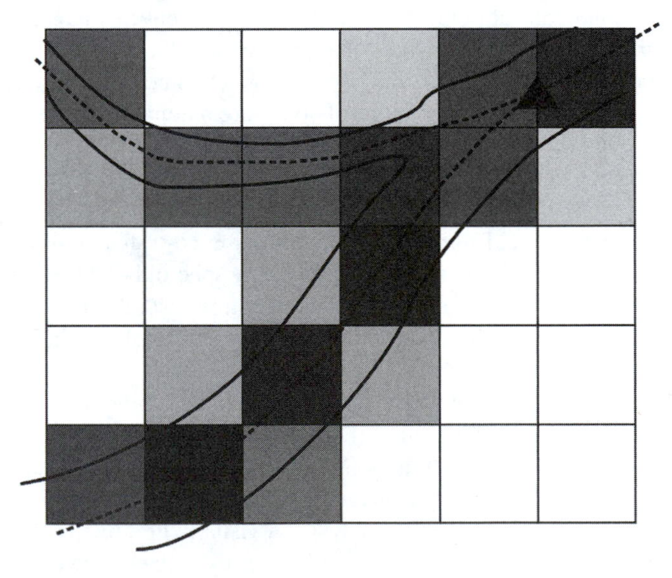

FIGURE 4.21 Pixel tones and their relationship to ground detail.

the expected range of values. The square root of the (sums of squares/number of observations) should be about ± half pixel in size. If they are significantly larger than this or if there appears to be a pattern to the residuals, then the analyst should look for a cause. The first place to look at is the pattern in the residuals. Consider the example shown in Figure 4.22. The least squares method will minimise the sums of squares of the residuals. The square of a large number is much larger than the square of a small number, so least squares will give a number of small residuals rather than one large residual. Figure 4.22 shows that a large error at one control point, in this case centred in the figure, will result

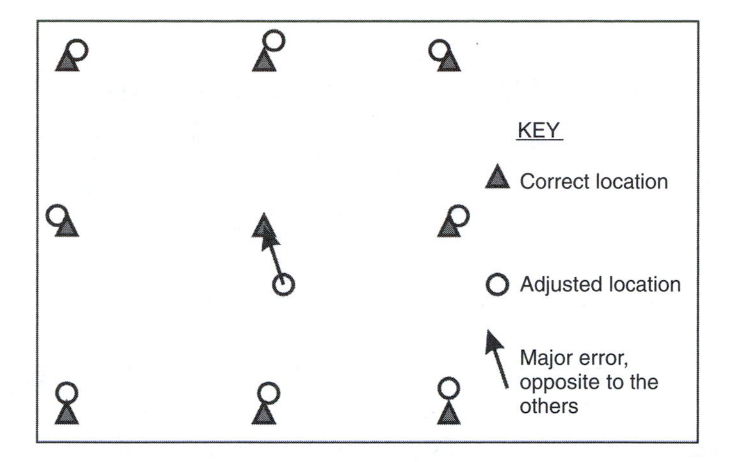

FIGURE 4.22 Distribution of residuals due to an error in the location of one control point.

in a relatively large correction at that point and smaller corrections that are opposite in sign at the other control points in the vicinity of this control point. There is thus a pattern of one large correction in one direction, countered by a number of smaller corrections at adjacent points.

If the analyst finds such patterns, then the offending control point should be checked for errors in identification and in the co-ordinates used in the adjustment, and the fitting conducted again. There are a number of causes of errors in the co-ordinates of control points:

1. *Misreading or mis-transferring map co-ordinates* into the fitting process.
2. *Misidentification of the control point in the image.*
3. *Misidentification of the ground control point.*

The analyst should avoid just removing control points from the fitting process. If the analyst cannot resolve the problems that seem to beset a control point, then it is possible that the problem is in one of the other control points. It may be necessary to reject that point, and identify another point. For example, an area used may appear to be the same point in both the image and the map, but they may be different points entirely. Road junctions can be moved, and river intersections can change.

Once a satisfactory fitting has been achieved, then the analyst conducts the resampling. In general, cubic convolution should be used for image data, whilst nearest neighbour is usually adequate for classified data. Many analysts prefer to classify the source image, fit using the source image and then resample using the classified image.

4.3.3 Radiometric Calibration

The energy incident on the sensor passes through the collecting optics to the detector. The detector converts the incident energy into an analog electrical signal in accordance with its response function, and subject to a lag or delay. The detector response function will normally be non-linear. This electrical signal is then sampled to create digital data in an analog to digital (A/D) converter. The A/D converter accepts input data over a power range and converts these to digital values over a second range. This conversion is usually a linear process. The second range depends on the number of bits that are used to store the digital data.

In all sensors the detector sensitivity varies with environmental conditions, particularly temperature, and age of the detector. For this reason, sensor calibration needs to be conducted on a regular basis, and the calibration of image data needs to use the relevant calibration data.

TABLE 4.5
In-Flight Calibration Constants for Landsat TM 4

Band	α [$Wm^{-2}sr^{-1}\upsilon m^{-1}R^{-1}$]	β [$Wm^{-2}sr^{-1}\upsilon m^{-1}R^{-1}$]
1	0.602	11.4
2	1.174	7.50
3	0.806	3.99
4	0.816	3.82
5	0.108	−0.38
7	0.057	−0.15

The owner of a satellite system should supply the information necessary to calibrate image data to radiance incident on the sensor, where this relationship will be of the form:

$$\text{Radiance} = \text{function(digital count, calibration factors)}$$

The shape of the calibration curve and its parameters should be supplied as part of the calibration factors. For example, Landsat TM data is calibrated using the linear relationship:

$$\text{Radiance} = (\text{digital count}) \times \alpha + \beta \tag{4.56}$$

where α and β are constants that can change over time, and will change from sensor to sensor. The values for α and β will be similar to those given in Table 4.5. With most image processing systems the user can create scripts that will take the image data and pass it through a user created filter or transformation to compute the required response values. This linear transformation is simple to accomplish in such script languages on a band-by-band basis. With most of these languages, the table could be input, and all of the bands transformed in the one step to compute radiance at the sensor. This should be done prior to atmospheric calibration. Since it does not involve resampling of the pixels, but simply changes each pixel value independently of its neighbours, it can be done either before or after rectification.

4.3.4 ATMOSPHERIC CORRECTION

The purpose of correcting for atmospheric absorption and scattering is to provide data that is capable of being used to provide estimates of surface condition, and to provide temporal image data sets that are internally consistent. As we know, variations in image data values are translated in estimation into estimates of surface condition. It follows that errors in the satellite data will give errors in the derived estimates.

The most useful basis for estimation is the use of reflectance data, rather than radiance, since reflectance is independent of variations in the incident radiation budget, atmospheric path length and the effects of the atmosphere on the incident radiation, and it can be estimated from ground instruments. Reflectance is not independent of non-Lambertian effects.

The amount of energy incident on a defined area such as a pixel, is a function of the angle of elevation of the sun as shown in Figure 4.23. As the angle of elevation, θ, increases, the cross-sectional area projected in the direction of the sun decreases relative to the ground area and, in consequence, the amount of energy available to illuminate the pixel area is lower. The relative dimensions of these two areas are a function of the cosine of the solar elevation, so that satellite data can be corrected for this effect by dividing the radiance data incident on the sensor by cos(solar elevation). Whilst this

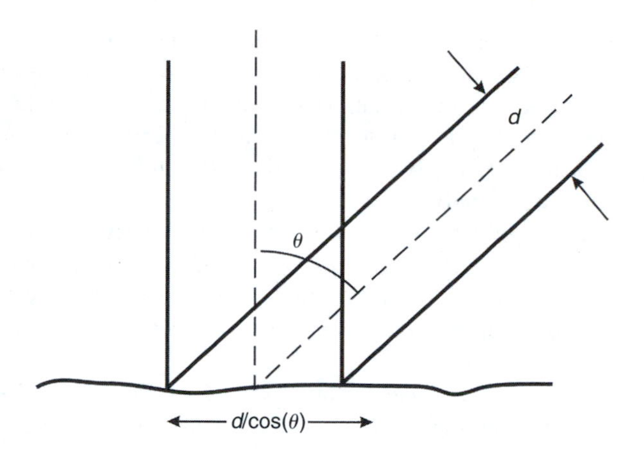

FIGURE 4.23 Relationship between solar elevation and energy incident on the target area.

transformation will correct for variations in incident solar radiation due to changes in elevation, it does not correct for variations in:

1. *Radiance due to changes in atmospheric path length.*
2. *Absorption and scattering in that path length.*
3. *Bi-directional reflectance variations*, introduced as a function of the solar incident angle.

The effects of variations in atmospheric conditions on satellite data, and the potential impact on image processing, particularly of temporal data sets, can be very great. In Chapter 2 we showed that the energy incident on a scanner:

$$L_{d,\lambda} = \left[e_\lambda t'_\lambda t_\lambda \rho_\lambda E_{s,\lambda} \cos(\theta) + e_\lambda t'_\lambda \rho_\lambda E_{A,\lambda} + e_\lambda E_{a,\lambda}\right]/\pi \tag{2.15}$$

which can be simplified by using radiance incident on the surface rather than radiance at the top of the atmosphere, to give

$$L_{d,\lambda} = e_\lambda\left[t'_\lambda \rho_\lambda E_{i,\lambda} + E_{a,\lambda}\right]/\pi \tag{2.16}$$

Assuming $e_\lambda = 1.0$, Equation (2.16) can be rearranged to give

$$\rho_\lambda = \left\{\pi L_{d,\lambda} - E_{a,\lambda}\right\}/\left(t'_\lambda E_{i,\lambda}\right) \tag{2.18}$$

Now the scanner response values, R_λ, should be calibrated prior to correction for atmospheric effects so that $L_{d,\lambda} = f(R_\lambda)$. The functional relationship between incident radiance at the sensor and satellite response values is usually linear, so that in this case we can say:

$$L_{d,\lambda} = k_1 R_\lambda + k_2 \tag{4.57}$$

If t'_λ, $E_{i,\lambda}$, $E_{a,\lambda}$, k_1 and k_2 are assumed to be constant, then Equation (2.18) can be simplified to:

$$\rho_\lambda = c_1 R_\lambda + c_2 \tag{4.58}$$

where ρ_λ is surface bi-directional reflectance, R_λ is the calibrated sensor radiance data and c_1 and c_2 are constants related to radiance from the surface and the atmosphere, respectively.

Equation 4.58 shows that a linear model is valid for atmospheric correction if the atmosphere is assumed to be constant across the image. In practice, t'_λ and $E_{a,\lambda}$ are rarely constant across the image, although the other factors are usually constant. They will vary due to variations in proportions of the atmospheric constituents, particularly water vapour and suspended solids or particulate matter. Variations in absorption and scattering by atmospheric gases and ozone generally follow global models, as local fluctuations are usually small.

Water vapour can exhibit significant local variations. There are significant moisture absorption bands within the wavelength ranges that are used in remote sensing systems. Sensors that record data in these bands, and on their shoulders, can be used to estimate the absorption effects of atmospheric moisture at each pixel. This data can be used to improve the corrections that are applied by the global models.

Variations in particulate matter are usually due to local sources. These local sources can be smoke, organic and chemical emissions and dust. In addition to the range of spectral conditions that can exist among these different sources, the particulate matter may undergo changes in the atmosphere, as the particles absorb moisture, changing their reflectance, and agglomerate into larger particles with different shape characteristics, further modifying the reflectance characteristics of the aerosols. As a consequence, the strategy of using an absorption band to estimate aerosol density in the atmosphere is not practical, unlike the case of water vapour. At present there is no established method for the correction of the effects of atmospheric aerosols on image data, other than the use of standard atmospheric conditions, or the adoption of aerosol loads based on physical measurements as being representative across the image. Such physical estimates of aerosol load can be based on sun photometer observations. A sun photometer records the spectral characteristics of the radiation incident at the surface, by accurately tracking the sun and recording the incident radiation in a set of very specific and highly calibrated wavebands of data. Since the radiation at the top of the atmosphere can also be accurately measured, the differences between the two can be attributed to atmospheric absorption and scattering in the path length. Once these data are corrected for the effects of the atmospheric chemical constituents, then the data can be used to estimate ozone concentrations, water vapour density and aerosol load.

Currently there is a network of sun photometers distributed around the globe, from which estimates of aerosol load are being estimated. Such data are an essential component in the development of a strategy for the correction of image data for the effects of aerosol absorption and scattering. However, by providing point observations of a spatially dynamic phenomenon, they are insufficient for the accurate correction of image data.

There are three avenues for correcting image for atmospheric effects:

1. *The use of the linear model*
2. *The use of global atmospheric correction models* with field data or the application of standard atmospheres
3. *The use of sensor data* to estimate the impact of atmospheric components, for use in global atmospheric models

4.3.4.1 The Use of a Linear Model for Atmospheric Correction

The linear model is of the form given in Equation (4.58):

$$\rho_\lambda = c_1 R_\lambda + c_2 \tag{4.58}$$

It has been shown that a linear model is a reasonable approximation for the relationship between calibrated image radiance and reflectance at a point, but that this linear relationship can vary across an image and between images as atmospheric conditions change. If the satellite response values are linearly related to radiance, then a linear model is a reasonable approximation of the relationship between response and reflectance at a point.

The optimum solution would be the ability to develop such a linear relationship for each pixel in the image. Since this is currently impractical, the next best solution is the establishment of a network of linear models across the image and interpolation for the linear model parameters between these points so as to derive linear model parameters for each pixel.

To derive the two parameters of the linear model, it is necessary to use two surfaces of very different reflectance characteristics that are in close proximity to each other, to derive the linear model at that location in the image. The image response and surface reflectance of these two surfaces are used to estimate the linear model parameters. The selected surfaces thus need to be large enough to be recorded in the image data, and they need to be stable in reflectance characteristics, since it is critical that they do not change significantly in reflectance between the time of field observation of reflectance and satellite overpass. Ideally the surfaces should be near to the extremes of reflectance in the wavebands to be corrected. There are three approaches to the determination of surface reflectance values:

1. *Image-to-image correction*
2. *Correction using field spectrometer observations*
3. *Correction using panels of known reflectance*

4.3.4.1.1 Image-to-Image Correction

A minimum of two points, with low- and high-response values in the image data, and considered to be of constant radiance in both the control image and the image to be corrected, are selected. Data values for these two points are determined in both images, and used to compute the gain and offset values in a linear transformation. The linear transformation is applied to the second image so that the data for both images occupy the same data space. The control image can be a standard image that is nominated as the control, or may be an image calibrated to reflectance using the techniques described below.

This method of correction is relatively simple to use. However, it assumes that sufficient points can be found and that the radiance of these points has not changed significantly between the images.

4.3.4.1.2 Correction Using Field Spectrometer Observations

From prior images, pairs of high- and low-reflectance targets that are both large enough and stable in reflectance are chosen. The reflectance of these sites is determined using field spectrometers to sample across the extent of the sites, using techniques described in Chapter 5. From this, average site reflectance values for the dark and the light surfaces are determined. Training areas are used in these sites to determine the satellite response or radiance values for each surface type; the size of the training area is discussed in Chapter 5. The interval between the field observations and the satellite overpass depends on the stability of the reflectance characteristics of the sites. If the sites exhibit significant diurnal variability, then the field data will also need to be collected at the same solar time as the satellite overpass. Both sets of data are then used to derive the linear correction model at that location.

4.3.4.1.3 Correction Using Control Panels of Known Reflectance

Conceptually this is identical to the above method. The difference is only that instead of using natural features, control panels of known reflectance are placed out in the image. The advantage of this is that these panels can be placed out in a regular pattern, can include high- and low-reflectance surfaces at each location and that there is no need to take spectrometer readings during or near to overpass. The disadvantage of the approach has been the large size panels required. The panels need to be much larger than pixel size, for the reasons discussed in Chapter 5. However, the advent of high-resolution satellite image data means that this approach is becoming more realistic.

The density and distribution of the chemical constituents of the atmosphere are relatively stable and uniform across an image. In addition, the atmospheric water vapour can be estimated using satellite data, so that if this information is available, then they can be most readily used in atmospheric models. The linear model thus has a role when this data are not available, the provision of a means of correcting for aerosol load, recognising that the method cannot adequately deal with point emissions, and the verification of the accuracy of other correction methods.

The major issue here continues to be that of aerosol absorption and scattering when there are significant local or introduced loads that cannot be specified sufficiently for them to be corrected by the use of atmospheric models. Under these conditions the linear model approach provides a mechanism for both correcting the data and arriving at an estimate of the quality of the corrected reflectance data.

4.3.4.2 Atmospheric Correction Using Atmospheric Models

Equation 2.15 can be solved if the various atmospheric parameters can be either estimated or assumed. The atmospheric effects that need to be considered include:

1. *Molecular absorption and scattering*
2. *Ozone absorption*
3. *Water vapour absorption and scattering*
4. *Aerosol absorption and scattering*

Of these, the effects of molecular absorption and scattering are quite stable and so they can be corrected quite readily using standard models. Ozone is also relatively stable, and the density of ozone in the atmosphere can be measured by satellite; these measurements can be used as the basis for estimation of the ozone absorption effects. Atmospheric moisture and aerosols are much more variable both spatially and temporally. Water vapour has constant refraction and reflection characteristics; however, the effect of water vapour on electromagnetic radiation depends on the density of the water vapour in the atmosphere and the size distribution of the droplets. Both can be highly variable.

Atmospheric moisture is dependent on the atmospheric conditions, the drying potential of the air and the availability of moisture. It can also thus change very dramatically with both time and space. However, this problem is easier to solve than that of dust as the reflectance characteristics of the moisture are known and are stable. Since the moisture absorption bands are known, the normal strategy to correct for atmospheric moisture is to select a part of the spectrum where there are little variations due to most known ground surface types or atmospheric aerosols, but which contains a (preferably narrow) moisture absorption band.

The energy is recorded in three wavebands, two on the shoulders of the moisture absorption band and one in the centre of the moisture absorption band. If three bands, as indicated in Figure 4.24 are measured as R_1, R_2 and R_3 respectively, in the NIR to mid infrared, then aerosols and ground reflectance will have a similar effect on all three wavebands so that the trough is almost entirely due to moisture absorption. Given this, the transmissivity in the central or trough waveband can be estimated using Equation (4.59):

$$t_2 = (R_2 - R_1)/(R_1 - R_3) \tag{4.59}$$

Once the transmissivity has been estimated in this waveband, then inversion of normal atmospheric correction software yields an estimate of the total moisture in the air column. Once this is

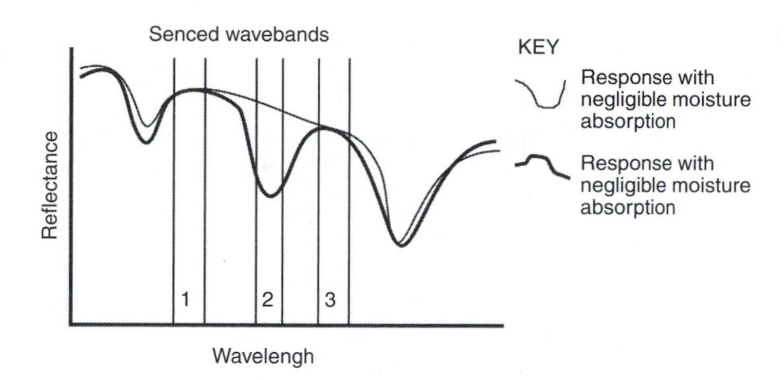

FIGURE 4.24 Sensing in three narrow wavebands to estimate the moisture content of the atmosphere.

known, then it can be used to estimate the moisture transmissivity in the other wavebands that are of interest to the analyst.

Conventionally, aerosols are dealt with by the use of standard atmospheric values for aerosol load. A better solution is to use sun photometer data to derive an accurate estimate of the aerosol load, at least at one location in the image. Whilst this is better than assuming standard conditions, it may still incur significant errors when there are local variations in aerosol load.

4.4 THE ENHANCEMENT OF IMAGE DATA

Image enhancement is the process of transforming image data to enhance certain features within the image. Enhancement selectively enhances some information in the data and may suppress other information. An analyst may well decide to use a number of complimentary enhancements in the conduct of the one interpretation task, where each enhancement is developed to assist with the interpretation of a specific aspect of the project.

Enhancement can be conducted in a number of domains:

1. *Radiometric enhancement*. The recorded radiometric data are enhanced to provide better grey tone or colour rendition in the viewed images.
2. *Spectral Enhancement*. The spectral channels are combined or transformed so as to enhance certain features in the data.
3. *Spatial Enhancement*, to extract information on spatial inter-relationships or the texture and patterns in the data.
4. *Temporal Enhancement*, to compress a temporal sequence of images, yet to extract information on the dynamic changes that are recorded in the image data.

4.4.1 RADIOMETRIC ENHANCEMENT

Image data usually contains either 256 (2^8 bits) grey levels for byte data, 65536 (2^{16} bits per pixel, per band) grey levels for integer data or many more potential levels for float data. In contrast to this resolution, the human eye can discriminate about 20 grey levels when viewing an image. Sensors are not only much more sensitive than the human eye, but they can usually detect energy at a much lower and higher levels than can the human eye, so that the image data covers a wider range of energy levels, and at finer resolution than can be detected by the human eye. Digital data will therefore contain a lot of information that will be hidden to the human eye. Radiometric enhancement is the process of enhancing image data so that the interpreter can view information in the data that is useful to him.

The process of deciding what is useful is very subjective. It means that the analyst must do much of the process of defining suitable enhancements. In doing this he or she will use experience as a guide, but must also be prepared to explore options different to those learnt from experience. Most image processing software systems provide the capacity to explore different ways to enhance image data; however, the different software systems provide different capacities, just as they provide different capacities in other segments of image processing. Ultimately the analyst needs to select the software system that is most suitable for the purpose on the basis of the suitability of the different systems to meet the needs. For image enhancement, where similar approaches may be adopted in different projects, the approach taken in some software systems of automatically recording the steps implemented in the enhancement of image data, as a recipe can be very convenient. With this approach, the recipe is retained by the system. When the analyst wishes to use this recipe he or she can insert different images into the recipe, and the system will process them in the way specified in the recipe.

Image enhancement, by suppressing irrelevant data and information, and emphasising that which is relevant, assists the interpreter to extract the required information in the most economical, reliable and accurate manner. Visual interpretation involves both physiological and psychological processes; different interpreters will often prefer different enhancements for their interpretation. The individual interpreter can only select the enhancement type and more particularly the transformation parameter values empirically. For this reason, facilities that are used to create enhanced images for visual interpretation must include a high quality colour video monitor to rapidly and cheaply display enhanced image data.

It is important that the interpreter be familiar with the reflectance characteristics of the surfaces that have to be imaged, as have been discussed in Chapter 2, when selecting a suitable method of enhancement. These reflectance characteristics will create different effects for the different surfaces with each enhancement. The interpreter thus has to also be aware of the effects of the selected enhancement. In this way he or she is less likely to misinterpret the imagery or suppress important information. One tactic that the interpreter may employ would be to conduct a preliminary interpretation on imagery that has a broadly based enhancement that suppresses little information, but does not reveal more subtle information, before focusing on detailed interpretation using imagery that has been optimised for this purpose.

There are infinite numbers of ways that imagery can be enhanced, but experience has shown that there are a few that are used for the majority of visual interpretation work. In addition, each time a new enhancement is used the interpreter has to learn the significance of the displayed colours, and if other interpreters are to be involved, then the same will apply to them. For this reason it is best to use a small armoury of enhancements wherever possible, even though this might cause some small loss in information content in the display.

4.4.1.1 The Display of an Image

One of the first things that you will want to do is display an image. Whilst there are a number of things that need to be specified to display an image, only two are essential, the others usually being given default values by most image processing systems. When you activate the display system in Idrisi, the window shown in Figure 4.25 is displayed. In this display it is essential that you:

1. *Specify the name of the image file.*
2. *Select a lookup table to use with the image*, for some file types.
3. *Check that the system is expecting the correct type of file.*

In addition to these, the following specifications are set to default values that can be overwritten at this stage or later:

1. *The scale of the display*
2. *The area to be displayed*
3. *The wavebands to be displayed* with each of the three colour guns in the video monitor.
4. *Display of information about the image* including a title and legend.

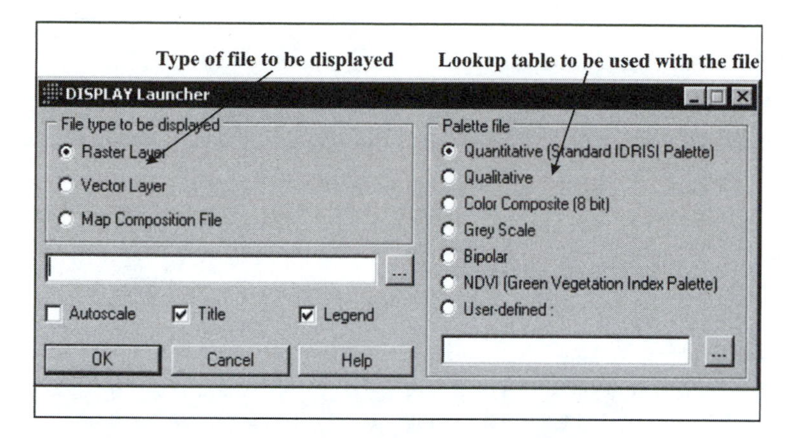

FIGURE 4.25 The Idrisi prompt window for the display of an image.

With multispectral images, the response values in the three selected channels in the image are converted to tonal values in accordance with the selected enhancement, where this is usually set to the default enhancement at the start of a session. Most default enhancements use a lookup table to convert the image data values to output or display values; the concept of lookup tables is discussed in the next section. If a single channel image is to be displayed, then this can be conducted by means of a grey scale enhancement, similar to that for a single channel of a multispectral image, or the image can be displayed as a pseudo-colour image. The display of pseudo-colour images is discussed in Section 4.4.1.2.

Most image processing systems set most parameters to default values when the system is started. With many systems these defaults can be changed to more suitable settings either at the time of first display or later. Once the image has been displayed, it is also possible to change most of the settings for the image, either interactively, or before re-displaying the image.

Most software systems implement enhancements used to display data within the computer display system itself, so that the software neither modifies the file data values nor creates new files. However, with some more computer-intensive transformations and enhancements, the software may write the transformed data to a file, for subsequent display using the display software.

4.4.1.2 The Pseudo-Colour Density Slice

While the human eye can only detect about 20 levels of grey tone, it can detect many more levels of colour. This facility is exploited with single band images by displaying them as pseudo-colour images. In a pseudo-colour image individual data values or ranges of data levels are colour-coded to different colours so as to convey more information to the analyst.

To create a pseudo-colour image, the single channel image values are used to assign values to the red, blue and green colour guns in accordance with a lookup table. Table 4.6 shows the values in both a black and white and a colour lookup table. In this table the values in the image are shown in the left-hand column, the tone or colour as will be seen by the user in the second or seventh columns, and the colour gun assignments are shown in the other columns. In practice the lookup table just consists of the three columns for the three colour guns. The row address in the table is the data value and the values under the colour guns are the output values that are sent to the display device. Table 4.6 shows just two examples of lookup tables that accept source data values in the range 0 to 15. Tables are also made to accept character and integer data with 256 and ±32,768 values respectively. If data are in a range different to the range of the lookup table, then the data are either linearly transformed into the correct range, or sub-ranges are defined so that the full data range is covered by the levels in the

TABLE 4.6
Typical 16 Level Grey Tone and Colour Lookup Tables

Data value	Tone	Blue	Green	Red	Data Value	Colour	Blue	Green	Red
0	Black	0	0	0	0	Blue	255	0	0
1		17	17	17	1		255	63	0
2		34	34	34	2		255	127	0
3	Light Grey	51	51	51	3		255	191	0
4		68	68	68	4	Cyan	255	255	0
5		85	85	85	5		170	255	0
6		102	102	102	6		85	255	0
7	Mid Grey	119	119	119	7	Green	0	255	0
8		136	136	136	8		0	255	85
9		153	153	153	9		0	255	170
10		170	170	170	10	Yellow	0	255	255
11	Dark Grey	187	187	187	11		0	170	255
12		204	204	204	12		0	85	255
13		221	221	221	13	Red	0	0	255
14		238	238	238	14		125	0	255
15	White	255	255	255	15	Magenta	255	0	255

lookup table. Very few systems have float lookup tables, and so when float data is to be displayed in a way that requires the use of a lookup table, then the data have to be transformed into the appropriate data range.

Pseudo-colour density slicing can also be used to display classification (Section 4.5) results. With many image-processing systems, classification results are created as a one channel image file, with the classes numerically coded from one upwards. These can usually be best viewed by conducting a pseudo-colour density slice.

4.4.1.3 Linear Enhancement

The simplest way to enhance an image is to use a linear enhancement. In a linear enhancement the selected subset of the image data values are used to define a linear relationship between the source image data and the display values in the form:

$$\begin{aligned} \text{Display} &= A + B \times \text{Image} && \text{when } 0 <= (A + B \times \text{Image}) <= M \\ &= 0 && \text{when } (A + B \times \text{Image}) < 0 \\ &= M && \text{when } (A + B \times \text{Image}) > M \end{aligned}$$

The reason why a subset of the image data response values is used is because the image histogram will show that the image data normally covers a subset of the possible values, say values in the range n–m. This data is then enhanced to cover the range 0–M. If there are source data values outside of the range n–m, then they will be assigned to 0 if they are below n–M if they are above m.

The subset range from 4 to 116 (n–m) in the source data shown in Figure 4.26 has been stretched to cover the range 0 to 255 (0–M). In this case the cumulative histogram of the source data was used to find the data values at which 1% and 99% of the total counts occurred, 4 and 116 in this case. These were then used as the lower and upper cut-off values and assigned to 0 and 255, respectively, in the transformed data. The linear transformation was then used to find the transformation gain and

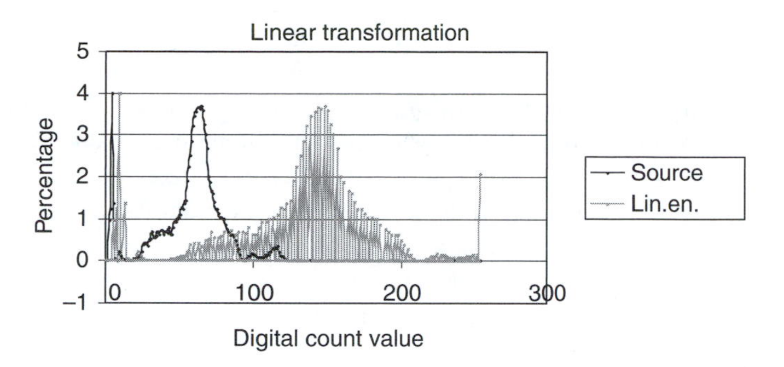

FIGURE 4.26 Linear enhancement of the data shown in Figure 4.1.

offset by solving the two equations:

$$0 = 4 \times \text{gain} + \text{offset}$$
$$255 = 116 \times \text{gain} + \text{offset}$$

With such an enhancement, all data values below 4 stayed at zero, and all data values above 116 were allocated the value of 255. Thus the new bin 0 contained all data that had values of 0, 1, 2, 3 and 4 in the source data, and bin 255 contained all of the data with values higher than 115. Since there are many more steps between 0 and 255 than between 4 and 116, it means that the data had to be allocated to every alternative or third bin value. This is why the histogram contains gaps. It will also be noticed that the last bin has a high-count value compared to the immediately prior bins. These artefacts are typical of the linear enhancement.

The gain and offset parameters of the linear transformation can be determined in a number of ways:

1. *Specification of the gain and the offset by the user.* The difficulty in this approach is the time involved in determination of the two values by the user. It is not a commonly used approach.
2. *Specification of the percentage of the data in the tails of the enhancement.* This specifies the percentage of the data that will have values of 0 or M as a result of the linear transformation. Often the cut-off percentages are set as default values of 1% or 2%, but can be overridden by the operator if required. The total pixel count in the histogram is used to calculate the number of pixels that should be in the lower and upper tails. The histogram values are summed upwards and downwards until this value is exceeded, and these histogram values, at the lower and upper ends, are the maximum and minimum histogram bin values that will be assigned 0 and M respectively in the transformed data. The resulting two simultaneous equations are solved for A and B as has been shown above.
3. *Specification of a pair of original data values and the corresponding transformed values.* The gain and offset are computed by solution of the two simultaneous equations created by these two pairs of values. The two original data values are usually selected by taking the mean values from training sets of the data established for this purpose. The operator then selects the values to which these will be transformed. This approach can be used when the reflectance values of various surfaces are observed during acquisition of the satellite image data, to convert the satellite data to reflectance. It can also be used to convert the range of values in one image to the comparable range in a second image, by using training sites in both images that are assumed to not change in reflectance between the acquisitions of the

two images. To be representative, the two sets of values must be near the extremities of the data values in the histograms.

4. *Change of the gain or offset by means of a digitising tablet, arrow keys on the keyboard or mouse.* Movements on the tablet, keyboard or by the mouse, in one direction, say the x-direction, changes one of the two parameters, say the offset value, and movement in the other direction changes the second parameter, say the gain. As the operator moves the mouse, key or digitising tablet cursor, the position and gradient on the linear enhancement changes causing a change in the enhancement of the displayed image. It is a very interactive way of establishing and modifying linear image enhancements.

The histogram for the transformed data shows that many pixels will have black to near-black tones (low digital count values) whilst the vast majority will be moderate grey in tone. There are relatively few pixels with moderate to dark grey tones and few with light tones (high digital count values). These disadvantages arise when the histogram is multimodal, as is the case here. These limitations of linear enhancement have led to the search for better ways to enhance image data.

The advantages of the linear transformation are that its effects are easy to visualise by the interpreter, and the transformed values are linearly correlated with the original response values. However, if the cumulative histogram exhibits a number of steps, as occurs when the histogram is multimodal, then the linear stretch will not provide a good enhancement across the whole of the data range. When this occurs, other stretches are often preferred.

4.4.1.4 Non-Linear Enhancements

Criticisms of linear enhancements are that they:

1. Do not match the logarithmic response of the human eye.
2. Create sharp cut-offs at the extremities of the enhancement.
3. Do not provide a good enhancement with multi-modal data, particularly if the peaks in the histogram are at both ends of the histogram, as can frequently occur in images that contain significant areas of water and green vegetation. Multi-modal data create steps in the cumulative histogram.

These limitations of the linear enhancement can be corrected by the use of non-linear enhancements that are usually a function of the form:

$$\text{Digital Number } (DN) = f(\text{Old})$$

where $f(\text{Old})$ is a non-linear function of the Old parameter such as:

$$\text{DN} = A \times \log(\text{Old}) + B \text{ or } DN = A \times \sin(B \times \text{Old}) + C, \text{etc.}$$

The response of the human eye is a logarithmic function of the radiance incident on the eye, of the form:

$$\text{Eye response} = \ln(\text{Radiance}) \tag{4.60}$$

This means that the eye is much more sensitive to changes in response at low light levels than at high light levels as shown in Figure 4.27. This response characteristic of the human eye means that equal changes in response at the darker and lighter ends of the enhancement will not be perceived as having equal tonal changes. A unit change in response at the darker end will elicit a larger visual

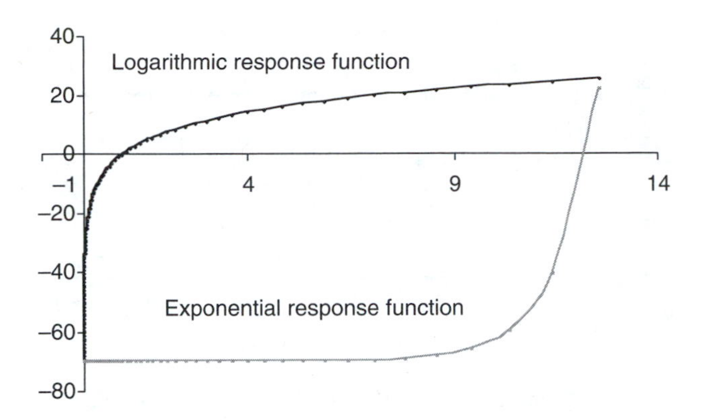

Figure 4.27 Logarithmic response function of the human eye.

response than will a unit change at the lighter end, since the human eye will detect smaller tonal changes at the darker end than at the lighter end.

An enhancement can be introduced, by use of an exponential function, to reverse the logarithmic response function of the human eye, so that perceived changes in response match the original changes in response. Such an enhancement would be of the form:

$$\mathrm{DN} = A + (b \times \mathrm{e}^{\mathrm{Old}}) \tag{4.61}$$

The effect of this transformation is that equal intervals in the original data at low, moderate and high values are seen as having equal changes in intensity in the transformed data by the observer, as long as the constants A and b are properly set.

Other forms of non-linear transformation can be applied in the same way as described above. The disadvantages of these forms of non-linear transformation are:

1. They can create difficulties for the analyst during interpretation as to the significance of the variations in tone, relative to conditions on the surface.
2. The transformation does not necessarily provide a good enhancement as it does not take into account the specific characteristics of the data itself.

Two types of non-linear enhancement that do take into account the characteristics of the actual data as shown in the histogram are the piecewise linear stretch and the histogram equalisation stretch.

4.4.1.5 Piecewise Linear Stretch

The piecewise linear enhancement or stretch is a variation on the linear stretch that goes some of the way towards reducing the disadvantages of the linear enhancement. In this enhancement there are a number of linear enhancements, each with different gain and offset values that are used to transform parts of the range of response values in the original data. The transformed data is usually made to fit the cumulative histogram of the actual data, so as to produce good quality enhanced images.

The usual way to compute the piecewise linear enhancement is to construct the cumulative histogram, identify the breakpoints in the gradients of the histogram and fit individual linear enhancements to the segments of the histogram between these breakpoints, as shown in Figure 4.28.

The piecewise linear enhancement gives a much better result than the linear enhancement when the data exhibit a multi-modal distribution across the histogram of the data values. The cost as with all non-linear enhancements, is that the simple linear correspondence between the image data and the original data is lost in the transformation.

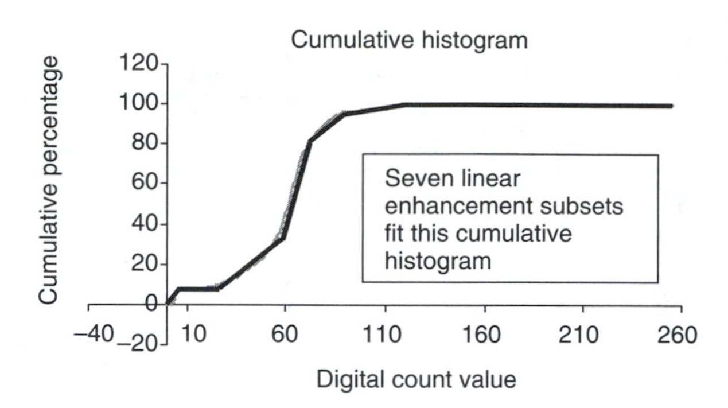

FIGURE 4.28 Typical piecewise linear stretch for data, showing the seven linear enhancements used for different subsets of the data.

4.4.1.6 Histogram Equalisation

The objective of histogram equalisation is to create images that have approximately equal numbers of pixels, or pixel counts at each transformed density value, that is, the transformed data will have an approximately uniform distribution. The enhancement does this by determining the optimum number of pixels to be assigned each value:

$$\text{optimum count} = \text{total number of pixels/number of values in the data.}$$

The algorithm starts at the lowest *old* value, sums the value counts in the old histogram until this optimum count is exceeded. All of the *old* values used to accumulate to this total count are assigned to the first *new* value. The count of the *old* values in excess of the *optimum count* are retained and carried forward as a starting count for accumulating *old* value counts, from further values in the *old* histogram, until the *optimum count* is again exceeded. The *old* values used to accumulate this second total count are assigned to the second *new* value, and the excess is again carried forward as a starting count to determine which *old* values are to be assigned to the third *new* value.

If the histogram count for an *old* value (and the total count that includes this histogram count) is larger than the *optimum count* then that one *old* value will be assigned to the *new* value. If the total count exceeds $n \times optimumcount$ where $n > 1$, then all of the *old* values will be assigned to the one *new* value, but the next $(n - 1)$ *new* values will have zero counts. In this way the average count in each bin or value is kept the same. Once the mapping from *old* values to *new* values is known through this process, then a lookup table can be constructed and the image transformed by the use of this lookup table.

Figure 4.29 shows the histogram equalisation transformation for the same source data as that used to show the linear transformation in Figure 4.26. The figure still shows a peak, since the bin of counts at a specific value cannot be sub-divided into a number of new bins, but all of the same old values have to go in the same new bin. However, the new bins are further apart where the bin counts are large, so that similar total counts will occur at low, middle and high values, and thus the enhancement will tend to look as though the digital counts are spread over the whole range in a more even way than in fact can be achieved.

4.4.2 SPECTRAL ENHANCEMENTS

Spectral enhancements usually transform the data by combining subsets of the wavebands in some way. The derived data may be used in the form, or it may be further processed and then either used directly or transformed back into the original spectral space, or some version of it. A special application of spectral enhancement is to use the frequency domain by means of the Fourier transformation.

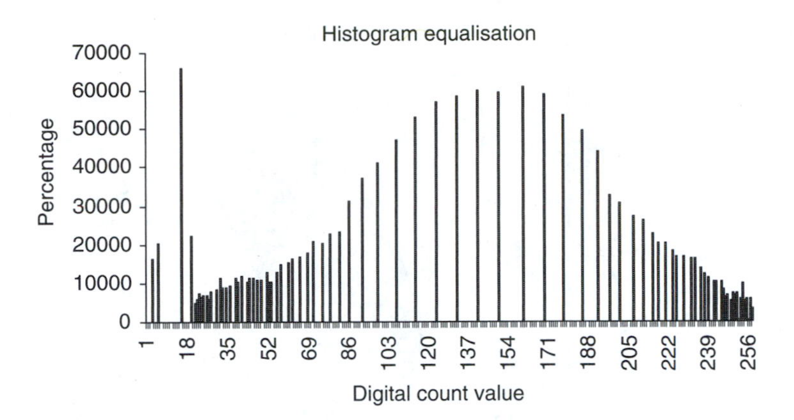

FIGURE 4.29 Histogram equalisation of the same data as has been used in the other enhancement transformations (Figure 4.26 and Figure 4.28).

4.4.2.1 Ratioing

Ratioing is the procedure of forming a ratio of one function of a subset of the N bands in the data, to another function of another subset of the N bands in the data.

$$\text{Ratio} = \Sigma(\text{Const}_i \times \text{Band } i)/\Sigma(\text{Const}_j \times \text{Band } j) \tag{4.62}$$

The parameters Const_i and Const_j are integer or real constants. A further constant may be added to the denominator to ensure that division by zero does not occur. In practice the functions are normally quite simple ones, consisting of only one band in each of the numerator and denominator to give ratios of the form:

$$\text{Ratio} = \text{Band } i/\text{Band } j \tag{4.63}$$

Derived ratios are normally highly non-linear relative to the source data. Consider a ratio of the form (Band i/ Band j) for data in the range 0–255. Add a value of 1 to each band so that they are in the range 1–256. The ratio will have values in the range (1/256) to 1 when Band $i <$ Band j and 1–256 when Band $i >$ Band j. A simple linear enhancement of the ratio will severely compress the values less than one relative to the values greater than one. Ratios are normally enhanced using a non-linear transformation. Suitable enhancements could be a piecewise linear or a logarithmic stretch as described earlier.

Assessment of the usefulness of ratioing requires an understanding of the reflectance characteristics of the different surface types as described in Chapter 2. Most discrimination between green vegetation, soil and water occurs in the red and infrared regions, so that a ratio of (NIR/Red) will yield high values for green vegetation, values slightly greater than unity for soils and less than one for water.

A form of ratioing that enhances the relative intensities in the different bands, called normalisation, is of the form:

$$\text{New}_i = (n \times \text{Old}_i)/(\Sigma\text{Old}_j,\ j = 1, n) \text{ for } n \text{ wavebands} \tag{4.64}$$

This transformation retains the relative intensities in the bands but eliminates variations in total intensity. Another special group of indices are the vegetation indices, designed to enhance green vegetation, and discussed below in Section 4.4.2.3.

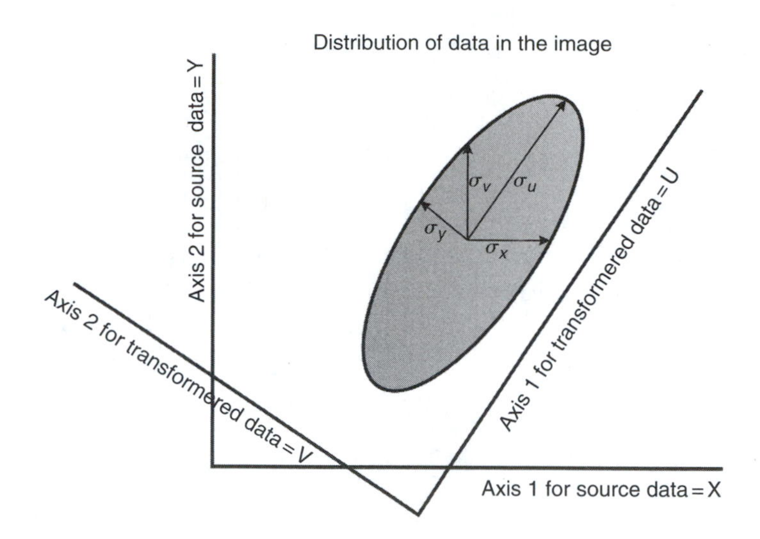

FIGURE 4.30 The principal components transformation.

4.4.2.2 Orthogonal Transformations

Remotely sensed data, in N bands, can be transformed by a rotation, to give values relative to a new set of axes that are at a specified orientation to the original axes. The most common orthogonal transformation is the principal components transformation. It uses the correlation matrix for the data to compute rotations designed to transform the data to create uncorrelated data sets. It achieves this by transforming the data onto the axes that coincide with the primary axes of the ellipses describing the distribution of the data. Consider data in two dimensions as that shown in Figure 4.30. The covariances can be depicted as an ellipse, the orientation of which indicates the correlation that exists between the two bands of data. Transforming the data onto the axes of the ellipse yields uncorrelated data, as discussed in Section 4.2.3.

The same principles apply to imagery in N-dimensional space. The first primary axis, coinciding with the first semi-major axis of the N-dimensional ellipsoid defined by the covariance array, will contain the maximum amount of variation in the sample data used to derive the covariance array. With image data the greatest variation is usually due to illumination so that component one is often termed the brightness component. Subsequent principal components will then depend on the cover types within the scene. Thus scenes that contain a lot of green vegetation may yield second principal components that are related to greenness. Since each transformation uses the data from that image, each transformation will be different. In addition, since the transformation uses all of the scene data, features that cover only a small part of the scene are likely to have a negligible effect on the transformation. They will only have an effect if their response values are very different to the remainder of the features in the scene. Features that do have response values that are very different will have a significant impact on the transformation.

By creating transformed data that coincides with the principal axes of the ellipses in the data, most of the common and very unique information in the data is concentrated in the first few components. The remainder of the components contains noise and the effects of surface types with statistically small samples that are not significantly different to the other surfaces types.

Preparing the orthogonal transformation given in Table 4.7 and Table 4.8 results in principal components images in which the variance in the components, and the percentage of the variance in each component are shown in Table 4.9. This data shows that 76.6% of the variance, that is, of the "common" information in the data, is contained in the first component, 92.0% in the first two

TABLE 4.7
Summary of Image Statistics Used to Create a Principal Components Transformation

	Covariance Array						Correlation Array					
Mean Array	Ch1	Ch2	Ch3	Ch4	Ch5	Ch7	Ch1	Ch2	Ch3	Ch4	Ch5	Ch7
74.47	156.35	89.53	157.07	56.47	208.61	123.50	1.000	0.925	0.900	0.166	0.571	0.761
28.38		59.93	104.98	89.56	176.52	88.59		1.000	0.972	0.424	0.780	0.882
29.60			194.62	151.61	326.88	164.13			1.000	0.399	0.802	0.912
61.80				743.63	603.72	184.75				1.000	0.758	0.528
54.61					854.22	349.26					1.000	0.921
21.10						168.35						1.000

TABLE 4.8
The Principal Components Transformation Parameters and the Contribution of the Variance to Each of the Components

Component	Ch1	Ch2	Ch3	Ch4	Ch5	Ch7
1	0.3693	0.4146	0.4175	0.2362	0.3893	0.4177
2	0.4199	0.1538	0.1583	−0.7776	−0.3697	−0.0527
3	0.3847	0.3355	0.1990	0.2132	−0.1283	−0.1890
4	−0.2903	−0.1023	0.1580	−0.4835	0.4170	0.5005
5	0.5956	−0.3168	−0.6152	0.0555	0.0665	0.3885
6	−0.0769	−0.3206	0.3727	0.2392	−0.6535	0.5167

TABLE 4.9
Contribution of Components to the Variance in the Data

Component	PCA Variance	Percent of Variance
1	706.3	76.6
2	142.1	15.4
3	44.8	4.9
4	19.7	2.1
5	6.0	0.7
6	3.0	0.3

components and 96.9% in the first three components. The first four components contain the most significant common information, the variance in the last two components appearing like noise.

One of the main difficulties with using the principal components transformation is the need to learn the significance of the colours in each transformation. One attempt to eliminate this problem has been the search for a standardised transformation that operates in a manner similar to the principal components transformation. One version of this approach developed by Kauth and Thomas (1976)

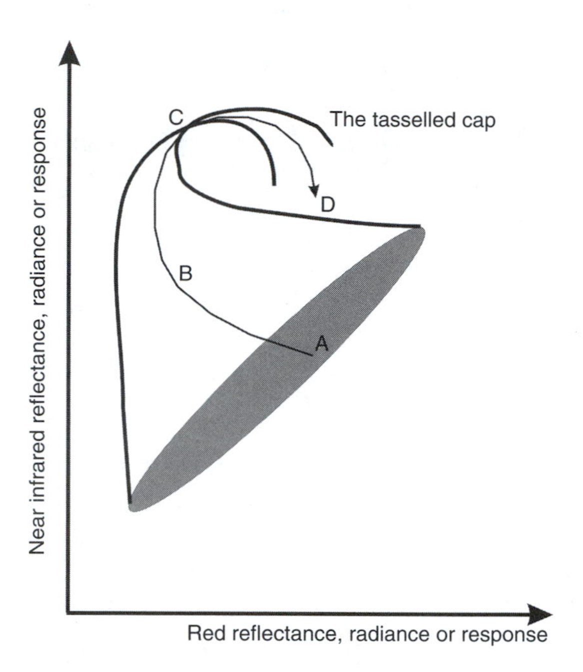

FIGURE 4.31 The tasselled cap concept (formulated by Kauth, R.J. and Thomas, G.S., *Proceedings, Symposium on Machine Processing of Remotely Sensed Data*, Purdue University, West Lafayette, Indiana, USA. With Permission) 1976.

for Landsat MSS data creates the transformed images:

$$\begin{aligned} \text{SBI} &= \text{Soil Brightness Index} \\ &= 0.433 \times \text{Band4} + 0.632 \times \text{Band5} + 0.586 \times \text{Band6} + 0.491 \times \text{Band7} \qquad (4.65) \\ \text{GVI} &= \text{Greenness Vegetation Index} \\ &= -0.29 \times \text{Band4} - 0.562 \times \text{Band5} + 0.60 \times \text{Band6} + 0.491 \times \text{Band7} \qquad (4.66) \end{aligned}$$

In this transformation the SBI and GVI correspond to the first two components of the principal components transformation for conditions that the authors believed were typical for Landsat MSS data. These two axes coincide with the line of soils or brightness and towards the response of green vegetation, respectively. These transformations were part of the development of the tasselled cap concept of crop growth maturity and decay, as depicted in Figure 4.31. In this concept, a crop starts from the soil colour of the actual pixel, in the plane of soils. As it grows, its reflectance progresses up the cap part of the tasselled cap, through growth, flowering until it comes to the tassels at maturity of the crop. Eventually the tassels will return to the plane of soils.

The sensitivity of principal components to surface conditions that are very different to most of the surfaces in the image is exploited in change detection studies using images at different dates. In these studies, an area of change may only represent a small percentage of the total area, yet they will usually have quite distinct spectral data over the set of images being used. They thus have an impact on the transformation that is out of proportion to their coverage. Principal components usually enhances changes, but the need to learn the significance of the colours in the derived images remains.

4.4.2.2.1 The Minimum Noise Fraction Transformation

The minimum noise fraction (MNF) Transformation was originally developed to transform multispectral image data, recognising that the data contained both information and noise. The MNF transformation is a linear transformation that is of a form similar to the principal components transformation, but is implemented in two steps. The first step is designed to minimise the noise in the higher order components and the second step is designed to maximise the remaining signal variance in the first through the other components, as is done in principal components (PCs) analysis.

Image data contains noise and information that can be put in the form:

$$Z(x, y, b) = S(x, y, b) + R(x, y, b) \tag{4.67}$$

where x, y and b are the co-ordinates and the band in the data, $S()$ is the information component of the signal and $R()$ is the noise component.

The noise component is minimised first. The noise in the data can be estimated using:

1. *Data external to the image.*
2. *Identification of areas in the image that just contain noise.* Such areas could be fields that are considered to be of a uniform landcover.
3. *Adjacent pixels could be assumed to have the same values* so that the difference between the adjacent pixel values is used to compute the noise covariance array.

This noise data is used to derive a noise covariance array that is used in a PC-type transformation so as to shift this noise to the last components and remove correlation between the bands due to the noise component or to "whiten" the data. Once this transformation has been completed, the "whitened" data are used to compute the image covariance array that is used as in the PC transformation to compute a rotational transformation. The "whitened" data are transformed using this transformation in the same way as is done with PC transformations.

Principal components, in focusing on the source data correlation, introduces a transformation that will take little account of the noise unless its statistical distribution matches that of the information component of the image data, so that the resulting components will usually retain the noise distributed through the components. The MNF transformation deals with these first, and then conducts the normal PC transformation, so that the derived higher order components will exhibit less noise than a standard principal components transformation, and accordingly contain the maximum amount of information.

4.4.2.3 Vegetation Indices

Vegetation indices transform a number of bands of data into an index that is correlated with the amount of green vegetative material in the canopy. There are many different indices but only a small number need to be considered for most analysis tasks. Each index uses specific bands in its computation. They are used both to create enhanced imagery and as a basis for the estimation of various parameters related to the green plant matter in the vegetative canopy. By concentrating most of the information about the green vegetation in the one image, the analyst's task of image interpretation is often simplified. Various vegetation parameters are correlated with the different vegetation indices where these physical parameters include above-ground biomass, green leaf area index (GLAI), the fraction, absorbed photosynthetically active radiation (FPAR) and chlorophyll density.

The reflectance of green vegetation shows the effects of strong chlorophyll absorption in the blue and red parts of the spectrum, the moderate absorption in the green part of the spectrum and the negligible absorption but significant reflectance and transmittance in the reflected Infrared (IR) (see Chapter 2). Thus, chlorophyll absorption creates quite unique reflectance spectra for vegetation in comparison with other surface types. Vegetation indices try to exploit this uniqueness, with the different indices trying to do this in different ways. The early indices exploited the large difference in red to near IR reflectance to enhance the effects of vegetation on the image data. These indices are affected by the background soil reflectance, primarily because of the transmissivity of vegetation in the near IR. The second approach was the attempt to develop indices that were less sensitive to

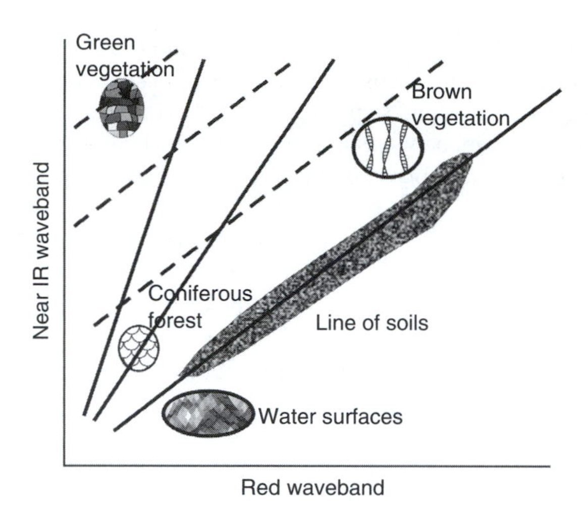

FIGURE 4.32 Typical response values in broad red/near IR waveband space for the common cover types and the shape of typical broadband vegetation indices.

this background reflectance. A third approach was the attempt to exploit the shape of the reflectance curve, particularly in the vicinity of the red edge, using hyperspectral data.

4.4.2.3.1 Broad Band Indices of Red and Near-IR Wavebands

These indices quantify the distance from the line of soils in the direction of the point of green herbage as indicated in Figure 4.32. As the amount of green herbage cover increases within a pixel from no herbage, the response of the pixel moves from the line of soils towards the point of green herbage as discussed in Chapter 2. The distance from the soil line, towards the point of green herbage is thus correlated with the amount of green herbage in the top of the canopy. These indices usually measure this distance as an angle relative to the soil line, as is typical for the ratio indices, or as distances from the soil line.

The more commonly used ratio vegetation indices include (Rouse et. al., 1974)

$$\begin{aligned} \text{RVI} &= \text{Ratio Vegetation Index} \\ &= (\text{Near IR})/\text{Red} \end{aligned} \tag{4.68}$$

$$\begin{aligned} \text{NDVI} &= \text{Normalised Difference Vegetation Index} \\ &= (\text{Near IR} - \text{Red})/(\text{Near IR} + \text{Red}) \end{aligned} \tag{4.69}$$

The more common distances vegetation indices include:

$$\begin{aligned} \text{PVI} &= \text{Perpendicular vegetation index} \\ &= (\text{NIR} - \text{red} \times a - b)/\sqrt{(1 + a^2)} \end{aligned} \tag{4.70}$$

$$\begin{aligned} \text{DVI} &= \text{Distance vegetation index} \\ &= \text{Near IR} - \text{red} \end{aligned} \tag{4.71}$$

Both the RVI and NDVI act like ratio indices with the values radiating out from the origin. The PVI and DVI values vary in proportion to their distance from the line of soils. All of these indices are affected by the background soil colour, since the Near IR waveband reflects and transmits approximately equal amounts of the incident energy. The distance vegetation indices have been found to perform somewhat better than the ratio vegetation indices at low vegetation levels, but the ratio vegetation indices perform better at moderate to high vegetation levels.

The most commonly used of these indices is NDVI as this has been shown to provide as good as or better estimates of green vegetative status than the other broadband indices. NDVI is highly linearly correlated with FPAR, and RVI is highly linearly correlated with GLAI; this provides them with an advantage for both image analysis and for the subsequent estimation of vegetation parameters.

4.4.2.3.2 Broad Band Indices that are Less Sensitive to the Soil Reflectance

A number of soil adjusted vegetation indices have been developed. Most of these indices recognise that the soil line does not go exactly through the origin, and does not lie at an orientation of 45° to the red and Near IR axes. The transformed soil adjusted vegetation index (TSAVI) and the adjusted TSAVI, or ATSAVI were developed with this idea in mind:

$$\text{TSAVI} = \frac{(\text{near IR} - \text{gain} \times \text{red} - \text{offset}) \times \text{gain}}{(\text{gain} \times \text{near IR} + \text{red} - \text{gain} \times \text{offset})} \tag{4.72}$$

$$\text{ATSAVI} = \frac{(\text{near IR} - \text{gain} \times \text{red} - \text{offset})}{(\text{gain} \times \text{near IR} + \text{red} - \text{gain} \times \text{offset} + x \times (1 + \text{gain}^2))} \tag{4.73}$$

where the gain and offset are derived from the actual soil line in the area being analysed, or standard values are used with somewhat less reliable values in the derived indices.

4.4.2.3.3 Measures at the Red Edge

A number of researchers have noted that the red edge, the sharp change in response of vegetation from the high absorption in the red to the high reflectance in the NIR, moves slightly in position and orientation with condition of the vegetation canopy. This has led to efforts to use the shape of the reflectance curve in the vicinity of this red edge as an indicator of vegetation status or condition. The goal of deriving better indicators of the status of green vegetation has been one of the driving forces behind hyperspectral remote sensing, or the use of image data with very many, typically more than 128, wavebands. A significant advantage of the use of the red edge is that it has been shown that the location of the point of inflection in the red edge, known as the red edge inflection point (REIP) is relatively insensitive to variations in illumination conditions and to the reflectance of the soil background but is highly correlated to vegetation greenness parameters. A number of approaches have been taken to the analysis of the red edge; to model the red edge and then derive parameters from that model, and to the rate of change in the red edge.

The first approach has been to exploit the characteristics of the REIP. The location of the REIP is usually found by computation of the first derivative from the spectral data:

$$\rho'_\lambda = \mathrm{d}\rho/\mathrm{d}\lambda = (r_{\lambda(i+1)} - r_{\lambda(i-1)})/(\lambda(i+1) - \lambda(i-1)) \tag{4.74}$$

The REIP occurs at the steepest gradient in the reflectance curve and so the curve of the first derivative will show a peak or a trough at this point. A commonly used model to find the REIP is the inverted Gaussian model so that the peak of the Gaussian curve occurs at the red trough and the standard deviation measures the location of the point of inflection, or the REIP:

$$\rho_\lambda = \rho_s - (\rho_s - \rho_o) \times \exp\{(\lambda - \lambda_o)/2\sigma^2\} \tag{4.75}$$

where ρ_s is the shoulder reflectance in the near IR, ρ_o is the trough reflectance in the red part of the spectrum and λ_o is the wavelength at which this minimum occurs.

An alternative is to start with the method of Guyot & Baret (1988) which assumes that the reflectance red edge can be simplified to a straight line centred around a midpoint between the

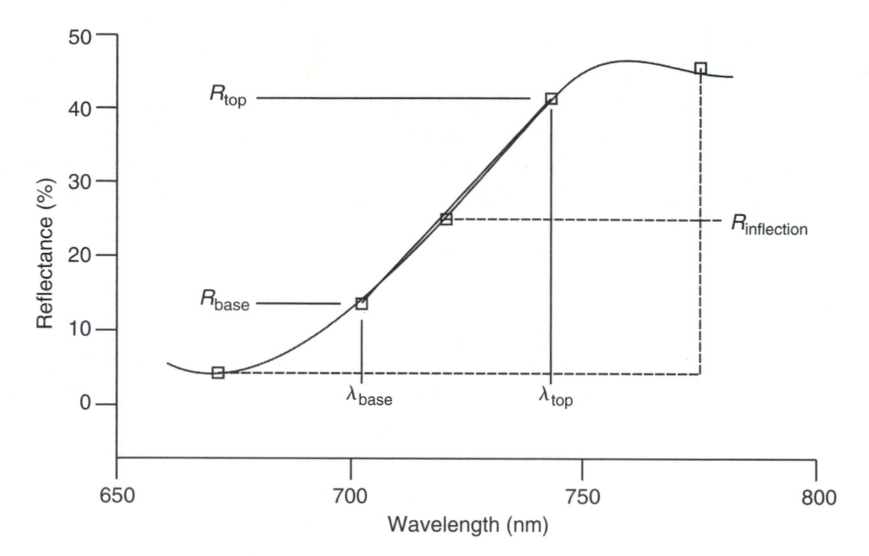

FIGURE 4.33 Illustration of the "linear method" (Guyot G. and Barret, F. *Proceedings of the 4th International Colloquium on Spectral Signature of Objects in Remote Sensing*, Aussois, France, January 18–22, 1988 (ESA SP-287, April 1988), pp. 279–286. With Permission) 1988.

reflectance in the NIR at 780 nm and the reflectance minimum of the chlorophyll absorption feature at about 670 nm (Figure 4.33). The method is described in detail by Clevers in De Jong and Meers (2004). It involves two steps:

1. Calculation of the reflectance at the inflection point, $R_{inflection}$

$$R_{inflection} = (R_{670} + R_{780})/2 \tag{4.76}$$

2. Calculation of the wavelength at this inflection point, $\lambda_{inflection}$

$$\lambda_{inflection} = \lambda_{base} + \frac{\delta\lambda\,(R_{inflection} - R_{base})}{(R_{top} - R_{base})} \tag{4.77}$$

where R_{670}, and R_{780} refer to the measured reflectances at these wavelengths; R_{base} and R_{top} are the reflectances at the base and the top of the linear part of the slope, $\delta\lambda = \lambda_{\tau op} - \lambda_{base}$ or the wavelength difference between the top and the base of the linear part of the slope and λ_{base} is the wavelength at the base of the linear part of the slope. The wavelengths at the base and top of the linear part of the slope are normally at 700 and 740 nm, respectively. They can vary from these values, in which case the actual values should be used. Thus, $\lambda_{base} = 700$ nm and $\delta\lambda = 40$ nm are the usually applied values.

Another approach to the use of the red edge has been to measure the area contained in the triangular space created between the green peak, the red trough and the near IR shoulder, or the area in the first derivative curve between the wavelengths of the green and the near IR.

4.4.2.4 The Fourier Transformation

The Fourier transformation has been introduced in Section 4.2.6. It transforms data into the frequency domain to derive information on the frequencies of the contributing signals to the image data, and the strength of these contributions. The Fourier transformation creates real and imaginary components.

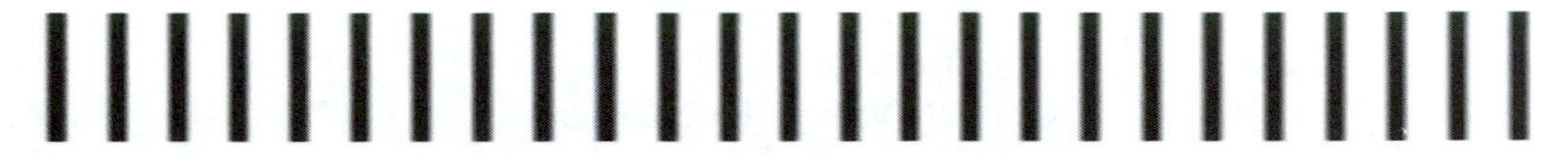

FIGURE 4.34 Image of a sine wave.

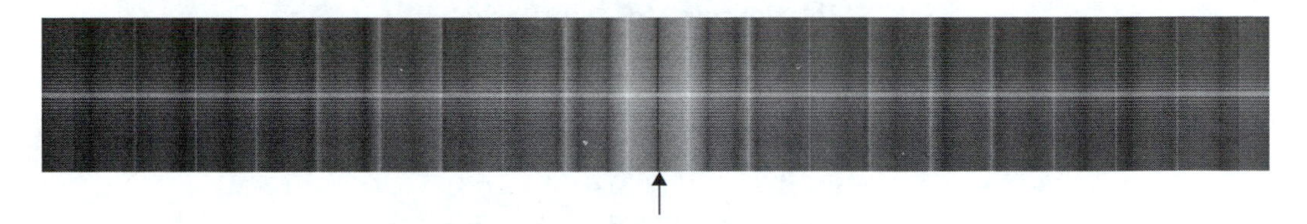

FIGURE 4.35 Fourier power spectrum of the image shown in Figure 4.34.

Let $g(x, y)$ be an image and $F(u, v)$ be its Fourier Transformation. $F(u, v)$ can be shown to be of the form:

$$F(u, v) = R(u, v) + j \times I(u, v) \tag{4.78}$$

where $j = (-1)^{1/2}$

From this can be derived the Fourier spectrum $|F(u, v)| = [R(u, v)^2 + I(u, v)^2]^{1/2}$

And the phase angle $\phi(u, v) = \tan^{-1}[R(u, v)/I(u, v)]$

Most image processing systems show the Fourier spectrum or its square function, called the power spectrum or the spectral density function. Two-dimensional image data is usually transformed into two-dimensional frequency data.

The interpretation of Fourier images is not straightforward. Figure 4.34 shows an image of columns constructed of identical sine waves progressing from left to right, and Figure 4.35 shows the Fourier power spectrum of this image.

The central value in the power spectrum (PS) image represents a signal of infinite wavelength and zero frequency. It thus represents the average value in the source image. As one moves away from the central pixel in the PS image, the pixels represent larger and larger frequencies, and hence shorter and shorter wavelengths. The outermost pixels represent the shortest wavelength that can be represented by the data, or the Nyquist frequency. The shortest wavelength is twice the pixel size. The x-axis depicts changes in frequency along the x-direction, and the y-axis changes in the y-direction. The top left-hand quadrant represents data with +ive x and y values, and the other quadrants contain −ive values in either or both x and y. In practice, the images in the four quadrants are usually mirror images of each other. The source image will contain N pixels in M lines. The derived PS image will contain n pixels in m lines where n and m are the next power to the 2 value higher than N or M, respectively. Thus, with this data there are 1000 pixels in the source data and the PS image contains 1024 pixels.

If there are N pixels in the image, then the value at the $\{(n/2) + 1 + x\}$ pixel in the PS image represents cosine wave components that repeat every N/x pixels in the source wave.

In this image there are 25 sine waves covering a distance of 1000 pixels, so that the sine wave has a wavelength of 40 pixels. The PS image shows a series of regular vertical parallel lines. The centre of the PS image is the vertical arrow. The first vertical white stripe in the PS image is the highest intensity away from the centre. It is at a distance of 26 pixels from the centre pixel. It represents $1000/26 = 38.46$ as the wavelength giving this strong feature. In practice, the peak may not exactly match the pixel and so these estimates are always somewhat approximate. The other vertical lines

TABLE 4.10
Pixel Values for Vertical Lines, Their Distance from the Centre and their Relationship to the Basic Wavelength

512 (Centre)	0	
538	26	λ
589	51	λ/3
640	51	λ/5
691	51	λ/7
742	51	λ/9
794	52	λ/11
845	51	λ/13
896	51	λ/15
947	51	λ/17
999	52	λ/19

are at distances given in Table 4.10. The main vertical line thus indicates the basic sine wave shape defined in the image in Figure 4.34. The other lines are harmonics of this line, introduced because the generated image is not a perfect sine wave, being assigned discrete integer values in the image.

If there are cyclic features at some orientation, α in the image, then these features will exhibit features at right angles to the source features in the PS image.

The power spectrum shows the relative contributions of the different frequencies to the source image data. It follows that dominant cyclic patterns in the source image will have strong spikes or areas of response in the power spectrum. One example of such a pattern could be anomalies caused by differences in calibration of the different detectors within the sensor, leading to striping in the image. Another example could be regular features in the image due to human activities, such as the rows in crops. The interpreter may find that such patterns are inhibiting interpretation and analysis. Once the expression of these regular features has been identified in the power spectrum, they can be removed by the use of a filter. There are a number of different filters, but they all remove selected frequency ranges in the data. The differences between them are usually related to the sharpness of the filtering effect from fully off to fully on.

Alternatively, the analyst may believe that the data are very noisy, and may wish to suppress random noise components in the image data by filtering out those frequencies that just express the background level of contribution to the power spectrum.

Filtering in these ways by use of the Fourier transformation is a global filtering process. The process cannot distinguish between noise that is to be removed and information that is to be retained, if both have the same frequency characteristics in the data. It thus applies the filter globally to the data. For example, the dominant frequency of image striping will coincide with the use of that frequency in other features in the image data, such as roads, buildings or fields. Filtering out that frequency means that its contribution to these other features is also affected. It may mean that these other features are smoothed to some degree, depending of the frequency that has been filtered out.

4.4.3 SPATIAL TRANSFORMATIONS OF IMAGE DATA

The spatial characteristics of image data are an important means of extracting information about the surface. Spatial information has always been particularly important in the visual interpretation of images. This interest has not usually been matched with the digital image processing of satellite-borne remote sensing data, primarily because of the apparent ease of spectral analysis of image data, and because the coarseness of scanner data has not encouraged the use of spatial techniques. The spectral data contains high overheads in calibration and correction if it is to be used to extract significantly more detailed information than that currently extracted. In addition, the spectral dimension has been extensively explored, so breakthroughs are less likely to occur in the exploitation of the spectral data than in other dimensions of the data. The advent of high-resolution image data has encouraged investigation into the use of spatial characteristics of the data.

Historically, most of the focus on spatial data in remote sensing has focused on:

1. *Filtering of image data*
2. *Derivation of texture* to be used as another channel in image classification, or as an aid in image interpretation
3. *Segmentation* of an image into homogeneous sub-areas as part of the classification process

The advent of high-resolution image data is prompting investigation into the identification of objects in the image data through their characteristics of size, distribution, spatial response characteristics and texture.

4.4.3.1 The Measurement of Texture

Texture has been defined in Chapter 3 as the degree of roughness or smoothness of an object or surface in the image. Pattern is the regular repetition of a tonal arrangement across an image. Texture can be of interest of itself when different land covers exhibit different amounts of texture. For example, turbid water, commercial centres and some soil and rock conditions can be displayed as cyan colour in standard false colour composites, but the three will exhibit quite different textures. Texture is a local phenomenon and as a consequence it can be measured within the local environs of the pixel. Thus, many measures of texture operate within a window around a pixel, and the derived textural values are assigned to the central pixel of that window. The window is of $n \times m$ pixels around the central pixel, where it is usual for $n = m$ and for each to be odd numbers so that the window is centred on the pixel of interest.

There are three basic approaches to the measurement of texture: statistical, structural and spectral. Statistical approaches derive the moments within the window used about the pixel. Structural approaches seek to detect the effect of small-scale patterns in the image space. Spectral techniques utilise the properties of the Fourier transform to detect regular patterns within the image space.

4.4.3.1.1 Statistical Techniques

The simplest techniques to measure texture are to derive the second, third and fourth moments within the window about the pixel where:

$$\text{Second moment} = \text{variance} = \Sigma(g(i,j) - m)^2 \times p(g(i,j)) = \sigma^2 \tag{4.79}$$

where $g(i,j)$ = picture function, m = mean value and $p()$ = probability of occurrence.

The variance is an important measure of the variability in the data in the window or sample area, with $\sigma^2 -> 0.0$ for smooth areas and with increasing values for areas of increasing roughness. Sometimes this measure can be normalised by:

$$R = (\sigma - \sigma_{\min})/(\sigma_{\max} + \sigma_{\min}) \tag{4.80}$$

where R takes a value near zero for smooth areas and approaches one for rough areas, and $\sigma_{\min}$ and $\sigma_{\max}$ are the minimum and maximum local variance values found in the image.

2	3	3	1	0
0	2	3	1	0
0	0	3	3	2
2	0	1	3	3
3	2	0	1	3

FIGURE 4.36 A 5 × 5 window of pixels and their image values.

The third moment is a measure of the skewness of the histogram:

$$\text{Skewness} = \Sigma(g(i,j) - m)^3 \times p(g(i,j)) \tag{4.81}$$

If the data in the window is clustered about a value, and there is a tail of a few high, or low values, then the histogram will be skewed.

Then fourth moment is a measure of the kurtosis or peakedness of the histogram.

$$\text{Kurtosis} = \Sigma(g(i,j) - m)^4 \times p(g(i,j)) \tag{4.82}$$

Measures of texture that are based on these statistics suffer the disadvantage that they provide no information on patterns in the data. Structural measures of texture attempt to provide some information on these characteristics.

4.4.3.1.2 Structural Methods

The most common structural method is the construction of co-occurrence matrices for each pixel using the data in a window around that pixel. Consider the image data shown in Figure 4.36. Choose a position operator that will detect the features of interest. Typical position operators are given below:

1. *Detection of edges at* $+45°$. For this position operator, create an array of rows i and columns j such that a location $[i,j]$ in the array is the sum of the number of occurrences of a pixel of value i being located one down and one to the right of a pixel of value j. Thus the first location in this array, $[0,0]$, is the count of the number of occurrences of 0 in the window down one line and to the right by one pixel from another value of 0. In the data shown in Figure 4.36 this location will have a value of 3. The full array for this data and with this position operator is

3	1	0	0
1	1	0	1
0	1	2	0
0	0	1	5

   Note that there are a possible 16 combinations of this type in a 5 × 5 array, so that the values in the above array must sum to 16.
2. *Detection of edges at* $-45°$. This position operator is similar to 1 above, but the first pixel position is one down and to the right of the second pixel. The values in the array for this

position operator and this window of data are

0	1	1	2
0	0	1	2
1	1	0	1
2	2	1	0

3. *Detection of horizontal edges*. For this position operator, create an array of rows i and columns j such that a location $[i, j]$ in the array is the sum of the number of occurrences of a pixel of value i being one down from a pixel of value j. The values in this array must sum to 20:

3	0	3	0
1	1	0	1
2	0	0	2
0	2	1	4

4. *Detection of vertical edges*. This position operator is similar to 3 above, but value i is to be to the right of value j. The result of this analysis for this window of data is:

1	2	2	0
2	0	0	2
2	0	0	2
0	2	2	3

These arrays are then divided by this total, 16 for 1 and 2, and 20 for 3 and 4, to give the co-occurrence arrays $C[i, j]$. For the first position operator:

0.1875	0.0625	0	0
0.0625	0.0625	0	0.0625
0	0.0625	0.125	0
0	0	0.0625	0.3125

For the second position operator:

0	0.0625	0.0625	0.125
0	0	0.0625	0.125
0.0625	0.0625	0	0.0625
0.125	0.125	0.0625	0

For the third position operator:

0.15	0	0.15	0
0.05	0.05	0	0.05
0.1	0	0	0.1
0	0.1	0.05	0.2

And the fourth position operator:

0.05	0.1	0.1	0
0.1	0	0	0.1
0.1	0	0	0.1
0	0.1	0.1	0.15

The co-occurrence matrix can then be used to derive various indices, the usual ones being:

1. Maximum probability $\max[C(i,j)]$
2. Element-difference moment of order k $\Sigma\Sigma(i-j)^k \times C(i,j)$
3. Inverse element-difference moment of order k $\Sigma\Sigma C(i,j)/(i-j)^k$ (i not equal j)
4. Entropy $-\Sigma\Sigma C(i,j) \times \log(C(i,j))$
5. Uniformity $\Sigma\Sigma C(i,j) \times C(i,j)$

The maximum probability indicates the strongest response to the position operator. The element-difference moment has lower values when high values in C are on the diagonal since the $(i-j)$ values are lowest here. The inverse element-difference moment has the opposite effect. Entropy is an indicator of randomness, achieving its highest value when all elements of C are the same. The uniformity indicator is lowest when all elements of C are the same.

It can be seen that the size of the arrays A, and hence C depend on the range of radiance values in the source data. If the range is great, then most of the array locations will be empty unless the samples used to create the arrays are also large. This negates one important application of texture, unless the range of values in the source data is reduced. In general the number of grey levels used should be the same as the number of rows or columns in the windows. It is thus important to be very careful in compressing the values in the source data, to ensure that important information is not lost. There are two commonly used options:

1. *The use of data values in the window to derive the compression.* This approach will give the best compression for the individual window, but it means that the compression used will vary from window to window, so that the values derived will not be consistent, window-to-window.
2. *The use of image values to derive the compression.* This approach will give consistent values from window to window, but will not give the best individual compression in a window.

One approach to the use of these indicators is to determine typical indicator values for different conditions, and then find other areas that are similar, and those that are different.

4.4.3.1.3 Frequency Domain Techniques

These approaches to texture analysis depend on the Fourier transform or analysis of the variogram. As such they are suited to the analysis of large areas as discussed in Section 4.4.2.4.

4.4.3.2 Edge Detection

Edge detection is a specific application of texture and pattern analysis in image data. Pixels usually overlap edges, giving them response values that are due to the various surface types that they sense. Their resulting response values can be derived using the concepts discussed in mixture analysis. As a consequence, edges rarely provide sharp discontinuities in image data, but rather provide a blurred

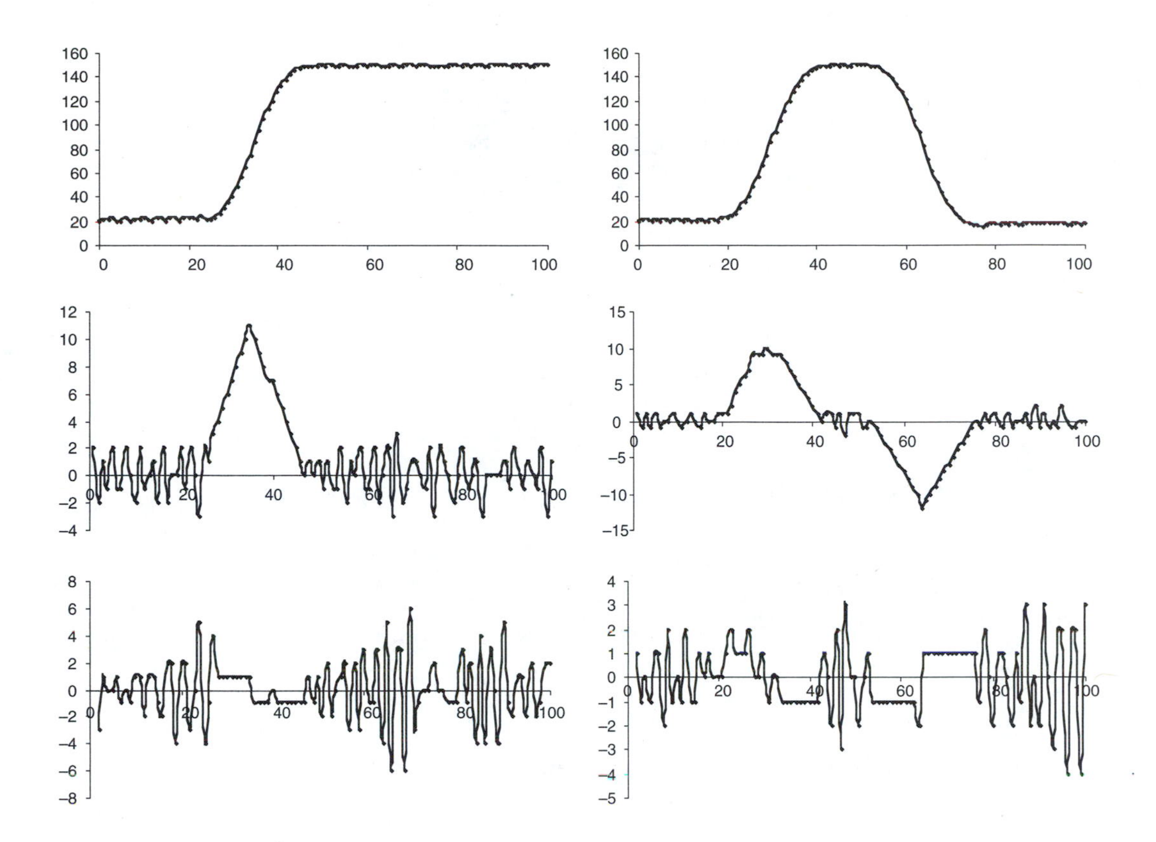

FIGURE 4.37 Response profiles through an edge and a road, their first and second derivatives.

edge. If the picture function is given as $G(i,j)$, then the first and second derivatives are $(\delta G/\delta x)$, $(\delta G/\delta y)$ and $(\delta^2 G/\delta x^2)$, $(\delta^2 G/\delta y^2)$ and $(\delta^2 G/\delta x \delta y)$, respectively. The first derivative measures the slope in the image and the second derivative measures the rate of change in the slope. Two important properties of the derivatives in the x and y directions mean that they are sufficient to describe the characteristics of the slope in an image. These properties are:

1.

$$\text{A vector,} \quad V\{G(i,j)\} = \begin{vmatrix} \delta G/\delta x \\ \delta G/\delta y \end{vmatrix} \tag{4.83}$$

points in the direction of maximum gradient or slope in the image.

2. The magnitude of $V = v = \{(\delta G/\delta x)^2 + (\delta G/\delta y)^2\}$ is the maximum gradient or slope, in the image.

If the picture function, $G(i,j)$ consists of an ideal picture function, $S(i,j)$ and a noise function $N(i,j)$ such that:

$$G(i,j) = S(i,j) + N(i,j) \tag{4.84}$$

then an ideal edge detector will determine the maximum gradient of $S(i,j)$ and be unaffected by variations in response due to $N(i,j)$. This can be done in a number of ways as given below.

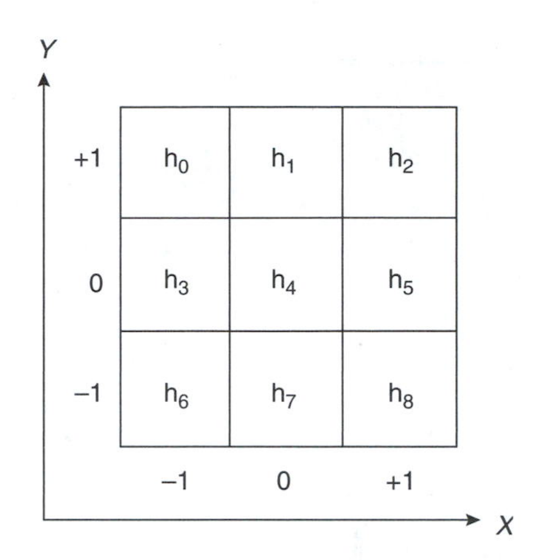

FIGURE 4.38 The local co-ordinate system used in computation of texture.

4.4.3.2.1 Fitting a Surface

If a surface is taken to be of the form:

$$A_o + A_1x + A_2y + A_3xy + A_4x^2 + A_5y^2 = R \tag{4.85}$$

the gradient at any point in this surface is the differential of this equation, that is:

$$\begin{aligned} \text{Gradient}_{x\text{-direction}} &= \mathrm{d}G/\mathrm{d}x = G_x = A_1 + A_3y + 2A_4x \\ \text{Gradient}_{y\text{-direction}} &= \mathrm{d}G/\mathrm{d}y = G_y = A_2 + A_3x + 2A_5y \end{aligned} \tag{4.86}$$

and the maximum gradient will be

$$G_{\max} = [G_x^2 + G_y^2]^{0.5} \tag{4.87}$$

The maximum gradient in a window of say 3 × 3 or 9 pixels can be made by using the co-ordinates and response values of these 9 pixels to solve Equation (4.85) by least squares. The easiest way to do this is to use local co-ordinates as shown in Figure 4.38. The equations are then of the form:

$$\begin{array}{rl} \text{Point} & \\ 1,1 & A_o + A_1 + A_2 + A_3 + A_4 + A_5 = R_{1,1} \\ 1,0 & A_o + A_1 + A_4 = R_{1,0} \\ 1,-1 & A_o + A_1 - A_2 - A_3 + A_4 + A_5 = R_{1,-1} \\ 0,1 & A_o + A_2 + A_5 = R_{0,1} \\ 0,0 & A_o = R_{0,0} \\ 0,-1 & A_o - A_2 + A_5 = R_{0,-1} \\ -1,1 & A_o - A_1 + A_2 - A_3 + A_4 + A_5 = R_{-1,1} \\ -1,0 & A_o - A_1 + A_4 = R_{-1,0} \\ -1,-1 & A_o - A_1 - A_2 + A_3 + A_4 + A_5 = R_{-1,-1} \end{array} \tag{4.88}$$

which can be solved to find A_1 and A_2, the gradients in the x and y directions at the centre pixel of co-ordinates $(0, 0)$:-

$$\begin{aligned} A_1 &= (R_{1,1} + R_{1,0} + R_{1,-1} - R_{-1,1} - R_{-1,0} - R_{-1,-1})/6 \\ A_2 &= (R_{1,1} + R_{0,1} + R_{-1,1} - R_{1,-1} - R_{0,-1} - R_{-1,-1})/6 \end{aligned} \tag{4.89}$$

Hence $G_x = A_1$, $G_y = A_2$ at point $(0, 0)$ and $G_{max} = (A_1^2 + A_2^2)^{0.5}$ or $G = |A_1| + |A_2|$ can be used as a computationally simpler derivation of the gradient parameter.

4.4.3.2.2 Using Window Transform Functions

Equation (4.89) gives the gradient in the x and y directions at the pixel centre, assuming a second order polynomial surface. It does not give the maximum gradient in the range ±0.5 co-ordinate units in the x or y directions, nor does it allow for variations from the second order polynomial. Because of these limitations many other forms of edge-detecting algorithms have been developed. Most of these other algorithms use a window transform function. The window transform function is multiplied by the image data in an equivalent window around the central pixel, to compute a value for the central pixel. The simplest window transform function for a (3, 3) window is:

$$\text{Window transform function} = \begin{matrix} 0 & 0 & 0 \\ 0 & 1 & 0 \\ 0 & 0 & 0 \end{matrix}$$

When this window transform function (WTF) is multiplied by the image data, the result is the same as the original image. A WTF of the form:

$$\text{Window transform function} = \begin{matrix} \frac{1}{9} & \frac{1}{9} & \frac{1}{9} \\ \frac{1}{9} & \frac{1}{9} & \frac{1}{9} \\ \frac{1}{9} & \frac{1}{9} & \frac{1}{9} \end{matrix}$$

yields the average of the eight surrounding and the central pixel. Such a WTF is a smoothing WTF. There are clearly many WTF that can be constructed. For edge detection G_x and G_y are estimated using the Sobel WTFs:

$$\begin{matrix} -1 & -2 & -1 \\ 0 & 0 & 0 \\ 1 & 2 & 1 \end{matrix} \qquad \begin{matrix} -1 & 0 & 1 \\ -2 & 0 & -2 \\ -1 & 0 & 1 \end{matrix}$$

The second order derivatives are estimated by the Laplacian WTF:

$$\begin{matrix} 0 & 1 & 0 \\ 1 & -4 & 1 \\ 0 & 1 & 0 \end{matrix}$$

The Laplacian WTF is very sensitive to noise and hence it is rarely used as an edge detector by itself. The Prewitt WTFs are another model for estimating the first and second derivatives:

$$\begin{matrix} 1 & 1 & 1 \\ 0 & 0 & 0 \\ -1 & -1 & -1 \end{matrix} \qquad \begin{matrix} 1 & 0 & -1 \\ 1 & 0 & -1 \\ 1 & 0 & -1 \end{matrix}$$

and

$$\begin{matrix} -1 & -1 & -1 \\ 2 & 2 & 2 \\ -1 & -1 & -1 \end{matrix} \qquad \begin{matrix} -1 & 2 & -1 \\ -1 & 2 & -1 \\ -1 & 2 & -1 \end{matrix}$$

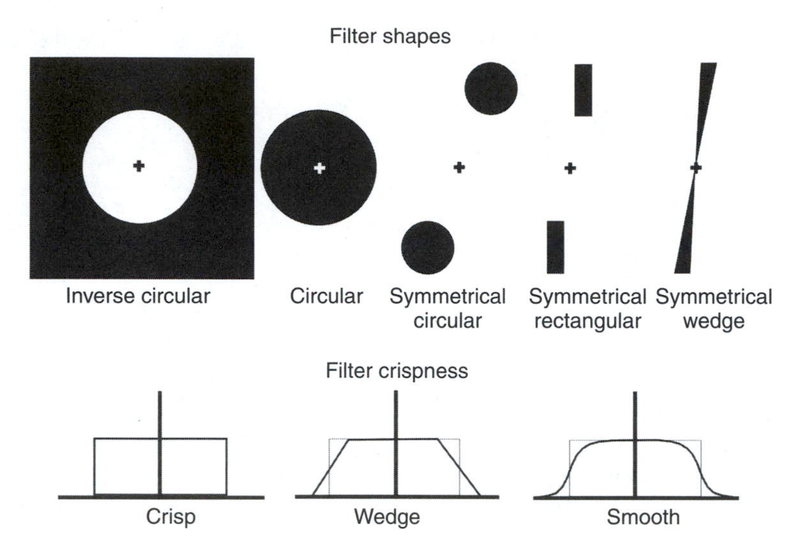

FIGURE 4.39 The filter shapes and crispness determinations typically used in filtering frequency domain data based on image data.

4.4.3.3 REMOVAL OF REGULAR NOISE IN IMAGE DATA

There are a number of sources of regular noise in image data. These sources include slight differences in detector response in the linear arrays of detectors in some scanners, introducing stripping into the optical imagery and speckle in radar data, due to the resampling of the source synthetic aperture amplitude and frequency data to create the high-resolution amplitude image data.

Regular noise will introduce a strong signal at the frequencies that correspond to the noise in the image. The most common method of removing such artefacts is to convert the image to the frequency domain by the use of the Fourier transform, identify the frequencies that correspond to the noise and then use a filter to remove these frequencies from the frequency data. Once this is complete to the satisfaction of the analyst, the frequency data is inverted to create the filtered image.

The common filters include high and low pass filters as well as those shown in Figure 4.39. High and low pass filters remove frequencies above or below the threshold set by the filter, suitable for edge detection and smoothing of an image. To remove noise, it would be more usual to find one of the filters shown in Figure 4.39 to be the most suitable, where these filters are based on either circular, rectangular or wedge-shaped figures. In addition to the shapes of the filters, the user can select a degree of crispness in the filter. The ideal filter is a crisp filter, providing a sharp transition between filtered and non-filtered data. The Bartlett filter provides a linear transition from totally filtered to non-filtered, whilst the Butterworth, Gaussian and Hanning filters provide a smooth transition in accordance with a specific transformation:

$$\text{Butterworth } H(u, v) = \frac{1}{1 + [D(u, v)/D_0]^{2n}} \tag{4.90}$$

$$\text{Gaussian } H(u, v) = e^{-(D(u,v)/D_0)^2} \tag{4.91}$$

$$\text{Hanning } H(u, v) = \tfrac{1}{2}\left(1 + \cos\left(\pi D(u, v)/2D_0\right)\right) \tag{4.92}$$

where $H(u, v)$ is the degree of filtering applied and $D(u, v)$ is the frequency value relative to the start frequency, D_0.

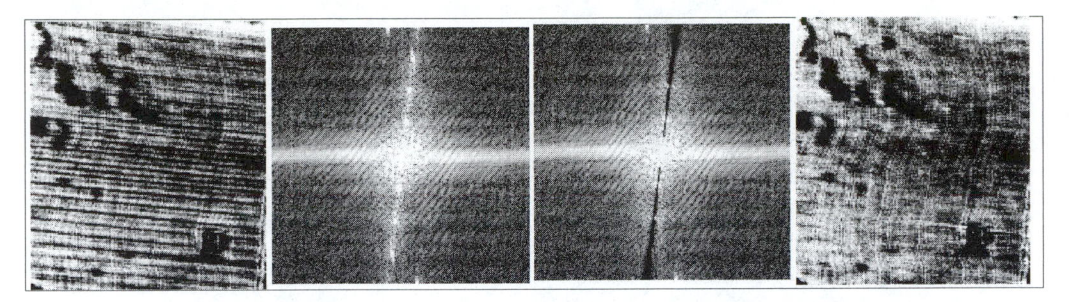

PHOTOGRAPH 4.1 A source airborne scanner image of an agricultural site, its frequency power spectrum, the filtered power spectrum and the derived image by inversion of the filtered power spectrum.

Photograph 4.1 shows an image of an agricultural area that exhibits strong striping due to cultivation. Such striping can interfere with visual interpretation and so it has been removed to facilitate the interpretation of the soils in the area. Adjacent to the image is its Fourier power spectrum, showing the strong horizontal signal due to the majority of features in the image, and the sloping vertical features due primarily to the cultivation effects that are to be removed. The third filter shows the Wedge filter used with Hanning transition applied, and the resulting image is shown to the right of the photograph. The filtering has removed the effects of the cultivation on the image, and revealed sensor striping that was not obvious in the source image. The sensor striping is wavy since the source data was airborne scanner data that has been rectified. The filtering has significantly improved the images for visual interpretation.

It should be recalled that the Fourier filtering process is a global filter. This means that removal of these frequencies will remove the offending noise in the image, but it will also remove these frequency components from other types of spatial patterns in the image. Most of these other patterns will be information, so that the global filter will interfere to some extent with the information content of the image.

4.4.3.4 Analysis of Spatial Correlation : The Variogram

The variogram has been introduced in Section 4.2.5, where the equation for the variogram was given as:

$$\delta(k) = \frac{1}{2(n-k)} \sum_{i=1}^{n-k} (z_i - z_{i+k})^2 \tag{4.17}$$

The variogram value at lag, l, is a measure of the variance between pixels at this separation. As adjacent pixels will usually have higher correlation than those further apart, so the variance can be expected to be less for lower lag values. The variogram value usually increases with lag distance whilst the level of correlation decreases. Once the pixels are uncorrelated, then the variance can be expected to become stationary. The variogram reaches the sill value at this range or lag value. This can be seen in the variogram plotted in Figure 4.40. If the variogram has increasing values for all lag values, then correlation is assumed to exist for all lag values used in computation of the variogram. In this situation, all the lines of data fall within the range value. Either larger or more pixels would need to be used to go beyond the range value in the variogram. If the variogram is flat, then the separation between adjacent pixels is greater than the range value so that the range value cannot be determined from the data. Either smaller pixels or data at a closer interval, would need to be used to derive information within the range of correlation of a point.

The variogram value cannot be computed at lag = 0 as it will give zero value. However, extension of the sloping line will usually cut the lag = 0 location on the graph at a value, z_o. It is usually

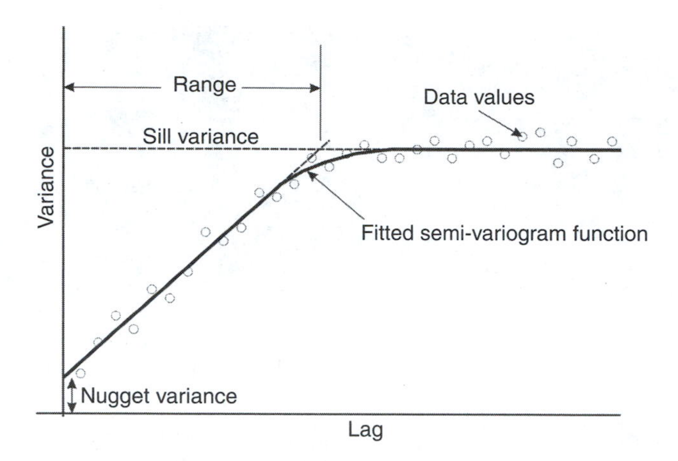

FIGURE 4.40 A typical variogram and its components.

assumed that this value is a measure of the inherent error level, or noise, in the data. The variogram thus provides information on the noise level in the data, the range of correlation between pixels, and the variance between uncorrelated pixels, as well as the variance between pixels at any specific separation. Equation (4.17) shows that the variogram is a sums of squares term. This means that differences are enhanced in the variogram relative to those differences in the data. The variogram is also an average value, so that these enhanced differences are smoothed relative to the source data. Variograms are thus often easier to interpret than the source data because of these characteristics of enhancement and smoothing.

The variogram is used to form the basis for influencing the interpolation between point observations in co-kriging (Chapter 6). Co-kriging is the process of interpolation between observations where another source of data is used to influence this interpolation. Remotely sensed images are a potentially important source of data to use in co-kriging. For this purpose, a mathematical model describing the shape of the variogram is required. The variogram is used to compute the parameters of this mathematical model, and the mathematical model is then used to provide weights to influence the interpolation. The variogram also enhances regular features in image data, such as the regular patterns that exist in cultivation. Such features affect the use of the variogram for interpolation. It may be necessary to filter out these features prior to the use of the variogram to derive the mathematical model to be used in the co-kriging. One way to conduct this filtering is by use of the Fourier transform (Section 4.5.3.3).

Variograms can be produced for two dimensions, and as such they are often very valuable in relation to image data. The regular features discussed above show up in two-dimensional variograms as regular parallel lines in the variogram.

$$\delta(k_x, k_y) = \Sigma(z(i,j) - z((i + k_x), (j + k_y)))^2/((n_x - k_x)(n_y - k_y)) \tag{4.93}$$

for $i = 0$ to $(n_x - k_x)$ and $j = 0$ to $(n_y - k_y)$

In this formula, when $i > 0$ and $j = 0$, then the result is the same as for variograms across the rows, and when $i = 0$ and $j > 0$, gives variograms down the columns. Note that the formula uses all of the rows and the columns in the computation, hence the product in the denominator.

In practice variograms do not always exhibit the shape shown in Figure 4.40. There are two main reasons for this:

1. *The existence of spatial structure at large intervals in the data.* Such structure is often caused by variations in the geomorphic characteristics of the area, such as variations in soil

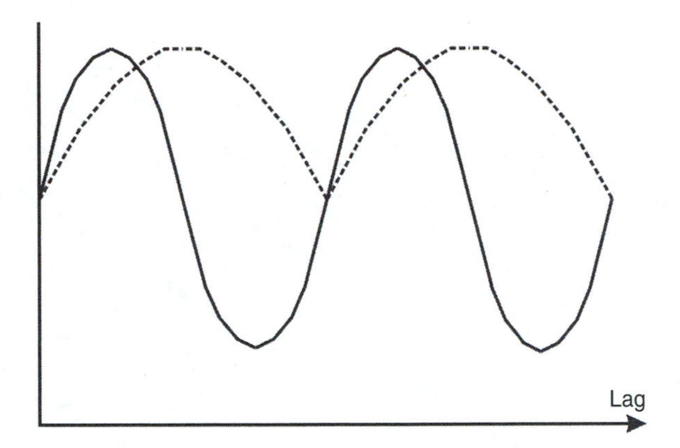

FIGURE 4.41 Sinusoidal and arch variograms derived from imagery dominated by repetitive linear features in the landscape or from long sequences of image data that is dominated by the cyclic seasonal growth of vegetation.

colour across the area of the image. This structure may show gradual changes, in which case the variogram will exhibit a gradual but constant increase in values with increasing lag. Alternatively, the structure may show sharp changes to another plateau level of values. When this occurs, the variogram will exhibit flat areas, and then sharp rises to another flat area, so that the variogram appears to exhibit steps in the variogram values.

2. *The existence of regular features in the image data.* Typical regular features are those made by man. These are very common in agricultural and forestry areas. Regular repetitive features in the image space create oscillatory variograms. If the regular features are approximately sinusoidal or square wave in response then the resulting variogram will be approximately sinusoidal in shape. If the features have marked differences in the duration of the high and low values, then the variograms will appear to be more like a viaduct or arch than a sinusoidal shape. Arch-shaped (Figure 4.41) variograms can also arise if the periods at the peak and troughs are very short compared to the periods at intermediate stages.

   Regular features will normally not be parallel to the rows and columns in the image data. They will exhibit regular sinusoidal or arch features that will have different frequencies in the two directions. The frequencies in the two directions can be used to compute both the orientation of the features in the image data, and the frequency at this orientation, as shown in Figure 4.42.

Temporal variograms are constructed in the time domain. A variogram is created for each pixel through a long series of images or hyper-temporal image data. In these variograms the lag is not a spatial distance, but an interval in time. For long-term datasets that cover a number of years, temporal variograms will usually be dominated by the seasonal fluctuations as recorded in the image data.

4.4.3.5 Image Segmentation

In remote sensing, segmentation is the process of partitioning an image into areas that are homogeneous in terms of response or some other criteria that will be used in the segmentation process, such as texture data. In computer vision, segmentation may mean the delineation of objects in the scene, where these objects may vary in terms of response, but have sharp boundaries. The goal of segmentation in remote sensing is to identify objects of interest, usually fields or stands of different cover types. These fields usually have relatively uniform response within the field, but different response values to the surrounding fields, and they may have clearly identifiable boundaries.

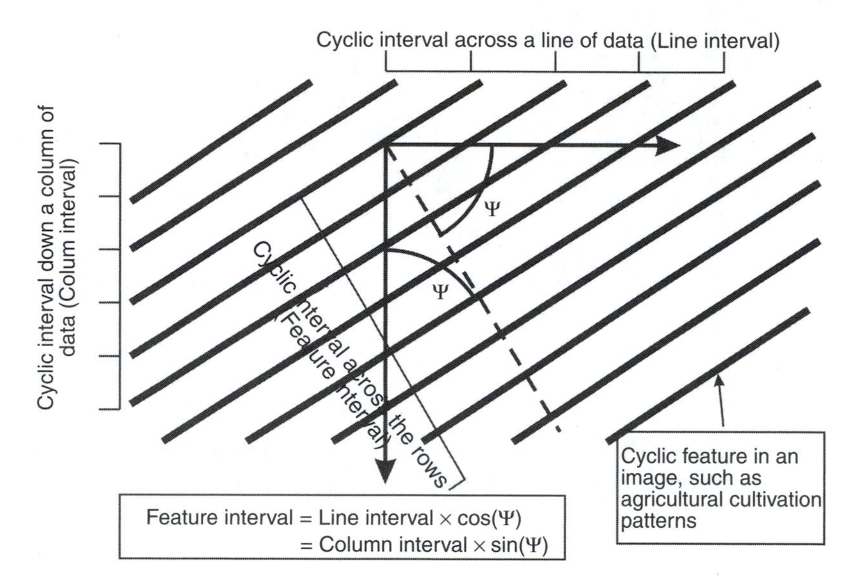

FIGURE 4.42 Sinusoidal features in image data, their orientation to the image pixels and lines and the frequency of the features.

Fields or areas that are managed in a common way usually have similar response values across their area. However, their responses can vary due to:

1. *Variations in cover type or condition across a field.* For example, a crop field can show areas that have not germinated, or in which the crop is in poor condition. In pasture fields there may be areas where the animals congregate, or where the cover conditions vary for other reasons.
2. *Isolations of other cover types within the field.* Examples are a small stand of trees within a field, a small lake or pond, or just the occurrence of farm machinery whilst the image was acquired.

Similarly, the boundaries between these areas can arise from different causes:

1. *The edge between the cover types.* When this occurs, the edge will only be obvious if the cover types have different response characteristics. If they do have different response characteristics, then the response across the edge will obey the linear mixture algorithm discussed in Section 4.5, and will appear similar to that shown in Figure 4.43(a).
2. *A linear feature may follow the edge.* Such features include roads, creeks, hedges and so forth. Often they contain a mixture of different cover types, such as the road and trees along the road. Both these types of edges are usually confined to a narrow strip between the adjacent field areas.
3. *Areas of mixed cover type.* Areas of mixed cover conditions can define boundaries. These can cover a small to a large area and may not be linear in extent.

With natural resources, the field locations tend to remain stationary over a number of seasons. This characteristic can also be used if sufficient and suitable data is available for this task.

Segmentation algorithms are generally based on one of these two basic properties of areas in the image:

1. *Detection of edges or discontinuities*
2. *Identification of similar areas (similarity)*

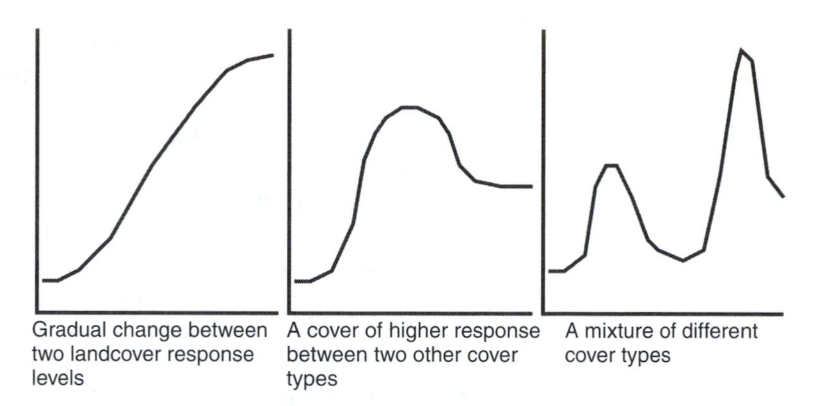

FIGURE 4.43 Types of boundaries that can be met in image data, being (a) mixtures of adjacent field cover types, (b) a different linear feature between the fields and (c) areas of mixed cover type.

4.4.3.5.1 Detection of Edges or Discontinuities

The detection of edges would normally be conducted using concepts like those discussed in Section 4.4.3.3 to identify the edge pixels. When adjacent pixels are found to contain edges, they are connected so as to create edge segments. Edge segments get broken up into edge arcs when joints are identified, as occurs when branching occurs in the edges.

In practice there are a number of difficulties with this approach:

1. *Breaks will occur in the edges*, simply because the edge has stopped at that location or because noise has made detection of the edge difficult.
2. *Multiple choices at junctions are not always easy to resolve.*

For these reasons, segmentation is rarely accomplished by the sole use of edge detection. It is most often used in conjunction with the analysis of similarity.

4.4.3.5.2 Similarity

Discontinuities represent areas of high variance, whilst similar areas represent low variance. Areas of similar response are found by region growing and splitting. This process occurs in accordance with the criteria:

1. *Completeness*. The whole image is to be partitioned into segments.
2. *Connectivity.* The pixels in a segment must connect in a reasonable way.
3. *Disjointness.* The pixels in one segment cannot belong to another segment.
4. *Segment Properties.* The pixels of a segment must meet user-defined criteria.
5. *Uniqueness.* The pixels in different segments must violate one or more of the above criteria. For example, the pixels in different segments may meet criterion (4), but they may not meet criterion (2).

The process starts by finding kernels of areas of low-variance in the image. Once a low variance kernel has been located, it is converted into a segment kernel and the addition of pixels to the segment in accordance with set criteria increases the size of the segment. This process is called region growing. This normal criterion is that the pixel will be added as long as its response values fall within a set tolerance of the segment mean, as a function of the segment variance. This is usually done by computing the Mahalonobis distance (Section 4.2) from the pixel to the class mean and testing that this falls within a specific value or range. Once the decision is made to accept a new pixel into the segment, then the segment mean and covariance statistics are re-computed by including the new pixel values in the computation. In this way it can be seen that the segment statistics will tend to change with the addition of more pixels to the segment, so that process is called the method of migrating means.

In this way the segments are grown around the kernels of low variance. When two adjacent segments connect by having a common edge, then it is necessary to determine whether the two segments should be joined into one larger segment, or whether they should remain as two segments. The normal way to do this is to use the JM distance between the class means and see if it is less than a set threshold value. If the segments are to be joined using this criterion, then the new segment statistics have to be re-computed using all of the pixels preent in both of the original segments.

The capacity of the means to migrate in this way can have disadvantages. Sometimes pixels may be rejected early in the process, because their response values were not within the set tolerances, whereas later they may be accepted, because the means might had migrated so as to fall within the tolerances. The same can happen in reverse, and so for both of these reasons, segmentation is usually conducted in an iterative manner, using the means from a previous iteration as the starting kernels for the next iteration.

If very tight constraints are applied to the criteria for inclusion in a segment, then many pixels will not fall in any of the established segments. When this occurs, the relaxation of constraints may be justified, in order to find out how the distribution of rejected pixels gets reduced with changes in the level of relaxation that are set. Making this type of judgement is a very intuitive process, requiring some machine–operator interaction.

Sometimes the merging process causes the mean and variance to diverge significantly from their starting values. When this occurs, it may be valid to investigate whether the class should be split into two. Find the two pixels in the class that are most different. Use them as new sub-class seed pixels, and investigate growing about them. Other criteria may also be imposed on the merging process. For example, there may be a lower limit set on the segment size, and there may be a limit set on the segment shape, such as the ratio of length to width.

Once region growing is complete, the process will be left with:

1. *Edges*. The edges are dealt with by means of edge detection, to define the edges.
2. *Areas of variable response, in a noisy or random sense*. These could be dealt with by means of region growing, but using more relaxed criteria than in the first pass. This step cannot overwrite segments that have been found in the first pass through the data.
3. *Areas of systematically changing response*. These areas could form segments using the first criterion. The resulting segments will be narrow, and the segments formed will be entirely dependent on the starting pixels. When this situation occurs, it may be resolved to join them using only some of the *Segment Properties* criteria, such as texture.
4. *Isolated anomalies*. These may be added to their surrounding segments.

The advantage of segmentation is that the segment statistical properties are more stable than are the statistical properties of individual pixels. This means that segmentation leads to more stable statistics for classification, with the result that classifications are often improved by the use of segmentation. The segment statistics are more stable because the pixel means are more stable, since the variance of the mean is smaller than the variance of the individual pixel:

$$\text{Segment Variance} = (\sigma_m)^2 = \text{Pixel Variance}/n = \sigma^2/n \tag{4.94}$$

4.4.3.6 Object Patterns and Object Sizes: The ALV Function

If an image contains objects that form a pattern in the image, then the average size of the object elements can be determined by the use of the average local variance (ALV) function. An object element is defined as the object and half of its immediate area between it and the adjacent objects.

An alternate way to consider it is as the average distance between objects, rather than as the size of the objects themselves. The ALV function computes the average local variance using a window of pixels to compute individual local variance values. It then moves this window by a selected step of pixels, and later lines, to compute the next local variance value, and all of these local variance values contribute to the average local variance value for that resolution in the data. The steps are thus to:

1. *Compute the image pixel values at the specific resolution*. Each coarsening of the resolution creates pixels that are larger than those at the prior resolution by 1 pixel and 1 line, and the average response within these areas are computed. Thus successive resolutions will use (1×1), (2×2), (3×3), etc, windows in the source data to compute each coarsening of the resolution in the data.
2. *Compute the local variance values*. For a window of the selected size, (2×2) is the best to use, compute the variance in the window as a local variance value. Shift the window by the selected step, the usual step is 1 pixel at a time, and later 1 line, to compute the next and subsequent local variance values.
3. *Compute the average local variance*. Use all of the local variance values to compute the average local variance value for that resolution:

$$ALV_{\mathrm{r}} = \Sigma_o^{(n-w)\text{step } k} \Sigma_o^{(m-w)\text{step } l} \sigma_{i,j}^2 / \{(n \times w)(m - w)/kl\} \quad (4.95)$$

where n = number of pixels at this resolution; m = number of lines at this resolution; w = window size; k = sampling interval along a line, usually set to 1 or w; l = sampling interval down a row, usually set to 1 or w; $\sigma_{i,j}^2$ = Window variance at sample location (i, j) and $(n \times w)(m - w)/k \times l$ = number of local variance values that will be computed.

The ALV function is constructed by deriving ALV values at each resolution, to give ALV functions similar to those shown in Figure 4.44. The general explanation for this shape is that when the pixels are very small compared to the size of the object elements, many pixels will either lie within objects, or within their surroundings, and they will thus give many low local variance values. As the pixels approach half the object size, most pixels will contain either object or background, and so the local variance will rise. The local variance reaches a peak at half of the object element size. As the pixels

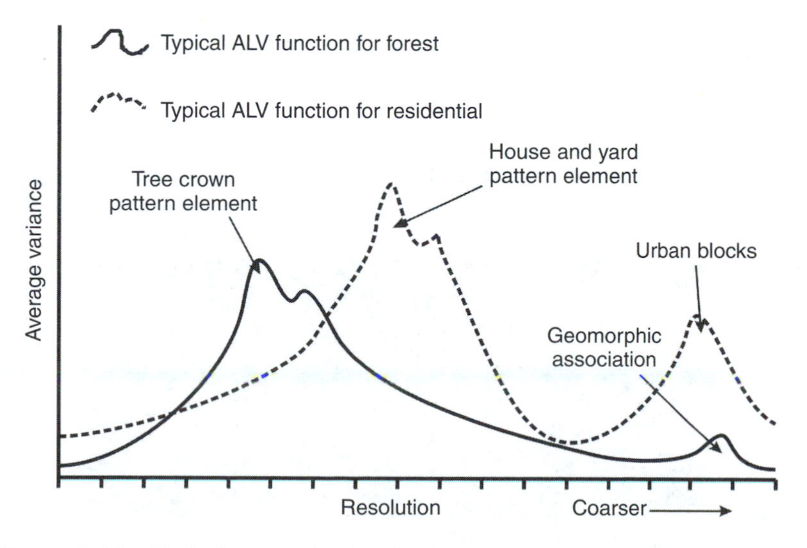

FIGURE 4.44 Typical average local variance (ALV) functions for two cover types.

continue to get larger, they will tend to contain more objects and their background, and so the average local variance will tend to decrease.

With normal images there is always a hierarchy of object sizes in the image. Within vegetation there are the leaves at the lowest level of concern in this discussion. Arrangements of leaves make up a plant, and arrangements of plants make up an ecological zone, a forest or an agricultural field. So, in vegetation there are at least three levels to be considered, and there may well be more if the terrain creates other levels. When this occurs, then peaks will occur that are related to each object pattern size. Since objects in nature are never exactly of the same size, the ALV function will exhibit spreading compared to that which would occur for an image consisting of a set of ideal object elements. With ideal object elements it can be shown that a very sharp and distinct peak will occur at the size of the object elements and a lower, broader peak will occur at half the size of the objects. In practice, the narrow and sharp peak at the object size is usually lost but the broader peak remains at half of the object pattern size. The narrow and sharp peak is usually lost because the image pixels do not exactly coincide with the edges of the objects, so that some mixed pixels occur along the edges, and because the source image pixels are not at fine enough resolution for the coarsening of the pixels to detect the fine peak. Or to put it another way, the broader peak at half the object size will always occur in the ALV function, but the narrow peak will only occur when the resolution of the source data is much smaller than the object size and the object edges coincide with the pixels in the image.

Typical ALV functions for an urban area and a forest area are shown in Figure 4.45 for the images depicted in Photograph 4.2. The dominant peaks in both functions are related to the size of the object elements in each image. The dominant object elements in the forest image are the tree crowns, so that in this case the objects are the same size as the object elements, although this will not be the case

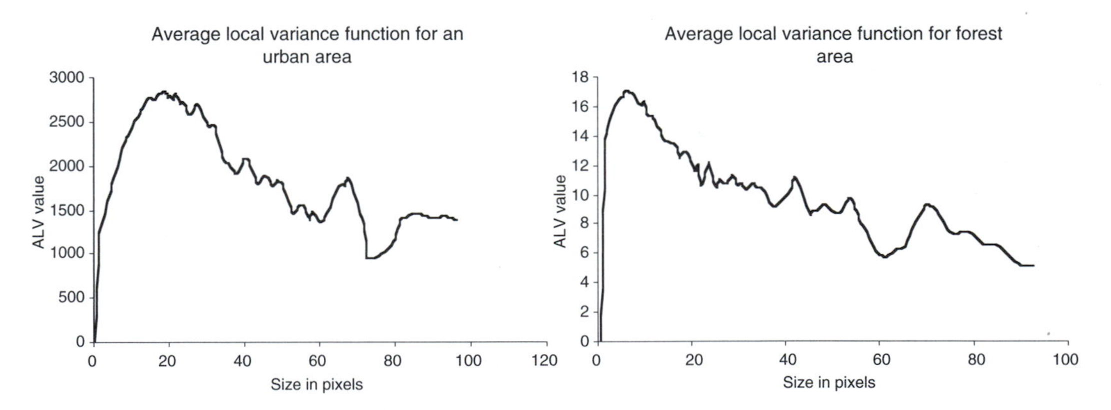

FIGURE 4.45 Average local variance functions for the two images shown in Photograph 4.2.

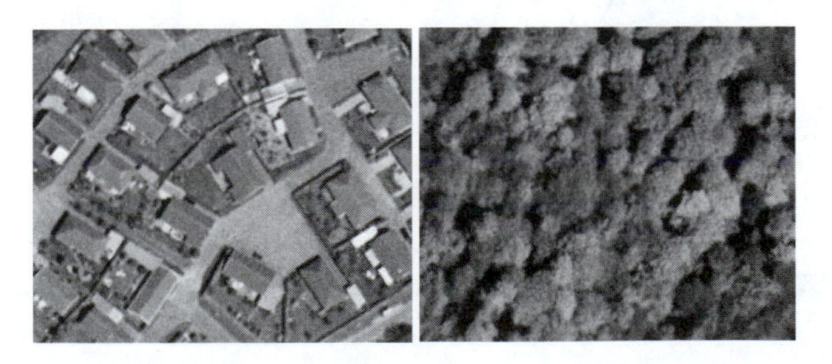

PHOTOGRAPH 4.2 The images used to derive the ALV functions shown in Figure 4.43.

when the forest has an open canopy. In the urban image the dominant object elements are the houses and their surrounding areas, so that the peak is related to half the size of the urban blocks, or to the separation between the houses rather than to the size of the houses themselves. As this separation is quite variable, so the peak is also much broader than is the peak for the forest area.

4.4.4 Temporal Enhancements

Hyper-temporal image data has been defined as a long temporal sequence of regularly acquired image data. Not only does the data capture information within each image, but the hyper-temporal data also captures information on how the covers change over time. Analysis of these changes can lead to a better understanding of the dynamic processes that are at work in the area and how they inter-relate with surface conditions.

A fundamental problem with the analysis of hyper-temporal image data is to be able to meaningfully understand the complexity of the data. With all transformations of the data, certain information is enhanced and some is lost from the display, but not from the source data. The same applies to the analysis of hyper-temporal data. Transforming the data will incur surrenders in the selected enhancement.

To provide reliable analysis of hyper-temporal imagery it is essential that the data set be internally consistent both spatially and spectrally. The minimum requirement is that the images be rectified to each other using image-to-image rectification and that the response values be consistent from image-to-image by image calibration, and then a radiometric transformation so that the images exhibit similar response values for similar reflectance conditions. However, it is often better to correct the data so that it is externally consistent, as in this way the data can be related to other types of data, and the results can be related to other types of information. To achieve this requires that the rectification be conducted using ground control for the rectification and proper correction of the radiometric values for atmospheric effects.

Temporal image analysis is likely to have one or more of the following purposes:

1. *To better understand the dynamic processes* that are recorded in the hyper-temporal image data. This focus is the display of the required information in the image data. The goal here is to reduce the effects of localised phenomena, yet to retain the effects of longer-term dynamic phenomena. These enhancements reduce the hyper-temporal image sequence into a few temporal images that are more amenable to visual analysis.
2. *To quantify aspects of these dynamic processes*, for example, to measure changes in the onset of greenup in spring and to quantify changes in the growing season of vegetation.
3. *To compare actual changes over time with those predicted by models*. This focus is interested in comparing two sets of hyper-temporal data and in reducing the data set into a smaller set of images that are more amenable to visual analysis.

These different foci of hyper-temporal image analysis are likely to use different analysis tools. This section will introduce four ways to condense and enhance hyper-temporal image data for visual analysis. Section 4.11 will focus on analytical methods that can be applied to the analysis of hyper-temporal data sets.

Four types of hyper-temporal transformation are currently available:

1. *Temporal Enhancement*
2. *Principal Components*
3. *Temporal Distance Images*
4. *Fourier Analysis*

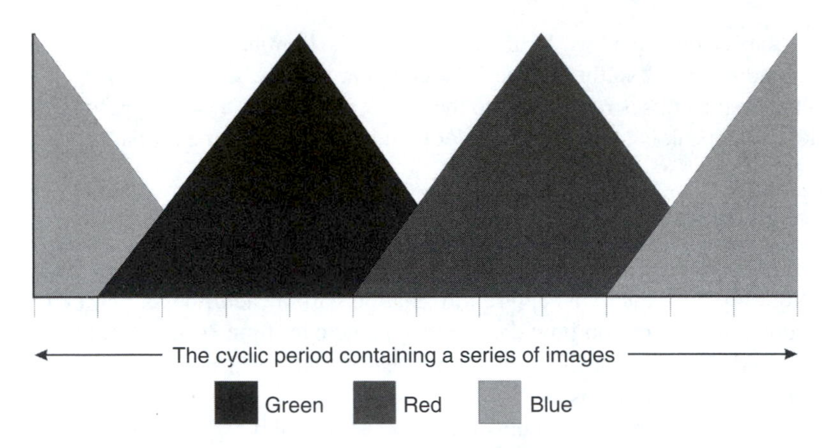

FIGURE 4.46 Colour assignment over time using the temporal enhancement.

4.4.4.1 The Temporal Enhancement

The temporal enhancement gives an image in which the strongest colour at each pixel is indicative of the period of maximum response within the cycle, and the magnitude of that response. Figure 4.46 indicates the intensities of the three primary colours at each stage in the cycle. The intensity in each primary colour assigned to a pixel is the sum of the products of the image values at that stage in the cycle, divided by the intensity at that stage. If the full cycle is a twelve-month period, then Figure 4.44 shows that January would be assigned blue, February — blue-green, March — cyan, April — green, May — green, June — yellow-green, July — yellow, August — red, September — red, October — magenta, November — purple and December — blue. If an area had maximum intensities in June, then that area would appear yellow-green in the temporal image. With annual data, the enhancement can be applied separately to each twelve-month period to give an enhanced image for each annual cycle, or to the average of all twelve monthly periods in the image data to give an average annual enhancement.

Clearly, the shape and position of the colour assignments shown in Figure 4.44 can be changed to suit the needs of the analyst. The enhancement can also be used for hyperspectral data, but to show the wavelength region of maximum intensity in the image data.

4.4.4.2 Principal Components

The concepts of principal components were introduced in Section 4.4.2.2. The concepts are exactly the same except that the data set contains the additional temporal dimension. The additional images are built onto the first image as a large multi-temporal image stack. If your image processing system can only deal with three-dimensional image data for the principal components transformation, then either each image will need to be transformed into a vegetation index, and then all of these vegetation index images used to create the one three-dimensional image stack, or a stack will need to be made of the images in a single waveband of the data.

4.4.4.3 Temporal Distance Images

The concept behind temporal distance images is that the analyst has a conceptual model of how a cover type will respond in the hyper-temporal image sequence and the temporal distance image depicts how close each pixel is to this conceptual model.

The normal way to construct a temporal distance image is to define a training area that represents the cover type of interest to the analyst. Mean and covariance arrays are created for this training

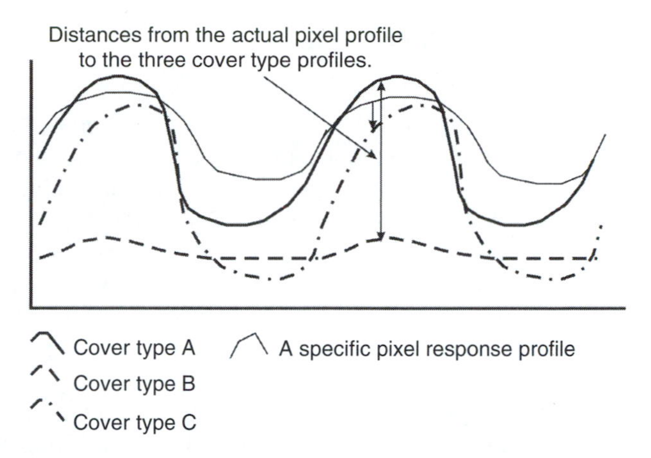

FIGURE 4.47 Construction of temporal distance images relative to each of the three cover types A, B and C.

area in each image in the sequence, and these are used to compute a selected distance metric for each pixel, in each image, from the mean and covariance values. The values for one image create a distance image for that single image. These single distance images can then be combined in some way to create a temporal distance image. The simplest method of combination is to derive the average distance across the temporal sequence. Three such temporal distance images can be combined in colour to provide a colour composite depicting the distances of each pixel from three different cover types. The concept is illustrated in Figure 4.47.

However, one could derive other forms of distance image that may be more useful, including:

1. The variance and the average distances
2. The fit of a function to the distances so as to deduce trends and cycles in the distances

Once each temporal distance image has been constructed, the images can be viewed individually or combined so as to view them in colour.

4.4.4.4 The Fourier Analysis of Hypertemporal Data

Fourier analysis has been introduced in Section 4.2.6. In temporal analysis the complex temporal signal is converted into a power spectrum displaying information on the dominant frequencies in the data, and their contribution to the overall temporal trajectory. The ability of Fourier analysis to convert a temporal trajectory into values of amplitude and phase for each frequency means that the complex temporal image sequence can often be simplified for analysis.

In most situations, the Fourier power spectrum that is derived from Fourier analysis will be dominated by two or three frequencies:

1. *The annual cycle*. Most vegetation and most climatic regions around the globe show an annual seasonality, although the magnitude of the annual cycle can vary a lot from place to place.
2. *A six-monthly cycle*. This cycle occurs where there are areas of bi-annual agriculture, as occurs in parts of India, China, Bangladesh and other tropical areas. Bi-annual cycles can also occur if the weather patterns exhibit a bi-modal annual distribution as occurs in parts of Africa.
3. *A bi-annual or tri-annual cycle*. In many places agriculture operates on a bi- or tri-annual cycle, and so these patterns can be exhibited in the Fourier power series as cycles of this duration.

4.5 ANALYSIS OF MIXTURES OR END MEMBER ANALYSIS

Mixture or end member analysis is the analysis of the spectral characteristics of pixels that either contain two or more types of landcovers or exhibit variations in the condition of the one landcover. The classic application of mixture analysis is for situations of pixels containing a mixture of two or more landcovers (mixels). This application is illustrated in Figure 4.48 for two cover types. It will be shown that when a pixel contains a mixture of two or more cover types, then the resulting mixels lie on a linear surface (a line with two cover types, a plane with three, etc.), the ends or nodes of which are defined by the response of the parent surfaces that are contained in the pixel. The linear end member model estimates position in this figure in co-ordinates that are relative to the nodes, or proportion co-ordinates.

Of equal importance is the second situation, which can occur quite frequently in nature, for example, variations in the turbidity or depth of the water in lakes, variations in the condition of vegetation and variations in surface soil colour. An example of this is illustrated in Figure 4.49, containing the mean response values for nine lakes in western Victoria (Australia) using the MSS sensor. The lakes contain water with varying levels of turbidity. The response values for the lakes show strong linear correlation. The nine lakes could be considered as all belonging to the one, turbid lake class, which has a long linear probability distribution that is correlated in some way with

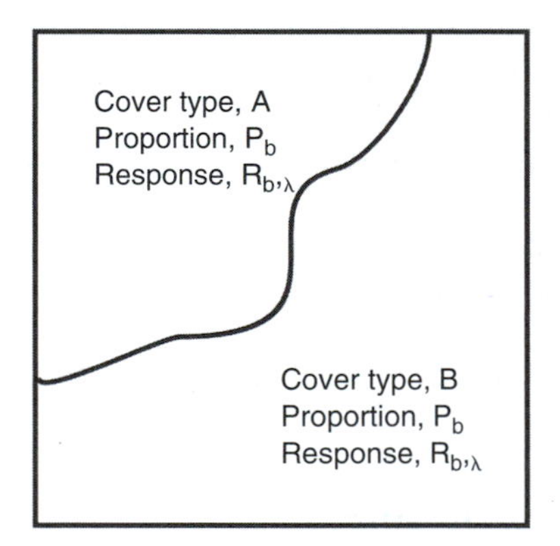

FIGURE 4.48 Mixture of two cover types in a pixel.

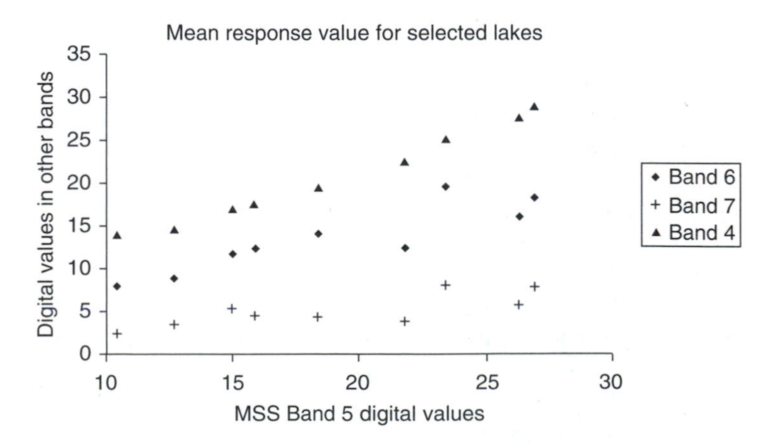

FIGURE 4.49 The response characteristics of ten lakes in Western Victoria (Australia), illustrating the linear change in response that can occur across certain landcovers.

turbidity. When natural situations demonstrate strong linear correlation of this type, then the linear end member model could be used to map the class and to estimate the position along this linearity as proportion co-ordinates relative to the end members or nodes. These proportion co-ordinates are then correlated in some way with the physical attributes that are driving the change in response, in this case turbidity. If this relationship is determined, then the proportion co-ordinates can be used to estimate the physical parameter for each pixel across the surface, or in each lake in the example discussed here.

All image pixels can be considered to be mixtures. With some conditions the mixing occurs at very fine resolutions or there are negligible changes in condition, and the pixel can be taken as imaging a homogeneous surface. At the other extreme are those pixels that straddle the interface between two or more land covers where there is a very large change in condition across the pixel. Most pixels sense a surface area that is somewhere between these extremes but which none the less contains proportions of various types of vegetative, soil and water cover conditions. For this type of situation, the analyst is interested in estimating the percentages or the condition of each landcover that is represented in each pixel. Thus, for example, the analyst may be interested in deriving more accurate statistics on the areas of crops within a field, more accurately identifying the area of dams, estimating soil colour or estimating the green and brown leaf area index within an area.

Changes of a condition occur in a landcover when, for example, there is a change in the soil colour, a change in the turbidity of the water, or some other type of change. With this type of change the cover can have a range of values in the changing attribute. Thus soil colours can cover a range of values, just as the turbidity of the water can cover a range of values and the condition of vegetation can cover a range of conditions. When this occurs, the analyst is interested in the average value of that condition in the individual pixel: that is, to derive an estimate of the soil colour, or the level of turbidity within the individual pixels.

There are thus two situations that occur with mixtures:

1. *Mixtures of finite areas of different cover types*, or pixels that consist of two or more landcovers that exist on either side of a definite interface. In this situation the mathematical surface defined by the response of the individual landcovers will be a planar surface so that the linear mixture model will be suitable to use.
2. *Variations in surface response within the one cover type*, where the different components of the surface, such as soil, green and brown herbage and water, have different densities and distributions in the different parts of the surface. If the response values lie close to a planar surface then the linear mixture model can be used. If, however, the surface defined by these response values is non-linear, then a suitable non-linear mixture algorithm will need to be applied. If there is significant multiple reflection between the different components, then the linear mixture model may not work satisfactorily.

It should be noted here that the emphasis here on the planar surface concerns the radiometric data values for the pixels. Even when the pixel response values lie on a planar surface so that the linear mixture model is suitable for use, the relationship between position on this surface and physical condition will often be non-linear. There are thus two activities that are often conducted in mixture analysis: the analysis of the data using the end member or mixture analyser to derive co-ordinates in the surface relative to the end members that are defining the surface, and relating these co-ordinates to physical conditions.

4.5.1 The Linear End Member Model

Consider a pixel consisting of two cover types, A and B, as depicted in Figure 4.50, with radiance values L_{A} and L_{B} and areas $_{\mathrm{A}}$ and A_{B}, respectively. The pixel with area A_{T}, will have a radiance

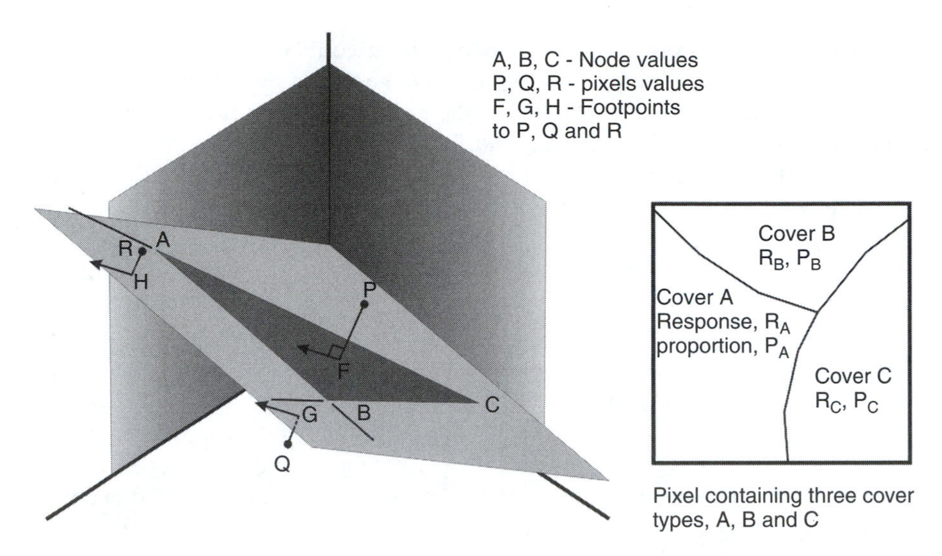

FIGURE 4.50 The domain of the linear end member algorithm for three end members in three-dimensional space. The end member response locations are at A, B and C, defining the vertices of the triangle. The relationship between the end members and proportions is shown in the figure.

value, L_T given by:

$$L_T A_T = L_A A_A + L_B A_B \tag{4.96}$$

so that

$$L_T = p_A L_A + p_B L_B$$

where p_A and p_B are the proportions of the pixel area covered by surfaces A and B respectively, and

$$p_A + p_B = 1.0$$

Now, reflectance is the ratio of radiance to incident irradiance; thus, it can be readily shown that:

$$\rho_T = p_A \rho_A + p_B \rho_B \tag{4.97}$$

This is the linear mixture model. Whilst it has been developed for just two surfaces, it can be just as easily extended to more than two surfaces, giving the general form:

$$\rho_T = \sum_{i=1}^{n} p_i \rho_i \tag{4.98}$$

If the satellite response values, R, are linearly related to radiance and hence reflectance, then this model can be applied to the data directly:

$$R_T = \sum_{i=1}^{n} p_i R_i, \quad \text{and} \quad \sum_{i=1}^{n} p_i = 1 \tag{4.99}$$

Equation (4.99) can be expanded to include n cover types in m wavebands:

$$\begin{aligned}
R_1 &= p_1 \times R_{1,1} + p_2 \times R_{2,1} + p_3 \times R_{3,1} + \cdots + p_n \times R_{n,1} \\
R_2 &= p_1 \times R_{1,2} + p_2 \times R_{2,2} + p_3 \times R_{3,2} + \cdots + p_n \times R_{n,2} \\
&\vdots \\
R_m &= p_1 \times R_{1,m} + p_2 \times R_{2,m} + p_3 \times R_{3,m} + \cdots + p_n \times R_{n,m}
\end{aligned} \tag{4.100}$$

where $\Sigma p_i = 1.0$, so that the equations can be restated in the form shown below:

$$\begin{aligned}
R_1 - R_{n,1} &= p_1 \times (R_{1,1} - R_{n,1}) + p_2 \times (R_{2,1} - R_{n,1}) + \ldots + p_{n-1} \times (R_{n-1,1} - R_{n,1}) \\
R_2 - R_{n,2} &= p_1 \times (R_{1,2} - R_{n,2}) + p_2 \times (R_{2,2} - R_{n,2}) + \ldots + p_{n-1} \times (R_{n-1,2} - R_{n,2}) \\
&\vdots \\
R_m - R_{n,m} &= p_1 \times (R_{1,m} - R_{n,m}) + p_2 \times (R_{2,m} - R_{n,m}) + \ldots + p_{n-1} \times (R_{n-1,m} - R_{n,m})
\end{aligned} \tag{4.101}$$

Equation (4.101) can be stated in the form of arrays:

$$A \times U = P \tag{4.102}$$

where A is the $(n-1, m)$ rectangular matrix derived from $(R_{i,j} - R_{n,j})$; U is the $(n-1)$ column matrix of unknown proportions, p_i and P is the m column matrix of $(R_j - R_{n,j})$.

Equation (4.101) is a matrix of m simultaneous equations in m wavebands containing $(n-1)$ unknowns where there are n nodes that can be solved if $(n-1) <= m$. The least squares method of solving these equations, when $(n-1) <= m$, finds a "best fit" solution of all equations that satisfies the criteria that the sums of squares of the residuals will be a minimum. This could be visualised as choosing the solution that minimises the Euclidean distance between the actual pixel response in m-dimensional space (the pixelpoint), and the linear surface defined by Equation (4.101) at a point called the footpoint. Using this criterion means that the solution finds a footpoint that is on the surface and closest to the pixelpoint in a Euclidean sense. Such a point is a normal from the pixelpoint onto the plane surface defined by the nodes.

Equations (4.101) can be solved by least squares to give:

$$\begin{aligned}
&A^{\mathrm{T}} \times A \times U = A^{\mathrm{T}} \times P \\
&\text{Let } A^{\mathrm{T}} \times A = B \text{ and } B^{-1} = C, \text{ so that we get } U = C \times A^{\mathrm{T}} \times P
\end{aligned} \tag{4.102}$$

where $C \times A^{\mathrm{T}}$ is a $(n-1, m)$ rectangular matrix and T represents the transpose of the matrix.

The values in array $(C \times A^{\mathrm{T}})$ will be constant if the values in array A remain constant. This is the normal situation. The values of array A may vary from pixel to pixel if:

1. *The values of a node vary from pixel to pixel.* This can occur, for example when the soil background is used as a node, and the soil colour changes.
2. *The values of a proportion are known.* This can occur when the proportions of a cover type are stable over time in comparison with the other cover types. If the proportions of the stable cover type are computed at one date, then they may be used at other dates. An example could be tree cover, for processing a number of images during a year, where the proportion of trees is considered to be stable throughout the year.

If $(C \times A^{T})$ is constant, then computation of the proportions at a pixel requires the determination of P for that pixel, where $P =$ (pixel response – response of the nth node) and the matrix multiplication $C' \times P$, where $(C \times A^{T}) = C'$.

The shape of the figure defined by Equation (4.103) is linear in m-dimensional space. The vertices of that figure are n nodes or end members with response values as given by $R_{i,j}$ for the ith end member with response value in the jth waveband. A typical triangle formed for three end members in three dimensions is shown in Figure 4.50. The solution, U, in Equation (4.103) gives the footpoint (F) on the figure, found by minimising the distance between the pixelpoint (P) and the figure. The footpoint is the Nadir point from the pixelpoint onto the planar surface of the figure. In Figure 4.50, point P represents a pixel with response values that would occur for a pixel containing just proportions of the three covers, A, B and C. Ideally P will lie directly on the plane of the triangle ABC, but this will happen rarely in practice, because of errors in the data. The footpoint-to-pixelpoint distance, PF, found by the least squares solution of Equation (4.103), is a measure of the closeness of the pixel values of P to the triangle, ABC. The derived proportions, p_{A}, p_{B} and p_{C}, can be considered as co-ordinates in the figure relative to the node points A, B and C, respectively.

From this description it can be seen that the derived proportion co-ordinates of pixels with response values identical to an end member point will all be zero, with the exception of the proportion of that actual end member, which will be one. If a pixel has response values that lie on or in the figure defined by the end members, then Equation (4.103) will derive proportions that will solve all equations in Equation (4.103) without residuals. Usually pixels will lie outside of the figure defined by Equation (4.103) in which case the solution will derive proportions for a point on the surface of the figure that is at the shortest Euclidean distance from the pixelpoint or the footpoint or FP distance. The FP distance can be used as a measure of the likelihood that the pixel belongs to the class by setting a threshold value. Below this value the pixel is considered to belong to the class and above this value it is considered to not belong to the class.

If the footpoint is on the figure defined by the end members then all of the derived proportions will be positive and will sum to one. If, however, the footpoint is on an extension of planar surfaces defined by subsets of the end members as shown in Figure 4.50 that is beyond one or more of the end members, then some of the proportions will be negative. Since they will still sum to unity, some of the proportions may be larger than 1.0. When one or more of the proportions are negative, then the least squares process has identified the shortest distance to the nearest planar surface defined by subsets of the end members, beyond the locations of the end members that define the planar surface. In Figure 4.50, pixels with values giving locations Q and R have footpoints at G and H respectively, that lie outside of triangle ABC. At G, beyond a node of the triangle, the proportion of that node will be greater than 1 and the proportions of the other two nodes will be less than 0. At H, beyond an edge of the triangle, the proportion of the opposite node, C in this case, will be negative and the proportions of A and B will sum to greater than 1. These extensions are not part of the figure defined by the end members, and so it is necessary to find a footpoint on the figure rather than on surfaces that are extensions of the figure.

This is done by the approximation:

1. Assign zero to all negative proportions and sum the remaining proportions.
2. Compute new proportions that will sum to unity by dividing the individual proportions after step 1, by the sum computed in step 1.

The effect of this process is to identify the closest point on the figure as the footpoint: for G, the closest point will be the node B; for H, the closest point will be on the edge of the triangle.

The footpoint distances are automatically calculated for pixels that have a footpoint that lies within the figure defined by the nodes, and can be calculated for all other adjusted footpoints, creating a distance image that is indicative of areas that are close to the domain defined by the nodes. This image is valuable in allowing the analyst to evaluate the suitability of the process for the task, and to identify areas that are suitable.

4.5.2 The Characteristics of the Linear End Member Model

The intrinsic characteristics of the linear end member algorithm are that it:

1. *Assumes that the end members are normally distributed.*
2. *Defines an $n-1$ dimensional linear surface* connecting n operator defined limiting conditions or end members.
3. *Transforms response co-ordinates into proportion co-ordinates* relative to the n end members.
4. *Tests for inclusion in the class by use of the shortest Euclidean distance* between the pixel values and the surface defined by the end members.
5. *Cannot be solved if the equations are numerically unstable*, as occurs when the n end members define, or approximately define an $n-i$ dimensional surface where $i > 1$. It may be possible to solve this problem by using less end members.

The extrinsic characteristics of the linear end member algorithm are that:

1. Determination of the end member response values can be done either from statistically adequate training sets or by extrapolation from the distribution of the response data in the response domain. For mixtures of monocultures there are usually satisfactory opportunities to collect adequate training data to determine mean and covariance array values for each end member. In conditions of environmental change there are rarely opportunities to identify statistically valid training data sets. Frequently there are no pixels representing the desired end members and indeed the class data distribution may be bounded by mixed pixels. In consequence, it is often impossible to define the full end member domain from training data in situations of environmental change. The alternative is to estimate the end member response values by extrapolation from analysis of the data response distribution.
2. The end members identified by analysis of the spectral distribution of data may not be true end members themselves, but mixtures of other end members (parent cover types). These end members can be used to process the data by means of the linear end member algorithm, but the resulting proportions are proportions of proportions.

   The mixtures involved in the end members used in the analysis would need to be determined by use of field data. Equation (4.100) gives the end member response of a pixel:

$$R_\lambda = p_1 R_{1,\lambda} + p_2 R_{2,\lambda} + p_3 R_{3,\lambda} + \cdots + p_n R_{n,\lambda}$$

   If $R_{1,\lambda}$, $R_{2,\lambda}$, etc are themselves mixtures of parent cover types, then, dropping the wavelength for convenience:

   $R_1 = p_{a,1} R_a + p_{b,1} R_b + \cdots$. which means that surface 1 is a mixture of surfaces a, b, etc., and

   $R_2 = p_{a,2} R_a + p_{b,2} R_b + \cdots$

and so on, so that by substitution into Equation (4.100):

$$\begin{aligned} R = & R_a\{(p_1 p_{a,1}) + (p_2 p_{a,2}) + (p_3 p_{a,3}) + \cdots + (p_n p_{a,n})\} \\ & + R_b\{(p_1 p_{b,1}) + (p_2 p_{b,2}) + (p_3 p_{b,3}) + \cdots + (p_n p_{b,n})\} \\ & + R_c\{(p_1 p_{c,1}) + (p_2 p_{c,2}) + (p_3 p_{c,3}) + \cdots + (p_n p_{c,n})\} + \text{etc.} \end{aligned} \quad (4.103)$$

3. The end member algorithm can be used to calculate proportions 1, 2, 3, etc. as long as the proportions *a*, *b*, *c*, etc. are deduced from field data and the response of surfaces A, B, etc. are observed in the field. To do this, identify the end member pixels in the field. Measure the responses of the parent cover types that constitute these end members. Use this data in the end member algorithm to compute the proportions of the parent cover types. The reliability of the classification depends upon the data used to define the end members being representative of the surface and at the extremities of the surface. The derived end members will apply to one area, at one time, and in a specific image response domain, unless the data have been calibrated for atmospheric and sensor effects.
4. The surface defined by the algorithm is a linear surface, and variations from this linearity in the class data distribution may incur errors. If the pixelpoint-to-footpoint distance is calculated, then analysis of the distribution of these distances will readily show if the data are a close fit to the planar surface or not.
5. All pixels falling within the class are assumed to be proportions of the specified end member surfaces. Situations will occur where either the response from a quite distinct surface type or a mixture of other surface types coincides with a location in the class, causing classification errors.
6. The end members define a sub-domain of the scanner data domain so that a change of one response value in the image data may result in large changes in the estimated proportions. The size of the sub-domain controls the meaningful resolution that can be achieved in transforming the data.
7. The more end members being processed then, in general, the shorter the distance between the end members and the more sensitive the processing to changes in response value. With seven end members in Landsat TM data, the inherent noise level in the data often affects the derived proportions. This sensitivity to small changes in response values also applies to end member locations: slight changes in end member position can significantly affect the derived proportions.
8. If a pixel response is affected by more end member surfaces than are used in the processing then the resulting proportions will be erroneous. However, if more end members are used than is necessary, then the derived proportions for the unrepresentative end members should be close to zero.
9. The proportions of a cover type that do not change significantly, in area, over a period of time, can be determined using one image and then applied as a known proportion in subsequent images so as to improve the analysis of the proportions of the other surface types.
10. One or more end member values can be changed from pixel to pixel, at the cost of extra processing time, as long as the end member values are stored in the appropriate form. Soil response data could be used in this way.

4.5.3 Identification of End Members

End members can be identified either by use of training data, or by analysis of the response distribution of the image data. It is difficult to ensure that training set data are in the vicinity of the extremities of the response distribution of the data of interest to the analyst. Whilst the analyst must strive to have them as close as possible to the optimum position, by accepting that the selected nodes are themselves proportions, the analyst can use field data to convert the calculated proportions to actual proportions. If the image data is in four or less bands (or contain four or less significant components from principal components analysis), then displaying all band pair combinations as two-dimensional histograms, and using training data mean values, can be used to select optimum node positions. The use of band pair histograms requires a good understanding of how to read the histograms.

4.5.4 Implementation of the Linear End Member Algorithm

The linear end member algorithm can be implemented in three ways:

1. *Standard form* in which the response values of all of the end members are constant and known and the proportions are to be determined for each pixel.
2. *Known proportion form* in which the proportion of one type of surface is assumed to be known at each pixel, having been derived using earlier data. Thus, the area of water surface, or woody cover may be assumed to be constant over a period of time. In this case, both the proportion and the response of the surface are known and can be subtracted from the response, reducing the number of unknowns.
3. *Variable response form* in which the response values for an end member vary from location to location in the image. An example of this situation is the response of soils that vary across an image. In this case the response of this surface type has to be determined separately and used as input to the processing.

4.6 IMAGE CLASSIFICATION

4.6.1 Principles of Classification

Classification is the process of partitioning an image data set into a discrete number of classes in accordance with specific criteria that are based, in part, on the individual image point data values. The image data set is the set of data values for all of the pixels in the image and the image pixel data values is the set of data values associated with an individual pixel. Classification is the process of using the pixel data values to allocate that pixel to one of a discrete number of classes. The pixel data set can contain other data, including data on the texture that uses the data for the surrounding pixels, and is added to the data set for the pixels as additional dimensions. The data values used in image processing are usually numerical values, but this is not essential in classification in the more general sense where coded and logical data can form the data set for classification.

It is often the case that pixel response values are such that the pixel could belong to a number of classes. Since each pixel is usually assigned to one class out of the alternatives (hard or sharp classification), some criteria has to be established on which the assignment criteria are to be based. These assignment criteria set decision surfaces that partition the data domain into class sub-domains. A pixel with data values on one side of a decision surface will be assigned to one class; pixels on the other side will be assigned to another class. Some classifiers do not force such an assignment, but derive probabilities of membership in the different classes, giving fuzzy classification.

Consider the case of the weights of 100 animals of two species, A and B, with all of the animals weighing between 0 and 60 units of weight. From this sample of animals it has been determined that the probability that an animal is of species A can be determined from the weight of the animal in accordance with the following equation:

$$P_r(A) = 0.95 - 0.019 \times \text{weight } (0 <= \text{weight} <= 60)$$

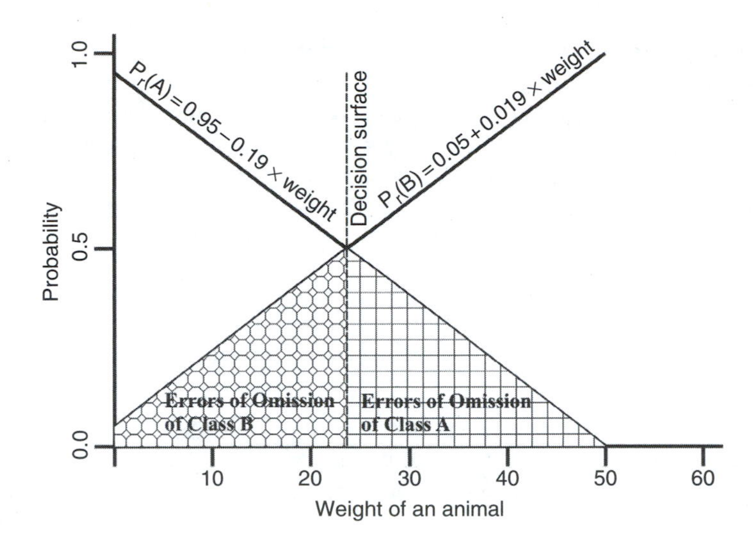

FIGURE 4.51 Probability distributions that an animal is either species A or B from its measured weight, with errors of classification (a) and (b) for the decision surface set at $W = 23.684$.

This probability distribution is shown in Figure 4.51. All of the animals are either of species A or B and the probability distribution of B is:

$$P_r(B) = 0.05 + 0.019 \times \text{weight}$$

If $P_r(A) = P_r(B)$ then there is equal probability of the animal being species A or B. Below this weight, found to be 23.684 by simultaneous solution of the above equations, $P_r(A) > P_r(B)$, the probability of the species being A is greater than the probability that it is B, so the decision is made to assign the animal to species A. In consequence, all animals below a weight of 23.684 will be assigned to species A. Above this weight the probability of the species being B is greater than the probability that it is A and so all animals will be assigned to species B. This value, where the decision changes, is called the decision surface.

The decision surface is the surface in the data domain where the decision assignments change from one class to another.

If a population of animals of species A and B are classified using this probability distribution, then some animals will be assigned to the wrong species class since some species A animals are likely to weigh more than 23.684, and some animals of species B will weigh less than 23.684. This gives rise to errors in classification. These errors of misclassification, shown in Figure 4.51 as the patterned areas, are of two types:

1. Errors of Omission when an area of one class is assigned to another class, or omitted from its correct class. In Figure 4.51 the errors of omission are shown as the two patterned figures. The errors of omission for class A is the triangle to the right of the decision surface and the errors of omission for class B are the rectangle and the triangle patterned to the left of the decision surface.
2. *Errors of Commission* when an area of another class is included with the class of interest.

In the case of the animals being classified into species A or B, the categories of errors of omission and commission depend on your perspective. From the perspective of species A, the errors of omission are all animals that are species A but that weigh more than 23.684 and hence are classified as being species B. The errors of commission are all of those animals that are species B but weigh less than 23.684. From the perspective of class B, the errors are reversed. Most situations are more complex than this, involving more than just two classes, in which case the errors of omission and commission do not usually exactly match and counter each other. With more complex situations, involving many classes, the errors of omission and commission can be constructed as an error matrix as is discussed in Chapters 5 and 6.

Misclassification incurs costs and the costs of misclassifying one class may be different to the costs of misclassifying another class. Consider in the case discussed above, that it may be more expensive to misclassify animals of species A than animals of species B. If the costs of misclassification are to be considered then the decision surface needs to be changed so that the costs of misclassification are minimised. Consider the situation where it is more expensive, by a factor c, to misclassify A than B. Consider Figure 4.51. If the costs of misclassification are identical, then equal patterned areas will represent equal numbers misclassified, and hence equal costs. In the above case, the areas under each curve for a decision surface at weight = 23.684 can be found:

$$\text{Proportion of A misclassified as B} = \tfrac{1}{2} \times (50 - \text{weight}) \times (0.95 - 0.019 \times \text{weight}) = 6.58$$

$$\text{Proportion of B misclassified as A} = (0.05 \times \text{weight}) + \tfrac{1}{2} \times \text{weight} \times (0.05 + 0.019 \times \text{weight} - 0.05) = 6.51$$

So, using a decision surface at 23.684 will give more errors of misclassification of class A than of class B. If the costs of misclassification are the same, then it would be better to set the decision surface so that the areas under the curve that represent errors should be set to be the same size. That is:

$$\tfrac{1}{2} \times (50 - \text{weight}) \times (0.95 - 0.019 \times \text{weight}) = (0.05 \times \text{weight}) + \tfrac{1}{2} \times \text{weight} \times 0.019 \times \text{weight}$$

This equation can be solved to give a weight of 23.75 as setting the same error areas under the curve. To take the general case, if the costs of misclassification are not equal then the areas (A) and (B) need to be adjusted so that:

$$\text{Misclass. Count of A} \times \text{Cost}_{\text{A}} = \text{Misclass. Count of B} \times \text{Cost}_{\text{B}}$$

Since the counts are proportional to the area under the curves, then:

$$\text{Area}_{\text{A}} \times c = \text{Area}_{\text{B}} \tag{4.104}$$

where c is the ratio of the costs of misclassification, [Cost A/Cost B]. If the areas (A) and (B) are adjusted so that:

$$\text{Area}_{\text{A}}/\text{Area}_{\text{B}} = 1/c$$

then the costs of misclassification will be minimised. To achieve this in the above example requires the decision surface to be shifted to make area (A) smaller and area (B) larger so that the ratio of these areas reflects the ratio of the costs of misclassification. Consider the example discussed above:

$$\text{Area } A = \tfrac{1}{2}(50 - \text{weight})(0.95 - 0.019\ \text{weight}) \tag{4.105a}$$

and

$$\text{Area } B = (0.05 \text{ weight}) + \tfrac{1}{2} \text{ weight} \times 0.019\text{weight} \tag{4.105b}$$

So that $c[\frac{1}{2}(50 - \text{weight})(0.95 - 0.019\text{weight})] = (0.05\text{weight}) + \frac{1}{2}\text{weight}0.019\text{weight}$

If $c = 1$ then this equation can be solved as above to give weight $= 23.75$, different to the point of intersection at 23.684 since the two lines, of $P_r(\text{A})$ and $P_r(\text{B})$ are not symmetrical and hence the two areas under the curve do not have the same area. If $c \neq 1$ then the equation can be put in the form of a quadratic equation in W for solution:

$$0.019(c - 1)\text{weight}^2 - (1.90c - 0.1) \times \text{weight} + 47.5 \times c = 0$$

In the case where $c = 2$, the equation can be solved to give $W = 30.43$ as the decision surface.

Classification contains a number of important concepts, the underlying assumptions and philosophies of which need to be understood to achieve reliable and consistent results. The concepts are:

1. *The use of numerical models are used to establish decision surfaces.* Partitioning the data sets into class domains is usually done by means of a classifier algorithm. The shape and position of the decision surfaces depend on the characteristics of the algorithm. If the algorithm makes assumptions about the distribution of class data that do not accord with how the data are actually distributed in the data set, then the decision surfaces will be wrongly positioned and errors will occur in the classification. It is therefore essential to understand the theoretical model on which the classifier algorithm is based, and be sure that the actual data is a satisfactory fit to this model.

In the earlier example of the weights of 100 animals of two species, the probability distribution was shown for that sample in Figure 4.51. However, a second sample of animals from the population may give a probability distribution of the weights of animals of species A as being normally distributed as shown in Figure 4.52. This figure shows species A as having a normal probability distribution, with a mean value of 17.0 and standard deviation of 4.0. If the probability that an animal is species B is the same as before, that is $p(\text{B}) = 0.05 + 0.019 \times W$, then animals weighing less than 8 and more than 23 will be classified as being in species B, assuming that the decision surface is set at $P_r(\text{A}) = P_r(\text{B})$.

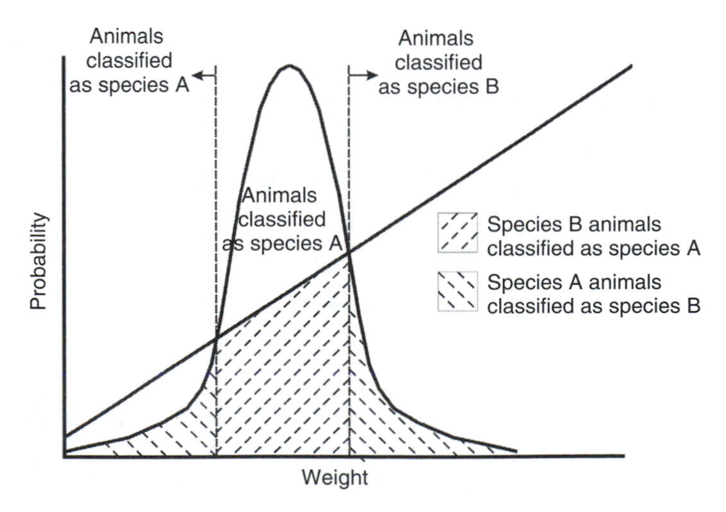

FIGURE 4.52 Probability distribution for a second sample of animals from the population sampled to construct the probability distributions in Figure 4.51.

If the analyst is in doubt as to the suitability of the theoretical model on which the classifier algorithm is based, then the goodness of fit of the actual data to the model that forms the basis of the classifier should be determined using appropriate statistical procedures such as the chi-square goodness of fit test.

2. *Use of parameters for those models that are usually derived from field data.* The values used for the algorithm parameters significantly affect the shape and position of the decision surfaces. In the example above, both of the algorithms discussed had two parameters. In the first algorithm the parameters were a gain and offset whilst in the second algorithm the parameters were a mean and variance. Changing any of these values affects the probability distribution, the position of the decision surfaces and hence the accuracy of the classification.

Estimation of parameter values is usually done by use of a sample of data selected for that purpose, or training data. The sample of data must be representative of the class, that is, it needs to represent the range of conditions that apply to the class, and it must be of an adequate size to give representative values for the class. For example, in the case of the two samples of animals from the one population as discussed above, there is clearly something wrong if two samples can result in such different probability distributions. The samples may be too small, or they may not truly represent the class conditions. Discussion on methods of collecting field data to establish class parameters is included in Chapter 5 on field data collection.

An important factor to consider in selecting the size of the sample is the sensitivity of the algorithm to slight variations in the derived parameter values. For example, the mean of a set of data is relatively insensitive to the inclusion of a few wild or erroneous pixels within a large sample data set, whereas the variance is very sensitive to such data values. If the algorithm is sensitive to the inclusion of such data then care needs to be taken to ensure that erroneous data are filtered out of the data set before determining the class parameters.

3. *Decisions are based on "Spectral" data.* Classifier algorithms allocate a pixel to a class either on the data set belonging to the pixel, or they can utilise additional data about the pixel or data derived from the surrounding pixels. Most classifiers limit classification to the use of the spectral data for the pixel and as such are called spectral classifiers. However, there are many situations where other data sets can significantly improve classification accuracy. For example, certain classes of tree species may have similar spectral values in the image data, but they may grow in distinct climatic regimes reflected in differences in elevation, gradient and aspect or temperature and rainfall, or a combination of these. In a situation like this, inclusion of additional data sets, effectively as another band of data, can improve the classification accuracy.

Another type of information that may improve classification accuracy under some conditions is the inclusion of texture as a source of discrimination. This can be done by processing the image data to create a data file of texture which can then be used as an additional band or dimension of data in the classification process.

4. *Partitioning this spectral domain into discrete classes.* Once the classification parameters are defined they create decision surfaces within the data domain such that all pixel values within the one sub-domain, as specified by the decision surfaces, will be assigned to the one class. The resulting classification is uniform within the class sub-domains.

In some circumstances this partitioning is a valid reflection of actual conditions, such as the mapping of landcover into major classes of cultural, herbage, woody cover, water, snow and ice, and soil when these land covers are quite discrete. In other situations, when the different conditions are not so discrete, then partitioning the image in this way introduces artificialities that can introduce errors in the classification and hence into the derived maps and statistics. Consider the case of pastures that can consist of varying amounts of soil, green and brown herbage, woody vegetation and water. The one class of pasture can consist of a wide variation in conditions and contributions by the different covers where these variations can occur in quite gradual and subtle ways. The variations, rather than being discrete steps from one cover type to another, are a gradual change from one to another as part of a continuum of conditions within the one pasture class. Many of the spectral classifiers cannot handle these within-class variations in a satisfactory way for a number of reasons:

(a) *Inadequate training data*. Because the response values are changing from pixel to pixel due to changes in surface conditions, it is often very difficult to define areas that are physically homogeneous and can therefore act as a training area of adequate size to define class parameters.
(b) *Poor fit of the data to the normal distribution*. Whilst the whole of the class area can be defined to establish training areas, the data within this area may be a poor fit to the normal distribution due to variations in response caused by variations in physical conditions. One solution is to cluster the data (Section 4.7), or create a number of spectral classes within the separate physical classes, in the hope that the clustering process will create classes that are a reasonable fit to the normal distribution.
(c) *Inadequacy of spectral classes* for estimating condition parameters. Classification of the data into discrete spectral classes provides a basis for estimating conditions in the physical class by assigning average conditions to all pixels in each class on the basis of field or other data. But actual conditions may not accord with the average weighting, introducing errors into images of conditions, and into statistical estimates derived from the image data.
(d) *The existence of mixed pixels* creates problems for most of the spectral classifiers.

The most commonly used spectral classifiers are:

1. *Discriminant function classifiers*
2. *Fuzzy classifiers*
3. *Neural network classifiers*
4. *Hierarchical classifiers*

4.6.2 Discriminant Function Classifiers

Consider an image containing a number of identifiable classes, from which the analyst has determined the mean and covariance arrays for each class. If a pixel has values that coincide with the mean values for a class then the pixel could reasonably be expected to belong to that class. It could be said to have a high probability of belonging to the class. As the pixel values move away from the mean value for a class, the probability that the pixel belongs to the class will decrease. If the pixel is simultaneously moving towards another class mean then the probability that the pixel belongs to the second class are increasing. At some stage these probabilities, weighted by the cost of misclassification, will become equal. The decision surface between those two classes is set at this point. On one side of the decision surface the pixel will be assigned to one class and to the other class on the other side of the decision surface.

The simplest decision surfaces are when the Euclidean distances from the pixel values to the class means are calculated, and the pixel assigned to the class with the shortest Euclidean distance. This type of algorithm is called the minimum distance classifier. It is a special case of the full discriminant function classifier where all of the classes are assigned the same variance in all of the bands, and the covariance is assumed to be zero. The class probability distributions are thus all spherical and of the same size or radius and the decision surfaces are planar surfaces in the n-dimensional space (Figure 4.53).

In practice, the covariance arrays for the different classes are usually quite different, indicating different variances in the different wavebands in the different classes, and different amounts of correlation. Furthermore, if one class occurs more frequently in an image then there is a higher probability that a pixel is of that class rather than other classes that occur less frequently. Finally, there may be costs associated with misclassifying a pixel as one class relative to misclassifying the

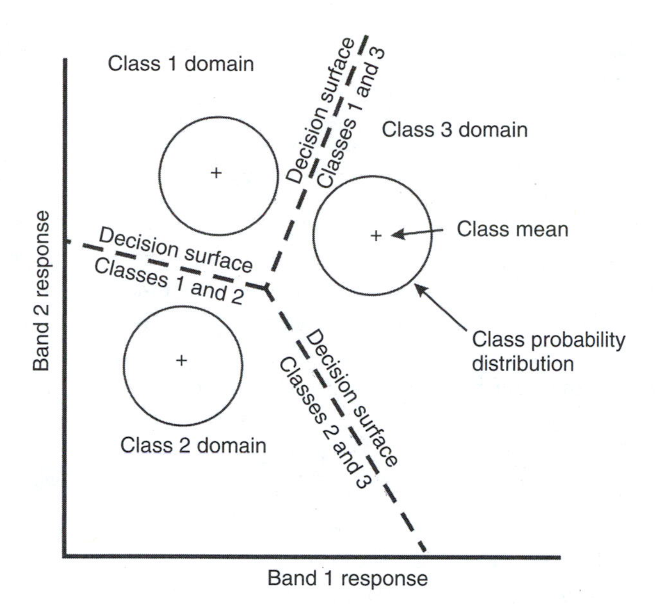

FIGURE 4.53 The probability distribution and decision surfaces for three classes in two dimensions for the minimum distance classifier.

pixel as another class. Incorporation of these concepts into classification has lead to the development of the maximum likelihood version of the discriminant function classifier.

4.6.2.1 Development of the Maximum Likelihood Classifier

Consider that there are k classes, $W_1, W_2, \cdots, W_k$ and a pixel with values in array X_p is to be assigned to one of these classes. The probability of the different classes being in the area sensed by the pixel is $P(W_1), P(W_2), \cdots, P(W_k)$ for each class in turn where:

$$\Sigma P_r(W_i) = 1.0 \tag{4.106}$$

If the area covered by all classes is equal, or assumed to be so, then there is equal probability of each class coinciding with the area of the pixel. In this situation these probabilities are all equal to:

$$P_r(W_i) = 1.0/k \text{ for } i = 1 \text{ to } k \tag{4.107}$$

In practice it is usual to assume equal probabilities for all classes even though this assumption is rarely true. It would be much better to estimate the area of each class within the area and use these values to calculate the probabilities of the occurrence of the classes for use in the subsequent classification.

The probability density function for a class, $P_r\{X_p/W_i\}$ is the probability that a point, p with values in array X_p is in the class W_i. The function assumes a distribution associated with the data for the class, and the parameters for that distribution have to be determined before the classification can take place. Usually the normal distribution is accepted as the probability distribution of the data and the parameters for this distribution are the mean vector, U_i and the covariance array, Σ_i.

The parameters of the distribution are usually determined either using training data or by the use of clustering algorithms (Section 4.7). The first results in supervised classification and the second in unsupervised classification; the relative merits and applications of both are discussed later in this section.

Use of the normal distribution means that the probability density function for $P\{X_p/W_i\}$ is multivariate gaussian of the form:

$$P_r\{X_p|W_i\} = \frac{e^{-0.5\times(X-U)^T\Sigma^{-1}(X-U)}}{2\pi|\sum|} \tag{4.108}$$

where $(X - U_i)$ is the linear array of pixel response values minus the mean response values for class "i", $(X - U_i)^T$ is the transpose of this matrix, Σ_i^{-1} is the inverse of the covariance matrix for class "i" and $|\Sigma_i|$ is the determinant of the covariance matrix. For simplicity the "i" will be dropped whenever it does not lead to a loss of clarity.

Bayes rule states that the probability of class i occurring for a given value of X_p is given by:

$$P\{W_i|X_p\} = \frac{P_r\{X_p|W_i\} \times P_r\{W_i\}}{\sum P_r\{X_p|W_j\} \times P_r\{W_j\}} \tag{4.109}$$

Bayes also states that there are costs incurred in misclassification, and that these costs might not be the same in the misclassification the different classes. Construct these costs as a loss function, $L\{W_i/W_j\}$ which is the loss or cost incurred when a pixel is assigned to class i when it is in class j. The loss function is a two-dimensional array of size $k \times k$ when there are k classes. If the loss function specifies unit loss when $i\#j$ and zero loss when $i = j$, that is, for correct classification, then the loss function is:

$$\begin{aligned} L\{W_i/W_j\} &= 1 \qquad i\,\#\,j \\ &= 0 \qquad i = j \end{aligned}$$

so that:

$$L\{W_i/W_j\} = \begin{matrix} 0 & 1 & 1 \\ 1 & 0 & 1 \\ 1 & 1 & 0 \end{matrix} \quad \text{when } k = 3 \tag{4.110}$$

Consider an example where a pixel is to be mapped into one of three classes. The probabilities that the pixel is of class i, given the actual pixel values, X_p are given by $P\{W_i/X_p\}$, and calculated to be 0.3, 0.5 and 0.2 for classes 1, 2 and 3 respectively. Now the weighted probability, with the weighting given by the loss function in Equation (4.109) above, is:

$$R\{W_i/X_p\} = \Sigma L\{W_i/W_j\} \times P_r\{W_j/X_p\} \text{ for } j = 1, k \text{ and } j\#i \tag{4.111}$$

and in the above example:

$$\begin{aligned} R\{W_1/X_p\} &= 1 \times 0.5 + 1 \times 0.2 = 0.7 \\ R\{W_2/X_p\} &= 1 \times 0.3 + 1 \times 0.2 = 0.5 \\ R\{W_3/X_p\} &= 1 \times 0.3 + 1 \times 0.5 = 0.8 \end{aligned}$$

As the intention is to minimise cost, the class with minimum $R\{W_i/X_p\}$ will be chosen, in this case class 2. It will be noted that for the pixel values used in this case, stored in array X_p, the $P\{W_2/X_p\}$ has the highest value. The zero loss function may not always be the most suitable to use. Consider, for the case above, the use of a loss function of the form:

$$L = \begin{matrix} 0.00 & 0.33 & 0.67 \\ 1.00 & 0.00 & 1.50 \\ 0.33 & 0.50 & 0.00 \end{matrix}$$

then

$$R\{W_1/X_p\} = 0.00 \times 0.3 + 0.33 \times 0.5 + 0.67 \times 0.2 = 0.31$$
$$R\{W_2/X_p\} = 1.00 \times 0.3 + 0.00 \times 0.5 + 1.50 \times 0.2 = 0.60$$
$$R\{W_3/X_p\} = 0.33 \times 0.3 + 0.50 \times 0.5 + 0.00 \times 0.2 = 0.35$$

so that class 1 would be chosen. The procedure then is to compute $[R\{W_i/X_p\}, i = 1, k]$ and select the class with the minimum value of $R\{W_i/X_p\}$. Because the selection is done by comparison or relativities, components of the equation that have an equal effect on all values of R can be ignored, thereby eliminating unnecessary computation. This is done by defining a discriminant function, $G_i(X)$ for class i as:

$$G_i\{X\} = -R\{W_i/X_p\} \tag{4.112}$$

Then the pixel will be assigned to the class with the largest value in the discriminant function.

$$G_i\{X\} = -\sum_{j=i} L\{W_i \mid W_j\} \times P_r\{W_j \mid X_p\} \tag{4.113}$$

so that

$$G_i\{X\} = -\sum_{j=i} L\{W_i \mid W_j\} \times \frac{P_r\{W_i\} \times P_r\{X_p \mid W_i\}}{\sum P_r\{W_j\} \times P_r\{X_p \mid W_j\}}$$

If the normal distribution is adopted as the probability density function for $P_r\{X_p/W_j\}$, then:

$$G_i\{X\} = \frac{-\sum L\{W_i \mid W_j\} \times P_r\{W_i\} \times \mathrm{e}^{-0.5\times(X-U)^{\mathrm{T}}\Sigma^{-1}(X-U)}}{2\pi \times \left|\sum\right| \times \sum P_r\{W_j\} \times P_r\{X_p \mid W_j\}} \tag{4.114}$$

The constant components of the denominator can be ignored without affecting the relative discriminant function values between the classes. The generalised form of the discriminant function is:

$$G_i\{X\} = \frac{-\sum L\{W_i \mid W_j\} \times P_r\{W_i\} \times \mathrm{e}^{-0.5\times(X-U)^{\mathrm{T}}\Sigma^{-1}(X-U)}}{\left|\sum\right|} \tag{4.115}$$

It can be seen that there are k products $P_r\{W_j\} \times P_r\{X_p/W_j\}$ to be computed and then k multiplications of $L\{W_i/W_j\} \times P_r\{W_j\} \times P_r\{X_p/W_j\}$ for each discriminant function so that there are $k \times k$ computations for each pixel. This is a very considerable processing cost that means that the zero loss function is usually adopted to reduce the processing load to k computations per pixel. When the zero loss function is adopted the discriminant function becomes:

$$G_i\{X\} = \frac{-1 \times P_r\{W_i\} \times \mathrm{e}^{-0.5\times(X-U)^{\mathrm{T}}\Sigma^{-1}(X-U)}}{\left|\sum\right|} \tag{4.116}$$

The -1 is common to all of the discriminant functions so that it can be ignored without affecting the relative values of the functions. The discriminant functions can be further simplified by taking the logarithm of both sides:

$$g_i\{X\} = \ln(G_i\{X\}) = \ln(P_r\{W_i\}) - 0.5 \times (X - U_i)^{\mathrm{T}}\Sigma^{-1}(X - U_i) - \ln\left(\left|\sum_i\right|\right) \quad (4.117)$$

The form of the discriminant function in Equation (4.117) is inherently quadratic because of the $(X - U_i)^{\mathrm{T}}(X - U_i)$ term. The decision surface can be circular, elliptical, parabolic, hyperbolic or linear depending on the parameter values for the different classes. Further assumptions can be made which will result in simpler forms of discriminant functions. Consider the case where all variances are assumed to be equal to S, and there are zero covariances.

$$\Sigma_i = S \times I \quad (4.118)$$

where I is the identity matrix. Then:

$$g_i\{X\} = \ln(P_r\{W_i\}) - \frac{(X - U_i)^{\mathrm{T}}(X - U_i)}{2S} - \ln(S) \quad (4.119)$$

Since S is the same for all classes, the value $\ln(S)$ will be the same for all classes and so will not affect the relative values of the discriminant functions. It can thus be discarded. If the remainder of $g_i\{X\}$ is expanded then the $X \times X$ product will be the same for all discriminant functions and can be discarded. Removal of these two components gives discriminant functions of the form:

$$g_i\{X\} = \ln(P_r\{W_i\}) + XU_i/S - (U_i)^2/2S \quad (4.120)$$

In this equation the first and last components are independent of the values in the individual pixels and so they can be computed once at the start of the classification to give a constant, K_i for each class so that the discriminant functions simplify to the form:

$$g_i\{X\} = K_i + X \times U_i/S \quad (4.121)$$

This form of the discriminant function is a linear function in X resulting in linear decision surfaces. Since $(X - U_i) \times (X - U_i)$ is the square of the Euclidean distance between the pixel values and the class means, this algorithm is called the minimum distance classifier, introduced at the start of the section.

A second simplification of the main form of the discriminant functions as given in Equation (4.121) is to assume that all classes have the same covariance array, Σ. From Equation (4.121):

$$g_i(X) = \ln(P\{W\}) - \frac{X^2\Sigma^{-1}}{2} + XU\Sigma^{-1} - \frac{U^2\Sigma^{-1}}{2} - \ln\left(\left|\sum\right|\right) \quad (4.122)$$

in which $X^2\Sigma^{-1}$ and $\ln(|\Sigma|)$ are the same for all classes and can be deleted as they will not affect the relative values in the discriminant functions. The discriminant functions simplify to:

$$g_i(X) = \ln(P\{W\}) + UX\Sigma^{-1} - \frac{U^2\Sigma^{-1}}{2} \quad (4.123)$$

This form of the discriminant function is also a linear equation in X so that it also gives linear decision surfaces.

If the probabilities of the classes occurring at a pixel are assumed to be equal, that is, $P_r\{W_i\}$ is the same for all classes, then this can also be eliminated. This assumption is usually made in classification without good justification. Often the reverse of applying different probabilities $P_r\{W_i\}$ to the different classes is more valid, either as constant values across an area, or varying as a function of conditions. If a pixel is surrounded by pixels of a specific class then there is a higher probability that the pixel belongs to the same class. In the simplest case, the probabilities $P_r\{W_i\}$, can be taken as the same as the proportions of the classes within the area being mapped even if these proportions are estimated using a prior classification.

4.6.2.2 Summary

The four versions of the discriminant function classifier are:

1. *Full maximum likelihood classifier*

$$G_i\{X\} = \frac{-\sum L\{W_i|W_j\} \times P_r\{W_i\} \times e^{-0.5\times(X-U)^{\mathrm{T}}\Sigma^{-1}(X-U)}}{2\pi \times \left|\sum\right| \times \sum P_r\{W_j\} \times P_r\{X_p \mid W_j\}} \qquad (4.114)$$

   This form of the algorithm is rarely used because of the high processing costs that would be incurred.

2. *Maximum likelihood classifier*

$$g_i\{X\} = \ln(G_i\{X\}) = \ln(P_r\{W_i\}) - 0.5 \times (X - U_i)^{\mathrm{T}}\Sigma^{-1}(X - U_i) - \ln\left(\left|\sum_i\right|\right) \qquad (4.117)$$

   Adopts the zero loss function, but still provides a very powerful classification algorithm. This is the commonly used version of a "full" maximum likelihood classifier.

3. *Minimum distance classifier*

$$g_i\{X\} = ln(P_r\{W_i\}) + XU_i/S - (U_i)^2/2S \qquad (4.120)$$

4. *Common covariance classifier*

$$g_i\,(X) = \ln\,(P\,\{W\}) + UX\Sigma^{-1} - \frac{U^2\Sigma^{-1}}{2} \qquad (4.123)$$

4.6.2.3 Characteristics of the Discriminant Function Family of Classifiers

The characteristics of the maximum likelihood family of classifiers are:

1. *The classes are weighted by the probability of occurrence, $P_r\{W_i\}$, of the class.* Usually all classes are assumed to have the same probability of occurrence; using better estimates of this probability will have a significant effect when the probabilities of all classes are dissimilar.
2. *The classes are weighted by the cost of misclassification.*
3. *The classifier assumes that the remotely sensed data for all classes are normally distributed*, so that deviations from normality in the data will incur errors in the classification. If the

classes are of surfaces that are fairly homogeneous in their landcover characteristics, such as agricultural crops, then the data are usually a good fit to the normal distribution. If the classes contain significant variation in landcover condition, as can occur in other landcovers including forested areas, water surfaces and rangelands then the data may not be a good fit to the normal distribution.

The goodness of fit of the data to any specified distribution can be tested statistically. The data should be so tested if the analyst is in doubt about the goodness of fit of the data to the statistical model being used. This analysis can only be done after the collection of field data, and not during the planning stage. To make reasonable judgements during the planning stage requires the analyst to consider the reflectance characteristics of the surfaces in question, and how changes in surface conditions might affect reflectance.

A method of handling the problem of the data not being a good fit to the assumed distribution is to have the domain covered by a landcover class be represented by more than one spectral class in the classification. Partitioning the training data into a number of spectral classes is usually done by means of clustering algorithms (Section 4.7).

4. *The decision surfaces of the maximum likelihood classifier are curvilinear whereas the simplifications of the discriminant function algorithm have linear decisions surfaces.* The magnitude of the covariance arrays can significantly affect the classification when using the maximum likelihood classifier. For example, a class with a small covariance array near a class with a large covariance array may have decision surfaces that define an island within the domain of the second class.

4.6.2.4 Implementation of the Maximum Likelihood Classifier

Implementation of the linear simplifications of the discriminant function classifier can be done by simplifying some of the steps in the implementation of the full classifier. The steps in implementing the full classifier are:

1. *Set class parameters,* $N\{U_i, \Sigma_i\}, i = 1, n$ for the n classes. Normally this is done by defining training areas for all of the information classes using the image data as a base map. If the spectral classes for the individual information classes are not a good fit to the normal distribution, or if the analyst considers that they may not be, then clustering techniques are used to create a number of spectral classes within the information classes of concern.
2. *Decide on class weights,* $P_r\{W_i\}$. Often these are made to be equal for all of the n classes, that is $[P_r\{(W_i\} = 1/n, i = 1, n]$.
3. *Set classification threshold.* The classification process uses Equation (4.116) to derive values for each $g_i\{X\}$ at each pixel. The pixel is then assigned to the class for which $g_i\{X\}$ is a maximum (or the smallest negative value). Adoption of this practice will assign each pixel to a class since each pixel will give values for each $g_i\{X\}$. However, the response of some pixels may be such that their response is a long way from all of the classes, that is $g_i\{X\}$ will be large negative for all classes. In this case the probability that the pixel belongs to any of the classes is low. The operator may decide to set minimum probabilities in the classification in which case pixels that have $g_i\{X\}$ values larger negative than this threshold would not be assigned to any class.
4. *Implement the classification.* The classification may be done for the whole area of the image, or for portion of the image, for example, for all areas that are held freehold, or all areas on specific soil types. If the classification is to be constrained to certain conditions in this way then the classification will be conducted within a mask of the areas to be classified. The source of the mask will depend on its purpose, but it would normally be derived either from maps or from previous image processing.

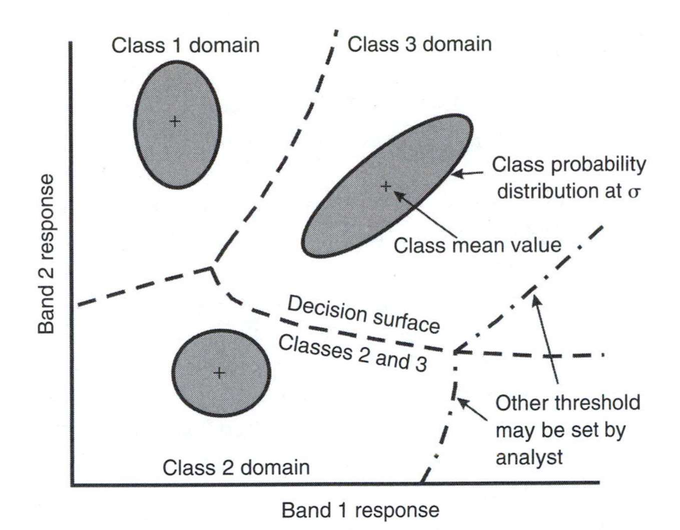

FIGURE 4.54 The maximum likelihood classifier for three classes in two dimensions showing the decisions surfaces for these three classes and the capacity to set a threshold or limit to the classification so as to reject wild pixels.

5. *Conduct accuracy assessment*. The conduct of accuracy assessment is an essential component in any classification task. Accuracy assessment must be conducted in accordance with strict statistical rules, as discussed in Chapter 5.

4.6.3 Fuzzy Classifiers

Probabilistic classifiers, like the discriminant function classifiers, give estimates of the probability of full membership of each class, and then assign each pixel to the class, which has the highest probability, or to an unclassified class if the operator sets a minimum probability threshold for assignment to a class. Thus a pixel is forced into being assigned to one of the classes, or to a waste bin for unassigned pixels. This gives what is called as "crisp" or "hard" classification, as is illustrated in Figure 4.54. The problems with this approach are that sometimes pixels do not contain just the one landcover or landuse, and so the assignment distorts the actual situation by not allowing the existence of proportions of more than one class or classes in each pixel.

Fuzzy classifiers, on the other hand, determine the grade of membership of a pixel in each class, where this grade of membership corresponds to the level of similarity of the pixel response values to the class mean values as a function of a measure of the distance between the pixel values and the class mean values. Typical measures of distance that can be used are either the Euclidean distance or the Mahalonobis distance.

There are many different methods of implementing fuzzy classifiers, but the most common is called the Fuzzy c-means (FCM) classifier. The typical way to implement the FCM classifier is as an unsupervised classifier. The first step in the FCM classifier is to develop the cluster mean and covariance arrays for the k classes identified by the clustering algorithm in the data set. The clustering is done in the conventional way. Once the clustering has been completed, the image is classified on a pixel-by-pixel basis by computing the distance of the pixel from each class mean and then computing the membership grade, $u_{i,c}$:

$$u_{i,c} = [(d_{i,c})^2]^{-1/(q-1)} / \Sigma^k [(d_{i,j})^2]^{-1/(q-1)} \tag{4.124}$$

where $d_{i,c}$ = the distance of pixel i from class c, k = the number of classes and q = the amount of fuzziness or overlap that is allowed in the classification. When $q = 1$ the classification will be crisp, with no overlap. As q increases in value so the fuzziness and overlap increases until when $q \gg 1$ the overlap is complete. A value of $q = 2$ is typically used.

The output of the fuzzy classifier is a vector of membership grades for each of the k classes, so the vector is a k-dimension vector. The operator can readily convert this to a crisp classification by selecting that class with the highest membership grade. However, the operator can just as easily pass the classification through a filter that considers the pixels surrounding the central pixel, and by giving weights to the central pixel and the surrounding pixels, allocate the pixel to a class based on the weighted membership from the pixel and its surroundings. As part of this process the operator may also choose to determine the relative strength of the chosen membership value and have this as another summary vector value. Thus, if the membership value is high, then higher confidence can be placed on the allocation, and, conversely, lower membership values will lower the confidence that one can place in the assignment. The operator may also set a minimum acceptable threshold on the membership allocation value (Figure 4.55).

The preceding discussion is assuming that the operator wishes to end up with an allocation of each pixel to a class. One advantage of this approach over the discriminant function classifiers as developed in Section 4.6.1 is that the surroundings, and other data, can be readily taken into account in setting the allocation. Another advantage is that the membership grades can be readily used to assign a cost or risk of misclassification. Thus pixels with a high membership grade will incur low risk or costs, whilst those with low membership grades in a number of classes will incur high risk or costs. Such costs can influence class allocation, but they can also be used in subsequent decision analysis in a GIS where the costs incurred are a factor influencing alternative decision options.

4.6.4 Neural Network Classifiers

The artificial neural network (ANN) classifier is an information-processing model based on a simplification of our perception of how the neurons and synapses in the human brain process data. The model consists of a number of highly inter-connected processing elements that are analogous to neurons. These are connected by weighting junctions that are analogous to synapses. In the two-layer ANN shown in Figure 4.55, the first layer represents the input data, the intermediate or hidden layer represents synapses and the final layer represents the output neurons. It is called a two-layer ANN because the input layer is not included as neurons. An ANN can have more than one hidden layer; it has been shown that all problems that are likely to be met can be addressed using an ANN with three hidden layers. However the processing costs of an ANN increase significantly with the number of neurons in a layer, and with the number of layers, and so both should only be as large as is necessary to solve the problem at hand.

The sum of the products of the input values and the synapse weights in the first hidden layer are transformed using a thresholding function into output values from that hidden layer. These output values of a hidden layer are then used as input values in a similar process for second and third hidden layers, if they exist in the ANN. The results of the output of the last hidden layer in the ANN are sent to the output neurons.

The ANN classifier has to train its synapses so that its output values match the actual class values. Learning in a biological system involves comparing the input with the output, and adjusting the weights until the output matches the input. The same applies in an ANN classifier. A set of training sites with known pixel data values (the inputs) and class assignment are used to train the ANN classifier by adjusting the weights until the output class assignment from the ANN matches the actual class value.

Consider that there are n inputs, representing the n input channels in the data, $X_{i,j} = \{x_1, x_2, x_3, \ldots, x_n\}$. Consider an ANN with one hidden layer with assigned weights $W_1 =$

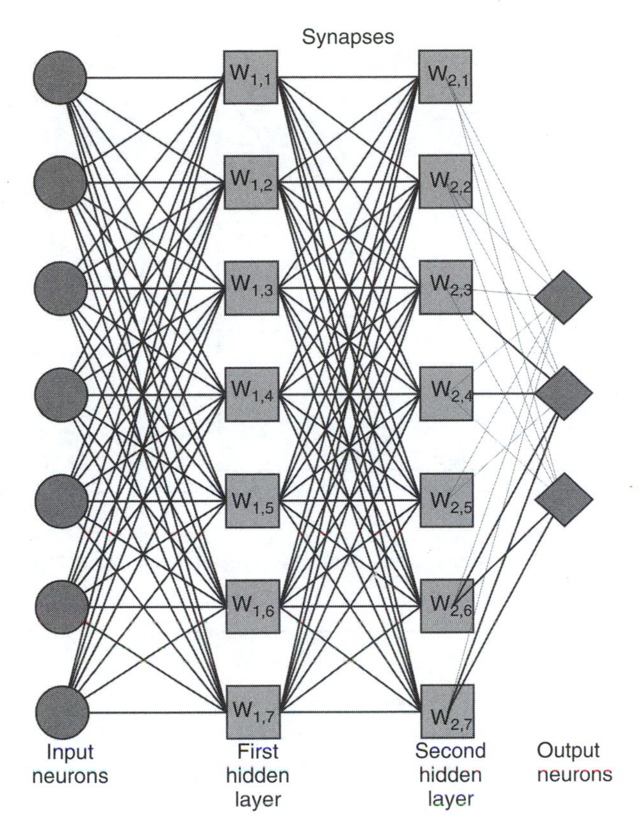

FIGURE 4.55 The architecture of a typical artificial neural network classifier consisting of input neurons, two hidden layers of synapses and the output neuron.

$\{w_{1,1}, w_{1,2}, w_{1,3}, \ldots, w_{1,n}\}$. The activation is then computed as:

$$a = \Sigma^n w_{1,i} \times x_i \tag{4.125}$$

The output is then found by thresholding the activation, a, by the use of a thresholding function. A simple thresholding function is the binary step function:

$$\begin{aligned} y &= 1 \text{if } a >= \text{threshold} \\ &= 0 \text{if } a < \text{threshold} \end{aligned} \tag{4.126}$$

In practice, the logistic function is a much more common thresholding function, although there are a number of thresholding functions that are used. The logistic thresholding function sends most values to the extremes, but it provides a softening of the transformation in comparison with the binary step function (Figure 4.56):

$$y = 1/(1 - \mathrm{e}^{-a}) \tag{4.127}$$

In a two-layer system the y values are then submitted to the second hidden layer in the same way to get an answer z. Consider a simple case of two input values that can take the value of 0 or 1. Initially set the two weights to unity and the threshold to 1.5. The results of the computation of the activation and the assignment of values to y are shown in Table 4.11. If the data space for the data in Table 4.11

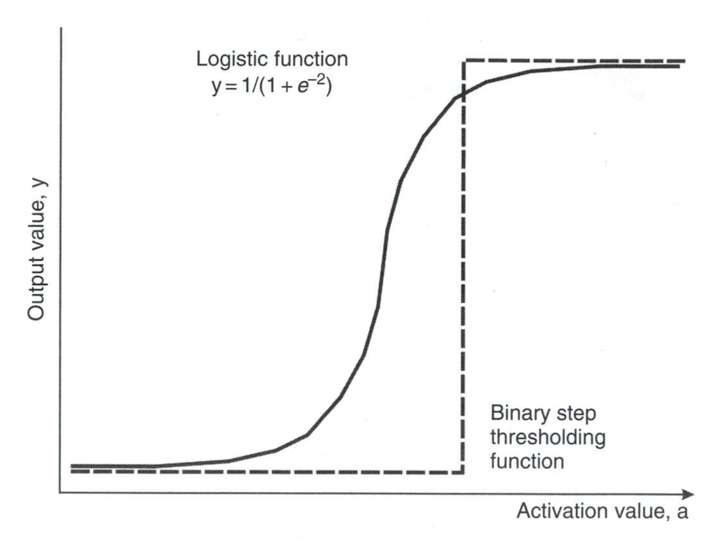

FIGURE 4.56 Two threshold functions used to transform the activation to output values at a synapse.

TABLE 4.11
The Input Values, Activation and Output Value for a Simple Two-Channel Artificial Neural Network

x_1	x_2	Activation, a	Output, y
0	0	0	0
0	1	1	0
1	0	1	0
1	1	2	1

is drawn as shown in Figure 4.57, then it can be seen that the threshold gives a linear decision surface. This can be readily verified. In this simple case the sum of the products at the decision surface will have the equation:

$$x1 \times w1 + x2 \times w2 = t \quad (4.128)$$

Equation 4.128 is easily seen to be a linear equation. It turns out that all ANN of this type give n-dimensional linear decision surfaces, so that the decision surfaces in n-dimensional space can be viewed as facetted n-dimensional planar surfaces between the classes. Because of this characteristic, this type of ANN is called a linear classifier.

The process of training the ANN is usually done by back-propagation. Back-propagation is conducted in the sequence:

1. The initial weights are set, either as random values, or as some values drawn from earlier experience.
2. The training data are shown to the ANN, with the input data values and the result or the correct assignment for that set of values.

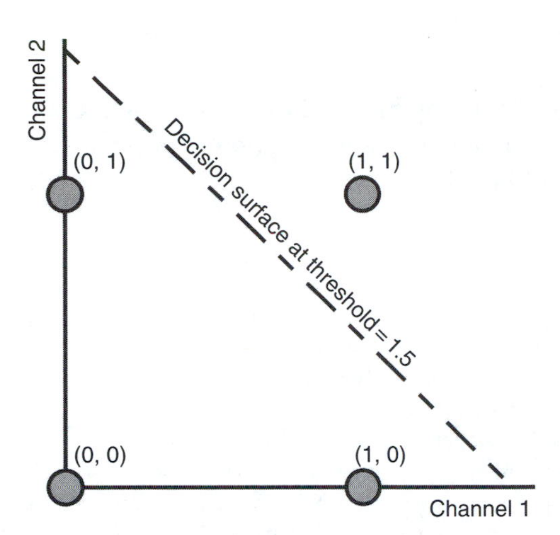

FIGURE 4.57 The data space and the decision surface for a simple two channel, single hidden layer ANN.

3. The ANN computes the output value using the method described above. The difference between the output and the result is the error, where the total error, $E = (1/N)\Sigma^N \Sigma^O e^2(n, i)$, N = the number of patterns used in the training dataset, O = number of neurons in the output layer and $e(n, i)$ = the error at a neuron on the output layer when the nth pattern is applied.
4. Back propagation is applied to determine how to change the weights so as to reach closer agreement between the output and the expected results. There are two common learning rules used in the back propagation; the perceptron learning rule and the delta rule. The perceptron learning rule adjusts the weights by a fraction of the error level. The delta rule adjusts the weights so as to minimise the error level.

The process of training will normally be conducted iteratively until the error level is below an acceptable level. If this cannot be achieved, then modifications have to be made to the architecture of the classifier or to the learning algorithm to achieve acceptable error levels. Modifications to the architecture include changing the number of hidden layers or changing the number of synapses in those layers. Most ANN applications use one or two hidden layers, and it has been shown that in principle any image classification problem can be solved by the use of three or less hidden layers.

Modifications to the learning algorithm include changing the learning rule and changing the threshold function.

The artificial neural network classifier has the advantage that it does not depend on a statistical model for the distribution of its data, and so if the data does not fit such a statistical model, then the ANN may be far superior in classification. However, the disadvantage is that sufficient training data has to be used to adequately train the classifier if accurate results are to be achieved. The process of training must always use data that covers the full range of conditions that can be expected to be met in the full data set, otherwise significant errors may be incurred. More than the discriminant function classifiers, the ANN cannot be used for extrapolation, beyond the conditions used to train the classifier.

4.6.5 Hierarchical Classifiers

The concept behind hierarchical classifiers is that different physical criteria characterise the classes at the different hierarchical levels and thus different criteria should be used for discrimination of the classes at these different levels. In general, the hierarchy proceeds from the most general to the

TABLE 4.12
Hierarchy of Classes That May be Used for Hierarchical Classification, and Working Towards the Development of Classification Criteria at Each Level to Guide the Development of the Classifier

Heirarchical Level	Classes	Criteria	Comments
1	1-land	High NIR values	Use NIR and TIR to
	2-water	Low NIR values	discriminate these classes
	3-cloud	Low TIR values	
1.1	1-green vegetation	High in Veg. indices	Use VI's, NIR and
	2-brown vegetation	Low in VI's, high in NIR	other visible bands
	3-soil/rock/artificial	Low in VI's, high in NIR	to discriminate these classes.
	4-shadow	Low in VI's, low in NIR	
1.2	1-open water	Low NIR, low VI's, uniform	Use NIR, VI's and texture to
	2-swamp and marsh	Low NIR, moderate, variable VI's	discriminate these classes
1.3	1-none		
1.1.1	1-herbaceous	Seasonal phenology	Temporal dynamics
	2-woody	No phenology except deciduous	And VI's
	3-desert	No phenology	
1.1.2	1-none		
1.1.3	1-rough	Texture	Texture
	2-smooth		
1.1.4	1-none		
1.2.1	1-sediment load	Response values in the	End Member analysis
	2-chlorophyll load	Blue-green part of the spectrum.	

most detailed. If the hierarchical levels are set as in Table 4.12, then the criteria for establishing the classes at each level can be seen to be quite different, suggesting different classification strategies may be useful at the different levels. Table 4.12 is not meant to provide a definitive description of the hierarchical levels and classes that should be used; these are likely to change from user to user, with the application and with the environmental conditions. Table 4.12 is meant to be indicative of the process. Perusal of Table 4.12 shows:

1. As the hierarchical level goes down, the differences between the classes often become smaller. This will often mean additional complexity and difficulty in the classification process, increasing the cost and reducing the accuracy.
2. The criteria for discrimination change with the prior classes in the hierarchy. Thus the criteria for discriminating between the water covers are quite different to those for discriminating between the different green vegetation types.
3. The criteria tend to be similar to classical classification hierarchies at the higher levels, but they tend to change as the levels drop towards estimation criteria.

Implementation of hierarchical classification thus currently requires a considerable design and development effort before implementation.

4.6.6 CLASSIFICATION STRATEGIES

The analyst must not only know the characteristics of the tools that are available to him, but must develop a plan to extract the required information from the data. The objectives of this plan are to extract the information required in the most cost-effective manner. This section analyses the factors that must be considered in the development of such a plan.

4.6.6.1 Types of Classes

There are two important types of classes to consider in classification: informational and spectral classes. Informational classes are those classes of interest to the user. They will be based in the physical attributes of the real world that are the interest of the user. The spectral classes are response classes derived from the image data that meet the statistical requirements of the classifier. It is clear that the criterion upon which the user separates the real world into informational classes are quite different to the criterion used to create the spectral classes in the computer. An assumption in classification is that one or more spectral classes can and will represent the informational classes unambiguously. At present, the only way that can be used to determine whether this assumption is being met is by accuracy assessment after the conduct of the classification, although it would be much better to incorporate these criteria into the classification process right from the start. There are thus two sets of criteria that have to be met: separation of the spectral classes in the spectral domain, and the suitability of these classes to represent the informational classes. Both criteria have to be met if the classification is going to provide an accurate classification in terms of the informational classes that are required from the classification.

Classification can be conducted by means of supervised and unsupervised classification techniques. Supervised classification uses operator-specified training data to form class statistics as the basis of the classification. Unsupervised classification clusters the image data to create spectral classes that form the basis of the classification. Supervised training means that the operator will start with the informational class criteria in defining these training sites. However, supervised training does not, of itself, ensure maximum spectral discriminability between the classes. If, however, all of the training site data were then subjected to clustering, then the analysts would be able to see whether the spectral classes derived from the clustering were also discrete in terms of the informational classes, or whether some of the spectral classes overlapped more than one informational class. Unsupervised classification derives classes based purely on spectral criteria, with no reference to the required informational classes. With unsupervised classification, the smaller the classes, and the more of them that are created, the more likely that each spectral class can be assigned to a specific informational class. However, ideally the unsupervised clustering should take into account the informational class needs in the creation of the spectral classes. At present there is no method that allows this to happen.

4.6.6.2 Selecting Classes and Classifiers

We have seen that there will be optimum decision boundaries in the spectral data for the information classes, but that it is often difficult to define them adequately. This problem can be made more tractable if we choose imagery in wavebands and at a time such that these boundaries are most clearly discernable in the imagery. It is important to choose imagery when there is maximum spectral discriminability between the information classes. It is not important to ensure good discriminability between the spectral classes within the one information class because this does not affect the finally derived information.

The imagery must be chosen in anticipation, from assessment of physical conditions in the field. This can be done if the analyst has a good appreciation of the distribution of data for the different information classes in the spectral data. The procedure requires the following:

1. *Analysts need to be familiar with the spectral characteristics of the information classes to choose the most suitable imagery.*
2. *Imagery must be chosen in appropriate wavebands and at the best time to maximize the discriminability of the information classes.*

    The information classes in the imagery may not be a good fit to the parametric classifier algorithms. In this situation it may be necessary to partition the information class domain into spectral class sub-domains that are a satisfactory fit to one of the classifiers, or it may be better to consider using another type of classifier algorithm. For both of these situations, create information class training areas that are both large enough and distributed across the area so as to be representative of the class. It is poor statistical practice to take one training area as a sample to represent a class. It is much better practice to select training areas for each class distributed across the whole area to be classified.

3. *Selection and acquisition of training data for each class across the area to be classified.*
   Once this data is accumulated it should be analysed for goodness of fit to the assumed distributions of the parametric classifiers that may be used. If the data are a good fit to one of the classifier distributions then that classifier should be used for the classification.
4. *Testing goodness of fit of the data to the various classifier distributions and adoption of the most suitable classifier.*
   If the data for an information class is not a close fit to the distributions assumed by the classifier, then either the class data should be partitioned into spectral classes using clustering techniques or a parametric classifier or a non-parametric classifier should be used.
5. *Partition of information class data into spectral classes using clustering techniques or use of a non-parametric classifier.*

4.6.6.3 Improving Interclass Discriminability

The more discrete the information classes in the spectral data, the more reliable and accurate will be the classification. If the discrimination is particularly good then simpler, and cheaper algorithms may be used, yet the required accuracy standards will be acheived. The use of imagery acquired in appropriate wavebands and taken at the right time increases discrimination. It is also increased if the confusions between different landcovers are reduced. Confusions can be reduced by:

1. *The use of masks.* Specify areas that are not to be classified, or are to be processed in different ways, by identifying the boundaries of these areas on a map or image and digitising this information to create an image mask. Because this image controls the classification process it is called a mask. Thus, in mapping irrigation it would be logical to mask out dams, rivers and other water surfaces that are clearly not irrigation areas, but which will respond very similarly to irrigation areas in remotely sensed imagery in some seasons. A mask can be used to specify areas that are not to be classified, to reduce both confusions and the cost of the classification. A mask can also be used to define areas that are to be classified using different techniques, for example, a different algorithm, different spectral classes or even different data sets. For example, localised rainfall may have caused quite different surface conditions, and management response, in some parts of the image than in others. It may be appropriate to partition the images on the basis of differences in condition and conduct quite different classifications in both areas. Topographic effects may control different landcovers or landuses. Inclusion of appropriate topographic data (slope, height, aspect) may significantly improve classification accuracy. Cloud cover may also convince the analyst of the advantage of using different data sets for different sub-image areas.

The use of masks is very much an integration of local and environmental knowledge in the classification process. By use of the masks the analyst is creating greater control over the classification process.

2. *Use of temporal change.* Different landuses are represented by different landcovers and landcover conditions, at different stages in the season. Careful selection of temporal imagery might assist in discriminating between the various landuses, such as crops and pasture. The temporal changes that occur in spectral data for the one landuse will often vary over the area being classified, creating different classes for the different sub-areas. Consequently the more temporal images that are used in the classification, and the larger the area, the more spectral classes are likely to be required, significantly increasing processing time and costs.

3. *Incorporation of additional data sets into the classification.* Classification is generally taken to use spectral data, controlled by masks as appropriate. However, other types of data, useful in increasing interclass discriminability, can be included in the classification process. Some examples of other types of data are:

(a) *A measure of texture.* Some landcovers will exhibit higher textural variability, the use of which may improve discriminability between two classes.
(b) *Information* on other parameters that may influence the class distribution such as topographic height, slope, aspect, temperature and rainfall.
(c) *Classification of adjacent pixels.* If a majority of adjacent pixels are classified into one class, then that may suggest the classification of the central pixel. Whilst this approach needs to be used with care, as there are instances of stands of trees in the centre of cropping fields and dams in pasture fields, the probability of classifying these adjacent pixels should often be a factor in considering the classification of the central pixel.

4.6.6.4 Cost Minimisation

A number of the techniques discussed above will improve classification accuracy, and will also reduce costs. There are other techniques that reduce costs, but their use needs to be assessed, particularly as they may affect the accuracy of the derived information. These other techniques include:

1. *Utilisation of only the essential data set*, and not data values that do not contribute significant information on the classes. Contribution is usually seen as a function of correlation. If the data values in two bands are highly correlated, as indicated in the correlation matrix, then one of the bands will contribute nearly as much information about the surface as will the two bands. However, computation of the correlation matrix using large data sets can mask divergences between classes with small populations. The best solution is therefore to compute correlation matrices using sample populations for each information class that are of similar size for each class, and reject a band or bands on the basis of analysis of the correlation in these matrices. The final decision on which bands to use in the classification should be based on evaluation of repetitive classifications of an area of test field data, with highly correlated bands being eliminated at each repetition. Comparison of the classification accuracies and processing costs for each classification of the test data will indicate the cost/benefit trade-off of using various band combinations.

2. *Utilisation of the simplest algorithm* that will achieve the desired accuracy. The accuracy required for any type of management information has to be considered in relation to the accuracies of the other types of information on which management decisions are to be made, as well as the costs incurred by improving the accuracy. There is usually no benefit in having information that is more than an order of magnitude more accurate than the other information being used because errors in the information are going to be swamped by the errors in the other information sources. The important aspects are the influence of the information on the decision-making process, and propagation of error through that process. In general the more accurately the information is required, the more costly it is to get that information.

3. *Utilisation of sampling techniques* to achieve the information required. Sampling is not applicable in some situations, for example, where it is required to locate every area of specific land covers. Sampling can be used when the information required is at resolutions that are coarse relative to the resolution of the data, for example, if the information required is the area under crop in the catchment area for a grain silo. Sampling can significantly reduce processing costs for little loss in accuracy. There are a number of approaches to sampling:

(a) *Pixel skipping.* Classify only one pixel every n pixels, and classify along one line every m lines. This approach is effectively sampling the data by taking a regular sample of $1/(m \times n)$ part of the total data set. Different sampling fractions could be adopted for different parts of the classification if required. For cropping a sampling fraction of between 1% and 5% has been found to give almost identical results to classification of the whole of the data set.

(b) *Property sample*. Classify a sample of the properties within the area to be classified, with the classification controlled by a mask that contains the required properties. The sample of properties could be chosen either as a random or a stratified random sample. This approach usually takes longer than pixel skipping because each pixel and line of the data has to be interrogated, even if only briefly.

4. *Stratification of the area* and use of one of either different data sets, different algorithms or different spectral classes in each strata. Stratification is justified if there are variations in either the class distributions, the within-class quality, or the management practices across the area, that suggest different strategies for classifying the different strata to get the desired information.

4.6.6.5 Field Data in Classification

The need for field data is considered in detail in Chapter 5. Collection of field data is likely to be one of the most expensive, and logistically difficult aspects of any mapping or monitoring task. There are significant benefits in using procedures that reduce the need for field data. Field data is required for two purposes:

1. *To train the classifier algorithm*. The more parameters in the classifier algorithm, and the more spectral classes, then the more field data that is likely to be required. In addition, the poorer the fit of the spectral data to the classifier probability distribution, the more field data that may be required to ensure that all class conditions are adequately represented. Design of classifier algorithms that have decision surfaces in close accord with those of the classes of interest can significantly reduce the amount of field data required for training purposes. This is particularly so if the algorithm exploits our understanding of the reflectance, phenology and the other physical properties, of the surfaces of interest in making a decision.
2. *To evaluate the accuracy of classification* into the information classes. The field data required for this is dependent on the number of information classes, and the reliability required from the accuracy assessment. This field data must be quite separate and independent of that used for training the classifier.

With any mapping it is essential to analyse the accuracies achieved using the field data collected for that purpose, and display this assessment on the map, or with the statistics produced. Only in this way can users of the information make a judgement on the weight that they will give to this information, relative to other information, knowledge or intuition in making a decision.

4.6.6.6 The Use of Different Classification Strategies

Classification of data to produce landcover or landuse information can be done using either parallel or sequential classification strategies.

In a parallel strategy all of the data are used simultaneously to arrive at a decision. Multi-temporal image classification, in which the data for a number of images are used in the one classification process is the classic illustration of the approach. Parallel classification strategies suffer the disadvantage that they are expensive for each classification and they duplicate processing costs when updating the classification, as will routinely occur in the monitoring of vegetation, water and other resource conditions. With the maximum likelihood classifier the cost of classification is a function of the square of the number of dimensions, or wavebands, in the data. Multi-temporal classification of five Landsat MSS images (twenty dimensions) would have costs that are a function of 20^2 for parallel classification instead of being a function of 5×4 in the case of sequential classification. If new data are to be added to the classification, either within the existing data set range of dates to improve classification accuracy, or after the existing data to effectively update the classification, then parallel classification will incur much greater costs than will sequential classification.

Sequential classification strategies process sets of data relevant to a date, to produce interim information, such as landcover information, for that date. These interim sets of information are then used to update or revise earlier provisional information, for example, of landuse, to map the landuse as at the date of the imagery. Sequential classification can be done by the use of the following strategies, either individually or together:

1. *Utilisation of phenological rules to control the updating process.* Classify each image in turn to derive landcover information. Use these images of landcover within the framework of environmental and phenological rules to derive landuse information. One way to do this would be to start with a landuse mask, however simple, and use the environmental and phenological rules to revise and update the landuse. In some situations the updating rules can be simple logic rules. Consider mapping flooded rice in an area that contains water surfaces and other agricultural crops that are regularly flood irrigated throughout their growing season, coinciding with the growing season of rice. Only rice, and swamp, of all of these landuses, is put under permanent flood and may develop a green canopy as the season progresses, although all of these landcovers can look like rice on individual images. The following phenological rules will separate the various landuses:

(i) Only rice, swamp and water bodies have the same landcover of water to green canopy on two or more sequential images in the growing season. Other land uses will look like rice on individual images, often more than one, but not sequential sets of images. Therefore if a pixel looks like rice on two or more sequential images then it can only be rice cultivation, swamp or water land uses.
(ii) Of these three only rice and swamp may develop a green canopy during the growing season. Therefore, a pixel must exhibit an increase in green canopy cover during the growing season for it to be rice cultivation, but it may still be swamp.
(iii) Swamps are normally flooded before the season starts, and may be known beforehand. Swamps can be deleted either by exclusion from the classification, or by mapping them prior to flooding of the rice bays.

Similar sets of rules are known for other conditions. The process can be built as an expert system designed to monitor the physical parameters of interest.

2. *Hierarchical classification* in which the data are partitioned into broad classes using one combination of data and processes, and then each of these broad classes is analysed in more detail using a different set of data and processes. Consider the above example of monitoring rice in which method (ii) uses increasing green canopy to discriminate rice and swamp areas from other water bodies. The first level of image classification is to discriminate the "looks like rice" from the "does not look like rice" classes, which is adequate for method (i). For method (ii), however, more information is required, by analysing only those pixels that fit into this category to determine whether they are exhibiting increasing green canopy. This second level of information satisfies method (ii).

Sequential classification strategies provide the basis for a more powerful and flexible approach to the establishment of a classification procedure, as well as providing interim information, than do parallel strategies.

Classification, by training on selected areas and then using the derived class statistics to assign pixels at other locations to one of the spectral classes, is extrapolation. Extrapolation means that the process is being conducted without a means of checking the results during the process of extrapolation. Interpolation, by contrast, is constrained to provide a result that is set within very specific bounds. Extrapolation is thus a risky way to operate. This is an important reason why accuracy assessment should always be conducted on a classification, as this is the only means open to the end user to be aware of the quality of the analysis. However, it would be better if classification strategies

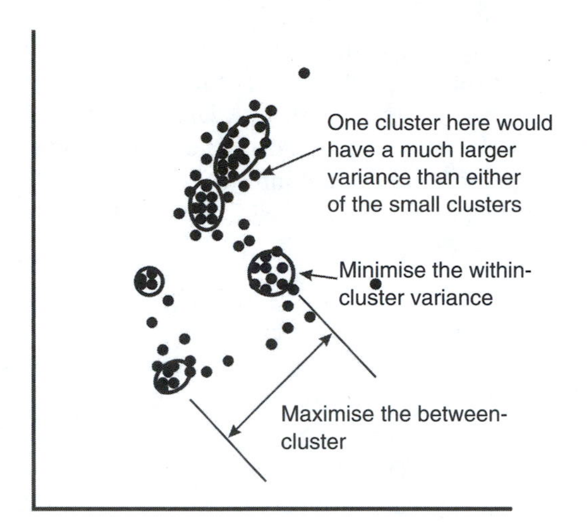

FIGURE 4.58 Minimise the within class spread and maximise the between-class distances.

were developed that move away from the extrapolation approach, and move towards the interpolation approach.

4.7 CLUSTERING

The concept behind clustering is that similar surface conditions will have similar response values, but which are different to the response values of other surface conditions. If these typical response values for each type of surface condition can be found, then it provides the basis for discriminating between the different surface conditions.

Clearly there are surface conditions that are of interest to users, and conditions that are not of interest. Equally clearly, the conditions of interest will vary in some criteria, but not in other criteria, for example, they may vary in some spectral bands but not in others. They may vary in other criteria, and they may vary over time. The success of clustering depends on the selection of data that are appropriate to the surface conditions that are to be discriminated and clustered, and which are suitable for separating the different conditions into different clusters.

4.7.1 CLUSTERING CRITERIA

The notion of clusters implies the notion of groups of data such that the average distance between values in a group is much smaller than the distance between groups. The usual criteria in clustering is thus to maximise the difference between the "between class" and "within class" distances (Figure 4.58). Most clustering algorithms also assume that the required clusters need to be a reasonable fit to the normal distribution. The usual methods of implementing this criteria are:

1. *Minimisation of the variance*
2. *Maximisation of the Divergence*

4.7.1.1 Minimisation of the Variance

Let n_i be the number of data values in cluster i of k clusters, m_i be the cluster mean value and $x_{j,i}$, $j = 1, n_i$ be the data values in the cluster. Hence:

$$m_i = (1/n_i) \times \Sigma x_{j,i}, j = 1, n_i \tag{4.129}$$

and the sums of squares is:

$$S = \Sigma\{\Sigma[(x_{j,i} - m_i)^2], j = 1, n_i\} i = 1, k \tag{4.130}$$

The optimal assignment of a new data value, or pixel value, is to that class which minimises S. If the new data value is shifted from one class to another and reduces S in the process, the pixel is assigned to the second class. This process is repeated until the pixel is assigned to that class which minimises S. Once a pixel is assigned to a class its value is used to recompute the mean for that class, to ensure that the mean will minimise the sums of squares in the class. In consequence, the means for the classes will gradually shift as new data values are assigned to the different classes. It is for this reason that this approach to clustering is called "*migrating mean*" clustering procedures.

Since the means migrate in a pass through the data, some of the early data might well be better assigned to other classes than to that class to which they were initially assigned. Iterating or processing the data a number of times solves this problem. With each iteration the changes in the class statistics become smaller until they tend to oscillate from one pass to the next. At this stage the iterations will result in no improvement in the results and the iterations should be stopped.

The minimum variance criteria works well if the data form compact clouds, which are well separated from each other. This criterion can, however create artificial clusters when a cluster has a large scatter and hence the sums of squares may give lower values of S by splitting the cluster into a number of smaller and tighter sub-clusters.

When considering multi-dimensional data the above approach needs to be expanded to consider the covariance arrays of the data. If m_i is defined as the mean for a cluster in Equation (4.130) and:

$$M = \{\Sigma[n_i \times m_i], i = 1, k\}/\{\Sigma[n_i], i = 1, k\} \tag{4.131}$$

as the mean for all of the data values, then:

$$S_t = \{\Sigma[\Sigma(x_{j,i} - M) \times (x_{j,i} - M)^{\mathrm{T}} j = 1, n_i] i = 1, k\} \tag{4.132}$$

is the total sums of squares for all data points about M. Expand Equation (4.132) to give:

$$\begin{aligned} S_t &= \Sigma\Sigma[(x_{j,i} - m_i) - (M - m_i)] \times [(x_{j,i} - m_i) - (M - m_i)]^{\mathrm{T}} \\ &= \Sigma\Sigma(x_{j,i} - m_i) \times (x_{j,i} - m_i)^{\mathrm{T}} + (M - m_i) \times (M - m_i)^{\mathrm{T}} \end{aligned} \tag{4.133}$$

since the cross products in each class will sum to zero. Hence:

$$S_t = S_w + S_b \tag{4.134}$$

where S_w is the within-class scatter and S_b is the between-class scatter. Now the total scatter matrix does not depend on the allocation of the individual pixels to the classes, but both S_w and S_b do depend on this allocation. In consequence, if one of the matrices is reduced then the other matrix will be increased. The normal criterion is thus to reduce the within-class scatter, S_w, which automatically increases the between-class scatter, S_b. This criterion is met by computing the within-class variance after adding the pixel values to each class in turn and allocating the pixel to that class that minimises S_w.

The size of the matrices, S_w, is measured using either the trace or the determinant, where the trace is the sum of the diagonal elements of the covariance array.

4.7.1.2 Maximise the Divergence

Consider maximising the ratio $(S_w)^{-1} \times S_b$ of the between- to within-class distances. The eigenvalues of an array are the variances along the semi-major and semi-minor axes of the data distribution depicted in the covariance array. The eigenvalues thus measure the maximum value of this ratio and so the criterion is to measure the trace of the eigenvalues derived from the ratio matrix $(S_w)^{-1} \times S_b$.

4.7.2 Clustering of Training Data

Training areas, whether they are regular or irregular in shape, can include pixels that do not properly belong to the class. Inclusion of these pixels will usually have little effect on the class means, as they will represent only a small proportion of the pixels in the training data set. Their inclusion will, however, have a large impact on the covariance array for the class and hence on the location and shape of the class decisions surfaces.

One way to minimise this problem is to cluster the data in the training area using clustering processes. Another way is to assume that the class in the training area is normally distributed, and reject pixel values that diverge from the class mean by more than some specified limit. For this approach, compute the Mahalonobis distance, D_{m}, for each pixel from the class mean, and reject those pixels with a D_{m} value larger than a set threshold value. A normal threshold value would be in the range 2 to 4.

4.7.3 Strategies for Clustering

Remotely sensed data provide significant challenges to clustering, simply because the physical size of the pixels ensures that an image contains many pixels that are mixtures of various cover types, and these pixels contribute to the smoothing in the waveband histograms, the two-dimensional scattergrams and the higher order equivalent of these. This means that classes rarely have sharp boundaries in the data distributions in the spectral space, but rather tend to blend and merge. One of the implications of this is that the class boundaries that are found in clustering may be very soft, in that slight changes in the clustering criteria can change those boundaries. This is a significant problem in derivation of statistical classes from the clustering, but may not be so much of a problem in the subsequent grouping of the statistical classes into informational classes. Clearly data should be chosen so as to maximise the discrimination between the informational classes. If the data does this, then the clustering may produce somewhat different statistical classes, depending on the parameters used, but the grouping of these into informational classes should proceed satisfactorily. It would be an advantage if the data chosen also made the clustering more robust in the sense that varying the clustering parameters gave little impact on the derived classes. If this is the case, then the subsequent steps of allocating the statistical classes to informational classes will proceed more quickly, and accuracy assessment is likely to be more straightforward. However this is not always possible, and so most clustering algorithms adopt internal strategies for maximising the accuracy and reliability of the clustering process, as well as providing the analyst with the opportunity to vary a number of clustering parameters.

Most internal clustering strategies designed to improve the accuracy and reliability of clustering focus on:

1. *Reducing the cluster variance and covariance values.*
2. *Increasing the between-cluster distances.*

4.7.3.1 Reduction of the Cluster Distributions

The discussion on the ALV function in Section 4.4.3.6 showed that the variance increased as the pixel size increases to about half of the object spacing. After this point, increases in pixel size are matched by a reduction in the ALV function value, until the next largest objects in the scene start to dominate the ALV function values, at which point the ALV function values start to increase again, until the pixels are again about half the size of these larger objects. An image with pixels at half the field size is clearly too coarse to accurately classify the fields and estimate their area. If we follow the pixel size to smaller values, the ALV function values decrease until the next smallest objects start to influence the ALV function values. It is at this trough that ALV function values are the smallest and the within-class variances will be the least, and so this is the optimum resolution to classify the image.

4.7.3.2 Increase of the Between-Cluster Distances

The best way to maximise the between-cluster distances is to select wavebands with this goal in mind, and to select the time in the season to also maximise the separation.

In addition to these strategies, clustering algorithms often allow the analyst to:

1. *Change the method of selection of the seed positions.* Clustering has to start with seeds for the clusters, as well as with techniques for the subsequent merging and splitting of clusters. The starting positions adopted vary from clustering algorithm to algorithm, but can include selection on the spatial diagonal of the image, a random scattering across the spatial image, as well as on the diagonal and random scattering in the spectral/radiometric space. All algorithms start with one of its options as default, and the operator can chose one of the other options.
2. *Change the number of classes.*
3. *Change the number of iterations and/or the class threshold criteria.* Clustering proceeds through a number of iterations, so changing the number can improve the clustering. In the same way the criteria for joining pixels to clusters, splitting and merging of clusters need to be set in the clustering procedure, and some clustering algorithms allow the analyst to set higher or lower criteria. As the criteria become harder, for example, by setting narrower probability or variance criteria on the joining, splitting and merging, so the number of classes will usually increase, and the number of iterations may need to also increase.

4.8 ESTIMATION

4.8.1 Introduction

Estimation is the process of deriving an estimate of specific resource parameters from remotely sensed data, often when used in conjunction with other data. It is a transformation of the remotely sensed and other data into estimates of specific resource parameters by the use of numerical mathematical algorithms or models. It provides information on the condition, in terms of the parameters that are estimated, of the sensed surfaces at the time of the acquisition of the remotely sensed data. Classification provides information on *WHAT* the nature of the surface landcover or landuse is and estimation provides information on the *CONDITION* of that surface.

Estimation is thus a modelling task involving transformation from some physical variables, such as the response values of image data into estimates of other physical variables, but it is not a prediction task. Estimation can be used to provide input data to drive models that may subsequently be used to provide predictions, and estimation may also provide data that can be compared with model outputs, so as to assess the accuracy of the models.

Since estimation uses the response or reflectance value at a pixel as the basis for estimating the physical attribute that is of interest, errors in these values may introduce errors into the derived estimates. For regression models, such errors will usually give errors in the result, whereas this may not be the case for the interpolation models. For this reason, imagery that is to be used for estimation would normally be calibrated and corrected for atmospheric effects.

Remotely sensed image data can contribute to the mapping of specific resource parameters when the image data, or parameters derived from the data, are correlated to some degree with the resource parameters. If there is no or negligible correlation between the image data and the parameters being derived from it then the image data contributes nothing significant to the estimation process and should not be used.

Estimation models are often created for a specific landuse or landcover. As a consequence, estimation needs to follow image classification so that the covers are known at each site in the image, so as to apply the correct estimation model, or to not process the image data into an estimate as is appropriate.

Estimation can be conducted in two ways:

1. By means of *regression models*
2. By means of *interpolation models*

With either type of estimation model, input to the model may be either response, radiance or reflectance in the selected wavebands, or it may be an index derived from the data. With the estimation of vegetation parameters, one of the vegetation indices are often derived from the image data and this is used in the estimation model to derive the required physical attributes.

4.8.2 Development of Regression Estimation Models

A regression estimation model is an empirically developed model that typically takes a set of observations of the physical attributes of the cover of interest, over the range of conditions that are likely to be met, and simultaneously takes spectral observations of that cover with a spectrometer. The spectrometer data are calibrated and converted into reflectance. The spectral reflectance data are then either used directly, or they are converted into an index for analysis of the relationship between this data and the field data. This analysis will involve fitting the two sets of data to a suitable regression model using least squares and statistically testing the goodness of fit of the derived model. The methodology follows that discussed in Chapter 5 on field techniques.

With all regression models, the higher the level of correlation between the physical parameters and the remotely sensed data, the better the model will be at providing estimates of the physical parameters from the image or remotely sensed data. In addition, the regression models should only be used under conditions that are similar to those in which the model was constructed. If the model is used under other conditions, then the relationship between the physical parameters of interest and the remotely sensed data may be different to that assumed in the model, so that the derived estimates will be in error, possibly by a long way.

4.8.3 Application of Regression Estimation Models

One common application of regression models is in the estimation of vegetation parameters from remotely sensed image data. With green vegetation, the dominant cause for the high absorption in the red part of the spectrum is due to the density of chlorophyll in the leaf tissue. Thus, parameters that are highly correlated with this characteristic will give the best models, and other parameters that are not so highly correlated will give less accurate or reliable models. Thus, the FPAR and leaf chlorophyll content can be expected to be highly correlated and thus provide a good model. The GLAI is the ratio of green leaf area to the ground area supporting that green vegetation. Since the level of chlorophyll in a green leaf can change with physical conditions in a leaf and the leaf reflectance and leaf orientation can affect reflectance but not the chlorophyll concentration, the correlation will not

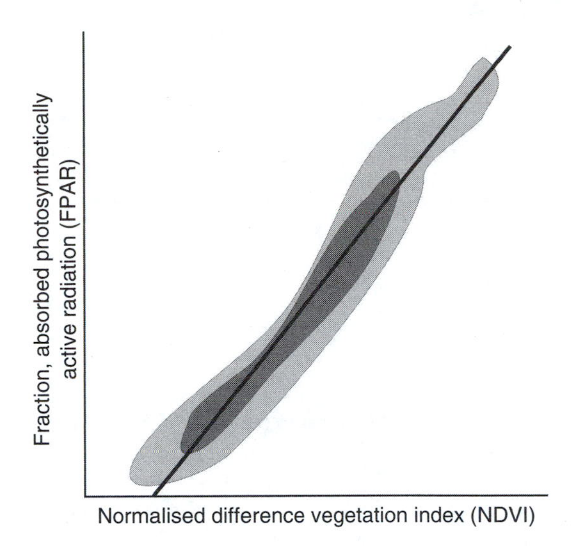

FIGURE 4.59 Typical relationship between FPAR and NDVI.

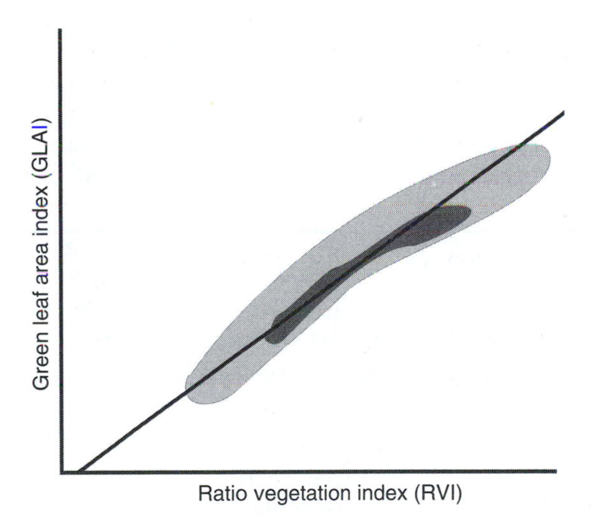

FIGURE 4.60 Typical relationship between GLAI and RVI for various crop types.

be as high, although the level of correlation is usually significant. Other parameters, such as biomass and yield have lower levels of correlation.

Extensive studies have shown that FPAR is approximately linearly correlated with NDVI, whilst GLAI is approximately linearly correlated with RVI, as shown in Figure 4.59 and Figure 4.60 respectively. RVI and NDVI are non-linearly related as can be seen in Figure 4.61 and as a consequence FPAR is non-linearly related to RVI and GLAI is non-linearly related to NDVI:

$$NDVI = \frac{(\text{NearIR} - red)}{(\text{NearIR} + red)} = \frac{((\text{NearIR}/\text{red}) - 1)}{((\text{NearIR}/\text{red}) + 1)} = \frac{\text{RVI} - 1}{\text{RVI} + 1}$$

In practice, RVI is very sensitive to errors in the red values, since the red values for green vegetation are very low, and as a consequence errors in red values will create large errors in the

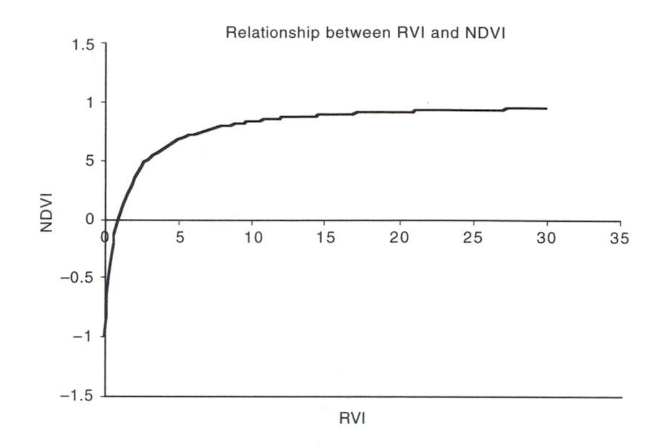

FIGURE 4.61 Relationship between NDVI and RVI.

derived RVI values. NDVI is not as sensitive to errors in the red values, and so it provides a more robust indice for estimation.

The main disadvantages of regression models are that: they depend on high correlation for the development of an accurate model, they can only be used under similar conditions that were met in the construction of the model, the model can change with changes in some conditions and they need accurately calibrated and corrected image data. Thus, regression models estimating crop conditions in one season may not be that accurate for another season if weather or management conditions were strikingly different between the seasons. Another problem is that the models are different not only for each species, but also between varieties, so that use of this approach commits the user to the construction of models for each species, and often for varieties as well.

4.8.4 Development of Interpolation Estimation Models

The conduct of interpolation is covered extensively in Chapter 6 on geographic information systems, so the topic will just be introduced at this stage. It has been seen in the previous section that empirically based regression models suffer a number of disadvantages:

1. *Individual models have to be built for each species, variety* and possibly for different growing conditions. This is an impractical demand.
2. *They require full atmospheric correction* of the image data, and even then errors in that correction will lead to errors in the estimates.
3. *A high level of correlation is necessary* between the parameters.

Because of these significant limitations, other approaches to estimation need to be developed. The use of interpolation is seen as an alternative approach. Interpolation has classically been used in the derivation of contour maps from a mesh of height observations across an area. The field data are observed for a mesh of points across the area of interest, and then intermediate values are estimated by linearly interpolating between the observations as shown in Figure 4.62. Whilst there are a number of methods of estimating the value of pixels, the use of a surface-fitting model is a common method of interpolation.

All such methods of interpolation are limited by the assumptions of the interpolation method and the suitability of the data for that method. With elevations, and creation of contour maps of the terrain, the surveyor has the ability to see the surface when setting out the field sites, thus having the chance of improving the distribution and density of points observed. So the surveyor should take observations along major changes in grade, so that interpolations are not conducted across such changes in grade. With many other parameters this luxury does not exist. However, methods

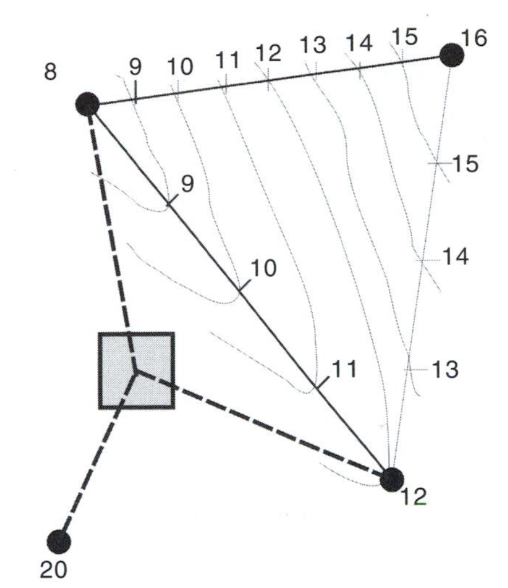

To produce contours

Linearly interpolate on each line joining field data to mark the crossing point of a counter line. Join the relevant contour line points on the different lines, to create the contours.

To estimate pixel values

Fit a geometrical surface to the observed field data points and use this surface to estimate the value of each pixel within the geometrical figure.

FIGURE 4.62 The interpolation between field data to create contour maps or estimate digital elevation and for other types of surface modelling.

have been developed that use auxiliary data that provides the equivalent type of information for the interpolation. These methods are called co-kriging. In co-kriging the auxiliary data, such as image data, are used to influence the interpolation process. In co-kriging, the image data are converted into as close an estimate of the final parameter as is possible and then used to influence both the setting of field data sites and the interpolation process. The image data may thus be converted into a vegetation index, and this in turn converted into an estimate of FPAR or GLAI using a standard regression model. The reason for this is to remove as much of the non-linearity as possible prior to the conduct of the co-kriging interpolation. The distribution of FPAR or GLAI across the image is then used to influence the number and location of the field sites. Field data is then collected at these sites and co-kriging conducted using the image-based estimate of FPAR or GLAI to influence the interpolation.

The method is tied down by the field data, whilst the image data is used to influence the interpolation process. If the level of correlation between the image data and the physical parameter being estimated is poor, then the method can still be used, but at the cost of using a denser net of field data to control the interpolation. Consequently, as the level of correlation increases, so the level of field data can be reduced.

In addition, if the image data is not corrected for atmospheric effects, this is likely to reduce the accuracy of the interpolation, requiring more field work to maintain similar standards as could be achieved using corrected data. As a consequence the method is more robust than the regression method, but it does require extensive field data.

4.8.5 Estimation Based on Physical Models

Physical models are based on our understanding of the physical basis of a component of the environment. Such models can be used to estimate reflectance, given specific physical conditions. As such they can assist us in better understanding the relationship between physical conditions and the derived reflectance values. Sometimes these models can be inverted so as to estimate physical attributes from the reflectance data. This will only be the case if the reflectance properties are dominantly derived from the effects of a few physical attributes, since there are a limited number of wavebands that can be used, and because of correlation between many wavebands. However they offer a way of deriving estimates that are based on sound physical principles.

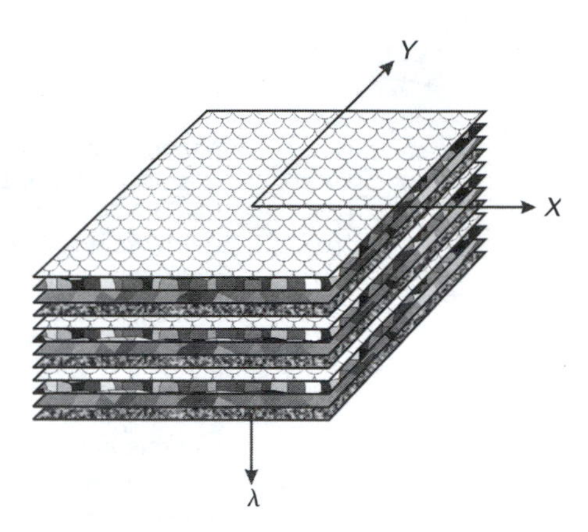

FIGURE 4.63 The concept of a hypercube of image data.

Most of the work on using physically based models in remote sensing has been done in relation to green vegetation. The use of models in this application is discussed in Chapter 7.

4.9 THE ANALYSIS OF HYPER-SPECTRAL IMAGE DATA

4.9.1 INTRODUCTION

Hyper-spectral image data is defined as image data that is recorded in sufficient wavebands to provide a reasonable estimate of the spectral profile for each pixel in the image data. Hyper-spectral imagers are thus also often called imaging spectrometers, recording the data in many wavebands across the spectral range of the data. The data thus forms a hypercube of data in three dimensions of (x, y, λ) as shown in Figure 4.63. All wavebands of the ideal hyper-spectral data will have the same spectral range and interval between wavebands and the same IFOV and thus the same spatial resolution. In practice this does not always occur, so that some ranges of wavebands, such as in the visible and NIR may have smaller waveband ranges and smaller pixels than in the mid to thermal IR regions.

In many aspects of hyper-spectral image analysis the data can be treated in the same manner as for multispectral image data. Thus the data may need to be rectified, calibrated and corrected for atmospheric effects, and these processes are conducted in exactly the same way as for multi-spectral image data. The data may be used in image classification, and again the principles are exactly the same as applies for multi-spectral image data. Although the principles are the same, it will sometimes be found that image processing software may not allow classification to be conducted with image data in so many wavebands. Some software has been adapted to deal with hyper-spectral data. Whilst hyper-spectral data can be used to compute the classical vegetation indices, they will usually be computed using narrow wavebands, representing a waste of a lot of data. Hyper-spectral image data are more likely to be used to derive vegetation indices based on the red edge characteristics of vegetation. The other application of hyper-spectral data is to use libraries of spectra derived from other sources, and look for similar spectra in the hyper-spectral data. Clearly the image data needs to be accurately calibrated and then corrected for this application to be successful.

4.9.2 VEGETATION MAPPING

One of the most potentially useful parts of the spectra of vegetation to detect condition information is the red edge. The position and slope of the red edge has been found to vary with age and condition in the leaves of plants. Hyper-spectral data provides the capacity to model the red edge so as to derive more than one parameter from the data, such as the location and the slope of the red edge.

4.9.3 Fitting of Spectra

The geological sciences have found that many rocks have absorption spectra that fall within the spectral range of image data. A hyper-spectral imager has the capacity to map the occurrence of these materials if it has wavebands that are sufficiently narrow and cover the appropriate spectral range. To do this, the geological community has constructed a number of libraries of the spectra for different materials. These spectra are then compared with the spectra that are derived from the image data so as to identify areas that have spectra that are similar to those in the library. These areas can then be assigned to the class to which that spectra belongs in the library.

4.10 THE ANALYSIS OF DYNAMIC PROCESSES

4.10.1 Introduction

Dynamic processes are defined as those physical environmental processes that are the result of the impact of forcing mechanisms on physical components of the environment. The effect of these forces is to induce changes to the status or condition of those and other physical components of the environment. These changes can in turn contribute to forcing mechanisms that then affect other components of the physical environment. The physical environment thus consists of many interlocking dynamic processes. Each process is driven by a number of forcing functions. Each process provides outputs that are either part of the environmental store, or that act as forcing functions for other processes. In this way each dynamic process can affect both the store of resources and other processes.

By their nature dynamic processes are affected to some degree by the fundamental forces that affect the globe. These forces include the orbital characteristics of the Earth around the Sun, the radiation from the sun and its cyclic patterns, the gravitational, magnetic and other characteristics of the Earth itself and the forces induced by the rotation of the Earth on its axis. These fundamental forces act on the Earth and its constituents, producing changes in these constituents that then affect other physical attributes of the environment. Most dynamical processes thus include cyclic components that are controlled by these fundamental forces. The annual or seasonal cycle is the best known and most dominant of these cycles, although there are a number of other known cycles, and there are probably a number of unknown cycles as well. However, individual dynamical processes can be affected by subsidiary cyclic and non-cyclic patterns that are derived from feedback or feedforward loops operating on other dynamic processes.

The characteristics of dynamic processes are always shown as changes in the temporal dimension and often as changes in the spatial dimension. Imagery recorded over a suitable period of time is a powerful way to capture information on these processes.

> Analysis of dynamic processes using image data requires the use of hyper-temporal image data.

Shannon's sampling theorem states that a sinusoidal cyclic process can be completely specified if at least two observations are taken within a cycle of the process. This theorem sets the minimum sampling rate to recover cyclic temporal patterns from temporal data. In practice a higher frequency is required because errors in the data may make estimation of the system characteristics very problematic if only two observations are used per cycle.

> Temporal image data can be used to depict temporally dynamic processes if the data is acquired at a frequency that meets Shannon's sampling theorem, in the absence of noise in the data, or at a higher frequency when there is noise in the data.

Hyper-temporal image data is defined as a dense temporal set of image data of a region, matching the definition of hyper-spectral image data. It follows that each image in that sequence of images captures information about the surface being imaged. This set of information includes, but is not exclusive to, information about the effects of the dynamic processes on the features in the area that have some of their characteristics captured in the image data. An image thus captures information on both some of the dynamic processes that operate in the area as well as information on other characteristics of the area. This means that the dynamic information that we seek will be embedded in other information in the image. Now the other information that is captured will include some that is stable over time, and some that is not, and so there are three types of information that are captured in a sequence of images:

1. *Stable information that is not reflecting the effects of the dynamic processes*. Such information can include the landuse of an area and some types of landcover condition information. Such information may be included in just a subset of the whole hyper-temporal data set.
2. *Variable information that does not reflect the effects of the dynamic processes*. Such information can include the effects of the atmosphere on the image data, and the condition of some landcovers, for example, due to the effects of fire and the activities of man on the surface. Such information may be captured in a subset of the hyper-temporal data set, albeit a different subset to that of other types of information.
3. *Variable information that derives its change in value from the effects of the dynamic processes on the covers in the area*. Such information will show changes over time and space that will reflect the influence of the dynamic process on those covers, where the nature of these changes can vary from location to location, and from time to time. They are none the less linked in a systematic way to the dynamic process, even if this systematic connection does not lead to predictable changes in the features that can be detected in the image data.

One of the characteristics of hyper-temporal image data is thus the high level of dimensionality of the data and another is the complex intermeshing of the different types of information that are captured in the data. This leads to significant challenges in the extraction of information about the temporal processes that are impinging on the image data from the other information that is captured in the data, or in sub-sets of the data. Most techniques of image analysis that are thus focussing on the extraction of dynamic information will thus have the twin goals of reducing the dimensionality of the data as well as identifying the relevant information from all of the other information that is captured in the image data.

In the conduct of hyper-temporal data analysis for the goal of extracting information on the dynamic processes, it is important not to ignore the use of classification and estimation. Indeed, these tools should be seen as prerequisites to this level of analysis. Classification is a very valuable way of reducing the dimensionality of the data as well as extracting information that is not directly relevant to this analysis, but which may be essential to this analysis. Thus the analysis of dynamic processes will often focus on specific landcovers or landuses, or may indeed operate in different ways on different landcovers or landuses. Estimation is also a valuable way of reducing the dimensionality of the data, and in the process can lead to dynamic analysis using attributes that are more useful in that analysis. Thus, conversion of image data into a vegetation index, or into an estimate of FPAR, or GLAI will significantly reduce the dimensionality of the data set to be analysed and may lead to better insights than will be achieved in using the source wavebands of data.

Dynamic analysis, because of its need to use image data captured over time and space, must use data sets that are internally consistent. This means that the data sets need to be rectified, so as to provide spatial consistency, and calibrated and atmospherically corrected so as to provide temporal radiometric internal consistency. Internal consistency is essential whilst external consistency will be essential for some purposes but may just be desirable for other purposes. Internal consistency means that all of the images are spatially and radiometrically related in a consistent way, although they may not be related to absolute co-ordinates or reflectance. External consistency means that the data are accurately related to external co-ordinates for spatial consistency and to reflectance for radiometric consistency. External consistency means reflectance since radiance is itself subject to dynamic processes, and it may eventually be found necessary to define some standard reflectance if BRDF is shown to significantly interfere with this level of image analysis.

4.10.2 Time Series Analysis of Image data

Hyper-temporal time series of image data can be analysed by means of the classical methods of time series analysis. In such an analysis the time series is considered to consist of:

1. *The seasonal cycle* — S. The seasonal cycle is defined as the cyclic data derived from known causes, so that the period of the cycle is known, even though the magnitude of the effects may not be known. The most common seasonality is derived from the annual cycle as the name indicates, although other cycles may also exist of this form.
2. *Long term secular movements or trends* — L . These refer to a general direction or trend that is taken by the dataset.
3. *Cyclic movements* — C. These are due to processes of unknown duration and intensity. They are periodic, but their causes are not be known.
4. *Irregular fluctuations* —I. These arise due to valid causes, but they are not cyclic, or not obviously so. They can appear to be similar to quasi-random fluctuations in the data, but their causes are quite different, and they are of considerable interest to the analyst.
5. *Random fluctuations* — R, due to noise and errors in the data.

A time series thus consists of the five components in either the additive or multiplicative forms:

$$V = S + L + C + I + R$$

$$V = S \times L \times C \times I \times R$$

Whilst most of the techniques that will be discussed in this section are based on the first form, both forms have some validity. The preferred form should be that which yields the most success in the analysis.

An important concept in time series analysis is that of stationarity.

Strict stationarity means that the statistics of two equal length time intervals of the time series will be statistically the same. This means that the differences between the means and the other moments as derived for both segments are not statistically significant. In many cases it is impractical and indeed it may be impossible, to measure all of the moments and so it may be impossible to test for strict stationarity. A weaker statement on stationarity is called the weak stationarity. A time series is weakly stationary if the differences between the means, variances and covariances of the two segments of the time series are not statistically significant.

The importance in defining stationarity is that some techniques in time series analysis assume that the series being analysed is stationary. It follows that a time series that violates this assumption may give unreliable results when using techniques that depend on this assumption.

4.10.2.1 Removal of the Seasonality

The period of the seasonality is known, but its magnitude is not known. It can be removed in either an additive or a proportional manner. The seasonal cycle consists of a set number of intervals (12 in monthly data) separating data acquisitions. The methods to remove seasonality are:

1. *Additive seasonal adjustment* At each pixel in turn, and in each of the known cycles, compute the mean value. Within each cycle, take its mean value from each acquisition value to give difference values for each acquisition in the cycle. Repeat this within each cycle. Use the differences at each interval to compute the average difference at each interval in each cycle. Such an analysis is illustrated with the data in Table 4.13, and the results shown in Figure 4.64.

TABLE 4.13
Additive Seasonal Adjustment of Three Seasonal Cycles for a Pixel in a Temporal Sequence of Monthly Image Data

Cycle 1		Cycle 2		Cycle 3		
Observed Values	Adjusted Values	Observed Values	Adjusted Values	Observed Values	Adjusted Values	Average Differences
26	67.28	24	65.28	24	65.28	−41.28
28	66.61	26	64.61	28	66.61	−38.61
31	64.94	30	63.94	35	68.94	−33.94
48	65.94	45	62.94	51	68.94	−17.94
79	65.61	75	61.61	84	70.61	13.39
125	71.28	106	52.28	128	74.28	53.72
135	67.94	120	52.94	144	76.94	67.06
129	67.28	116	54.28	138	76.28	61.72
86	60.61	79	53.61	109	83.61	25.39
42	60.61	37	55.61	63	81.61	−18.61
31	62.61	29	60.61	43	74.61	−31.61
26	65.28	25	64.28	29	68.28	−39.28
Average 65.5		Average 59.33		Average 73.00		

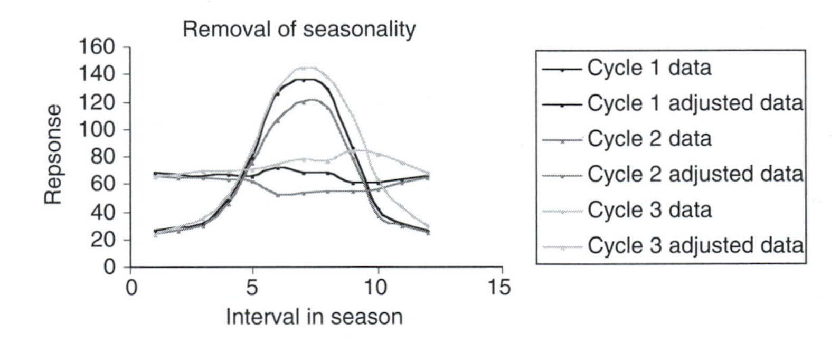

FIGURE 4.64 The additive seasonal adjustment of the data in Table 4.13.

2. *Proportional seasonal adjustment* At each pixel derive the seasonal mean in each cycle as in the first method. Compute the proportion of this that each acquisition represents in each cycle, using the cycle mean. Average the proportions and then adjust them to sum to 12.00. Divide the acquisition value by the average proportion for that acquisition period within the cycle to derive adjusted seasonal data. The data displayed in Table 4.13 has been used for this form of adjustment as shown in Table 4.14 and Figure 4.65. The two techniques derive somewhat different results as can be seen in comparing the results in the two figures. The analyst should find out which form of seasonality adjustment works best under the different conditions that are likely to be met, and then use the appropriate form of adjustment.

The average seasonal adjustments that are derived in this way should be kept as they represent that component of the data. They can be used to create an image that can then be analysed so as to better understand the characteristics of the seasonal component in the data. Seasonal adjustment in this way

TABLE 4.14
Proportional Seasonal Adjustment of Three Seasonal Cycles for a Pixel in a Temporal Sequence of Monthly Image Data

Cycle 1		Cycle 2		Cycle 3		
Observed Values	Adjusted Values	Observed Values	Adjusted Values	Observed Values	Adjusted Values	Average Differences
26	69.01	24	63.71	24	63.71	0.38
28	67.24	26	62.44	28	67.24	0.42
31	63.77	30	61.71	35	72.00	0.49
48	65.76	45	61.65	51	69.87	0.73
79	65.45	75	62.14	84	69.60	1.21
125	68.83	106	58.37	128	70.48	1.82
135	66.87	120	59.44	144	71.33	2.02
129	66.55	116	59.85	138	71.20	1.94
86	62.36	79	57.28	109	79.03	1.38
42	59.22	37	52.17	63	88.82	0.71
31	59.96	29	56.09	43	83.17	0.52
26	64.17	25	61.70	29	71.57	0.41
Average 65.5		Average 59.33		Average 73.00		1.00

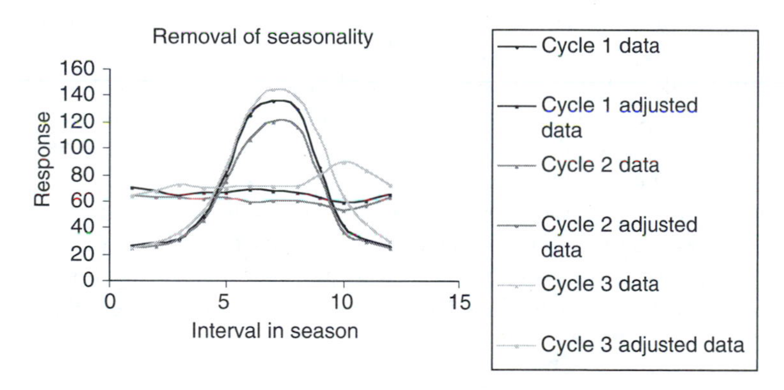

FIGURE 4.65 The proportional seasonal adjustment of the data in Table 4.14.

does not remove the long-term trends in the data, nor does it remove other cyclic components from the data, unless they have a cyclic period that matches the season. If they do match the season, then of course they are removed as part of the seasonal cycle. Finally, the seasonality represents a very large part of the variation in the data, and the sums of squares due to the seasonality will dominate the total sums of squares for many parts around the globe. If the seasonality is not removed prior to trend analysis, then the seasonality will remain after the trend analysis, and the seasonality will contribute to the sums of squares that are remaining after applying the regression.

4.10.2.2 Analysis

4.10.2.2.1 Analysis of the Trend

The long-term trend through the temporal pixel data is removed by the use of least squares to fit a curve to the data, as discussed in Section 4.2. The analyst needs to select an appropriate shape curve for the fitting. This shape may be suggested by the analysis of the actual data for a number of selected pixels in the image area or from an understanding of the expected shapes of the trends derived from an understanding of the processes that are involved.

The normal practice in the conduct of such a curve fitting is to analyse the statistical significance of the derived parameters of the curve, as introduced in Section 4.2, and to only use those curves that have statistically significant parameters. The problem with this approach is that the application of the curve fit to some pixels, and not to others will introduce discontinuities in the image data that may be large at the boundary between pixels that have been affected by the curve fitting and those that have not been affected. It may, for example, introduce large discrepancies into the residuals data after application of the curve fit, and these discrepancies will tend to influence the analysis of the residuals data. The analyst has the choice of following this approach, and then having to deal with the introduced anomalies later, or he may decide to apply more relaxed constraints on the use of the derived curves, accepting all curve fits with a confidence higher than, say, 50% rather than 90%, 95% or 99% or he may ignore the significance analysis and apply all curve fits, and in this case recognize that some of the corrections will introduce errors into the residuals. In practice it is a good idea to try the alternatives and then use that combination that the analyst thinks is most appropriate for the work being conducted.

4.10.2.2.2 Analysis of Cyclic Patterns

The cyclic patterns of interest are of unknown duration and intensity. The best way to address this issue is by means of the use of the fourier transform. The Fourier transformation has been introduced in Section 4.2 in relation to spatial data, and it has been shown that significant spikes in the data indicate the existence of that spatial frequency in the data. The same applies to temporal data. A Fourier transformation of the data in the time dimension will contain spikes at those frequencies that represent significant cyclic processes in the data. It was further shown in Section 4.4 that the Fourier Transform could be filtered to remove these spikes. Once the spikes are removed in this way, then inversion of the Fourier transform yields the temporal record, but without the cyclic component in the temporal dimension. If the filtering is conducted in a proper way, then the residual data that is left contains just the irregular and the random components.

Since the fourier transformation is being conducted on each pixel in the time dimension, it is not realistic to analyse each of these transformation manually to filter out the cyclic component. It is necessary to remove them using an automated process, and storing both the cyclic components and the residual image data as separate images that can then be subjected to separate analysis.

4.10.2.2.3 Irregular Patterns

Irregular patterns are those caused by legitimate activities or processes, but which appear to be quasi-random in form in a temporal profile of data for one pixel. They differ from actual random processes in that they are due to actual events in the environment and these are quite non-random in character. However, when considering just one pixel, they would appear to be quasi-random. One of the characteristics of such processes in comparison with random processes is that they will usually affect more than one pixel. Thus, if a farmer implements an action of a field that is inconsistent with the remainder of the fields in the area, the result may look quasi- random, but it is not. These irregular processes thus will normally affect an area of pixels, so they will create a level of correlation between adjacent pixels. This correlation is thus an indication of the existence of irregular patterns in the data. Construction of the variogram for the data remaining after removal of cyclic patterns will show the existence of this local correlation, and the nugget will show the level of random processes in the data. Thus the course of action is to compute the variograms for the data processed to remove cyclic processes, so as to identify the magnitude of the irregular and random components in the data. Once the time series analysis has separated the five components of the signal, each can be analysed in turn.

4.10.3 Comparison of Two Time Series

An important application of time series analysis is the comparison of two time series. This comparison may be between two time series of image data, or more usual, a comparison of a time series of image data and a time series derived from a model. With this sort of comparison, both time series can be considered to be a reasonable fit to a function of the form:

$$X(t) = f(t)$$
$$Y(t) = f'(t)$$

where $X(t)$ and $Y(t)$ are the two time series. If t can be removed from these equations then you get an equation that is of the form:

$$X(t) = F(Y(t))$$

Any functions can be used as long as the data is a reasonable fit to the model, and t can be removed from both functions to form a functional relationship between the time series. If we consider vegetation over annual and longer periods, then we can consider the two time series as sine functions of the form:

$$X(t) = b_0 + b_1 \times \sin(b_2 t + b_3)$$
$$Y(t) = d_0 + d_1 \times \sin(d_2 t + d_3)$$

For vegetation, the mean and amplitude values (b_0, d_0 and b_1, d_1, respectively) can be expected to vary, the phase shifts (b_3 and d_3) may vary and the wavelengths are likely to be either annual or of six-monthly duration, that is $b_2 = d_2$ or $b_2 = 2d_2$. If the two series only vary in mean and/or amplitude values then $b_2 = d_2$ and $b_3 = d_3$ and it can be shown that the relationship between $X(t)$ and $Y(t)$ is a line:

$$(X - b_0) \times d_1 = (Y - d_0) \times b_1$$

or

$$X = Y \times (b_1/d_1) + (b_0 - d_0 \times (b_1/d_1)) \tag{4.135}$$

The figure resulting from Equation (4.135) is a line with the gain setting equal to the ratio of the amplitudes and the offset being a function of the mean and amplitude values. If the amplitudes are about the same then the gain will be about 1 and the offset will be the difference between the mean values. If the mean values are about the same then the gain and offset tend to be mirror images of each other since the gain $= (b_1/d_1)$ and the offset will have values of about mean $\times (1 - (b_1/d_1))$.

If there is a difference in phase between the two signals then the resulting equation is an ellipse or a circle. When there is a difference in the wavelength of the two signals, then a variety of higher order figures can evolve. In practice, since the dominant cycles of nature are the annual and six monthly cycles, only this situation will be considered. In the simple situation the resulting figure is a figure of eight. The different situations considered here are shown in Figure 4.66.

If the two time series just vary in amplitude and mean, then the formula relating the two time series is a line with the gain and the offset a function of the differences in the mean and amplitude values. When the two time series differ in phase, then the derived figure is an ellipse or a circle. If an artificial phase shift is introduced into the data that is opposite to the actual phase shift, then the relationship can be reduced to a line at one phase shift that matches the differences in the phase of the signals. Thus, if the analysis has monthly data to compare, and linear regressions are fitted to

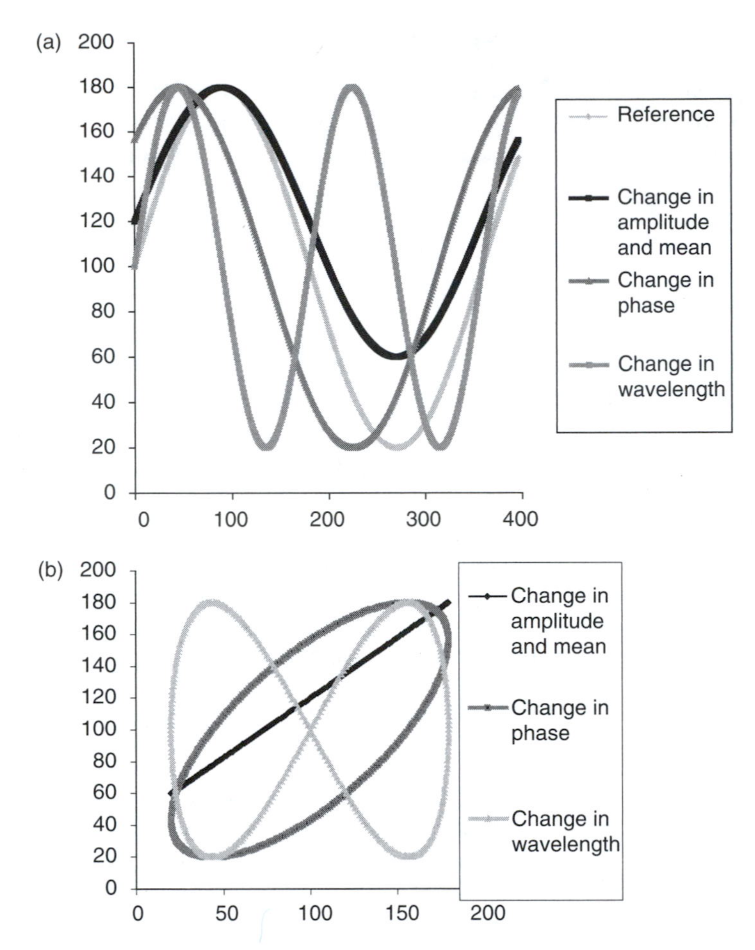

FIGURE 4.66 Comparison of a reference sine wave with a second sine wave. (a) the various source signals for the comparison, (b) the comparison of each signal with the reference signal.

the two data sets, at each of the 12 possible combinations of phase shift, then one of these will be a close fit to the linear regression, if the series only vary in their means, amplitudes or phase shifts.

The maximum R^2 values are then used to indicate the best fit between the data sets, and this phase shift is then taken as the phase shift that exists between the data sets. When this is done, then the significance of the regression gain and offset values are the same as for the case where there are just differences in the mean or amplitudes. Thus, changes in mean, amplitude and phase shift can be expressed as variations in the regression gain and offset, and the phase shift between the signals. These parameters can be taken to be a valid solution to the analyses when the value of the regression coefficient is above an acceptable threshold value.

In the physical world, four types of cyclic pattern dominate actual data sets, but are by no means the only cyclic elements in those data sets. These four types of patterns are:

1. *Different mean values.* The signal amplitude, phase shift and wavelength are similar, but the signals have different mean values. This situation exists in say a comparison of NDVI signals from rainforest with that from desert conditions. Both conditions have very small amplitudes but a large difference in the NDVI values.
2. *Similar wavelength and phase shift, different means and amplitudes.* The average signal value and the amplitude of each data set may be different, but the wavelength and phase

shifts are similar. This is consistent with the comparison of NDVI values for say rainforest and temperate forest, or between many temperate landcovers.

3. *Similar wavelength, but different phase shift, mean and amplitude*. This is consistent with greening up and senescence consistently occurring either earlier or later in one data set than in the other data set, or with greenup occurring in a different season in one data set to the other data set. This can occur when comparing an area of natural vegetation with an agricultural area where the greenup of the agricultural areas is significantly earlier, or later, that that of the natural vegetation.
4. *Different wavelength, phase shift, mean and amplitude.* A doubling of the frequency or halving of the wavelength of one data set relative to the other will be considered of all those that are possible because this situation is consistent with double cropping each year and with bi-modal climates that occur in some places around the globe. As a consequence one data set will exhibit six-monthly cycles whilst the other data set exhibits annual cycles.

It can be seen that if a comparison of two data sets yields a linear relationship, then the features of that relationship are readily explained in terms of the two series, if the series are assumed to be sine waves. It is not, however, necessary that the time series need to fit a sine wave. They can fit any other form of equation, the main requirement being that the equations used can be manipulated so as to remove t and leave a relationship between the time series. Indeed, it is not even necessary that the two equations be of the same form, so that, for example, one could have two time series that fitted a line and a second order polynomial respectively:

$$X(t) = b_0 + b_1 \times t$$
$$Y(t) = d_0 + d_1 \times t + d_3 \times t^2$$

which gives a parabolic fit between the two time series. For annual data the sine wave is quite robust, however for data that may occupy only part of an annual cycle, some other form may be more suitable.

4.10.4 The Analysis of Spatio-Temporal Dynamic Processes

Many environmental processes not only change over time, but they can also change in location. Thus, the distribution of plants and animals can both change in intensity and location over time, where some of these changes can be very dramatic, as with the case of animal migration patterns. However, many other processes show the same sorts of characteristics such as the spread of disease in crops, and animals.

At present no techniques have been developed to address the need to understand and analyse spatio-temporal patterns of dynamic processes, and this is a gap that I hope gets filled in the future.

4.11 SUMMARY

In the thirty or so years that the image processing of image data has been implemented, there have been some significant developments and there have been cases of our inability to resolve difficult, maybe even intractable problems. The most spectacular developments in image processing have come from improvements in the technology and our improved understanding of how light is reflected in a vegetation canopy. The spectacular improvements in computer technology have been matched by better image processing software. Scientifically the development of models of canopy reflectance have allowed the development of better sensors and better ways to extract information on vegetation from image data.

The most intractable problem has been that of providing internally and externally consistent image data. Internal geometric consistency has been achieved. External consistency can be achieved in an expensive way through the use of ground control and rectification. However, there is going to be a proliferation of image analysis, including the analysis of temporal image data and current methods of rectification are too expensive. There is a need for the development of automated rectification processes. For most parts of the globe, where there is adequate ground identifiable points to use auto-correlation techniques, this should be achievable.

The largest problem has been our inability to achieve internal and external radiometric consistency. The single largest problem is the impact that the atmosphere has on image data. Whilst there are atmospheric correction models and software, the problem arises primarily because of the impact of water vapour and aerosols on image data, and the difficulty of correcting for these processes due to the very fast changes that can occur in their status in the atmosphere, both spatially and temporally. Water vapour is the least of these problems, since the absorptance characteristics of water vapour do not change, but the density does, and so the issue is that of estimating the density and then knowing this, being able to correct for that density at all the wavebands of interest. The best way of achieving this so far is by the use of moisture absorption bands in the atmosphere, assuming the surface does not have its own absorption features at these narrow wavebands. There are now satellite sensing systems that can correct for this, and so, if these prove to be effective, then this problem may have eventually been solved.

Aerosols are a different mater again. Aerosols can have different reflectance characteristics when they went aloft, and their reflectance characteristics can change whilst they are aloft. This means that the assumptions on which the correction of water vapour is based, cannot be met. At present there are two approaches that seem to have some merit: assuming the low and constant reflectance of water to estimate aerosols over the oceans and the multiple path image approach for estimation over the land. Unfortunately, both approaches make some fairly significant assumptions that are not likely to be met in practice. It is likely that this problem has not yet been satisfactorily solved.

Our inability to create externally consistent image data has meant that applications that require this characteristic have been hampered in their implementation of remote sensing, because of our inability to provide consistent information to the resource manager. The classical example of this is the estimation of vegetation status in terms of such things as green biomass and green leaf area index.

Even in classification, our achievements have not been as impressive as one would have hoped. The classes that we wish to map are such that there is significant spectral overlap between them in the image data, and this means quite simply that they cannot be completely discriminated using classical statistical techniques. If we could routinely achieve accuracy in classification that would be acceptable to a court of law, then the way we use image data would be revolutionised. This should be the goal of classification. Given this goal, we need to think more about how to achieve this goal. If spectral data alone, using current techniques cannot give us this accuracy, then how do we either improve the spectral data, improve the techniques, or incorporate other information into the process. The use of hyper-spectral data is an attempt to improve the data, and this has, so far, not shown a lot of promise in achieving this goal. There may be some potential with improved radar, or some combination of the two, but nothing has suggested that this will be the case. There has been a lot of work in the development of alternate classification strategies, but none of them have yet given us the breakthrough that we are seeking.

The answer, I believe, almost certainly will come from taking a fresh approach to classification that incorporates other dimensions of data and knowledge into the classification process. Some approaches to this issue are discussed in relation to vegetation in Chapter 7. In addition, we hardly take the spatial dimension into account (only by segmentation), and I think that there is a lot more to be made from this dimension. One of the things that has come out of the work on the ALV function is that local variance is a function of the image resolution relative to object sizes. Since classification is improved by minimising the within-class variance and maximising the between-class differences, spatial analysis tools may tell us a lot about how to reduce the within-class variance whilst the spectral data tells what we can do to maximise the between-class distances.

Finally, the analysis of long sequences of image data, to better understand dynamic environmental process, is starting to take off in a big way. This is likely to be an area of fruitful research and development for a number of years, but of course, it fundamentally depends on the derivation of externally consistent data.

FURTHER READING

There are a number of texts that deal in more detail with digital image processing, and the reader is referred to those by Richards (1992) and Gonzales and Woods (1993). However, even these texts do not deal with specialist subjects in full, and so the reader may wish to go further afield should they want to gain deeper insights into specific topics. For an understanding of Fourier theory, the books by Korner (1989), Weaver (1983) and Rees (1981) are a good starting point. Goovaerts (1997) provides a very good and practical understanding of geostatistics and its role in the natural resource evaluation. Should you want a more theoretical treatment, then Cressie (1993) is suggested as a starting point. Finally, should the reader wish to read further about the least squares method, then Lawson (1974) provides a good basic text, whilst Collins (1992) is more up to date and focuses on spatial data as is relevant to our interests.

REFERENCES

Collins, W.A., 1992. "The Analysis of Observations with Applications in the Atmospheric Sciences", *The Advanced Study Program Colloquium Series*, National Centre for Atmospheric Research, Boulder, CO.

Cressie, N.A.S., 1993. *Statistics for Spatial Data*, John Wiley and Sons, New York.

Gonzales, R.C. and Woods, R.E., 1993. *Digital Image Processing*, Addison Wesley, MA.

Goovaerts, P., 1997. *Geostatistics for Natural Resources Evaluation*, Oxford University Press, Oxford.

Guyot, G. and Baret, F. 1988. Utilisation de la haute resolution spectrale pour suivre l'etat des couverts vegetaux, *Proceedings 4th International Colloquium on Spectral Signatures of Objects in Remote Sensing, Aussois,* France, January 18–22, 1988 (ESA SP-287, April 1988), pp. 279-286.

Kauth, R.J. and Thomas, G.S., 1976. "The tasselled cap — A graphic description of the spectral-temporal development of agricultural crops as seen by Landsat", *Proceedings, Symposium on Machine Processing of Remotely Sensed Data*, Purdue University, West Lafayette, IN.

Korner, T.W., 1989. *Fourier Analysis*, Cambridge University Press, Cambridge.

Lawson, C.L. and Harrison, R.J., 1974. *Solving Least Squares Problems*, Prentice Hall, NJ.

Rees, C.S., Shah, S.M. and Stanojevic, C.V., 1981. *Theory and Application of Fourier Analysis*, Marcel Dekker, New York.

Richards, J.A., 1986. Digital Image Processing, Springer-Verlag: New York.

Rouse, J.A., Haas, R.H., Deering, D.W., and Schell, J.A., 1974. "Monitoring the vernal advancement and retrogradation (The Greenwave Effect) of natural vegetation," ERTS Final Report, Remote Sensing Centre, Texas A&M University, College Station, Texas.

Van der Meer F.D. and S.M. de Jong (Eds.), 2001, Imaging Spectrometry: Basic Principles and Prospective Applications. *Book series Remote Sensing and Digital Image Processing* Vol. 4, 403, Kluwer Academic Publishers, Dordrecht.

Weaver, H.J., 1983. *Applications of Discrete and Continuous Fourier Analysis*, John Wiley and Sons, New York.

5 The Use of Field Data

5.1 THE PURPOSE OF FIELD DATA

5.1.1 Definition and Description of Field Data

Field data are defined as verified, or verifiable, data collected using proven, relatively close range techniques such that the resultant data are much more accurate, consistent and at a higher resolution than the information derived from remotely sensed data with which the field data are to be used. Remotely sensed data are used to provide consistent, spatially extensive information. Field data are used to improve information extraction from remotely sensed data, to calibrate either the data or the information, and to assess the accuracy of the derived information. Field data are point or localised data.

Field data can be of many types: data collected in the field, derived from statistical data in the office, from aerial or ground photographs, or from aerial observation. The key differences between field data and the information derived from remotely sensed data are that the field data are localised, more accurate and will often be more detailed, either spatially or informationally.

There are a number of aspects of this definition that require further consideration:

1. *Verifiable*. The field data must be capable of independent verification, duplication or replication using similar or different techniques. This is in contrast to the information derived from the remotely sensed data that can be duplicated using the same data sources but which can only be verified using field data.
2. *Proven techniques*. The techniques used in acquiring field data must be well established, of known characteristics, accuracy and reliability. These techniques can include:
   (a) Physical site visits to collect specimens of the physical characteristics of interest, for example, observations on geomorphology and samples of vegetation, rocks and soils.
   (b) Measurement of parameters in the field, for example, the measurement of radiometric, gravimetric and magnetic data using remote, but close sensors; the measurement of temperature and rainfall using direct contact measurement techniques.
   (c) Measurement of parameters using established remote sensing techniques such as the measurement of field areas on aerial photographs and the use of contours derived by photogrammetry, as long as the measurements meet the criteria set above.
   (d) Sampling surveys to extend field data to cover larger areas, for example, to estimate the biomass for a pixel area.
3. *Accuracy, consistency and resolution*. Information derived from both field data and image analysis contain errors. The magnitudes of both sources of error are of concern to the analyst. In this context the accuracy of information derived from remotely sensed images is defined as the closeness of the estimated values to the values determined by the use of field data. As field techniques are not perfect, the estimate of accuracy is not an estimate of the closeness of the extracted information to "true" values but rather to the field data values. If the accuracy of the field data is much higher than that of the image analysis, then the errors in the field data will introduce negligibly small errors in the derived information, and the derived accuracies will be a good estimate of the accuracy of the image analysis relative to "true" values. There is extensive evidence from the physical sciences that there

is no such thing as a "true" value, since at very fine resolutions the methods of measurement start to interfere with the measurement itself.

5.1.2 Role and Types of Field Data

Field data have a role to play in all facets of information extraction from remotely sensed data, including:

1. Correction for atmospheric effects
2. Support for both visual interpretation, and image processing in information extraction
3. Development of estimation models and in estimation processing
4. Accuracy assessment in both classification and estimation

The techniques used in collecting field data are often relevant to more than one of these facets. These techniques include:

1. *Spectrometer observations* for atmospheric corrections, in the development of estimation models and improving the conceptual models used by the interpreter to relate physical conditions to remotely sensed image data.
2. *Collection of physical field data* on surface conditions. These types of field data use all the standard techniques for the conduct of surveys and the collection of data including botanical surveys, collection of botanical, geologic and other samples and the observation of geomorphic conditions. These field techniques, being well covered in texts on the relevant disciplines, will not be covered in this text.
3. *Observation of close range measurement of parameters* including gravity, magnetic and electromagnetic induction surveys as well as the identification and mapping of surficial features from aerial photographs. In general, these techniques are also well covered in the relevant texts and will not be covered in this text. Some of the simpler techniques that can be used in mapping features from aerial photographs have been covered in Chapter 3.
4. *Random sampling surveys* in both estimation processing and accuracy assessment.
5. *Conduct of accuracy assessment.*

Whatever method, or methods, is used in collecting field data, the analyst must start with:

1. *A clear definition of the purpose* of the field data.
2. *A specification of the criteria that the field data must meet*, including the types, resolution, timeliness and accuracy of the data to be collected.

The purposes of field data are best understood by analysing the nature of the information extraction process. A remotely sensed image, as a record of the energy reflected or radiated from the surface in selected wavebands at a moment in time, is not the information on characteristics of the surface that are of interest to resource managers, but are indicative of these characteristics to some degree. They may also be indicative of other characteristics to a greater or lesser degree. In consequence, there is considerable potential for confusion and conflict, resulting in the drawing of wrong inferences or conclusions, during extraction of information from the data. Often a process of extrapolation rather than interpolation is used to draw these inferences in image analysis, where:

Interpolation is defined as the estimation of intermediate values between two or more known boundary conditions by the use of a linear, or some other kind of model. During interpolation, the boundary or limiting conditions are known, and the estimation is to derive values between these boundary conditions. Thus, errors will arise when the model used as the basis for the interpolation is different to the actual surface being estimated. For example, contours are often drawn by interpolation between points of known height using a linear interpolation model. If linear interpolation is used then errors will occur when the shape of the surface deviates from this model. But the resulting errors are usually small, particularly if the locations of the spot heights are carefully chosen and usually will only occur within a known range of values that is controlled by the boundary values.

Extrapolation is defined as projecting from known information to estimate unknown information outside the known conditions. The classical extrapolation situation is to say that conditions last year were $(x + y)$, this year they are $(x + 2y)$ and therefore they will be $(x + 3y)$ next year. Extrapolation is going outside the bounds of the known or observed data. In the analysis of remotely sensed data, it is assumed that the same response values represent the same conditions across the image, and that these response values are sufficiently unique so as to discriminate the various classes. Unfortunately, this is a very flawed assumption. The remotely sensed data, representing just a few attributes of the surface can have similar response values for quite different conditions. In addition, similar surface conditions can have different response values in the image data, because of the effects of atmospheric attenuation and scattering and also due to the variations in other physical attributes of the different sites, such as variations in soil colour, topographic shape of the land surface and the effects of variations in management. In extrapolation there is no check in the processing to tell the analyst that the analysis is giving results that contain large errors. As a consequence, large errors can occur, and the analyst will not normally be aware of their existence from the analysis itself.

During information extraction the known attributes about the features of interest are the image data itself. Other data, such as field and training data; maps and other auxiliary data and information, such as soil, topographic and land ownership data; and knowledge, including heuristic, numerical and other models, form the basis for the development of specific relationships between the image attributes and the information being sought under particular conditions or locations. The information extraction process, by identifying the known attributes and using the available accumulated data, information and knowledge, selects the most valid of these relationships to extrapolate from the known image attributes to estimate the causal physical characteristics. Often the assumptions linking physical characteristics to causal attributes on which this process is based are valid and the extrapolation will lead to correct results. However, extrapolation is fraught with risk because it is often very difficult to know whether the assumptions used in the extrapolation are valid in the specific conditions pertaining to the particular image and surface. Auxiliary information and knowledge are invaluable in assisting in determining whether the assumptions are being, or are likely to be, met. Despite the use of such information and knowledge there are always likely to be circumstances where a different conjunction of environmental conditions than those considered by the analyst will create the specific data characteristics as depicted in the imagery. When this occurs the imagery will often depict conditions quite differently than those expected by the analyst, or to put this another way, the depiction of the surface in the image may match a set of conditions that the analyst may expect to occur, but due to quite different physical conditions than actually exist. The analyst, by using his or her constructed relationships or models may draw the wrong conclusions from the data, leading to errors in the information extraction process. Not only may large errors arise, but also since there is no other data or criteria to use as a check, the existence of these errors will not be noticed by the system or the analyst.

With extrapolation the errors can be very large if the data analysed leads to assumptions that create a model a long way from actual conditions. Consider the case of estimation of leaf area index (LAI) from a vegetation index image using a numerical model developed for the purpose. The numerical model would have been developed to operate over a range of species and growing conditions, where these will include variations in canopy structure and soil conditions. If these conditions are replicated in subsequent mappings, then reasonable estimates of LAI can be expected.

However, other conditions may be met than those covered in the creation of the model, including, but not exclusive to:

1. Changes in growing conditions may have an effect on the canopy geometry or reflectance that have not have been met in the construction of the model.
2. Moisture on the leaves due to rainfall will affect the reflectance, particularly in the Near-infrared (NIR), significantly reducing the vegetation index values from the crop.
3. Windy conditions will change the canopy geometry and may change the canopy reflectance during image acquisition, causing large increases or decreases in the vegetation index values derived from the crop.
4. Surface soil moisture may reduce the NIR soil reflectance, again reducing the vegetation index values.

When one or more of these situations arise, then the model will take the vegetation index values and derive estimates of LAI that may be a long way from the actual values. However, the system, and the analyst, not being aware of the variation away from conditions met in the construction of the model, will not be aware that errors are being made in the estimation.

It is therefore essential that analyst, in conducting an image interpretation task, consider:

1. *The assumptions being made* during each extrapolation task.
2. *Identify alternative, potentially conflicting, sets of assumptions*, so as to determine which are the more valid set to apply under the actual image conditions. For example, a map of LAI may be constructed using a model, of the form LAI = f(NDVI), with its own set of assumptions. Alternatively, field data can be acquired across the area at an appropriate sampling of points and then the map of LAI constructed by interpolation. Both methods are based on their own sets of assumptions and both will have cost and accuracy structures associated with them.
3. *Ensure the validity of the assumptions*, as practically as possible during the analysis. If there is a discrepancy between the physical conditions interpreted from imagery and other evidence available to the analyst, then the analyst has to be very cautious in accepting that set of assumptions as an appropriate extrapolation model.

Field data have an important role to play in meeting these requirements during image analysis. An important assumption usually made in using field data is that the field data are significantly more accurate than the information being derived from remotely sensed data. For this reason the collection of field data should not contain the same processes of inference making by extrapolation as occurs in image analysis. If similar processes are used in collecting the field data as for information extraction, then the observations must be taken under tightly controlled conditions so that all the significant boundary conditions are known. In the example given above, of estimating LAI for a pixel, sampling can be used as long as a sufficiently large and representative sample is taken to keep the errors well below those that might occur in the image processing.

Because field data are not error free they cannot be called "ground truth." This unfortunate term is both misleading and incorrect; it should never be used.

All users of information derived from remotely sensed data should and usually are aware of the errors that exist in the information. This realisation is likely to prompt one or more of three responses from users:

1. A demand that the information derived from remotely sensed imagery undergoes rigorous accuracy and reliability assessment in accordance with established or agreed criteria.
2. That they have access to the accuracy assessment results during their use of the information.

3. Downgrading of the information by using it simply as a pointer or indicator of specific conditions, and then using field visits to verify or reject the information supplied by the pointer. In this way the user is protected from legal action arising out of errors in the information. But the cost is in the need to implement a more extensive and detailed programme of field visits. This downgrading also changes the information from being essential to valuable to the user.

All resource managers, in using any information, weigh that information in accordance with their perceived expectation of the accuracy and reliability of the information. In the absence of reliable, understandable and relevant accuracy assessment, a manager will downgrade information so as to reduce the risk of a mistake. If information is to be used properly, it is essential that a statistically valid and relavant accuracy assessment be conducted on the information. The results of this accuracy assessment should also be conveyed to the user in a readily understood manner.

5.1.3 Accuracy and Reliability

A recurrent theme throughout this chapter is that of accuracy and reliability. This theme occurs because the role of field data is to either improve or to determine the accuracy and reliability of the information that is being derived from the remotely sensed image data. It is therefore necessary to be very clear about what is meant by accuracy and reliability, and their importance in the analysis of image data.

Accuracy is defined in Chapter 1 as a measure of the closeness of the estimated values to the "true" values of an attribute. This is an ideal definition in that it is impossible to measure the true value of any attribute. As devices become more accurate and sensitive, the values measured approach the true values, until such a stage is reached when the act of measurement itself interferes with the attributes being measured. So, in practise, accuracy is the closeness of the estimated values to those estimated using significantly more accurate methods of measurement, or reference methods. If the errors associated with the reference methods are significantly smaller than the errors in the attributes as derived from remotely sensed data, then the errors in the reference methods will have an insignificant effect on the derived estimates of accuracy.

Consider the case where the accuracy is derived using the equation:

$$A = \frac{O}{E} \tag{5.1}$$

where O is the observed data and E is the expected data. By the Law of the Propagation of Error and assuming that O and E are independent:

$$\sigma_A^2 = \frac{1}{E^2}\sigma_O^2 + \frac{A^2}{E^2}\sigma_E^2 \tag{5.2}$$

So that if $(A^2\sigma_E^2 \ll \sigma_o^2)$ then σ_E^2 will have a negligible effect on the variance of the accuracy. This is usually not difficult to achieve, remembering that the accuracy, A, is always less than 1.0, and hence that A^2 can be much less than 1.0.

Reliability, or consistency, is defined as a measure of the closeness of a set of related accuracy assessments. Reliability may refer to the accuracy assessments made of a temporal sequence of estimations of the attribute, such as the reliability of landuse mapping conducted on an annual basis. Reliability can also be concerned with spatially extensive estimates of an attribute, for example the reliability of a method of estimation of LAI across a country covered by a number of image areas.

Both concepts are important, and indeed in some tasks, one concept may be more important than the other, depending on the purpose to which the information is to be used. Thus, a grain merchant, concerned with the sale of grain, may want to know the production that can be expected in the current season. To this user the reliability is of prime importance, since he will use the estimate and multiply it by an adjusting figure so as to derive what he considers to be an accurate estimate of production. If the information he gets is reliable, then he can use the one factor, and get a good estimate of the total production each year. The level of accuracy is not that important. If, however the reliability is poor, then he cannot use the factor with confidence. On the other hand, a manager may wish to use landuse maps to identify individual land managers for a promotional campaign, or for some other purpose. He does not want to contact those who cannot use his product. In this situation accuracy is more important than reliability. As a general rule, accuracy is more important when the information is to be linked to individual and specific users or locations, and reliability is more important when the information is to be used as a generalisation for a region.

The measures of accuracy and reliability are also important. There are at least three different types of accuracy that need to be considered:

1. *The accuracy of classification.* In Chapter 4 we discussed the derivation of landuse maps by the classification of image data. Field data need to be collected in the area mapped so as to assess the accuracy of the classification. What forms of accuracy can be tested, and which should be tested?
   (a) Will the analyst test the accuracy of the classification of individual pixels?
   (b) Is it more relevant to test the accuracy of assignment of fields to classes?
   (c) Is the analyst more interested in the accuracy of the areas of the fields?
   There are many different things that can be tested. If the classification were conducted on a pixel-by-pixel basis, then it would seem reasonable to test the accuracy of classification of the pixels. However, if the classification is then subjected to filtering to remove isolated pixels that have been classified into a different class than its surroundings, and particularly if the classification is subject to further aggregation rules so as to form segments, then testing the accuracy of the derived segments is likely to be more useful to the end user, just as classification that is conducted post segmentation of an image may be subjected to this form of accuracy assessment. On the other hand the end user may be interested in the area of the land covers within the fields and the ability to detect the boundary conditions, in which case assessment of the accuracy of the area estimates may be more useful. This is the situation when the end user wants the area of a crop, so as to estimate production. In this situation ignoring edge effects will result in an over-estimate of the crop area. Clearly, the form of accuracy assessment that is to be conducted depends on the use to which the derived information is to be put, and the form of accuracy assessment will usually affect the design of the accuracy assessment process. The basic components of this form of accuracy assessment are discussed in Section 5.6.
2. *The accuracy of estimates of a parameter.* In Chapter 4 we also discussed the use of remotely sensed images to derive estimates of one or more physical attributes of the environment in the area imaged. The form of accuracy assessment that is appropriate for estimation is usually more straightforward in terms of the goal of the accuracy assessment, although it may be more labour intensive in its execution. A major challenge with this form of accuracy assessment is that the image derives estimates in the finite area of a pixel, whilst the fieldwork used to conduct the accuracy assessment is conducted on a point-by-point basis. In this situation the individual field sample is likely to be such as to ensure that a single pixel is contained within the area, and a number of samples need to be collected so as to provide a statistically valid sample. This form of accuracy assessment is also discussed in Section 5.6.
3. *The accuracy of location.* For some purposes the accuracy of delineation of boundaries are important, where this may be the borders of fields or the locations of borders of landuse or landcover classes. This form of accuracy assessment is also covered in Section 5.6.

Clearly, more than one form of accuracy assessment may be necessary to address the needs of the single, or the group of end users of the information being derived from the image data.

The reliability of a mapping, or of a series of mappings, are usually stated in the form of a variance associated with the derived accuracies, or with a confidence interval (CI) that is derived from the variance, and using a set probability level. Each section dealing with accuracy assessment will also deal with reliability.

5.2 COLLECTION OF FIELD SPECTRAL DATA

5.2.1 Purpose of Collecting Field Spectral Data

Spectral data are collected in the field using spectrometers or scatterometers. Spectrometers are instruments that record the electromagnetic radiation entering the optics of the instrument in set waveband ranges between about 300 nm (ultraviolet) and 15 μm (beyond the thermal infrared). They may be designed to collect the energy in many, but very narrow wavebands, or they may sense a much smaller number of broader wavebands that will often match the wavebands of a specific sensor.

Scatterometers are instruments that generate microwave radiation in one or more wavebands, and receive the return signal where both the generated and recorded returned signals may have like or opposite polarisation. The wavebands used in scatterometers would normally correspond to one or more of the standard radar wavebands.

Spectral data are usually collected to:

1. *Optimise bands used in data acquisition*. The spectral data will be used to identify the spectral bands, phenological stages and seasonal conditions that optimise discrimination among the different physical features or conditions of interest to the analyst. Once the optimum wavebands and conditions have been determined then imagery acquired in accordance with these criteria should provide the best opportunity to discriminate the features in the imagery of interest to the analyst, at the minimum cost and with maximum accuracy. The ability of the imagery to discriminate the covers of interest should make the information extraction process faster, easier, more accurate and consistent, in this way reducing processing costs and increasing usability.
2. *Better understand interaction mechanisms*, or the relationship between surface conditions, environmental processes and the reflectance characteristics of the surface. A better understanding of these interactions will facilitate the development of better sensors and sensing systems, lead to improved analysis techniques and develop more robust and consistent strategies to extract information from the data.
3. *Develop estimation models*, to estimate the values of physical parameters or attributes of the surface from spectral data acquired of that surface.
4. *Facilitate the incorporation of environmental rules* based on our environmental knowledge into the information extraction processes so as to improve their accuracy and reliability or reduce their costs.

5.2.2 Measurement of Field Spectral Data

The radiance emanating from a surface depends primarily on the amount of energy incident on the surface, and the reflectance characteristics of that surface. The radiance emanating from a surface is not constant for a given surface condition because of atmospheric and path geometry effects. Reflectance is much more consistent for given surface conditions. Reflectance has been defined in Chapter 2 as:

$$\rho_\lambda = \pi L_\lambda / E_\lambda \tag{5.3}$$

where L_λ is the radiance reflected from the surface in waveband λ and E_λ is the irradiance incident on the surface.

The reflectance is normally determined in either of two ways:

1. *The direct measurement of reflectance*. Measure incident and reflected radiation to calculate the reflectance directly from the pair of observations.
2. *The measurement of reflectance factors*. Measure the reflected radiation from the target and from a control surface to derive the reflectance factor of the target, or target reflectance relative to the control surface. If the reflectance of the control surface is known then the target reflectance can be calculated.

5.2.2.1 Method 1: The Direct Measurement of Reflectance

There are two ways of implementing this method. The first uses field instruments that can simultaneously measure incident solar irradiance and reflected radiance from the surface. The second is used for instruments that can only measure one of these at a time.

If a spectrometer is used that can simultaneously measure irradiance and radiance, then the procedure is to:

1. Set the instrument up so that the incident radiation sensor is pointing directly upwards, with its diffuse objective set horizontal, ensuring that it is away from any shadow or stray reflected light effects.
2. Set the reflected sensor so as to sense a pre-defined field of view (FOV) within the target surface, ensuring that the sensor is set at the correct angle, usually vertically downwards, and that the instrument casts no shadow in this FOV.
3. Ensure that the instrument, or other equipment, does not project stray light into the sensor.
4. Acquire and record the observations in accordance with the specific instrumental procedure.
4. Calibrate the observations to radiance and compute reflectance.

The main advantage of this method is that it provides true reflectance as long as the sensors are properly calibrated, it is simpler to implement and that it can be used under sunlit conditions as well as conditions of scattered cloud as long as they are a long way from the sun. If clouds are close to the sun, then they can introduce very fast changes in irradiance that may introduce errors into the observations since the two sets of observations (upwards and downwards) are not taken at exactly the same instant in time.

With one or more instruments that can only measure either irradiance or radiance, then the second method has to be used. Either one, or sets of instruments are used to measure the incident irradiance on, and the reflected radiance from, the target surface. Both observations are calibrated and the reflectance calculated as the ratio of the reflected to incident radiation. The general observational procedure is to:

1. Measure the incident irradiance by means of a spectrometer with opaque diffuse objectives set horizontal so as to receive all incident (hemispheric) radiation.
2. Measure the reflected or target radiance using objective lens with a defined FOV.
3. Under cloud free atmospheric conditions, groups of observations of various targets can be observed between observations of incident irradiance, as long as irradiance observations are taken before and after observing the targets, and not more than 1 h separates each observation of irradiance.
4. Repeat the procedure until all the required observations have been taken.

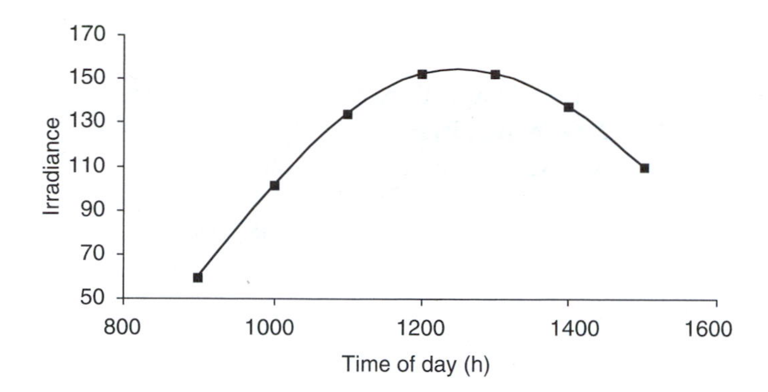

FIGURE 5.1 Interpolated graph of irradiance from regular irradiance observations taken throughout one day.

5. Calibrate the irradiance and radiance observations using calibration curves for the individual instruments. It is important that the instruments be regularly calibrated to maintain accuracy.
6. Compute the reflectance for the different surfaces. How the reflectance is calculated depends on the manner in which the observations were taken as discussed below.

If two or more instruments are available then the optimum situation of taking simultaneous observations of irradiance and radiance can be implemented. This is essential if there is any cloud as reflectance off the cloud can cause local fluctuations in irradiance that will be unnoticed by the observer. In this case the irradiance and radiance observations can be calibrated and the reflectance calculated as the ratio of the two observations.

If one instrument is used, or if one instrument is set up to record the irradiance at a regular interval then the irradiance observations will not correspond, in time, to the radiance observations. The solution is to use the calibrated irradiance observations to fit a surface, such as a sine curve, between the observations knowing that the irradiance rises to a peak at solar midday and then decreases (Figure 5.1). Use the fitted curve to interpolate a value of irradiance at the time of observation of radiance of the target and use this value with the target radiance to calculate target reflectance. To use this technique the irradiance observations must be taken:

1. *At a regular interval*, of not more than hourly separation between the observations, from before until after the radiance observations.
2. *At the same location* as the radiance observations.
3. *Under cloud-free conditions*.

The irradiance observations can be fitted to either a linear (Equation [5.4]) or sinusoidal (Equation [5.5]) curve by regression in a computer. Generally, linear interpolation will only be used when only two observations of irradiance have been taken, before and after the radiance observations of the target surfaces. Since the irradiance does not change linearly, a linear model should only be used over short time periods, for example, when taking radiance observations of the one target. The derived regression curve is then used to estimate irradiance at the time of observation of target radiance, to calculate target reflectance. The linear curve is of the form:

$$E_\lambda = A_0 T + B_0 \tag{5.4}$$

TABLE 5.1
Irradiance Observations at Orange on 15 June 1986, Estimated Irradiance Values by the Use of a Sine Curve, and the Residuals

Time	Observed Irradiance (W/sr/m^2)	Sine Regression Irradiance	Sine Regression Residuals
0900	59	59.48	−0.48
1000	101	102.64	−1.64
1100	134	134.41	−0.41
1200	152	151.64	+0.36
1300	152	152.61	−0.61
1400	137	137.22	−0.22
1500	110	107.00	+3.00
R^2			0.9981

And the sinusoidal curve is of the form:

$$E_\lambda = A_0 \sin\left(\pi\left(\frac{T - T_r}{T_s - T_r}\right)\right) + B_0 \tag{5.5}$$

where T is local time of the observation of irradiance in hours, T_r is local time of sunrise and T_s is the local time for sunset. Consider the observations plotted on Figure 5.1, and fitted to a sine curve, with the observed irradiance, the estimated irradiance values and the residuals included in Table 5.1. The regression curve derived using the data in Table 5.1 is:

$$E_\lambda = 165.9819 \sin\left(\frac{\pi(T - T_r)}{T_s - T_r}\right) - 11.75$$

where $T_r = 07.6$ h and $T_s = 17.5166667$ h for this particular location at this time of the year.

If the residuals are sufficiently small then the derived regression curve is used to estimate the irradiance at the time of acquisition of radiance observations of the targets, so as to calculate target reflectance.

The main disadvantages of this method are:

1. *Variations in irradiance* between the regression estimate and the actual irradiance primarily due to cloud.
2. *Differences in calibration* between the instruments.

5.2.2.2 The Measurement of Reflectance Factors

In this method the observations of incident irradiance are replaced with observations of a control surface that has stable and known reflectance characteristics. Normally, both the target and control surfaces are measured using the same instrument and identical optical systems so that calibration errors have an equal affect on both observations. If the target and the control surfaces are measured with different instruments then different calibration factors will apply to the different observations. However, simultaneous observations of the control surface with both instruments provides the necessary relative calibration of the two instruments as is required by this method. The ratio of the target to control observations (for the one instrument) or calibrated observations (for different instruments) is the reflectance of the target relative to the control surface reflectance, or reflectance factors. As long as the reflectance of the control is stable then the method readily provides consistent measurements of reflectance factors for the specified control surface.

The reflectance of the targets can be calculated from:

Reflectance of the control surface, $r_{c,\lambda} = \pi L_{c,\lambda}/E_{i,\lambda}$

Reflectance factor of the target, $f_{t,\lambda} = L_{t,\lambda}/L_{c,\lambda}$

Reflectance of the target, $r_{t,\lambda} = \pi L_{t,\lambda}/E_{i,\lambda}$

$$= [L_{t,\lambda}/L_{c,\lambda}] \cdot [\pi L_{c,\lambda}/E_{i,\lambda}]$$

so that

$$r_{t,\lambda}=f_{t,\lambda} \cdot r_{c,\lambda} \tag{5.6}$$

With some instruments the radiance of the target and the control are measured simultaneously. This practice avoids errors due to variations in the incident radiation.

The main advantage of the method is that it can provide good values of target reflectance factors. The main disadvantage of the method is its dependence on the control surface having consistent reflectance values for all the observations being taken, and the need to either maintain or recalibrate the control surfaces.

The control surface needs to have the following characteristics:

1. Be Lambertian or nearly Lambertian.
2. Not significantly degradable, in terms of reflectance value, with time or exposure to sunlight.
3. Be resistant to wear and tear in its use.
4. Have either known reflectance values, that are similar, or slightly greater to the reflectances expected, or reflectance values near to 100%. The advantage of "grey" control surfaces is that they create similar electronic responses as the target in the instrument, thereby minimising one source of instrumental error. However, the derived reflectance factors will be quite different to the actual reflectance of the targets, and so the reflectance factors will need to be converted into reflectance before they can be used. The advantage of "white" control panels is that the reflectance factors are similar to actual reflectance and so they are easily checked for errors. Disadvantages are that users might be tempted to use these reflectance factors as actual reflectance values and that the differences in radiance from the two surfaces may elicit different instrumental error responses that are not fully dealt with in instrumental calibration.

Various types of control surfaces are used, but the two most common are:

1. *Halon powder*. Halon powder is compressed under constant pressure of 14 Lb/in.^{2} to form a uniform, reasonably Lambertian reflectance surface with typical reflectance values as shown in Figure 5.2. The main disadvantage of the powder is that it gets dirty very easily, so that it has to be frequently replaced. To ensure that the surface formed is flat and of consistent compression, the halon has to be filled into a mould and compressed using a matt surface that will not significantly distort during application of the pressure. Thick matt glass is suitable for smaller control surfaces.
2. *Halon paint*. Halon paint is used to cover a reasonably smooth surface to a depth of at least 0.1 mm. Suitable bases must be matt and have reasonably Lambertian reflectance characteristics. One solution is to create a sandwich of heavy-duty particle board between

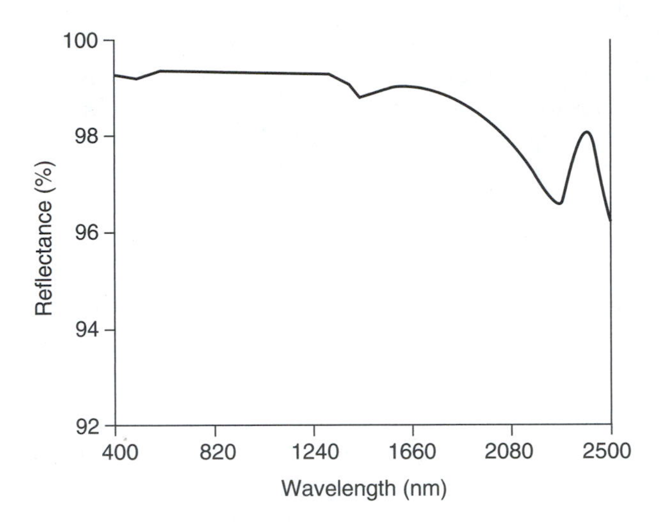

Figure 5.2 Typical reflectance characteristics of Halon powder (from Grove et al., 1992 *Jet Propulsion Laboratory Publication 92-2*, Pasadena, CA. With permission).

aluminium plates. The surfaces of the aluminium plates are etched with hydrochloric acid to provide the necessary matt surface, which are then repeatedly painted to accumulate the necessary depth of paint on the surface.

5.2.3 Considerations in Collecting Field Spectral Data

5.2.3.1 Need for Representative Data

The reflectance of a vegetative canopy depends upon the reflective, transmissive and absorptive characteristics of all of the components in the canopy, and their area, location in the canopy and orientation as well as the reflective, transmissive and absorptive characteristics of the background soil or water layer. Since the distribution of vegetative components in a canopy will never be identical in different FOVs, and in fact may change over time due to such factors as wind and orientation of leaves towards the sun, the measured reflectance will never be identical at all points in the canopy. It is necessary to correct for this variability by using the mean of a sample of observations as being representative of the cover type.

5.2.3.2 Need for Lambertian Control Surfaces

Most of the reflectance data required in remote sensing is bi-directional in form, that is the dominant source of energy is from one direction, and the energy is recorded in another direction. Such reflectance is very dependent upon the Lambertian characteristics of the target and the control surfaces. The Lambertian characteristics of the control must be measured so as to determine the range of solar zenith angles over which the control can be used. This is best done by the use of an artificial light source to represent the source, and measure the reflected radiation at set zenith angles, at a constant distance from the target. Variations in the measured radiance from the control are indicative of variations from the Lambertian norm by the surface.

5.2.3.3 The Effect of Non-Lambertian Surfaces on Reflectance

Many surfaces are non-Lambertian to some degree. Most of these surfaces exhibit significant variations in their BRDF with variations in the sun–surface–sensor geometry and with waveband.

If the BRDF is to be determined, then observations need to be taken over a range of angles in the sun–surface–sensor geometry, for each selected waveband as discussed in Chapter 2.

5.2.3.4 Impact of Clouds on Measurement of Reflectance

Clouds block and reflect solar radiation where the level of effect can vary from negligible (1 to 2%) to complete interference (100%). Negligible to small effects occur when the cloud is very thin or very small in which case they will not be detected in the image data. When this occurs then the irradiance will decrease due to absorption and scattering and increase due to reflection over affected areas relative to that which would occur solely due to the direct solar beam. The reflection typically occurs off the sides of the clouds, whilst the absorption and scattering occurs during transmission through the cloud. The changes in irradiance can create significant spikes and troughs in the irradiance recorded at the surface, as shown in Figure 5.3. It is therefore safest to take field spectral observations when there are no, or only a few clouds that are a long way from the sun–surface–sensor geometric figure.

The data in Figure 5.3 contains other Interesting information. The irradiance and the radiance curves should have maximum values that follow a curve that peaks at solar midday, decreasing away from this point. The curves both show spikes, due to reflection from the clouds, and troughs due to absorption and scattering. The reflectance curve should be flat throughout the day. The reflectance curve shown in Figure 5.3 has peaks at low solar elevations indicating that the diffuse hemispherical objective acquiring the irradiance data is not collecting all the incoming radiation at these low angles. The curve is not flat at higher angles, either because the surface is not truly Lambertian or because the objective is still affecting the incoming radiation. The ratio curves show that the ratio of the reflectances is consistent throughout the day, as is expected from the definition of reflectance. The ratio of the radiances exhibits significant changes when there was cloud cover, indicating that the proportions of red and NIR energy changed under these diffuse light conditions.

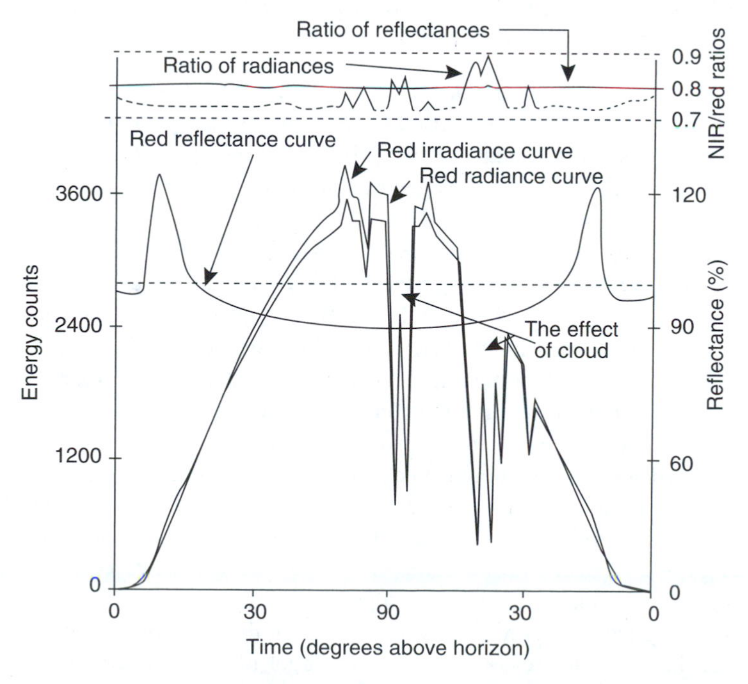

FIGURE 5.3 Radiance, irradiance and reflectance curves for a red waveband recorded throughout the day, for a Lambertian Halon target, and curves of the ratios of (NIR/red) radiance and reflectance values using the same field data set for red and NIR wavebands.

5.2.4 Collection of Other Data With Spectral Data

If the spectral data are being collected to develop numerical models that are intended to estimate canopy physical parameter values then a number of matters need to be considered, including:

1. The ability of the spectral data to estimate the desired parameters
2. Need for representative spectral and physical data
3. Impact of variations in canopy reflectance on derived models
4. Conversion of field-based models to airborne or space-borne applications

5.2.4.1 Ability of Spectral Data to Estimate Selected Physical Parameters

The more closely linked the physical parameters are to canopy reflectance or radiance, the more likely that the reflectance or radiance data will provide good estimates of that parameter, as discussed in Chapters 2 and 4. Conversely, the more factors that affect the parameter of interest but do not significantly affect the reflectance or radiance from that surface, the less likely that the image data will provide good estimates of that parameter. Spectrometer data may be used to establish the level of correlation between selected spectral bands and the physical parameters of interest, and to develop a model to estimate the physical parameters from remotely sensed data in specific wavebands.

Similar considerations apply if the objective is to develop models estimating parameters that may be primarily influenced by thermal radiation, such as stress and moisture content, or used with radar imagery, such as biomass, moisture content and canopy structure.

5.2.4.2 Need for Representative Physical and Spectral Data

The physical and spectral data collected to investigate relationships between the two, must be compatible as follows:

1. Cover the full range of conditions that are likely to be met by the model. Consider the development of models designed to estimate pasture (grassland) canopy parameters. Canopies of green vegetation exhibit variations in reflectance in the NIR up to green leaf area index (GLAI) values of about 6 due to the transmissivity of green vegetation at these wavelengths. Green vegetation shows little variation in red reflectance above GLAI values of about 3, and brown vegetation shows little variations in reflectance in both wavelengths above LAI values of about 3, because of the opacity of vegetation under these conditions. The spectral and physical data needs to be distributed across a domain that varies between $0 \leq \text{GLAI} \leq 6$ and $0 \leq \text{BLAI} \leq 3$, similar to that shown in Figure 5.4. Similar considerations apply for estimating attribute values for other landcover types. In developing numerical models it is important that the samples are distributed across the appropriate domain, and sufficient samples are taken to ensure that the conclusions drawn are statistically valid. In general a minimum of 30 samples would be required for this purpose.

2. The samples used must be of sufficient area to include representative proportions of the different cover types that are likely to be met. In the case used above for illustration, the samples need to be large enough to include representative proportions of vegetation, soil, shadow and litter within the spectrometer FOV so that the observations of reflectance will be representative of that cover type. If the herbage contains individual plant clumps then the sample area must be large enough to be representative of cover conditions across the surface. It is for this reason that samples of herbage will normally be larger than 1 m^2 in area, and samples of larger plants will need to be correspondingly larger again. The FOV of the instrument is specified as a cone angle, ϕ, so that the area covered can be calculated as a function of the distance, d, from the sensor and the tilt angle, θ, as shown in Figure 5.5.

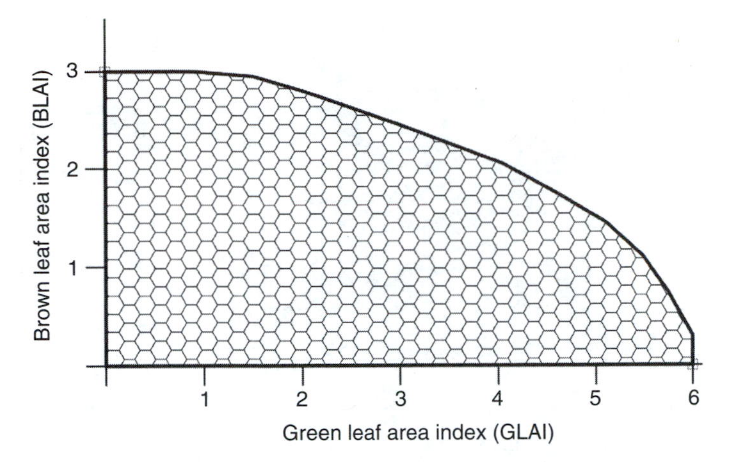

FIGURE 5.4 Distribution of data required to develop a model to estimate GLAI and BLAI for pastures from spectral data.

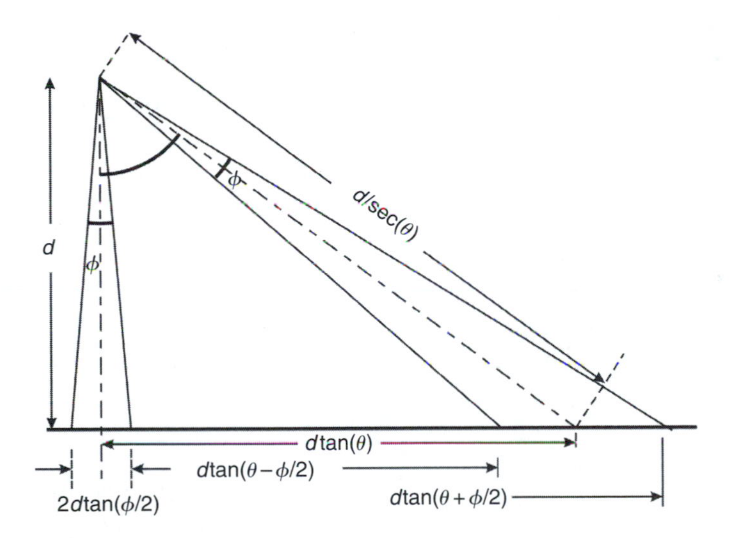

FIGURE 5.5 Calculation of the area covered by an instrument of specified IFOV at distance, d from the target.

If the distance from the target to the sensor is d then the diameter of the FOV in the nadir direction ($\theta = 0°$) is given by:

$$\text{Diameter, FOV} = 2d \times \tan(\phi/2) \tag{5.7}$$

where ϕ is the solid angle defining the instruments FOV.

$$\text{The area of the FOV} = \pi \left(d \times \tan\left(\frac{\phi}{2}\right)\right)^2 \tag{5.8}$$

If the spectrometer is tilted by θ then the footprint of the FOV is an ellipse with the semi-major axis in the direction of the tilt and the semi-minor axis at right angles to the direction of tilt.

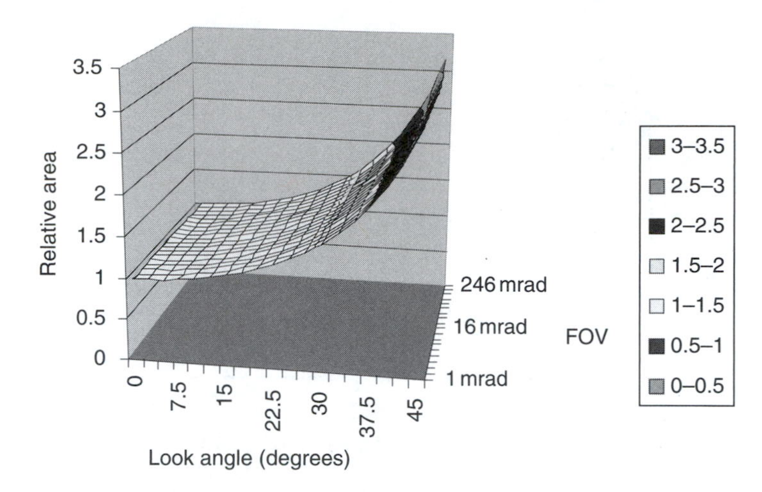

FIGURE 5.6 The relative area of a pixel at different look angles and different FOV values.

The semi-major and semi-minor axes are given from:

$$\text{Semi-major Axes (FOV}, \theta) = 2d \tan\left(\frac{\phi}{2}\right) \Big/ \left[\cos^2(\theta) - \sin^2(\theta)\tan^2\left(\frac{\phi}{2}\right)\right] \tag{5.9}$$

$$\text{Semi-Minor Axes (FOV}, \theta) = 2d \tan\left(\frac{\phi}{2}\right) \Big/ \cos(\theta) \tag{5.10}$$

The area of the FOV is approximately:

$$\text{Area} \approx \pi \times \left(d \tan\left(\frac{\phi}{2}\right)\right)^2 \times \left(\frac{1}{\cos^2(\theta) - \sin^2(\theta)\tan^2(\phi/2)} + \frac{1}{\cos(\theta)}\right)^2 \tag{5.11}$$

or:

$$\text{Area} \approx \pi \times \left(d \tan\left(\frac{\phi}{2}\right)\right)^2 \times \left(\frac{1 + \cos(\theta) - \sin(\theta)\tan(\theta)\tan^2(\phi/2)}{\cos^2(\theta) - \sin^2(\theta)\tan^2(\phi/2)}\right)^2$$

In this equation $\tan^2(\phi/2)$ is very small since the FOV is small, so that:

$$\text{Area} \approx \pi \left(d \tan\left(\frac{\phi}{2}\right)\right)^2 \left(\frac{1 + \cos(\theta)}{\cos^2(\theta)}\right) \tag{5.12}$$

The full Equation (5.11) has been used to create Figure 5.6, by dividing it by the area at the nadir Equation (5.8) to give relative area. Perusal of the figure shows that the size of the instantaneous field of view (IFOV) has very little effect on the relative area, although of course it will have a large impact on the actual area. The look angle also has a small effect for angles less than about 15°, but after this the effect rapidly increases.

3. The field parameters and spectral observations should be measured to similar levels of accuracy and resolution. The simplest field spectrometers can measure the radiance incident on the sensor to resolutions of about 1% of full-scale deflection, and standard errors of about 3% of the full-scale deflection. Many better-quality field instruments can measure radiance to better than 1% of full-scale deflection with standard errors of less than 1%. Before starting to collect field data take a set of observations on a standard surface, to determine the resolution and standard error that can be

achieved with the instrument. The field observations of canopy parameters should achieve similar resolutions and accuracies to that determined for the instrument. To achieve this measurement of a number of samples (or sub-samples) as discussed below may be required, and relate the average values to each spectral observation since:

$$\sigma_{\text{mean}} = \sigma_{\text{obs}}/\sqrt{n} \tag{5.13}$$

for n observations.

The usual method of collecting field data is to:

(a) Clip and weigh the whole sample for fresh weight.
(b) Shuffle or mix the sample and partition it into fractions, such as halves, and then halves again, or quarters. If this sub-sample is still too large then it is further shuffled and halved until the sub-samples are suitable for measurement.
(c) Sub-samples are selected to measure GLAI and the other relevant parameters. Different sub-samples may be used to measure the different parameters, to sort into species, etc.
(d) Each of the sub-samples used are weighed, dried and reweighed, with the weights of the sub-samples and the whole samples being used to extrapolate to estimate the parameters for the whole sample.

The standard measure of vegetation weight is of the vegetation after it has been dried in an oven for a minimum of 18 h at a temperature of 450°.

There are many factors that will cause variations in the physical parameters of the land cover of interest. Measuring these parameters usually involves disturbing that land cover in some way. If this is the case then the experimental design must incorporate sufficient training sites to cover this destructive collection of field data and ensure that an adequate temporal sequence of training sites is available.

In designing field data collection programs, it is usual that some trade-off among resolution, accuracy and cost has to be made. The analyst should carefully consider the implications of this trade-off in preparing the experimental design. If the analyst is exploring new ideas that may, or may not be successful then a slight reduction in resolution and accuracy in the field data being collected may allow him to more economically collect a mass of data that more exhaustively covers the full range of possible conditions. Such data sets will allow the analyst to quickly and economically verify or refute the hypotheses being tested, but will not necessarily allow the proposed model to be developed to the required level of resolution or accuracy. Having verified the concept, more expensive data sets can be used to fine-tune the model. This approach has the advantage that development of the crude model will provide considerable insights into the processes being studied, and how the model will actually be used; these insights may well influence future experimental design. If the analyst is conducting the work to prepare for an operational program then resolution and accuracy are of supreme importance. The analyst may choose to take very detailed observations over subsets of the conditions that will be met, and know that the analysis will be good across the whole range of conditions, but that it will be more accurate within the subsets than outside them.

5.2.4.3 Impact of Variations in Canopy Reflectance on Derived Models

The reflectance of a canopy will usually exhibit variations in response due to a variety of causes. Some of these causes, such as variations in canopy geometry and the impact of wind on the canopy, can be considered to be a random noise component that has little impact on the average taken for a set of observations. Other causes do not act in a random manner, but introduce a bias, either up or down, into the reflectance data. These causes include variations in atmospheric attenuation, and some surface conditions such as crop stress. Since these causes bias the reflectance values they will often reduce the accuracy of any derived models. It is important to either minimise or correct for their effects as far as possible.

Systematic errors should also be corrected in the data. In practice many systematic errors are too small to justify the significant amount of work necessary to establish a credible correction mechanism. In consequence, they remain in the data and reduce the accuracy associated with estimates derived from the data. However, the analyst cannot make a final decision on this matter without collecting sufficient field data to:

1. *Identify the significant sources of systematic error*, so as to correct for them prior to using the data in the estimation model.
2. *Indicate the magnitude of the residual errors*, as indicated by the variance of the data corrected for (1) above.

5.2.4.4 Implementation of Models in Airborne and Space-Borne Systems

Models developed on the ground generally estimate physical parameters from measurements of reflectance data. Satellite and airborne data are derived from the radiances received at the sensor. To use models that have been developed on the ground requires that the airborne or space-borne sensor data be corrected from a measure of radiance to reflectance at the surface. This requires an atmospheric correction to the sensor data as discussed in Chapters 2 and 4. Errors in the atmospherically corrected data will translate directly into errors in the derived estimates.

5.2.5 Construction of Models from a Set of Related Spectral and Field Data

A set of field and spectral data may have been collected to support the building of process or deterministic models, or models based on a set of physical laws that govern the behaviour being modelled. When this is the situation then the sets of data will have been collected so as to address the specific needs of the model. However, in many situations it is not appropriate to use a process model, because:

1. Our knowledge of the underlying processes may be inadequate, that is, we may not be able to address questions like, "What is the nature of the actual underlying processes; how to quantify them or how to weigh or balance them in the proposed model?" sufficiently to construct a model of an acceptable accuracy, or likely to cover a satisfactory range of conditions.
2. The models developed may require an extensive suite of input parameters that are either impractical or too expensive to acquire.
3. The models may not be invertible. Models may have been built of the forward process, but these models may not be invertible. For example, a model may have been built to estimate canopy reflectance, given a range of canopy parameters and information on the incident radiation. However, it may not be possible to invert that model so that input values of reflectance can be used to estimate the canopy parameters.
4. Some processes are inherently stochastic. When this is the case then implementation of process models that cannot deal with these stochastic attributes will not provide a satisfactory solution.

For these reasons stochastic or probabilistic models are also used for a range of applications. Whilst some stochastic models are built using a stochastic version of the physical laws governing a process, many are built empirically by the use of regression techniques.

Regression analysis is the action of numerically fitting a numerical equation through two sets of matching data values so as to provide a model of the relationship between the parameters, for the conditions met in the acquisition of the data sets. With regression analysis it is essential that there be a significant level of correlation between the parameters for the model to give useful results. The higher the correlation, the more reliance can be placed on the model results, as long as the model is used under similar conditions and circumstances to those met in the collection of the data used to create the model.

The first step in the conduct of regression analysis is to decide on the type of relationship that holds between the parameters, and to ensure that there is an adequate level of correlation between the parameters. Graphing the data sets as shown in Figure 5.7 is the best way to do this. From analysis of this graphing the analyst will decide on whether the level of correlation is sufficient to continue, and the nature of the relationship between the variables. The data shown in Figure 5.8 is a close fit to a linear model, so that simple linear regression would be a suitable way to proceed with this data set.

Consider the situation depicted in Figure 5.8(a), in which the relationship between the data sets is approximately logarithmic. When this situation arises, it is often best to transform one or both sets of data so that the relationship between their transformed values is linear, as shown in Figure 5.8(b). It is often much easier to conduct linear regression than to conduct more complex forms of regression. Once the analyst has viewed the data and selected a model to be used for the regression, then the next step is the actual conduct of the regression analyses.

Regression analysis is normally conducted by the least squares technique as described in Chapter 4.

Consider the data listed in Table 5.2 and displayed in Figure 5.9. Simple least squares regression was conducted on the data in the manner described in Chapter 4 and shown in Table 5.2.

The purpose of fitting a curve in this way is to use that curve as a representation of the data, or to use it to derive estimates of the dependent parameter from other values of the independent parameter. If the curve is to be used to represent the data, then what confidence can you place in this curve as representing the data? If it is to be used to derive estimates from other values, or to make predictions, then what confidence can you have in the predicted values? We have seen that the mean

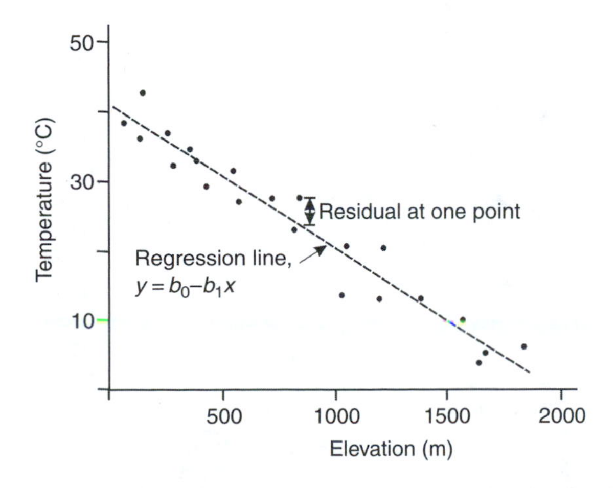

FIGURE 5.7 A set of observational data and a linear regression line fitted to the data.

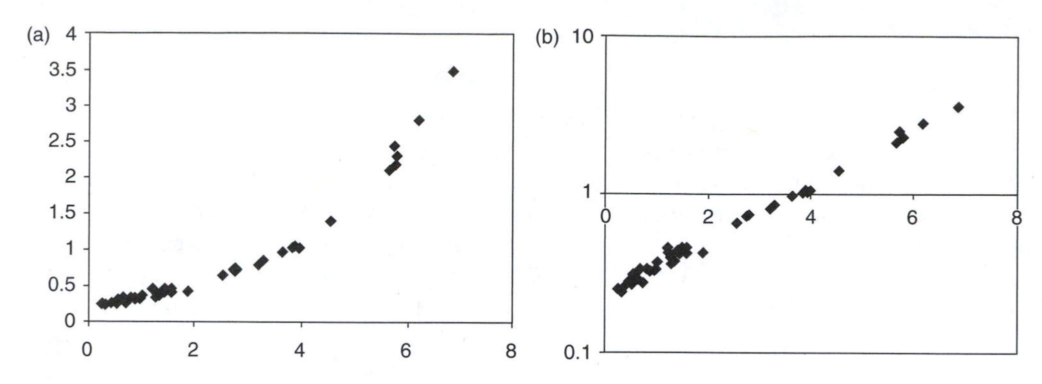

FIGURE 5.8 The non-linear relationship between two sets of variables (a) and their transformation into a linear relationship by transforming one of the datasets (b).

TABLE 5.2
A Set of 31 Observations of Temperature and Elevation used to Compute a Linear Regression Fitting the Data by Means of Least Squares

Elevation (m)	Temperature (°C)	Residual
8	35.9	0.049
9.5	36	0.168
8.9	35.7	−0.139
15.6	35.5	−0.254
100.8	35.4	0.738
125.7	34	−0.343
368	30.5	−0.740
1035	22.3	−0.397
514	28.4	−0.970
433	30.1	−0.307
79.6	34.8	−0.134
159	33.3	−0.617
295	32.1	−0.075
658	27.9	0.374
697	28	0.974
725	27.2	0.533
652	27.4	−0.202
687	26.8	−0.354
899	24.3	−0.139
38.9	35.4	−0.055
43.1	36.4	0.999
1947	11.6	0.584
1367	18.8	0.355
1427	17.1	−0.576
122	33.9	−0.491
287	33.2	0.923
242	33.4	0.546
156	33.9	−0.055
179	34.1	0.439
188	33	−0.545
1879	11.6	−0.287
Mean temperature	29.61	
SST	1457.375	
SSE	8.554	
SSR	1449.185	
Regression coefficient	0.9943	

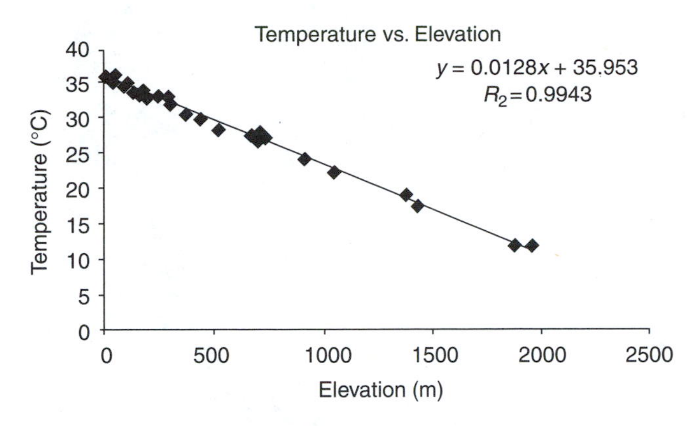

FIGURE 5.9 Data values, linear regression and confidence limits.

value of a normally distributed set of data is also normally distributed. In the same way, if the data used to create a regression are normally distributed, then the regression line will have residuals that are normally distributed about it. Graphing the residuals will show if this is the case, and if this can be tested statistically. If the residuals exhibit patterns, then the analyst should consider whether the form of the curve used for the regression, a straight line in the above case, is satisfactory, or whether the assumption of normality which is implicit in this method of derivation of regression parameters is valid.

Once the analyst is satisfied that the residuals meet this criteria, then he or she may wish to determine the significance of the regression; the confidence level that they can put in the regression terms themselves, or set a confidence level in the predictions made using the regression. Thus the analysts may wish to:

1. *Test the level of significance of the regression.* If the regression is found to be significant at a satisfactory confidence level, then it means that the regression curve can be used to provide meaningful information. Usually a 90, 95 or 99% confidence level is set for such tests. If the regression is found to not be significant, then it cannot be relied upon to provide meaningful information.
2. *Test the significance of the individual parameters.* In the above case, find the range of values for the regression parameters at a satisfactory confidence level. This is particularly important when conducting multiple regression as some of the parameters may not be contributing significantly to the regression. If this is found to be the case, then it may be better to drop these variables from the regression and re-compute the regression without them.
3. *Set confidence levels for estimates derived by the regression.* The normal distribution associated with the regression curve is not constant at all values, so that the confidence limits will vary with changes in the independent parameter or parameters.

It is usual to set a 90, 95 or 99% confidence interval and then find the limits at the selected interval. The percentage that is set defines the percentage of times that an observation will fall within the defined range. We have seen that 68% of all observations in normally distributed data fall within one standard deviation of the mean value. Thus, a 68% confidence interval has a confidence limit of one standard deviation either side of the mean. Analysis of the standard normal distribution shows that 90, 95 and 99% confidence limits set confidence intervals of ± 1.645, ± 1.96 and ± 2.575 standard deviations, respectively. As you would expect, if you are going to be more confident of the results, then you need to set broader limits, so as to catch more of the data. Thus, the higher the level of confidence that is demanded, the wider will be the range, or the larger will be the distance between the two limit curves in case (3) above, and similar considerations apply to both of the other cases.

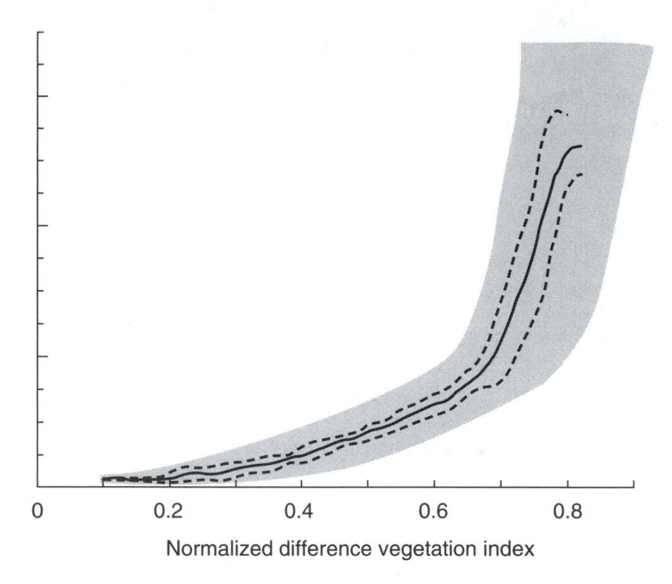

FIGURE 5.10 Typical relationship between GLAI and the normalised difference vegetation index (NDVI) (courtesy of Wolfgang Buermann, Boston University).

Figure 5.10 shows a regression curve fitting NDVI derived from field spectral data against Green Leaf Area Index (GLAI) data, and how the confidence range can vary for a constant confidence limit, with position along that regression.

5.3 THE USE OF FIELD DATA IN VISUAL INTERPRETATION

Our general knowledge of the physical environment, its processes and how they are manifested in remotely sensed data, are largely based on conceptual models that are constructed from a mix of scientific and human observations. These models are made more relevant to specific conditions by the use of factual information on aspects of the environment. Not always is the process error free; errors are made in the collection of data, its analysis and in making inferences to amend or build on existing conceptual models. The process is therefore very much an empirical one of model building and evaluation, with modifications to the models only being accepted after exhaustive evaluation.

This process of conceptualisation and quantification of models form the basis of all decisions made in extracting information by visual interpretation. The process involves two tasks:

1. Identification
2. Interpretation

5.3.1 IDENTIFICATION

Identification is the process of identifying objects and features in the image or images. It is the simpler of the two tasks, only requiring sufficient information in the imagery to first discriminate the feature from other features and then identify the general geometric shape and characteristics of the feature. If the imagery and facilities used do not allow the routine identification of features then errors will occur in the interpretation. These errors of non-identification are called *CLASS A* errors.

They require consideration of the:

1. Scale of the imagery
2. Resolution characteristics of the imagery
3. Contrast of the object relative to its surroundings
4. Photographic processing
5. The interpretation facilities used and the physiological condition of the interpreter
6. The knowledge levels of the interpreter and the physiological aptitude of the interpreter for the task

An important component of project design is to ensure adequate scale and resolution of the objects in the imagery, at the contrast that will exist in that imagery, that the processing is sufficient and that the interpretation facilities are of an adequate quality. What is always more difficult to assess is the physiological and psychological aptitude of the interpreter. However, the importance of these factors should not be underestimated, and procedures should be established, and interpersonal relationships maintained that are conducive to a good working environment.

The most reliable way to assess the suitability of an interpreter for a task is through accuracy assessment. However, accuracy assessment is conducted last so that all the work has been done before this stage cannot be used to influence the allocation of tasks to interpretation staff, although it can be used for this purpose in future years. It is important, therefore, not to depend entirely on this check but to build a positive interpretation environment for the interpreter.

5.3.2 Interpretation

The interpretation of imagery can be at different levels of complexity as discussed in Chapter 3. Whatever the level of complexity, all interpretation tasks may contain the following types of errors:

1. A feature may be identified in the image, but not recognised as being important to the interpretation, or errors of mis-identification. Errors of this type are called *CLASS B* errors.
2. The feature may be identified and described to the satisfaction of the interpreter, using established conceptual models in the interpretation, but be erroneous. These errors of mis-interpretation are called *CLASS C* errors.
3. The features may be identified, their importance recognised, but this importance is not quantified to the satisfaction of the analyst. In this case the analyst may either use field or auxiliary data to satisfactorily describe the feature, or he may isolate the feature as an unresolved object, creating *CLASS D* errors, or errors of non-interpretation.

A major reason why analysts should have considerable knowledge of the disciplines associated with the interpretation is to minimise *CLASS B* and *CLASS C* errors during the interpretation. *CLASS B* errors, arising when the importance of features are not recognised, are minimised through the knowledge of the disciplines of interest and the consequent development of good conceptual models. Field data are of some use in reducing *CLASS B* error during the interpretation process, but are primarily used to minimise *CLASS C* and *CLASS D* errors. The field data does this by facilitating the inductive and deductive processes that strengthen and improve the conceptual models employed by the analyst, thereby improving the accuracy and efficiency of the interpretation. The field data does this by identifying errors in the interpretation (*CLASS C* errors) that will lead to a re-evaluation of that interpretation, and similar interpretation elsewhere in the image. The field data will also explain features (*CLASS D* errors), leading to a re-evaluation and re-interpretation.

Conceptually *CLASS C* and *CLASS D* errors differ in that *CLASS D* errors have been recognised as a limitation of the data, facilities, knowledge base or interpretation skill of the analyst whereas *CLASS C* errors have not been recognised. Since the *CLASS D* errors have been recognised they will be visited and resolved in the field visit. On the other hand *CLASS C* errors would normally be addressed only by driving through areas typical of the interpretation so as to assess the correctness of that interpretation and addressing errors, and their more global implications, when they are found.

To summarise the four types of errors:

5.3.2.1 *CLASS A* Errors or Errors of Non-Identification

These errors will be minimised by ensuring that the imagery used is of adequate resolution, the processing employed does not degrade the photography, the facilities used are of adequate optical and illumination quality and the interpreter is both interested and capable, physically and psychologically, of conducting the interpretation.

5.3.2.2 *CLASS B* Errors or Errors of Mis-Identification

Ensuring that the analyst has adequate knowledge levels of the discipline and the characteristics of the area being interpreted primarily minimises these errors. These errors can be further reduced by the proper use of field data.

5.3.2.3 *CLASS C* Errors or Errors of Mis-Interpretation

Minimised by ensuring that the interpreter has adequate knowledge levels of the discipline and the area being interpreted, has knowledge of and is skilled in the methods of extracting information from imagery, conducts adequate field work and is both physically and psychologically suited to the interpretation task. Psychological preparation for interpretation depends on many factors, but it can be affected by physical exhaustion, personal affairs and work conditions.

5.3.2.4 *CLASS D* Errors or Errors of Non-Interpretation

Minimised by conducting adequate fieldwork during interpretation and by the knowledge levels of the interpreter.

The amount and type of field data that are required as part of visual interpretation is thus dependent upon the knowledge levels of the interpreter, the status of the conceptual models constructed by the interpreter and their relevance to the interpretation task at hand. It is therefore impossible to specify the amount of field data that would be required in any given situation. In general the inexperienced analyst will require more field data than will the experienced analyst, and any analyst confronting new conditions will require more field data than when interpreting conditions that are similar to those met in previous tasks. The analyst has to judge for himself the amount of field data that are required, and the relative density of field data that is required in the different conditions being met in the interpretation. Some of this field data should be acquired before the conduct of the interpretation task, by moving through the area to gain an appreciation of the area and its conditions. This initial traverse through the area should not be restricted to too small a portion of the area, or too small a sample, as it is important for the analyst to sample the full range of conditions that might be met within the area. Sampling the full range of conditions allows the analyst to be confident that the conceptual models being constructed represent the range of conditions within the area. The general approach to the collection of field data is therefore to:

1. Drive or move through the area to assess the full range of conditions that will be met and to indicate the range of conceptual models that might need to be developed or used.
2. Conduct a preliminary interpretation of the imagery. This interpretation will readily identify many of the existing limitations of the conceptual models held by the analyst and identify

confusions that will require either new models or refinements of existing models. Sometimes this preliminary interpretation can be conducted as part of the full interpretation, by conducting the interpretation of some of the simpler, yet crucial components of the interpretation task as this phase. Such tasks could include the interpretation of the drainage in the area.

3. Conduct full interpretation of the area, proceeding from the easier yet more indicative tasks to the more complex and self contained tasks. Drainage interpretation is relatively straightforward, yet it is important as the drainage patterns and structures can tell a lot about the landforms and the geologic evolution of an area. Drainage can thus be very indicative of geology and soils in an area. The second interpretation task could well be terrain classification, since this both builds on the drainage classification and is indicative of geologic and soil conditions in an area.
4. Conduct a second field visit to the area, to establish other models as required, and to refine existing models.
5. Repeat steps (3) and (4) until the analyst is satisfied with the interpretation results being achieved. However, there will still be errors in the interpretation so that it is important to further assess the results.
6. Conduct a statistically based accuracy assessment of the interpretation results. If this accuracy assessment shows that the interpretation is not of an adequate standard then the analyst has to identify the dominating classes of errors so as to decide on an appropriate course of action.

5.4 THE USE OF FIELD DATA IN THE CLASSIFICATION OF DIGITAL IMAGE DATA

Classification was defined as partitioning the response domain into a number of discrete class sub-domains, where individual or groups of these sub-domains represent different physical or informational classes of landcover or landuse. In this section we will consider the role of field data in the two modes of classification that can be conducted with some classifiers (supervised and unsupervised) and then we will consider the use of field data when classifying using the mixture model.

5.4.1 Classification with the Normal Classifiers

With most classifiers the process of classification can be conducted in either of two ways:

1. Classification then field verification.
2. Collection of field data then Classification.

5.4.1.1 Classification then Verification

Classification then verification is normally adopted with unsupervised classification. The unsupervised classification produces a classified image containing a number of statistical classes based on the spectral and radiometric image data. The analyst then uses visual interpretation with field visits so as to allocate single and groups of the statistical classes to individual informational classes. Ultimately, each informational class will contain one or more statistical classes.

If in this process, the analyst finds that the one statistical class can represent more than one informational class, then the unsupervised classification may be unsatisfactory. This situation can arise either because the statistical characteristics of the two informational classes show so much overlap in the data being used that it is not realistic to try and separate them, or because the clustering process partitioned the data in an unsatisfactory manner. The first situation can arise because:

1. The radiometric characteristics of the two informational classes may show significant overlap in the data being used. If this is the case then only the use of other data, either alone or integrated with some or all of the existing data sets, has the possibility of solving the problem. This additional data may be in other wavebands or at other stages in the year or phenological stages of the plant communities that are involved in the overlapping classes.
2. Mixed pixels may be creating a significant problem, particularly if the image resolution is too coarse to map these informational classes. Mixed pixels will give a blob in the radiometric domain, made up of both land covers and the mixtures of the two cover conditions.

The second situation is made worse when the above conditions apply, since the sharp boundaries between class areas will be lost, as the classes tend to merge and form blobs in the spectral domain. Clustering starts with seed positions for the classes, and then uses the spectral data in a clustering technique, as discussed in Chapter 4, to find the actual clusters. If the class distributions have merged in the source data, then the clustering may not identify two clusters but one, and it may also be the case that this class will be centred on some mixture of the two classes rather than on the classes themselves. In addition, clustering should give similar results, independent of the method of selection of the starting seed positions. This will usually occur when the classes are quite discreet in the spectral domain. However, if the classes have a lot of overlap so that the boundary between them is not clear, then the selection of the starting position can have a significant impact on the classes that are selected. Therefore, the different options for selection of the starting seed positions should be tried, just as the options on the number of classes and the number of iterations should also be explored. It should be noted that increasing the number of classes or iterations in clustering increases the workload during the clustering and in the subsequent assignment of the spectral classes to informational classes, but with the advantage that the possibility of spectral classes overlapping informational classes is reduced. It is thus better to err on the side of using to many classes in clustering rather than too few.

5.4.1.2 Field Data Collection then Classification

This approach is normally adopted with supervised classification. In this approach field data are collected for known samples of all physical classes so as to define physical class training data. The boundaries of these training areas are digitised so as to create training areas in the image data, to determine class statistics and analyse the goodness of fit of the data to the model that underlies the classifier algorithm when this occurs. If analysis of the spectral distribution in the classes shows that the data are not a good fit to the probability distribution that is assumed by the classifier, when this occurs, then the data needs to be split into a number of spectral classes that meet the classifier assumptions using clustering techniques. It should be noted that the maximum likelihood family of classifiers assumes that the class data are normally distributed, and all current clustering algorithms make the same assumptions about the data. After the spectral classification the groups of statistical classes are assigned to their relevant informational classes in preparation of accuracy assessment. The training data must therefore:

1. *Be representative of all of the physical classes.* To be representative it is necessary to ensure that the field data covers the full range of conditions that are likely to be met in each class. The safest way to do this is to ensure that the field data adequately samples the whole area covered by each class.
2. *Be sufficient to derive the parameters for all spectral classes* contained in each physical class. If the data for a physical class are a poor fit to the probability distribution that underpins the selected classifier algorithm then the data for the physical class may need to be partitioned into a number of spectral classes, as discussed in Chapter 4. If this occurs then collection of field data will need to reflect the increased use of that data by more classes.

5.4.2 Classification Using the Mixture Model

Unlike the usual classifiers, the End Member method of classification assumes a linear domain for the class data, where changes in response within that domain reflect changes in physical conditions within the class. The class domain is separated from the domain of other classes by a threshold surface at some distance from the linear surface defined by the response values for the nodes used in the classification. The End Member classifier is suitable for use where these criteria, as can occur in the natural environment, are met. Field data are therefore required to:

1. Verify that the class data occupy a linear domain
2. Assist in defining node response values
3. Develop models to estimate physical parameters from the proportion co-ordinates derived by the End Member classifier

5.4.2.1 Verify that the Class Data Occupy a Linear Domain

Collect field data that is representative of the full range of class conditions that are likely to be met. Analyse the statistical characteristics of the response or reflectance data for these field sites to see if they are a close fit to a linear surface. If the data occupies a linear domain, use the observed data to assist in identifying node response values. If the analyst considers that two or three nodes represent the linear domain, then plotting the data and simple regression analysis will show whether or not the data are a close fit to the linear model. If the analyst considers that four or more nodes are necessary, then simple graphical analysis will not show whether the data fits the linear model or not. In this case, the analyst should collect data that varies in two or three parameters, and not in the other parameters, for different combinations of the nodes and analyse this data for linearity. These data will show whether the data are linearly related between the subset of nodes, giving different linear relationships for the different combinations of nodes.

5.4.2.2 Assist in Defining Node Response Values

Collect field spectral data for conditions that represent node conditions such as full green or brown herbage canopy conditions, or if these limiting conditions are not met, field data should be collected in the vicinity of the nodes that are to be used, so as to estimate the mixtures that exist at these nodes.

Develop models to estimate physical parameters from the proportion co-ordinates derived by the End Member classifier.

Use field data to develop models that relate variations in physical conditions to the proportion co-ordinates derived by the End Member classifier. Examples of this are to convert proportion co-ordinates into estimates of herbage LAI and biomass.

5.5 THE STRATIFIED RANDOM SAMPLING METHOD

An important technique in assessing the accuracy of information that has been derived from remotely sensed data is by the use of a stratified random sample survey consisting of the following components:

1. Delineation of strata
2. Determination of the strata sampling fractions and the number of samples
3. Selection of samples
4. Measurement of samples
5. Comparison of the samples with derived information

5.5.1 Delineation of Strata

The simplest way to conduct a sample survey is to select a set of unbiased samples across the area to be assessed. The unbiased set of samples is selected as a random sample from a population using a specific frame of reference that describes and contains the population in the random selection process. This frame of reference might be the range of map co-ordinates covering the area (in this case the population for the sampling is the range of map co-ordinates that can be chosen), a telephone listing of all property owners (the population in this case is the telephone book listing of numbers) in the area, a graphical list (as a map) of all cadastral land parcels (which are the population in this case) numbered sequentially from 1 to *n*, and so forth. In this process the whole frame of reference, whether it be the co-ordinates for the whole area, the complete telephone listing or the set of numbers for all cadastral land parcels, is considered as the one strata or domain, which is treated as a unit. This is suitable if all the classes to be assessed represent similar proportions of the total area, so that the random sample will select samples from each class.

Generally, however, these conditions are not met. Often some classes only represent a small fraction of the total area, the mix of classes varies significantly in different parts of the area classified, or different processing is employed in different parts of the area. When this occurs then the random sampling process will usually not select representative samples for all areas or classes. The solution is to partition the total area into a number of strata, and conduct random sampling within each of these strata.

Strata are often chosen on the basis of geomorphic, climatic, land using and land parcel size conformity, with each stratum representing areas that are relatively homogeneous in the chosen characteristic(s). In general, the strata will not be chosen on the basis of classes within digital classification since errors in the classification may introduce bias in the stratification. However, if different processing procedures are used in different parts of the area then these parts form strata boundaries because of differences in the classification accuracy in the different parts.

Once the strata have been selected they must be mapped over the frame of reference that will be used to select the samples. If co-ordinates are to be used as the basis for selecting the samples then the strata need to be mapped on an appropriate map. If the frame of reference is the listing of telephone numbers then the population of telephone numbers needs to be partitioned into the different strata and so forth for different populations.

Within the chosen strata the samples need to be selected in an unbiased way. The simplest way is to select the samples using a random sampling process. The random samples can be chosen from various frames of reference within the strata, as long as the chosen frame of reference includes, but does not duplicate, all members of the strata population. Suitable frames of reference include those discussed previously for random sampling.

5.5.2 Determination of the Sampling Fraction and the Number of Samples

The sampling fraction (SF) is the fraction of the whole area, or the frame of reference, that will be sampled. If the whole area, or the whole frame of reference is sampled (i.e., the SF = 1.0) then the samples will collect all the information in the area and will represent the "correct" situation. As the SF is reduced from 1.0, the proportion of the area being sampled is reduced, the possibility that the sample will not be representative of the population increases and the cost of the sampling decreases (Figure 5.11). The objective is to select an SF so that the samples are a reasonable representation of the population, whilst the cost of the sampling is minimised. The most suitable SF depends on the size and number of samples as well as their nature or shape. The SF could all be contained in the one, large sample, but this will lead to errors in the sampling. Generally the more, but smaller the individual samples, the better, as discussed below. For remote sensing an SF of 1% is the largest that is usually required as this SF has been found to provide samples that are representative of the whole population

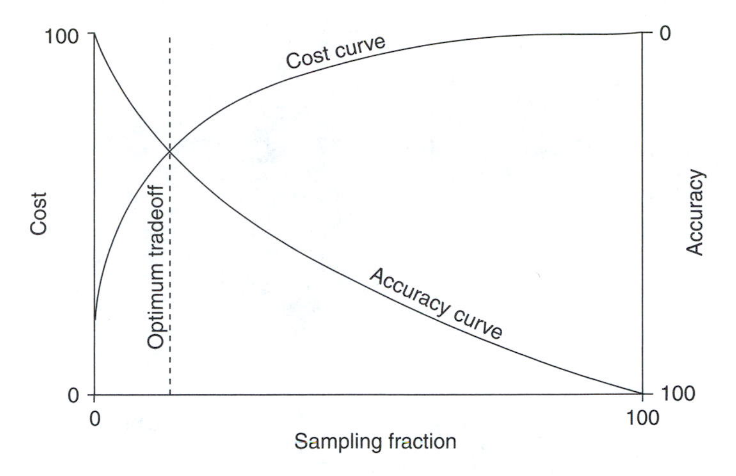

FIGURE 5.11 Relationship between sampling fraction, cost of field data collection and accuracy.

at a realistic cost, when the individual samples are of 1 ha in size or smaller. The smallest SF to be used should be larger than 0.01% as this SF has been found not to provide a representative sample of the population in some studies, at this sample size.

Once the SF has been selected this indicates the size of the area to be sampled. This area is then divided amongst the number of samples that are to be taken. In general, the more, and the smaller the samples, the better the results. In practice, the more samples that are used, the higher the cost; it is not realistic to have the samples smaller than the resolution of the mapping that is being assessed. With small samples it is often very difficult and expensive to locate the samples in the field and in the mapping, hence the analyst needs to be very carefull while considering the sample size.

If point or area samples are chosen that are to be of a specific size, then the minimum size is controlled by the spatial registration and resolution of the data. If the image data have been registered to a map with a standard error of $\pm s$, then 68% of the time the actual pixel will be within $\pm s$ of the nominal or known position of the pixel, and 95% of the time the actual pixel will be located within a distance of $\pm 2s$ of its nominal position. This means that the actual pixel position is within an area, at the 95% confidence level, that is, the pixel size $\pm 2s$ on all four sides, as shown in Figure 5.12. Consider a Landsat TM image that has been rectified to a resolution of 30 m with a standard error of ± 20 m in both directions. A sample for a pixel will need to be of size 30 ± 40 m at the 95% confidence level, or an area of 110×110 m^2. Clearly, whilst the actual pixel is likely to be within this area, so will be other pixels as well, the actual location of the pixel is not known within this area. The samples of this size therefore must be chosen so that they are reasonably homogeneous in the physical parameters that are being measured.

The accuracy of the rectification has less impact on larger samples. Consider samples that are to be $n \times n$ pixels in size, then the sample area will need to be ($n \times$ pixel dimension $\pm 2 \times$ standard error) at the 95% confidence level. Consider Landsat TM data where the pixel sizes are 30×30 m^2, for a sample of 10×10 pixels, rectified to an accuracy of ± 20 m standard error. The sample size will be a square of side:

$$\text{Length of side} = [n \times \text{pixel size}] + 2Cs \tag{5.14}$$

where n is the number of pixels along, or across the side of a sample, C is the multiplication factor at the set confidence interval ($C = 1$ at 68%, 1.645 at 90%, 1.96 at 95% and 2.575 at 99%) and s is the standard error associated with rectification of the image. For the example given above and using

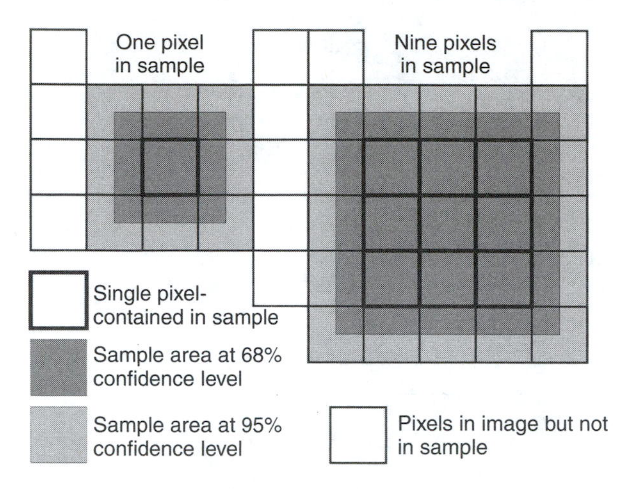

FIGURE 5.12 Sampling for a pixel in scanner data.

a 95% confidence interval:

$$\begin{aligned}\text{Length of side} &= 10 \times 30 + 4 \times 20 \\ &= 380 \text{ m}\end{aligned}$$

Landsat MSS is somewhat different since the separation of the pixels along a scanline is only 60 m, whilst the pixel size is 80 m. If the samples are to be 10 × 10 pixels (of size 80 × 80 m^2) and the rectification accuracy had a standard error, s, of ±50 m, then at the 95% confidence Interval the sample size will need to be: Along the scanlines:

$$\begin{aligned}\text{Size} &= [\text{pixel size} + (n - 1) \times \text{pixel interval}] + 2Cs \\ &= [80 + 9 \times 60] + 4 \times 50 \text{ at the 95\%CI} \\ &= 820 \text{ m}\end{aligned}$$

Across the scanlines:

$$\begin{aligned}\text{Size} &= [n \times \text{pixel size}] + 2Cs \\ &= [10 \times 80] + 4 \times 50 \text{ at the 95\%CI} \\ &= 1000 \text{ m}\end{aligned}$$

Please note that when the rectification standard error is given in pixel units and the pixel interval coincides with the pixel size, then it is simple to derive the distance in metres for the standard error. If the pixel or line intervals are different to the pixel size, then the pixel or line interval should be used to estimate the standard error in metres, rather than the pixel size. For the 10 × 10 pixel sample with TM data only 80 m is due to the errors in position, representing 38% of the sample area in comparison to the situation when sampling for one pixel where 93% of the sample is due to registration errors, for the condition that the standard error is ±0.68 of the pixel size.

For TM and SPOT data the ratio of the field area sampled to the image-based sample is given by:

$$\begin{aligned}\text{Area ratio} &= [nd + 2Cs]^2/(nd)^2 \\ &= [1.0 + 2Cs/nd]^2 \end{aligned} \tag{5.15}$$

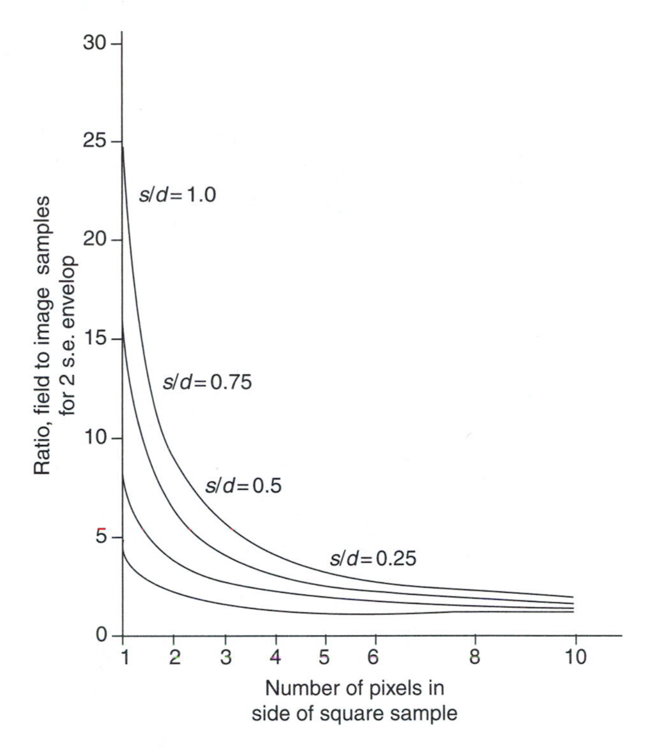

FIGURE 5.13 Ratio of the field sample area to image sample size for different numbers of pixels in the image samples, for different rectification accuracies (different s/d values).

where C is factor to set confidence range, s is the standard error, d is the pixel dimension and n is the number of pixels in the sample.

The relationship between this ratio and the proportion (s/d) is given in Figure 5.13 when the envelope about the image sample is $\pm 2s$ (i.e., a 95% confidence interval). This figure shows the inefficiency introduced into the collection of field data when (s/d) is large, due to low rectification accuracies, and when n is small. This is a particularly good example of the use of high-resolution data that is registered to the coarser multispectral data being used in the analysis, to define field sample areas that are not much larger than the actual pixel size.

If only a small number of samples are chosen, say 10, then the number of errors discovered by those samples would also be small, say 0,1 or 2. However, the achievement of zero errors does not imply that the total population is error free as this result may occur by chance since the number of samples is so small. Let the proportion of the classification that is in error be specified by p (or $p\%$). The probability of discovering no errors when taking x samples from this population with $p\%$ actual errors is given by the binomial distribution as $(1-p)^x$, and the probability of discovering y errors in x samples is ${}^{x}C_y \times p^y \times (1-p)^{(x-y)}$.

Table 5.3 shows the probability of scoring no errors in samples of varying size taken from a population with a range of real error proportions, p. This table indicates that sampling with zero errors can quite easily arise with small sample populations, even when the actual error rate is quite high. Taking the conventional probability level of 0.95/0.05, or 95%/5%, the table can be divided into two parts as has been done with a stepped line. Above and to the left of the line, the probabilities of obtaining error-free sampling results are high, even when the actual errors are present in significant proportions. Below and to the right of the line there is a small to negligible chance of getting an error-free result from the sampling when there are actual errors in the population.

Thus, if the permissible error rate in the image interpretation is set at about 85 to 90%, as suggested in the U.S. Geological Publication on landuse mapping (USGS Circular 671) or as required in an

Table 5.3

Probability of Scoring No Errors in Samples of Varying Sizes from a Population with a Probability, *p*, of Real Errors in Each Pixel. Stepped Line Indicates Approximate 0.05 Probability Level

Probability of real errors	$n = 5$	$n = 10$	$n = 15$	$n = 20$	$n = 25$	$n = 30$	$n = 35$	$n = 40$	$n = 45$	$n = 50$	$n = 55$	$n = 60$
0.01	0.9510	0.9044	0.8601	0.8179	0.7778	0.7397	0.7034	0.6690	0.6362	0.6050	0.5754	0.5472
0.05	0.7738	0.5987	0.4633	0.3595	0.2274	0.2146	0.1661	0.1285	0.0994	0.0769	0.0595	0.0461
0.10	0.5905	0.3487	0.2059	0.1216	0.0718	0.0424	0.0250	0.0148	0.0087	0.0052	0.0030	0.0018
0.15	0.4437	0.1969	0.0874	0.0388	0.0172	0.0076	0.0034	0.0015	0.0007	0.0003	0.0001	0.0
0.20	0.3277	0.1074	0.0352	0.0115	0.0038	0.0012	0.0004	0.0001	0.0	0.0	0.0	0.0
0.25	0.2372	0.0563	0.0134	0.0032	0.0008	0.0002	0.0	0.0	0.0	0.0	0.0	0.0
0.30	0.1681	0.0282	0.0047	0.0008	0.0001	0.0	0.0	0.0	0.0	0.0	0.0	0.0
0.40	0.0778	0.0060	0.0005	0.0000	0.0000	0.0	0.0	0.0	0.0	0.0	0.0	0.0
0.50	0.0313	0.0010	0.0000	0.0000	0.0000	0.0	0.0	0.0	0.0	0.0	0.0	0.0

TABLE 5.4
Probability of Scoring Errors in Samples of Varying Sizes from a Population with Real Errors of 15%, that is an Interpretation Accuracy of 85%. Stepped Line Indicates Approximate 0.05 Probability Level

	Number of Errors					
Sample Size	0	1	2	3	4	5
15	0.0874					
20	0.0388	0.1368				
25	0.0172	0.0759	0.1607			
30	0.0076	0.0404	0.1034			
35	0.0034	0.0209	0.0627	0.1218		
40			0.0365	0.0816		
45			0.0206	0.0520	0.0963	
50				0.0319	0.0661	0.1072
55				0.0189	0.0434	0.0781
60					0.0275	0.0544
65						0.0365

TABLE 5.5
Probability of Scoring Errors in Samples of Varying Sizes from a Population with Real Errors of 10%, that is an Interpretation Accuracy of 90%. Stepped Line Indicates Approximate 0.05 Probability Level

	Number of Errors			
Sample Size	0	1	2	3
15	0.2059			
20	0.1216			
25	0.0718	0.1994		
30	0.0424	0.1413		
35	0.0250	0.0973		
40		0.0657		
45		0.0436	0.1067	
50		0.0286	0.0779	
55			0.0558	0.1095
60			0.0393	0.0844
65				0.0636
70				0.0470
75				0.0297

operational programme specification, then the sample size in each category or strata necessary for 85% interpretation accuracy is at least 20 samples, and at least 30 samples for 90% accuracy. The tables can thus be used to determine the minimum number of samples that are required for any given interpretation accuracy. Tables 5.4 to 5.6 inclusive provide more detailed calculations of the

TABLE 5.6
Probability of Scoring Errors in Samples of Varying Sizes from a Population with Real Errors of 5%, that is an Interpretation Accuracy of 95%. Stepped Line Indicates Approximate 0.05 Probability Level

Sample Size	Number of Errors 0	1	2
45	0.0994		
50	0.0769		
55	0.0595		
60	0.0461	0.1455	
65	0.0356	0.1219	
70		0.1016	
75		0.0843	
80		0.0695	
85		0.0572	
90		0.0468	0.1097
100			0.0812

probabilities of scoring errors in samples of varying sizes with the specific interpretation accuracy levels of 85, 90 and 95%, respectively. It is noted in Chapter 8 that the standard level of accuracy and reliability required for topographic maps is 95/95 or 95% accurate, 95% of the time. This level of accuracy and reliability is likely to be legally acceptable, whereas lower levels of accuracy and reliability may not be accepted at court. Given this, it should be the aim to achieve accuracies such that the derived information can stand on its own which means that it needs to be of an acceptable level of accuracy and reliability to be accepted at court.

5.5.3 Selection of Samples

It is necessary to consider both the types of samples to be used and how they will be selected. There are at least two types of samples that can be used:

1. *Uniform size*, without the consideration for existing physical or administrative boundaries. These samples are all of one size and shape for a particular project or task. The minimum size of the samples will need to consider the influence of data resolution as discussed earlier. They can be circular, centred on a location, or rectangular as is suitable to the task. The samples are selected as a random sample within each stratum.

   A major difficulty with this type of sample is to identify the location and boundaries of the sample in the mapping being assessed, and on the ground. The samples may contain information on more than one class, or the condition of the landcover within the sample may vary across the sample. If the samples contain more than one class then the proportions of the classes need to be determined and carried through the accuracy assessment. If the assessment is on condition of the covers then variation in condition can create difficulties in the assessment. It may be necessary to select other samples that do not contain too much variation in condition within the individual samples.

This form of sample is very appropriate for assessing the accuracies of:

- Condition parameters at the resolution of the data, for example, to assess mapping of vegetative biomass, cover, water quality, etc. For these assessments the samples should lie within the one cover type for the assessment, and be reasonably uniform in cover condition to avoid, as much as possible, discrepancies between the actual and the measured cover due to discrepancies between the areas mapped and the area field sampled. Each sample would normally be designed to match one or more pixels in the mapping being assessed.
- Thematic classifications, such as soils, land systems and landuse mapping, particularly where the mapping does not follow, or may not be adequately represented by samples that are constrained by, administrative or physical boundaries. The samples would often be large relative to the resolution in the data, and may contain more than one class within each sample so that the samples need to be located and then mapped for comparison with the classification.

Samples of this type would normally be selected as a random sample from a regular grid across the area to be assessed. The random selection of sample sites can be done with the use of a table of random numbers, or by use of random numbers generated in a computer. The gridlines, both northings and eastings, are both numbered and a pair of random numbers specifies each co-ordinate point of a sample.

2. *Irregular samples* in which the sample boundaries are defined by physical or administrative boundaries, such as property or paddock areas. The main advantages of this type of sample are that the samples can be unambiguously located in the mapping and on the ground, and the sample can be readily revisited if required. Another advantage is that the individual fields will often contain the one landuse or landcover, but the boundary of the field will contain mixed pixels that need to be considered. Irregular samples may have the disadvantage that they may be very large relative to the sampling fraction, and so the area sampled, to ensure that the number of samples is adequate, may be larger than is necessary.

The first step in selecting the samples is to decide on the sample units to be used (e.g., paddocks or properties). Once these have been chosen, they are identified on a map and numbered in a regular way. Random numbers are then generated, either from a table or in the computer, to cover the range of these field numbers, so that each field with the same number as the generated random number, becomes a sample. In this way successive generation of random numbers and identification of the relevant field selects all the samples. The paddocks or properties can be numbered in any way that is convenient, unbiased and the numbers on each sample unit are unique. It may be convenient to use telephone numbers if all properties have the telephone. With samples of this type there may be two distinct forms of error to be considered:

- Errors in classification itself, that is, errors in assigning pixels that are of one class to that class. The frequency of this type of error will be determined by comparing the classifications within the sample. Accuracy assessment for this purpose can be done by defining the sample boundaries within the fields so as to avoid the mixed pixels around the border of the field.
- Errors due to mixed pixels along the boundary of areas of landcover or landuse types. The errors associated with mixed pixels are likely to be quite different to the errors associated with the centres of fields, so that it may be necessary to consider these two sources of error quite separately. Digitisation on the border of fields may not be suitable because slight errors in the location of this border in either mapping may introduce significant "apparent" rather than "real" errors into the accuracy assessment. A solution is to create samples that are larger than the fields and compare the areas of the field as stated in the field and the mapping.

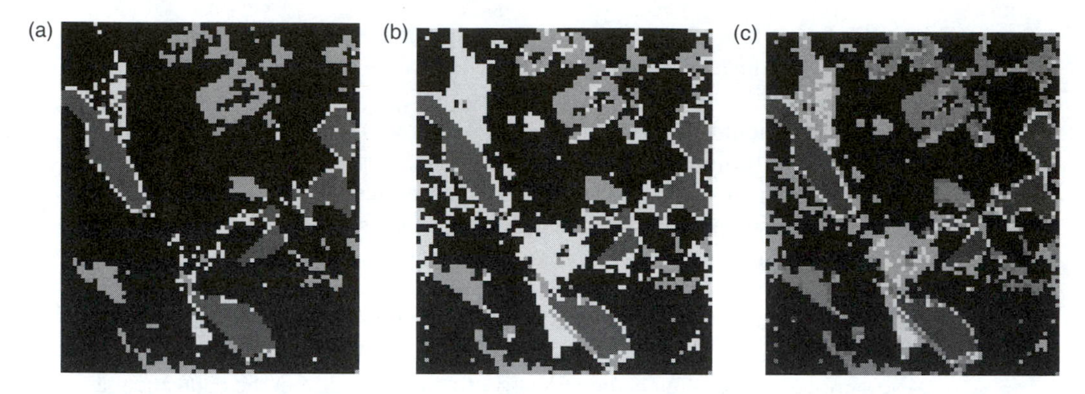

FIGURE 5.14 (a) Classification of an image, (b) field data map and (c) cross-tabulation of the two classifications. Black = pasture, dark grey = water, light grey = cropping and white = forest.

5.5.4 MEASUREMENT OF SAMPLES

The samples are located in the field and the parameters of interest are measured within the sample area. The method of measurement depends on the parameters to be measured:

1. *Landuse and landcover*. A typical mapping of landcover or landuse within a rectangular sample is shown in Figure 5.14. For landuse or landcover accuracy assessment the most accurate method is to map the sample areas on the ground, aided by current, high-resolution aerial photographs. Visual interpretation of moderate scale aerial photographs, supported by extensive field visits would be the second best choice. The mapped field information for a sample is then compared to the same area as mapped by the remotely sensed data, using either of the two methods:
    - Overlay one mapping on the other as shown in Figure 5.14, and a dot grid is laid on top. Count the dots within each sub-sample area and for the total sample to compute the proportion of the sample area that is covered by each sub-sample area. These proportions are then used in accuracy assessment (Section 5.6).
    - Digitise the field mapping to create a theme image file of the field data. Compare, on a pixel-by-pixel basis, this theme file with the classification theme file derived from the remotely sensed data, using a Geographic Information System (GIS), as the tool, as discussed in Chapter 6.

    The classification (a) and field maps (b) depicted in Figure 5.14 have been compared in a GIS to produce the cross-tabulation image in (c), and the table of cross-tabulations are given in Table 5.7 as count values, and as proportions in Table 5.8. A cross-tabulation table shows the field classes across the top and the classes being compared from the image-based classification down the side.
2. *Condition parameters*. The sample areas are defined and the physical attributes of interest are measured for a random distribution of points across each sample, so as to derive estimates of the parameters for the sample. The set of data for all of the samples is used in the accuracy assessment.

Perusal of the classifications shows that the water areas (dark grey) are consistent, and that the cropping areas are reasonably consistent. There are considerable differences between the two classifications in assignment of forest areas. This is primarily due to differences in definition of forest from woodland from pasture with tree cover, but there may be some confusion between forest and water, primarily along the edges of the water bodies.

TABLE 5.7
Cross-Tabulation of Field Classification Against Image Classification as Total Counts

Image Classification	Field Classification				
	Pasture	Water	Cropping	Forest	Total
Pasture	3898	0	174	771	4843
Water	0	544	0	43	587
Cropping	2	0	502	3	507
Forest	5	19	3	228	255
Total	3905	563	679	1045	6192

TABLE 5.8
Proportional Cross-Tabulations of the Data in Table 5.7

Image Classification	Field Classification				
	Pasture	Water	Cropping	Forest	Total
Pasture	0.6295	0.0	0.0281	0.1245	0.7821
Water	0.0	0.0879	0.0	0.0069	0.0948
Cropping	0.0003	0.0	0.0811	0.0005	0.0819
Forest	0.0008	0.0031	0.0005	0.0368	0.0412
Total	0.6307	0.0909	0.1097	0.1688	1.0

5.6 ACCURACY ASSESSMENT

5.6.1 THE ROLE OF ACCURACY ASSESSMENT

Much has been made of the opportunities that exist for errors to arise in extracting information from remotely sensed data. Resource managers, being well aware of this characteristic, will want to weigh the value of this information relative to other information, in making decisions. They will want to decide when to use this information as the basis of final decisions, and when to use it as an indicator, to be supplemented by other information on which to base final decisions. At present, information derived from remote sensing is more usually used in the second mode, making the information valuable but not essential. As the accuracy and reliability of information derived from remote sensing improves, so it will increasingly be used in the first role, in which case the information becomes an essential component of the resource management process.

If resource managers weigh the value of this information too lowly then they will not be making the best use of the information available to them; it will represent a waste of expensive processing time and field data collection, or worse, it will not be collected at all. If they weigh the information too highly then they may make erroneous decisions that might reflect on them, and on this source of information. It is therefore important to provide the users of this information with an objective means of properly weighing the accuracy of this information relative to other types of information that they might use.

A very good basis for weighting information is the accuracy assessment done on the information, as long as that accuracy assessment is:

1. Relevant
2. Representative of the population
3. Easily understood and useable
4. Can be related to other information

5.6.1.1 Relevance

Accuracy assessment can assess many facets of the information derived from remotely sensed data. The resource manager, wishing to use the information for a specific purpose or purposes will find that some forms of accuracy assessment are more relevant than others. For example, in digital classification the manager may be interested in the accuracy of areas estimated from the classification. If this is the case then the accuracy assessment should be of the accuracy in determining areas rather than the accuracy in identifying fields. Alternatively, the manager may be interested in the locational accuracy of the classification so as to relate this information to other types of information or data. In this case the accuracy assessment should be conducted on the accuracy of identifying pixels or fields. In visual interpretation the manager may be interested in the positional accuracy of the interpreted class boundaries, or in the accuracy of classifying specific locations.

5.6.1.2 Representative of the Population

The easiest way to ensure that the accuracy assessment is representative of the population is to ensure that the accuracy assessment is conducted using a statistically valid and sufficient sample to conduct the accuracy assessment. The sample must:

1. Cover the area of interest
2. Represent all classes
3. Be distributed in an unbiased way
4. Be of sufficient size
5. Select appropriate sample types

5.6.1.3 Easily Understood and Usable

It is essential that accuracy assessment results be stated simply and in an unambiguous fashion. Simplification of accuracy assessment is often likely to be achieved by some compromise between theoretical precision and purity on the one hand and practical simplicity on the other. Consider the solution used in topographic mapping where the position of contour lines are stated as being "$x\%$ accurate $y\%$ of the time" where a contour is considered to be accurate if it is within $\pm$ (half a contour interval) from its "true" position. Statistically, the contours are plotted with accuracies that are related to the $\pm$ (half contour interval) and the "$x\%$ accuracy $y\%$ of the time" rule gives a confidence interval.

5.6.1.4 Can be Related to Other Information

The accuracy of information derived from remotely sensed data often needs to be related to the accuracy of other information. This is necessary to:

1. Allocate relative weights to the different sets of information in making decisions
2. To derive the accuracy of information extracted from combinations of the parent information types

The issues raised in relating one set of information to others have not been adequately addressed in the literature. The discussion in Section 5.1 identified two complimentary indicators of the quality of information derived from remotely sensed data, accuracy and reliability.

Accuracy assessment of information extracted from remotely sensed imagery depends upon whether the information is concerned with classification or estimation. Accordingly they will be dealt with separately.

5.6.2 Comparison of Field Data with Classification Results

There are three different aspects of classification that may require accuracy assessment:

1. *Classification accuracy*. The accuracy of classification of individual pixels or image elements.
2. *Accuracy of identifying production units* such as paddocks. A management unit can be delineated even though some of the pixels contained within the unit might be wrongly classified. The important criteria is that sufficient pixels are properly classified so as to assign the paddock to the proper class. It is thus a much less rigorous criterion for accuracy assessment than (1), yet it is adequate for many purposes.
3. *Boundary definition*, the accuracy of delineation of the boundaries of management units such as paddocks. The accuracy of delineation of the mixed boundary pixels will usually be different to the accuracy of classification of the internal pixels in an area of the one land class. Yet, this accuracy is important if the resource manager is concerned with areas, or with an accurate definition of the boundaries of the paddocks.

Selection of the size and number of samples within the sampling fraction of the area classified depends on the form of accuracy assessment that is to be conducted. If the accuracy assessment is to be conducted for:

1. *Classification accuracies at the pixel level*, then point samples are adequate, and these are related to the pixel at that location. Since there are errors in rectification, the population of co-ordinates that are used as the basis for selection of the samples should be restricted to the co-ordinates of the pixel centres. Thus, if the image has been rectified to 25 m resolution, then the co-ordinate grid at this resolution, and matching the pixel centres, should be used as the basis of selection for the samples.
2. *Classification accuracies at the field level*, for which the samples need to be fields themselves. Point samples cannot be taken, since the point locations may be in error in the classification, but ignored, as they were part of a scattering of errors in the classification that were below the accepted threshold for accepting that the field was of the one class. Although, the field is the sample, the field sampler only needs to view the field to ensure that they can assess the cover type that is typical for the field.
3. *Setting of boundaries or determination of areas*. For these the fields will again be the sample, and the fields need to be mapped as part of the field data collection so as to be able to assess the accuracy of location or of area determination.

5.6.2.1 Classification Accuracy

Accuracy assessment of an image-based classification against field data for pixels or fields will create a cross-tabulation table as depicted in Table 5.7 or Table 5.8. Initially, the table will contain count values as in Table 5.7, but these need to be converted into proportions of the total population as shown in Table 5.8. Cross-tabulation tables can be constructed for each sample, similar to Table 5.9(a), for all of the samples in the one stratum (Table 5.9[b]) and for all samples in all strata (Table 5.9[c]).

TABLE 5.9
Comparison of the Pixel Counts for the Combinations of Classes in Two Classification Theme Files in Which One File is from Image Processing and One is Derived from Field Data

	Field-Based Classes					
Image-Based Classes	**Cropping**	**Pasture**	**Forest**	**Bare Soil**	**Water**	**Total**
(a) *Cross-tabulation table for one sample*						
Cropping	82	0	0	0	0	82
Pasture	6	15	0	0	0	21
Forest	0	0	2	0	0	2
Bare soil	0	3	0	0	0	3
Water	0	0	0	0	0	0
Total	88	18	2	0	0	108
(b) *All samples in the one strata*						
Cropping	1421	43	0	5	0	1469
Pasture	247	1360	0	81	1	1689
Forest	2	104	288	0	1	395
Bare Soil	0	29	0	494	0	523
Water	0	1	3	0	104	108
Total	1670	1537	291	580	106	4184
(c) *For all samples in all strata*						
Cropping	3694	106	0	74	0	3874
Pasture	432	11872	106	234	6	12651
Forest	36	327	1096	0	1	1460
Bare Soil	59	221	0	736	0	1016
Water	0	1	14	0	429	444
Total	4221	12437	1216	1044	436	4184

The first analysis that should be conducted of the cross-tabulation table is to assess the agreement between field-based and imaged-based classifications that is often conducted by the use of the kappa statistic:

$$\kappa = \frac{p - e(\kappa)}{1 - e(\kappa)} \tag{5.16}$$

where p is the overall agreement between the classifications and so it is the sum of the diagonal proportion terms (Table 5.9) and $e(\kappa)$ is the sum of the probabilities of each classification independently classifying pixels into each class. It can be found for a proportional cross-tabulation table with n entries or classes and a total proportion column and row:

$$e(\kappa) = \sum_{i=1}^{n} (\text{Table}(i, \text{Total}) \times \text{Table}(\text{Total}, i)) \tag{5.17}$$

Consider the data shown for all samples in Table 5.9(c), and converted into proportions in Table 5.10 in which the columns have been given upper case designations and the rows lower case as shown in

TABLE 5.10
The Proportions of the Total Population for the Data in Table 5.9(c)

Image-Based Classes	Field-Based Classes					
	Cropping, C	Pasture, P	Forest, F	Bare Soil, B	Water, W	Total, T
Cropping, c	0.18998	0.00545	0.00000	0.00381	0.00000	0.19924
Pasture, p	0.02222	0.61057	0.00545	0.01203	0.00031	0.65059
Forest, f	0.00185	0.01682	0.05637	0.00000	0.00005	0.07509
Bare soil, b	0.00303	0.01137	0.00000	0.03785	0.00000	0.05225
Water, w	0.00000	0.00005	0.00072	0.00000	0.02206	0.02283
Total, t	0.21708	0.64426	0.06254	0.05369	0.02242	1.00000

the table. Using Equation (5.16) and Equation (5.17) for the data in Table 5.10:

$$p = \text{Table}(\text{C}, \text{c}) + \text{Table}(\text{P}, \text{p}) + \text{Table}(\text{F}, \text{f}) + \text{Table}(\text{B}, \text{b}) + \text{Table}(\text{W}, \text{w}) = 0.9168$$

$$e(\kappa) = \text{Table}(\text{T}, \text{c}) \times \text{Table}(\text{C}, \text{t}) + \text{Table}(\text{T}, \text{p}) \times \text{Table}(\text{P}, \text{t}) + \text{Table}(\text{T}, \text{f}) \times \text{Table}(\text{F}, \text{t}) + \text{Table}(\text{T}, \text{b}) \times \text{Table}(\text{B}, \text{t}) + \text{Table}(\text{T}, \text{w}) \times \text{Table}(\text{W}, \text{t}) = 0.4704$$

$$\kappa = \frac{p - e(\kappa)}{1 - e(\kappa)} = \frac{(0.9168 - 0.4704)}{(1 - 0.4704)} = 0.8430$$

The kappa statistic indicates the agreement between the classifications after chance agreement has been removed. Kappa values of about 0.0 indicate that any agreement between the classifications is due to chance, and a kappa value of 1 would indicate perfect agreement, without chance. Kappa values higher than 0.7 indicate that there is good non-chance agreement between the two classifications that are being compared. The value of 0.843 derived above thus indicates that there is good, non-chance agreement between the field data and the classification data. This is valuable as long as the field data is collected in a statistically valid way.

Once the correspondence between the classifications has been conducted using the kappa statistic, the analyst may wish to evaluate the errors of omission or commission. Errors of omission are the errors associated with classifying areas mapped from the field data as being in the one class, that are mapped from the image data into other classes. They have been omitted from the correct class. The errors of omission are the column values, other than the cell of correct classifications, in the cross-tabulation table (Table 5.10) and are depicted in Figure 5.15. The errors of commission are the errors associated with mapping other classes in the field data into the one class. They are thus additional areas that have been included in the class. Errors of commission are the row values in the cross-tabulation table, other than for the cell of correct classifications (Figure 5.15).

Consider the crop class given in Table 5.9(c) and in Table 5.10. The accuracy of classifying crops is given by:

$$\Pr\{\text{Crop}|\text{Crop}\} = \frac{3694}{4221} = \frac{0.18998}{0.21708} = 0.875 \tag{5.18}$$

The errors of omission, E_0, of classifying crops and an another class, such as pasture is:

$$E_0\{\text{Pasture}|\text{Crop}\} = \frac{432}{4221} = \frac{0.02222}{0.21708} = 0.102 \tag{5.19}$$

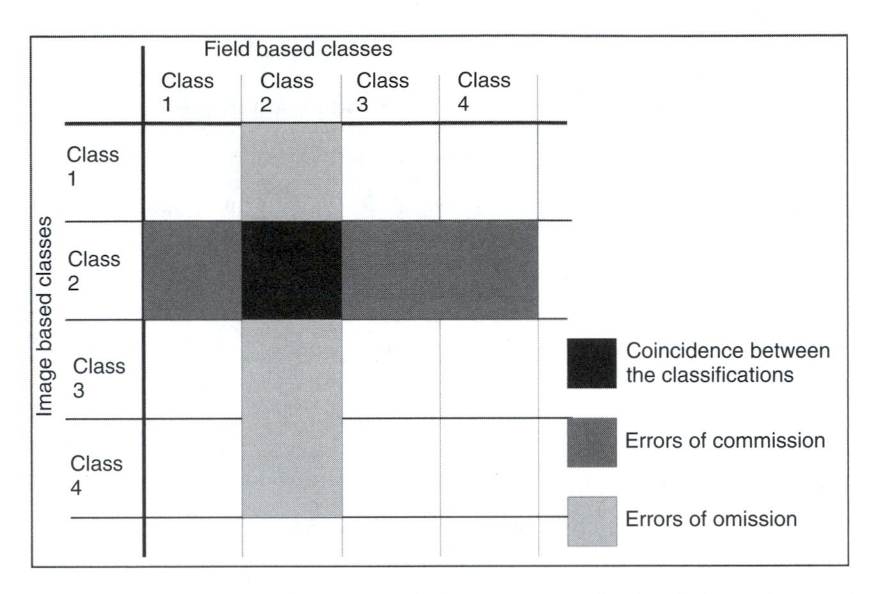

FIGURE 5.15 Accuracies and errors of omission and commission in relation to the one class.

TABLE 5.11
Accuracies (Black) and Errors of Omission (Gray) from the Data Given in Table 5.9(c).

Image-Based Classes	Field-Based Classes					
	Cropping	Pasture	Forest	Bare Soil	Water	Total
Cropping	0.87514807	0.008462	0	0.070881	0	0.954491
Pasture	0.10234542	0.947713	0.087171	0.224138	0.013761	1.375129
Forest	0.00852878	0.026104	0.901316	0	0.002294	0.938242
Bare Soil	0.01397773	0.017642	0	0.704981	0	0.7366
Water	0	7.98E−05	0.011513	0	0.983945	0.995538
Total	1	1	1	1	1	5

This assessment can be conducted for each column in the cross-tabulation table as shown in Table 5.11.

Perusal of this table shows that a significant proportion of the soil area has been wrongly classified as pasture (22%) and that 19.4% of the urban areas have been misclassified as water. This information can then be used by the resource manager to assess the reliance that he should place on the information, and by the analyst to identify weaknesses in his analysis, and rectify them if this is justified. Similar analyses can be conducted on the errors of commission as shown in Table 5.12.

In any classification, the class boundaries can be adjusted so as to change some of the errors. Usually, such adjustments do not have a lot of impact on the overall error levels in the classification, but rather reductions in one error level are matched by increases in other error levels. Such adjustments can thus have a big impact on the individual error levels, and thus the location of the decision surface, and the subsequent error levels depends on the purpose of the classification and how the resulting information is going to be used by the resource manager. The most common effect of shifting the decision surfaces is to change the errors of omission in one direction (say a decrease), and change the errors of commission in the opposite direction (increase). Since the errors of commission for one

TABLE 5.12
The Errors of Commission are Shown in the Grey Areas, for the Data Given in Table 5.9(c)

Image-Based Classes	Field-Based Classes					
	Cropping	Pasture	Forest	Bare Soil	Water	Total
Cropping	0.9535364	0.027362	0	0.019102	0	1
Pasture	0.0341502	0.938498	0.008379	0.018498	0.000474	1
Forest	0.02465753	0.223973	0.750685	0	0.000685	1
Bare Soil	0.05807087	0.21752	0	0.724409	0	1
Water	0	0.002252	0.031532	0	0.966216	1
Total	1.07041499	1.409604	0.790596	0.762009	0.967375	5

TABLE 5.13
Variances Associated with the Classification Accuracies and the Errors of Omission

Image-Based Classes	Field-Based Classes				
	Cropping	Pasture	Forest	Bare Soil	Water
Cropping	0.00543864	0.008897	0	0	0
Pasture	0.014583	0.002043	0.027399	0.027261	0.047561
Forest	0.01532613	0.008817	0.009009	0	0.047836
Bare soil	0.01528395	0.008855	0	0.01681	0
Water	0	0.008934	0.028511	0	0.006068

class are errors of omission for other classes, then such a shift will change the errors of omission for these other classes in the same direction as the errors of commission for the first class (increase).

Consider the situation where the crops are being mapped in an area for the purpose of verifying that the farmers under contract are growing the contracted crop. In this situation the contractor wants to be sure that the farmers are growing sufficient area of the crop; they can grow more and hope to sell this additional produce on the open market, but that is not of concern to the contractor. In this situation the contractor will be most concerned with errors of commission, since a mapping that shows that everything is satisfactory when in fact it is not, is a problem to him. The reverse situation, of errors of omission will show that some farmers do not have sufficient area, but these will have been identified in the classification and field visits will verify whether the problem lies with the classification. In such a situation the cost structure in the Maximum Likelihood classifier, if that is used, can be adjusted so as to minimise errors of commission for that cropping class. The accuracy assessment can then show whether the change in the decision surfaces is having the desired effect, or whether it is too stringent, or not stringent enough.

In many applications the reliability of the classification is also important. If the accuracies are conducted for each stratum in the classification, then this gives a population with which to assess the reliability of the classification. Alternate populations would be mappings of a number of areas, and mappings over time. The accuracies as given in Table 5.13 for all the members of the population can be used to derive mean and variance values for the accuracy and errors of omission of each class. The variances can then be used to set confidence intervals around the mean value at some set confidence level, usually 95%.

If this is not possible, then an alternative is to use the data in Table 5.10 to derive confidence intervals. The Central Limit Theorem states that the proportions in this table are average proportions and as such they are approximately normally distributed as long as more than 30 samples have been used to derive the proportions. The standard error associated with each proportion, p, or σ_p can be calculated from the equation:

$$\sigma_p = \sqrt{\frac{(N-n)}{(N-1)}} \times \sqrt{\frac{p(1-p)}{n}} \tag{5.20}$$

for a finite population, N, with a sample population, n. When the whole population is infinite then the equation can be simplified to:

$$\sigma_p = \sqrt{\frac{p(1-p)}{n}} \tag{5.21}$$

To use Equation (5.20) or Equation (5.21), the total population, N, can be taken from the classification results and the sample population taken from the cross-tabulation table. As N gets very large, the value derived from Equation (5.20) approaches that coming from Equation (5.21). These variances associated with each proportion can be determined and used in the analysis of the propagation of error. The variances associated with the proportions in Table 5.10 are given in Table 5.13.

The cross-tabulation table can also be used with the classified areas of the different cover types from the image classification to derive better estimates of the area of each cover type within the total area mapped. The errors of omission analysis as given in Table 5.11 showed the accuracy of classification and the errors of omission. If, in this table we use the convention Pr{Column|Row}, a_i is the image classified area of class i, (equivalent to the total across the row for that class) and A_i is the more accurate estimate of the area of class i, then:

$$A_i = \sum_{j=1}^{n} a_j \times \Pr\{i|j\} \tag{5.22}$$

So that

$$A_1 = a_1\Pr\{1|1\} + a_2\Pr\{1|2\} + a_3\Pr\{1|3\} + \ldots$$

It should be noted that the sum of the new areas should be the same as the sum of the old areas, as long as the classes cover the whole of the area.

5.6.2.2 Accuracy of Identifying Production Units

For some purposes the resource manager may be interested in the accuracy of classifying production units, such as fields. In this situation the analyst will have conducted the classification, and then allocated the parcels a landcover by:

1. Removing the salt and pepper effect.
2. Using the known field boundaries to find out the proportions of the different classes within the field. If one class dominated the field area above a designated threshold level, then the field will be assigned that class.

For this situation the classification after the conduct of this form of cosmetic changes should be used in the accuracy assessment, using field samples as the basis of the accuracy assessment.

5.6.2.3 Boundary and Area Accuracy Assessment

Mapping of boundaries on an image creates errors in the accuracy of location of that boundary that is due to a combination of interpretation, identification and cartographic errors. These errors are usually larger when attempting to delineate boundaries that are poorly defined on the image. Thus, the accuracy of defining the boundaries of soil groups on a soil image will usually be less accurate than will be the accuracy of defining the boundaries of cropping fields. The accuracy of defining boundaries can be assessed by:

1. Plotting the locations of the boundaries in the field on one or more random traverses across the area, and comparing this with the intersections of the traverse with the mapped boundaries. The difference between the two, as a distance, can be computed as an average and a variance, taking differences in one direction as positive and as negative in the other. Traverses can require considerable fieldwork and it can be difficult ensuring that they are representative.
2. Taking a set of sample areas along the various boundaries, and measuring the areas of each landcover in each sample, both from the classification and from another source such as field mapping. The differences between the two, taken negative in one direction and positive in the reverse, are used in a similar manner to the distances to compute an average and a variance. The areas calculation will tend to average small idiosyncrasies that may occur in either of the mappings.
3. Overlay the mappings in a GIS and extract the areas that are within the overlap as errors of omission and commission, and the areas that are classified the same, in both mappings, as being correctly classified.

The end user may be more interested in the areas estimated by the classification, so that the accuracy assessment of the mapping of areas may be an important component of the accuracy assessment. To conduct this form of accuracy assessment, the areas of the samples need to be measured from the image classification and in the field using more accurate techniques such as field survey techniques or by measurement on high-resolution orthophotographs. The two sets of area measurement can then be analysed using regression techniques to derive the relationship between them, and to set confidence ranges on the estimates.

5.6.3 The Use of Field Data in Estimation

Accuracy assessment in estimation would normally be conducted by the use of a distribution of regular area samples matched to the pixel size and rectification accuracy as previously discussed. These samples would be distributed across the area by means of a stratified sampling methodology, as previously described. The physical parameters of interest are then measured at a sample of points across each sample, so as to derive reliable estimates of the parameters for each sample. The derived sample field data values are then compared to the image-based estimates by the use of regression techniques so as to establish the relationship between the field and the image-based estimators of the physical parameters, and to set confidence ranges on the image-based estimators.

5.7 SUMMARY

The collection of field data will always remain a crucial and expensive component to the conduct of high-quality image analysis tasks. It is crucial because the imagery cannot stand on its own, but it needs field data to link the image data reliably to field conditions. The analyst should not consider that image analysis removes the need for field data, but rather the use of image data extends the reach of the analyst, or enables him to derive more accurate and more detailed information than would be possible without the use of the image data.

The costs of field data mean that it behoves the analyst to ensure that the field data component is conducted in as efficient and effective way as possible, and that most amount of information as possible is drawn from the field data. At present field data collection and analysis usually represents about one third of the costs of a program using remote sensing, whereas the costs of data and analysis are each about one third of the costs as well. Whilst the costs of data and analysis may come down, the costs of field data are less likely to be reduced. It is therefore important to continue to look for ways to reduce these costs whilst maintaining the validity of the whole process of information extraction.

FURTHER READING

For those who wish to delve into the assessment of accuracy of information derived from remotely sensed images, then Congalton (1999) is recommended, whilst Raj (1968) provides a good treatment of Sampling Theory.

REFERENCES

Congalton, R.G. and Green, K., 1999. *Assessing the Accuracy of Remotely Sensed Data: Principles and Practices*, Lewis Publishers, MI.

Grove, C.I., Hook, S.J. and Paylor, E.D. II, 1992. "Laboratory reflectance spectra of 160 minerals, 0.4 to 2.5 micrometers," *Jet Propulsion Laboratory Publication 92-2*, Jet Propulsion Laboratory, NASA, Pasadena, CA.

Raj, D. 1968. *Sampling Theory*, McGraw-Hill, New York.

6 Geographic Information Systems[†]

6.1 INTRODUCTION TO GEOGRAPHIC INFORMATION SYSTEMS

Consider the situation where a resource manager has a range of spatial information. The information may include elevation, slope and aspect data, soil classes, individual site soil tests, current landownership, drainage and water features, current and historical landuse, the road and rail transport corridors in the area and the power and telephone lines in the area, some of which are depicted in Figure 6.1. Different subsets of this information are used to address different issues. To promote the use of a particular fertiliser requires information on landuse history, soils and land ownership, to locate those farmers that are the target client group. Alternatively, the manager may wish to select a route for a road, power or telecommunications line through an area, where the cost of construction is a function of soil type, topography, water obstacles and land values, as could be assessed from landuse. Integration of these factors will indicate potential corridors that can then be assessed in more detailed analysis. Clearly, each management issue will use a different subset of the available information but there will often be overlaps in the subsets required to address the different questions posed both within the one area of resource management, and across different resource management disciplines.

Each type of spatial information could be conceptually viewed as a layer of information on that aspect of the environment. The set of information layers create a set of overlays for the area, much like a "Dagwood sandwich." In some of these layers, the information varies continuously across the area, like elevation, slope and aspect. In other layers the information is consistent across sub-areas, and then the information changes abruptly for another sub-area. Thus, the layer of landownership may be consistent in its information content for each farm, but will be different from farm to farm. In some of these layers, each unique sub-area may have only one attribute value, for example, an elevation layer may have only the one attribute value of elevation for each location in the overlay. In other layers there may be multiple attribute values. The layer of land ownership may include, but not be exclusive to the attributes listed in Table 6.1.

The layer of soil classes may have attribute values within each sub-area of soil class namely, soil association, typical soil colour and soil chemistry information. Additional information that may be required to characterise the spatial variability of soils could include texture and the thickness of the different soil horizons.

The whole set of information is a set of databases, that include one or more spatial and attribute databases. Such a database set that contains spatial information is called a geographic information system (GIS). The mix of spatial and attribute data in the database set will vary from situation to situation. The GIS can be constructed as a database in the classical database sense, however, in general they are constructed differently. They are usually constructed as one or more spatial database files and one or more attribute files. The spatial database files contain the spatial information on each layer and a method of connecting the spatial entities both to each other and to the attribute data in the relevant attribute databases. The attribute databases are conventional databases that can be connected to the spatial databases. They contain all the attribute information for each area or spatial entity as contained in the spatial databases. The power of the GIS comes from the use of the spatial

[†]Chapter written in collaboration between Keith McCloy and Susanne Kickner.

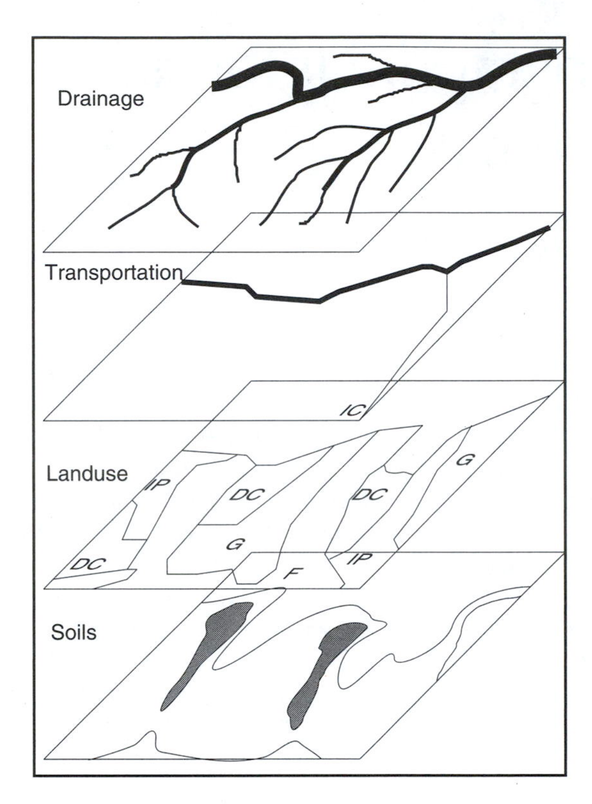

FIGURE 6.1 Layers of data on attributes of the environment that may be used in addressing some specific questions in resource management.

TABLE 6.1
Possible Fields in an Attribute File Describing Important Aspects of the Ownership of Each Land Parcel

Field Number	Name of Field	Format of Field	Comments
1	Parcel number	20 char string	Key field
2	Name of owner	50 char string	
3	Parcel address	120 char string	
4	Parcel area	Float	
5	Parcel valuation number	10 char string	
6	Parcel valuation	Integer	

and attribute databases to analyse spatio/temporal relationships and thus provide better information for spatio/temporally related resource management issues and tasks.

A GIS is defined as a computer-assisted system for the acquisition, storage, analysis, modelling and display of spatially related sets of resource data. Analyses applications across apparently different layers yield a better understanding of the physical, social, economic and other types of environmental relationships within the area of interest (AOI). Modelling applications provide management information and decision support.

There are many situations where the geographic relationships between different types of information are crucial in making good decisions. In these situations, a GIS is a critical tool in the proper management of these types of resources. Consider these questions:

1. Where in this county can I buy a farm that I can then use for raising pigs, if pigs require a certain quality and quantity of feed that will be produced on the farm, and they produce sewage wastes that can create serious pollution problems. A solution to this question not only requires information on the soils, topography, climate and landownership across the county, but also information on existing piggeries and models to show the amount of wastes that will be produced by each piggery so as to map areas that can absorb more wastes.
2. How can we best assess the biodiversity status of the state, which environments are threatened and assess the impacts of remediation? Existing landuse, landcover, climate, soil and topographic data can be used to identify those conditions that may be colonised by the different ecological associations in the area, that is to build a model of the potential for an ecological association based on the main drivers and historical landcovers. Once we have conducted this analysis, then we can use this model to show the areas that could have been originally been occupied by the different environments, and we can use this information, with existing cover information, and models of movement and migration to assess which environments are most at risk and the best ways to protect these environments.
3. What are the costs of the alternative routes for the proposed new road? The costs of a new road include the costs of the land used, the costs of construction of the road foundation, including excavation and embankment costs, the costs of the pavement and finally the costs associated with the impact of the road on the human activities in the area and on the environment. Often there are different costs and benefits associated with the alternatives. Thus, for a transport corridor, the lowest cost may well be through a wilderness area, but the corridor may be of benefit to but few in the community. Often spatially related decisions have to deal with cost/benefit ratios in finding an optimum solution. As with the costs, the benefits in terms of access and usage often has to be assessed using spatial data.

In all these cases, and many more, the final answer depends upon the spatial relationships that exist among some or all the parameters that affect the resources that are the focus of the question being posed. In each case you cannot just take the statistics and derive a satisfactory answer to the question. The answer depends on the spatial relationships that exist among some or all the resources that influence the solution or solutions. For these sorts of questions, a GIS will provide a much better basis for the formulation of a response than can be done using data that does not take the spatial inter-relationships into account.

Computer-based GIS have their origins in the early days of computers. Many of the systems developed in the 1960s were designed to address a specific task. For example, the Universities of Minnesota and Michigan both developed GIS to analyse environmental data, and to assist in the management of their states' natural resources. Both these systems were started as punch card, batch oriented, systems that depended on grid cell or raster data as the form of the spatial data that they processed. These systems suffered from the difficulty they had of creating large files since each layer had to be stored as a grid consisting of *n* lines, each of *m* cells. Very fine resolution raster files (that is small cell size), meant a large number of cells, incurring high storage and processing costs and creating difficulties in printing hardcopy products because of the finite number of lines, and cells per line, that could be handled by printing and display devices. Reducing the storage and processing costs to acceptable levels meant sacrificing resolution in raster files by adopting a larger cell size.

Other early developments were more concerned in drawing architectural or engineering drawings, or in the production of line maps, such as topographic maps. Systems with this type of focus started with linear features that were converted into line or vector data, where lines are retained in the computer as strings of (x, y) co-ordinates. Storage of vector data is more accurate, and at a higher resolution than that of raster data because there is little limit on the recorded (x, y) values. Vector oriented systems evolved into computer-aided design (CAD), computer-aided mapping (CAM) or GIS systems such as the Canadian Geographic Information System, which is acknowledged as being the first system to use the term GIS. Generally, these systems were developed most quickly in the 1960s and 1970s because the vector data was much less demanding of storage space, and easier to display.

A third important component in the evolution in GIS came from the concepts of the landscape as being made up of various layers, each representing an aspect of the environment. The concept then was that the environment in an area could be described by some combination of these basic layers. This concept has been incorporated to some degree in most existing GIS systems.

In the 1980s the cost of display and storage devices became significantly less. This technological development has allowed the raster-based systems to compete with the vector-based systems on more equal terms. In consequence, the 1980s saw the re-emergence of a number of raster-based systems for the analysis of spatial relationships.

To this day the GIS world is still affected by the cleavage between the raster and the vector-based forms of GIS, although most GIS software systems have attempted to bridge this gap in one way or the other. The reality is that both approaches to GIS have their advantages and disadvantages and so both are best for particular types of analyses. Raster GIS usually have an advantage where some of the more significant attributes can be described as being a continuous function of spatial location, such as soil physical and chemical attributes, topography, climate parameters and species distributions. When these parameters are used in the analysis, then the derived information will usually also be a continuous function with spatial location. Often the attributes vary in different ways across an area. Thus, soil pH, or soil acidity, may change in a somewhat different way to soil clay content, and both may change in somewhat different ways to the humus and silt contents of the soil. In general, each raster layer contains values for a single attribute. It is not essential that this be the case, and attribute files can, and are, used with raster GIS systems, in which case the raster layer contains values in one attribute from the attribute database, and the GIS can map the other attributes onto the raster layer when this is required by the analyst. It does mean that there needs to be a fixed relationship between the value in the key attribute and the other attributes in the table, that is, every possible combination of the attributes requires an entry into the table.

In a vector GIS the spatial features are of points, lines and two or three-dimensional polygons. Vector data can be very good where the data are of features that are defined by sharp boundaries or the features are either linear or point features such as landownership parcels, roads and other transportation corridors, power, water and other linear features, water and soil test or monitoring sites and other point features. Vector GIS are superior when the analysis is based on features that have simple geometric forms, such as lines and polygons but which may contain a lot of attribute information. Thus, questions concerning the optimisation of a power grid or a transportation corridor, upgrading of the water or sewage systems and so forth can be conducted much more readily in vector GIS than in raster GIS. In all cases of this type, the input and removals from the actual system are site specific, whilst the losses occurring throughout the system can be calculated, where this information is the basis of optimisation of the system. Unlike raster-based data, which are spatially rich but are usually attribute poor in each layer, vector-based data are spatially simple in each layer but they are often attribute rich. Not only may each layer contain attribute data in a file, but also these attribute files may be connected to other attribute data in other files. Thus, the attribute table for land parcels given in Table 6.1 could be linked to other databases, including:

1. A property valuation database
2. A taxation database for rates and taxes
3. Landuse and land zoning databases

4. Voter address databases for local elections
5. Databases associated with individuals living at the address

These combinations of information can form a very powerful tool for the analysis of specific issues.

Thus, many questions to do with landownership, population statistics and economic issues can be well addressed using a vector-based GIS. Vector GIS is also very good at creating zones around features, even though subsequent analyses may be done using raster-based techniques. Thus, if there is interest in the land uses in buffer zones around nature areas, then vector GIS is very good at defining such buffer zones, even though they can be readily converted into raster zones for analyses, if that is appropriate.

Of course there are many spatial issues that include both types of features, and set questions where some parts of the question are best addressed in one way, and other parts in another way. Because of this, it is likely that GIS will continue to include both types of data, and the analysis will focus more and more on the easy integration and analysis of data sets that contain both data types and deriving data that is in the most appropriate form for the subsequent analysis. The components of a GIS are shown in Figure 6.2 and include:

1. The spatial databases
2. The attribute databases
3. Data input systems
4. Data display and output systems
5. Data management
6. Data creation and transformation
7. Data analysis and evaluation
8. Modelling in a GIS
9. Connections to the Web

A GIS holds data for a section of the earth's surface. This means, amongst other things, that the data must bear a known relationship to the actual features on the surface, and all the data sets that are to be used together must have the same relationship. This relationship is known as a map projection, as discussed in Chapter 3.

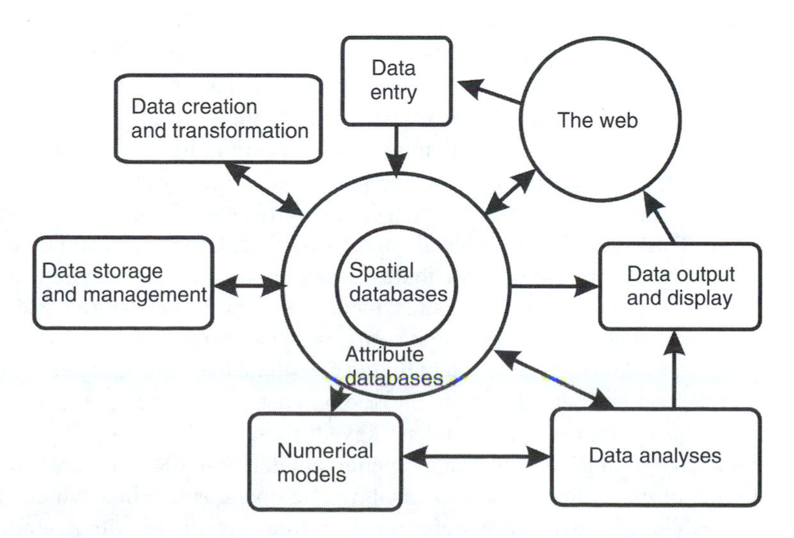

FIGURE 6.2 The components of a geographic information system.

6.2 DATA INPUT

Data in a GIS file is usually one of four types:

1. *Binary.* Binary data is either on or off. The digital values are either 0 or 1. A typical GIS layer that could contain data of this type would be a mask defining the limits of a study area, administrative region, etc. Each cell or pixel in the binary data can be stored in one bit of computer storage.
2. *Octal or char.* Octal data is stored in 8 bits, or a byte of computer storage. All characters in a computer are stored in a byte. A byte can hold one of 2^8 or 256 values, usually in the range 0 to 255.
3. *Integer.* Integer data is stored in 2 bytes, so that the data can take one value in the range 0 to 2^{16} or 65,536. However, storage of 2^{16} possible values does not allow the use of a sign, so that data are called unsigned. If the data can take a positive or negative sign, then the sign takes one bit leaving $\pm 2^{15}$ or $\pm 32{,}768$ possible signed values.
4. *Float.* Each data value is stored in 4 bytes.

It can be seen that the data storage demand increases from the binary data through each type to the float that is the most expensive in terms of data storage demand. Thus, float data should only be used where absolutely necessary.

There are two types of data that need to be entered into and which will be used in a GIS: spatial and attribute data. The spatial data are those data that define geographic locations and at least one attribute of the data, where that attribute may be a key attribute that then points to other attributes. The attribute data describe or quantify a condition or quality of the object or the mapped surface. Both types of data are held in databases, spatial and attribute databases in turn.

In practice, it is usual to form the spatial databases first and then build the attribute databases. However, since they all use databases, we will deal with the nature of these databases, and the attribute data first. After that we will deal with the creation of the spatial databases.

6.2.1 DATABASES AND ATTRIBUTE DATA

The database structure of raster and vector-based data layers are typically quite different in a GIS. In a raster GIS each layer consists of a rectangle of $n \times m$ cells or pixels, like in an image, consisting of a set of n lines, each containing m columns. The data are normally stored line by line for the n lines, with each line an array of m data values representing the cells in that line. As is the case with image data, the line and column co-ordinates are stored implicitly in the structure of the data. The header information for the layer will contain information on the projection, the co-ordinates of a corner of the layer, the number of rows and columns and the pixel size in the layer, so that the co-ordinates can be computed for any pixel in the layer. The value recorded in each raster cell is the cell value for one specific attribute. Thus, a raster layer on soil colour would contain the cell soil colour as the cell attribute value. One way to store different attributes is to store a layer for each attribute. However, when there is a relationship between the attributes, then an alternative is to store these relationships in a table or a database and create one layer that contains as its attribute value a key into either the table or the database. Consider the data shown in Figure 6.3 where the soil classes have been itemised in a database, with the soil attributes for each class being entered into the appropriate fields in the database. The spatial file contains just the database key attribute.

GIS systems can take such a combination of spatial and database files, and paint the spatial file with one of the other attributes for analysis or display. The operator specifies which attribute is to be used in the analysis or display. The system, on accessing a pixel, gets its key attribute value, finds the relevant value of the specified attribute from the database or table and then paints this value

12	15	6	6	6	6
12	12	6	6	6	6
12	12	12	6	6	6
1	12	12	6	1	6
4	1	12	1	1	6
4	1	12	12	1	6
4	1	12	12	12	6

Spatial file containing key attribute values

Key	pH	%Sand	%Clay	Next field
1	07.2	25	15	
2	07.4	22	17	
3	07.4	20	16	
4	06.9	25	12	
5	07.0	25	13	
6	06.5	18	24	
7	07.5	20	23	
8				
9				
10				

Relational database file showing some fields and some entries

FIGURE 6.3 A spatial file containing the key attribute to a database of soil attributes.

into the pixel attribute value. This new spatial layer can be displayed, used in analysis or saved as a separate layer from the source layer.

With vector GIS data, there exists at least one spatial and several attribute databases. The spatial database contains all the point, arc and polygon data. This data is connected to the attribute database through a key attribute that is common to both databases.

The attribute data, whilst being connected to the spatial data through one field, or the key field in the database file, is of much more conventional form. Databases are normally of one of the forms:

1. *Tabular or flat file*. These are simple lists of data in a file, without any information on their relativity or connectivity.
2. *Hierarchical file*. In a hierarchical file structure, each higher-level value has one or more entries below it, that respond to it. There are no lateral connections in a hierarchical file. Typical hierarchical files are the species hierarchies in botany, and landownership information (country–county–sub-division–land parcel details). In all these cases there are no connections or links necessary in the lateral directions across from, say one land parcel to another.
3. *Network file*. In a network file the lower-level entries can connect to more than one higher-level entry. Thus, in a school the students academic records may be required by the different faculties, as well as by the student administration.
4. *Relational file*. This is the most flexible of the forms of database and is the form normally used in a GIS. A relational database consists of: (1) one or more files, (2) linkages between these files, (3) queries that can be made on the files and forms that can be made and printed from the files. Each database file consists of a variable number of records and each record contains a fixed number of fields. Each field is of a specific type set up when the file is designed, and it will contain data on that attribute as is relevant to that instance or example of the class or feature about which information is stored in the file. The set of fields thus contain the information that is stored on those instances that are recorded in the file, each instance being recorded in one record of the file. Thus, the information in a database file on landownership may contain the information given in Table 6.2.

When the user needs to add information to the file, the information is added for each field in the record. The record is then automatically appended to the file. The file structure shown in Table 6.2

TABLE 6.2
Contents of Two Interlinked Relational Database Files

Field Number	Description of Contents	Type of Information
Data on cadastral parcel of land		
1	Reference Number (Key Field)	Text (50 characters)
2	Parcel number	Text (50 characters)
3	Owners name	Text (50 characters)
4	Owners address #1	Text (25 characters)
5	Owners address #2	Text (25 characters)
6	Postcode	Integer (5 characters)
7	Area of parcel	Float (8.3 characters)
Database on the attributes of the postcode area		
1	Postcode	Integer (5 characters)
2	Area	Float (10.3 characters)
3	Number of farms	Integer (8 characters)
4	Number of businesses	Integer (8 characters)
5	Number of residents	Integer (8 characters)

The first contains data on a cadastral parcel of land, the second is a linked database on the attributes of the postcode area.

is an attribute file as it contains information describing some characteristics of a legal parcel of land. It does not contain a lot of other information about the parcel, for example, the soil distribution across the parcel, the elevations across the parcel, or how the land has been used within the parcel. These other types of information may well be the subject of separate attribute files. If this is the case, then they could be connected to this file through a common parameter. The common parameter that would normally be used is the Reference Number (Field Number 1), although other parameters can be used. Thus, a file on fields would logically use Field 1 as the connection, whereas information on the financial status of the owner may well use the owner's name as the connecting field. It is good database design to use the same field as the connection between all files wherever possible.

This attribute database file is also connected to the spatial database through the Key Field. This key field number will appear in the polygon file of the spatial database. The polygon database will contain this Reference Number as its Field #1, the numbers of the arcs that surround the polygon and the co-ordinates of a centroid to the polygon. The numbers of the arcs are then used as connections into the arc database file that will contain information on each arc and its co-ordinate pairs. The structure of the spatial database is discussed in more depth in the next section.

If the analysis is concerned with landuse and landcover information, then there will be an attribute file of landuse and landcover that will contain the Reference Number, the Polygon Reference Number and fields for landuse and landcover.

One of the strengths of databases is their ability to provide linkages between the databases and to query the databases. One form of linkage has been discussed. Other forms include the capacity to automatically update interconnected files when one file is updated. Queries allow the analyst to transform the data in the file and to address a range of questions that may require some form of analysis or summary of the information contained in the file.

Current trends in database design are towards the development of object oriented data models. In this approach each object type is defined as having a set of properties, behaviours and relationships. An object is an example of a specific object type, in which case the properties, behaviours and relationships are given specific values. For example, a house may be an object type, with properties of location, external dimensions, building material used, roof type and so forth as appropriate. It may have a price behaviour mechanism as well as water and power consumption behaviours. It may have relationships with other buildings, with its access road and with the services that come to it. When forming or using an object type, an object type database structure is established. When each object of the object type is created or declared, a specific example of that type of database is created. Since the behaviours and relationships are held in the databases, the relationships between objects, not only those of the one object type, can be established and maintained.

Object oriented data models have been adopted by a number of GIS systems, including ArcInfo. With most of these GIS a number of object types are predefined within the software, and the user can define others as required.

6.2.2 Creation of Spatial GIS Layers by Digitising from an Existing Map or Image

We will now deal with the spatial files in vector GIS, by starting with their creation by data acquisition and then with the formation and structure of the files. First we must consider the vector data model as is relevant to the collection and analysis of vector data, before we consider the acquisition of vector data.

At the simplest level, a vector GIS consists of sets of points or lines located in a co-ordinate system for each theme of data in the GIS, and the attribute databases associated with these spatial geometric entities. Lines and junctions have been given various names. They were called "edge" and "vertex" by graph theorists, "chain" is the word officially sanctioned by the U.S. National Standard and the terms "arc" and "node" are used by many GIS today. In contrast to graphics and CAD software, arc and node data in a vector GIS must be "intelligent." This means that the arcs and nodes are connected with topological information. For example, arcs have an attribute of the arc length, attributes which identify the polygons on either side of the arc as well as attributes that identify the nodes at the ends of the arc. An arc can represent a line object like a river or a boundary of a polygon. A point can be the digital representation of a point object like viewpoints or spot heights or they can represent objects such as houses and windmills. A point can also be a label point. A label point is a point positioned inside a polygon. In this case a polygon is defined by a series of arcs comprising its border and by a label point. Polygon identification number and attributes of polygons are assigned to the label point. Besides the topological features a vector GIS can store a lot of attribute data, which can be used to select and analyse spatial objects. GIS that store features in vector format, are preferred in urban applications where legal boundaries, high accuracy and the analysis of networks are important.

In comparison with raster systems, vector systems requires complex data structures but have the following advantages:

- Points, lines and areas are stored at an operator selected precision in the form of co-ordinates.
- The map output is of high quality.
- Vector data requires less storage space and the maintenance of topological relationships is easier.
- Areas are easy to overlay with other areas, points and lines.
- By overlay functions new objects can be formed.

The spatial data of arcs and points that are associated with a theme are saved in single layers called coverages in a vector-oriented GIS. The layer concept is a formal structure method for data with a spatial geometric reference. A thematic data layer in a GIS should be as homogeneous as possible. It is important that only objects of the same geometric type such as points or arcs and their polygons, are contained in the one layer. If a theme contains different types of spatial objects, then they need to be held in different layers. Consider a landuse theme that includes grassland areas as polygons and individual trees as points. The polygons need to be stored in one layer and the points are stored in a second layer. For the grasslands, a polygon object geometry is defined whilst a point object geometry is defined for the individual tree layer. These layers are usually put on top of each other in the cartographical representation.

All the layers must be stored in the same map reference system if they are to be analysed together or overlaid for display. This means that the following details must be identical for these layers:

- The spheroid and its origin, for example, Bessel or WGS84.
- The map projection type and details, for example, the UTM and its Euclidean co-ordinates.
- The units of distance, for example, degrees, meters, miles.

Database creation involves several stages: input of the spatial and attribute data and after this linking spatial and attribute data. Spatial data is entered via digitised points and lines, scanned and vectorised lines or can be imported from other digital sources. Once points are entered and geometric lines are created, topology must be "built." Topology in GIS practice means primarily to constitute the relations to other objects (neighbourhood relations). General basic principles must already be taken into account at the digitalisation to be able to build up a topology later:

- A node must be set at each intersection of two lines
- Lines which have a common node must meet it exactly
- Common borders of areas are digitised once
- Areas must be closed

A prerequisite to the construction of a correct topology is a fault-free recording (digitalisation) of the data. Prior to the construction of a topology you must check and clean up or edit any polygons that do not close. This can happen if lines were digitised too short (undershoot) or too long (overshoot) as shown in Figure 6.4. To ensure that the nodes — like start and end node of a polygon boundary — coincide exactly from digitisation, a snap distance is fixed in the GIS. If a digitised node is within the snap distance of another node then both nodes automatically "snap" together. The snap distance must be adapted to the density of the objects and window size, otherwise different nodes may unintentionally snap together, if the snap distance is too large as shown in Figure 6.4.

A border between two polygons must be digitised only once. After the construction of the topology it is known which arcs form the boundaries of which polygons, and how the arcs are arranged around the polygons. The topology is constructed after digitisation and editing by either:

- Directly online during the digitalisation, or
- By special functions in the GIS

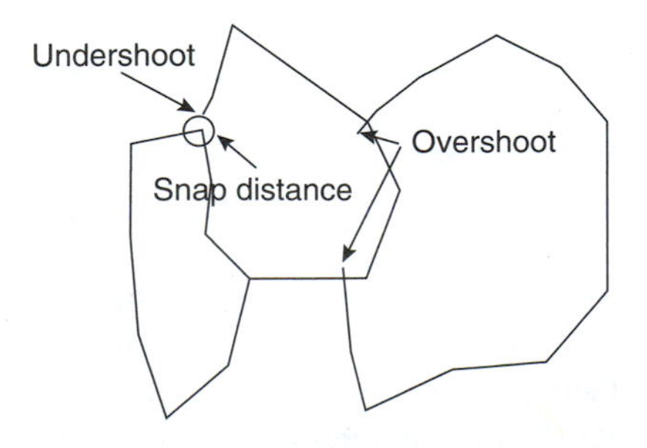

FIGURE 6.4 Typical faults at digitising.

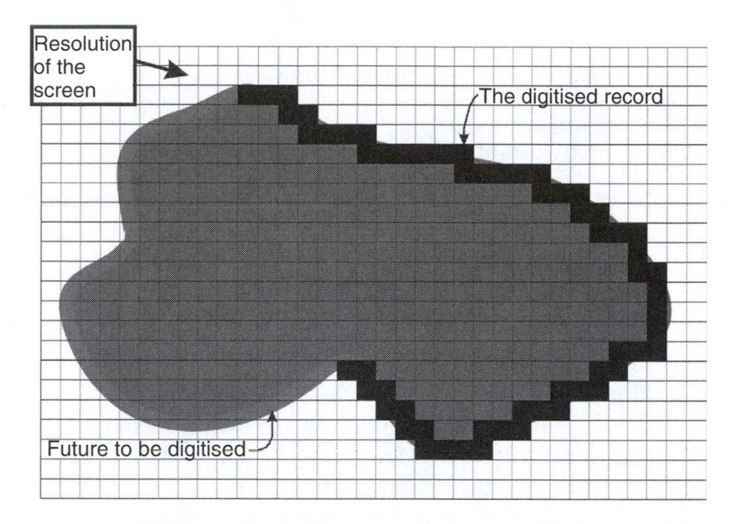

FIGURE 6.5 Digitisation of data on a video screen showing the relationship between the screen pixels, the feature to be digitised (grey) and the stored digitised data (black). The figure may be considered to be misleading since the digitising is always done at the finest spatial resolution that can be achieved on the screen. Even if the pixels are magnified so that they look like those depicted in this figure, the digitising will appear to be a continuous curve since the set of (x, y) co-ordinates that are stored at the finest achievable screen resolution. What is depicted in the figure is how the digitising would appear if it is digitised at one scale and then subsequently greatly magnified.

Spatial data can be input to a GIS in four ways:

- Digitisation from an existing map or image
- Field collected data and observations
- Processed results of image processing
- Drawn from existing spatial databases

Digitisation is the process of creating a digital spatial layer from an analog map or image. It can be done by:

- Manual digitising on a viewing screen
- Manual digitising by the use of a digitising tablet or table
- Scanning the map or image to record a digital image of the scanned map or image

6.2.2.1 Manual Digitising on a Viewing Screen

This is the simplest and easiest method of manual digitising. The source map or image will often be a digital image, such as an orthophotograph or rectified satellite image. An orthophotograph is a photograph that has been rectified and corrected for height distortions so that it represents the areas of the earth's surface as a projected plan at a constant scale. This image is displayed on the screen. The cursor is used to follow the features that are to be digitised. As the cursor moves, the software records its position creating a vector spatial file of the digitised data (Figure 6.5).

Figure 6.5 shows that the video screen consists of a rectangular area of cells or pixels of a finite resolution. When the feature of interest is displayed on such a screen, that feature is shown at the resolution of the screen. The operator, in following this feature attempts to do so as accurately as possible, but of course errors will occur. Finally, the screen resolution and the scale of the image as depicted on that screen control the maximum resolution or accuracy that can be achieved in the digitisation. Thus, higher resolution can be achieved by magnifying the image on the screen.

The data recorded in this way is in the same projection as the source image or map. It is therefore important to use rectified images for this type of digitising.

6.2.2.2 Manual Digitising by use of a Tablet or Table

A digitising tablet or table consists of a table and controller attached to a computer. The table contains a fine electrical graticule or grid of parallel and vertical wires embedded within the table, usually at a resolution of about 0.05 to 0.2 mm, representing the finest resolution data that can be recorded from the table. The controller consists of a cursor and controller keys. The cursor is a glass or Perspex window containing an etched cross-hair and electrical loop. The controller handle usually contains between four and 16 cursor keys that, by sending different signals to the computer, can control the operation of the computer. As the cursor is made to follow a line by the operator, pressing the appropriate cursor button completes the circuit containing the cursor and table. The intersection within the graticule of wires in the table closest to the cursor is identified. This intersection is recorded in the computer as digitiser (x, y) co-ordinates. As the cursor is moved along the line, a series of co-ordinate sets are recorded, where the mode of recording the data can be chosen by the operator from the alternatives discussed below. As the data are digitised they are displayed on a graphics screen so that the operator can see how the work is progressing.

The digitiser co-ordinates recorded in this way need to be converted into map co-ordinates before use, since the digitiser co-ordinates will change if the map is positioned differently on the table. Recording the digitiser co-ordinates of at least two and preferably four control points starts digitising. The control point co-ordinates and the digitiser co-ordinates are related through a rotation, translation and change in scale if the control point co-ordinates are in a Cartesian co-ordinate system. In this case the transformation is of the form:

$$\text{CONTROL } X = A + B \times (\text{digitiser } x) - C \times (\text{digitiser } y)$$

$$\text{CONTROL } Y = D + C \times (\text{digitiser } x) + B \times (\text{digitiser } y) \tag{6.1}$$

Whilst this transformation requires a minimum of two control points because of the four unknowns (A, B, C and D), it is good practice to use four points located near the corners of the area being digitised so as to have redundancies in the data for error checking. The best control points are grid intersections as these are easily seen on the map, and their co-ordinates are precisely known. With most topographical map series the grid co-ordinates are Cartesian co-ordinates and so can be used in this way quite satisfactorily. The digitiser co-ordinates are transformed into the same projection as the map co-ordinates using the formulae above. When digitising a map, it is best to use either the base map on plastic film, or a new copy of the map that has never been folded. Take care to ensure that the map used does not contain distortions, as can occur during folding, or when a map is or has been wet or torn. These distortions will create errors in the digital record.

The strings of map co-ordinates (x, y), are then stored as vector data in which the vectors represent either points, linear features or the boundaries of area features.

Digitising can normally be done in a variety of modes. Three modes that are typical include:

1. Point mode
2. Stream mode
3. Continuous mode

6.2.2.2.1 Point Mode

Once the cursor is positioned over a point, the relevant cursor button is pressed to record the co-ordinates for that point. The method is slow but accurate. It is suitable for digitising control points, individual point features such as houses and linear features that consist of straight stretches, such as landownership boundaries, by digitising just the changes in direction along the boundary. Digitising by point mode leads to the most efficient and accurate storage of data.

6.2.2.2.2 Stream Mode

Once the cursor is correctly positioned the appropriate cursor button is pressed and points are digitised at a constant time rate or frequency. Points are thus recorded automatically at the set time interval as the cursor is moved around the features of interest. With some systems the time rate can be adjusted, to allow for differences in operator experience and variations in the complexity of the information being digitised. This approach often leads to excessive storage of data, but it allows the experienced operator to vary his digitising pace with variations in the complexity of the information being digitised. It is a good way for experienced operators to digitise irregular features such as creeks and contours.

6.2.2.2.3 Continuous Mode

Similar to the stream mode except that points are digitised at a set spatial interval rather than a set time interval. This mode does not allow for variations in the complexity of the information, but can reduce the amount of data stored by inexperienced staff.

Whilst the names of the modes used here may be unique to one GIS, most GIS have similar modes, even though they may use other names. Vector data can be stored in four ways (Figure 6.6):

1. Points
2. Lines
3. Areas

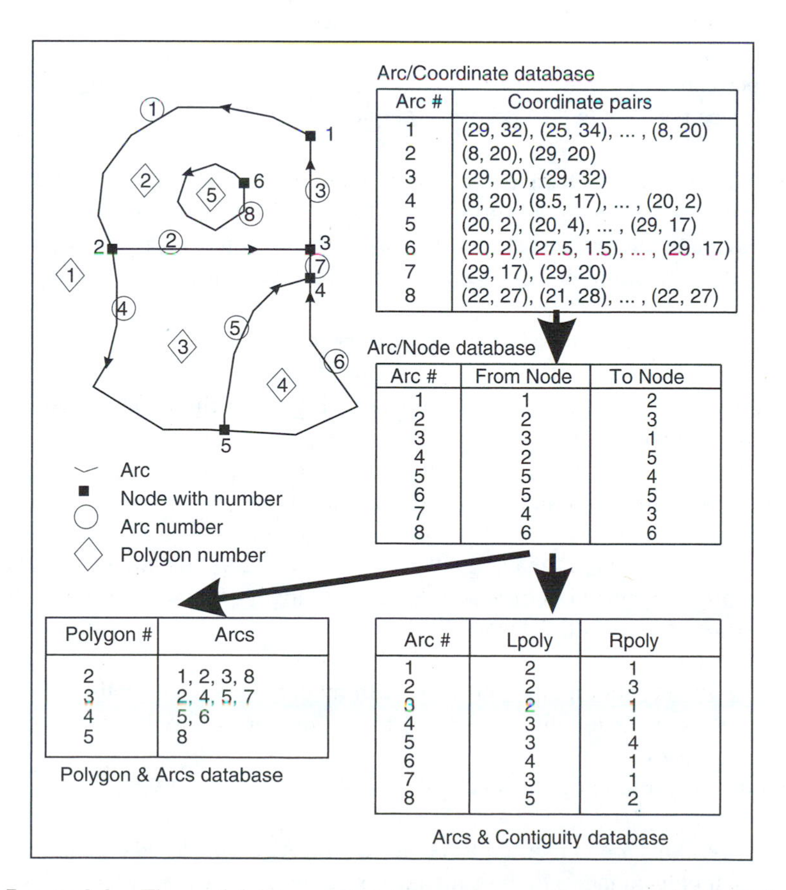

Arc/Coordinate database

Arc #	Coordinate pairs
1	(29, 32), (25, 34), ... , (8, 20)
2	(8, 20), (29, 20)
3	(29, 20), (29, 32)
4	(8, 20), (8.5, 17), ... , (20, 2)
5	(20, 2), (20, 4), ... , (29, 17)
6	(20, 2), (27.5, 1.5), ... , (29, 17)
7	(29, 17), (29, 20)
8	(22, 27), (21, 28), ... , (22, 27)

Arc/Node database

Arc #	From Node	To Node
1	1	2
2	2	3
3	3	1
4	2	5
5	5	4
6	5	5
7	4	3
8	6	6

Polygon #	Arcs
2	1, 2, 3, 8
3	2, 4, 5, 7
4	5, 6
5	8

Polygon & Arcs database

Arc #	Lpoly	Rpoly
1	2	1
2	2	3
3	2	1
4	3	1
5	3	4
6	4	1
7	3	1
8	5	2

Arcs & Contiguity database

FIGURE 6.6 The spatial data and its databases that provide topology in the GIS.

6.2.2.2.4 Points

Each point is independently stored in the data file. Attribute value(s) may be either recorded with each pair of co-ordinates in the file, or they may be recorded once for a set of site co-ordinate pairs, for example, the sites of all houses. Point data is appropriate for many types of point observations, such as soil tests, climate station records and so forth.

6.2.2.2.5 Lines

The arcs between segment joins or nodes are digitised once and stored as separate entities. Each arc thus constitutes the nodes at either end and the intervening digitised points or vertices. It is necessary to specify the relationships between the arcs, between the arcs and the polygons, and between the polygons to provide full topology to the data. Topology in a GIS is discussed further below. Polygon information is stored either as a set of nodes that specify the arcs, or as lists of arcs required to construct the polygon. This approach does not incur the errors or costs of double digitising and minimises storage costs, but it incurs some additional processing costs in reconstructing the polygons. Arc-node storage is suitable for storing all three types of data, point, line and boundary. It is thus the best of the currently used methods of storing vector data.

6.2.2.2.6 Areas

Polygons can be created either by direct digitising or from arc-node data. The first approach is inefficient, since there is duplication of digitising, and will contain inconsistencies because these digitisations will not cover exactly the same path. It is thus a poor method to adopt in practice. The second approach avoids these disadvantages.

Topology is the process of establishing the relationships between sets of data. Topology thus defines the way the spatial data items are interconnected and how these are interconnected to the attribute data. Topological data and information is crucial to some GIS processes and it provides the GIS with the capacity to conduct more intelligent analyses that require information on the relationships between objects in the GIS, for example:

1. To find all farms larger than 10 ha in size that are bounded by major rivers, requires topological information on the attributes of the arcs and the arc-polygon topology.
2. To find all landholdings of between 0.1 and 5 ha within 10 km of all towns requires topological information on the landuse attributes of polygons.

To establish topology during digitisation it is necessary to:

1. *First digitise all features.* This is done by first digitising the points of an arc and then specifying its connections to other arcs and to the polygons that it bounds.
2. *Construct the topology* by specifying the point and arc connections as well as the relationship between the arcs and the polygons, particularly the polygons on either side of an arc.
3. Identify and rectify errors such as gaps that have occurred in the digitising. The best way to ensure proper connection of arcs is to use a snapping facility to snap to the adjacent feature, or to the node of that feature.

Whilst this is a strict description of the process, with modern GIS the establishment of topology is automatically established as the digitising takes place, significantly reducing the costs associated with the digitising process.

There are three major topological concepts:

1. *Connectivity.* Arcs are connected to other arcs through nodes, where an arc is a vector segment that only connects to, or is intersected at, its end, that is, at the nodes.

2. *Areality*. Areas are bounded by polygons that consist of one or more arcs.
3. *Contiguity*. Arcs have direction and a left and right-hand side.

6.2.2.2.7 Connectivity

Each arc consists of an arc number and a series of (x, y) co-ordinate pairs that defines the shape and the length of the arc. These are written into a database file, similar to that shown in Figure 6.6. This figure shows a very simple situation, with eight short arcs of up to nine vertices in each arc. The end vertices are called the node points to which this arc connects. The nodes are also stored in an arc-node database that is created during the construction of the topology (Figure 6.6). The common field to the arc-co-ordinate and the arc-node databases is the arc number. The arc-node database enables the analyst to link the arcs together and to see which arcs connect to which other arcs.

6.2.2.2.8 Areality

The polygons can be stored either as a set of (x, y) co-ordinates extracted from the arcs, or as a set of arcs. The second method is the most favoured because it involves the storage of less data and it is less error prone in amending the data. Figure 6.6 shows this component as the polygon–arcs database, which may contain other attribute data about the polygons, such as the polygon area.

6.2.2.2.9 Contiguity

Contiguity enables the system to identify which polygons are adjacent to each other. Contiguity is established by creating a left polygon — right polygon database. The database shows that polygons 2 and 3 are at either side of arc 2, hence they are adjacent polygons. The same applies to polygons 3 and 4 around arc 5. The label for polygon 1 is outside the area. This polygon is the external polygon representing the area outside of all the polygons in the digitised area.

A digitised spatial layer is conventionally called a coverage. So far we have seen how to create the spatial boundaries of points, arcs and polygons in a spatial database, but we have not assigned attributes to those databases. Each type of spatial feature represents different types of features in the environment, and as a consequence they will have different attribute databases associated with them. Typical types of information that are represented by the different types of spatial features are listed in Table 6.3. This table listed another type in addition to those mentioned earlier, called labels. A label is a type of point location that is defined as a location in a polygon, and it is used to link the polygon with the attributes of that polygon.

6.2.2.3 Scan Digitising

A scanning device senses the density in a strip or scanline across the photograph or map, and converts that density into a digital value for each of a finite number of elements, or pixels, across the line. A series of lines are recorded to scan digitise the whole of the image. Conceptually, the process is the

TABLE 6.3
Spatial Features and Typical Examples of Their Use

Spatial Feature	Typical Uses of This Feature
Point	Wells, soil test sites, accident sites, crime sites
Arc or line	Roads, power and other transmission lines, rivers
Polygon	Land parcels, soil types, landuses, administrative regions
Label	Polygon identification
Nodes in an arc	Supply and demand points for a service, impedance in street networks

same as scanning the environment with an imaging scanner. In scanning a photograph, the density within the field of view (FOV) controls the amount of energy that enters the scanning system. This density depends on the range of densities in the photograph and the illumination of the photograph. Most scanners these days are flat bed scanners, although some drum scanners exist.

The scan-digitised data is in raster form with the size of the grid cells being set by the scanning. The raster data produced by a scan digitiser has pixels of a specific size that may not be at the desired location or orientation. The scanned data needs to be resampled as discussed in Chapter 4 so as to create raster data at a suitable resolution, position and orientation. If vector data are required then the resampled image can be interrogated using line following software to identify the linear and point features in the scanned data. Line following routines can be confused, particularly if the scanned data contains a lot of noise. In consequence, the line following is usually followed with editing to clean up the final record, and to add text and other details not included in the line following phase of the work.

6.2.3 Field Collected Data and Observations

Point observations must include co-ordinate and time data as well as other types of data, or attribute data. The attribute data will depend on the nature of the point data but it may be weather data (rainfall in units of time or accumulated rainfall, temperature, humidity, wind direction and velocity), soil data (soil pH, humus content, moisture content and friability) or some other type of data.

When recording field data, it is important to include:

1. *The co-ordinates of the location*. It is normal to acquire the co-ordinates by means of global positioning system (GPS) instruments that will be discussed below. These co-ordinates should be recorded in a consistent co-ordinate system and preferably the one that is being used for the entire project mapping in the GIS.
2. *The time of the observation.*
3. *The attributes to be recorded.*

This data can be entered into the GIS, in a number of ways:

1. *Directly into the spatial database.* If the data to be entered into the GIS simply includes the co-ordinates and the value in one attribute, such as height above sea level (ASL), then this is an appropriate method of entry. It is a common approach when a GPS is linked into the GIS so that the GPS co-ordinates can be entered directly into the spatial database file. It can be very convenient when, for example, the operator is moving from place to place recording just landcover or landuse, and he/she requires that the location and the information are stored in the GIS.
2. *Directly into the spatial and attribute databases.* This is suitable when a number of attributes are to be recorded at a location, such as elevation, time of day, landcover type and condition. However, it requires more work by the operator in the field.
3. *Entered into a database in the field, and subsequently into the GIS spatial and attribute databases.* This may be the most appropriate approach when the data will require additional processing prior to incorporation into the GIS, or where the databases require the addition of other fields that may come from other sources. Thus, field soil samples will need to be tested in the laboratory prior to the incorporation of the additional laboratory results with the other data in the database.

Global positioning system is a method of determining the (x, y, z) co-ordinates of the GPS instrument above the ground surface. The GPS instrument is usually a hand-held instrument that uses trilateration based on the known co-ordinates of a number of satellites, at least three, at the time of observation,

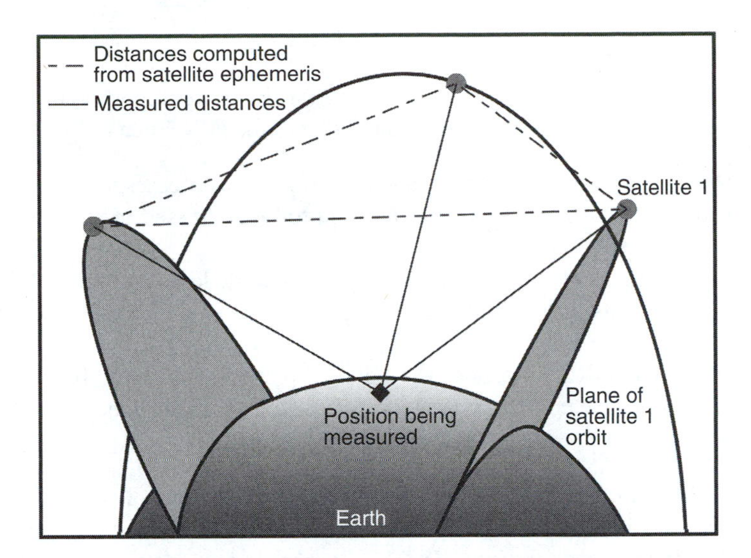

FIGURE 6.7 The relationship between a location on the Earth's surface and the satellites being used to fix its location using GPS. Only three satellites are depicted, whereas at least four should be observable before taking observations of location. The satellite ephemeris is the satellite orbit characteristics. The three distances between the satellites are calculated using the calculated ephemeris data and the three distances to the ground stations are observed prior to calculation.

and the measured distances from the satellites to the GPS receiver, as shown in Figure 6.7. The satellites are travelling on orbits whose characteristics are known very accurately, using a network of ground stations for the continuous monitoring and measurement of their path.

The orbits are all at about the same distance above the earth, but are at various orientations. Even though there are more than 24 such satellites in orbit at any one time, because of earth curvature, and the need to have the line of sight between the satellite and the ground station more than 10° above the horizon, there are rarely more than 6 to 8 satellites that are accessible at any one time. The signal from the satellite to the ground station travels at the speed of light, but this speed varies slightly with atmospheric density, hence small errors are introduced when the satellites are close to the horizon. In addition, the solution of the triangles is best when the triangles are well conditioned, by which is meant that the internal angles to the triangles should be greater than about 30° and preferably about 60°. In addition the satellites should not be distributed in a plane, but define volumetric triangles, with the base in the position of the satellites and the apex at the GPS receiver.

The distances between the satellite and the ground station are measured simultaneously to at least three satellites, and preferably between four and six satellites. The distances are measured by the length of time that it takes a pulse to travel from the satellite to the ground station. Knowing that the radio waves travel at the speed of light enables the distance to be computed. The satellites are travelling at very high speeds, so the observations are taken, virtually simultaneously, and many observations are taken so that they can be averaged. Knowing the position of the satellites, and having the three or more distances measured from them to the ground stations means that the triangles involved can be solved to find the location of the ground station relative to the satellite co-ordinates. With four or more satellite observations, this can be done by least squares. The solution has to be performed in spherical trigonometry, and the derived spherical co-ordinates then have to be converted into the selected map projection co-ordinates or latitude and longitude.

With most GPS the co-ordinates and other details can be recorded in the GPS, in a connected computer or in a data logger. The data can thus be directed into a GIS as it is recorded, or this can be done later. The accuracy of most GPS is given for those systems and is usually satisfactory for most GIS work.

6.2.4 Information from Image Data

Whilst images can be used as a backdrop in a GIS, by far the more important use of image data is the incorporation of information derived from that data into the GIS. Such information could be:

1. Classification of landuse or landcover. Such information addresses the question, "What is that?" It is the first and most important question to be addressed in relation to most land resource management issues.
2. Estimation of surface conditions, where these can include leaf area index (LAI), biomass or soil moisture status. Information of this type addresses the second important question of "What is its condition?"
3. Analysis of Dynamic Response, or "How does it respond to forcing functions?" This is the third important question to be dealt with in resource management, and is relatively undeveloped in remote sensing at this stage.

Information derived from image data, by its nature, relates to the time of acquisition of the image, the season of the cover or landuse or to a period of time if dynamic analyses have provided the information that is being incorporated into GIS. Thus, remote sensing data can provide very extensive yet consistent information at a date, when acquired from one image (for example, LAI or biomass data), over a season (such as landuse classification that uses a series of images through the season) or over a longer period (when assessing the dynamic processes as recorded in image data). The derived information will only be accurate when it is carefully integrated with the appropriate field data and knowledge. Thus, image data extends the reach of the professional environmental officer, agronomist, soil scientist or other professional groups concerned with the management of natural resources, rather than replacing them. Only by the use of image data can much of this type of information be made available for use in a GIS.

The information derived by image processing from remotely sensed images is in a raster form, but it can be converted into vector data either in the image processing software, or in the GIS. However, the resolution of the image data limits the realistic resolution of the data in the GIS. Raster files derived from remote sensing have to be rectified and geocoded before they can be used in a GIS as this allows the different data sets to be registered one with the other. The standard resolution and coverage of GIS files is given in Table 6.4, although other resolutions may be adopted for specific purposes.

Whilst standard map coverages need not be used as such, none the less for many routine tasks partitioning and maintaining the data in this way will lead to easier and more efficient storage and retrieval of GIS data from the data management system. GIS raster files can also be very large. Since there is often a lot of redundancy in the information contained within the GIS raster files, there is considerable scope to reduce data storage needs by the use of data compression techniques.

TABLE 6.4
Relationship Between Scale, Pixel Size, Map Coverage Area and the Number of Pixels in the Image

Map Scale	Typical Map Coverage	Pixel Size	Number of Pixels in a Coverage
1:1,000,000	6° longitude by 4° latitude	1×1 km^2	260,000
1:250,000	1.5° longitude by 1° latitude	100×100 m^2	1,600,000
1:100,000	30′ longitude by 30′ latitude	50×50 m^2	1,100,000
1:100,000	30′ longitude by 30′ latitude	25×25 m^2	4,300,000
1:50,000	15′ longitude by 15′ latitude	25×25 m^2	1,100,000
1:25,000	7.5′ longitude by 7.5′ latitude	10×10 m^2	1,700,000

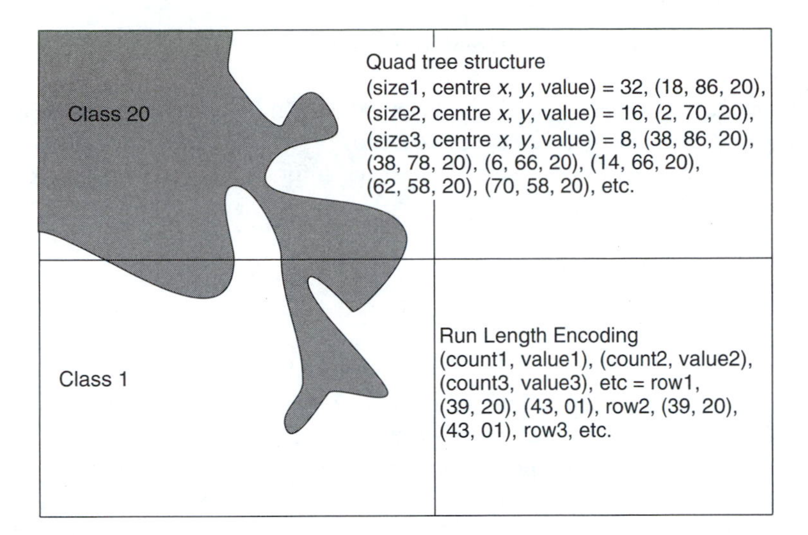

FIGURE 6.8 Quad tree and run length encoding techniques for spatial data compression.

Inspection of the GIS raster data files in Figure 6.8 will show that many of the cells in the GIS layers have the same values, creating data redundancies since the data can be reconstituted from a much smaller data set. Compression of the data to eliminate data redundancies will reduce the storage requirements for the data. Whilst there are many techniques for data compression, two are commonly used within GIS raster systems:

- Quadtree data structures
- Run length encoding

6.2.4.1 Quadtree Data Structures

In the quadtree data structure, the raster data are analysed to find the largest square that is a multiple of 1, 2, 4, 8, 16, . . . etc. cells within each area of the same values in the layer. Once the largest square has been found it replaces all the equivalent individual cells. The area that remains around this first (and largest) cell is then analysed again on the same basis. In this way an area of the same values is stored as a series of cells of varying size, but with sizes that are related to each other. This approach significantly reduces data storage without affecting the quality of the information.

6.2.4.2 Run Length Encoding

If a line of data contains a run of contiguous cells with the same values then storage savings can be made if the run of values is replaced with the value, and the number of occurrences. Storage of the data includes the number of occurrences of a value, and the value, at each change in value along the line of data. It can save a lot of space when there is high redundancy, but will cause higher storage costs if there is little data redundancy.

6.3 SIMPLE RASTER DATA ANALYSIS IN A GIS

One of the first things that you are likely to want to do is display some GIS data. After that you will start to analyse that data in various ways. In this section we will consider:

- *Displaying data.* The types of data that can be displayed and the selection of a layer, the scale of the data and the colour table to use.

- *Overlaying layer*. Overlaying different forms and types of data of an area for display and analysis.
- *Combining layers* Numerical and logical methods of combining layers to create new layers.
- *Filtering and neighbourhood analyses*, to remove "salt and pepper" effects and to derive information that is a form of summation of the information in a neighbourhood set of pixels.
- *Distance and contextual analysis*.
- *Cost distances and least cost pathways*.
- *Statistical analyses*.

By convention, in an equation, $y = a + b \times x$, then x is an independent function variable, or variable in short, which may have a variable value or value, and a and b are function parameters that may have constant values. Whenever an actual software command is written as a command in the text, such as **PRINT**, then it is written in upper case bold. When it is used as an English word in the text, for example, as a heading, then it will adopt normal written text conventions. Command parameter and output parameter values are written in lower case bold, for example, **ADD(image1, image2**).

6.3.1 Displaying Data

There are usually four questions that need to be answered for a GIS to display a specific layer on the screen:

- The name of the layer to be displayed
- The type of data to be displayed
- The colour table to be used for that display
- The scale of the display or the portion of the whole layer that is to be displayed

6.3.1.1 The Name of the Layer

It is strongly recommended that your data be well structured, since it is easy to create new layers in a GIS, and easy to forget what they mean, or their relevance. If you have a directory for projects, then each project should occupy a separate sub-directory. Often a number of projects may share some files. In this case you need to decide whether to duplicate those files in each project sub-directory, or whether to have the common files in their own sub-directory.

Within the project sub-directory you may then create a further layer or layers of sub-directories to reflect sub-tasks being conducted in that project. Thus, a project may have a landuse mapping task, a soil mapping task and a hydrologic modelling task. Each would be best placed in their own sub-directories.

In some GIS, the layers associated with a project are grouped into a project folder; indeed with some it is necessary to create this project folder before you can display any data. It is logical to have the directory structure closely follow the project structure that you establish within the GIS.

Once the name has been selected then the geographic characteristics (projection, etc.) will control the display of subsequent data. With vector data, or any data with an associated attribute database, you can also choose which attribute is to be displayed once the associated attribute table is linked to the spatial file.

6.3.1.2 The Type of Data to be Displayed

Some types of data can be displayed on their own, whilst with others it is necessary to have data already on display before they can be displayed. The types of data that can be displayed on their own usually include:

- Raster data
- Vector data
- Annotation data

Annotation data are descriptive graphics data that are often created in the GIS to highlight specific features or items of interest. For example, the analyst may create an annotation layer to show the locations of field sites, or to name the towns and other key features in the area.

Some types of graphics data, such as Areas of Interest (AOI's), used for derivation of statistics, may not be able to be displayed without first displaying the layers to which the AOI layer refers.

6.3.1.3 The Colour Table to be Used

A GIS layer has digital numbers associated with the features in the layer, and these numbers need to be allocated to a colour or a tone for display on the screen. This is usually done by means of a lookup table, very much like those discussed in Chapter 3. All GIS have a number of standard lookup tables. Examples of 16 shade grey tone and colour lookup tables are shown in Table 6.5. In this table the first column represents the data value in the pixel location in the layer. The second column is the colour seen by the analyst and the last three columns are the intensities as sent to the three colour guns in the screen. In practice, the lookup table only contains three columns, one for each of the colour assignments. The pixel value is used as the row address into the table. Thus, a pixel data value of 7 will give (blue = 119, green = 119, red = 119) in the grey tone table and (blue = 0, green = 255, red = 0) in the colour lookup table. These values are then sent to the colour guns of the screen. In a lookup table, either each possible value that can be taken by the data, in the case of 8 bit or integer data, or range of values in float data, is allocated a grey tone or values in blue, green and red. With float data, the full range of float values is divided by the number of entries in the table (usually 16, 64 or 256) and then each value in the table is allocated a sequential sub-range of the

TABLE 6.5
Typical 16 Level Grey Tone and Colour Lookup Tables

Data Value	Tone	Blue	Green	Red	Data Value	Colour	Blue	Green	Red
0	Black	0	0	0	0	Blue	255	0	0
1		17	17	17	1		255	63	0
2		34	34	34	2		255	127	0
3	Light grey	51	51	51	3		255	191	0
4		68	68	68	4	Cyan	255	255	0
5		85	85	85	5		170	255	0
6		102	102	102	6		85	255	0
7	Mid grey	119	119	119	7	Green	0	255	0
8		136	136	136	8		0	255	85
9		153	153	153	9		0	255	170
10		170	170	170	10	Yellow	0	255	255
11	Dark grey	187	187	187	11		0	170	255
12		204	204	204	12		0	85	255
13		221	221	221	13	Red	0	0	255
14		238	238	238	14		125	0	255
15	White	255	255	255	15	Magenta	255	0	255

size as computed, within the full range. With some GIS it is also possible with vector data to display the layer with various patterns associated with different attribute values. It is very easy to make new lookup tables so as to change the way the actual values are to be displayed. These, with the colour attributes of the screen will affect the resulting display.

Once the data to be displayed and the colour lookup table have been selected, the individual map data pixel values are used as the array address in the lookup table to find the colour assignments to send to the colour guns. If the layer data range is larger than the range of the lookup table, then the input data are first linearly transformed into the same range as the lookup table before using the transformed value as the address into the lookup table (Chapter 4).

6.3.1.4 The Scale of the Display and/or the Portion of the Whole Layer that is to be Displayed

Most GIS display the selected layer within a window established for the purpose. The size of this window can be changed. However, the resolution of the screen will have some control over the area that can be displayed. The layer area may be larger than the sized area in the screen, in which case you will need to decide whether to display only a part of the layer area, or whether to reduce the display scale so as to display the whole layer. With most systems the choice of one of these options at the start does not preclude changing the option later.

With most GIS there are defaults established for most or all these questions except for the name of the layer to be displayed.

Figure 6.9 shows the enquiry that is launched in the Idrisi image-processing software when the display launcher is started. The figure shows that the operator needs to specify the type of file, the file name and the colour lookup table when this is appropriate. In this system the scale and portion of the layer that are initially displayed are defaulted by the system.

6.3.2 Overlaying Layers of Data on the Screen

With most systems the procedure described above for the display of a layer can be repeated a number of times, each time to display another layer. When this occurs, then one layer will be the top layer,

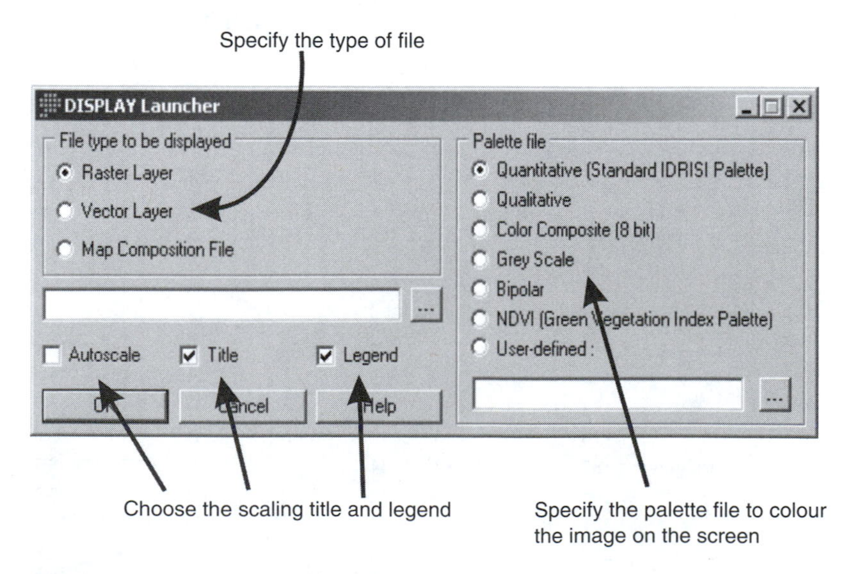

FIGURE 6.9 The interactive window used in Idrisi to enable the operator to define what layer is to be displayed, the type of file and the lookup table to be used, if appropriate.

and the other layers will be beneath this layer. The layers beneath the top layer can only be seen if the top layer contains gaps in its data. Thus, it is often best to first display a raster layer and then place on top of this the vector layers, and finally the annotation layers. However, with most systems, the sequence of layers can be changed, and in this way the user can view the different layers. With some systems it is also possible to turn some layers on or off with the click of the mouse. With such systems, the list of layers is shown, as well as their current status.

6.3.3 Combining Layers, Numerically and Logically

One of the main uses of a GIS is to combine layers in different ways, so as to address questions of the type:

- Which areas have sandy soil on slopes greater than 5%?
- Which farms have cropping on soils with slopes greater than 5, 10 and 15%?
- Which areas are most at risk from soil erosion, and what is the magnitude of the risk in these areas?

With questions of this type, certain attributes in one layer are to be displayed when certain attributes in another layer or layers meet certain constraints. When these types of questions are posed, the analyst would normally identify the attributes of interest in the first layer, and create a layer with just these activated. The same is then done with each other layer. The new layers are then combined either logically or numerically to create a second generation layer for display or further analysis. This is illustrated in Figure 6.10 for the first and second of the above questions.

If more than one attribute are required in a layer, as is the case with the second question that is set above and shown in Figure 6.10 in the second flowchart, then all three conditions would be shown in the one new layer, but assigned different values. Once this is complete the analyst combines the two layers, displaying only those areas that met the required conditions in all layers. In these examples, the information in each layer is identified using a simple **if** test and then the combined layer is created using logical tests. Consider the steps to be implemented for the first example in Figure 6.10:

Step 1. The soil layer attribute database shows that there are two soil types that meet the sandiness condition (say types 2 and 6), so, set an **if** test to be implemented for each pixel in the image of the form:

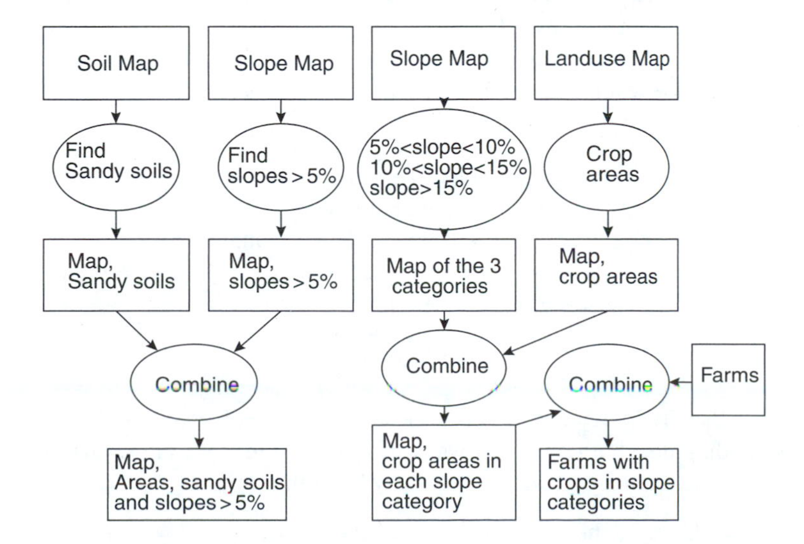

Figure 6.10 Flowcharts illustrating the steps taken to combine layers using logical combinatorial techniques.

IF ((pixel == soil_type_2) OR (pixel == soil_type_6)) then assign 1 to the new image otherwise assign 0. This creates a new layer with values of 1 where soil classes 2 or 6 occur and 0 where all other soil classes occur. Since the process is to be conducted on each pixel, most systems just use the image name instead of pixel, as the default is to process each pixel in this way.

Step 2. The slope map is of type float with fractional proportion values. Find the areas with slopes greater than 0.05 using an **if** test.

IF(pixel >= 0.05) then assign 1 otherwise assign 0.

This again creates a new layer with just two values. This layer could easily be a binary or octal layer, to save space. However, if the actual slope were to be used later in a calculation, then it would be better to carry the slope values through if they are greater or equal to 0.05.

Step 3. Combine these layers. This can be done with an **if** test of the form:

IF((layer_1 == 1) AND (layer_2 == 1)) then assign 1 to the new layer, otherwise assign 0. However, often the interrogation can be done just as easily with numerical combinations, for example:

New_layer = layer_1 × layer_2, which will give exactly the same result as the IF test above. The reader can find logical and numerical ways to combine these layers so as to display the four possible combinations of values that can come from these two layers as different colours.

Once this analysis is complete, then the analyst will need to add navigational and possibly other forms of information to the layer prior to display.

With the second example, similar logic applies. With all GIS, a range of options exists to enable the analyst to conduct filtering, branching, combining and calculation options as described above. These options include:

1. *Reassignment*. Each value in the input layer is assigned a new value in the output layer, where groups of input values can be assigned to the same output value. It can be seen that this technique could have been used in the above example.
2. *Logical tests*. A logical test is a test to determine whether a condition exists. If the test result is TRUE, then it returns a 1, otherwise it returns FALSE or zero. Usually, the tests are constructed so that they cause one action to be taken if the test result is TRUE, and a second action if the test result is FALSE. The typical tests are:

$a == b$	a has the same value as b
$a > b$	a is greater than b
$a >= b$	a is greater than or equal to b
$a < b$	a is less then b
$a <= b$	a is less than or equal to b
$a \mathrel{!=} b$	a is not the same as b

3. *Branching*. Usually a logical test is conducted and then either of two alternatives can be taken depending on whether the test was true or not. Many GIS allow chains of branches to be taken in the branching, where these may be of the form:

If (layer_1 == layer_2) then do action 1,
Else if (layer_1 > layer_2) then do action 2,

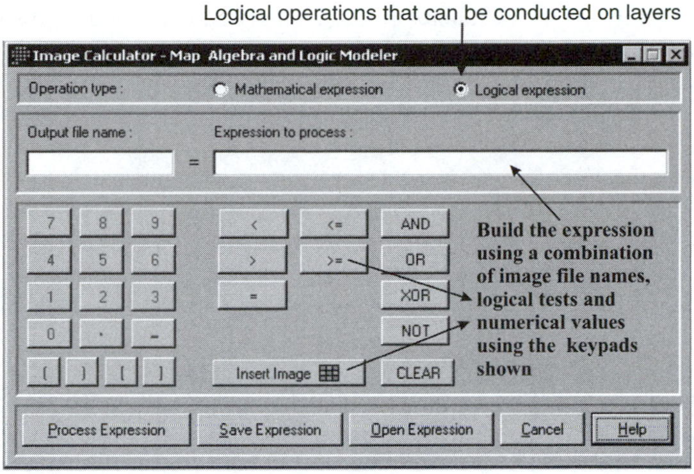

FIGURE 6.11 The Idrisi window for the construction of logical expressions.

Else do action 3. Note the action 3 takes place when layer_1 < layer_2. Also note that this process is conducted on each pixel, so that it is essential that layer 1 and layer 2 are of the same size.

Of course, such IF tests need not be restricted to just two layers. Thus, an application may require different actions that depend on the landcovers, different soil types, different slope zones, access and distance to market.

The Idrisi window for the conduct of logical operations of the type that may be required for some of the above tasks is shown in Figure 6.11. In some systems the branching choices are explicit, in Idrisi they can be achieved by the use of multiple brackets in the expression.

There are many cases where layers need to be combined numerically. Take the Universal Soil Loss Equation (USLE), which is commonly used to estimate soil loss through water erosion, as would be a typical approach to take with the third example. The USLE is of the form:

$$A = 0.224 \times R \times K \times L \times S \times C \times P \tag{6.2}$$

where A is the soil loss per unit area, R is the rainfall and runoff factor, K is the soil erodibility factor, L is the slope length factor, S is the slope steepness factor, C is the cover and management factor and P is the conservation practices used factor.

Calculation of the soil loss per unit area, A, from the USLE requires the creation of layers for each parameter contributing to the calculation, and then combining the layers as a product of those layers and 0.224. The layers are often created as follows:

1. *Soil erodibility, K*. Soil mapping units can be categorised in terms of soil erodibility, usually based on field visits for the purpose, or experience. Thus, a soil map will form the base data for this analysis. Field or other data will be used to assign a soil erodibility value to each soil mapping class. In a raster GIS the soil class layer will then be used with these transformation values in a **REASSIGN** process to assign soil erodibility values to each pixel based on their soil class. In a vector GIS the attribute table would be used to create a layer of the relevant soil erodibility values derived from or replacing the soil class values.
2. *Slope factor, LS*. Slope maps are derived by the analysis of digital elevation data (DEM). The drainage lines and watersheds in the area can be found semi-automatically in some

GIS. Alternatively, they can be interpreted manually and entered into the GIS. The GIS can then compute the slope distance from the pixel to the drainage line, and compute the slope steepness on this line. The two are then combined to derive a slope factor, *LS* by a COMBINE operation that combines two or more layers together in accordance with a set numerical formula.

3. *Cover and management factor, CP.* Both management and climatic factors affect landcover changes. It is important to assess the amount of cover that can be expected at the time when the area is most at risk from erosion. Satellite imagery can be used to assess cover at the time of acquisition, and landuse in that season. The management factor is often set as a function of the landuse. Thus, the landuse information can be used in an ASSIGN operation to derive a *CP* layer.
4. *Rainfall and runoff factor, R.* The rainfall erosivity can be estimated knowing the typical rainfall intensity and duration in the study area. Usually, *R* will be constant across an area. However, the rainfall duration and intensity may be a function of elevation, in which case the DEM, may be used in an ASSIGN operation to convert the DEM values to *R* values.

The USLE can now be computed as the product of the four parameters by 0.224. This would typically be done in a COMBINE operation.

Most GIS include a number of different operations to combine layers numerically or logically. Many also include a capacity to write other operators using map algebra in a simple form of computing programming language that will be discussed later. The typical operators will include the following, although the name given to these operators, and how they work, varies from one system to another:

1. **TRANSFORM.** To transform the values in a layer in accordance with a numerical function. Such functions would typically include addition, subtraction, multiplication, division, remainder, power, logarithm, exponential and the trigonometric functions.
2. **COMBINE.** To combine two or more layers using the addition, subtraction, multiplication, division, remainder, power, logarithm, exponential and trigonometric functions. This is the function that would be used to derive Erosion Risk once the four factors have been derived using the above operators.

In using all these operators it is important to always keep in mind that systems cannot:

1. *Divide by zero.* If you attempt to do this then you will get an error. It is always best to structure your formulations to ensure that a divisor can never have a zero value, if it does then a suitable logical **if** test is conducted on the data, and when the divisor is zero, then a different process takes place than when the divisor is not zero. For example, when a divisor is zero, the result can be a value that would normally never occur, so that these pixels can be readily identified.
2. *Compute the square root of a negative number.* Again this cannot be solved numerically. The comments made above also apply in this case.
3. *The value limits of different data types.* The range of values that can occur in character, integer and float data have been discussed. If a calculation creates values that are outside the range of the data type, then an error will result. For example, if a value is calculated to be 257 in character (char) data, then the value derived will be 1 and this value is what will appear in the results file. It is essential to check the range of values that may occur in all calculations, to ensure that this does not occur in your analysis. This source of error can easily arise when character or integer data are combined numerically and are assigned to character or integer constants or images. For example, if *A* and *B* are of type char with values of 65 and 4, respectively, then the product of the two is 260. If this is assigned to a character image or constant, then it will overflow the possible range of 0 to 255 that can be handled

by this type of data, and will take the value 4 (new value = old value − 256 × integer value such that the resulting new value is in the range 0 to 255). This will, of course, introduce errors, that may sometimes be difficult to detect.

6.3.4 Filtering and Neighbourhood Analyses

Neighbourhood operators operate on a window of pixels surrounding the central pixel, so as to derive a value for the central pixel that is in some way an estimate of the pixel values in the window. There are various applications of such operators including:

- To reduce or remove the *salt and peppering* effect that can often be seen in classification derived from remotely sensed image data.
- To find areas in a classification or a layer where there is the maximum level of variation, for example, this may indicate areas of high biodiversity.
- Find the classes or values in a layer that most frequently occur as isolated pixels, as these areas may indicate vulnerable nature sights.

Neighbourhood operators operate on a window around the central pixel. The results of the analysis are applied to the central pixel. The size of the window can be changed by the operator, with (3 × 3), (5 × 5) and (7 × 7) being the commonly used sizes. There are a number of different types of operators that can be used including:

- **Sum** — Compute the sum of the values in the window.
- **Majority** — Assign the most commonly occurring value to the central pixel. Typically used in cleaning up *salt and pepper* effects in classification.
- **Minority** — Replace the central value by the least commonly occurring value in the window. Highlights those layer values that most frequently occur in isolation.
- **Maximum** — Assign the highest value to the central pixel.
- **Minimum** — Assign the lowest value to the central pixel.
- **Diversity** — Replace the central pixel value with the number of different values that occur in the window. Another way to estimate variation.
- **Density** — By the number of pixels with the same value as the central pixel.

Another form of neighbourhood analyses is where the neighbourhood is used to calculate the layer gradient and orientation at each centre pixel. Consider a 3 × 3 window, in which the local co-ordinates are as given in Figure 6.12. If the surface is considered to be a close approximation to a second order

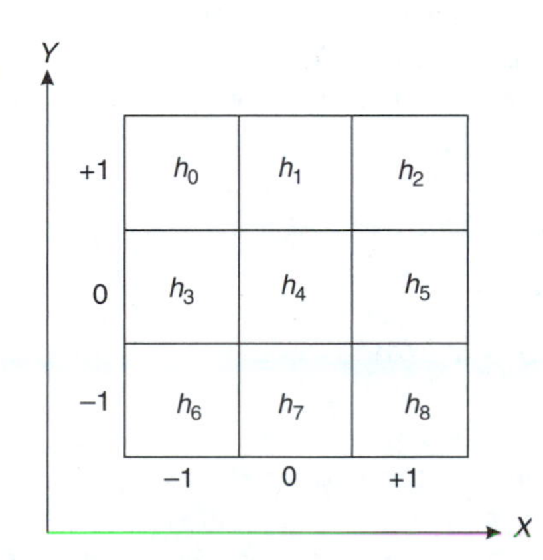

FIGURE 6.12 Local (x, y) co-ordinates and pixel values to solve for gradient, and rate of change in gradient.

polynomial, then the equation for that surface is of the form:

$$h = a_0 + a_1 \times x + a_2 \times y + a_3 \times xy + a_4 \times x^2 + a_5 \times y^2 \tag{6.3}$$

The equation can be solved by using the local co-ordinates as shown in Figure 6.12 and pixel values for these nine pixels using Equation (6.3) to give the equations:

$$\begin{aligned} a_0 - a_1 + a_2 - a_3 + a_4 + a_5 &= h_0 \\ a_0 + a_2 + a_5 &= h_1 \\ a_0 + a_1 + a_2 + a_3 + a_4 + a_5 &= h_2 \\ a_0 - a_1 + a_4 &= h_3 \\ a_0 &= h_4 \\ a_0 + a_1 + a_4 &= h_5 \\ a_0 - a_1 - a_2 + a_3 + a_4 + a_5 &= h_6 \\ a_0 - a_2 + a_5 &= h_7 \\ a_0 + a_1 - a_2 - a_3 + a_4 + a_5 &= h_8 \end{aligned} \tag{6.4}$$

These equations can be solved for the unknowns, a_0 to a_5 by means of least squares techniques. The gradient is the first derivative of this equation, that is:

$$\begin{aligned} \text{Gradient in the } X \text{ direction} &= \mathrm{d}h/\mathrm{d}x = a_1 + a_3 y + 2a_4 x \\ \text{Gradient in the } Y \text{ direction} &= \mathrm{d}h/\mathrm{d}y = a_2 + a_3 x + 2a_5 y \end{aligned} \tag{6.5}$$

We are interested in the gradient at the centre pixel, which we have given local co-ordinates of (0, 0), as can be seen in Figure 6.12, in the above equations, so that:

$$\begin{aligned} \text{Gradient in the } X \text{ direction (at } x = 0) &= a_1 = [(h_2 + h_5 + h_8 - (h_0 + h_3 + h_6)]/6 \\ \text{Gradient in the } Y \text{ direction (at } y = 0) &= a_2 = [(h_0 + h_1 + h_2) - (h_6 + h_7 + h_8)]/6 \end{aligned} \tag{6.6}$$

Now it would be possible to construct Equation (6.6) using the numerical tools available in most GIS. However, a quicker way is to construct a template table as shown in Figure 6.13, and use this template with the image data by multiplying the template value in each cell of the template by the image values in the corresponding image cell and summing the results. The resulting figure is the same as would be derived by computation of Equation (6.6).

The rate of change in the gradient can also be found since the rate of change in the gradient is the second derivative, that is:

$$\begin{aligned} &\text{Rate of change (} X \text{ direction)} = \mathrm{d}^2h/\mathrm{d}x^2 = 2 \times a_4 = [(h_0 + h_2 + h_3 + h_5 + h_6 + h_8) \\ &\quad - 2 \times (h_1 + h_4 + h_7)]/3 \\ &\text{Rate of change (} Y \text{ direction)} = \mathrm{d}^2h/\mathrm{d}y^2 = 2 \times a_5 = [(h_0 + h_1 + h_2 + h_6 + h_7 + h_8) \\ &\quad - 2 \times (h_3 + h_4 + h_5)]/3 \\ &\text{Rate of change on the diagonal} = \mathrm{d}^2h/\mathrm{d}x\mathrm{d}x = a_3 = [(h_2 + h_6) - (h_0 + h_8)]/4 \end{aligned} \tag{6.7}$$

These equations can also be implemented as a neighbourhood template where the sum of the products gives the rate of change.

Template to compute gradient in the X direction

−0.1667	0.0	+0.1667
−0.1667	0.0	+0.1667
−0.1667	0.0	+0.1667

Template to compute gradient in the Y direction

+0.1667	+0.1667	+0.1667
0.0	0.0	0.0
−0.1667	−0.1667	−0.1667

FIGURE 6.13 Templates that can be used to compute gradients in the X and Y directions using Equation (6.7).

6.3.5 DISTANCES, COST SURFACES, LEAST COST PATHWAYS AND CONTEXTUAL ANALYSIS

Distance operators calculate the distance from some feature or set of features and assign the pixel this distance value. There are many types of distances that can be used, but the two most common in a GIS environment are Euclidean or geometric distance, and cost distance. The Euclidean distance between two points (x_1, y_1) and (x_2, y_2) is:

$$E = ((x_1 - x_2)^2 + (y_1 - y_2)^2)^{0.5} \tag{6.8}$$

The cost distance between two points is equal to:

$$C = \Sigma(d_i \times C_i), \quad (i = 1 \text{ to } n) \tag{6.9}$$

where C_i = the cost associated with the ith sub-distance and d_i is the Euclidean distance in the ith sub-distance.

All decisions involve choices. One way to assist a manager is to reduce complex interactions to a simple choice based on costs. To base decisions on costs requires that at least two steps need to be implemented, the actual number depending on the task at hand. The two that constitute the minimum are:

1. Create cost surfaces
2. Use these cost surfaces to either delineate choices in landuse or to construct least cost pathways

Cost surface modelling is a way of accounting for the cost of moving through space, where the cost of movement is a function of the standard costs associated with movement and with the forces or frictions that facilitate or impede movement. For example, the base costs of moving produce to market include loading and unloading costs and the base costs associated with the hire and running of the vehicles used to conduct the transport of the products. Forces and frictions that may act on these vehicles in a way that influence these costs may include road conditions, road gradients and the density of traffic.

Costs can be isotropic, in which case the costs have the same magnitude independent of direction, or costs can be anisotropic, or the directions influence the magnitudes of the costs. In the above example, the road surface conditions may be considered isotropic, whereas the gradient is anisotropic.

Cost surface modelling starts from an origin located within the data set. If the modelling is to be conducted with isotropic forces or frictions, then the analyst will derive a layer for each force

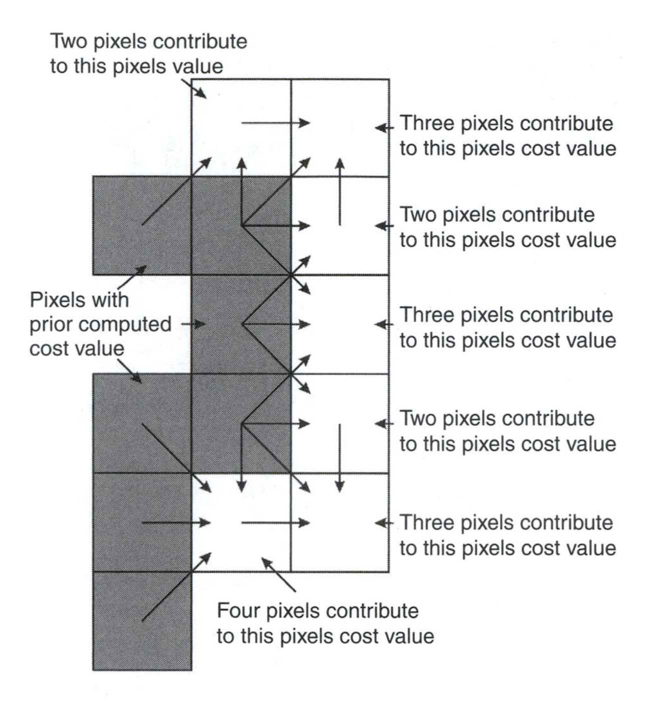

FIGURE 6.14 Pixels used to compute the cost value associated with pixels in different positions.

or friction, with values in that layer representing the magnitude of the force or friction per unit distance travelled. The cost surface is then modelled incrementally out from the origin, by means of concentric rings or pixels. In general, each pixel in a concentric ring can be reached by three pixels of computed cost, by the pixel immediately adjacent and the two diagonal pixels. In general, if a pixel has one common edge with a pixel of known cost value, then two or three pixels will be used to find the cost value of the pixel. If the pixel has two edges that are common with pixels of known cost value, then three or more pixels will be used to compute the pixels' cost value. There remain pixels with no common edge, but a common point. These pixels are not dealt with until they have at least one common edge with pixels of computed cost value (Figure 6.14). All costs from prior costed pixels (usually three as stated) should be computed and then weighted by the distance, so that the adjacent pixel has unit weight and the diagonal pixels will have a weight of $1/\sqrt{2}$. The minimum of the computed costs should be adopted.

Isotropic cost modelling thus needs one layer to provide the distance information and one or more layers to provide the parameters to the cost model so as to compute the cost per unit distance information. In some instances the distance layer contains the simple horizontal distance, whilst for other applications, the slope distance may be necessary.

Anisotropic forces or frictions act with a maximum positive impact in one direction, negative impact in the opposite direction, and with variable impact between these directions, such that they will give zero impact for directions at right angles to the direction of maximum impact. For anisotropic forces or frictions it is thus necessary to use a distance image and two images for each anisotropic force or friction, representing the maximum value of the force or friction, and its direction. The image pairs thus specify a vector magnitude and direction at each pixel. For anisotropic cost surface modelling at a pixel, the same incremental costs have to be calculated from the prior costed pixels, but now the costs need to be computed in each of the directions to the prior costed pixels using the vector image pair and then the incremental costs are computed prior to selection of the lowest cost.

In general, the cost force or friction in a direction is computed as the product of the maximum force or friction by the cosine of the angle between the direction of the maximum force and the direction from the prior costed pixel and the pixel for which the cost value is being computed. However, some forces do not operate in this way, but may, for example, decrease much more rapidly away from the direction of the force or friction. With some GIS the operator can specify the anisotropic function to be used to derive the force or friction in any specific direction from the source information.

6.3.5.1 Least Cost Pathways

A least cost pathway is the lowest cost route from the source feature or location to the target feature or location, where the cost surfaces have already been computed relative to the target. The cost surface will thus have a trough at the target, and the source will be at a point of higher relative cost. The construction of a least cost pathway is then based on two principles:

1. The shortest Euclidean distance is the lowest cost pathway if all other things are equal. By constructing a cost surface, we are saying that not all other things are equal, but the Euclidean distance is the benchmark against which other paths will initially be tested.
2. The pathway has, on average, to travel downhill from the source feature to the target feature, since the target feature is in a trough in the cost surface.

The process thus starts by constructing the Euclidean distance between the source and the target and computing the summed cost across this route. However, whilst this is the route of shortest distance, it may not be the route of lowest cost. To find the route of lowest cost, at each pixel along the route, in turn, the algorithm:

1. Determines whether the gradient of the cost surface is in the direction of travel, or to either side.
2. If the gradient is in the direction of travel then the path is not moved.
3. If the gradient is to either side, then the path is moved one pixel to the side of lower cost and in the process one or more pixels may be added or removed from the path. The cost of this new path is then compared with the cost of the prior path, and if it is lower, then it is accepted, otherwise the original path is retained.
4. The process is iterated until all changes result in higher cost pathways.

6.3.5.2 Context Operators

Context operators derive the new pixel value based on the values in the surrounding pixels. We have seen in Chapter 3 that one class of context operators are used to derive textural information for image processing. In image processing, contextual operators are normally used with continuous data, to derive a value based numerically on that continuous data. The same operators can, of course, be applied to GIS data when appropriate, but it is much more usual in GIS analysis to use contextual operators with nominal or ordinal data, to which the rules of numerical techniques no longer apply. In GIS the following types of contextual operators are used:

1. *View or watershed operators*. These operators use a continuous surface to derive information on the areas that are inter-visible on the surface or to delineate watersheds in the surface.
2. *Context operators*. These operators return a value that depends on the operator's specifications in relation to the distribution of nominal or ordinal classes in a window around the pixel. For example, in classification, the operator may wish to reduce the *salt and pepper* effect by having a single pixel of one class, which is surrounded by pixels of a second class, converted to the second class. Context operators may return the dominant class, the number of classes, the dominance of the dominant class, etc., depending on the operator.

3. *Area operators*. These operators are used to identify clumps of pixels with the same value that meet specific criterion. This criterion would typically be area criterion; they need to be larger than or smaller than set criterion. In this case the software calculates the area of continuous pixels of the one class, computes the area and then decides if it meets the set criterion.
4. *Buffer operators*. These operators are used to define buffer zones around area or linear features in a GIS layer. The software uses specified distance criteria to set a buffer zone as a category around the specified features in the specified layer. Such a process may be used, for example, to identify buffer zones around environmentally sensitive areas.

6.3.6 Statistical Analysis

The most common forms of statistical analysis that can be conducted in most GIS systems includes:

1. Creation of histograms, scattergrams and the derivation of basic statistical data of the first four moments (mean, variance, skewness and kurtosis)
2. Creation of scattergrams and the analysis of correlation
3. Regression analysis

Histograms have been introduced in Chapter 4. Their x-axis represents the range of data values that can occur in the data, for example 0 to 255 for byte data. The y-value at each x-value is the count of how many times that value occurred in the GIS layer to which the histogram refers. The frequency distribution of the data is created by dividing the counts at each x-value by the total number of counts in the data. This also gives the probability distribution for the data in that layer.

A scattergram is a similar idea, but in two dimensions. To create a scattergram, simply count the number of occurrences of each combination of the two layers. The scattergram provides an indication of the level of correlation between the two layers.

The mean, variance, skewness and kurtosis represent the first four moments of a set of n data values, $x_1, x_2, \ldots, x_n$. These are defined as:

$$\text{Mean} = E(x_i) = (x_1 + x_2 + x_3 + \cdots + x_n)/n = \Sigma x_i, \quad (i = 1 \text{ to } n)/n = \mu \tag{6.10}$$

where $E()$ is the statistical expectation.

The mean value has the characteristic that the sum of the residuals about the mean will be zero and the sums of the squares of the residuals will be the minimum of any set of residuals, that is:

$$\Sigma(x_i - \mu) = 0 \quad \text{and} \quad \Sigma(x_i - \mu)^2 < \Sigma(x_i - x')^2 \quad \text{where } \mu \# x'$$

$$\text{Variance} = E(x_i - \mu)^2 = \Sigma(x_i - \mu)^2/(n-1), \quad (i = 1 \text{ to } n) = (\Sigma x_i^2 - (\Sigma x_i \times \Sigma x_i)/n)/(n-1)$$
$$= (\Sigma x_i^2 - (\mu^2 \times n))/(n-1) = \sigma^2 \tag{6.11}$$

The square root of the variance is called the standard deviation, and it indicates the distance from the mean to the point of inflection in the curve, as shown in Figure 6.15(a).

The skewness is a measure of the asymmetry of a set of data (Figure 6.15[b]). If the frequency curve of a distribution has a longer tail to the right of the peak than to the left, then the data is said to be positively skewed and vice versa.

The skewness can be computed from:

$$\text{Skewness} = \{\Sigma(x_i - \mu)^3/(n-1)\}/\{\Sigma(x_i - \mu)^2/(n-1)\}^{3/2} \tag{6.12}$$

The skewness of a normal curve is zero.

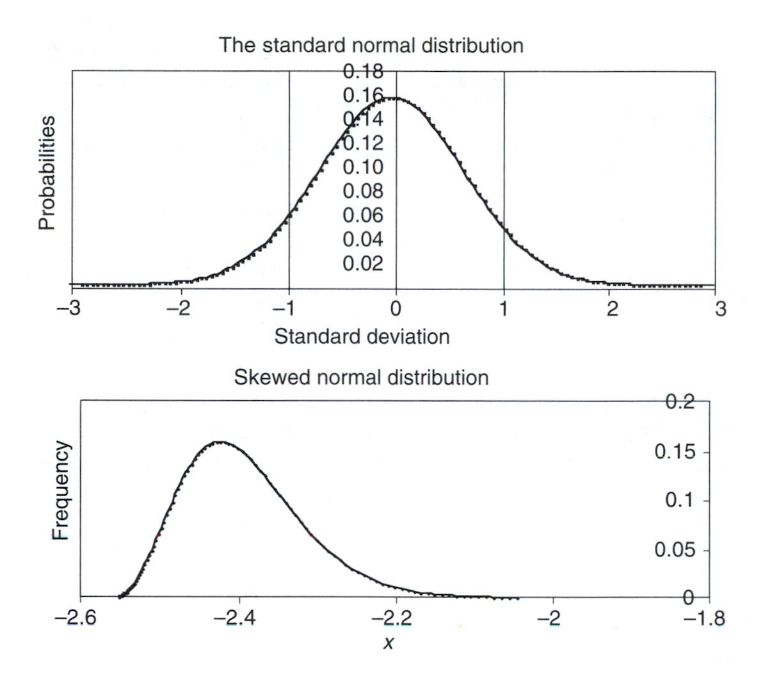

FIGURE 6.15 The normal distribution (a) and skewed normal distribution (b).

Kurtosis is the degree of peakedness of the frequency curve of a distribution, where:

$$\text{Kurtosis} = \{\Sigma(x_i - \mu)^4/(n-1)\}/\{\Sigma(x_i - \mu)^2/(n-1)\}^2 \tag{6.13}$$

The kurtosis of a normally distributed set of data is 3. As the frequency distribution becomes more peaked, the kurtosis increases higher than 3, and kurtosis is less than 3 for distributions that are flatter than a normal distribution.

The covariance between two variables is an indication of how changes in one variable are mirrored in variations in the other variable. If changes in one variable are matched by random changes in the other variable, then the covariance and correlation will be zero, whilst on the other hand if the changes in the second variable can be exactly predicted from changes in the first variable then the correlation between the variables is 1.0 and the covariance will be high. If we have a set of n data values in two variables, $(x_1, y_1), (x_2, y_2), \ldots, (x_n, y_n)$ then the means, variances, skewness and kurtosis of the individual variables can be computed in the same was as described above. However, these values, whilst telling a lot about the individual variables, tell us nothing about how they related to each other. For that it is necessary to compute the covariance array and after that the correlation array:

$$\text{Cov}_{x,y} = \Sigma(x_i - \mu_x) \times (y_i - \mu_y)/(n-1) \tag{6.14}$$

and

$$\text{Corr}_{x,y} = \text{Cov}_{x,y}/(\sigma_x \times \sigma_y) \tag{6.15}$$

where σ is the standard deviation or the square root of the variance. It should be noted that covariance and correlation show the relationships between the variables. They do not necessarily indicate a cause to effect or independence/dependence relationships between these variables. Whether they indicate a cause to effect or independence/dependence relationship is something that the analyst has to decide, based on this and other evidence.

The covariance and correlation are often stated in the form of arrays. This has been discussed in terms of the covariance and correlation between wavebands in remotely sensed image data in Chapter 4. However, the same concepts can apply to GIS data, in which case it can be advantageous to merge the data into overlay layers and then derive scatterplots or scattergrams as discussed above and in Chapter 4 in relation to image data.

In a GIS there is one application of scattergrams that is quite different to that in image processing. This has to do with the evaluation of the accuracy of a layer. If field data have been taken of, for example, landuse, then the data can be entered into a layer of landuse field data. This layer of typically sparse data is then compared with the classification of landuse layer to show the correlation between the field data and the classification. This application of correlation analysis is discussed at some length in Chapter 5.

The basis of regression analysis was discussed in Chapter 4, using one-dimensional linear regression as the main focus for that development. The same principles apply when conducting regression analysis in two or more dimensions. The differences are that there will be more independent variables, and this is likely to affect the number of unknowns to be found, as well as affecting the values of some parameters when constructing confidence limits. When dealing with two or more dimensions, the analyst also has a slightly more complicated task when deciding on the most suitable model to be used in the regression. When doing this the analyst can create plots of the marginal distributions, that is, if the data is of the form, $v = f(x, y)$, then the marginal distributions are of the form $v = F(x)$, where each x value includes all the data values with this x value, and of course having a range of y values. Such marginal distributions may be of help, but they may also exhibit such a range of values that it is difficult to select an appropriate model to use. The analyst can also take slices through the data, of the form, $v = F(x, y_{\text{set}})$ in which only the y values used are those within a set narrow range of values. Such an approach obviously does not use all the available data in each slice, but it will give a reasonable idea on the appropriate shape to use at that part of the data domain, as long as sufficient data was used to give a credible set of data in the plot. It is most likely that some combination of the two will be found to be the most useful. The analysts can also use three-dimensional plots when just two independent variables are involved, but this form of display rapidly becomes unusable once more than three dimensions become involved.

6.4 VECTOR GIS DATA ANALYSIS FUNCTIONS (Susanne Kickner)

Typical functions used in routine vector data analysis include layer overlay operations, selection of layer subsets or sub-themes, neighbourhood operations or connectivity functions and so on. Consider the need to overlay different layers, for example, landuse, soil and slope data. A vector GIS can create new spatial data by the creation of the spatial overlay. Spatial overlay reconstructs the topology of spatial objects, when they are represented by arc-node-structure. Overlay functions normally calculate the intersection of line segments and the area of polygons for the resulting layer.

Spatial analytical questions that may be answered with vector analysis are those as shown in Table 6.6.

6.4.1 SELECTION

Selection is an extremely important concept in GIS. It creates a subset of the thematic layer. The result of a selection can be stored as a new layer, can be highlighted or can be used to summarise some attributes. In the simplest case single objects can be selected with the mouse, however, selection is also possible using attribute values or location relative to the object as the criteria.

TABLE 6.6
Typical Tasks Conducted in Routine Vector GIS Analysis

Question	Kind of Layers	Function
What kind of landuses can we find in a certain region?	Polygon and polygon	Union
How many houses are lying within 100 m of a road?	Points and polygon	Buffer, identity
How long is the common border between two counties?	Polygon with arc topology	Selection
Where is a certain city?	Polygon or point	Selection
How long is the distance over a street network from A to B?	Arcs with network topology	Path finding
In what soil types can we find a certain plant?	Areas and points	Intersection
Which countries does a certain river pass through?	Areas and lines	Intersection

TABLE 6.7
Conditional Operators Used in Most GIS Systems

=	Equal
>	Greater than
<	Less than
>=	Greater than or equal
<=	Less than or equal
<>	Not equal to
cn	Contains (for character items)

6.4.1.1 Selecting by Attribute

Normally, for selecting by attribute, a relational database consisting of one or more tables is connected with the spatial objects. The spatial database data or information is stored in these tables. The SELECT statement is used to query the database and retrieve selected data that match the criteria that are specified by the operator. To select objects with certain attributes the typical tools of database queries are available. Typical simple conditional selections with comparison operators used are given in Table 6.7.

The returned result of a conditional operator is either **TRUE** (=1) or **FALSE** (=0). The operator will specify the different follow on actions that are to be taken depending on whether the result is **TRUE** or **FALSE**. For example, given the conditional test (**A** <= **B**) will return **TRUE** when **A** = **5** and **B** = **6** and **FALSE** when **A** = **6** and **B** = **5**.

More advanced queries are implemented by combining conditional and Boolean operators to select objects. The **AND** and **OR** operators can be used to combine conditions. For the **AND** operator the conditions on both sides of the **AND** condition must be **TRUE** for the Boolean **AND** to return **TRUE** and for the action specified for **TRUE** to be taken. With the **OR** operator, either or both sides must be **TRUE** for the appropriate action to be taken. If we consider a Boolean test to be of the form ([condition 1 test] Boolean operator [condition 2 test]), then the actions that result from the two Boolean operators is as given in Table 6.8.

Consider an example from Germany. Consider the case where one has a GIS layer with municipalities in Germany that contains attribute information on the populations of each municipality and one would like to select all smaller spas. Spas in Germany normally contain the word "Bad" (bath) in or before the town name. As a size limit 20,000 inhabitants will be fixed as the upper limit. The table of this layer looks like that shown in Table 6.9.

TABLE 6.8
Results of the Application of the Boolean Operators

Condition 1	Condition 2	Returned Result	
		AND	OR
TRUE	TRUE	TRUE	TRUE
TRUE	FALSE	FALSE	TRUE
FALSE	TRUE	FALSE	TRUE
FALSE	FALSE	FALSE	FALSE

TABLE 6.9
Some Municipalities in the GIS Municipal Layer, with their Attribute Information

ID	Municipality	Inhabitants	Federal State
1	Bad Reichenhall	16375	Bavaria
2	München	1210223	Bavaria
3	Göttingen	121132	Lower Saxony
4	Bad Kleinen	3896	Mecklenburg-Vorpommern
5	Bad Pyrmont	11336	Lower Saxony
6	Karlsruhe	278558	Baden-Württemberg
7	Badenhausen	2033	Lower Saxony
8	Bad Nauheim	30199	Hesse
9	Hamburg	1715392	Hamburg

TABLE 6.10
Data Records Selected as a Result of the Query Above

ID	Municipality	Inhabitants	Federal State
1	Bad Reichenhall	16375	Bavaria
4	Bad Kleinen	3896	Mecklenburg-Vorpommern
5	Bad Pyrmont	11336	Lower Saxony

The query then should be formulated as:

SELECT *municipality* **cn**'*Bad*' **AND** *inhabitants* <= 20000 (6.16)

For this query, only those records in the database that return **TRUE** for both conditions will be selected. As a result the data records shown in Table 6.10 are selected.

TABLE 6.11
Data Records Selected as a Result of the Second Query

ID	Municipality	Inhabitants	Federal State
1	Bad Reichenhall	16375	Bavaria
4	Bad Kleinen	3896	Mecklenburg-Vorpommern
5	Bad Pyrmont	11336	Lower Saxony
8	Bad Nauheim	30199	Hesse

TABLE 6.12
Typical Questions that can be Addressed by Selection by Location in Vector GIS Analysis

Question	Layer Type of Objects to be Selected	Layer Type with Locational Information
Which properties border on a certain street?	Area	Lines or areas
Which buildings lie within a postcode district?	Points or areas	Areas
Which street sections run through woods?	Lines	Areas
Which farms are attached to which water pipe?	Points or areas or lines	Lines
Certain plants are found on which soils?	Points	Areas

If either all spas shall be selected which have less than 20,000 inhabitants or all spas lying in Hesse, the query is as follows:

SELECT (*municipality* cn 'Bad' **AND** *inhabitants* < 20000) **OR**
(*municipality* cn 'Bad' **AND** *federal state* = Hesse)

As a result the data records listed in Table 6.11 are selected. It should be noted that a test of this form follows a particular hierarchical sequence. The test is evaluated first within brackets. If there are embedded brackets, then the most deeply embedded are evaluated first. Then conditions are evaluated followed by the Boolean operators. If you want to be absolutely clear on the sequence to be adopted, then use brackets to achieve this.

6.4.1.2 Selecting by Location

Objects in a layer can also be selected according to the location, the outline or the size of lines, points and areas in the same or in another layer in the vector GIS. The prerequisite condition necessary to be able to do this is the existence of topological information on the objects in the layers being interrogated. Typical questions for selection by location include those listed in Table 6.12.

There exist a variety of selection methods by spatial query in the different GIS. Some of the methods available in ArcGIS from Environmental Systems Research Institute, Inc. (ESRI) are described in the next sections:

1. *Intersect: selecting lines or polygons that are crossed by lines*. This method selects the features that are crossed by the lines of another layer. For example, you can select roads in one layer which

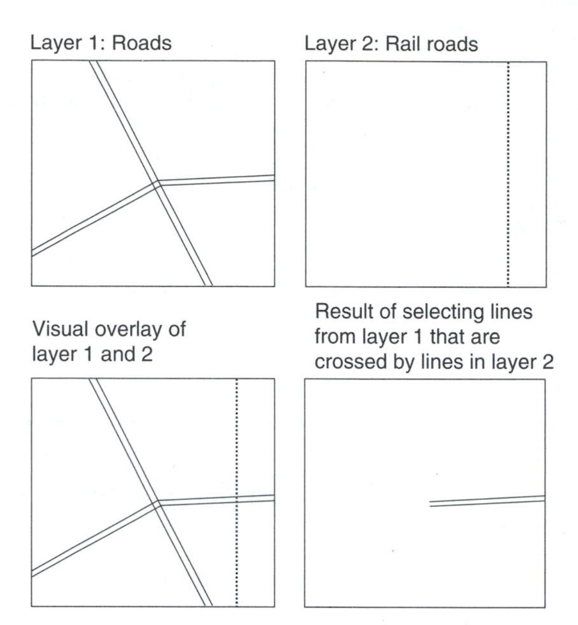

FIGURE 6.16 Selecting lines that are crossed by lines either in the same or in different layers of the GIS.

are crossed by railways (see Figure 6.16) saved in another layer or you can select areas in one layer intersected by rivers saved as lines in a second layer.

2. *Adjacency: selecting features that are within a certain distance or adjacent to other features.* This method selects lines, points or polygons within a certain distance of features in the same or in a different layer. If one wants to inform all people who live in the danger area of a fire for example, all houses, which are located within a defined distance of the fire source, can be selected with this function. In order to identify all houses in the potential flood area of a river, one can select all points that contain house attribute values within a specified distance of linear features that contain river attribute values (see Figure 6.17).

Selecting adjacent features can also be conducted by this method by selecting polygon features with a common line between them. The first step is to select the features you are interested in. To find the adjacent polygons, you would select the areas within zero distance of the features already selected. Another way to find adjacent polygons is to select by attribute. In this case you select all areas that are bounded on one side by arcs with an identical internal number.

3. *Enclosures: selecting polygon features that are completely within or completely contain polygons of another layer*. With this function the analyst can find areas that are completely contained within other polygons. For example, this function can be used to interrogate GIS layers to identify the vegetation on river islands or clearings within forests. To do this, areas of the vegetation layer without trees but which lie completely within the forest polygon or water areas of a layer with landuse information are selected (see Figure 6.18). It is also possible to select only those internal polygons that lie within, or beyond, a specified distance of the boundary of the water or forest by specifying a buffer distance that the contained polygons are to be within or beyond.

On the other hand, the analyst may be interested in the reverse case or identifying those forests that contain clearings. This analysis is conducted using the **COMPLETELY CONTAIN** function to select polygons in one layer that completely contain the features in another layer (see Figure 6.19).

If the areas inside the polygon of another layer partly touch the boundaries of this area, they can be selected by using the function **CONTAIN**.

4. *Commonality: selecting lines which touch the boundary of lines or polygons in another layer*. This selection method is used to identify lines or areas in one layer that share line segments, vertices or nodes with lines or polygons in another layer. For example, this function is useful in examining

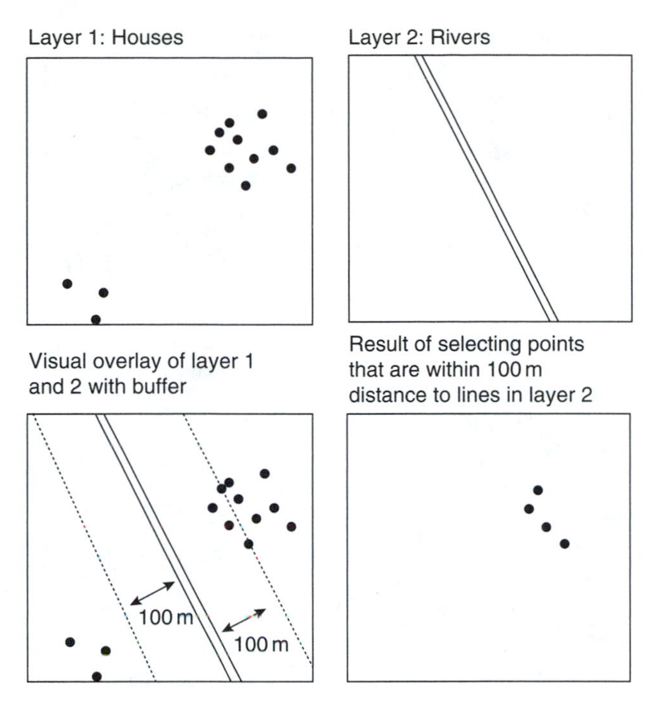

FIGURE 6.17 Selecting points that are within a certain distance to lines.

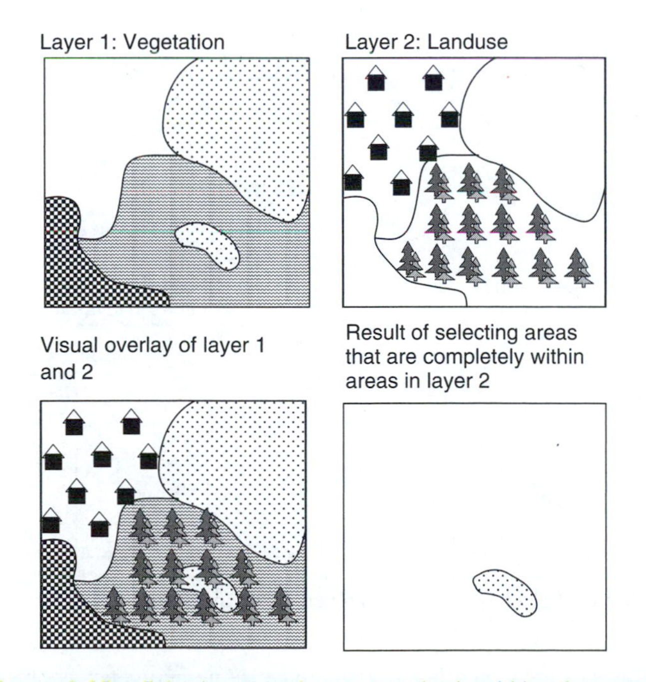

FIGURE 6.18 Selecting areas that are completely within other areas.

a cycle track network to identify those segments where the cycle track is on a road or street since these segments may be more dangerous for cyclists and thus may require additional treatment (Figure 6.20).

5. *Identicality: Selecting features that are identical to features in another layer*. In this case the feature types of both layers must be the same in their geometrical definition, but not in their attributes.

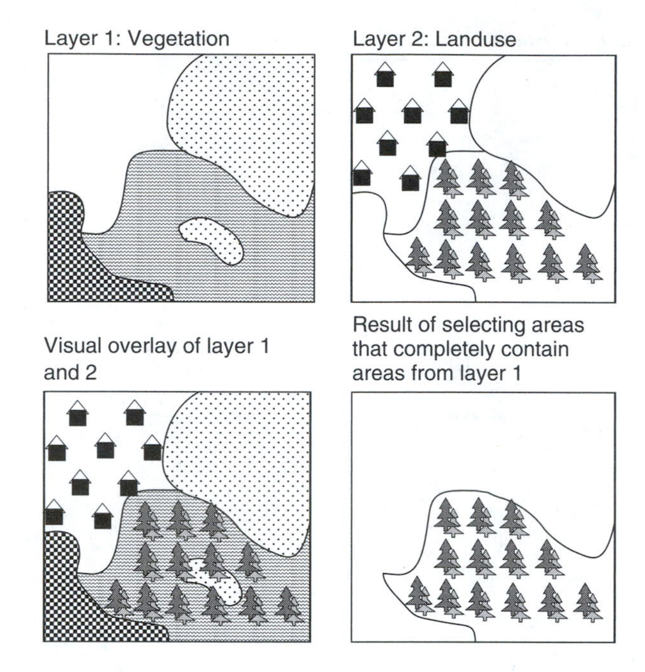

FIGURE 6.19 Selecting areas that completely contain other areas.

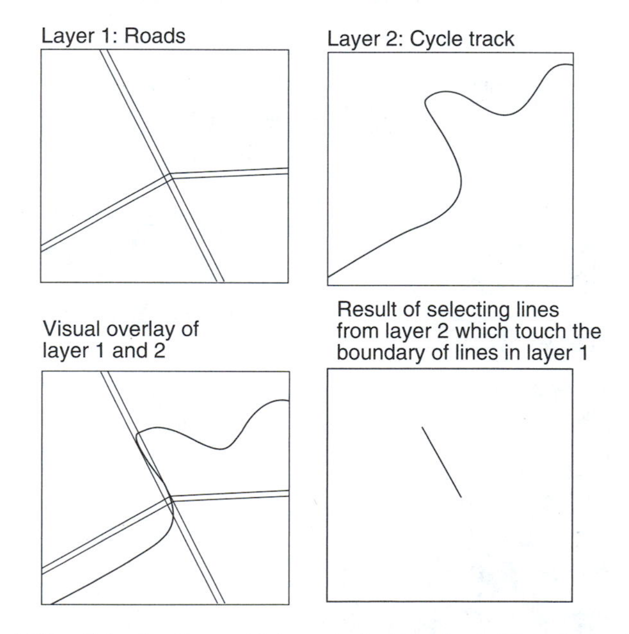

FIGURE 6.20 Selecting lines which touch the boundary of lines in another layer.

Given this, points, lines or areas in a layer, which have same geometry as objects in another layer can be selected (Figure 6.21).

6.4.2 Dividing and Joining Areas

Areas can be divided in the vector GIS into patches by digitising or importation of lines. Neighbouring areas with a common border can be united automatically into one area if one or more attributes of the adjacent areas are identical by use of the function **DISSOLVE** (Figure 6.22). This function

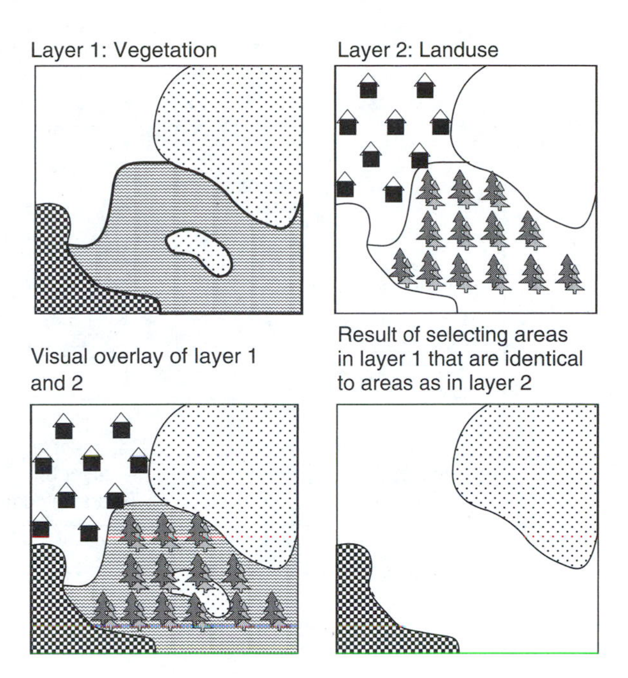

FIGURE 6.21 Selecting areas with identical geometry as areas in another layer.

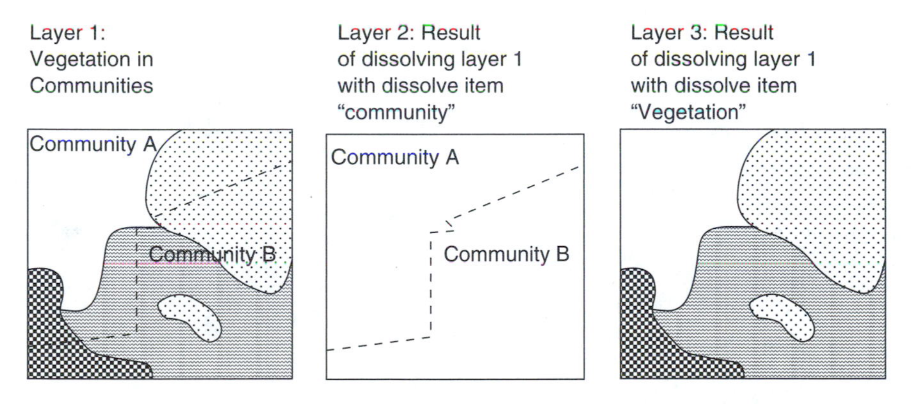

FIGURE 6.22 Example for the function DISSOLVE.

automatically removes borders between two adjacent polygons and rebuilds the topology of the new, combined, areas. If areas are shared or joined together in this way, then attribute values can be passed on. This passing on is steered through parameters which take the relative size of the original areas or the length of the common boundary into account and which are selected depending on need. For example, it can be specified that the combined areas are assigned the attribute of the largest polygon or the new areas will be attributed with the maximum (minimum) or the average of the original areas. It is also possible to summarise the values of the original areas as attributes for the newly merged areas. The question of how attribute data is to be transferred to the new areas is dependent from the kind and content of the attributes. This transfer must be carefully planned; otherwise it can lead to erroneous results.

6.4.3 TYPES OF SPATIAL OVERLAY

Overlay functions include clip, union or intersection operations and are described in the following sections in detail. In general, the different overlay operations produce different results from the same

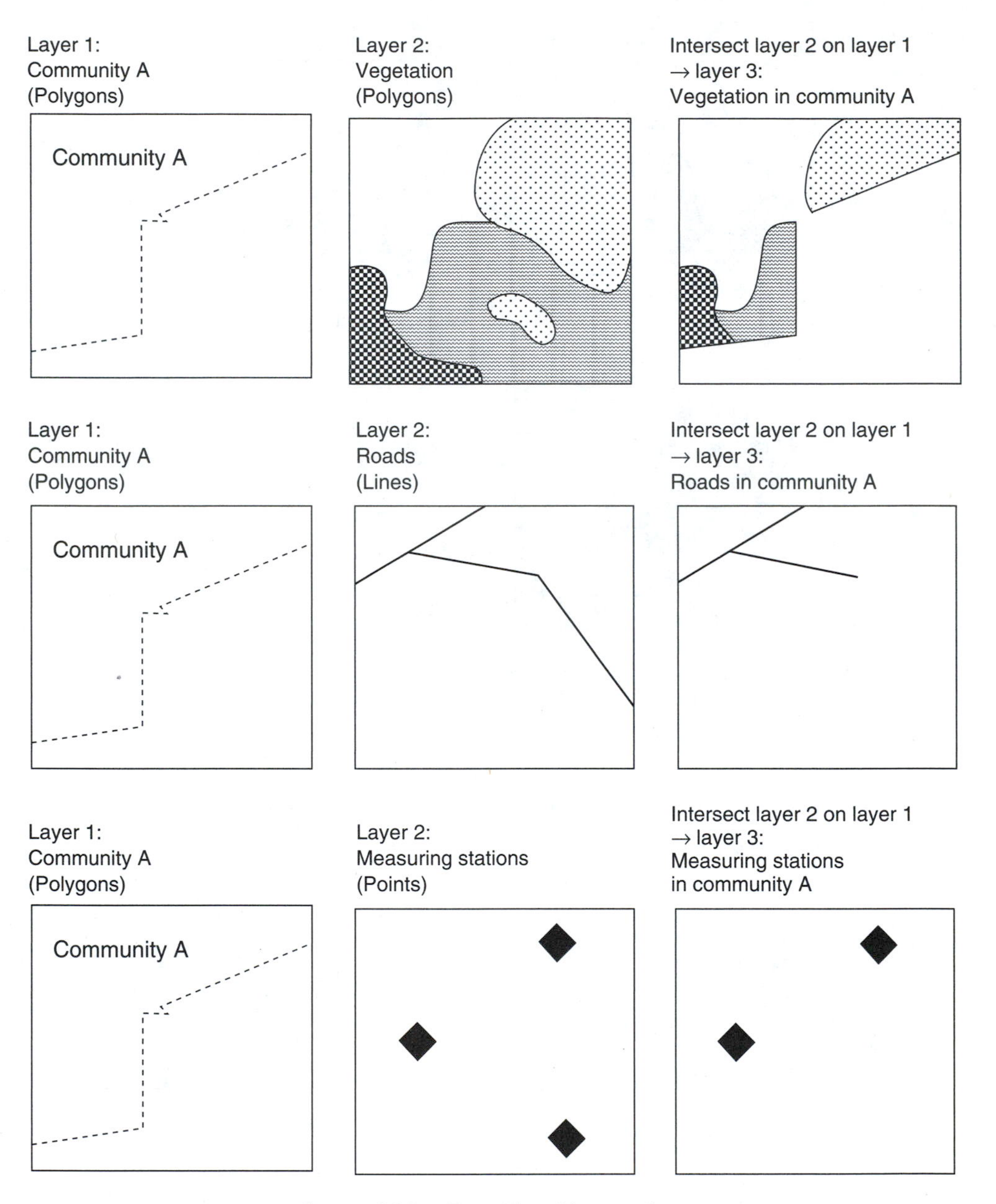

FIGURE 6.23 Examples of intersections.

input data. The result of an overlay operation is a new layer with new geometry and with all the items of the overlaid layers. The problems mentioned in Section 6.3.2 concerning the transfer of attribute data have also to be taken into account in this case.

6.4.3.1 Intersection

Intersection combines features of two layers and preserves the features in the common area and the attributes of both layers (Figure 6.23). In formal logic "Intersection" corresponds to the operator AND. Intersection means a polygon-on-polygon, a line-in-polygon or a point in polygon-overlay. The intersection layer must always have a polygon topology, but the output will depend on the kind

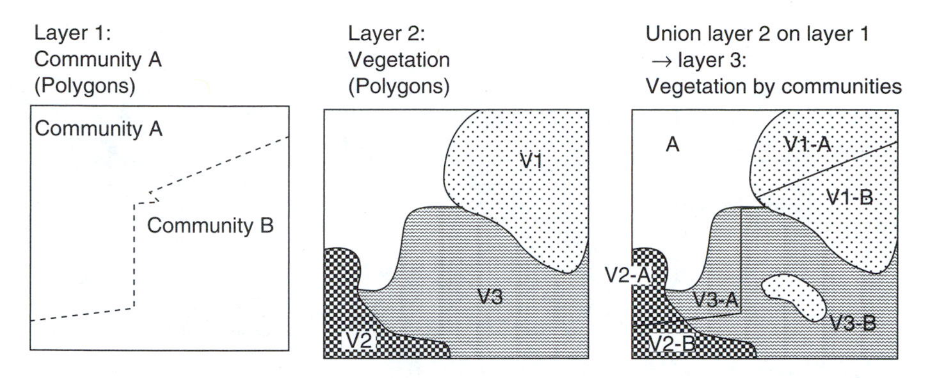

FIGURE 6.24 The UNION operation in vector GIS.

of features in the first layer. In the case of a line-in-polygon intersection the resulting layer would be a layer with line topology, a point-in-polygon intersection results to a point layer. A line-in-line or a point-in-point intersection can be done by copy and paste operations.

Typical questions, which can be answered by use of intersection include:

- On which grounds (in which polygons) is there vegetation?
- Which streets run through a certain area?
- In which municipalities are measuring stations located?

6.4.3.2 Union

Union also combines polygon features of two layers but preserves the entire area and all attributes of both layers. In formal logic union corresponds to the operator **OR** (Figure 6.24).

6.4.3.3 Identity

Identity also computes the geometric intersection of two layers. The order of two polygon layers does not affect the result of union or intersect. If we use identity for computing the intersection of two polygon layers, the first layer serves as the base layer, the second is the overlaid layer. There would be a different result, if we change the order of the two layers (see Figure 6.25). In the output layer are preserved all features of the base layer and those features of the overlaid layer that overlap the base layer. If there were two layers with identical spatial extents, the result of identity would be the same as using **UNION**. Identity also allows the geometric intersection of point-in-polygon or line-in-polygon.

6.4.3.4 Erase

Erase also creates a new layer by overlaying two layers. One layer defines the erasing region and all features of the other layer that are within the erase layer region are removed. The output layer will contain only those input features that are outside the erasing region. For example, if one would like to summarise the woods areas of the European Union (EU) states from a vegetation map of Europe then this command can be used. As the erase layer one could define the national frontiers of those countries that are not part of the European Union, such as Switzerland. The input layer would be the vegetation map of Europe. The result will show only those forested areas that lie within the European Union, whilst all other forested areas, and all other vegetation types, will be given a value of 0 in the output (Figure 6.26).

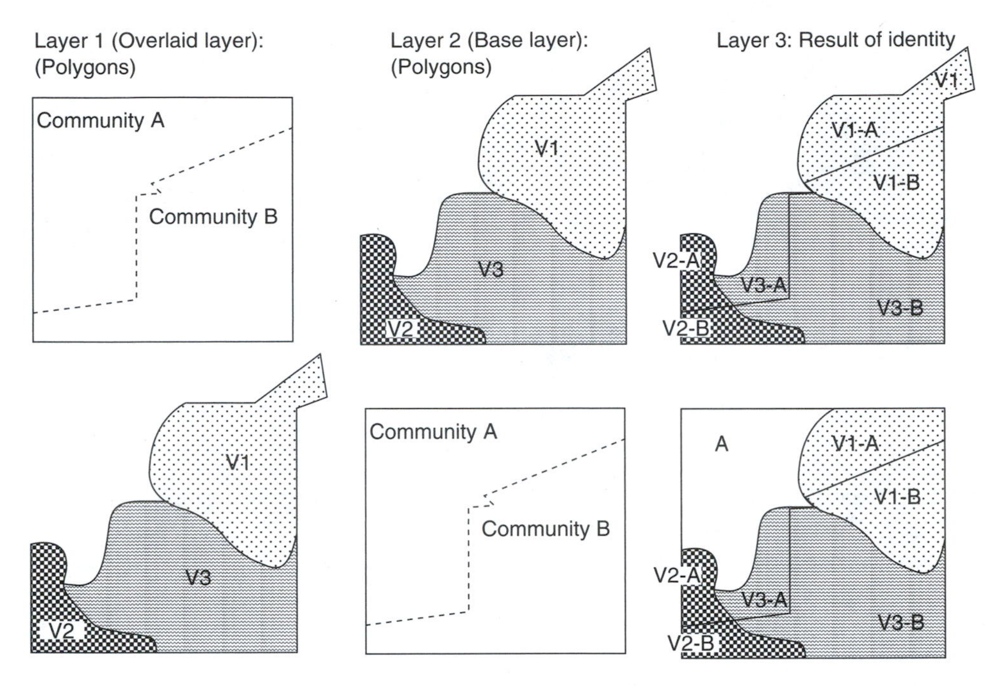

Figure 6.25 The IDENTITY operation in vector GIS.

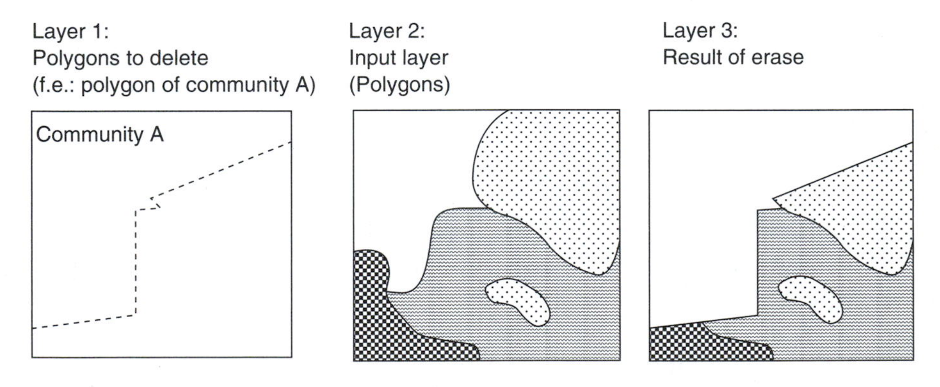

Figure 6.26 Examples of the application of the ERASE function.

Polygons will also be removed with the "delete" or "eliminate" functions, however, with these functions the conditions for the areas to be deleted are defined. With "delete" or "eliminate," all polygons, which satisfy specified conditions, are deleted. With the command "erase" on the other hand, all features which lie within a certain area, are deleted.

6.4.3.5 Update

Update replaces a part of the map layer by the data for that area from another layer using a cut and paste operation. In most GIS the user can decide whether to keep the boundaries between input and new layer or remove these boundaries after the update. If you choose to keep these boundaries and the input and update layers have identical item definitions then the result of update would be the same as if you use the **UNION** command. If you use, in a second step to **UNION**, the command **DISSOLVE** you will get the same result as if you used only **UPDATE**. This command is very useful

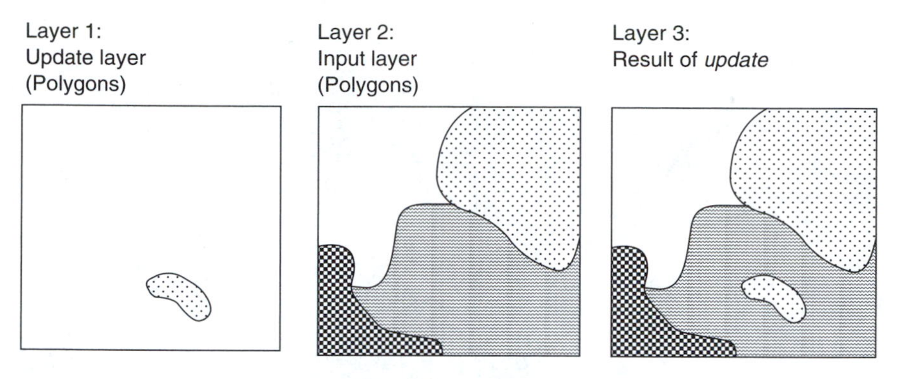

FIGURE 6.27 Examples of UPDATE.

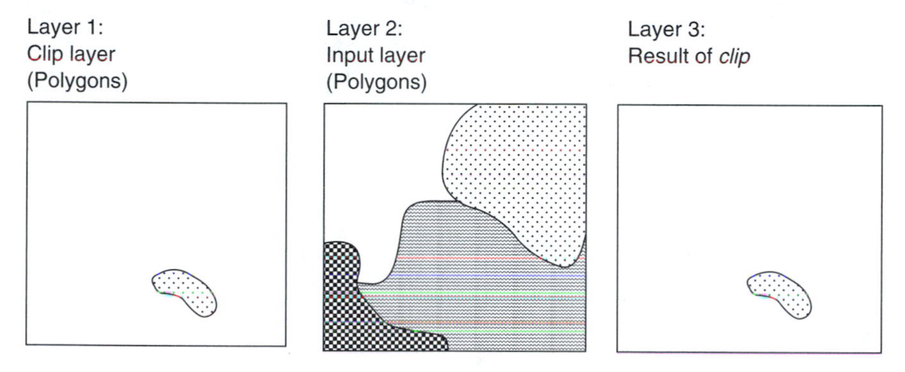

FIGURE 6.28 Examples of CLIP.

when new data exists for just part of a map area, or when more detailed data has become available for just part of the mapped area (Figure 6.27).

6.4.3.6 Clip

A clip layer defines the area on the input layer from which spatial objects are extracted. However, unlike the command **ERASE** the result contains only the unchanged features of input layer within the area of the clip layer. Consider the example given in the section on **ERASE** where the interest was in displaying the forested areas within the European Union, given a vegetation map of Europe. If one is interested in the reverse, the forested areas outside the European Union, then **CLIP** can be used. As a clip layer one could define the national frontiers of those countries that are not in the European Union, such as Switzerland and the input layer would be the vegetation map of Europe (Figure 6.28).

6.4.4 Proximity Analysis

Proximity analysis and buffer creation are essential features of GIS functionality. It is also possible to visualise buffer zones in various desktop publishing and CAD programs, even if this means using greater line widths. Only in a GIS will the result of **BUFFER** operations be new topological objects (polygons), which can then be used for further analysis. A typical question, which can be answered by proximity analysis is, "Which biotopes fall within a corridor around a planned new street and how big is the area share of the biotopes within the corridor?" Another question, which can be answered by proximity analysis is, "How close are certain specified features to other specified features, for

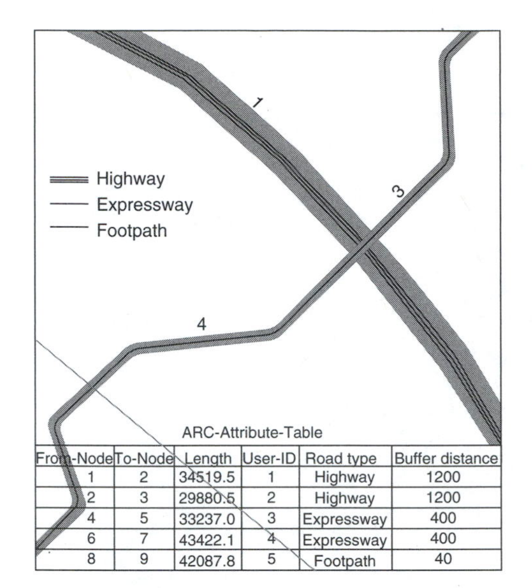

From-Node	To-Node	Length	User-ID	Road type	Buffer distance
1	2	34519.5	1	Highway	1200
2	3	29880.5	2	Highway	1200
4	5	33237.0	3	Expressway	400
6	7	43422.1	4	Expressway	400
8	9	42087.8	5	Footpath	40

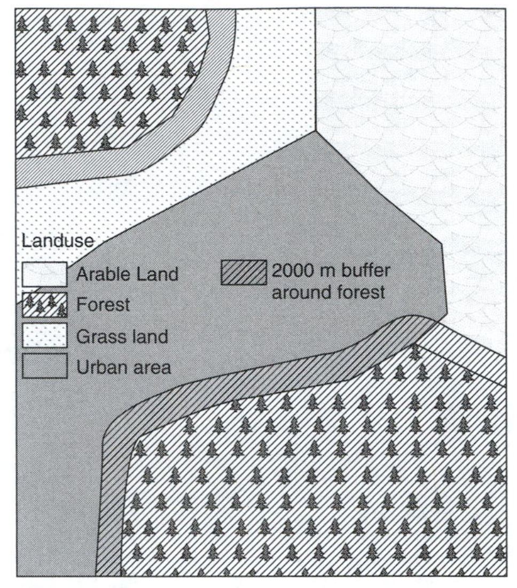

FIGURE 6.29 Examples of a line buffer with different buffer distances and a polygon buffer with the same distance around all forests.

example, what are the distances of nesting sites to rivers or roads?" These distances can subsequently be classified into distance classes (Figure 6.29).

6.4.4.1 Near

With **NEAR**, distances are calculated by points in the first layer to the nearest lines, points or nodes in a second layer. As a result of this command, the attribute table of the first layer is enlarged by appending two items, for each point the Euclidean distance and the internal number of the next object in the second layer. This information can be used as a base for further analyses or for distance classification.

6.4.4.2 Buffer Creation

Buffer functions create polygons around specified input layer features. A constant distance value can be indicated for all input features, or an attribute item, which contains different distances depending on the item value, and can be used as the buffer distance. For example, it can be necessary, for the calculation of noise zones, to produce different sized buffers for streets that are a function of the daily vehicle density on those streets. Buffer polygons can be created around points, lines or polygons.

6.5 DATA MANAGEMENT IN A GIS

In the analysis of images and GIS layer data it is very easy to create new layers, and it is also very easy to forget exactly what has been done, and the name of the resultant files. There is thus a significant need to establish rigour in the way files are created, named and stored, so as to maintain order in the system and allow the analyst to reclaim previously conducted work.

We have seen that the data in a GIS consists of spatial and attribute data. We have also seen that the spatial data can be of various types, including raster, vector and annotation data. Other types of spatial data include AOI data defining areas from which statistical information can be derived. In addition to this the formats of the data used in the software system available to the user will often be unique and different to the formats used in other systems, or the formats used for the standard storage of data. The normal way to solve this problem is to give each data type a unique but standardised extension. These extensions would normally be dealt with automatically in the software, so that the analyst does not have to consider them when conducting analytical work. However, it is important for the analyst to be aware of the different extensions and the type of data that they contain, as the extension used on a file will immediately tell the analyst the type of information that is contained in the file. It is obviously a good idea to maintain these conventions, and not to change the assigned file extension. All image-processing systems and GIS adopt this approach in relation to the files that are used internal to the software.

When file names were limited to 8 characters it was very difficult to make these names meaningful. Nowadays, most systems can accept longer filenames. Whilst the shorter the name, the easier to type and the less prone to errors during typing, there none the less are significant advantages in ensuring that the names give an idea of the type of information contained in the file, and the origins of the data. In doing this the analyst can also adopt certain conventions in the abbreviations used, the number of characters used for each segment of the name and the number of segments in the name. The analyst may decide to have four segments in the name, consisting of:

- The project or area code, a 3-character code that is unique to the project or the area of analysis within the project.
- The information type in the data as a two character code indicating the type of data, for example lu = landuse, lc = landcover, cr = rainfall, ct = temperature, vi = vegetation index, la = LAI values, etc.
- The year of the data as a two-integer code.
- The sequence of analysis as a single integer code. Often the same analysis is conducted a number of times, with changes to some parameters, before the results achieved are satisfactory. When this occurs, then having an analysis sequence code can be very valuable.

This approach will give an 8-character file name. However, it may not contain enough information to suit the analyst or the users. For example, it does not contain information on the processing steps that have been conducted, the source data sets used in the construction of the layer, or information on ownership of the data. In different situations, all these facets and others may be critical. When this occurs, then the analyst may need to select filenames that are longer than 8 characters.

Structure in the data can also be improved by the use of directories and sub-directories. It is often the case that source data may be used by more than one project. If this is the case, then the source data should be kept in its own directory, and separate from the unique project files. Each project should be allocated its own sub-directory, and if the project is divided into a number of activities than each activity should be given its own sub-directory within the project directory. Common data files should be kept at the level that is the parent to all of the applications of the data. A typical structure could look similar to that shown in Figure 6.30. The file structures that could be constructed to manage the data for these projects are represented. In this structure the information and data at a level is listed in the relevant ellipse. The other projects can access the higher and same level directories. Thus the BIOD and SOIL projects can access the landuse and LAI data in the leaf are index mapping (LULA) directory, but not the Private and Public sub-directories to the LULA directory. All three projects can access the main directory.

The public sub-directories are of data that are and can be made public. The Private sub-directory will contain unique data and analysis that may have been collected specifically for the project, that may contain private information that cannot be disclosed and intermediate results and analyses. Of course if there are sub-projects, for example, the LULA project could have landuse and LAI sub-projects, then a more involved directory structure would need to be made. Similarly, there may be activities within each sub-project that would be best managed with their own sub-directory. In general, it is better to have too many sub-directories than too few.

Once the directory structure has been designed, and every time that it is amended, then this metadata needs to be recorded and information on the metadata file structure, as well as the filename conventions should be kept in the top directory.

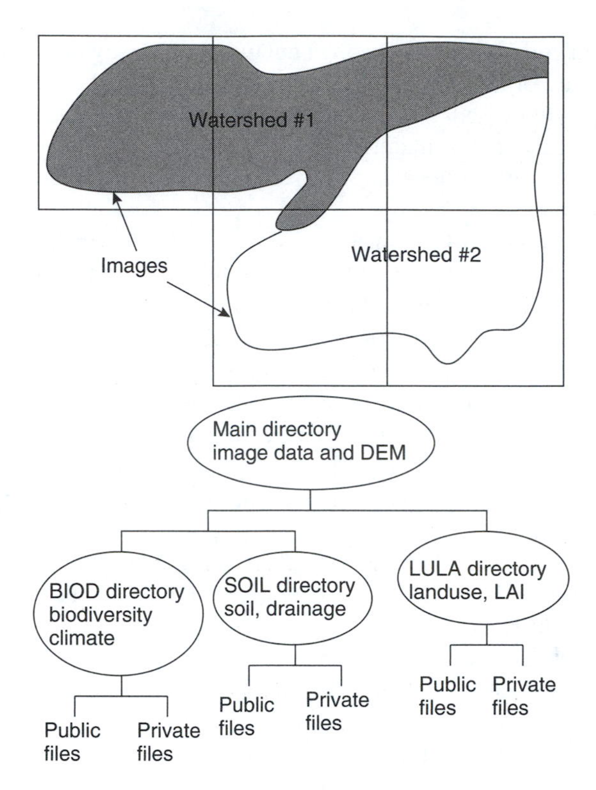

FIGURE 6.30 Example file structure for three projects; Biodiversity Investigation (BIOD) in Watershed #2, Soil Mapping (SOIL) in both watersheds and Landuse and LULA in both watersheds.

The next step is to manage the processes that have been conducted. The processing that has been conducted can be recorded in at least three alternative ways:

1. *As an entry in the data header file*. Thus the header file can retain a record of the history of that file. This is quite a handy way to record the history if the software allows this to happen, and even better if the software does this automatically. Thus, the data may record that the data was rectified and the filenames of the control points and residuals, calibrated, again with relevant filenames, classified, with details of the classes and the signature filename, etc. Clearly, this can lead to very large header files, particularly if the parameters used in each processing step are also retained.
2. *By retaining the record of processing conducted*. Many systems record the processing conducted in log files. These files can be retained. A difficulty arises with this if the analyst is handling a number of sub-tasks in the same session, and they all get lumped together in the log file. The log file will also retain a record of all of the analysis that has been conducted in that session, so that the log file contains a lot of irrelevant details.
3. *Recording the activities as they are conducted*. The operator holds a file for each project, sub-project or activity, and enters into this file the details of the actions taken as they are conducted. Whilst this is a slower process, it is none the less very reliable if conducted properly.

There are also a number of data management tasks that may require action within the GIS. These tasks include:

1. *Data merging and windowing*. Data merging is the process of joining a series of geographically adjacent layers together to form one layer of all of the areas represented in the

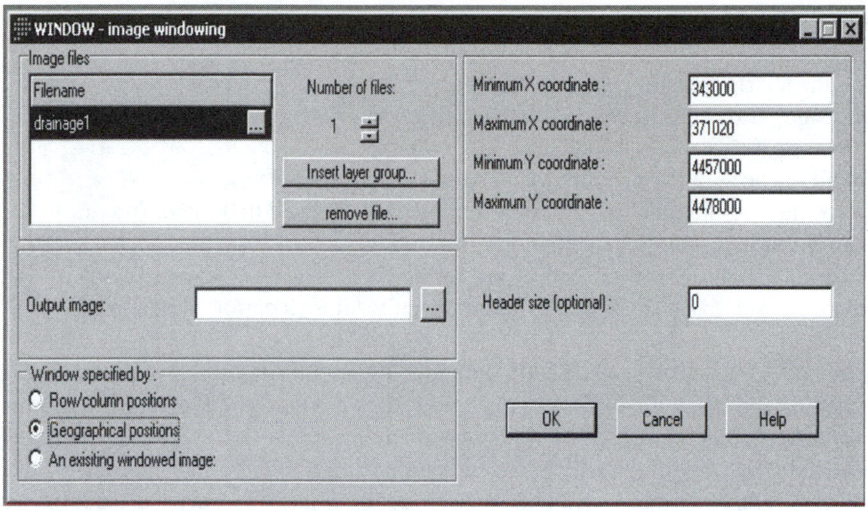

FIGURE 6.31 The window provided in Idrisi to window into an image file.

different source layers. The normal process is to specify a core layer, and then to specify the names of the adjacent layers and their locations relative to the core layer. With some systems, since the co-ordinates of the layer corners are known, the way that they are to be joined is taken automatically from the corner co-ordinates. If areas exist within the derived layer area that are not covered by any of the layers, then this area would typically appear as black. How duplication is dealt with varies from system to system, where some systems will not proceed when duplication is discovered. Windowing is the process of making a copy of a sub-layer area from within a layer area.

Windowing is thus the reverse of merging. An application of these techniques could well be to merge the layer information derived from image data, so as to form the one large layer that covers the AOI. Then windowing may be used to create a subset of the AOI. The window used to window in Idrisi is shown in Figure 6.31. The analyst specifies the source file, in this case drainage 1. The system automatically displays the co-ordinates of the boundary area of the layer. The operator then modifies these co-ordinates to specify the area to be windowed, and then provides a new filename for the system to act on the instructions and create a windowed file.

2. *Data masking*. Masking is the process of using a mask that shows the areas that are of interest to the analyst that is to mask out areas that are not of interest. Once the layer has been masked in this way, then areas masked out will not be analysed, nor will statistics be derived for these areas. Masking thus enables the analyst to focus himself and his clients on the AOI, it avoids wasting time conducting analysis on areas that are not of interest, and it means that any statistics that are derived are related to the AOI.

6.6 ADVANCED ANALYSIS TECHNIQUES IN A VECTOR GIS — NETWORK MODELLING (SUSANNE KICKNER)

The network data model is an abstract representation of the components and characteristics of real world network systems like rivers, roads, power grids or cable networks. In general, a spatial network is a system of interconnected linear features through which resources are transported or communication is achieved. A spatial network can help to solve the following problems:

- Find the shortest path through a network.
- Solve the "travelling salesman" problem (find the most efficient path joining a series of locations).
- Assign portions of the network to a location (allocation).
- Determine whether one location in a network is connected to another (tracing).
- Estimate the potential for interactions.
- Calculate distances between sets of origins and destinations (distance matrix).
- Determine site locations and to assign demand to sites (location-allocation).

To be able to analyse a network, some additional topological information is required. These features as well as details and methods for some application cases are described in the next sections.

A vector-based network model is more suitable than a raster model for analysing precisely defined paths. Raster network modelling uses a completely different approach to the topological-linked vector model. The grid cells of a raster-based GIS can only approximate the exact shapes of lines due to the resolution constraints of the cells. Direction is also only approximated with a raster-based system since there are only 8 possible directions from each cell. In addition, the line and cell attributes in a raster network must be stored as separate layers for each attribute, resulting in a vast number of layers. A raster-based network model on the other hand has advantages, when the problem is concerned with finding a path across terrain and where the criteria for finding the route are spatially variable parameters, such as slope, soil type and so forth.

6.6.1 Topology in a Network

Line data, for example, representing traffic roads, pipelines, power networks or rivers, requires additional topological information so as to show whether two lines (arcs, edges) are directly connected, whether one line crosses another or whether a change in direction may occur. This type of topological data is called network data. Topologically structured network vector data is based on graph theory. A graph is the complete set of nodes and edges (arcs, lines) in structured order and with topological data that defines the possible connections between all edges. This is called the nodes–edges model.

Topological network information can be generated from ordinary line data in GIS. It is necessary, however, to take some special features into account. One or more nodes are set at each crossing of two arcs. If the arcs intersect, as occurs at road junctions, then one node is set to an edge at the intersection. If the arcs cross over at some displacement, for example, a road going over a river, then nodes are set in each arc at the same horizontal position but separated in height. Sometimes intersections or turns cannot be followed, for example, a one-way street may intersect with a two-way street, in which case access against the traffic in the one-way street is forbidden. When this situation occurs, then the node may be given a turn impedance or resistance (normally a minus value). These nodes are entered at the digitising process manually or designed by GIS routines afterwards.

For a network as a real-world model, edges have an associated direction and a measure of impedance, determining the resistance or travel cost along the network. The To–From and the From–To directions are distinguished at the edges. The direction of a graph is always important. Different feature or attribute values must be assigned to edges depending on the direction of movement along this line. The attribute or feature values will also depend on the load or carrying capacity of the system. For example, there is only one flow direction in a river, but its velocity and volume of water can vary. With a road, there may be two lanes in one direction, but only one in the other direction, and the volume of traffic will vary over time. These types of graphs are called a directed graph or digraph.

The resistance value, called impedance or cost, must also be fixed for the lines and nodes. The impedance values are resistance factors that determine the time, length or monetary costs of the movement along a line or of the turn or crossing of a node. Impedances could be derived by several line or node attributes, such as length, speed limit, delay at traffic or in taking turns, road class, slope or flow properties.

6.6.2 Address Geocoding

Address geocoding is not a network function in the true sense. However, companies and institutions frequently have address data of customers or business partners who shall be supplied or visited. There are tools available in most GIS to enable the analyst to use these addresses as a destination or starting point of a route. Address geocoding is used to find locations on a map and provides the means to combine the geometric topological relations of the road network with the address data like customer names, street names or street numbers. For each address the co-ordinates will be computed so that the address can be used in spatial analyses.

6.6.3 Path Finding in Vector GIS

The search for the shortest or least cost route or connection is one of the most frequently conducted network analysis tasks. Consider a traffic network: What is the shortest route between A to B; which is the fastest route from A to B, or which is the cheapest route from A to B when locations C, D and E have to be passed en-route? For the shortest distance, only the travel distance is considered. For the fastest route, the distance and the legal and safe velocities have to be considered. For the cheapest, the distance, velocity and the costs, including tolls need to be taken into account. One approach is to examine all possible paths between the passing points and choose the path that best meets the set criteria. However, it would be nearly impossible to examine all possible paths between the two points in a larger network and so path finding algorithms have been developed. These algorithms fall into two main categories: matrix algorithms and tree-building algorithms. Tree-building algorithms are most commonly used in GIS network analysis.

Matrix algorithms work in iterative steps by sequentially eliminating the least favourable nodes. Tree-building algorithms find the shortest path from an origin node to all other nodes, producing a spanning tree. A spanning tree of a graph is any tree that includes every node in the graph. Formally, a spanning tree of a graph G is a sub-graph that is a tree and contains all the nodes of G. An edge of a spanning tree is called a branch. It functions by constructing a shortest-path tree from the initial node to every other node in the graph.

The most implemented algorithm is that originally developed by Dijkstra (1959). The Dijkstra algorithm solves the single-source shortest path problem in a weighted, directed graph. Edges in this graph represent possible connections and are weighted after the path length or cost. The complete graph is investigated starting out from the start node and in a complete search the shortest way to the destination node will be searched. Dijkstra's algorithm partitions vertices in two distinct sets, the set of "unsettled" vertices Q and the set of "settled" vertices S. A node is considered settled or reached, and moved from Q to S, once its shortest distance from the source has been found. Initially, all vertices are unsettled, and the algorithm ends once all vertices are in the settled set.

Steps to solve the problem:

1. At the origin node or the initial point as the first "reached" node. The origin node is the first node in set S with a cumulative impedance of 0. Starting from one origin node, the search tree builds branches in all directions.
2. For all adjacent nodes to the last reached nodes (the nodes at the other end of the arcs connected to the "reached" node or nodes) except the already reached nodes are typed with the respective cumulative time, length or cost of movement into a temporary event list, the set Q. The vertex with the current least cost is extracted from the queue and added to set S as a "reached" node.

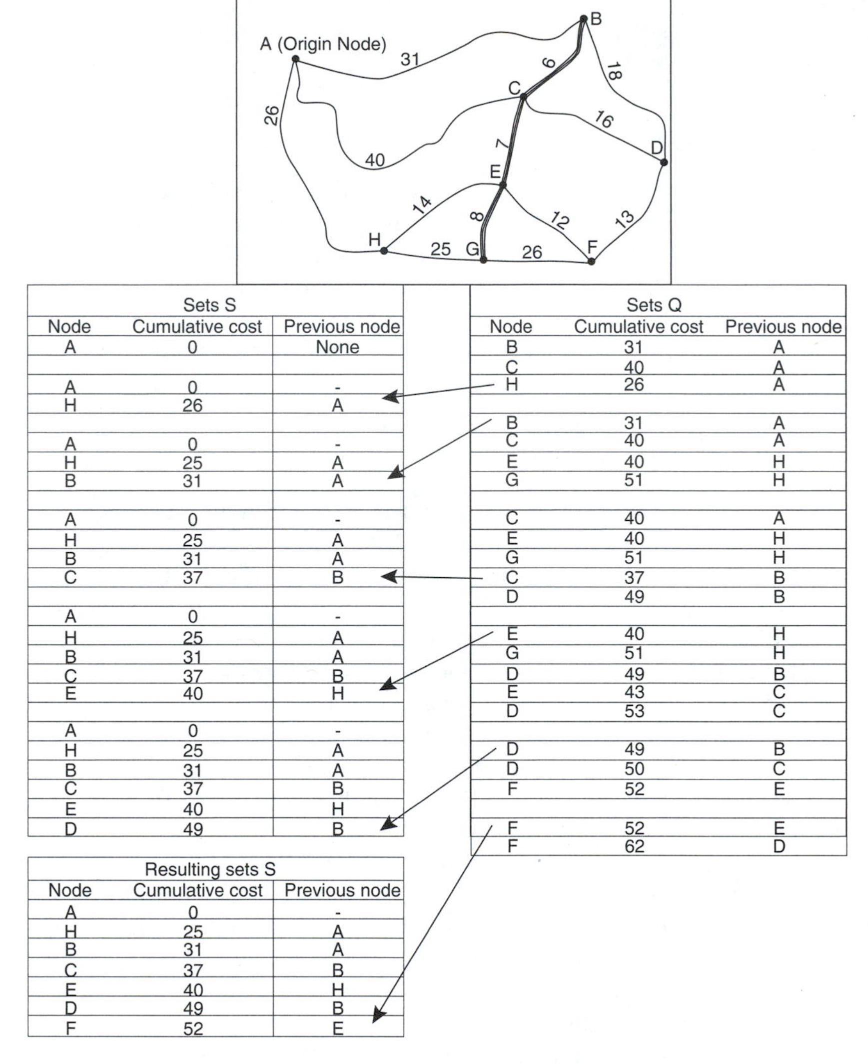

Sets S		
Node	Cumulative cost	Previous node
A	0	None
A	0	-
H	26	A
A	0	-
H	25	A
B	31	A
A	0	-
H	25	A
B	31	A
C	37	B
A	0	-
H	25	A
B	31	A
C	37	B
E	40	H
A	0	-
H	25	A
B	31	A
C	37	B
E	40	H
D	49	B

Sets Q		
Node	Cumulative cost	Previous node
B	31	A
C	40	A
H	26	A
B	31	A
C	40	A
E	40	H
G	51	H
C	40	A
E	40	H
G	51	H
C	37	B
D	49	B
E	40	H
G	51	H
D	49	B
E	43	C
D	53	C
D	49	B
D	50	C
F	52	E
F	52	E
F	62	D

Resulting sets S		
Node	Cumulative cost	Previous node
A	0	-
H	25	A
B	31	A
C	37	B
E	40	H
D	49	B
F	52	E

FIGURE 6.32 Application of the Dijkstra algorithm for finding the shortest route from A to F.

3. Update set Q, by retaining all of the nodes in Q from the previous step, except for the node that was "reached," and adding the nodes connected to this newly reached node to set Q.
4. Repeat steps 2 and 3, until all nodes are added to set S as "reached nodes."

An example is given in Figure 6.32 of a street network with different road types and travel times as impedances.

For many resource management questions it is not sufficient to have information on how fast one can reach point A from B. For example, the time required loading the cargo onto the truck or the order of delivery must also be taken into account. These factors can be taken into account by storing information on them in separate files, for example, locations that need to be visited for delivery or pick up, the delay that will occur at each of these staging points and so forth. In such a file the delay times involved in each visit, the purpose of the visit (e.g. refueling) and the sequence of visits are entered into the file as attribute data to the locations of the stops. A good vector GIS can then determine the minimum-impedance path between each stop and every other stop and then reorder the path, within the constraints imposed so as to find the minimum impedance from the start to the end nodes. The classic problem of finding a path through a weighted graph which starts and ends at the same node, includes every other vertex exactly once, and minimises the total cost of travel is called the Traveling Salesman problem. Less formally this means finding the lowest cost path for a salesman to visit every listed city in his itinerary. For very large networks, approximate or heuristic solutions are normally used because it is not currently possible to implement mathematically correct solutions on current computers or within realistic times frames.

6.6.4 Location–Allocation

For a variety of reasons, there are often advantages in providing services from a few centralised sites. Sometimes, the number of sites is known in advance, in other cases the optimum number of sites has to be found. The question of where to place these central facilities can be solved by location models, whilst allocation models can be used to determine the subsets of the total demand that should be met by each of the proposed sites. Allocation models are used to find the best solution when, for example; (a) there are limited sites to collect and store agricultural products in a farming area; (b) there is a need to define service or trade areas in a residential area (an example of doctors services is given in Figure 6.33) or (c) there is a need to assign streets within a 10 min radius to the nearest fire station under consideration of the road net.

In a vector GIS this is done by assigning arcs or nodes in a network along the least impedance paths to the closest facility, until the facility limit or each arc's limit of impedance is reached or in other words, allocating a region in space to a point facility.

Depending on the nature of applications all or part of the following information must be specified:

1. *A centre file*, usually a point layer that contains information about the supply capacity of the centres and the maximum impedance of allocated arcs or nodes, for example, the most reasonable travel time or the maximum acceptable cost.
2. *Impedance items* and assigned values in the network layer, reflecting the nature of the problem being addressed.
3. *Demand items* and assigned values in the network layer, specifying the level of demand on the arcs or nodes, for example, population, number of students or the quantity of the produced agricultural goods which shall be delivered to a central place.

Figure 6.33 Example for allocation of doctors services in a township area.

Figure 6.34 Example of a location–allocation solution to the provision of medical services in a town.

Consider an example. The provision of medical support in a town is to be reviewed in light of changing population numbers and distributions. For this analysis it is assumed that:

1. It is reasonable for citizens to travel up to 5 km to the doctors surgery for a consultation.
2. A single doctor can provide care for at most 3000 patients.

The sites of three doctors and the number of the inhabitants per street section are known. Analysis shows that the inhabitants of large portions of the town must go further than 5 km to reach a doctor. The assumptions suggest that more doctors' offices are required. However, analysis also shows that the service area of the different doctors is very different. Doctor 1 services a large area since the population density in this area is low. The population density is greatest in the area serviced by Doctor 3, who has only a spatially small service area.

Location–Allocation models can be used to provide the population with the minimum number of doctors yet which meet the criteria specified above. There are a large number of alternative heuristic algorithms that can be used to solve such a problem (e.g., the Global Regional Interchange Algorithm [GRIA] or the Teitz-Bart heuristic model [Teitz and Bart 1968]). For the given example the best solution using the GRIA algorithm is ten doctors in the solution shown in Figure 6.34.

6.6.5 Gravity Models

Gravity models assume that the influence of phenomena or populations on each other varies inversely with the distance between them. Gravity models help in solving the following problems in a GIS:

- To compute the potential of a site under consideration given the attractiveness of all other locations to the users of the site being considered, for example, to select a site for a hotel in a tourist area.
- To compute levels of interaction (flows) between pairs of locations taking into account the properties of the origin (this is usually a function of the population at that origin) in producing trip and destination properties (for a consumer this can be warehouses or for a worker this can be offices) that contribute to the occurrence of trips between the pair of locations.

The population potential of a region is a measure of the possibility of spatial interactions. The larger the attainable population and the lower the effort needed to pass through an area, the higher the population potential. For example, a high market potential is an advantage in site selection for a business that needs to be accessible to the buying public. The population potential represents a dimension of concentration. It is analogous to physical gravity, by assuming that the "gravitational attraction" of a place increases with its size and density, and decreases in proportion to its distance from other sites. In early work, the population potential was calculated using Newton's gravitation approach of the sum of the populations of the reached areas, divided by the exponentially weighted direct distance between the two places of interest. More recent studies have shown that negative exponential functions that weight the travel path lengths or times give better results since they better reflect the action spaces of interacting individuals. Other measures of attractiveness can be used instead of population when they are appropriate. Attractiveness means to combine in one number the factors that make a centre attractive; this could be the retail floor space, the number of the employees or the income of the population.

The accessibility or the regional potential of an area by using a exponential distance–decay function is calculated with the following formula:

$$P_i = \sum_{j=1}^{n} A_j \times e^{\beta d_{ij}} \tag{6.17}$$

where P is the accessibility or regional potential, A is the attractiveness, e is 2.7183, n is the number of locations in the region, β is the distance exponent and d_{ij} is the impedance between i and j, for example, time for travelling over the road network from i to j.

6.7 ADVANCED RASTER ANALYSIS TECHNIQUES IN A GIS

The activities conducted using the techniques described in Section 6.4 will often be used in most projects as steps in the derivation of the final data. They are commonly used techniques in GIS analysis. There are a number of other techniques that are very powerful but which may not be used as often as these techniques, but which nevertheless are very important in the analysis of GIS data. These techniques include:

- *Geostatistics* or the analysis of spatial interdependency.
- *The use of map algebra and script languages.*
- *Interpolation techniques.* Interpolation techniques take a distribution of point observations and interpolate between them to create a surface or image layer. There are a number of different methods of conducting interpolation.
- *Analyses of cross-tabulation tables* or the correlation between two GIS layers.
- *Analysis of clumps* or areas of contiguous class coverages.
- *Zonal analysis* or the analysis of layer attributes within vector coverages.

6.7.1 Geostatistics

Geostatistics have been covered in Chapter 3 in relation to image processing. The reader will find the details of the semi-variogram, its computation and interpretation in that section.

The use of the semi-variogram in a GIS context has much to do with its use in interpolation as will be discussed in Section 6.7.3, or for spatial analysis as has been discussed in Chapter 4.

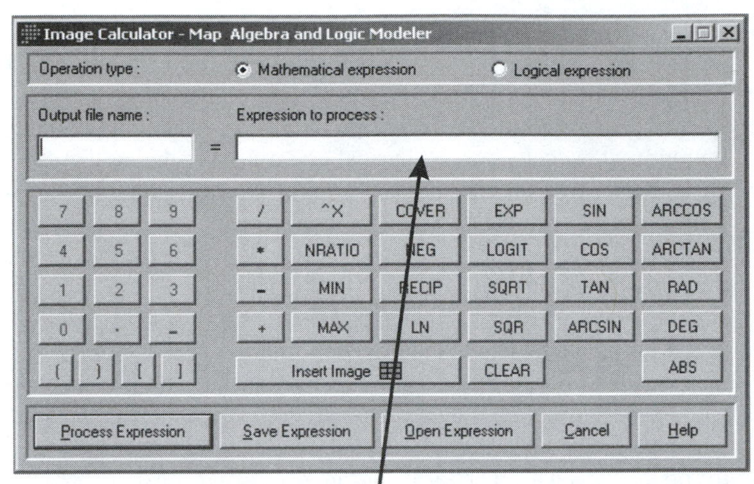

FIGURE 6.35 The map algebra window in the Idrisi software for numerical analyses.

6.7.2 MAP ALGEBRA AND SCRIPT LANGUAGES

Many software systems include script language facilities and map algebra that allows the user to construct sophisticated analysis tools not available from the standard features offered by the software. Such facilities also allow the user to customise certain practices so as to achieve higher levels of automation than that existing in the standard software. The map algebra window in Idrisi is shown in Figure 6.35 for numerical analysis. The operator can select either numerical or logical processes, with different options available on the keypad for logical operations than those shown in the figure. The operator specifies the output image name on the left-hand side and then constructs the function on the right-hand side. This function will include raster data, specified using the Insert Image key, and a combination of the function elements given on the keypad.

In other software, such as Imagine, map algebra is incorporated into a graphical script language. Many of the functions in Imagine are created using this script language, using a set of basic modules. All these script language modules are given with the software. When one opens the modeller, one is given a blank work pad on which they can draw the graphics depicting the processing to be conducted and option menus as shown in Figure 6.36, and a toolbox. The user creates a graphic similar to that shown in Figure 6.36 using the toolbox of graphic elements that include input and output (raster, vector, matrix, table and scalar) objects, direction of processing arrows and function objects. To build a process, the operator simply activates the selected graphics in the toolbox window and then double clicks on the workspace pad to have copies of them deposited on the pad. Arrows are used to show the flow in the process. Double clicking on each input or output object deposited in this way on the work pad opens a window. The operator sets parameters in this window about the data that will be represented by that window. For example, the data set can be specified during construction or during running of the process. The data can be input, a temporary file created during the process, or output. The operator can also specify if the data will overwrite an existing file, if it exists. The n1_landcover (Figure 6.36) indicates that the landcover data is specified during construction of the process and that it is available for subsequent processing in the process as variable n1. The n2_PROMPT_USER indicates that the second set of data is to be specified during running of the process, and that it will be available for subsequent analysis in the process as variable n2. If multi-band image data is selected then both the whole image and individual bands are available for analysis in the process.

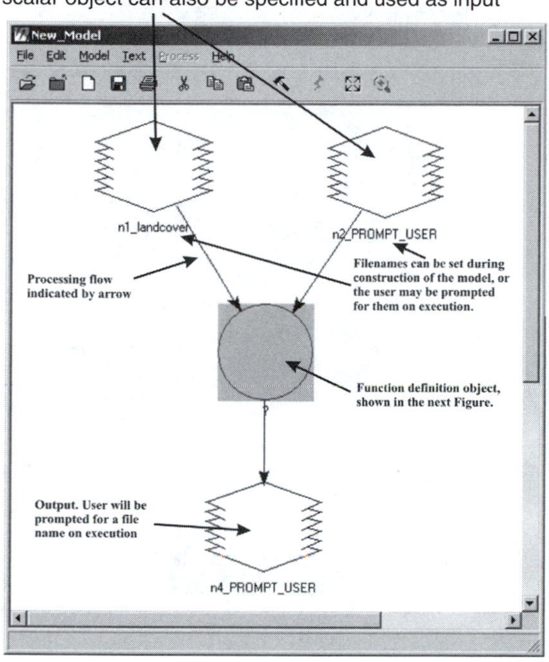

FIGURE 6.36 The modelling window in Imagine showing some of the graphical script language ikons and the flow directions for processing.

Activation of the function object creates a window as shown in Figure 6.37. The window shows the variables that are available to it, in this case landcover and the data specified by the user at run time. To the left is a list of the libraries of functions that are available, with the full list of the available libraries shown in the figure. Each library contains a suite of functions, and some of the functions that are available from the Analysis library are shown. The functions can be grouped into the following broad categories:

1. *Mathematical and trigonometrical functions* to conduct numerical operations on the data specified from that available, including matrix functions.
2. *Statistical functions*, to derive layer, local, global and stack statistics. Layer statistics are the statistics for the layer of data. Local statistics are statistics within windows in the data. Global statistics are to derive statistics across the whole suite of layers in the data, and stack statistics are to derive statistics at a pixel, but across the layers in the data.
3. *Testing, branching and evaluation* functions to test conditions and status and to choose between processing options in the analysis. This allows the development of "what . . . if" type processes and other types of choice actions.
4. *Colour transformations.*
5. *String operations.*
6. *A number of unique analysis transformations*, including principal components and distance operations.

The function is constructed using the appropriate function elements, where these may include:

1. *The data to be available to the process.* These include those that are to be used as source data, those to be created as an intermediate step and those that are to be created and saved

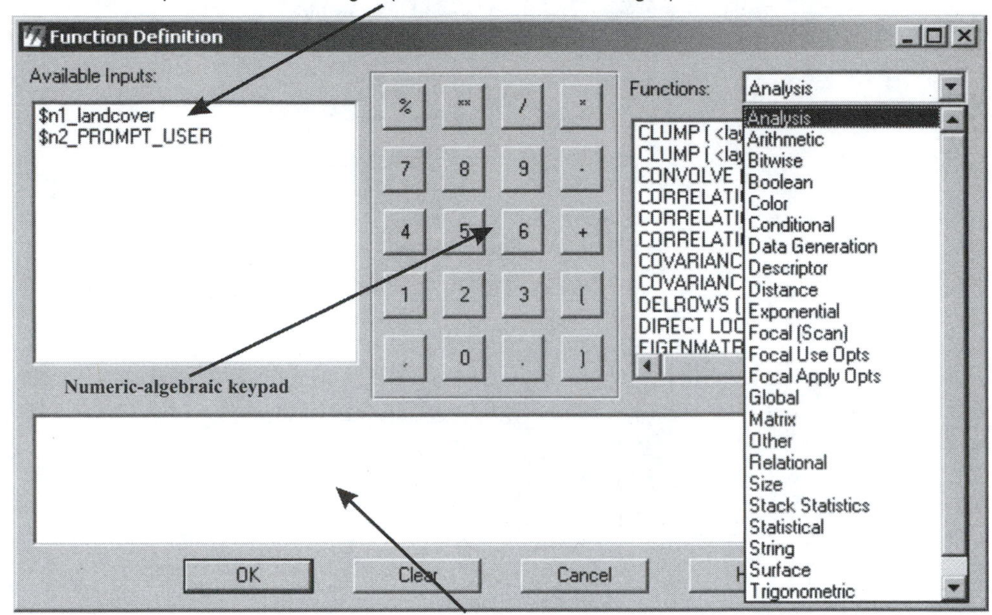

FIGURE 6.37 The Imagine function definition window used to create the function to be used in the model being constructed in Figure 6.38.

for subsequent analysis. The intermediate files are destroyed on completion of the process. Various types of data can be used in the process, including images, matrices, tables and scalar values and each have their unique icon. When the selected icon is double clicked in the tools palette, a copy is made and this is transferred to a location in the graphical display (Figure 6.36) using the mouse. Once the data icons have been transferred to the graphical display, or after the framework of the process has been constructed, then each data icon can be given parameter values. Double clicking on the icon in the graphical display creates a window containing options to do this. These options will include whether the data file name is specified now or during run time and the nature of the data. Other options can exist for different situations.

2. *The analysis to be conducted needs to be specified.* The analysis can be split into a number of steps, with each step separated by existing, temporary or new data files of the appropriate types. Each analysis step requires that a copy of the analysis icon be imported from the tools palette into the Graphics display space. Each analysis step can consist of various analysis options as previously introduced.
3. *The analysis flow* is specified by the user using the arrow icon. When this icon is imported it automatically clicks onto the nearest data and analysis icons. If these are incorrect, then the arrow icon needs to be moved to show the correct flow.

Consider two simple applications. The first will be to compute a normalised difference vegetation index (NDVI) image from image data where the reader will recall that NDVI = (NearIR − red)/(NearIR + red). In the second, the NDVI image is converted into an estimate of the fraction, photosynthetically active radiation (FPAR). FPAR is approximately linearly related to NDVI, where this relationship is known, but varies among different landcover classes.

6.7.2.1 Compute an NDVI Image

- Display the graphics workspace and the toolbox of icons. Set two image icons into the workspace, one for input and one for output.
- Set one analysis icon into the workspace.
- Connect the input image to the analysis icon and the analysis icon to the output image icon.
- Activate the input image icon, to reveal a window. In this window specify the name of the input image file. The characteristics of the image are displayed.
- Activate the analysis icon, to display an analysis function definition window (Figure 6.37). In this window are displayed the available inputs, the whole image and the individual bands in the input image. Set a bracket into the Function construction window, click on the Near-IR band image, a minus sign, the red band and then a closing bracket. You have now constructed the numerator. Insert a divide symbol, and then construct the denominator. The analysis function is now complete. It will look something like this ($n1_image[3] − $n1_image[2])/($n1_image[3] + $n1_image[2]) where band[3] is the Near-IR band and band[2] is the red band. Note that if both $n1_image[3] and $n1_image[2] are zero then a division by zero will occur and this will create an error. If this is possible with the data being used, then it is necessary to protect the process from this type of error. Protection can be in the form of an IF test, however, the easiest way is probably to add a unit value to both bands, to give the equation ($n1_image[3] − $n1_image[2])/(2 + $n1_image[3]+ $n1_image[2]).
- Specify the output image, or allow it to be specified at run time. Be sure to specify that the output image file will be of type float for this particular process. If you want to convert float data to "int" or "char," then first consider multiplying the data by a constant value. For example, NDVI typically has values in the range 0.1 to 0.9. If these are converted to "int" or "char" then they will all return with a value of 0 or possibly with 1 depending on how the conversion is applied, so that all or most of the information in the data is lost. This can be avoided by multiplying the data by a constant value first.

6.7.2.2 Compute an FPAR Image

The relationship between FPAR and NDVI is of the form $\text{FPAR}_i = a_i + b_i \times \text{NDVI}_i$ for each of the i landcovers. Figure 6.38 shows this construction:

- Set into the workspace two input image icons, for the source image and for the landcover layer. Also set in an intermediate image icon and an output image icon.
- Set an analysis icon between the source image icon and the intermediate image icon, to be used to hold the computed NDVI as a temporary file. It could be made a permanent file. Set a second analysis icon between the intermediate image, the landcover layer, and the output image icon.
- Connect an arrow from the input image to the first analysis window, and from the first analysis window to the intermediate image. Connect other arrow icons from the intermediate window and the landcover layer icon to the second analysis icon, and from this icon to the output image icon.
- Specify the two input images, and characteristics of the intermediate image (type float) and the output image.

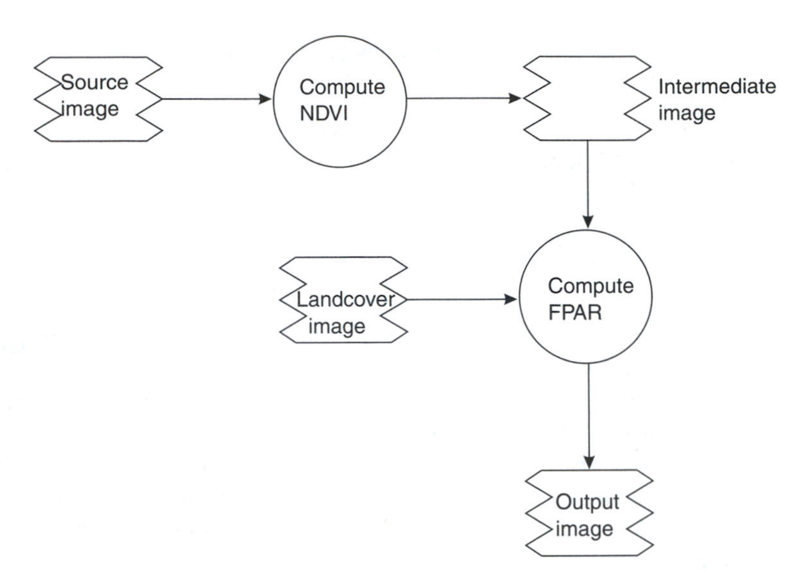

FIGURE 6.38 Flow chart for the computation of FPAR image using the Imagine script language.

- Activate the first analysis window and implement the algorithm to compute NDVI.
- In the second analysis window, select a multiple-choice function. Set the first test and the choice if it is correct, where this is likely to be of the form, **if** ($n2_land_cover == 1) then $(a_1 + b_1 \times \$n3)$
  else **if** ($n2_land_cover == 2) then $(a_2 + b_2 \times \$n3) + \cdots$ where $n2_land_cover is the landcover data and $n3 is the intermediate (NDVI) data. The linear constants could be specified by the operator, or they could also be input data in the form of a table.

The script models used in Imagine are available for use by the analyst. Thus, the analyst can also take one of these that closely match the work that the analyst wishes to do, and adapt it to his specific needs.

The script language approach provides a more powerful set of tools than the approach of using map algebra from a window, since the scripts can be saved, used and easily modified, they can include a number of steps in the one model, thereby reducing time, as well as facilitating an understanding of the action taken.

6.7.3 Surface Analysis

A GIS may use point data simply as a network of point observations. However, such data are often used as a distribution of data points for interpolation between the point observations to either create contours as vector data files, or to calculate DEM grid cell values for raster files. Whilst elevation data is commonly used in this way, and the terminology comes from this application, in fact any data can be processed in this way, including rainfall, temperature, soil acidity, soil colour and LAI data. There are three alternative approaches that can be used to the conversion of a mesh of point observations to a surface:

- Interpolation
- Surface Fitting
- Kriging or weighted interpolation

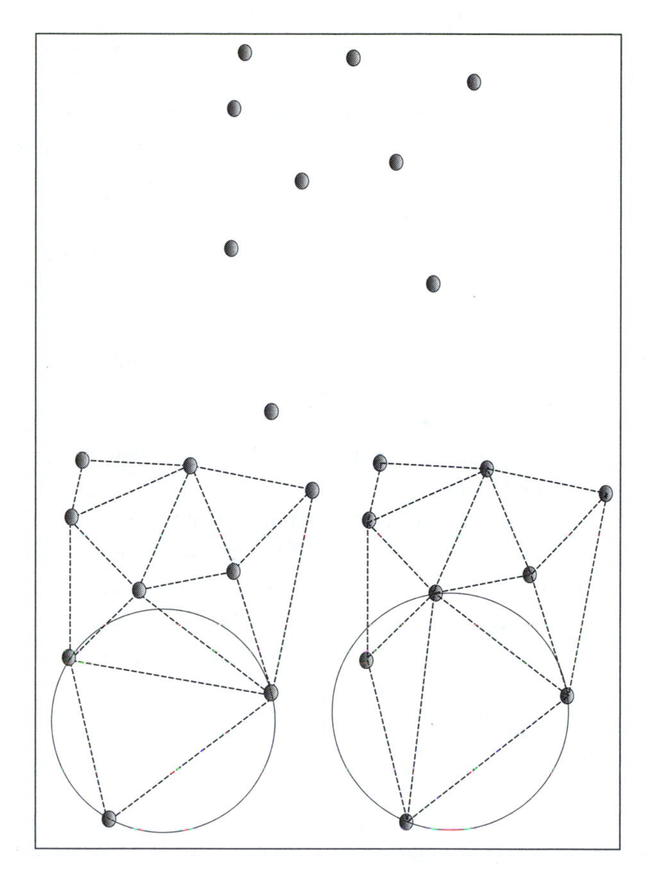

FIGURE 6.39 A network of data points and two triangulations of them; Delaunay triangulation (left) and non-Delaunay triangulation (right) showing how other points fall within a circle containing the three points of some triangles.

6.7.3.1 Linear Interpolation Using Triangulated Irregular Networks

Triangulated Irregular Networks (TIN) analysis is a commonly used way to interpolate spatially from point source data in which the data points are used to define triangles that are then used as the basis of interpolation within each triangle. The Delaunay triangulation process is commonly used to define the triangles to be used in TIN interpolation. Delaunay triangulation uses three criteria for the definition of the triangles that will be used in the interpolation:

1. A circle passing through the three points of a triangle will not contain any other point.
2. The triangles do not overlap.
3. There are no gaps between triangles.

A natural result of the Delaunay triangulation is that the minimum angle in any of the selected triangles is maximised. Figure 6.40 shows a distribution of points and two triangulations of them, one meeting the Delaunay criteria and one that does not and Figure 6.39 shows an example of observed data values and a selection of triangles to use for interpolation that meet the Delaunay triangulation criteria.

Interpolation is conducted by creating a mathematical surface through the known points and interpolating to estimate values at the pixel centres within the area covered by the surface, or finding the locations where the vector contour line passes through the TIN and joining these to form the final contours of the parameter being mapped.

The simplest surface is linear interpolation as a plane of the form:

$$\text{Estimate} = A_0 + B_0 \times X + C_0 \times Y \tag{6.18}$$

Such a surface can be constructed within triangles of points across the surface. Planar models of this type can be suitable if:

1. The points are observed on changes in gradient, as can be done for parameters that can be observed, such as ground elevations. It may be impossible to identify such changes in gradient in some parameters, such as rainfall.
2. The distribution of points forms well-conditioned triangles, which means that none of the internal angles should be less than 30°.
3. The unknowns (A, B, C) are solved to the number of decimal places necessary to ensure that rounding off errors do not significantly affect the final values.

Within each triangle, the three nodes of the triangle are used to solve this figure. The value at any other point in the triangle can then be computed by using the co-ordinates of the point to determine its estimated surface value. Since the co-ordinates of the pixel centres in the raster data are known, the raster cell values can be computed in this way.

Delaunay Triangulation breaks down when:

1. The observations get further apart than the feature being mapped; that is, the features in the surface are not represented in the point data.
2. The surface becomes more anisotropic. Isotropic surfaces are surfaces that are similar in all directions, and anisotropic surfaces are those that are different in some directions to other directions. For landforms, linear geologic structures such as dykes and formations created by erosion of metamorphic rock formations often create anisotropic surfaces.
3. The data do not match the shape of the features in the surface. Hilltops and valleys are two features that often create this type of problem, so that some types of interpolation techniques will not adequately deal them with. The simplest way to deal with these features is to have observations taken at all major changes in grade, so that many of the observations will be found along ridges and valleys.

Another disadvantage of interpolation techniques are that they do not provide an estimate of how well the fitted surface is to the actual surface and they cannot be used for extrapolation outside the triangles.

The concept of the Delaunay triangulation as shown in Figure 6.40 was discussed in Section 6.2.3. Once the triangulation has been completed, then the interpolation can be implemented. For this the surface is assumed to be a planar surface in each triangle of the form:

$$\text{Height} = a_0 + a_1 \times x + a_2 \times y \tag{6.19}$$

The three nodes of the triangle are used to solve this figure. The value at any other point in the triangle can then be computed by using the co-ordinates of the point to determine its estimated height value. Since the co-ordinates of the pixel centres in the raster data are known, the raster cell values can be computed in this way.

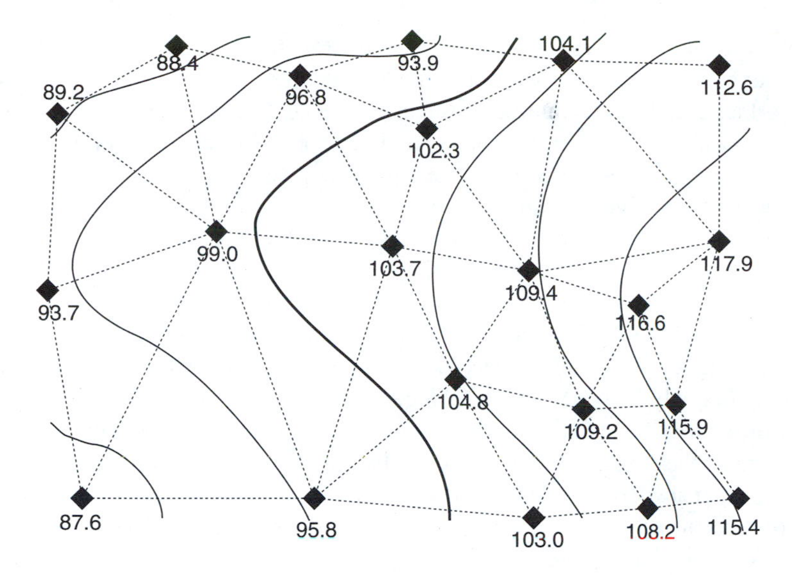

FIGURE 6.40 Distribution of spot heights used to create a digital terrain model, and estimated grid cell values using linear interpolation within triangles of points.

Whilst this is a common interpolation technique, it suffers a number of disadvantages. The main disadvantage is that it cannot adequately deal with areas that do not obey the linear assumption, such as would occur at the top of hills, in valleys and often for intermediate small features as well. The only way to deal with these anomalies is to make them as small as possible by ensuring that the field data points are always located at changes of grade. Thus, many points should occur along ridgelines, in valleys, and in changes in grade along the sides of valleys, when this occurs. The other main disadvantages of the techniques are that it does not provide any estimate of how well the fitted surface is to the actual surface and it cannot be used for extrapolation outside the triangles. Linear triangulation cannot adequately deal with anisotropic surfaces, unless the observed points lie along the lines of changes in grade, if this is the cause of the anisotropy. If rather a busier form of difference with direction causes the anisotropy, then this becomes impractical. Thus, in areas of extensive sand dunes, linear triangulation would require observations, and triangle edges, to be along the ridges of the dunes, and in the troughs. This may be acceptable if the ridges are large enough to justify this density of observations. However, if the ridges are smaller, but still significant in terms of the interpolation required, then this constraint would require too many observations.

6.7.3.2 Curve Fitting

Curve fitting has the advantage that if more data points are used than the minimum number required for solution by means of least squares then the residuals that are derived can be used to estimate how close the surface is to the data values. Usually, the surface is to be modelled using either a second or third order polynomial.

$$\text{Height} = a_0 + a_1 \times x + a_2 \times y + a_3 \times xy + a_4 \times x^2 + a_5 \times y^2$$

$$\text{Height} = a_0 + a_1 \times x + a_2 \times y + a_3 \times xy + a_4 \times x^2 + a_5 \times y^2 + a_6 \times xy^2 \qquad (6.20)$$

$$+ a_7 \times x^2 y + a_8 \times x^3 + a_9 \times y^3$$

A global fit will use all the data values to derive one equation for the whole area. Such a solution will give a continuous surface across the whole area, but it will often incur large residuals or discrepancies between the surface and the data values. It will also produce a smoother surface than would a local curve fitting. A local fit must use sufficient points to solve for the unknowns in the equations, necessitating 6 or 9 points, respectively, in the equations given above. However, it is better to use additional points so as to solve the equations by least squares. Once the unknowns have been found, the pixel co-ordinates are used to estimate the height value at the pixel location.

A local fit will result in discontinuities at the boundaries between the different local fits. If the data are dense enough, then these discontinuities will be small and probably not noticeable. However, one of the advantages of interpolation is to reduce the amount of field data collection, hence this characteristic is a significant disadvantage. One way to attempt to reduce this effect is to use the derived interpolation only within the inner triangle used, and then average the pixel values that are in the vicinity of the triangle sides using the two fits from either side of the edge.

Curve fitting has the advantage that if excess than the minimum number of data points is used then the residuals that are derived can be used to estimate how close the surface is to the data values. Whilst in theory the surface could be extrapolated beyond the observed data values, in practice the derived surfaces can rapidly diverge from the actual surface, so that extrapolation is not a good idea. Another problem is that the (x, y) values can cover a large range of values and so the derived observation or normal equations may be ill conditioned and produce a poor solution. Great care needs to be taken with curve fitting to ensure that this problem does not arise.

Spline curves exactly fit a curve to a small number of data points, creating new curves for each subset of a few points, but having the curves form a continuous join between each pair of spline curves. This means that it is possible to re-compute a section of the spline curve without affecting the remainder of the curve, something that is impossible to do with a least squares curve fit. Since only a few points are used, the order of the polynomial used in the spline curve can be manageable, whereas this may not be the case with least squares curve fitting. However, there are no residuals, so it is impossible to estimate the accuracy of the curve, without the use of additional points. However, this should also be done with least squares curve fits, as the residuals from the points used in the curve fit will usually be smaller than the residuals from other points.

6.7.3.3 Kriging

Kriging is a method of interpolation based on the use of geostatistical methods of defining spatial correlation to drive the interpolation mechanism. Its underlying philosophy is that points closer together are more likely to be correlated than are points that are further apart, and that after some distance, points are effectively independent. Only points that are correlated with the point of interest should influence the interpolation of that point, and the influence should be functionally related in some way with the level of correlation. The level of spatial auto-correlation can be measured using the semi-variogram or the spatial auto-correlelogram. The semi-variogram has been introduced in Chapter 3 and the reader is referred to this section for a detailed discussion on the computation, shape and interpretation of the semi-variogram.

In that Section, it was shown that a semi-variogram could be computed in any number of dimensions, but that its computation becomes prohibitive above three dimensions. Most semi-variograms are computed in one or two dimensions. In one dimension a semi-variogram may look like that shown in Figure 6.41. A two-dimensional variogram for an image may look like that shown in Photograph 6.1. In this photograph, the origin for the variogram is the top left-hand corner. This corner has relatively lower values, and hence appears as dark. The two-dimensional variogram appears to be relatively uniform in all directions, suggesting that the variogram is isotropic. There are large variations towards the extremity of the variogram. A one-dimensional variogram has been taken from this two-dimensional variogram along the transect shown in Photograph 6.1 and is shown in Figure 6.42. This variogram shows the classical variogram shape, with a nugget value, a range of about 13 and a sill value of about 0.0035.

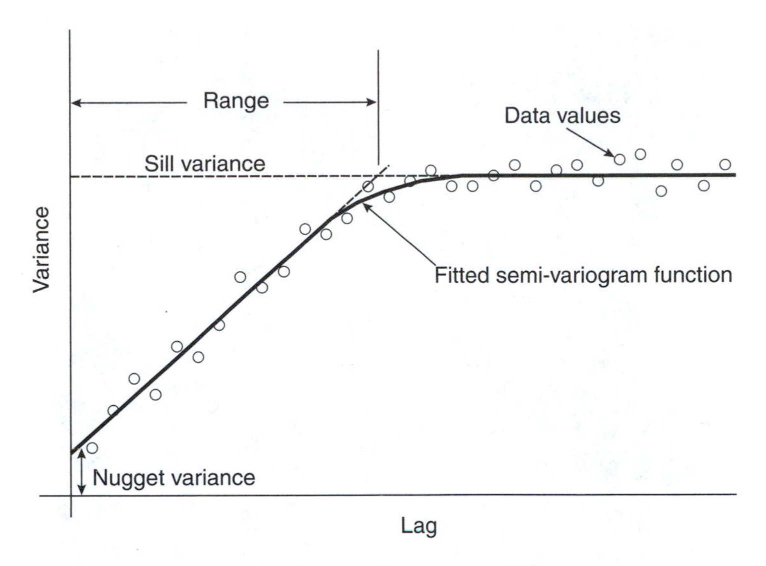

FIGURE 6.41 The semi-variogram and its elements.

With most semi-variograms some variation is exhibited with direction; that is, most surfaces are anisotropic to some degree. The anisotropy will have an effect on the semi-variogram, and on the interpolation. For interpolation, two-dimensional semi-variograms should be computed. If the data are isotropic then the two-dimensional semi-variogram will look the same in all directions from the origin of the semi-variogram. If this does not occur, then the data are anisotropic to some degree, as is usually the case. For the interpolation a one-dimensional semi-variogram will be used. If the data are isotropic, then a profile through the two-dimensional semi-variogram will provide a suitable one-dimensional semi-variogram for the next stage in the analysis. If, however, the data are anisotropic, then the analyst needs to find the direction in the two-dimensional semi-variogram that provides a profile that best represents the data, and this profile is used as the one-dimensional semi-variogram for the next stage in the analysis.

The semi-variogram shows the range over which a point will influence adjacent points. For interpolation, points should only be used that are within the semi-variogram range of the point to be found. The value to be found should be influenced more by closer points than by more distant points. The degree of influence should be proportional to the level of correlation, and this can be found from the semi-variogram. The semi-variogram is thus used to compute weights when interpolating points from a number of adjacent points. To do this, the discrete nature of the actual semi-variogram (the experimental semi-variogram) has to be replaced by a continuous function so that semi-variogram values can be computed for fractional lags. There are four types of function that meet mathematical constraints and provide the necessary shape, as shown in Figure 6.43.

The equations for the four functions are given below and depicted in Figure 6.43. The linear function has not been depicted, but it would be used when the variogram is approximately linear, as can occur either before the range distance has been reached in the data, or where the range distance is less than the lag interval.

1. The spherical function

$$\gamma(h) = c_0 + c_1[(3h/2a) - \tfrac{1}{2}(h/a)^3] \tag{6.21}$$

where c_0 is the nugget variance, $(c_0 + c_1)$ is the sill variance, h is the lag and a is the range.

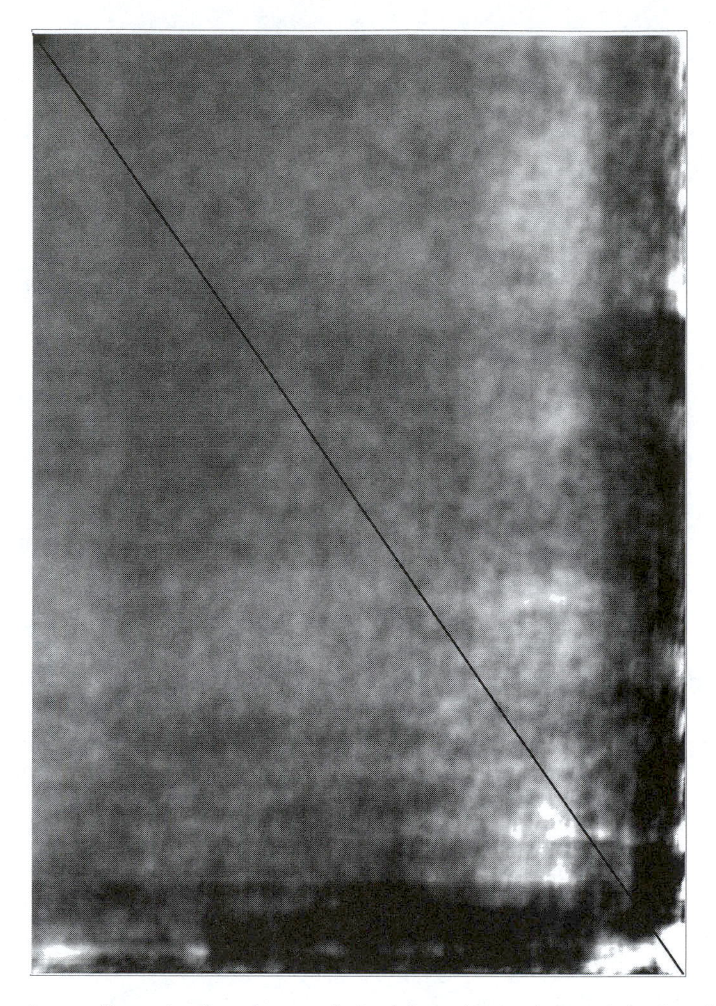

Photograph 6.1 A two dimensional variogram derived from image data and the location of the transect depicted in Figure 6.42. The origin for the variogram is the top right hand corner.

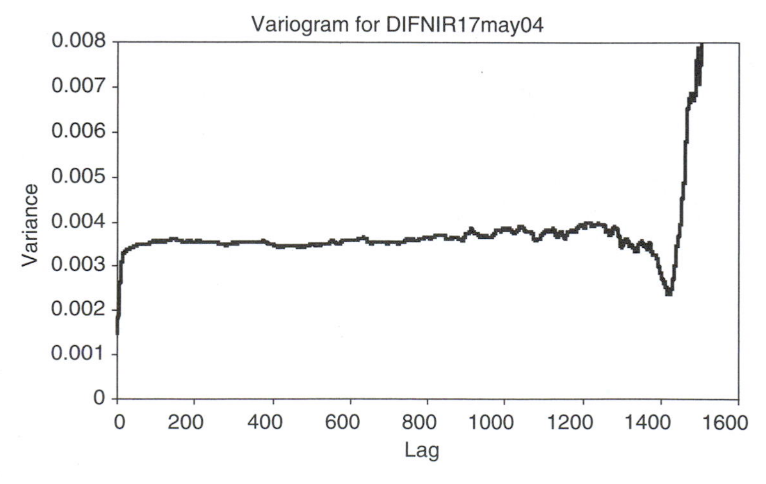

FIGURE 6.42 Transect one-dimensional variogram taken from the two-dimensional variogram depicted in Photograph 6.1.

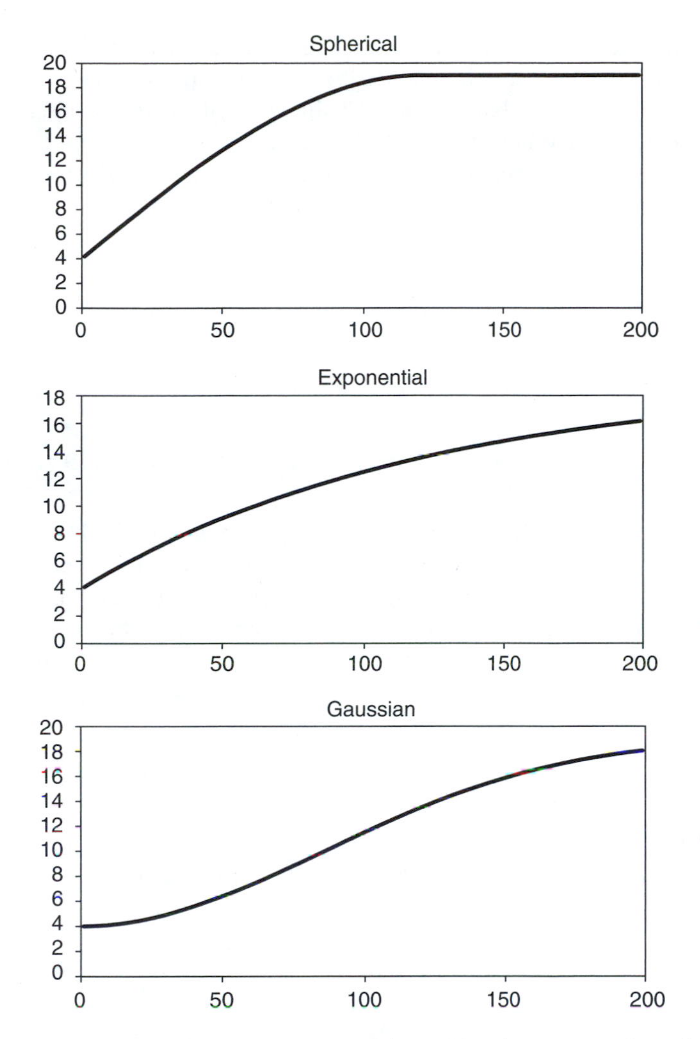

FIGURE 6.43 Examples of three of the variogram models in which the nugget variance is 4, the sill variance is 19 and the range is 120.

2. The exponential function

$$\gamma(h) = c_0 + c_1[1 - e^{(-h/a)}] \tag{6.22}$$

with the parameters as described above.

3. The Gaussian function

$$\gamma(h) = c_0 + c_1[1 - e^{(-h/a)(h/a)}] \tag{6.23}$$

with the parameters as described above.

4. The linear function

$$\gamma(h) = c_0 + bh \tag{6.24}$$

where b is the slope of the line.

The choice of function should come from inspection of the data, and from fitting the different functions to the data if the analyst is in doubt as to which one to use. The first three functions fit different shaped semi-variograms, but all reach the sill within the range of observations. The linear model is used where the sill has not been reached.

In standard interpolation, as discussed above with the use of the Delaunay triangulation, the linear distance between the point and each of the observed points gives the weights for the interpolation, so that effectively:

$$z_0 = \Sigma(d_i \times z_i)/\Sigma d_i, \quad i = 1, \ldots, n \text{ for } n \text{ points being used in the interpolation} \tag{6.25}$$

In Ordinary Kriging the weights are derived from the semi-variogram:

$$z_0 = \Sigma(\lambda_i \times z_i), \quad i = 1, \ldots, n \quad \text{where } \Sigma\lambda_i = 1.0 \tag{6.26}$$

It can be seen that Equation (6.25) and Equation (6.26) are of the same form, the difference being in the estimation of the weights, λ_i, and the number of points that can be used. In standard interpolation the number of points used is three, representing the three nodes of the surrounding triangle. In Kriging, all points should be used that are within the semi-variogram range of the point to be estimated. In Ordinary Kriging the minimum variance of $(z_0^{\wedge} - z_0)$, the difference between the estimate of z and the actual value of z, is given by:

$$\sigma^{\wedge 2} = \Sigma\lambda_i \times \gamma(x_i, x_0) + \phi, \quad i = 1, \ldots, n \tag{6.27}$$

where $\sigma^{\wedge 2}$ is the Kriging variance, λ_i is the weight given to point i, $\gamma(x_i, x_0)$ is the semi-variance between point i and the unknown point and ϕ is the Lagrangian multiplier. This is achieved when:

$$\Sigma\lambda_i \times \gamma(x_i, x_j) + \phi = \gamma(x_j, x_0) \quad \text{for } j = 1, \ldots, n \tag{6.28}$$

and

$$\Sigma\lambda_i = 1.0$$

If there are four known points being used, as shown in Figure 6.44 then this gives the equations:

$$\begin{aligned}
&\lambda_1 \times \gamma(x_1, x_1) + \lambda_2 \times \gamma(x_2, x_1) + \lambda_3 \times \gamma(x_3, x_1) + \lambda_4 \times \gamma(x_4, x_1) + \phi = \gamma(x_1, x_0) \\
&\lambda_1 \times \gamma(x_1, x_2) + \lambda_2 \times \gamma(x_2, x_2) + \lambda_3 \times \gamma(x_3, x_2) + \lambda_4 \times \gamma(x_4, x_2) + \phi = \gamma(x_2, x_0) \\
&\lambda_1 \times \gamma(x_1, x_3) + \lambda_2 \times \gamma(x_2, x_3) + \lambda_3 \times \gamma(x_3, x_3) + \lambda_4 \times \gamma(x_4, x_3) + \phi = \gamma(x_3, x_0) \\
&\lambda_1 \times \gamma(x_1, x_4) + \lambda_2 \times \gamma(x_2, x_4) + \lambda_3 \times \gamma(x_3, x_4) + \lambda_4 \times \gamma(x_4, x_4) + \phi = \gamma(x_4, x_0) \\
&\lambda_1 + \lambda_2 + \lambda_3 + \lambda_4 = 1.0
\end{aligned} \tag{6.29}$$

The last equation is both necessary as a condition of the function, and to solve the equations. Equation (6.29) can be put in the form of matrix notation:

$$A \times U = B \tag{6.30}$$

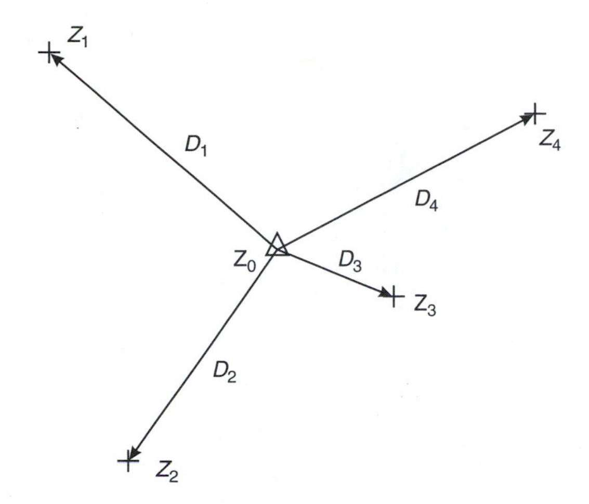

FIGURE 6.44 Example relationship between four known points, $(z_1, \ldots, z_4)$ and the unknown point to be interpolated, z_0.

where

$$A = \begin{bmatrix} \gamma(x_1,x_1) & \gamma(x_2,x_1) & \gamma(x_3,x_1) & \gamma(x_4,x_1) & 1.0 \\ \gamma(x_1,x_2) & \gamma(x_2,x_2) & \gamma(x_3,x_2) & \gamma(x_4,x_2) & 1.0 \\ \gamma(x_1,x_3) & \gamma(x_2,x_3) & \gamma(x_3,x_3) & \gamma(x_4,x_3) & 1.0 \\ \gamma(x_1,x_4) & \gamma(x_2,x_4) & \gamma(x_3,x_4) & \gamma(x_4,x_4) & 1.0 \\ \lambda_1 & \lambda_2 & \lambda_3 & \lambda_4 & 0.0 \end{bmatrix}$$

$$U = \begin{bmatrix} \lambda_1 \\ \lambda_2 \\ \lambda_3 \\ \lambda_4 \\ \phi \end{bmatrix} \quad \text{and} \quad B = \begin{bmatrix} \gamma(x_1,x_0) \\ \gamma(x_2,x_0) \\ \gamma(x_3,x_0) \\ \gamma(x_4,x_0) \\ 1.0 \end{bmatrix}$$

Matrices A and B are found by computing the distances between the observed points, and between each observed point and the unknown point. These distances are then used in the semi-variogram function to compute the semi-variogram value or weight. Note that there are four distances of zero relevant to array A, so that the semi-variance of these four distances will be the nugget variance. The equations are solved by dividing both sides of the equation by matrix A, since there are no residuals:

$$U = A^{-1} \times B \tag{6.31}$$

Ordinary Kriging is exact in that there is exactly the minimum number of equations to solve for the unknowns. The solution cannot contain residuals. However, the accuracy of the Kriging can be estimated by computing the Kriging variance from Equation (6.29) after the components of this equation have been found from the matrix multiplication. The advantage of being able to estimate the variance is that it enables the user to compute the accuracy at a user-specified confidence level.

As the reader can see from the above description of ordinary kriging, there are a number of steps that need to be implemented. Some of these steps are open to interpretation as to the best way to proceed. This particularly applies to the selection of the kriging function among the four available, and the setting of the function parameters in the presence of anisotropy. It will be found that taking

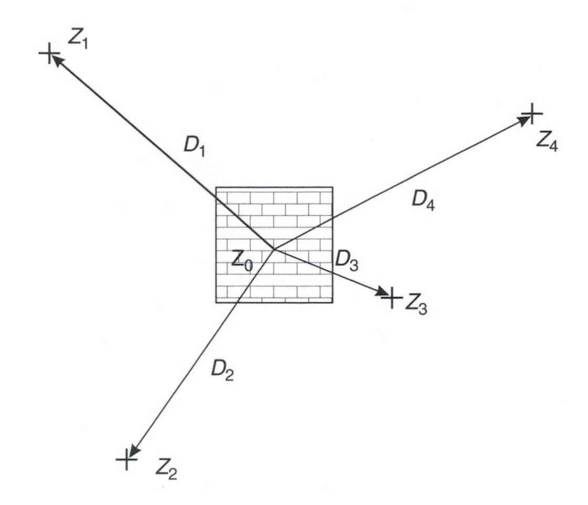

FIGURE 6.45 Block Kriging to find the average value within a block or pixel area.

different choices can give quite different results. The analyst should try the various alternatives and then select the solution that the analyst considers best represents the surface being fitted.

Ordinary Kriging assumes that all interpolated values relate to the support area of the source data. With images the support is the size of the pixel, but with most field data the support is a point if the data are taken as point observations. The data thus represent not an area, but a point, and with many types of field data the resulting observations can show sharp anomalies. As a consequence, Ordinary Kriging may show sharp spikes or troughs at individual data points. To overcome this problem Block Kriging is used to give an interpolated value over an area, such as a pixel (Figure 6.45). Block Kriging would be used, for example, when pixel values are to be derived from the source data.

$$\text{The area } z_0(B) = Iz_0(x)\mathrm{d}x/\text{area}B \tag{6.32}$$

$$\text{Estimated by } z_0^{\wedge}(B) = \Sigma\lambda_i \times z(x_i), \quad i = 1, \ldots, n \tag{6.33}$$

and the minimum variance is now given by the equation:

$$\sigma^{\wedge 2}(B) = \Sigma\lambda_i \times \mu_\gamma(x_i, B) + \phi - \mu_\gamma(B, B), \quad i = 1, \ldots, n \tag{6.34}$$

and this is obtained when:

$$\Sigma\lambda_i \times \gamma(x_i, x_j) + \phi = \mu_\gamma(x_j, B) \quad \text{for } j = 1, \ldots, n \tag{6.35}$$

and

$$\Sigma\lambda_i = 1.0$$

Equation (6.35) is similar to Equation (6.28) except that the right-hand side is the average semi-variance across the block. This can be calculated as the average of the semi-variance values for the four corners and the centre.

Often data may be held of two correlated parameters, one of which may have abundant observations, and one of which may have few observations. This can readily occur when one type of observation is either easy or cheap to acquire and the other is either difficult or expensive. It can also occur when image data provides a cheap estimate of a parameter and relatively sparse field data may be held of the actual parameter. When this situation occurs then the abundant data can be used to improve the interpolation of the parameter with sparse values in Co-Kriging. Consider two variables, U and V, where the cross semi-variograms at lag h between them is computed from the formula:

$$2\gamma_{UV}(h) = [\Sigma(z_U(x_i) - z_U(x_i + h)) \times (z_V(x_i) - z_V(x_i + h))]/n \tag{6.36}$$

It should be noted that cross semi-variograms can increase or decrease in value with increasing lag, depending on the nature of the correlation between U and V.

6.7.4 Analyses of Cross-Tabulation Tables

An important area of activity in GIS analysis is to evaluate the accuracy of data and information. The evaluation of the accuracy of a classification or of GIS information is normally conducted by the use of field data and comparing this with the estimates derived from the analysis, as discussed in Chapter 5. Our interest here is in how this work is conducted within a GIS, since the concepts have been well covered in Chapter 5. To conduct this work the field data is entered into the GIS to create a raster layer with the pixels corresponding to the sample areas showing the field data class codes for classification assessment or physical attribute data values for estimation assessment, and the remainder of the pixels being empty or being assigned a zero or default value representing an empty value. This is compared with a raster layer containing the equivalent information as derived from the analysis, ensuring that the field data is made the independent parameter. The analysis is then conducted as a comparison of the two layers, pixel by pixel, to create a table that contains the count of the number of occurrences of each combination of values in each layer. Thus, if the image data layer contains 36 themes and the field data layer contains 47 themes, then a table will be created that contains 47 columns and 36 rows. The derived table can then be exported from the GIS software for more detailed statistical analysis.

In Idrisi this task is conducted using CROSSTAB (Figure 6.46). CROSSTAB performs two operations. The first operation is cross-classification which shows all combinations of the logical AND operation. The result is a new layer that shows the locations of all combinations of the categories in the original layers. A legend is automatically produced showing these combinations with the left-hand categories referring to the layer listed on the left in the title and those on the right referring to the layer listed on the right in the title. Cross-classification thus produces a map representation of all non-zero entries in the cross-tabulation table.

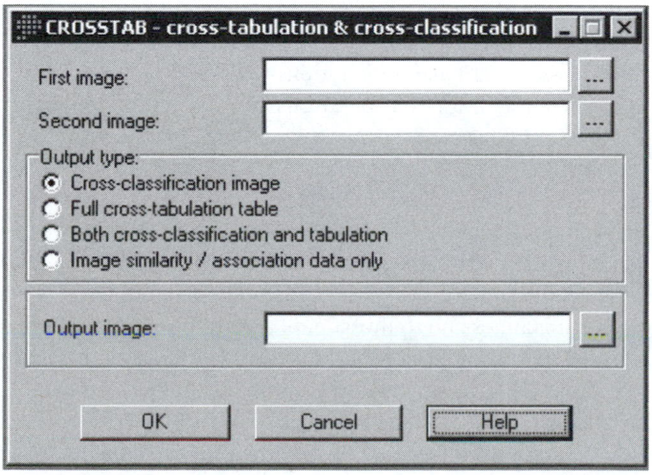

FIGURE 6.46 The Idrisi window for cross-tabulation of two layers.

The second operation is image cross-tabulation in which the categories of one layer are compared with those of the second layer and the results stored as a table of the occurrences of each combination of values in the two source layers. The result of this operation is the tabulation table as well as one and possibly two measures of association between the images, such as the kappa statistic discussed in Chapter 5.

6.7.5 Clump Analysis

Clumping analysis is concerned with deriving information on areas of contiguous pixels in a raster layer. For example, if the analyst is interested in biodiversity in an area, then he may be interested in identifying areas of specific classes that are large enough to support specific species or communities. To do this he would first conduct a clumping of the layer, to identify the clumps that exist. Then he would sieve the image to eliminate those clumps that are smaller than a specified size. He is left with those clumps that meet his specifications.

6.7.6 Zonal Analysis

In zonal analysis the analyst is interested in the statistics of certain attributes within zones defined in a vector polygon file. For example, he/she may be interested in the statistical distribution of classes in the different soil classes, or the statistical distribution of soil characteristics in the different landuse classes. A version of this is to create buffer zones and to then derive statistics for within these zones. For example, it may be considered desirous to limit certain landuse practices within a buffer zone of rivers, waterways, lakes and swamp areas. A buffer zone can be placed around such areas, and then statistics derived for the land classes that occur within this zone, so that the impact of the regulation can be assessed.

6.8 MODELLING IN A GIS

6.8.1 What is Modelling?

A model is an abstraction designed to represent some aspects of the real world. There are physical, conceptual and numerical models:

1. *Conceptual models*. Conceptual models are descriptive models that may be retained in the mind of the owner of the model, or they may be set down in words so as to share them with others. We retain conceptual models of many of those things that are important to us, such as a model of the route to work and our relationship with other people. Conceptual models can also be constructed as the basis for discussion on, and subsequent development of other forms of models. In this case the conceptual model will be set down in a descriptive manner.
2. *Physical models*. A model car, plane or boat are classical examples of physical models, where these models may simply be made for enjoyment, or that may be made to test some components of the eventually planned actual machine. The purpose of the model will influence the details and the components included in the model. Physical models were used to test various machines and to investigate certain processes. Thus, models of planes were used to test the aerodynamics of the proposed design, and of boats were used to test their capacity to move through water. Physical models have also been used to investigate things like the action of waves along a beachfront and the effect of this action on the deposition and erosion of sand from a beach. However, physical models suffer a number of significant disadvantages, and as such they are not used as much as they used to be. These disadvantages include the inability to use them to investigate many conditions, their

inflexibility and the fact that they do not allow numerical quantification of the process of themselves. Whilst physical models can be used for many purposes, for many others, in particular many environmental situations, physical models are not realistic to construct.

3. *Numerical models*. Numerical models are based on numerical formulae that are implemented in accordance with a specific sequence or set of rules established within software, and run on a computer. Many numerical models are based on a good understanding of the characteristics of the process being modelled and hence are called process models. Process models can be based on numerical formula, in which case they are called deterministic models, or that can be based on probabilistic formulae, in which case they are called stochastic models. The alternative to this understanding is to use data on the process to build an empirical model that is usually constructed by the use of regression techniques. Regression models are based on a data set that is extensive enough to develop a model that will cover all the expected conditions, so that the analyst deriving the model may develop a good understanding of the characteristics of the processes, even though the underlying principles may not be obvious to them. However, the fundamental lack of understanding of the underlying principles of the process that characterises regression models means that the analyst can never be sure that the situations being met during development of the model is to be used, is within the bounds for which it is desired to use the model, or the range of conditions that may be met in using the model. Process models, on the other hand, which are based on the underlying principles governing the process, have the big advantage that the range of conditions in which the model can be used comes directly from analysis of these underlying principles. Numerical models have the advantage over physical models that they are flexible; they can be adapted easily and many of them can be adapted to provide predictions. Thus, they can be used to test scenarios in a way that would be difficult to do with physical models.

There are many different models that have been developed and used to varying degrees in the various disciplines that are relevant to the management of resources, including economics, environmental science, hydrology, physics, geography and biology. The ones that are relevant to this book are those that are representing aspects of reality that contain spatial and temporal dependencies. Models that do not contain spatial and temporal dependencies do not need to be implemented in a GIS, and as such they are not of interest to this discussion. Thus, a simple model of the influence of gravity on an object:

$$\text{Velocity at time } t = \text{Start velocity} + \text{Acceleration due to gravity} \times \text{time}.$$

or $V = U + at$ is not likely need to be implemented in a GIS on its own, although it may form a component of other models. However, most processes that involve the physical environment involve spatial and temporal dependencies. Consider just two examples:

1. *Bush and grass fires.* The vigour and direction of burning of bush and grass fires depends on the availability of dry fuel, the temperature and wind velocity and the shape of the terrain. The amount of fuel and its dryness, the shape of the terrain and the velocity of the wind all contain spatial dependencies, whilst the weather and the dryness of the fuel contain temporal dependencies. As a consequence, a fire model designed to assist fire fighters plan the control of fires needs to take these spatial and temporal dependencies into account, and this can best be done within a GIS environment.
2. *Water-borne soil erosion.* The amount of soil eroded during rainfall depends on the rainfall intensity and duration, the impact velocity of the rain on the soil, the velocity and volume of the water flowing across the landscape. The rainfall intensity and duration, as well as the

volume and velocity of the water flow all contain temporal dependencies, whilst the impact velocity of the rain on the soil, the velocity and volume of flow have spatial dependencies.

In both these cases, and in many other cases concerning the physical environment, the aspects of the environment being modelled contain spatial and temporal dependencies, and as such they should be implemented in a GIS. There is an increasing trend to do this, and the result will be an improvement in the accuracy and reliability of the models.

All processes can be considered to contain three components:

$$\text{Process} = \text{Predictable component} + \text{Unpredictable component} + \varepsilon$$

The *predictable component* can be modelled using a numerical formula or set of formulae that are driven by a number of independent parameters. Once such formulae are known, then input of the parameter values into the model will give an estimate of the response. A prediction can be made.

The *unpredictable component* is implemented by repetitively iterating a set of relatively simple rules to each item of interest and to its neighbours in turn, where these rules will often contain a random component. The unpredictable component operates in a quasi-random manner within a moderate time/spatial period, but it produces a pattern over longer time/space periods that could not be predicted from the driving parameters that are used to create each step in the process. Since it operates in a quasi-random manner over short to moderate time periods, and because each starting value, and each step operate in quasi-random manner, these models cannot be used to make specific predictions.

ε is the actual random errors that occur in any system.

Let us consider a simple example of Newton's Law of Gravity, which we have discussed earlier. The velocity of the object depends on the force of gravity and the effect of wind resistance. Such a model can be formulated and it forms the predictable component of the process (Figure 6.47). Unpredictable components of this process include:

1. Variations in the force of gravity
2. Variations in the density of the atmosphere
3. The effects of wind velocity

In practice, these other components are very small in comparison with the dominating two components in this particular model. Thus, in modelling this process, the predictable components dominate the process, and as such the process can be modelled with satisfactory accuracy for many purposes using just the predictable component. When the predictable component dominates the process being

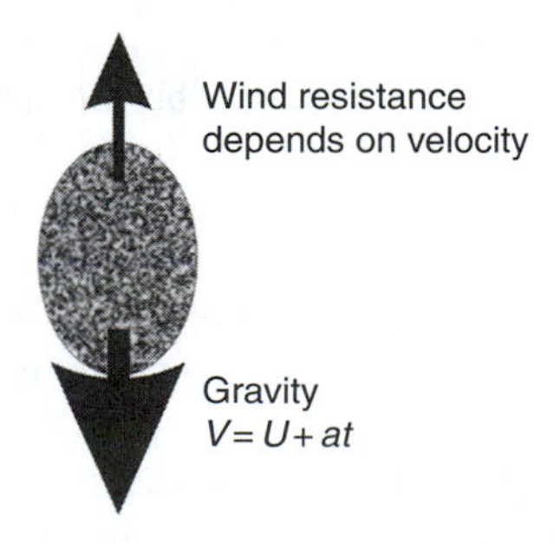

FIGURE 6.47 The dominating influences on a falling object: gravity and air resistance.

modelled, then the predictable component can be used to make reasonable predictions about the process.

In the physical environment, however, there are many processes which have an unpredictable component that is as large, or larger than the predictable component. When this is the case, then the predictable component cannot be used to derive reliable estimates or predictions. Consider the weather. There is a predictable component based on the annual cycle of the Earth around the sun, enabling us to predict the onset of the various seasons. But the day-to-day weather is dominated by the unpredictable components of the process such that it is impossible to provide reliable forecasts further than about 3 days in advance.

The predictable component of a process is modelled using a process or a regression model. To construct a regression model, a large data set is accumulated, designed to cover all the variations in conditions that might be expected to occur. A least squares best fit is made to this data, giving a regression model of the process. The regression model is a "global" model in the sense that the one model holds for all conditions for which the data was collected, or it is a "top-down" model. In such models the unpredictable and random components contribute to the residuals from the regression, however, we have seen that the unpredictable component gives a pattern over space/time, and as such it may contribute to the derived regression equation so that the regression equation will represent the predictable component and may partly represent the unpredictable component.

There are some significant limitations to the use of regression models, specifically:

1. Regression models do not, of themselves, help us to better understand the process being modelled.
2. Regression models can only be used within the range of conditions that are represented in the development of the model. Whilst the builder of the model will strive to cover as large a range as possible, if he/she is unaware of some of the drivers to the process of interest, then he/she may not adequately cover all the possible range of conditions.

Because of these limitations, there continues to be considerable effort expended on trying to better understand the processes of interest in the environment sufficient to build process models that describe these processes. To take the gravity analogy again, before Newton discovered and developed the formulation given above, experimental work may have given sufficient data to derive a regression model of the process. By the nature of the process, the experimental data may have given a similar model to the one developed by Newton, but it would have been somewhat different due to experimental error, and it may have been somewhat different due to the effects of wind on the experiment. However, Newton's model is much better in that it provides information on some of the assumptions on which it is based and hence it better defines the range of conditions over which it can be used, and these extend well beyond those that would have pertained in the experimental data used to derive the relationship.

In attempting to describe the process numerically, many models use a mix of both approaches. For example, experimental work may have elucidated the key drivers to a process and some of the relationships between some of them. Such information can then be used in the experimental design to elucidate these relationships and subsequently to develop the regression model, and may be able to be incorporated into the actual regression model itself. For example, if it is known that there is a functional relationship of the form $\text{Parameter}_a = k \times \text{Parameter}_b$ then such a relationship can be built into the regression model.

In reality some process and regression models may be too complex to implement, may depend on parameters that are impossible, impractical or expensive to acquire data for, may be impossible to invert if this is necessary or may be excessively accurate relative to the other components in the process. If a component is excessively accurate, then this represents overkill, incurring costs and delays in the application of the system. When these conditions apply, then the best option may be to simplify the model for implementation. When models are simplified in this way, then they will often yield less accurate results, or they may limit the range of conditions over which the model can be used. However, finding a balance between the complexity and accuracy of the model on the one hand, and its usefulness and costs on the other, gives rise to many different modelling solutions.

However the "top-down" approach, on its own, whether coming from regression or from a process-based model is fundamentally flawed when the process contains an unpredictable component that is of similar or larger magnitude to that of the predictable component. In many environmental processes this is the case, and so this reality has driven the search for better ways to derive models of these processes. Two types of model have been developed to address this need: cellular automata (CA) and agent-based (AB) models.

6.8.1.1 Cellular Automata

Cellular automata are discrete dynamical systems whose behaviour is completely specified in terms of a local relation. A cellular automaton can be thought of as a stylised universe. Space is represented as a uniform grid, with each cell containing data. Time advances in discrete steps. Each cell operates in accordance with a small set of rules that respond to a set of drivers and to the local neighbours of the cell, to modify the state of the cell in some way. Thus, the system's laws are local and uniform.

A simple example of a CA model is "Life," developed by John Horton Conway, a young mathematician at Cambridge University (United States) in 1970. Conway's cellular automata had only two states; a cell would be filled or empty, "alive" or "dead." The rules that he set were:

1. *Stasis* — If, for a given cell, the number of "on" neighbours is exactly two, the cell maintains its status quo into the next generation. If the cell is "on," it stays "on," if it is "off," it stays "off."
2. *Growth* — If the number of "on" neighbours is exactly three, the cell will be "on" in the next generation. This is regardless of the cell's current state.
3. *Death* — If the number of "on" neighbours is 0, 1, 4 to 8, the cell will be "off" in the next generation.

If *Life's* rules said that any cell with a live neighbour qualifies for a birth and no cell ever dies, then any initial pattern would grow like a crystal endlessly. If on the other hand the rules were too anti-growth, then everything would die out — Conway contrived to balance the tendencies for growth and death. Simple rules can have complex consequences. Whilst the *Life* model is not affected by other stimuli, many other CA models do respond to other stimuli just as they also affect, and are affected by, their neighbours.

6.8.1.2 Agent-Based Models

Agent-based models are simulations based on the global consequences of local interactions of members of a population. These *individuals* might represent plants and animals in ecosystems, vehicles in traffic, people in crowds or autonomous characters in animation and games. These models typically consist of an environment or framework in which the interactions occur and some number of individuals defined in terms of their behaviours (procedural rules) and characteristic parameters. In an AB model, the characteristics of each individual are tracked through time. Thus, AB models are based on the actions of individual members of a population, whilst CA models are based on the responses of spatially fixed cells. However, some AB models are spatially explicit, that is, the responses of the individuals depend on their locations in a matrix or grid. Such models are appropriate for animals in their environment, where the animals may have different rates of mobility, and the plants are immobile.

An example of a simple agent-based model (ABM) is "boids," developed by Craig Reynolds in 1987. The individual flying birds are generated at random and then they fly in accordance with a set

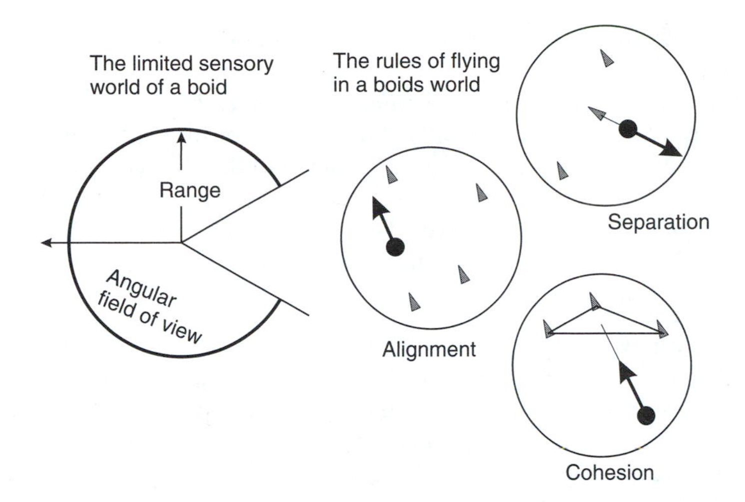

FIGURE 6.48 The world of Boids.

of three simple rules, influenced by other birds within their restricted visual range (Figure 6.48):

1. *Separation* — Steer to avoid crowding local flockmates
2. *Alignment* — Steer towards the average heading of local flockmates
1. *Cohesion* — Steer to move towards the average position of local flockmates

It can be seen in "boids," that if the range of visual perception of the individual "boid" is set too small, then they can only see the other "boids" sufficiently to avoid collision, and cannot respond adequately to the other laws. The effect of this would be to create non-collision random or chaotic behaviour. If, on the other hand, the range of vision was such that all "boids" could be seen, then each "boid" would respond such that stable patterns of behaviour would result. The resulting behaviour would not be typical of flocking birds. Thus, adjusting the perception distance will change the way that the birds flock in this model.

Both CA and AB models are based on the behaviour of individuals, and as such they are "bottom-up" modelling methods. It can be seen that the behaviour of individual cells in CA or individual objects in AB models are relatively predictable within one step of the modelling, but that they will rapidly become unpredictable. Despite this such models exhibit one of three types of behaviour, depending on the rules and the parameter values set into those rules:

1. *Stable or unchanging behaviour.*
2. *Patterns to their behaviour*, although these patterns will be very non-linear and probably impossible to describe using conventional numerical formulae on their own.
3. *Chaotic behaviour.*

Because of these characteristics, most current CA and AB models cannot be used, on their own, for predictive purposes, but they are very good at helping us understand many processes and objects in the real world. Thus, versions of CA models can construct very life-like versions of trees, leaves and the porosity of the soil or rocks, and AB models can provide very similar patterns of behaviour as animals in the environment and other processes.

A study of the weather, or most other natural systems, will show that these real systems:

1. *Exhibit predictable behaviour over some specified spatial and temporal scales.* The above discussion shows that such predictable behaviour is likely to be within the spatio/temporal scales of the individual interactions in that particular process. Since, the

spatio/temporal scales of different processes are quite different, the predictable component will also vary.

2. *Exhibit unpredictable patterns of behaviour over longer time intervals*, as long as the system is not chaotic. This seems to be the desired state for most systems as it enables the system to evolve in an "advantageous" manner to the objects that are the subject of the model.
3. *Can exhibit chaotic behaviour.* This would not seem to be the desired state for real systems, but CA and AB models suggest that if the system gets too far out of synchronisation, then this can be the result.

This suggests that real systems may exhibit different characteristics at different spatio/temporal scales, that is, the relative influence of the predictable and the unpredictable components may be scale dependent. The suitability of the different types of models for a task therefore depends on the purpose of the model, the scale of interest to the user, and the relative contribution of the predictable and the unpredictable components at that scale. Most of these issues have not yet been adequately resolved in modelling.

The uses of model in resource management include:

1. Provide more detailed and rigorous information than that available from other sources, so as to improve our management of those resources, by making better and timely decisions. Thus, models can be used with a network of field observations to provide better information at each cell in the GIS.
2. Better understand the processes that are influencing the resources being managed. This improved understanding may include a better understanding of the suite of drivers to the process of interest, their relative importance and how this can change with changes in the other drivers, and with changes in spatial and temporal scales. Whilst such knowledge does not directly lead to specific resource management decisions, it can be used to influence a range of decisions.
3. Compare and contrast alternative decision scenarios, so as to decide which decision it is best to make. Decisions almost always involve trade-offs between alternatives. By providing information on the alternatives, the resource manager can then assess the costs and benefits of the alternatives before making a decision. Of course, the costs and benefits of the various alternatives may not be the only criterion for making a decision. There may be social, political and future expectations criteria that need to be incorporated into the decision. What the modelling allows is that these components can now be objectively analysed and weighed in the decision process.
4. Extrapolate to a future so as to assess the possible impacts of the proposed action. One of the most difficult things for a manager to do is to assess the potential implications of a proposed action. By analysing the possible implications, the manager can factor the possibility of theses effects occurring into the decision process. Both (3) and (4) in this list involve spatial and temporal prediction.

In this section we have reviewed the status of modelling, using very simple examples to illustrate the principles on which all models are based. There are very many models that are relevant to students of this text, but they are all based on these principles, or combinations of these principles. In the following section we will meet a particular scaling issue that afflicts vector data, and then we will consider the USLE model in more detail.

6.8.2 The Modifiable Areal Unit Problem

For many purposes it is necessary to derive statistics from GIS layers, related to administrative boundaries of some sort. Thus, it may be of interest to determine the percentages of degraded land

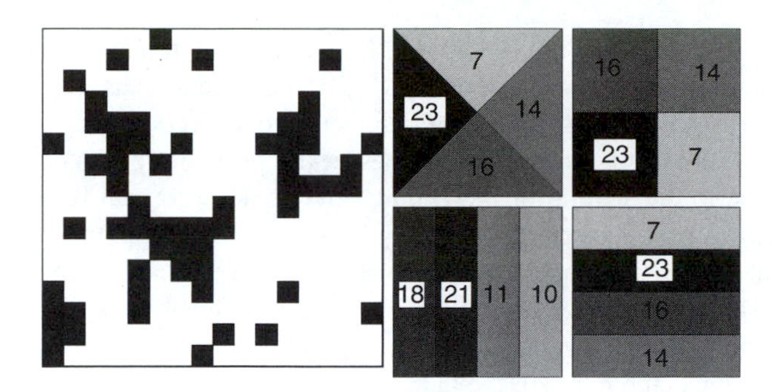

FIGURE 6.49 A map of two values in an attribute (left-hand side) and the derived statistical information using four different ways of partitioning the area into four equal sized sub-areas on the right-hand side.

by farm, by community or by larger regions or territories, just as it may be of interest to determine the percentages of forestland on similar administrative units. When statistical information is extracted from GIS layers, the modifiable areal unit problem (MAUP) may create statistical information that can be interpreted in a quite different way to that extracted using other administrative boundaries to extract and display the statistical information. The problem that arises is that the statistical information that is derived for different administrative boundaries can, when viewed across the whole area, suggest different interpretations for each types of boundary used. There are two aspects to the problem:

1. *The scale effect*. This is the tendency, within a system of modifiable areal units to derive different statistical results from the same data when the information is grouped at different levels of spatial resolution. The modifiable areal units here are the administrative units used to extract the statistical information, such as farms, communities and regions.
2. *The zoning effect*. This is the variability that can come from the use of a different distribution of modifiable areal units of the same size.

The problem arises because the source data contains specific information at specific locations, but when this is generalised into other areas then the size and position of the new area units will affect the values derived in the new statistical information. Consider the simple situation shown in the map on the left-hand side of Figure 6.49. This map contains pixels of either of two values of the attribute, whether that is landcover, soil condition or whatever, and is shown in black on a white background. This map is then partitioned into four areas in four different ways as shown on the right-hand side of the figure. The four different methods of partitioning the area are used to derive statistics on the frequency of occurrence of the black class, and these counts are shown as grey tone shading over the maps and as counts within the maps. If the analyst was to view the shading, then each of the partitions could be easily interpreted in a different way as to the distribution of the black class within the source layer of information.

The problem is not with the statistics that are derived, they are correct. The problem is that the shading that comes from them, even though they are all consistent in the relationship between count and shade, none the less give different interpretations as to the density of the black class across the whole of the area. This ability of statistical data to suggest differences is called MAUP.

Of the four methods of partitioning the area, all of them suggest quite different distributions of the black class. The MAUP decreases in severity as the classes become more uniformly distributed across the area, and as the size of the individual areas of each class become smaller, relative to the administrative areas being used to derive the statistics. Conversely, the MAUP can become worse as the areas become larger and as they become less uniformly distributed across the area. It should be remembered that it is usual to only derive statistics for those areal units that are of interest. What the MAUP shows is that if the areal units are of the wrong size, then they will give a distribution of the statistical data that may lead to errors in the interpretation that the analyst will not be aware of. There is thus a critical need to be aware of when MAUP may occur so as to protect the users of the derived statistical information from misinterpretations of the derived statistical data.

The general rule enunciated above applies whether the problem that is causing the differences in distribution are due to the scaling or the zoning effects. This general rule also suggests a way to either avoid MAUP, or to be aware when it may cause problems with the statistical data that are to be derived from it. MAUP will occur when the statistical areal units are of a size such that they contain data that is spatially correlated within the areal units. Since the semi-variogram shows the extent of spatial correlation within a data layer, computation of the semi-variogram will show the extent of this correlation as the range in the data. Modifiable areal units that are of a size similar to or smaller than the range in the semi-variogram may be affected by the MAUP and consequently may create statistics that will imply distributions that are different to that which should be implied. If the areal units are larger than the area supported by the range, than MAUP should not create a significant problem, and the larger the area is relative to the region of support as specified by the semi-variogram, the less likely is the MAUP to create a problem. The response then is to compute the semi-variogram for the data, determine the range and the area of support for this range ($\pi \times \text{range}^2$). If the statistical areas to be used are similar in size or smaller than this, then a problem may arise due to MAUP.

It should be noted that derivation of coarse pixel data from finer layer pixel data would incur the same problems with MAUP as will any other form of derived statistical data. If layer raster data are to be coarsened, then MAUP may cause a problem within the range of the source data.

A particular aspect of MAUP occurs when the analyst is interested in deriving information on attributes of individual members of a population when he has aggregated area data in those attributes, such as census or administrative district data. Numerous studies have shown that the plot of area-level versus individual-level correlation coefficients gives an S shaped figure as depicted in Figure 6.50. It can be seen in the figure that the area-level coefficients cover a larger range than the individual-level coefficients, and there is not a one-to-one correspondence between them.

This means that area-level data are a biased estimator of individual-level attributes, such as incomes and the medical and sociological characteristics of individuals. The problem arises because

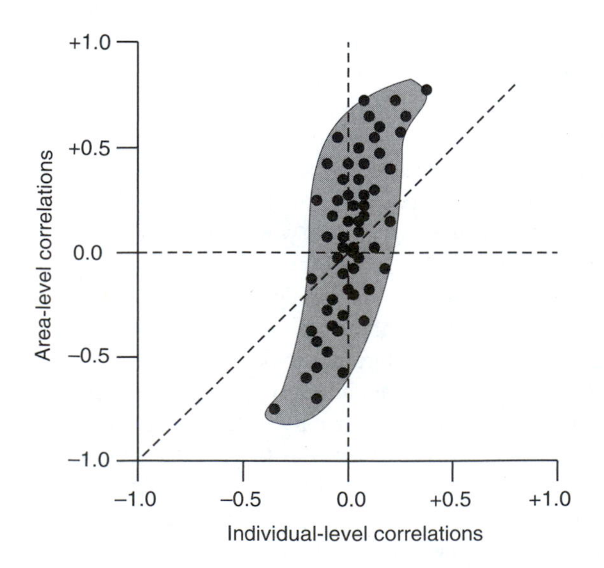

FIGURE 6.50 A typical S shaped plot of area-level correlation coefficients against individual-level coefficients.

area-level statistics represent those attributes for that area, not for the individual in that area. The same sort of problem will also arise if inappropriate temporal units are used in the analysis of temporal data.

Consider that there is a population of N individuals, each with a vector of attribute values, y_i. The whole population has a distribution with mean, υ_y and covariance matrix Σ_{yy}. However, this data is unavailable to the analyst.

The analyst has data in M districts into which the population of N individuals reside, with district populations of $N_j, j = 1, M$. The vector, c_i, contains the assignment of the individual to the district. The analyst has data, x_i for each district in the same attributes as for the individual, that is in y_i. The mean value for a district in these attributes are stored in x_i and $\varpi = (1/N)\Sigma N_j x_j$ is the overall mean for the districts. In the same way, the covariance array, $S_{xx} = (1/M)\Sigma N_j(x_j - \varpi)^2$ can be computed for the population of district values.

The MUAP arises because S_{xx} is a biased estimator of Σ_{yy} and as a result statistics that are based on the covariance matrix, such as regressions and principal components, will be biased and hence provide unreliable estimators of individual-level regressions and transformations.

6.8.3 Measures of Global and Local Homogeneity/Heterogeneity

The correlelogram and the variogram, discussed in Chapter 3, are normally used to provide global measures of heterogeneity in a layer or an image. However, most study areas are not consistent in their characteristics across that area, and so global statistics may be of limited value for some types of analyses where the analyst is interested in the characteristics within sub-areas of the image. The analysts can approach this problem from the perspective of finding the sub-areas through a form of analysis designed to identify homogeneous areas, such as image segmentation techniques, discussed in Chapter 4. He can then derive statistics in these sub-areas to describe the spatial characteristics within each sub-area. The correlelogram or the variogram can be used for this purpose.

An alternative approach is to use local measures, with the goal of identifying localised anomalies in the data. Most of these measures are based on variance measures or versions of these. One of the most useful is the pair of K statistics:

$$K_{1i} = \Sigma_j w_{ij} \times |x_j - x_i|$$

and

$$K_{2i} = \Sigma_j w_{ij} \times (x_j - x_i)^2 \tag{6.37}$$

6.8.4 An Example of a "Top-Down," Regression Based Model — The Universal Soil Loss Equation

The USLE is a regression model given in Section 6.3.3 as:

$$A = 0.224 \times R \times K \times L \times S \times C \times P \tag{6.38}$$

where A is the soil loss per unit area (kg/m^2), R is the rainfall erosivity factor, K is the soil erodibility factor, L is the slope length factor, S is the slope gradient factor, C is the crop management factor and P is the conservation practices used factor.

The USLE was developed as a method of estimating or predicting average annual soil loss from inter-rill and rill erosion. If this estimate or prediction produces estimates of unacceptably large annual soil losses, then the management response has to take actions that will reduce one or more of the factors so that the resulting soil loss is within an acceptable range. The only factors that can be affected in this way by resource managers are the land using practices (i.e., to affect the crop management factor, C and the conservation practices factor, P). The model can first be used with different values in these factors, to find those that give acceptable results. Having found a goal, the resource managers will need to work with landowners to find practices that will meet this goal and which are practical to implement at the farm level. Thus, the USLE may properly be used to:

1. Estimate or predict average annual soil loss from a field with specific weather, slope, soil and landuse conditions.
2. Guide the selection of crop management and conservation practices for specific weather, soil and slope conditions.
3. Predict the change in soil loss that would result from a change in crop management for a specific field, where these changes may include changes in conservation practices.
4. Determine how conservation practices may be applied or altered to allow more intensive cultivation.
5. Estimate soil loss from all lands.
6. Provide estimates of soil loss for conservationists to use in determining conservation needs in an area.

The equation was developed to estimate the average annual soil loss, hence its application in a particular year, or as a result of particular storm conditions may not be appropriate. The parameter values used must thus be representative of the whole year and not just one season or one rainfall event. The parameters that are affected in this way are those that change during the year, and from year to year, such as the rainfall erosivity factor, R, the crop management factor, C and the conservation practices factor, P.

The USLE factors were developed using an evaluation unit called the standard plot. A standard plot is 22.13 m long on a uniform lengthwise slope of 9%. Let us consider the factors in turn:

1. *The rainfall erosivity factor, R*. It is a measure of the erosivity of the rainfall events in an area. It is affected by two rainstorm characteristics: the kinetic energy created by the intensity of the rainfall on the surface and the duration of this intensity. These rainfall factor values for a large area can be presented as curves of equal erosivity (iso-erodents) on a map of the AOI, as shown for the United States in Figure 6.51.

2. *The soil erodibility factor, K*. It is a quantitative description of the inherent erodibility of a particular soil. Soils with different K values will erode at different rates, when the other factors are kept the same. The erodibility of soils is primarily affected by the percent silt, sand, clay and organic matter as well as the soil structure and permeability. For a particular soil the K value is the rate of erosion per unit of erosion index from a standard plot. The K value can be estimated from knowledge of the soil characteristics in the six physical parameters of percent silt sand, very fine sand, organic matter content, structure and permeability, where Table 6.6 shows the physical sizes of the particles in sand and silt in both the United States and Europe. Once values have been obtained for these parameters, they can be fed into the nomograph shown in Figure 6.52 to estimate the K value for that soil (Table 6.13).

3. *The slope length factor*, L *and the slope gradient factor*, S. The effects of slope length and gradient are presented in the USLE as L and S, respectively, however, they are often evaluated as a single topographic factor, LS. Slope length is defined as the distance from the point of origin of overland flow to the point where the slope decreases sufficiently for deposition to occur or to the point where runoff enters a defined channel. Irregular slopes are divided into n segments, where each segment should be uniform in slope and soil type. The soil loss for the entire slope is then computed using:

$$A = \frac{(0.224)RKCP}{x_c(22.13)^m} \times \{\Sigma(S_j x_j^{m+1} - S_j x_{j-1}^{m+1})\} \tag{6.39}$$

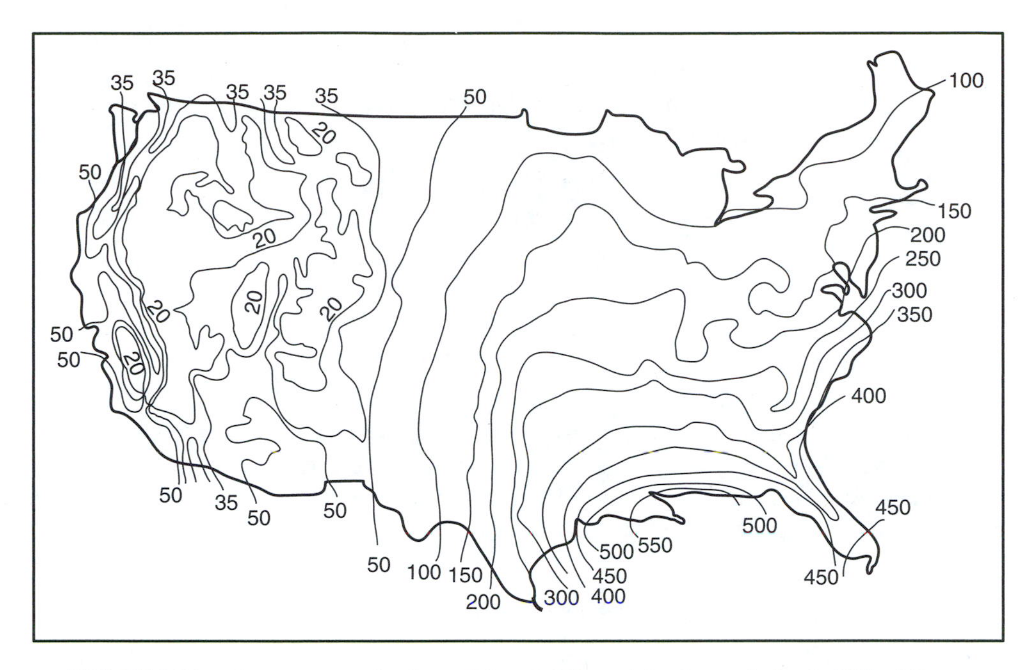

FIGURE 6.51 Average annual values of the rainfall erosivity factor, R, for the United States.

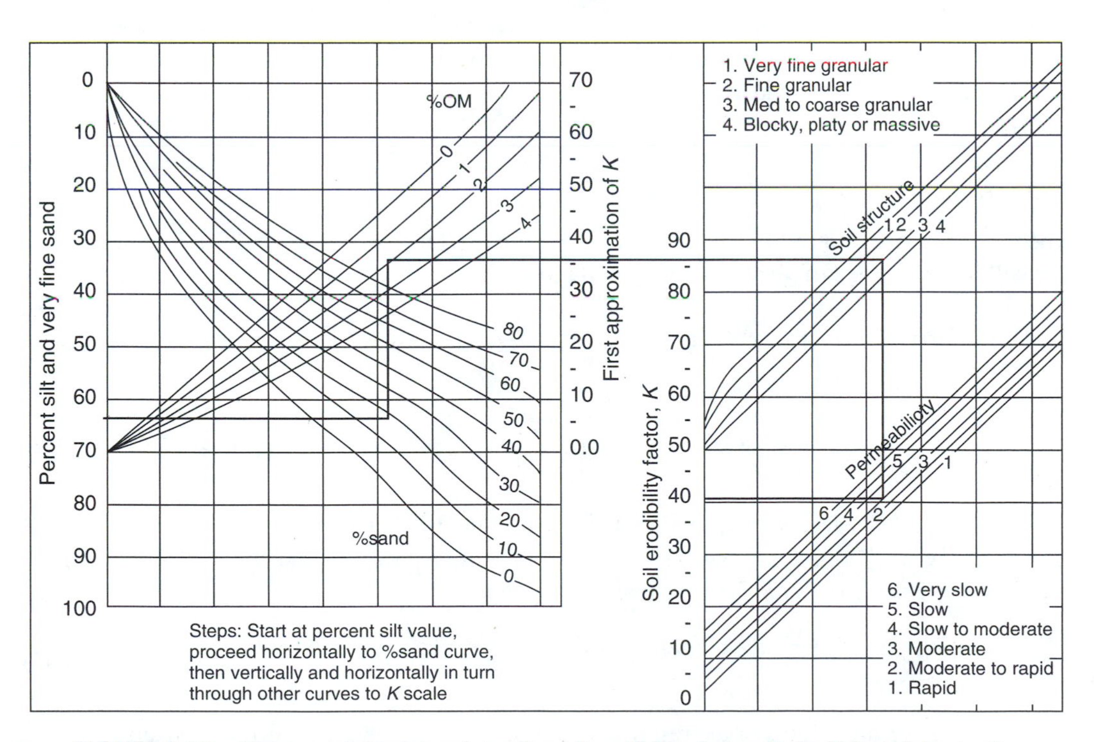

FIGURE 6.52 Nomograph for determining the soil erodibility factor, K, for U.S. mainland soils.

where x_j is the distance from the top of the slope to the lower end of the jth segment in meters, x_{j-1} is the distance from the top of the slope to the upper end of the jth segment in metres, x_c is the overall slope length in metres, S_j is the value of the slope gradient factor for the jth segment, m is the power that is a function of the slope, A, R, K, C and P have been defined previously.

TABLE 6.13
The Grain Size of the Different Categories of Soil Material

Grain size	Classes in the United States (mm)	Classes in Europe (mm)
Silt	0.002–0.05	0.002–0.063
Very fine sand	0.05–0.10	0.063–0.125
Sand	0.05–2.0	0.063–0.063

It would be reasonable to expect that whilst progressing down a slope the other parameters would also change. Changes in soil type may incur changes in the soil erosivity factor, K, and changes in landcover may result in changes in C and P. The USLE can be modified to allow for changes in all parameters down the slope, giving the modified equation:

$$A = \frac{(0.224)}{x_c(22.13)^m} \times \{\Sigma R_j K_j C_j P_j (S_j x_j^{m+1} - S_j x_{j-1}^{m+1})\} \tag{6.40}$$

This equation is almost suitable for use in a GIS as the values for R_j, K_j, C_j, P_j and S_j can be calculated for each cell or polygon in the GIS. However, the component $(x_j^{m+1} - x_{j-1}^{m+1})$ is dependent on both the distance of the point from the start of the slope and on the slope at the point x_j. The component $(x_j^{m+1} - x_{j-1}^{m+1})$ can be expanded:

$$(x_j^{m+1} - x_{j-1}^{m+1}) = x_j^{m+1} - (x_j - p)^{m+1} \tag{6.41}$$

where p is the size of a cell in the GIS layers in metres and the processing will be done in units of one cell in size in the GIS. Expanding Equation (6.41) gives:

$$\begin{aligned}(x_j^{m+1} - x_{j-1}^{m+1}) &= x_j^{m+1} - x_j^{m+1} + (m+1)x_j^m p - \tfrac{1}{2}(m+1)m x_j^{m-1} p^2 \\ &\quad + \tfrac{1}{6}(m+1)m(m-1)x_j^{m-2} p^3 - \cdots . \\ &= (m+1)x_j^m p - \tfrac{1}{2}(m+1)m x_j^{m-1} p^2 + \tfrac{1}{6}(m+1)m(m-1)x_j^{m-2} p^3 - \cdots \end{aligned} \tag{6.42}$$

For small watersheds the distance x_j will usually be less than $100p$, hence the range of values expected for x_j^m, x_j^{m-1} and x_j^{m-2} are as shown in Table 6.14. In consequence, the first term in Equation (6.42) is a reasonable approximation for Equation (6.41) giving:

$$(x_j^{m+1} - x_{j-1}^{m+1}) = (m+1)x_j^m p \tag{6.43}$$

Now x_j^m and p are in units of metres, whereas distances calculated in a GIS will usually be in cell units of size p m. This is most easily accommodated by the use of the conversion $x_j^m = X_j^m p$ where X is the distance from the start of the slope in units of cells in the GIS. Equation (6.40) can now be put in a more useful form for application within a GIS:

$$A = \frac{(0.224)}{x_c(22.13)^m} \times \{\Sigma R_j K_j C_j P_j S_j (m+1) X_j^m p^2\} \tag{6.44}$$

TABLE 6.14
Values of x_j^m, x_j^{m-1} and x_j^{m-2} Expected for the *m* Values Used and the Distances of One Cell, 10 Cells and 10 Cells from the Start of the Slope

		m											
		0.5			0.4			0.3			0.2		
		x_j^m	x_j^{m-1}	x_j^{m-2}	x_j^m	x_j^{m-1}	x_j^{m-2}	x_j^m	x_j^{m-1}	x_j^{m-2}	x_j^m	x_j^{m-1}	x_j^{m-2}
x_j	1p	1.0	1.0	1.0	1.0	1.0	1.0	1.0	1.0	1.0	1.0	1.0	1.0
	10p	3.2	0.3	0.0	2.5	0.3	0.0	2.0	0.2	0.0	1.6	0.1	0.0
	100p	10.0	0.1	0.0	6.3	0.1	0.0	4.0	0.0	0.0	2.5	0.0	0.0

for A in units of kg/m^2 per year, or:

$$A = \frac{(0.224)}{x_c(22.13)^m} \times \{\Sigma R_j K_j C_j P_j S_j (m+1) X_j^m\} \tag{6.45}$$

for A in units of kg/cell area per year.

4. *The Crop Management Factor, C*. It represents the ratio of soil loss from a specific cropping or covers condition to the soil loss from a tiled continuous fallow condition for the same slope, soil type and the same rainfall conditions. This factor includes the interrelated effects of cover, crop sequence, productivity level, growth season length, cultural practices, residue management and rainfall distribution. The evaluation of the C factor is often difficult because of the many cropping and management systems. Crops can be grown continuously or rotated with other crops; rotations are of various lengths and sequences; residues can be removed, burnt, left to decay in the field or incorporated into the soil, and the soil may be clean tiled or one of the various conservation practices may be adopted. Each segment of the cropping and management sequence must be evaluated in determining the C factor for a field. The C factor for each crop stage and then allocating weights to them based on the proportion of the year covered by that stage allows the computation of the final C factor for the field.

Sets of C factor tables have been produced for various countries and regions. An example of such data is given in Table 6.15, representing the cropping conditions in West Africa. In many situations the analyst may also be interested in estimating the soil loss from non-agricultural lands. Table 6.16 gives C factors for undisturbed lands in the Unites States.

5. *The erosion control practice factor, P*. It is the ratio of soil loss using the specific practice compared with the soil loss using up and down hill tillage practices. The erosion control practices usually included in this factor are contouring, contour strip cropping and terracing. Conservation tillage, crop rotations, fertility treatments and the retention of residues are important erosion control practices, but they are included in the cropping management factor discussed earlier.

The practice factors for the three major mechanical practices of contouring, contour strip cropping, irrigated furrows and terracing are shown in Table 6.17. Within a practice type, the P factor is most effective for the 3 to 8% slope class and values increase as the slope increases, and decease as slopes decrease away from this range. As the slopes decrease below 2%, the P factor increases due to the reduced effect of the practice when compared with up-and-down hill cultivation. The factor in Table 6.17 is for the prediction of the total off-the-field soil loss. If within-terrace interval soil loss is desired, the terrace interval distance should be used for the slope length factor L and the contouring P value used for the practice factor.

TABLE 6.15
The Vegetal Cover Factor and Cultural Techniques (*C* Factor) for West Africa

Practice	Annual Average *C* Factor
Bare soil	1.0
Forest or dense scrub, high mulch crops	0.001
Savannah and prairie in good condition	0.01
Overgrazed savannah or prairie	0.1
Crop cover of slow development or late planting: 1st year	0.3–0.8
Crop cover of rapid development or early planting: 1st year	0.01–0.1
Crop cover of slow development or late planting: 2nd year	0.01–0.1
Corn, sorghum, millet (function of yield)	0.4–0.9
Rice (intensive fertilisation)	0.1–0.2
Cotton, tobacco (2nd rotation)	0.5–0.7
Peanuts (as a function of yield and the date of planting)	0.4–0.8
1st year Cassava and yam (as a function of the date of planting)	0.2–0.8
Palm tree, coffee, cocoa with crop cover	0.1–0.3
Palm tree, coffee, cocoa with crop cover — burnt residue	0.2–0.5
Pineapple on contour (as a function of slope) — buried residue	0.1–0.3
Pineapple on contour (as a function of slope) — surface residue	0.01
Pineapple and tie ridging (slope of 7%)	0.1

TABLE 6.16
***C* Factor Values for Undisturbed Land in the United States**

		Mulch or Vegetation on the Ground (% Cover)					
Surface type	%Cover	0	20	40	60	80	95–100
None		0.45	0.24	0.15	0.09	0.04	0.01
0.5 m effective height	25	0.36	0.20	0.13	0.08	0.04	0.01
	75	0.17	0.12	0.09	0.07	0.04	0.01
2.0 m effective height	25	0.40	0.22	0.14	0.09	0.04	0.01
	75	0.28	0.17	0.12	0.08	0.04	0.01
4 m effective height	25	0.42	0.23	0.14	0.09	0.04	0.01
	75	0.36	0.20	0.13	0.08	0.04	0.01
Grass or compacted ground cover		1.00	0.83	0.67	0.46	0.30	0.27

Soil loss tolerance is the maximum rate of soil erosion that permits a high level of productivity to be sustained. In general, deep, medium textured, moderately permeable soils that have sub-soil characteristics favourable to plant growth are assigned tolerances of 1.1 kg/m^2/yr. Soils with a shallow root zone or other detrimental characteristics are assigned lower tolerances. Recommended soil loss tolerance values for soils in the United States can be obtained from Soil Conservation Handbooks.

The soil loss tolerance for a specific soil is used as a guide for soil conservation planning. The USLE is used to estimate the actual loss and to evaluate how changes in practices can be applied to reduce soil loss below the tolerance level.

TABLE 6.17
Erosion Control Practice, *P*

Land Slope (%)	Contouring	Contour Strip Cropping and Irrigated Furrows	Terracing
1–2	0.6	0.30	0.12
3–8	0.5	0.25	0.10
9–12	0.6	0.30	0.12
13–16	0.7	0.35	0.14
17–20	0.8	0.40	0.16
21–25	0.9	0.45	0.18

6.8.5 MODELING OF ECOLOGICAL SYSTEMS

The modelling of ecological systems has a long history. Lotka and Volterra developed the first population model in the 1920s. Since then many other process models have been developed and have evolved. It is beyond the scope of this text to discuss in detail the design, development and testing of models for which there are whole texts devoted to the subject. The reader is referred to these texts for advice and guidance in the selection and implementation of models. Most of the models that have been developed are suited to some research oriented tasks, but few are suited to operational and resource management tasks. This is because although the model formulations that are used in both may be the same, all models make simplifying assumptions about the spatial and temporal environments around them, but the level of simplification varies between the models. In addition to this, models designed for a research environment may require a wealth of input parameters, where many of these may be impractical or expensive to collect in an operational environment. Such an approach can be justified in a research environment if the model is to be used to investigate the process being modelled, but it is not acceptable for resource management implementations of a model. Thus in a research environment, the environment can be selected so as to fit the assumptions of the model, or the costs of describing the environment in an adequate way for incorporation into the model can be undertaken. However, this is not satisfactory for management models. For operational applications it is necessary to describe the environment in a cost-effective way that is not usually a constraint in a research environment.

All ecological systems are characterised by being extremely adaptive, having both the ability of self-organisation and a large number of feedback mechanisms. Ecological systems are irreducible, by which is meant that the system cannot be broken into its component parts and have those parts operate as they would when they are part of the whole. It is possible to study component parts of an ecological system, but the reactions that come from such studies are often different to that which occur when the part is part of the whole. Such reductionism thus has some value in better understanding the component part and how it operates internal to itself, but little value in understanding how that part interacts with other parts, and in turn influences those other parts. This is a classical example of where the whole system is more than the sum of the parts. This knowledge of how the component parts operate is essential in the construction of models of these systems, as the components must operate in accordance with their internal laws and constraints, but they operate in an environment of many outside influences, hence the internal understanding tells very little about how the components operate in the whole system.

Reductionism designed to describe the internal operation of components therefore needs to be matched with a capacity to conduct integrative science, and the only realistic way to do this for complex systems is by the use of numerical models. However, the models must meet some very

TABLE 6.18
The Hierarchy of Regulating Feedback Mechanisms in an Aquatic Environment (Jørgensen and Bendoricchio, 2001)

Level	Explanation of Regulation Process	Exemplified by Phytoplankton Growth
1	Rate by concentration in medium	Uptake of phosphorus in accordance with phosphorus concentration
2	Rate by needs	Uptake of phosphorus in accordance with intracellular concentration
3	Rate by other external factors	Chlorophyll concentration in accordance with previous solar radiation
4	Adaptation of properties	Change of optimal temperature for growth
5	Selection of other species	Shift to better fitted species
6	Selection of other food webs	Shift to better fitted food web
7	Mutations, new sexual recombinations and other shifts of genes	Emergence of new species or shifts of species properties

complex challenges. One of the main challenges has to do with the multiple levels of simultaneous operation that occurs in most ecological systems, and which are interdependent. Jørgensen and Bendoricchio (2001) has identified seven levels of simultaneous process that occurs in a wetlands environment, and these are listed in Table 6.18.

Similar tables can be constructed for other ecological situations, and indeed the table would be improved by consideration of the time and spatial scales that are relevant to each level in the table. It is clear that a model of an ecological system will need to sub-model some or all the levels that operate in the system in a simultaneous way, and that the linkages between these sub-models will need to be explicitly constructed in the model. Such forms of models are currently evolving in the fifth generation of models, and they are characterised by the incorporation of the principles:

1. Recognition that there are a number of levels of sub-process that operates within the ecological system in a simultaneous way.
2. The interdependencies that operate between processes at the same level, and between levels means that there are a huge number of feedback loops in a system, and that these provide for a very high level of adaptability of the system to changes in its environment.
3. Many processes are stochastic by their nature, and that the models that deal with these processes need to be able to deal with stochastic events. For example, all ecological systems are impacted by events outside the system, over which the environment has no control, and which are unpredictable by the system. Such events are stochastic by nature, and the system needs to be able to deal with such events. Stochastic events that impact on an ecological system include climate events, catastrophes such as fires, many forms of predation and the actions of other ecological systems, such as man.
4. Ecosystems display a high level of spatial and temporal heterogeneity. It has been shown that even under uniform conditions, ecological systems will display heterogeneity, and this will be exacerbated by heterogeneity in the underlying base conditions on which the model depends. Thus the terrain form affects the distribution of sunlight and may affect some or all the weather variables. Variations in the soils will affect other parts of the ecological process.
5. Units in an ecological system usually operate in accordance with economic principles that state that a system will make decisions on the basis of the best net benefit to the system. This means that the system will make the choice that gives $(\text{Total good} - \text{Total cost})_{\text{maximum}}$. In practice, there are different criteria for deciding on the value of a benefit to a system, and indeed the value of a benefit may change with changes in the condition of the system. Thus, a system that is facing starvation will place a higher value on something that ensures

survival over something that ensures reproduction, whereas a system that is healthy and well fed will place a higher premium on things that ensure reproduction rather than on survival.

6. Competition between species, and the fight for survival means that systems will tend to favour those individuals and species that are more fitted and that can better adapt to a higher level of fittedness. This means that systems will tend to evolve towards higher levels of complexity and individuals to a higher level of fitting for the environment in which they operate.
7. The rate of adaptation of natural systems is governed by the reproduction rate of the species in the system and their capacity to mutate. Since these factors will vary between species, systems that are under pressure to adapt are likely to see some species adapt well and others slowly, so that the slowly adapting species will become under threat before the adaptable species do.

6.9 UNCERTAINTY IN GIS ANALYSIS

Uncertainty is a factor in all decision making. Uncertainty in decision making involves lack of knowledge, ambiguity, errors and experience. There are other dimensions to the issue of uncertainty, but they can usually be reduced to some combination of these four factors. Thus, uncertainty arises when there is insufficient time to a make a decision, but that can be translated into a lack of experience or knowledge or some combination of both.

Just as there are a number of dimensions to uncertainty, there are also to the issue of lack of knowledge, where we will recognise the following dimensions to this issue:

1. *Lack of knowledge about a process*. This is the classical dimension of lack of knowledge, where we understand that our knowledge is inadequate simply because there are relevant questions about a process that we cannot answer, yet the answer may have relevance for the decision that we wish to make.
2. *Unknown knowledge gaps about a process*. In addition to the known gaps in our knowledge, there may well be gaps that we are not even aware of. Clearly, the less we know about a process, the more likely that this situation exists. At some stage we learn enough about a process to know that there are knowledge gaps, and of course this will often spur enquiry so as to identify and explain the discrepancies that have been revealed. Thus, for example, the gaps in the global carbon budget may be due to gaps in our knowledge about how the global carbon cycle operates.
3. *Lack of knowledge about relevance or criticality*. Sometimes we do not have sufficient knowledge about the importance of a process so as to decide on its relevance or criticality in relation to a decision that has to be made. This is rarely lack of knowledge about the process itself, but rather lack of knowledge about how it inter-relates to other processes or resources in the environment.
4. *Lack of knowledge about user dependence*. For some decisions it is important to understand the importance of the decision to the different resources managers who are dependent on the decision. Thus, managers who are not that dependent on the decision can have adverse decision made, in relation to this issue, without the decision having much, or any, impact on their operations. On the other hand, for other managers the decisions may have very serious implications for their production processes.

Ambiguity can arise through a lack of knowledge, but it can also arise through the difficulty of making decisions based on complex inter-relationships, particularly where there may be multiple choices and multiple goals that need to be achieved, and where the different choices may well have different levels of significance for the different parties who are dependent on the decision. The processes of Decision Support, discussed in relation to their use in a GIS environment in Chapter 8 are designed to assist in resolving ambiguities of these types. The nature of errors and their propagation in GIS, how to measure them and deal with them are the subject of the following three sections. The last section on this topic is that of decision making under uncertainty.

6.9.1 Errors within a Layer and Error Statements

Each layer in a GIS contains errors. These errors can occur in the identification or classification of features, in the assignment of them to a location in the layer, or in the estimation of the value of the feature. Thus, the types of errors include:

- *Errors in position and in the location of boundaries* in vector data. Such errors arise during rectification of image data, digitising of vector data and in the setting of decision boundaries in image classification where this causes a shift in position of the class relative to the correct position of the class.
- *Errors in assignment or allocation of cell values*. These errors arise in image classification and estimation and in setting attribute values for vector data.

Because of the way a lot of GIS data is created, the level of errors can vary from class to class and from location to location within a layer as well as between layers of the same type, but created using different data sets, different analysis techniques or at different times. Thus, in landuse classification, water is easily discriminated from land surfaces, and so it is likely to have a higher level of accuracy than of some of the land-based classes. In Chapter 5 we saw how the accuracy of classification and estimation can be determined from the use of field data and how the variance associated with classification and estimation can be derived.

With any mapping task, the map specifications will set down a confidence limit that is acceptable at a specified confidence interval. As at 2003, the U.S. topographic mapping specifications set down a locational accuracy specification that states (http://geography.usgs.gov/standards/):

1. *For large-scale maps* (larger than 1:20,000 scale) less than 10% of all well-defined points can have errors greater than 1/30th of an inch, or 0.85 mm.
2. *For small-scale maps* less than 10% of all well-defined points can have errors greater than 1/50th of an inch, or 0.5 mm.

If we consider the errors as being normally distributed, then the 10% that is referred to in the above specifications, is the area in the right-hand tail of the normal distribution (Figure 6.53) and not in both tails. This is because we are only interested in errors in position that are too large and not in those that are very small. The accuracy statement is saying that the cut-off value that leaves 10% of the area under the normal curve at higher values than this cut-off is set at 0.85 mm for large-scale maps and at 0.5 mm for small-scale maps. If the map scale is set at $1{:}x$, then these map distances represent ($0.85x/1000$ m) for large-scale maps and ($0.5x/1000$ m) for small-scale maps. From a table for areas under the standard normal curve we can see that this cut-off occurs at 1.28 standard deviations. Thus, it is stating that the confident limit is 1.28σ at an 80% confidence interval since 80% means 10% in each tail, even though we are only interested in the one tail in this situation. We can use this to work back and find out what variance we can accept in a sample of points, if the mapping is going to be accepted as meeting these specifications. The standard deviation, σ must be no larger than a value such that $1.28 \times \sigma \leq$ the set confidence limit:

$$1.28\sigma \leq 0.85x/1000 \text{ or } \sigma \leq 0.0006640625x \text{ for large-scale maps}$$

$$\leq 0.5x/1000 \text{ or } \sigma \leq 0.000390625x \text{ for small-scale maps}$$

If we consider 1:10,000 and 1:50,000 mapping, then σ must be equal or less than 6.64 and 19.53 m, respectively.

Similar considerations occur in relation to classification and estimation. The specifications for mapping accuracies will set a confidence limit on the classification. Thus, the specifications may require 90% accuracy at the 95% confidence level. This means that the accuracy must exceed 90% for 95% of the time. The construction of the cross-tabulation table and the derivation of the accuracy of classification and the variance associated with that accuracy has been discussed in Chapter 5. Thus, for each class there is a mean classification accuracy, μ and a variance, σ^2. At the mean value, 50% of the time the actual classification is more accurate, and 50% of the time it is less accurate. Thus, the mean accuracy value has a 50% confidence level. If this is not satisfactory then a new accuracy value has to be set such that it gives an acceptable confidence level. If a higher confidence level is required than 50%, then the boundary has to be set to the left, of the mean or at a lower accuracy value. Now we know that at one standard deviation from the mean, 34% of the observations will occur between the mean and this value, that is, there is a $(50+34)\%$ confidence level that the accuracy will be better then $(\mu - \sigma)$. If we wish to set a 95% confidence level, then this means that 5% of the chance will be left in the left-hand tail of the normal distribution. The standard normal distribution shows that this confidence level is achieved at 1.645σ, so that the user can be 95% confident that the accuracy exceeds $(\mu - 1.645\sigma)$. If this value is less than the specified accuracy (i.e., 90%), then the mapping meets the specifications, whereas if this value is higher than the specifications then the mapping is unsatisfactory (Figure 6.54).

Consider a classification that has an average class accuracy of $\mu = 0.948$ and a standard error, σ, of 0.000754726. It will be recalled from Chapter 5 that the standard error can be derived from the

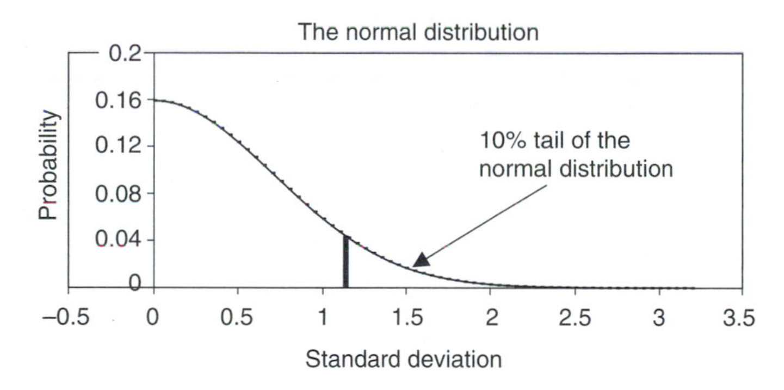

FIGURE 6.53 The normal distribution and the 10% tail.

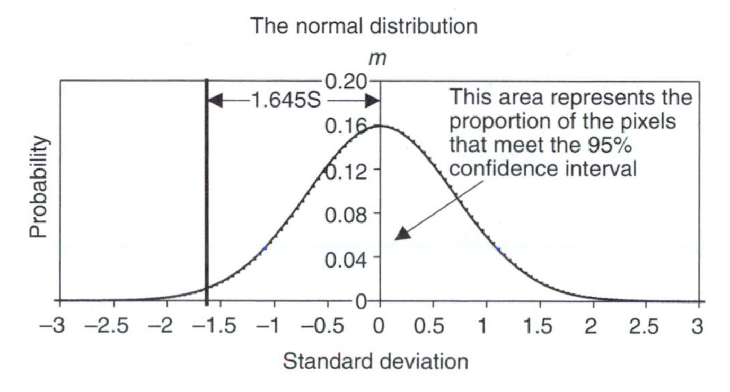

FIGURE 6.54 The normal distribution and the probability of the classification accuracy being satisfactory for more than 95% of the time.

equation:

$$\text{Standard error, } \sigma = [(N - n)pq/(n \times (N - 1))]^{0.5} \tag{6.46}$$

where N is the total population and n is the sample population, taken to be 10,00,000 and 79,650 in this example and $q = 1 - p$. For this class:

$$(\mu - 1.645\sigma) = 0.948 - 1.645 \times 0.000754726 = 0.947 > 0.90$$

and so this class meets the specifications but would just fall short of meeting the 95/95 criteria (95% accurate 95% of the time).

Similar logic applies to estimation accuracy. In this case the comparison of the field and the image estimates of the variable will yield a regression curve. This curve can be analysed in the same way as discussed earlier (Chapter 5) in relation to the linear regression.

6.9.2 Propagation of Error in Combining Layers

Not only are there errors in each layer of the data, but also these errors get propagated to new layers as they are created during analysis. For this discussion it is essential to differentiate between numerical and logical operations.

6.9.2.1 Propagation of Error in Numerical Analysis

Consider the case when a new layer, M is to be created as the addition of two old layers, P and Q. At each pixel the new layer value is computed as:

$$M = P + Q \tag{6.47}$$

Now let each of these contain an error, then:

$$M + \delta M = P + \delta P + Q + \delta Q \tag{6.48}$$

Subtract Equation (6.47) from Equation (6.48) to give:

$$\delta M = \delta P + \delta Q \tag{6.49}$$

Now in this case, the δ refers to one error. The variance is the sum of the squares of these errors divided by $(n - 1)$ where there are n observations. Since this will be identical in each case, it can be ignored, so the equation can be squared to derived variances:

$$\delta M^2 = (\delta P + \delta Q)^2 = \delta P^2 + \delta Q^2 + 2 \times \delta P \times \delta Q \tag{6.50}$$

If δM represents an error, the $(\delta M)^2$ is the square of that error. If these are summed across the n observation values in an image then we get the variance as:

$$\sigma_M^2 = \sum (\delta M)^2/(n - 1) \tag{6.51}$$

so that $(dM)^2$ is equal to $(n-1)$Variance(M). The same can be applied to the other variables in Equation (6.50), so that the $(n-1)$ is common and can be dropped to give:

$$\sigma_M^2 = \sigma_P^2 + \sigma_Q^2 + 2\sigma_{PQ} \tag{6.52}$$

In this equation, σ_{PQ} is the covariance between P and Q and is normally assumed to be zero, that is, zero correlation, even though there are many cases where this is not a valid assumption.

$$\sigma_M^2 = \sigma_P^2 + \sigma_Q^2 \tag{6.53}$$

It can be seen that the variances associated with the derived data are larger than the variances of either of the source data layers, and this is normally what occurs. Thus, as the analysis proceeds, the errors increase in the layers that are derived. This is one incentive for minimising the steps in an analysis. Instead of deriving the error propagation equations in this way, it is usually better to use the differential approach:

$$M = P + Q \tag{6.47}$$

Partially differentiate this equation:

$$\delta M = (dM/dP) \times \delta P + (dM/dQ) \times \delta Q \tag{6.54}$$

Square to give:

$$\begin{aligned}(\delta M)^2 &= ((dM/dP) \times \delta P + (dM/dQ) \times \delta Q)^2 \\ &= (dM/dP)^2 \times (\delta P)^2 + (dM/dQ)^2 \times (\delta Q)^2 + 2 \times (dM/dP) \times (dM/dQ) \times \delta P \times \delta Q\end{aligned}$$

Now $dM/dP = 1$ and $dM/dQ = 1$ for Equation (6.47). They are substituted into Equation (6.54) and the covariance is again assumed to be zero, to give:

$$(\delta M)^2 = \sigma_M^2 = \sigma_P^2 + \sigma_Q^2 \tag{6.55}$$

By similar logic it can be shown that the variances for other typical functions are:

If

$$M = P - Q \text{ then } \sigma_M^2 = \sigma_P^2 + \sigma_Q^2$$

$$M = P/Q \text{ then } \sigma_M^2 = (\sigma_P^2/Q^2) + (M^2 \times \sigma_Q^2/Q^2)$$

$$M = P \times Q \text{ then } \sigma_M^2 = (\sigma_P^2 \times Q^2) + (\sigma_Q^2 \times P^2)$$

$$M = \text{constant} \times P \text{ then } \sigma_M^2 = (\sigma_P^2 \times (\text{constant})^2)$$

$$M = c_1 \times P + c_2 \times Q + c_3 \text{ then } \sigma_M^2 = ((c_1)^2\sigma_P^2 + (c_2)^2\sigma_q^2)$$

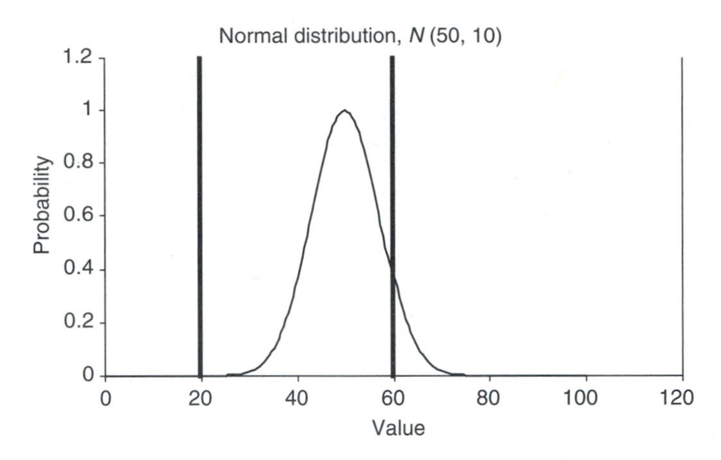

FIGURE 6.55 A normally distributed data value, d is $N(50, 10)$ in relation to set bounds of $a = 20$ and $b = 60$.

Similar formulae can be created for other forms of numerical analysis, although of course some of them can become quite complex. When conducting this sort of analysis, it would be normal to do so using the population mean and variance data, so that the analysis has to do with the reliability of the data rather than its accuracy at a point. The same sort of analysis can, of course be done on the analysis at a point, but then the sample mean and variance values at that point in both data sets need to be used.

6.9.2.2 Propagation of Error in Logical Operations

Logical operations are often of the form:

"If layer A has values in the range a to b, then accept layer B values, otherwise accept layer C values."

In such a process there are two potential sources of error:

1. *Errors in the criteria set for the allocation decision.* In this case the values set as the criteria (a and b) create errors, or the nature of the test creates errors.
2. *Errors in the transferred values.* In this case errors in the data in layers B or C.

Let us first consider the transferred data values. If the decision is to transfer the value from layer B, then the error transferred refers to that data value. The same applies to layer C. So, the errors in the new layer will contain errors that are sourced from the source layers, whichever layers they happen to be. Now we need to consider the allocation errors. The actual value being tested in layer A will have a variance associated with it, so we will assume that the value d from layer A is $N(d, \sigma^2)$, that is, it is normally distributed with a mean value of d and a variance of σ^2 that is the same as the variance of an observation in layer A. If d is a long way from either a or b, then its distribution will have a low probability at either a or b, and thus there will be a very low chance of a value d actually being a value outside the range, and hence negligible errors will be introduced. If, on the other hand, d is inside but close to either a or b, then its distribution will show a significant probability outside this range. This means that the value d has this probability of actually being a value outside a or b, and so this probability is of an error occurring in the allocation (Figure 6.55). In the case of $d = a$ or b, then there is a 50% probability of making a wrong decision for the allocation.

The probability of making a wrong decision can be determined by determining the probability. In the case here, take the upper bound $b = 60$ and find its position relative to d for a standard normal distribution by computing:

$$\text{Distance} = (b - d)/\sigma = (60 - 50)/10 = 1$$

So the point b is at exactly one standard deviation from the centre of the normal distribution for d. From the standard normal distribution we get the probability at this point of 0.3413. This is the probability from the mean value for the distribution, to the point of interest. However, we want the tail probability, so we want $(0.5 - 0.3413) = 0.1587$. For a value of $d = 50$, there is a 15.87% chance that we will make the wrong decision and thus make the wrong allocation. It can be seen that this probability decreases to negligible as d gets further and further away from the boundaries, and increases as it approaches one of the boundaries. The error propagated from the old data to the new is thus the same as the error in the old data when d is a long way from the assignment criteria, a and b. As d approaches one of these criterion, the propagated error will change in accordance with:

$$\text{Propagated variance} = \text{Variance in B} \times \text{Probability of assignment of B} + \text{Variance of C} \times \text{Probability of assignment of C.}$$

In the above case, $\sigma^2_{\text{new}} = (1 - 0.1587) \times \sigma^2_{\text{B}} + 0.1587 \times \sigma^2_{\text{C}}$

In many logical operations, the decision rule does not contain numerical data with a variance value, but may contain alphanumerical values. In such cases, the probability of error cannot be evaluated in the same way. What is normally done in these situations is to consider that the error propagated is simply that due to the errors in the source data, that is, the errors in layers B and C in the above case.

This method of analysis of error propagation can be conducted for all analytical operations. However, when the combinations become very complex, then derivation of the formulae can be difficult. If the equations cannot be differentiated, then it may be impossible to derive error propagation formulae. If this is the case, then an alternative is to use Monte Carlo techniques.

In Monte Carlo analysis, each input parameter in the formulae is allocated a range of values that would usually be normally distributed about a typical mean value, based on actual data collected for that parameter. The formula is then used with all combinations of these parameter values and the results used to derive a mean and variance for the output value. The derived variance is an estimate of the error propagated through the process. The affect of individual parameters can be estimated by using the mean value in all the input parameters except the parameter of interest, for which the full range of values are used. The resulting variance gives an estimate of the effect of the single parameter on the process.

Such a process will yield realistic estimates of variance when there is negligible correlation between the parameters. When there is correlation, then the correlation introduces a weighting of some combinations relative to other combinations of the parameters, and this weighting may introduce errors into the estimated variance.

6.9.3 Sensitivity Analyses

Sensitivity analysis is the analysis of the sensitivity of a numerical formula to variations in each of the input variables. In most equations, some variables have a much larger impact on the output than do other variables. The model is thus more sensitive to those variables that can produce large impacts on the formula than it is to the variables that have a small impact.

The purposes of sensitivity analysis are to:

1. Identify the relative sensitivity of the model to the values in the different input variables.
2. Determine the relative accuracy with which each input variable needs to be measured or estimated for use in the model.

3. To determine the resolution that needs to be used in estimating the input variable values.

Variations in sensitivity are due to:

1. *The magnitude of the constants associated with each variable.* A variable whose constant = 10 will have a much larger impact than one whose constant = 1.
2. *The formulation of the variable in the formula.* A variable that is raised to a power, such as $(\text{variable})^{3.14159}$ will have a much larger impact than a variable that is not raised.
3. *The relationship between the variables in the formula.* When you get a combination of the form $(\text{variable}_1)^{\text{variable}}$ then variable has a larger impact than variable_1.
4. *The range of values that can be taken by a variable.* A variable that can take a range of ± 50 will have a larger impact than one that only takes a range of ± 1, all other things being equal.

Sensitivity analysis can be conducted by means of the propagation of error for simpler models, such as the USLE model. However, most models are too complex to conduct sensitivity analysis in this manner. When this occurs then the best way to conduct sensitivity analysis is to use Monte Carlo techniques.

Consider the Levin's model of species propagation:

$$\mathrm{d}p/\mathrm{d}t = cp(1-p) - mp = \delta_p \tag{6.56}$$

Now if there is a small change in each factor, Δ, then we get:

$$\begin{aligned}\delta_p + \Delta_\delta &= (c+\Delta c)\times(p+\Delta p) - (c+\Delta c)\times(p+\Delta p)^2 - (m+\Delta m)\times(p+\Delta p)\\ &= cp + c\Delta p + p\Delta c + \Delta c\Delta p - cp^2 - 2cp\Delta p - c(\Delta p)^2 - \Delta cp^2 - 2p\Delta c\Delta p - \Delta c(\Delta p)^2\\ &\quad - mp - m\Delta p - p\Delta m - \Delta m\Delta p\end{aligned} \tag{6.57}$$

We can subtract the original equation from this equation to give:

$$\begin{aligned}\Delta_\delta &= c\Delta p + p\Delta c + \Delta c\Delta p - 2cp\Delta p - c(\Delta p)^2 - \Delta cp^2 - 2p\Delta c\Delta p - \Delta c(\Delta p)^2\\ &\quad - m\Delta p - p\Delta m - \Delta m\Delta p\end{aligned} \tag{6.58}$$

In this equation the Δ values represent small changes, so that the product of two or more of them are very small, and can be ignored. This gives:

$$\Delta_\delta = c\Delta p + p\Delta c - 2cp\Delta p - m\Delta p - p\Delta m = \Delta p(c - 2cp - m) + p\Delta c - p\Delta m \tag{6.59}$$

There are three input parameters for which we are interested in their sensitivities and each of them is multiplied by a factor. Both Δc and Δm are multiplied by the same factor, p, so that these two have the same sensitivity. The other factor, Δp, is multiplied by $(c - 2cp - m)$ and its sensitivity relative to the other parameters is a function of the size of $(c - 2cp - m)$ relative to the size of p. If, for example, $p = 0.3$, $c = 0.07$ and $m = 0.05$ then the sensitivities of p, c and m are 0.022, 0.3, 0.3, respectively. It can be seen that c and m need to be measured more accurately than p, in fact by the ratio $0.3/0.022 = 13.6$ times more accurately.

6.9.4 Decision Making Under Uncertainty

Decision making under uncertainty involves moving aware from hard or crisp decisions to soft or probabilistic decisions. With probabilistic decisions, a process provides probabilities that alternate solutions are correct. Clearly, if one has three alternate solutions, A, B and C, where the probabilities are given as 0.86, 0.12 and 0.02 for each of these in turn, then clearly one can place a lot of confidence in making the decision to adopt solution A. If on the other hand, the probabilities are found to be 0.35, 0.33 and 0.32, respectively, then one may adopt decision A, but one will have a low level of confidence in this choice. There are two main avenues for the implementation of probabilistic decisions: by the use of Bayes rule or by the use of Dempster–Shafer theory.

6.9.4.1 Bayes

The concept behind Bayes theory is covered under classification in Chapter 4 in detail. In that Chapter, Bayes rule was given as:

$$P_r\{X_p|W_i\} = \frac{P_r\{X_p|W_i\} \times P_r\{W_i\}}{\sum P_r\{X_p|W_j\} \times P_r\{W_j\}}$$

where $P_r\{X_p|W_i\}$ is the probability of class W_i given data values X_p, $P_r\{X_p|W_i\}$ is the probability of getting the response values X_p for class W_i, $P_r\{W_i\}$ is the probability of class W_i occurring independently of the evidence (data), and the denominator is the sum of these across all classes. Training data for class W_i gives the probability distribution of the evidence, or data, given the class. The numerator is the product of this probability by the probability that the class will occur and the denominator ensures that the probabilities sum to unity. So, Bayes theory can give probabilities that each pixel belongs to each class, even though in standard classification this is simplified and ends up with a crisp classification. These probabilities can be used as the basis of a soft classification.

6.9.4.2 Dempster–Shafer

In Bayes theory the probabilities arise not due to lack of knowledge, but rather due to poor evidence or data. Dempster–Shafer took this another step by saying that probabilities should also be assigned due to limitations on our knowledge. Thus in Bayes theory, with perfect data, the derived probabilities would come down to 1.0 for the correct class and 0.0 for each other class. With Dempster–Shafer, probabilities will still be assigned under conditions of prefect data, because there are gaps in our knowledge.

6.10 PRESENTATION IN A GIS

The result of all GIS analysis will be information in the form of tables or layers, or both. If the output includes layers as maps then the analyst needs to prepare these in a way that is convenient for the user to read and use. This means that the map will usually require ancillary information superimposed on it, and supplementary information of the map itself displayed around the map. The ancillary information that will need to be placed on the map depends on the use of the map, but it may include (Figure 6.56):

1. *Navigational details* such as roads, railway lines and the drainage network.
2. *Administrative details* such as farm boundaries, administrative district boundaries and some names, such as towns and locations on the map.

The supplementary information that needs to be displayed around the map includes:

1. *Map name* Standard topographic maps all have standardised names. If the map is of the same area as the topographic map, then it should have the same name, but include other information describing the main content of the map. Thus, a landuse map for the Culloden

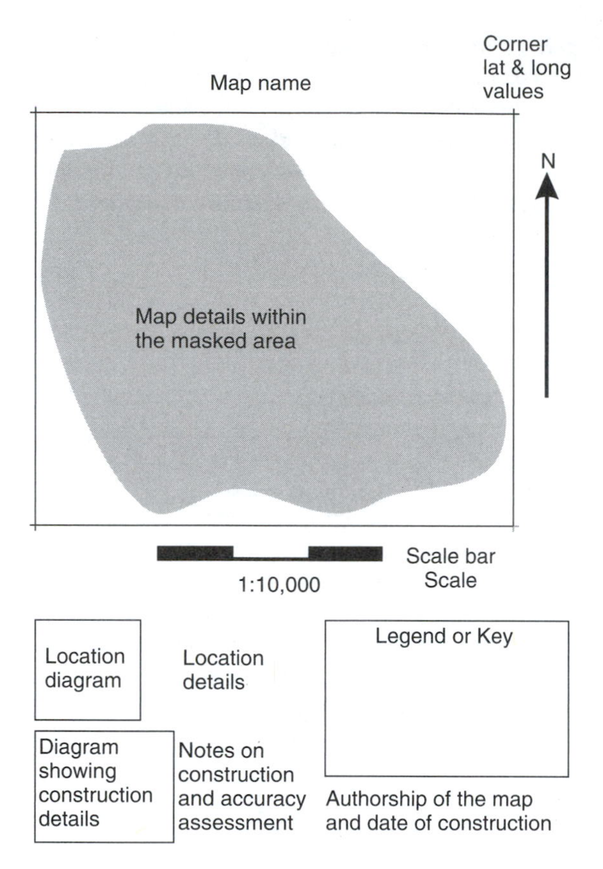

FIGURE 6.56 Supplementary information that should be placed on a map for the assistance of the user.

map sheet could be called Culloden Landuse Map. The map name should be at the top centre of the sheet, above the map itself.

2. *Scale details.* The scale details will normally include the scale of the map in the form 1:50,000 and a scale bar. It will normally be placed at the centre of the map and immediately below the map.
3. *North point* indicating the direction of grid north, and possibly magnetic and true north.
4. *Key* explaining the colours and symbols on the map.
5. *Locational boundary information* such as tick marks for the latitude and longitude for the corners of the map. This detail can be readily seen on all topographic maps. If the map is a special map, then this detail map not be included, but the location of the map relative to the topographic maps in the area may be displayed as a diagram at the base of the map.
6. *Description of the source of the map.* This map be in the form of notes or a diagram. Thus, if the map is constructed from one set of satellite data for part of the map, and from field data for the remainder of the map, then this can be shown as a diagram, possibly with adjacent notes. If the map is derived from analysis, then the main analytical steps need to be indicated.
7. *Accuracy and reliability of the map.* Often the analyst will only have the overall accuracy of the map, and this should be stated as the map accuracy. Of course, following our earlier discussion this is the same as the reliability of the map.
8. *Authorship and date of construction.* If some of the map details have been provided by another group or organisation, then this ownership should be acknowledged here, as well as the date that this information was constructed.

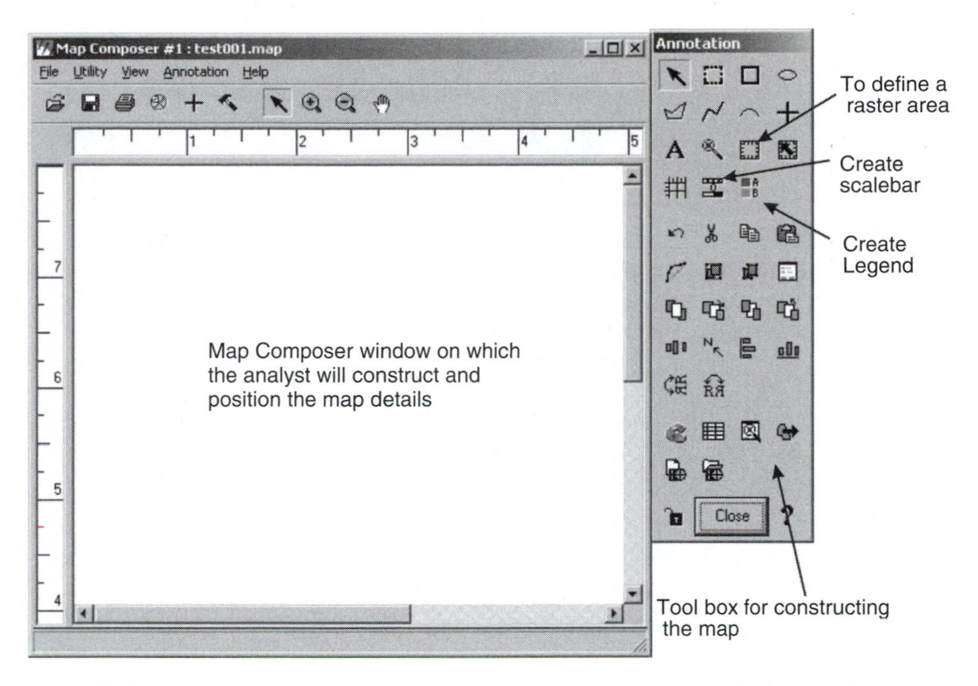

FIGURE 6.57 Tools for the construction of a map sheet as supplied by Imagine.

Most GIS systems provide the analyst with the capacity to construct such maps. Figure 6.57 depicts the tools provided in Imagine to construct a map. These tools consist of a sheet on which the analyst will draw the map composition. In doing this the analyst can control the size and location of the map details, and the surrounding details as well. The other tools consist of the toolbox shown in Figure 6.57. This toolbox allows the analyst to select images for inclusion in the map, and to then position and scale those images. It allows the analyst to place text in the map, and to construct a scale bar and legend. In this way the analyst can place the map in the image and then other maps depicting the location of the main map if this is appropriate.

6.11 THREE-DIMENSIONAL GIS

For some applications a three-dimensional GIS is a distinct advantage. Thus, three-dimensional GIS have been developed to depict geologic structures using seismic, magnetic and gravity data to construct these features. Another application is to depict the surface characteristics of groundwater relative to the geologic features in the area. Other applications could include depiction of features in the atmosphere and the nature of dynamic ecological processes, where the third dimension would depict time rather than depth into the soil, water or atmospheric surface.

The major difficulties with three-dimensional GIS are the costs of construction of the data, where (x, y, z, t) co-ordinates may need to be automatically recorded, and the depiction of the data in the GIS. Depiction can be achieved with a few parameters where the surfaces are displayed in a semi-transparent way so that features beneath the surfaces can also be seen. Finally, good navigational data needs to be shown in the display for orientation of the analyst.

ADDITIONAL READING

For those who wish to read more on GIS, the texts by Peter Borrough (Burrough and McDonnell 1998) and Robert Laurini (Laurini and Thompson 1992) are recommended. There is an extensive range of books dealing with modelling theory and its applications in hydrology, soil science and ecology. Few models currently focus on using GIS, that is, in taking the spatial dimension fully into account, but this situation is changing.

REFERENCES

Burrough, P.A. and McDonnell, R.A., 1998. *Principles of Geographical Information Systems*, Oxford University Press, Oxford, England.

Dijkstra, E.W., 1959. *A Note on Two Problems in Connection with Graphs, Numerische Mathematik*, Vol. 1, Springer-Verlag, Berlin.

Jørgensen, S.F. and Bendoricchio, G., (Ed.), 2001. *Fundamentals of Ecological Modelling*, Elseveir, Amsterdam.

Laurini, R. and Thompson, D., 1992. *Fundamentals of Spatial Information Systems*, Academic Press, London.

Teitz, M.B. and Bart, P., 1968. "Heuristic methods for estimating the vertex median of a weighted graph," *Operations Research*, 16, 955–961.

7 The Analysis and Interpretation of Vegetation

7.1 INTRODUCTION

Vegetation is one of the most critical components of the biosphere. Most forms of life on Earth depend on it either directly or indirectly for food, and some use it to provide safety or security. Man also depends on it for some forms of fuel and fibre. Vegetation has been crucial in creating and maintaining the atmospheric conditions that are amenable to other forms of life on Earth. In addition to these essential values of vegetation, it is valuable for many aesthetic reasons to the majority of people, such as for the provision of flowers, and the provision of forest and other recreational areas.

By capturing solar radiation, and using this with water, carbon dioxide and minerals to produce plant material and oxygen, vegetation lays the basis for the food chains that involve most species on Earth. Plants do this by absorbing carbon dioxide through spores on the leaves, water through the leaf spores and the roots, minerals through the roots and solar energy by absorption by the leaf chlorophyll pigments. They are the dominant way in which solar energy is captured and converted into material that can be used by other species in the environment or in the food chain.

By conducting this transformation, plants interact directly with a number of key biogeochemical (BGC) processes or cycles in the environment. They directly influence these processes by their consumption of some resources and the creation of products and of wastes that are then released into the environment. These processes thus directly influence vegetation. In addition to the impact of these biogeochemical processes, plants, by absorbing energy from the sun, affect the energy balance at the Earth's surface. Vegetation is thus one of the building blocks of life, and one of the major factors influencing the key BGC processes that ultimately control the conduct of biological activity on this globe.

There is evidence that man is having a significant impact on all these processes. Whether this impact is good or bad is at the heart of a significant global debate that is not likely to go away for some time. What all sides of that debate accept, however, is the need to understand what is happening to the global environment and from this develop reasonable strategies so as to ensure the long-term sustainability of our use of global resources.

The purposes for which we require information on vegetation from this global perspective are thus:

1. To improve our understanding of local, regional and global BGC processes so as to be able to live within the constraints imposed by the available resources and the dynamic processes that interact with those resources, as well as to develop ways of using the resources of the globe in a sustainable manner, including the constraints that this objective requires.
2. To more efficiently and effectively manage the production of food, fuel and fibre, so as to reduce the impact of this production on the environment and to use fewer resources in this production.

The above discussion has set very general goals or purposes for the collection of data about vegetation. However, it is now necessary to set a much more specific focus if we are to deduce more specific information needs about vegetation. A plant, by consuming resources from the environment including water, minerals and carbon dioxide, as well as absorbing some of the incident solar energy and converting them into plant products and plant wastes that include oxygen is being affected by the global dynamic forcing mechanisms, and in turn, is modifying them. The interactions that this implies with the global BGC processes thus include:

1. Removal of carbon dioxide from the atmosphere and deposition of this carbon as plant material. The removal of carbon dioxide from the atmosphere modifies the global warming processes of carbon dioxide as a greenhouse gas and places the carbon into a sink. The duration of that sink depends very much on the type of vegetation that has fixed the carbon. Plants thus contribute to the carbon cycle.
2. Removal of water from the soil and the surface, and its use in the conversion of carbon into complex organic compounds, as well as its excretion into the atmosphere through evaporation and transpiration. Plants thus shift water from the soil sink into the atmosphere, where it is then available for further precipitation, as well as converting some of the water into a sink in the plants as part of the plant tissue. Plants are thus influencing components of the hydrologic cycle.
3. Absorption of solar short-wave radiation with some of it being consumed in the conversion of carbon dioxide and water into plant tissue, some being used to heat the plants, and then being re-radiated as long-wave radiation and some being used to convert water to the gaseous form during transpiration. Plants are thus affecting the global energy budget, and the amount of energy that is available to other parts of the budget, such as that available for warming of the planet.
4. Absorption of minerals from the soil, depleting the soil of these minerals, but making them available to other parts of the biosphere.

In their interaction with these processes, plants operate in very different ways depending on the processes themselves, the climate regime in the area and the species involved. Each species interacts with the environment in a somewhat different way, at least at some stages throughout their phenological cycle. Thus, the impact of plants on the BGC processes depends to some degree on the species involved, and the climatic regime that is in place. It should be emphasised that some of these differences operate at a point in time, but many of them are linked to the phenology of the individual plants or their communities.

It is not the purpose of this discussion to get into the details of these interactions, as there are many excellent texts that cover this material including those by Chapin et al. (2002) and Butcher et al. (2000). What is the goal here is to start from a basis of an understanding of plant interactions with the BGC processes to work towards an understanding of the types of information that we need about vegetation so as to better understand their role in the BGC processes, and ultimately how we adapt to these processes so as to live in harmony with the major BGC processes that support life on Earth. In order to do this we need to consider these cycles in somewhat more detail.

7.1.1 The Energy Budget

Energy is stored as either sensible or latent heat. Sensible heat is the heat that can be measured using a thermometer. Latent heat is the heat that is absorbed or released as part of a chemical reaction or a change in state. Thus, when water is vaporised in evaporation or transpiration, it absorbs energy in the form of latent heat. Such heat is released when the gaseous form of water condenses into rain or snow.

The energy incident on the top of the atmosphere from the sun is either reflected back into space by the elements of the Earth's atmosphere or biosphere, or the atmosphere or the components of the biosphere absorb it (Figure 7.1). Some of this energy is converted into latent heat in the process of

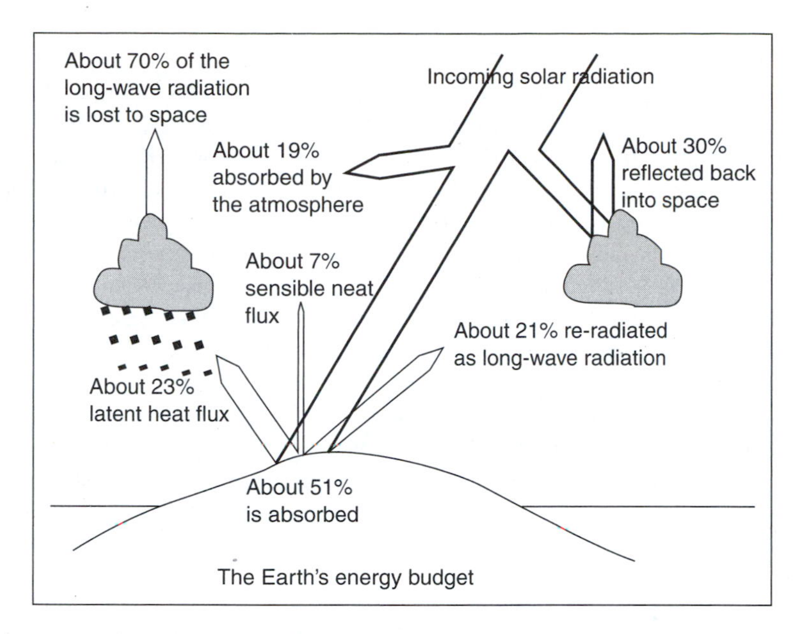

FIGURE 7.1 The components of the Earth's energy budget.

absorption, as occurs when it causes the evaporation of water and chemical reactions in the leaves of green plants, and some is stored as sensible heat. The sensible heat affects the temperature of the atmosphere and the Earth's surface and thus modifies the characteristics of the long-wave radiation that is radiated by the atmosphere and the surface of the Earth back out into the atmosphere and into space. It will be recalled from Chapter 2 that all bodies that are above absolute zero temperature radiate energy, the wavelength distribution and intensity of which is a function of the temperature and emissivity of the surface. Sensible heat transfers also occur between objects of different temperature, so that sensible heat is transferred from warmer to colder objects.

Albedo is defined as the ratio of light reflected back out into space to the light received. Since the amount of energy that is incident on the Earth's atmosphere is known accurately, knowledge of the Earth's albedo can be used to estimate how much of the incident radiation is lost directly back into space.

The major changes in state that occur in the biosphere are the conversion of chemicals and energy into plant matter, the growth, consumption and decay of plant and animal matter and the conversions of water into vapour and back again as part of the hydrologic cycle. Thus, physical attributes of vegetation that can be used to gain an appreciation of the components of the energy cycle are:

1. Green leaf area index (GLAI) is a valuable diagnostic as it is used to estimate the transpiration of moisture as part of the hydrologic cycle, as well as being a driving parameter for most plant growth models.
2. The fraction, absorbed photosynthetically active radiation (FPAR), is used in estimating the photosynthetic rate and hence the accumulation of plant material as a component of the energy budget and the carbon cycle.
3. The fraction, vegetative cover (FVC) is used to estimate the proportion of rain that will intercept the canopy before being intercepted by the ground surface, important in assessing the erosion impact of rainfall. With GLAI this can be used to estimate the proportion of the rain that remains on the leaves and is thus evaporated from the leaves.
4. The albedo is of use in improving our estimates of the total energy budget.

7.1.2 The Hydrologic Cycle

Plants modify the hydrologic cycle in a number of ways. By intercepting rainfall, they both reduce the erosive impact of the rain on the soil and they hold some of that rain on the plants where it is available for absorption by the plant material and for evaporation from the plants. Plants also tend to make soils more absorptive to rain, through their impact on the soil surface roughness and texture, by opening passageways into the soil, and by slowing the percolation of water across the soil surface due to the roots and litter. Thus, plants tend to increase the amount of rainfall that enters into the soil, and reduce the amount that is lost from that location by surface runoff. In addition to this, by drawing water out of the soil through their roots, plants take water out of this sink and through transpiration they excrete water into the atmosphere as moisture, that is then available for condensation and deposition as rainfall at some other time and location.

The Fraction, Vegetative Cover is important in estimating the proportion of the rain that will be estimated by the vegetative cover and thus not impact directly on the soil surface. FVC varies between 0 and 1. The total plant leaf area index (TLAI) is important in estimating the amount of moisture that may be held on the plant after rain, and thus is available for evaporation from the vegetative cover. Finally, GLAI is important in determining the transpiration that may take place from the plant.

Most of the moisture taken up by plants for photosynthesis is done so through the root system. The root systems of plants exhibit as much variation as do the canopy structures, with some species depending almost entirely on deep rooted tap roots to extract water and minerals from the deeper levels of the soil, others depending entirely on shallow root systems and some depending on a mix of the two. The effectiveness of either approach depends on the climatic regime, the availability of water in the different soil layers and the species or community mix that exists in the area. Therefore, information is required on the species/community mix in addition to the above parameters in the assessment of this component of the hydrologic cycle.

The hydrologic cycle is very dynamic. At some stages there are likely to be insufficient moisture to meet the needs of plants. A moisture deficit occurs when the potential evaporation, or the evaporation that would occur if there were adequate water, exceeds precipitation. Plants respond to this by closing their stomata, so as to reduce moisture loss from the leaves. Continuing or increasing water deficit causes a loss of leaf moisture, leading to a rise in leaf temperature and wilting of the leaves. These processes are, of course, accompanied by a reduction in transpiration and plant growth. If the deficit is only for a short period, then the plants can recover and resume growth. If, however, the period is too long, then the leaves die and eventually so will the plants. In addition to this general description of the process, different species have developed different strategies to deal with water deficits, which they can routinely expect, in their climate regimes. Thus, many grass species germinate, grow, flower, fruit and die all in the one season. Some tree species have adapted so as to shed their leaves when the soil moisture levels drop so that a moisture deficit occurs, forming deciduous species. Others, such as many Australian Eucalypt species, hang their leaves vertically so as to receive maximum radiation in the morning and evening, and very little during the heat of the day. Other species, not adapted to periods of moisture stress in their environments, are very vulnerable and can rapidly succumb to even low levels of moisture stress.

There are a number of implications of this complex response of vegetation to variations in the available water. These implications are that any modelling of the moisture status of the plants requires information on the landcover, linked to climate and soils data, to estimate soil moisture. Soil moisture information is an integral part of climatic, hydrologic and carbon cycle modelling because of the interdependence between these processes and the green vegetation.

7.1.3 The Carbon Cycle

Net primary production (NPP) is defined as the amount of carbon fixed and converted into plant material. Plants fix carbon during photosynthesis during the day and then transpire some CO_2 at night. Gross primary production is the total fixed during the day, and NPP is this figure less than that transpired at night. NPP is a fundamental figure for understanding the carbon cycle, the amount of carbon being fixed by plants and the amount of carbon held in vegetation. Estimation of NPP requires information on three important features of the canopy:

1. The architecture of the canopy
2. The optical properties of the canopy elements
3. The atmospheric and soil conditions as they influence respiration and transpiration

Thus, estimation of these parameters requires information on the landcover, from which the architectural properties can be deduced; FPAR, so as to deduce the relevant optical properties, soil and weather conditions.

The fixation of carbon by green plant material depends on the amount of energy that the plants absorb in the photosynthetically active part of the spectrum, as long at the plants have adequate water to conduct this chemical process. The first parameter of interest is therefore the proportion of FPAR that is used by the green plant material. When the canopy is entirely green and of high bulk, then this proportion is equal to the FPAR, however, it will decrease from this when there is a significant level of brown material in the canopy, or when there is a significant level of bare soil. The other part of this issue is the availability of adequate moisture. If plants do not have adequate water, then they loose turgidity in their leaves and thus wilt, changing the canopy geometry and thus the canopy reflectance. An important component of our assessment of the rate of carbon fixation is the derivation of information on the temporal history of wilting of green vegetation in an area.

The total amount of plant material is also important in assessing the magnitude of the sink, and the cover type is important in assessing the value of that sink, since much of the standing plant material from grasses and crops will be lost by decomposition, with much of the carbon being either returned to the atmosphere or converted into animal products. Much of the sink of carbon in vegetation is thus a short-term sink. The significance of the sink also depends on the way that the vegetation decays. If the vegetation decays into soil carbon pools, then the vegetation may be contributing to long-term sinks. Mapping of areas where these processes are occurring is thus an important component in building an accurate picture of the carbon cycle in an area.

The magnitude of this form of sink also depends on the below ground sink. Therefore, the landcovers are an important component of the information that is necessary in assessing the role of vegetation in the carbon budget.

7.1.4 Nutrient Budgets

In natural systems most of the nutrients get re-cycled in the one area, with the exception of that proportion that are lost as part of the natural erosion processes. Water-based erosion processes affect all land surfaces that have rainfall at some stage or another. Wind-based erosion also affects those areas that are subjected to strong winds and where the surface vegetative cover is not sufficient to slow the wind at the surface sufficient so that it does not have the power to lift particles off the soil surface. In both cases, the more fertile soils are the most vulnerable since they are usually lighter and more friable and thus easier to lift and carry. Some of the material lost through erosion is deposited in other areas, where they can then be used, but some is deposited in the sea and in areas that are inaccessible. Erosion means that nutrients are continuously removed from some areas and continuously deposited in other areas. This means that erosion causes some areas to become deficient in nutrients and some areas to be regularly supplied with new nutrient material.

Another cause of loss of nutrient is that due to solution of the minerals by the means of chemical action, for example, the erosivity of slightly acidic rain, or simply by the amount of rain under suitable conditions for the solution of the minerals, as is more likely to occur in warm climates. Thus, many soils in the tropical regions are very poor in nutrients because of solution of the minerals by the large amounts of rainfall that occur in these areas, with the nutrients then being carried downstream by surface water or through the soil by the groundwater.

Plants remove nutrients from the soil and incorporate them into their plant material. Some of this plant material is then consumed by herbivores and so enters the animal food chain. However, the majority of this plant material decomposes when the plants die, so that the majority of the nutrients held in this way are returned to the soil, where they are again available for user by other plants. The nutrients consumed by animals move up the food chain, but they also, eventually, return to the soil, where they are again available for use. They will not be returned from where they were removed, but on average the distribution of nutrients from this cause should be approximately in balance with the sources.

Agriculture is however, a quite different situation. In agriculture, the food, fuel or fibre products harvested from the vegetation are routinely removed from the area, so that there is a continual drain of nutrients from the soils of agricultural areas. The nutrients removed depend on the species, so that some species remove more of some nutrients and less of others. Creating a balance in the losses from agriculture is one of the rationales behind both crop rotation practices and the application of fertilisers. Nutrients come in two main forms. The trace elements are consumed in small amounts by the plants, but adequate supplies of them are critical to the proper growth of plants. Different species tend to consume differing amounts of the trace elements. The major common nutrients are phosphorus and nitrogen.

Losses in the trace elements can be remedied by their application, but in general crop rotation practices are used as a surrogate allowing the fields to recover from one crop, with its particular demand for trace elements, by the use of another crop with a different set of trace element demands. The common nutrients are generally replaced by the use of fertiliser being applied between one and three times to the crop during the growing season. However, the application of fertiliser is relatively inefficient, in part because of inadequacies in our detailed knowledge of the crop status and its specific nutrient needs, and in part due to the economic imperative to minimise the application time and cost. It is not economically practical to apply fertiliser in exactly the right amount, exactly when the plant needs it and as a consequence, the actual applications exceed the plants capacity to absorb the fertiliser for a period, and during this period some of the fertiliser can be removed by rainfall either into surface or sub-surface water flow.

Different landcovers have adapted to different nutrient regimes and make different demands on the nutrient supply, so that landcover mapping is an important component of information systems designed to deal with nutrients balances.

The above discussion on the BGC cycles shows that understanding, modelling and ultimately living in harmony with the major processes that control the global environment requires the following types of information on vegetation:

1. Species/community distributions. Landcover mapping is required at adequate spatial, temporal and informational resolution on the distribution of species/communities.
2. Total above ground biomass is important in assessing the magnitude of the carbon sinks in the vegetation component, as well as being a factor in assessing the distribution of rainfall in the hydrologic cycle. With information on the species/community, this information can be used to assess the total carbon pool in vegetation and to assess the significance of this pool.
3. Green leaf area index so as to assess transpiration for use in the carbon and hydrologic cycles.
4. The amount of energy used in photosynthesis is important in assessing the rate of fixation of carbon, the consumption of water in photosynthesis and the production of oxygen as a waste product from that process. This can be determined from information on the FPAR combined with information on the proportion of this energy that is absorbed by green vegetation.
5. Short-wave reflectance provides essential information on the amount of energy that is lost to the system through reflection from the surface. That such energy is not used for surface

warming has implications for evaporation of water from the surface as well as for the maintenance of other processes on the lands surface.

6. Long-wave radiation provides information on the amount of energy that is lost to the system in this way and not converted into evaporated moisture.

The reader should not interpret this discussion in too simplistic a manner. The processes introduced here are the subject of texts on them in their own right; they are complex subjects to deal with. What is being attempted here is to identify the major lines of information that are required, and which can only be realistically collected with the active use of remotely sensed data. In developing programs to collect information in support of an investigation into any one of these processes, the reader will become aware of the complexity of the issues involved in the cycle itself, and which may then require the collection of other forms of information. In addition, each of the parameters discussed here contains its own complexities in terms of its estimation. Some of these complexities will arise in subsequent sections as we address the issues involved in the measurement of some specific parameters. Consider the issue of GLAI. GLAI has been defined as the ratio of the area of single surface green leaf to the ground area beneath it. For a given GLAI there can be variations in canopy reflectance that arise from species dependent and environmental causes. The species-based causes give rise to differences in the reflectance, absorptance and transmittance at the leaf level, as well as to variations in the canopy bi-directional reflectance function (BRDF), where both can change both diurnally and throughout the season.

The environmental causes include:

1. Variations in chlorophyll densities due to variations in soil nutrient levels, meteorological history and moisture regime, species, slope and aspect, influencing in turn the reflectance, absorptance and transmittance of the canopy leaf elements.
2. Variations in canopy architecture as caused by wilting due to high levels of moisture deficit in the leaves.

Thus, a given GLAI can exhibit a range of reflectance values. The species dependent causes of this variability can be defined by the use of landcover mapping to delineate the areas of the different species mixes. Species distributions in nature tend to change much more slowly than does the condition of the species, and so landcover mapping can be used to delineate the important species groups. The environmental differences are much more difficult to deal with. Environmental causes are temporally variable so that the most promising way to derive information on environmental variability will be through the use of temporal image data. Some of the environmental causes also change the canopy architecture, so that integration of BRDF with temporal image data may be a valuable source of information on environmental effects on vegetation.

Figure 7.2 shows set of 16 temporal ratio vegetation index (RVI) growth curves for winter wheat on sandy loam soil at the Foulum Experimental Station in Denmark. Each curve indicates the temporal RVI profile observed for the mean of a set of samples given the application of fertiliser in kg/ha at three dates during the growing season as specified in the legend. The three vertical lines indicate the three dates of application, where these dates are 5th April (Degree Days [DD] = 24.0), 2nd May (DD = 266, 55) and 15th May (DD = 435, 2). The x axis of Figure 7.2 is the accumulated Degree Days (DD), calculated for winter wheat as the sum of degrees over 5°C after 1st March. The curves show that there is similar initial growth up to DD = 100, but then there is a steady divergence between the curves over the subsequent vigorous part of the growing season, that is to about DD = 280. After this crop growth depends on the amount of fertiliser still available to the crops. Those that had run out of fertiliser, such as those with early applications of fertiliser, plateau after DD = 280, and some of these resumed growth with the application of the third round of fertiliser. The maximum gradient of the envelope curve is correlated with nutrient level in the soil and with other predictable factors such as the soil type, slope, aspect and cover type, where the envelope curve covers the maximum RVI

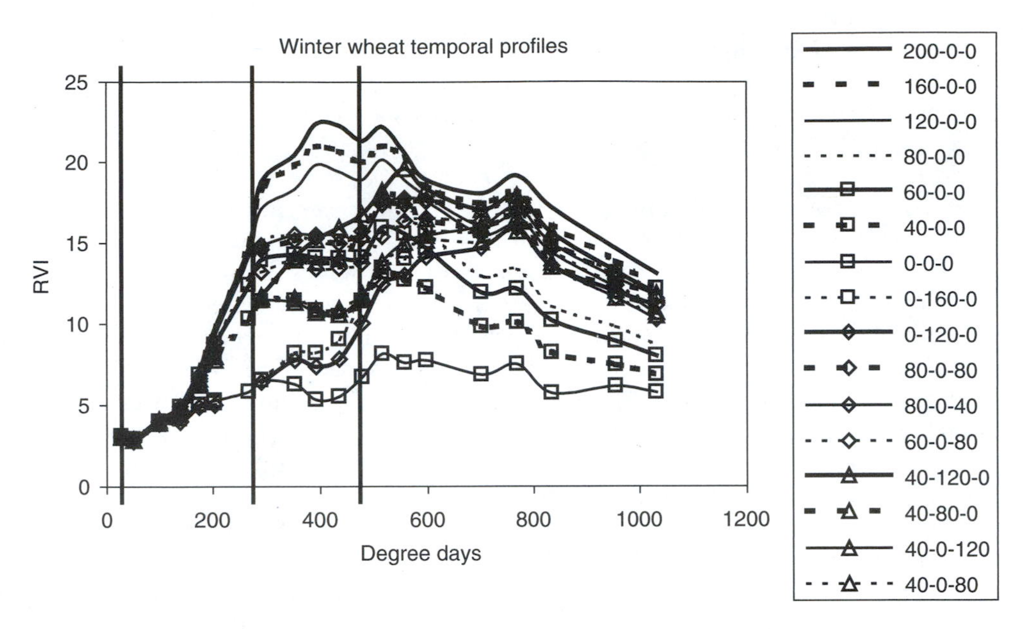

FIGURE 7.2 Temporal winter wheat crop growth curves for 16 different fertiliser treatments on the same sandy-loam soil in Denmark, and the dates of application of fertiliser.

values throughout the growing season. Unpredictable effects on the crop, such as moisture deficit, disease and predation, cause divergence or reduction from the envelope by the actual crop curves.

There are significant difficulties in accurately mapping most vegetation parameters without the use of remotely sensed data and as a consequence, any such alternate mapping will be very expensive and time consuming. It is essential that remote sensing evolve reliable and accurate techniques for the measurement of the key parameters of vegetation as are required for the monitoring and management of BGC process resources.

The above discussion has focussed on the global to broad regional issues that require information that can often be best acquired by the use of remotely sensed data, or indeed that can only be acquired by the use of remotely sensed data. These issues have to do primarily with the global environment, the processes that operate in that environment, and our adjusting to live within the constraints imposed by these systems so as to ensure the long-term sustainable use of the resources of the globe. In general, as we find issues here that require actions, those actions come down to the individual person or the individual unit of production. So, for example, with the ozone hole, the ultimate responses to the need to reduce the release of specific organic chemicals into the atmosphere became the responsibility of governments to set the rules and individuals to implement those rules. The same will apply to other issues, such as water quality, soil maintenance, atmospheric pollution and so forth. Thus, the global monitoring systems that are implied by the use of remote sensing in the above discussion, will ultimately require some form of regional monitoring system to support the implementation of the policies and programmes that are set in place.

The BGC processes and all forms of degradation have either point or area drivers. The drivers for vegetative growth are predominantly area drivers such as solar insolation and rainfall since they impact on the whole area, but to varying degrees, in contrast to deposition of pollution in a river, usually from point sources. The effects of these area drivers are spatially and temporally extensive. Thus soil erosion, which may be created at a point, or point source pollution, will usually be deposited over an area. However, even though most drivers for environmentally dynamic processes are spatially and temporally extensive, they can often be monitored satisfactorily using point measures. Thus, water quality can be satisfactorily monitored using point observations at critical places in the streams of the catchments. Point observations can thus be used to summarise the state of a resource in an area.

However, when these point observations show that there is a problem, then they are inadequate to explain the causes of the problem, or to suggest remedial action. Invariably, more detailed and extensive data is required for this purpose.

As the issues move from the broad region to the local area and the unit of production, the focus also changes from that of primarily focussing on the sustainable use of the land to the maintenance of productivity. Information systems that are put in place to maintain the resources of an area should also be used to improve the productivity of the area.

7.2 REGIONAL VEGETATION MAPPING AND MONITORING

Chapter 8 will show that the provision of information for regional resource management is the natural market for information derived from satellite-based remote sensing systems. In Chapter 8 a region is defined in terms of the management functions and not in terms of geographic extent. This is deliberate, since the geographic extent is a function of the regional management issues. In this context, concern with global warming and climate change, in dealing with global processes, involves a global region. In a more conventional sense, regional management deals with much smaller geographic areas. Thus, the geographic extent when an issue is concerned with surface water-borne processes will deal with a watershed or sub-watershed. Other geographic areas are appropriate when dealing with sub-surface flow, wind erosion, animal home areas or with vegetation distribution and so forth.

The issues being dealt with at the global level, including the global carbon cycle, the global hydrologic cycle, the global energy balance and global weather, will have an impact at the smaller regional level, but they are not the only factors and are unlikely to be the dominant factors. Important local factors driving regional management have to do with productivity and sustainability. Since the priority issues in each region are going to depend in part on the perceived importance of local issues, it is not realistic to attempt to identify the different sorts of information that is required, as was attempted in Section 7.1, in relation to global issues. Instead, this section will discuss the characteristics of two key types of information that are likely to be required for regional management, in the hope that these principles will apply to the other types of information that will also be required. The two types of information that will be considered will be landcover mapping and estimation of GLAI.

There are a number of critical differences between current global resource assessment and local regional resource management:

1. *Field data.* At present no effort is being expended on the collection of a statistically valid field data set to be integrated with the global monitoring and assessment programs, primarily because of cost. The global programs attempt to overcome this limitation by means of averaging, the use of algorithms that are as accurate as can currently be achieved and severely limiting the information derived from the image data. Thus, the spatial extent is averaged and the images used are averaged to derive an estimate representative of a longer time period. The algorithms are designed to derive estimates of those parameters that are highly correlated with the radiance values in the wavebands, using dedicated algorithms based largely on the physical modelling work that has been done. Finally, the information is used to derive estimates for regions that are relatively large compared with the source and derived image data, another form of averaging.

For local regional management, none of these conditions will apply, at least with currently available satellite systems. Regional management will require information down to the unit of production, which is the field as discussed in Chapter 8 in more detail. The size of fields varies considerably, but they will usually require 5 to 100 m resolution image data to provide adequate information. There is thus limited scope for spatial averaging. The management at the regional level will be dealing with local resource managers who know what is happening at the field level. The regional information will need to have a level of accuracy that is consistent with that of the local managers, and thus there is no scope for temporal averaging. A consequence of this is that local regional management will need to have a higher level of accuracy in the information as estimated at the pixel level, than what is required from the global information systems. It was seen in Chapter 5, that accuracies and reliabilities of the order of the 95/95 rule are likely to be required from operational systems. The only way to achieve this level of accuracy, without significant averaging, is by the integration of adequate field data into the analysis process, for the derivation of local regional resource management information.

2. *Frequency of repetition and data calibration.* All global monitoring systems start with daily data acquired using very cheap data acquisitions systems, covering the whole of the globe. Because of the resolution, the land/water interface is often used as the basis for the rectification of these data sets. Calibration and atmospheric correction of the data is conducted as far as is currently practically achievable, but does not currently include correction for variations in the water vapour column nor aerosols. For local regional management, data will be required at between 5 and 100 m depending on the information required, and ideally at daily intervals so as to provide reasonable temporal coverage during the growth seasons when significant changes in status can occur over periods of days to 1 week. At present this can only be achieved at high cost. In addition, the rectification of the data requires the use of ground control, incurring additional costs. Thus, the situation that exists for global coverage in terms of frequent, low cost data does not exist for local regional management. At present, the finest temporal frequency that can be achieved is 2 to 5 days, so that there is negligible scope for temporal averaging, and indeed, given cloud cover, it is more likely that interpolation between the images that could be acquired will have to be considered. In 2000, the French space agency, CNES, acquired all possible SPOT 1, 2 and 4 images of a site in Rumania, acquiring 39 images over a 10-month period. Many of these images had some cloud contamination and there were inconsistencies in the image set over time suggesting difficulties with correction for water vapour and aerosols. For many resource management tasks, that require temporal image data, this is not yet a satisfactory situation.

3. *Integration into a management hierarchy.* Local regional management is integrated into a management structure from the strategic level down to the local level that has political, economic and social dimensions as discussed in Chapter 8. One of the implications of this will be the need for consistency between the information used at the different levels; inconsistencies will reveal the existence of either errors or inconsistent ways of data acquisition between the levels, and neither will be accepted. Thus, the information derived for local regional management will need to be of an accuracy sufficient for it to have credibility at the interface to the local management information. It has already been stated that this accuracy is likely to need to be at the 95/95 level.

4. *Informational resolution.* The resolution in the informational classes is likely to be higher at the local regional level than at the global level. Agriculture is likely to be classified into the major crop classes for local regional management, or indeed into each crop species if that is possible, whereas this is not necessary for the global regional management information.

Given these differences between global and local regional management, let us consider how they are likely to impact on the collection of landcover and GLAI information at the local regional management level.

7.2.1 Landcover

Landcover information will be required at the field level and on a repetitive basis such that significant errors do not arise due to errors in the landcover classification used in the management of resources. Since a significant number of agricultural fields change in landcover, at the resolutions sought for this information, on a seasonal basis, agricultural areas are likely to need to be mapped on a seasonal basis. For most parts of the globe this means an annual basis, but not for some areas in Asia and Africa. Other land covers do not change this frequently, but they change less predictably. In areas where the images are acquired for agricultural landcover mapping, they can be used to determine whether the level of changes is sufficient to justify re-mapping the other landcovers. If the area is not covered with images for this purpose, then some other technique is required to assist managers decide whether the area needs to be re-mapped. This technique may be field visits or the use of coarse resolution image data to look for significant changes in response as an indication of changes in landcover.

Some of the limitations of current methods of landcover mapping are discussed later in this chapter, with some suggested ways of addressing these limitations.

7.2.2 Estimation

As with landcover mapping, estimates of GLAI will need to be at field resolution or better. This resolution is required so as to seamlessly connect to the unit of local management since there will need to be consistency between the information used at the regional level and the summary information that can be deduced from the local management level. Such a resolution is not of much use in local management, of course. But regional information will only have credibility with local managers, such as farmers, if the information at the field level is consistent with their information from within the fields. GLAI may also be required at a temporal frequency suitable for estimation of other parameters by means of models, or for the verification of other models. At present there are four approaches to the direct estimation of GLAI:

1. Empirical models
2. Empirical interpolation
3. Inversion of physical models
4. Linear approximations to physical models

The relative merits of these four approaches are discussed in a following section on estimation. First, let us consider the way vegetation is recorded in image data and then the modelling of vegetation reflectance.

7.3 THE SIGNATURES OF VEGETATION

Vegetation is recorded in image data in terms of its signatures. There are currently five recognised forms of signature that are characteristic of vegetation in image data: spectral, spatial, temporal, angular and polarisation. Although there has been some work conducted on the fluorescence of vegetation, so far this has not achieved results sufficient to justify its consideration although it is possible that this situation may change in the future. Currently, most methods of extraction of information on vegetation from image data focus on using one or at most two of these signature types. Yet there is information in all five forms of signature on the types or status of vegetation and so it is reasonable to expect that either more information on vegetation could be derived from image data if more types of signature are used in the analysis or the information derived would be more accurate and reliable.

It has been shown that some types of information are closely connected to particular types of vegetation signatures. Thus, FPAR is closely connected to the absorption characteristics of vegetation and so it is closely connected to the use of optical data in the appropriate wavebands. Derivation of FPAR may not benefit from inclusion of spatial information. It may be that the analysis of vegetation needs to be more sophisticated that is currently envisaged, using a hierarchy of analysis, with the signatures used at each hierarchy, and indeed within each hierarchy depending on the information that is to be derived. The broad hierarchy identified throughout this text is to address the sequence:

1. What is it that is in the image — classification
2. What is its condition — estimation
3. How does vegetation interact with dynamic processes — temporal analysis

Whilst these are the broad goals set for most work in remote sensing, the levels of accuracy and reliability achieved in most remote sensing work are far from the optimum. Given this situation, there is a need to see how to better integrate the various dimensions of the information space so as to derive information of an acceptable accuracy and reliability.

There is also the issue of accuracy and reliability. Accuracy and reliability have been discussed in Chapter 5. Accuracy can be defined as the closeness of the derived results to the true conditions, whilst reliability is the repeatability of accuracy. There has been little consideration in the remote sensing community of what levels of accuracy and reliability should be achieved in the information that is derived from image data, although it would seem that there is an implicit acceptance that current levels are not of an acceptable standard. One basis for consideration of this issue is what might be legally acceptable.

Legal decisions are made on the basis of "beyond reasonable doubt." The interpretation of what this phrase actually means is open to many subtle variations in interpretation and so it is impossible for this writer to make a judgement of what may be the position of any specific court in any specific situation. However, examples may be of help. The accuracy and reliability of topographic mapping has been accepted by most mapping agencies as 95% accurate, 95% of the time. Topographic maps are accepted as being valid evidence, when used within the limitations of their scale and information content in most countries, so that it would seem that information that meets the 95% accurate for 95% of the time rule is likely to be accepted by most courts, when used within the limitations of its time of creation, scale and so forth. It would seem that this level of accuracy and reliability should be the goal of any mapping or monitoring program, until such times as a better specification is set, either by resource managers themselves, or by the courts.

7.3.1 Spectral Signatures (λ)

The basis for the unique characteristics of vegetation signatures in the optical region is the absorption characteristics of the plant pigments. As we have seen, these operate at the leaf level. However, the absorption characteristics of the plant pigments are modified by the leaf cell structure and by the canopy geometry, so that the result is that these pigments give relatively broad absorption bands in image data of a canopy. This is in contrast to the narrow absorption features that can occur in geology. The fact that the same pigments are used in all plants and the relatively broad characteristics of the absorption bands means that these bands can be similar for different communities. There is a significant level of overlap in the signatures from different species and communities and as a consequence, most analysis based just on spectral criteria does not meet the accuracy and reliability specifications that have been set above.

Spectral signatures are also significantly affected by atmospheric attenuation, further reducing their information content. This will be a factor in the variation seen in spectral signatures, particularly across multiple images as the effects of aerosols and water vapour on the scattering and absorption of radiation is currently hard to predict and correct (Figure 7.3).

We have seen that spectral signatures are connected to both temporal and angular signatures. They are connected to the temporal signatures through the changes in the spectral signatures that occur with phenology, particularly of crops, but also of many other types of vegetation. They are connected to the angular signatures through the BRDF.

Spectral signatures in the thermal and microwave regions lack the unique features of vegetation signatures in the optical region. However, vegetation is a rather unique volume scatterer in the

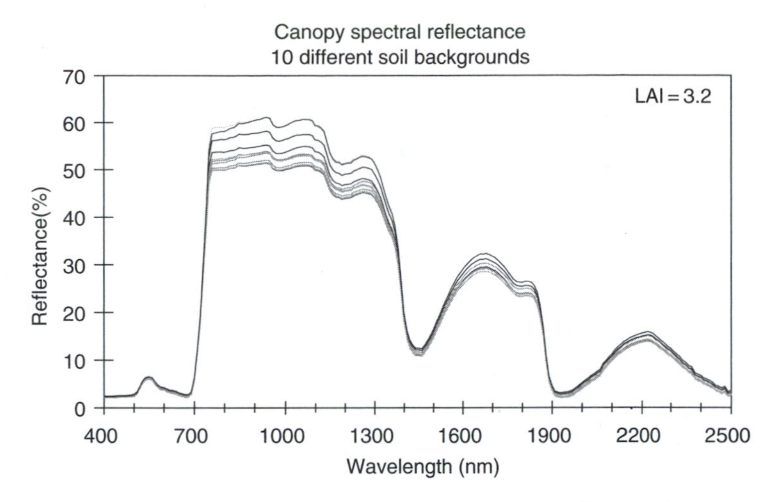

FIGURE 7.3 The reflectance of winter wheat over different soil backgrounds at an LAI of 3.2. The graphs illustrate the impact of the soil background on reflectance in the NIR, at even high LAI values, whereas there is negligible effect in the visible part of the spectrum. Figure supplied by Dr. Niels Broge.

microwave region and this can give it unique features relative to other surface types. Thus, vegetation tends to be a surface scatterer at shorter microwave wavelengths, a volume scatterer at the middle wavelengths and transparent at the longer wavelengths, although the actual interactions that occur depend very much on the nature of the canopy, particularly the amount of woody and dead material in the canopy.

7.3.2 SPATIAL SIGNATURES (x, y, z)

Spatial signatures occurs at least at three levels from vegetation: within the individual plant due to the arrangements in the canopy, within a community and within micro-climatic and geomorphic units. Within the plant there are some significant differences between species and families such as the differences between the needle leaf and broadleaf species. Discrimination based on this level of spatial information would require very fine resolution image data even though the canopy differences that result have a significant impact on the spectral signatures. The next level is the spatial mix of species and their distributional characteristics within a community, including the existence of gaps. Current resolution satellite images are getting very close to being adequate to provide information on this level of spatial signatures, particularly in forests and shrub communities.

There are a number of tools that are available for the conduct of spatial analysis including the use of variograms to analyse the spatial correlation, texture measures, the average local variance function to estimate the size of object elements in image data and segmentation. The reality is though, that the extraction of spatial information from image data has suffered from the relatively course resolution of the image data compared to the objects of interest and the complexity of the spatial features and extraction of information on them. This is an avenue that should benefit from the finer resolution image data that is increasingly becoming available, and the continued development of better tools and techniques (Figure 7.4).

One advantage of the use of spatial analysis is that spatial signatures are relatively insensitive to the degradation of image data due to atmospheric effects.

7.3.3 TEMPORAL SIGNATURES (t)

Temporal signatures are dominated by the processes that affect vegetation, and by quasi-random, often catastrophic events. Since the processes that affect vegetation are dominated by annual and

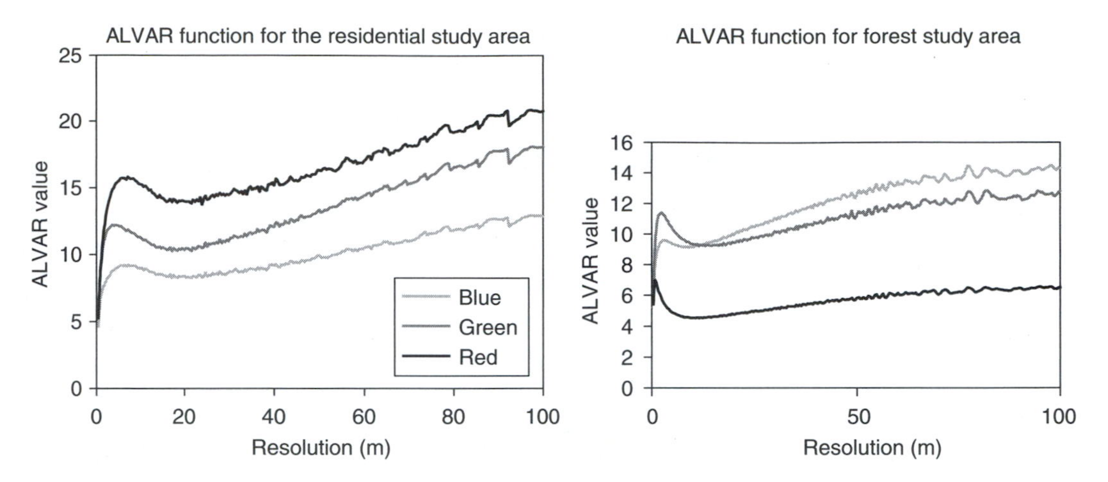

FIGURE 7.4 Average local variance functions for two distinct landcover conditions, derived from 0.4 m resolution colour orthophotographs, with the individual band ALVAR functions shown in the figure.

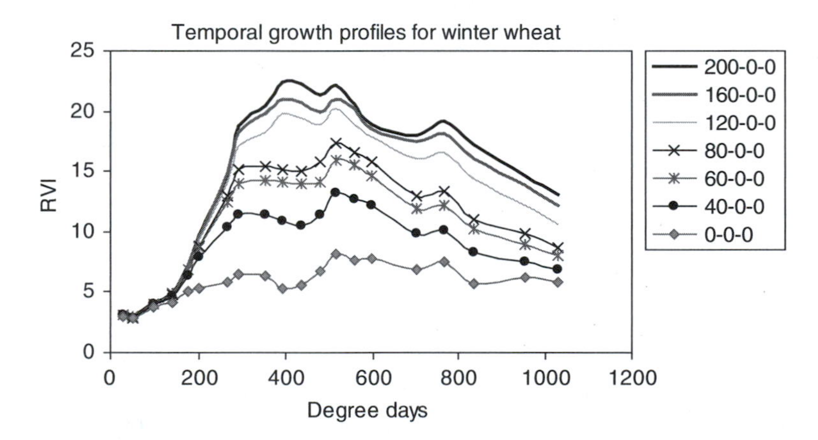

FIGURE 7.5 Temporal RVI profiles for seven different nutrient applications to winter wheat, varying from 200 kg/ha at the first application date early in the season to no application, and no applications at the two typical subsequent application dates in Denmark.

daily cycles, temporal signatures are dominated by the phenology of plants and their diurnal cycles. However, these cyclic processes can be significantly modified by catastrophic events such as floods, fires, drought, disease and predation and the activities of man. These quasi-random events tend to confuse the cyclic processes and so that analysis of temporal signatures will need to adequately deal with discrimination of the cyclic events themselves from the noise created by these events so as to be able to analyse both types of events (Figure 7.5).

Plant phenology is one of the major strategies used by plants to compete and to occupy unique ecological niches. For this reason, the phenology of plants is a valuable source of information on the plants themselves, and in reverse, on the state of the environment. Just as the phenology is an important driver for vegetation, so too can be the diurnal cycle. In the analysis of vegetation the diurnal cycle can be used to derive certain types of information, and it can confuse others if the image data is taken at different times during the day.

Temporal signatures are very vulnerable to image calibration issues. Whereas, a single image may have relatively satisfactory internal geometric, spectral and radiometric consistency, temporal images inherently have no internal consistency. They have to be given it through calibration and

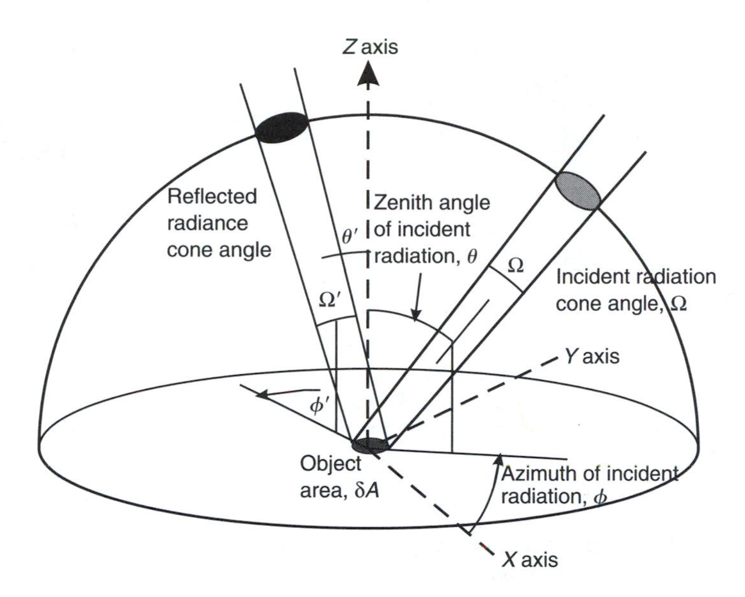

Figure 7.6 Geometrical relationships in the BRDF.

correction. Thus rectification, calibration and correction for atmospheric effects are crucial costs associated with the analysis of temporal signatures. This is a significant overhead with optical data and geometrically with radar data.

Temporal signatures are the only type of signatures that can provide significant information on the way that vegetation is responding to the dynamic forcing mechanisms that interact with that vegetation. To do this, though, requires the development of tools for the analysis and interpretation of temporal image sequences.

7.3.4 Angular Signatures

The non-Lambertian characteristics of vegetation mean that the canopy reflectance has an angular dependency that is given in the BRDF for the canopy, where the angular relationships in the BRDF are shown in Figure 7.6. Not only does the canopy BRDF vary with sun–surface–sensor geometry, but also it can vary during the day, due to the effect of environmental stresses on the plant, and also due to the capacity of some plants to have their leaf elements follow the sun. It has been shown in Chapter 2 that the angular distribution of reflectance is very dependent on the canopy architecture. Up until recently the non-Lambertian characteristics of canopy reflectance have been viewed as a problem to be overcome in remote sensing. In a sense this is correct, but the advent of multi-angle imagery may show that crucial information can be extracted from this dimension of the data, particularly on different canopy geometries. Examples of BRDFs for different covers are given in Chapter 2.

7.3.5 Polarisation Signatures

Polarisation has been shown to be valuable with microwave imagery because of the unique volume scattering characteristics of vegetative canopies. At shorter radar wavelengths, surface scattering dominates for most canopies. When this occurs then the polarisation tends to be maintained in the return signal, so that like-polarised images give a stronger response than the cross-polarised images. However, as the wavelength increases, the volume scattering increases and the surface scattering decreases. With volume scattering, the polarisation tends to be lost, so that the cross-polarised signal gains in strength at the expense of the like-polarised signal.

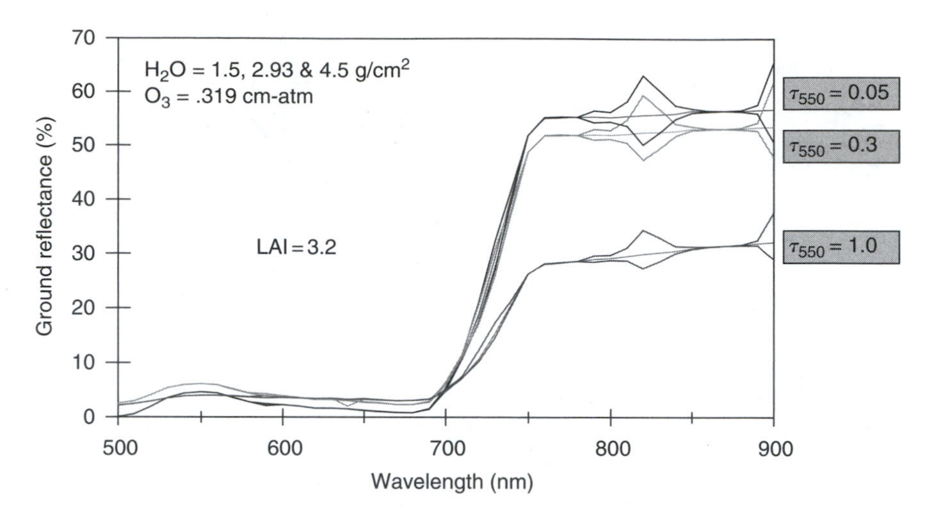

FIGURE 7.7 Simulated effect of variations of atmospheric moisture content on the sensed signal from a crop canopy with LAI = 3.4 at the top of the atmosphere, due to three different levels of water vapour. The curves have been calculated using a mid latitude summer atmospheric model for the atmospheric constituents. Figure supplied by Dr. Niels Broge.

The effects of polarisation on radar signatures have been discussed in Chapter 2. The reader is referred to that section for more details on the radar signatures of surfaces.

7.4 MODELLING CANOPY REFLECTANCE

7.4.1 INTRODUCTION

It has been shown in Chapter 2 that the reflectance from a canopy is a complex function that depends on a number of factors in each of the five component parts of the transmission of energy from a source to a sensor (Figure 7.7):

1. *Source effects* (s_i), that includes its spectral intensity, I_λ, orientation in terms of its azimuthal (θ_i) and zenith (θ_i) angles across the hemisphere that is incident on the surface. The intensity of the radiation from the solar source for optical energy varies in a predictable way with changes in the orbit distance between the Earth and the Sun due to the elliptical nature of the Earth's orbit and the tilt of the Earth's axis. It varies in unpredictable ways with Sunspot activity, but the effects of these are measured by satellite. The source of energy for thermal remote sensing is the Earth's surface and objects on it. The energy radiated by these is a function of their temperature and emissivity.

2. *Atmospheric effects* (a_i) of absorption and scattering, thereby modifying the distribution and intensity of radiation. Some of these components, such as the effects of the atmospheric gasses, other than ozone and water vapour, are relatively stable and hence predictable. Water vapour and ozone are less predictable in their density in the atmosphere, and thus their effects are more difficult to estimate, even though their effects on radiation are well known and can be modelled. As a consequence, they can be corrected as long as suitable data is available. Correction of the scattering and absorption due to aerosols is much more difficult. Aerosol particulates in the atmosphere have a reflectance that is a function of their source. The main sources are burnt organic compounds, organic materials and dust. The first two have relatively stable reflectance and absorptance characteristics, but the dust reflectance is related to that of the parent materials. The problem is however, made more complex due to changes in reflectance, particle size distribution and particle shapes that can occur. Particulates absorb moisture in the atmosphere, changing their reflectance and causing them

to aggregate, changing their size and shape in the process, further affecting their absorption and scattering properties. This complexity is compounded, of course, by the fact that aggregations may not be of the same material, but may come from different sources, each with their own reflectance characteristics. In addition to these difficulties, aerosols can only remain in the atmosphere whilst the wind energy is sufficient to do so, so that changes in wind energy may affect the characteristics of the particulate matter that remain in the atmosphere. Particulates can remain in the atmosphere from minutes, to, typically days, although some particulates shot into the atmosphere from volcanoes have been recorded in the upper atmosphere for months. Because of the multiple sources, highly dynamic nature of the winds that lift them and hold them in the atmosphere, aerosols are a highly variable component of atmospheric absorption and scattering. Accurate correction for their effects will have to be conducted using data acquired at the same time as imagery is acquired for other purposes (indeed the imagery may be used for both purposes), where the best methods that have been developed to date are the dark background method over oceans and the multiple angle approach discussed in Chapter 2.

3. The *vegetative canopy* (v_i). The canopy components include the spatial distribution and orientation of the different canopy components as well as their absorption, reflectance and transmittance characteristics. All these canopy characteristics can change in response to environmental pressures, with some capable of very rapid change, and others only capable of more gradual change. As has been noted, the reflectance, transmittance and absorptance characteristics of the canopy elements varies between species, just as the canopy reflectance varies between species, and the plant phenology introduces further temporal differences in reflectance between species.

4. The soil background (b_i) where variations in absorptance and reflectance with chemical composition, surface roughness and moisture content all introduce variations in the effect of the background on canopy reflectance.

5. Sensor effects (t_i) including its orientation (θ_s, ϕ_s) relative to the sensed surface, its spectral sensitivity, and its optical, geometrical and electrical efficiencies and distortions.

Different subsets of these cause different responses in the vegetation. Thus, one subset influences variations in phenological response, another subset causes variations in the diurnal response and other subsets cause noncyclic responses. For example, some of the non-cyclic variations in the vegetation canopy are caused by:

1. *Moisture stress*. When canopies go under moisture stress, the first major physiological effect is the wilting of the leaves that accompanies the loss of leaf water. This wilting changes the canopy geometry, changing the reflectance.
2. *Leaf age*. It has been shown that the reflectance characteristics of leaves change with age of the leaf due to physiological and chemical changes in the leaves, and this induces changes in canopy reflectance.
3. *Wind velocity*. A canopy is affected by the wind, inducing changes in reflectance that will change canopy reflectance as the canopy shape and distribution characteristics change.

As a consequence, the reflectance characteristics of vegetation are very complex. They are dominated by the seasonal variations in reflectance that accompany the phenological growth cycle of the plants. For most plants this phenological cycle is characterised by new leaf at the beginning of the cycle, where this new leaf rapidly increases in area in many species during this period. This increase is approximately exponential for cereal crops and pastures. This growth then stabilises and in fact may decay somewhat when the plants change from a growth strategy to a reproduction strategy by the production of flowers and then fruit. The phenological cycle is then finalised with death or dormancy in many species.

This dominant phenological cycle is modified by the diurnal cycle and non-cyclic events. Thus, some plants can change the orientation of their leaves as a function of the direction of the incident

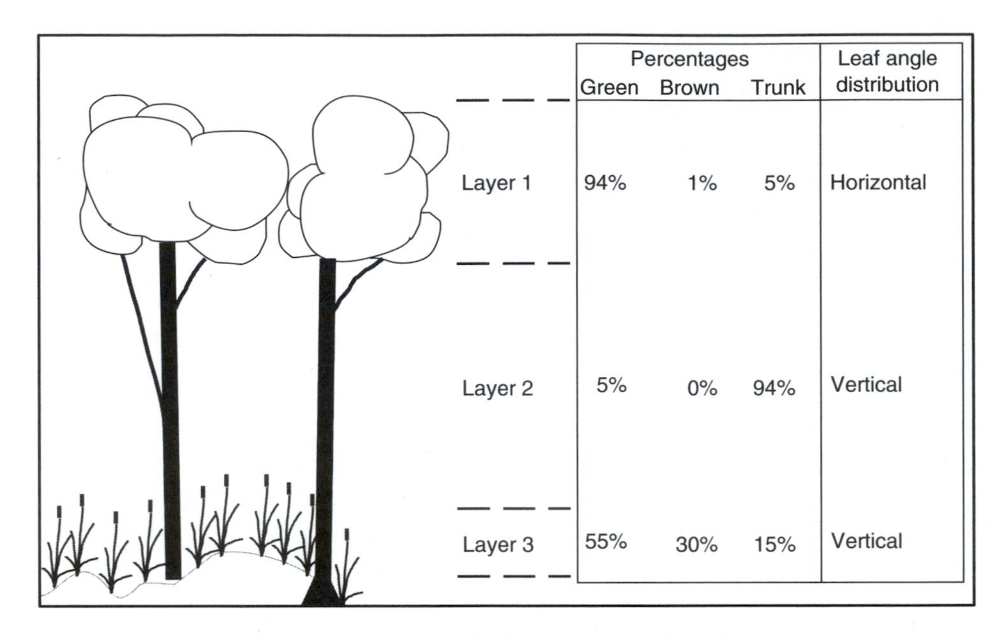

FIGURE 7.8 The spatial distributions that can be found in a canopy.

radiation, thus also changing their BRDF characteristics. There may be other non-cyclic diurnal process, such as the wilting of leaves that can also have a significant effect on the reflectance of vegetation. The non-cyclic events that can change the reflectance of vegetation include the effects of disease, predation, storm damage and the activities of man (Figure 7.8).

In addition to the very dynamic nature of vegetation and hence its reflectance, the canopy contains a number of different types of canopy elements, each with their own reflectance and transmittance characteristics, where these characteristics can change. The main types that can be expected in most canopies include green and brown leaf, leaf stem and bark or woody material, where there may be a number of sub-divisions in some or all of these. In addition to the different reflectance characteristics of these different elements, they will usually have different spatial and angular orientation distributions.

The complexity of the inter-relationships between the factors that affect canopy reflectance and the resulting effects on the distribution of reflectance is a major factor currently limiting the use of remotely sensed data for the accurate and reliable estimation of many vegetation conditions. Recognition of this limitation in our capacity to extract information from remotely sensed data has led to considerable effort to model the reflectance characteristics of the canopy so as to better understand the relationship between the canopy, the incident radiation and the sensor and hence to be able to derive more reliable and accurate information from image data.

The advent of high spectral, spatial and temporal resolution image data has created an even greater need for an accurate understanding of these processes, since the imagery is approaching, or even getting finer than the resolution of the plants in the field, so that small differences in individual plants can have a significant impact on the image data. The reflectance of a canopy can thus be characterised as a function of the form:

$$\rho(\lambda, \theta_s, \phi_s) = F(s_i, a_i, v_i, b_i, t_i) \tag{7.1}$$

where $\rho(\theta_s, \phi_s)$ is the reflectance in waveband λ, and in direction (θ_s, ϕ_s) being the zenith and azimuthal angles, respectively. The function $(F())$ parameters are the components that affect the transmission of energy from the source to the sensor as discussed at the commencement of this section. Some of the components in Equation (7.1) are known in that they will not vary as part of the sensing process. For example, the components to do with the source and the sensor can be categorised in this way, and so a more valid form of Equation (7.1) is:

$$\rho(\lambda, \theta_s, \phi_s) = F(a_i, v_i, b_i \mid s_i, t_i) \tag{7.2}$$

where $F(x \mid y)$ means x given y. One of the main goals of modelling the canopy is to build an understanding of the inter-relationship between a canopy and its reflectance, so as to improve our understanding of the impact of canopy characteristics on reflectance and hence to be able to design data acquisition systems and strategies that will yield more or better information on canopy status. A corollary goal is to use the models to estimate canopy parameters from image data, in which case Equation (7.3) needs to be inverted to give:

$$\bar{v}_i = F'(s_i, a_i, \hat{v}_i, b_i, t_i, \rho) \tag{7.3}$$

in which $\bar{v}_i$ represents estimates of selected vegetation parameters, such as GLAI or biomass, $\hat{v}_i$ represents either other known or independently estimated vegetation parameters, such as the species mix that contribute to the canopy structure and reflectance or preliminary estimates of the parameters being derived from the model, and the other parameters in $F'()$ are either measured or estimated.

These two activities represent different modes of operation:

1. *Forward modelling* is where one starts with a model of the physical canopy conditions and estimates the reflectance that will occur when this model interacts with, or is acted upon by electromagnetic radiation, that is solving Equation (7.2). Such models can be used to test our understanding of interactions of radiation within the canopy; if the derived reflectance is similar to that found experimentally across a range of conditions then we can deduce that we understand the processes that are occurring in the canopy. Such modelling can also be used to build BRDFs for a canopy across a range of geometric conditions, without the need to replicate all of them experimentally. Such models can also be used to test what reflectances may arise under differing conditions, and as such be used as part of the planning for data acquisition campaigns. Of course any such models need to be extensively validated and then only used under conditions over which the assumptions of the model are being met, or approximately so.
2. *Inverse modelling* is where one takes a set of reflectance values and uses these to drive an inverse model, such as Equation (7.3) so as to estimate specific canopy variable values. Such canopy parameters could include the FPAR, GLAI, biomass or yield. They are called inversion models because one of the classical ways to make such models is to take a forward model and invert it. The other classical way to make inversion models is by the use of regression techniques. There are a number of significant hurdles to overcome in the act of inverting a forward model of the form of Equation (7.2) so as to form the inverse model Equation (7.3). First, the nature of the forward model may mean that it is not mathematically invertible. Second, the model may be invertible, but it may have a number of local minima, so that the solution to the equation may go to incorrect local minima, incurring significant errors in the derived estimates. In this section we will deal with modelling as such and take up some of the more practical matters involved in inverse modelling in the next section.

The numerical modelling of canopy reflectance has evolved from relatively simple models to very sophisticated and complex numerical models that are beyond the scope of this text. There is an extensive set of literature on the subject, with some very good review articles, including that by Goel (1988) and Goel and Thompson (2000) to which the interested reader is directed.

A canopy consists of:

1. Canopy elements, including the green and brown leaves, stems and bark. Each type of element will have optical properties that are characteristic of that type of element, where

the actual optical properties of the elements can vary with species, their position in the canopy and on their age.
2. The canopy of a plant, consisting of a spatial and orientation distribution of the canopy elements in that canopy. Some of the canopy elements will have more strongly bound constraints on their distribution in the canopy than will have others. As has been seen the spatial and orientation distributions of the elements can change in both systematic and random ways throughout the day and throughout the phenological life of the plant.
3. The distribution of plants, forming the community canopy. For some communities, the distribution is quite regular, as is the case with agricultural canopies, however, with others the distribution can be either random or clumpy in which case the plants can form distinct clumps in the canopy.
4. The background of the soil or water beneath the vegetative canopy.

Attempting to model this complex environment has led to the investigation of many different approaches, and where some approaches have been found to be more suitable for some canopies than for other canopies. The more common approaches include:

1. *Turbid medium models.* In these models the canopy is assumed to consist of a number of horizontal, uniform, plane parallel layers above the background surface. Within each layer the scattering and absorbing elements of the canopy are distributed in a random way, but with a specified density and specified orientation and optical scattering and absorption distribution functions. The density is often expressed in terms of GLAI and the orientation distribution in terms of the Leaf Angle Distribution (LAD) function. The horizontally uniform assumption behind this class of model mean that they have been found to be useful for dense, relatively uniform canopies as can be found amongst some agricultural crops, but not for canopies that exhibit clumping, such as many forest canopies or row crops.

2. *Geometric models.* In these models the ground surface is covered by a distribution of canopy elements that have a defined geometric shape, size and reflectance characteristics. Each canopy element is illuminated by the source, and creates a shadow. The distribution of sunlit and shadowed areas depends on the size, shape, density and distribution of these geometrical canopy elements, as well as the source–surface–sensor geometry. The relationship between the canopy geometric components and the incident and reflected radiation leads to the four elements of sunlit canopy and background, and shadowed canopy and background. The various combinations of these give rise to the derived canopy reflectance. Whilst some of the early canopy reflectance models were geometric models, interest was renewed in this class of model when it was realised that the turbid medium models are significantly limited in application and accuracy when the canopy is patchy. Geometric models are most often used for forest and plantation conditions where each plant has a canopy shape that approximates a geometric shape and is separated from each other plant in the canopy.

A number of models use the geometric concepts to define the characteristics of the distribution of plants in the canopy, and the turbid model approach to define the optical properties within the individual plant canopies. This class of models have been categorised by some as hybrid models.

3. *Ray tracing models.* In these models a photon of energy is tracked from outside the canopy through the canopy until it is absorbed, either by the canopy or the background, or it emerges from the canopy, contributing to the canopy reflectance. A record of the history of each photon is kept and these are accumulated so as to construct the BRDF of the canopy. For each photon, at each stage of its travel through the canopy, there is a probability of interception with an element of the canopy and random number selection is used to decide whether an interception occurs or not. At an interception, there is a distribution function for the orientation of the canopy elements with the actual angle of incidence chosen using a random number generator. From this information, and the orientation of the ray, the angle of incidence of the ray is determined. The canopy elements also have their absorption,

reflection and transmission probability distributions and again random numbers are used to select whether the photon is absorbed, reflected or transmitted and then its angle of emergence relative to the leaf, if it is reflected or transmitted. The known orientation of the canopy element is then used to derive the absolute orientation of the ray.

Ray tracing models have been found to be successful in the construction of canopy BRDF functions even though they are computer intensive. However, by their nature they are not invertible and so cannot be used as the basis of methods to estimate canopy parameters from reflectance data.

4. *Flux or radiosity models.* In these models the canopy is constructed so as to consist of a distribution of geometric shapes in the canopy, with the elements having specified orientation and optical properties. The radiation on an element comes either from direct radiation from the source, or from illumination of the other elements in the canopy. A major activity in these models is thus to determine which elements are illuminating which other elements. It can be seen that one set of elements will illuminate each specific canopy element, and that each canopy element will then contribute to the illumination of another set of canopy elements. All the energy emitted or reflected by every surface in a scene is accounted for by its scattering or absorption by other surfaces. These models are even more computer intensive than the ray tracing models.

5. *Empirical models.* In these models the canopy reflectance is assumed to be simply related to combinations of the incident and reflected zenith and azimuthal angles. Observations are used to derive the relationship that is then used for the whole scene. The advantages of empirical models is that they are relatively easy to parameterise; the major disadvantages are their level of accuracy, that the parameters have no physical meaning and that they are limited to situations covered by the data used to parameterise the model.

6. *Linear semi-empirical adaptations from physical models.* The physical models are used as the basis for construction of linear models that depend on parameters readily estimated during the acquisition process, such as the solar–surface–sensor geometrical characteristics. Linearisation requires approximations that in turn create limitations on the use of the models and as a consequence they may be used as mosaic models, with different models being used in different parts of the mosaic. There needs to be an independent basis for selecting which model is appropriate at each location in the mosaic. For example, different models may be used for different landcovers, so that a landcover map becomes the basis for assigning models to mosaic elements.

The main thrust of the early modelling work was on the construction of canopy models. However, as these improved, it became obvious that a limitation of many of these models was the assumed reflectance characteristics of the canopy elements. As a consequence, significant effort was then directed at the estimation of the reflectance characteristics of the individual canopy elements, and using the results of this modelling within canopy reflectance models. We will thus first look at the modelling of leaf reflectance and transmittance and subsequently at the modelling of canopy reflectance.

7.4.2 Leaf Reflectance Models

A variety of approaches to the estimation of leaf reflectance have been explored. However, practically all of the more commonly used models are based on one of two different approaches: plate and the ray tracing models.

7.4.2.1 Plate Models

Plate models assume that a leaf element consists of one or more plane parallel plates, where each plate has its own unique refractive index, reflectance, absorptance and transmittance (Allan 1969, 1970). The absorptance of the layer depends primarily on the absorption characteristics of the chemicals within the layer, whilst the reflectance and transmittance distribution depends primarily on the cell structure within the layer. The Prospect model (Jacquemoud and Baret 1990; Jacquemoud et al. 1996) is a plate model that assumes that a leaf consists of N plates and $(N-1)$ airspaces. Let us first consider a plate. Each plate is assumed to be an absorbing plate with rough surfaces giving rise to Lambertian diffusion within a solid angle α. Within each plate there is absorption and transmission out of the plate at a specified index of refraction. The general formula for reflectance, ρ_α and transmittance, t_α within a layer is given by:

$$\rho_\alpha = [1 - t_{av}(\alpha, n)] + t_{av}(90, n)t_{av}(\alpha, n)\theta^2[n^2 - t_{av}(90, n)]/\{n^4 - \theta^2[n^2 - t_{av}(90, n)]\} \quad (7.4a)$$

$$t_\alpha = t_{av}(90, n)t_{av}(\alpha, n)\theta n^2/\{n^4 - \theta^2[n^2 - t_{av}(90, n)]\} \quad (7.4b)$$

where α is the maximum angle of transmission of the ray into the plate, n is the refractive index for the layer, θ is the transmission coefficient of the plate, $t_{av}(\alpha, n)$ is the average transmissivity of a dielectric plane surface given the incident radiation direction and $t_{av}(90, n)$ is the average transmissivity of a dielectric plane surface for all directions.

From Equation (7.4) it can be shown that:

$$\rho_\alpha = k_1 \rho_{90} + k_2 \quad (7.5a)$$

$$\tau_\alpha = k_1 \tau_{90} \quad (7.5b)$$

where the constants, k_1 and k_2 are $t_{av}(\alpha, n)/t_{av}(90, n)$ and $k_1(t_{av}(90, n) - 1) + 1 - t_{av}(\alpha, n)$ respectively, and the reflectance and transmittance through all directions, ρ_{90}, τ_{90}, are different functions of the refractive index, the transmission coefficient of the plate and the average transmissivity of a dielectric plane surface for all directions, $t_{av}(90, n)$. The average transmissivity of a dielectric plane surface is complex to derive and is shown in Figure 7.9. The figure shows that it remains approximately constant for $\alpha < 60°$ but above this it decreases as would be expected as grazing rays are much more

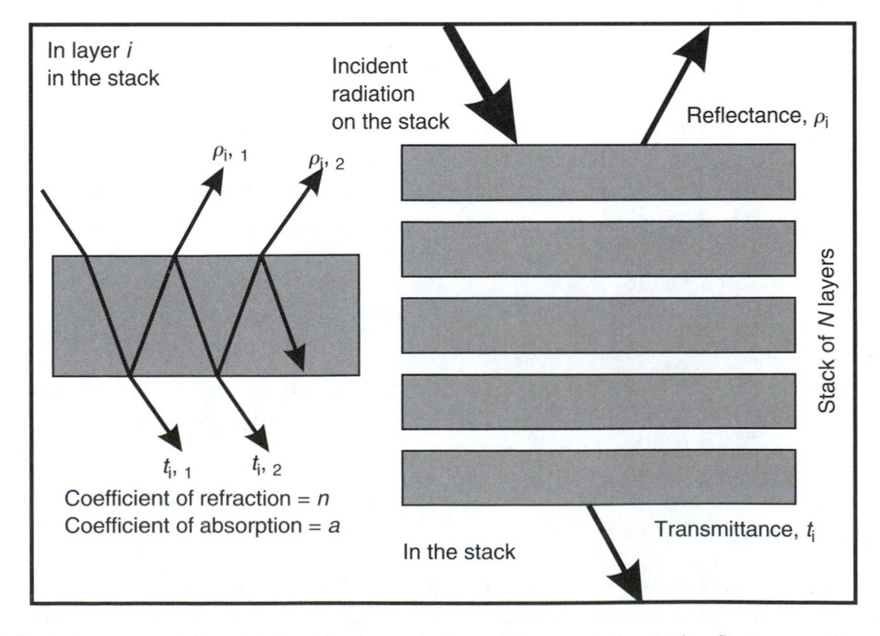

FIGURE 7.9 The rays contain multiple reflections within each layer and the stack reflectance and transmittance is the summation of all of the individual components.

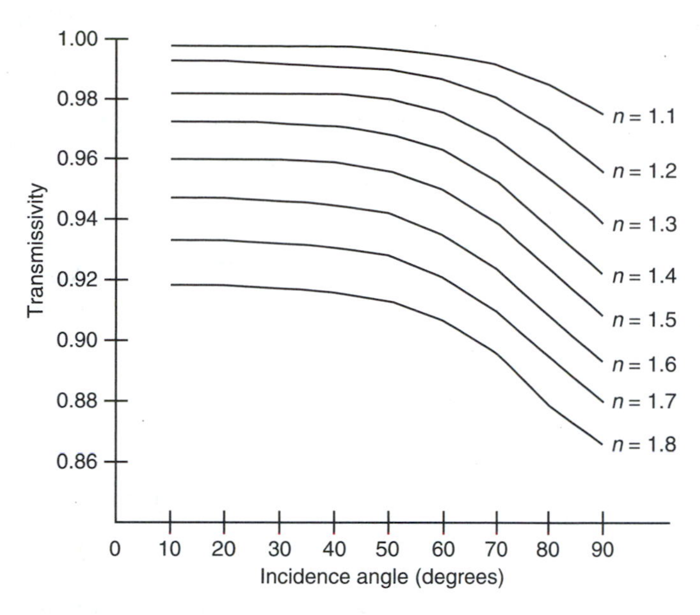

FIGURE 7.10 The average transmissivity of a dielectric plane surface, $t_{\mathrm{av}}(\alpha, n)$ as a function of the refractive index, n, and the incidence angle, a (from Jacquemuod and Baret, 1990).

likely to be reflected than transmitted into the leaf. It can be seen that it is dependent on the refractive index, n.

A single plate model of this form thus has three input parameters, the refractive index, n, the incidence angle, α and the transmission coefficient, θ. However, such a model only works for so-called compact leaves that are represented accurately as a single layer. This is not the case with many types of leaves and certainly not for dead leaves and so the solution adopted in the PROSPECT model is to consider the leaf as consisting of N layers and $(N - 1)$ air interspaces between the layers (Figure 7.10). The first layer has incident radiation incident upon it at incidence angle whilst the subsequent $(N - 1)$ layers have incidence radiation through all angles (incidence angle $= 90°$). It can be shown that the reflectance and transmittance of the N layers can be solved to give:

$$\rho_{N,\alpha} = \rho_\alpha + [\tau_\alpha \tau_{90} \rho_{N-1,90}/(1 - \rho_{90}\rho_{N-1,90})] \tag{7.6a}$$

$$\tau_{N,\alpha} = \tau_\alpha \tau_{N-1,90}/(1 - \rho_{90}\rho_{N-1,90}) \tag{7.6b}$$

From which ρ_α and t_α can be eliminated to give:

$$\rho_{N,\alpha} = k_1 \rho_{N,90} + k_2 \quad \text{and} \quad t_{N,\alpha} = k_1 t_{N,90} \tag{7.7}$$

where the constants, k_1 and k_2 are as described for Equation (7.5) above. This set of complex equations in N layers can be simplified by use of Stokes' equations to the form:

$$\frac{\rho_{N,90}}{b_{90}^{N} - b_{90}^{-N}} = \frac{t_{N,90}}{a_{90} - a_{90}^{-1}} = \frac{1}{a_{90}b_{90}^{N} - a_{90}^{-1}b_{90}^{-N}} \tag{7.8}$$

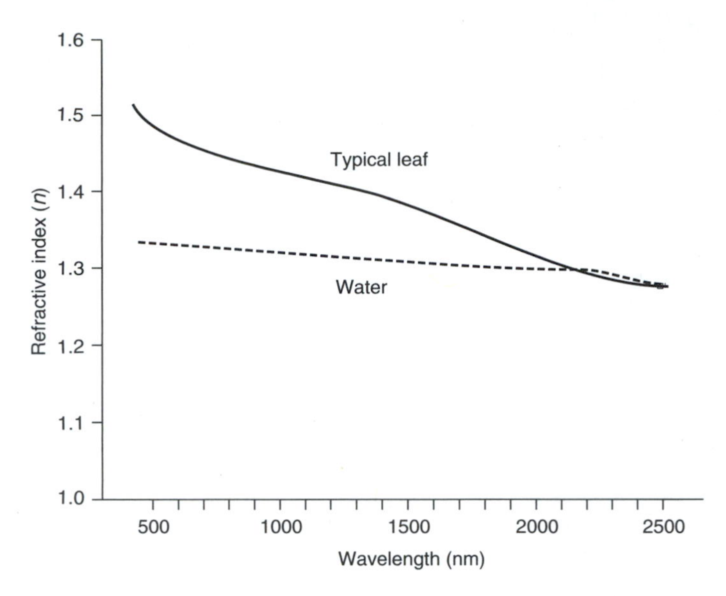

FIGURE 7.11 The refractive index of typical leaf material and water as a function of wavelength.

where

$$
\begin{aligned}
a_{90} &= (1 + \rho_{90}^2 - t_{90}^2 + \delta_{90})/2\rho_{90} \\
b_{90} &= (1 - \rho_{90}^2 + t_{90}^2 + \delta_{90})/2t_{90} \\
\delta_{90} &= \sqrt{(t_{90}^2 - \rho_{90}^2 - 1)^2 - 4\rho_{90}^2}
\end{aligned}
\tag{7.9}
$$

The Prospect model requires four parameters:

1. Scattering is described by the refractive index of leaf materials (n), by a parameter characterising the leaf mesophyl structure (N), modified by the incidence angle, α.
2. Absorption is modelled using pigment concentration and water depth and the corresponding specific absorption coefficients to derive the transmission coefficient, θ.

The refractive index of leaf material is about 1.4 in the visible part of the spectrum, and decreases in a regular way towards to Near-IR (NIR) as shown in Figure 7.11.

The mesophyl structure parameter (N) is assumed to affect the whole spectrum, however, in the visible part of the spectrum, the effects of the mesophyl structure is masked by the chlorophyll absorption, so wavebands are used in the NIR to estimate N. Three wavelengths corresponding to the maximum reflectance (λ_r), the maximum transmittance (λ_t), and the minimum absorptance (λ_a) are selected. For fresh leaves, these three wavelengths are located in the NIR plateau; for dry leaves, they are shifted towards longer wavelengths. The structure parameter of each leaf was adjusted at the same time as the three absorption coefficients by minimising:

$$
\sum_{\lambda} [\rho_{\text{meas}}(\lambda) - \rho(\lambda, N, a_\lambda)]^2 + [t_{\text{meas}}(\lambda) - t(\lambda, N, a_\lambda)]^2 \tag{7.10}
$$

where $\rho_{\text{meas}}(\lambda)$ and $t_{\text{meas}}(\lambda)$ are, respectively, the three reflectances and three transmittances measured at λ_r, λ_t and λ_a. For the same species, N estimated on dry leaves is higher than N estimated on fresh leaves. This is due to an increase of the multiple scattering resulting from the loss of water in dry leaves.

The wavelength independent mesophyll structure parameter N allows the inversion of the Stokes equations: using measured reflectance $\rho(\lambda)$ and transmittance $\tau(\lambda)$, the optical properties of the compact layer ($N = 1$) are easily calculated for each leaf, permitting the determination of a spectral absorption coefficient $a_0(\lambda)$. If the assumption is made that the leaf is a homogeneous mixture of biochemical components, this coefficient can be written as:

$$a_o(\lambda) = a_e(\lambda) + \sum_i \frac{C_i \times a_i(\lambda)}{N} \tag{7.11}$$

where λ is the wavelength, C_i the concentration of the constituent i, and $a_i(\lambda)$ the corresponding specific absorption coefficients, $a_e(\lambda)$ explains the nonzero absorption of an albino leaf under 500 nm. At this point either of two approaches can be considered. The first approach predicts the constituent concentrations and then compares these predictions with measured values. This approach presumes that the specific absorption coefficients are known, for example, deduced from optical measurements performed on pure substances. Whilst this has been done for the leaf pigments and water, the measurement of the absorption spectra shows shifts of up to 10 nm, attributed both to the influence of the solvent and to the fact that the chlorophylls inside leaf tissues are complexed with other pigments and proteins. The second approach uses the absorption coefficients $a_0(\lambda)$ and the measured concentrations to deduce the specific absorption coefficients of leaf biochemical components $a_i(\lambda)$.

7.4.2.2 Ray Tracing Models

The concept behind ray tracing is to generate a computer model of the three-dimension leaf, complete with its cells, and their contents, similar to that shown in Figure 7.12, containing the upper epidermis or guard cells, the palisade cells containing the chloroplasts, the spongy mesophyll cells and the lower epidermis. The cells are described be geometrical primitives (spheres, ellipses, cylinders) as are appropriate, and arranged so as to form a typical leaf cross-section.

The models include a detailed geometrical description of individual cell types and their unique arrangement within the leaf tissue. The optical constants of leaf materials such as the cell walls, cytoplasm, pigments, air cavities, etc. are defined, as is the moisture content of the leaves. Using the laws of reflection, refraction and absorption, it is then possible to simulate the propagation of individual photons incident on the leaf surface to their absorption in the leaf, reflection out of the leaf or transmission through the leaf. The orientation of the incident photon and the orientation of those that are reflected or transmitted are also recorded. Once a sufficient number of rays have been

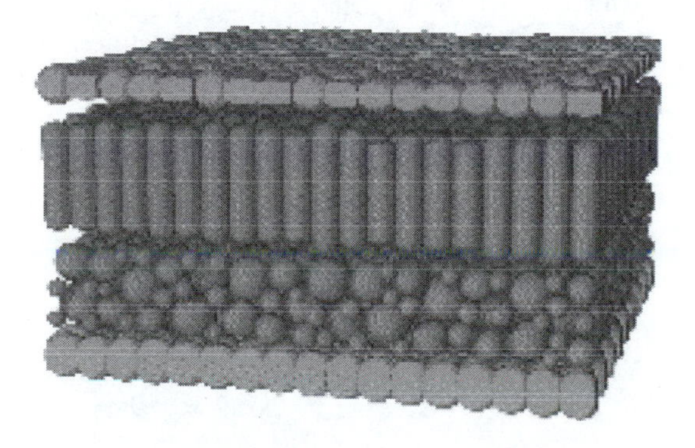

FIGURE 7.12 A computer generated depiction of a leaf structure used for ray tracing.

simulated through their absorption, transmission and reflection, statistically valid estimates of the radiation transfer in a leaf may be deduced.

Some studies have shown that the epidermal cells may have a shape that tends to focus the incident radiation onto the upper region of the palisade parenchyma, which contains many chloroplasts adapted to high light (Figure 7.13). This phenomenon has been mainly presented as an adaptation to the low light environment on the tropical forest floor but others have hypothesised that the epidermal lenses of *Medicago sativa*, a cultivated plant, could increase absorption of light at low sun angles.

The most current ray tracing model is the Raytran model that has been successfully used to simulated the optical properties of a virtual dicotyledon leaf (Govaerts and Verstraete 1998; Jacquemoud et al. 2000). This ray tracing code, requires a three-dimensional representation of leaf internal structure which fulfils the constraints of plant anatomy and physiology.

The advantage of the plate models is that they reduce the complex interactions in a leaf to a simple combination of a few parameters that can then be readily inverted to estimate leaf reflectance and transmittance characteristics from leaf physical data. They are thus invaluable components of more complex canopy reflectance models, so as to provide better estimates of the reflectance, transmittance and absorptance of canopy elements. However, their simplicity means that they do not predict the differences in reflectance of the upper and lower surfaces of leaves, as have been observed, nor do they account for the effects of epidermal lenses.

The ray tracing models are potentially more accurate, but they cannot, at present, be numerically inverted and hence they are a valuable tool in understanding the relationship between leaf structure and composition and the resulting leaf reflectance, but they cannot be used to directly estimate leaf properties from leaf reflectance.

7.4.3 Canopy Reflectance Models

The need to understand canopy interactions has led to the development of numerous numerical canopy reflectance models. Whilst many of these models are based on one or more of the approaches introduced in Section 7.4.1, a number of models have explored other theories of canopy reflectance. One of the first models developed to investigate canopy reflectance was that of Allan et al. (1970), who used the Kubelka–Munk (K–M) approximation of the turbid medium theory of radiative transport

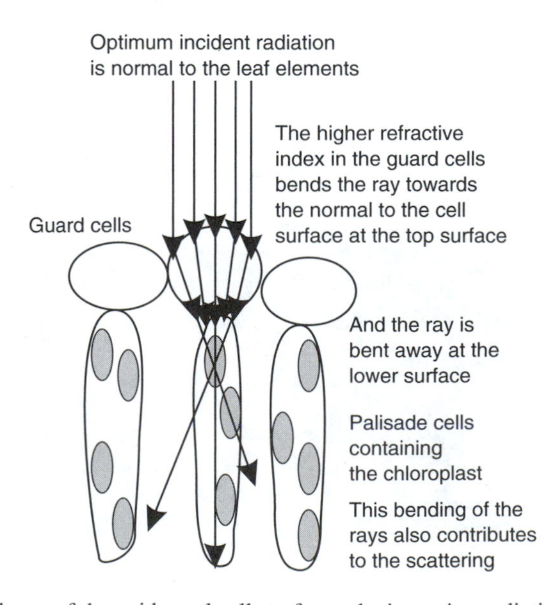

FIGURE 7.13 Tendency of the epidermal cells to focus the incoming radiation on the chloroplasts.

as the basis of their model. This model was adapted and improved to create the Suits model (Suits 1972), generally recognised as the first true canopy reflectance model that could give a reasonable approximation of actual reflectance, under some conditions. However, the limitations of this model led to the development of many refinements of the approach, of whish the SAIL model (Verhoef 1984) is the best known.

An early geometric model was developed by Egbert (1977), but the main activity in the development and evaluation of this type of model came with the work of Li and Strahler between 1985 and 1992 (Li and Strahler 1986). The other approaches to modelling canopy reflectance are more recent than this, mostly starting with the advent of cheap computer power in the mid to late 1980s.

In addition to the physical models, a number of empirically based models have been developed. As a consequence, most models can be categorised as being in one of six model types:

1. Empirical models
2. Turbid medium physical models
3. Geometric-optical physical models
4. Ray tracing physical-stochastic models
5. Flux or radiosity physical models
6. Linear, semi-empirical approximations on physical models

It is not realistic to trace the development of all of the different models, so that the main lines of development will be discussed and then some models discussed in more detail, to illustrate the main principles of that type of model.

7.4.4 Empirical Models

Empirical models attempt to describe reflectance from the surface by fitting a function to observed reflectance data in the form:

$$\rho = F(p) \tag{7.12}$$

where $F(p)$ is some arbitrary function that is not related to the physical properties of the canopy. One of the first empirical models developed to address issues in a vegetative canopy was that of Walthall et al. (1985) (Figure 7.14). In this model the BRDF is expressed as:

$$\rho = a\theta_{\mathrm{v}}^2 + b\theta_{\mathrm{v}}\cos(\phi_{\mathrm{v}} - \phi_{\mathrm{s}}) + c \tag{7.13}$$

where ϕ_{v} and ϕ_{s} are the view and solar azimuth angles, respectively, and a, b and c are constants to be determined empirically. The first term is meant to model the upward bowl shape that is often observed in BRDF data, the second term provides a linear dependence on zenith angle but which can allow for some of the observed anisotropy, such as that due to the Hotspot effect, and c is a brightness factor. The model has been shown to fit a number of measured BRDFs.

7.4.5 Turbid Medium Models

7.4.5.1 Theoretical Basis for These Models

In Section 2.2 we discussed the propagation of an electromagnetic wave through and absorbing medium. In that section we showed that Lambert had found that the radiance of energy traversing an absorbing medium suffers equal proportions of loss for equal distances travelled. That is:

$$\frac{\mathrm{d}L}{L} = -\beta\,\mathrm{d}z \tag{2.6}$$

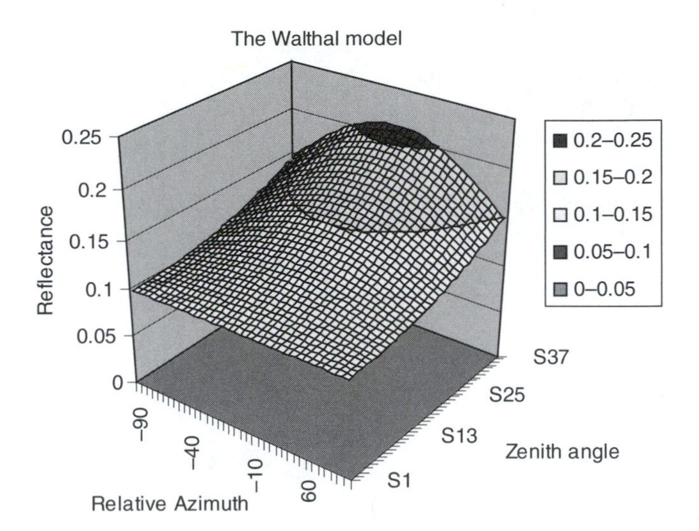

FIGURE 7.14 The Walthal Empirical Canopy Reflectance model (from Walthall, et al, 1985).

where β is the volume extinction coefficient. We can re-arrange Equation (2.6) into the form:

$$\frac{\mathrm{d}L}{\mathrm{d}z} = -\beta L \tag{7.14}$$

Chandrasekhar (1950) adapted this law to the problem of the transmission of radiation through gaseous clouds of interstellar dust. He assumed that interstellar clouds consisted of a randomly uniform scatter of small scattering and absorbing objects (dust particles) that were sufficiently far apart so that no mutual shadowing occurred between them. He developed the integral–differential equation to describe such transport:

$$-\mu\frac{\partial L(z,\Omega)}{\partial z} + \sigma_{\mathrm{e}}(z,\Omega)L(z,\Omega) - \int_{4\pi} \sigma_{\mathrm{s}}(z,\Omega' \rightarrow \Omega)L(z,\Omega')\,\mathrm{d}\Omega' = 0 \tag{7.15}$$

where $L(z, \Omega)$ is the specific radiance in direction Ω, at a height z within the turbid layer of total depth, T $(0 < z < T)$ as shown in Figure 7.15 after adaptation to a vegetative canopy. In this adaptation, the canopy is assumed to consist of one or more horizontal, plane parallel layers of infinite spatial (x, y) extent. Ω is the direction of the ray, consisting of a zenith angle, θ, and an azimuth, ϕ, that is $\Omega = (\theta, \phi)$. μ is the cosine of the zenith angle of the radiation, $(\partial L(z, \Omega)/\partial z)$ is the steady state radiance distribution function, or the way that the radiance varies within the canopy as a function of depth through the canopy. σ_{e} is the extinction coefficient of the canopy medium so that $\sigma_{\mathrm{e}}(z, \Omega)I(z, \Omega)$ is the rate of absorption of energy as it passes through the canopy layer. σ_{s} is the scattering coefficient for photon scattering from the illumination direction to the viewing direction so that $\int_{4\pi} \sigma_{\mathrm{s}}(z, \Omega' \rightarrow \Omega)I(z, \Omega')\,\mathrm{d}\Omega'$ is the accumulated radiation scattered in a specific direction within the canopy.

It can be seen that there are a number of assumptions made either by Chandrasekhar (1950) in his development of the original theory, or in the way that this theory has been adapted to vegetative canopies, that can lead to errors:

1. Very small scatterers and absorbers. The finite size and orientation of the canopy elements are a distinct divergence from the theory. This is one of the reasons why models based strictly on this theory do not explain the Hotspot effect.

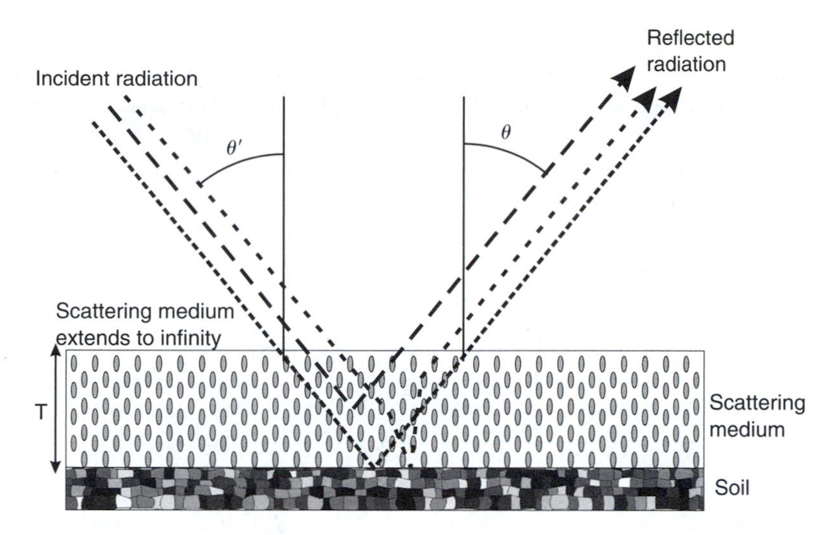

FIGURE 7.15 The turbid medium concept in canopy modelling, depicted for a single layer canopy.

2. Lambertian scatterers and absorbers. The theory assumes Lambertian scatterers and absorbers, even though some of the more recent adaptations of the theory allow for a relaxation of this assumption.
3. Random uniform distribution of scatterers and absorbers. Whilst some modestly dense canopies, such as some agricultural canopies, can approximate this assumption, it is far from the characteristics of most natural canopies and indeed many agricultural canopies. It is the reason why a number of adaptations have focussed on dealing with row crops, and the development of geometric models for forest canopies.
4. No mutual shadowing between the scatterers or absorbers. This assumption is violated when the canopy becomes dense, as often occurs within agricultural canopies.

Whilst there are many ways to solve the integral–differential Equation (7.15), all of them involve making simplifications of one sort or another when dealing with vegetation canopies. In the development of canopy reflectance models, a great variety of simplifications have been tried. One of the more common assumptions that has been used is the adaptation of the Kubelka–Munk theory to this problem. The Kubelka–Munk theory assumes that there are only four fluxes within the canopy:

1. Direct radiation that has not been intercepted by the canopy and flowing either up or down through the canopy; symbolised by F_+ and F_- for the upwards or downwards flux, respectively. The downward direct flux, F_- is the direct solar beam prior to interception with the canopy. Upward direct flux, F_+ is radiation that has not been reflected, but which is travelling upwards through the canopy. This flux is assumed to be zero at the base of the canopy in canopy reflectance models in the visible part of the spectrum.
2. Diffuse radiation in the upward and downward directions and symbolised by E_+ and E_-, respectively. Diffuse radiation is that scattered by the canopy, the soil background or the diffuse skylight that enters the canopy.

To solve Equation (7.15) using these assumptions involves two steps:

1. Solution of the phase function, $\sigma_s(z, \Omega' \rightarrow \Omega)$. The phase function is the probability that radiance coming from direction Ω' will be scattered into direction Ω. Solution of the phase function is the purpose of the leaf reflectance models discussed in Section 7.4.2.

2. Solution of the radiative transport equation for a given phase function and set boundary conditions. To solve this equation, the first assumption made is that the canopy consists of one or more plane parallel horizontal layers of infinite extent. This assumption means that spatial position in the canopy is only important in the z direction. Boundary conditions must be set that define the radiation incident on the canopy topmost layer, and to ensure that there is a consistent transfer of radiation from one layer to the next.

These assumptions allows Equation (7.15) to be put in the form:

$$\mu \frac{\partial L(z,\Omega)}{\partial z} = L(z,\Omega) - K(z,\Omega) \tag{7.16}$$

where

$$K(z,\Omega) = \frac{a}{4\pi}\int_0^{2\pi} \mathrm{d}\phi \int_{-1}^{1} \sigma_\mathrm{s}(\mu_\mathrm{i},\phi_\mathrm{i} \rightarrow \mu_\mathrm{e},\phi_\mathrm{e}) L(z,\mu_\mathrm{i},\phi_\mathrm{i})\,\mathrm{d}\mu_\mathrm{i} \tag{7.17}$$

where $\mu = \cos(\theta)$ and a is the albedo for single scattering so that $a = \sigma_\mathrm{s}/(\sigma_\mathrm{s} + \sigma_\mathrm{e})$. Equation (7.16) can be formally integrated to give:

$$\begin{aligned} L(z,\pm\Omega) &= L(z_\mathrm{o},\pm\Omega)\,\mathrm{e}^{-(\beta(z-z_\mathrm{o})/\mu)} + \frac{1}{\mu}\int_{z_\mathrm{o}}^{z} K(z_\mathrm{o},\pm\Omega)\,\mathrm{e}^{(-\beta(z-z')/\mu)}\,\mathrm{d}z \\ &= L(z_\mathrm{o},\pm\Omega)\,\mathrm{e}^{-(\tau_z/\mu)} + \frac{1}{\beta\mu}\int_{z_\mathrm{o}}^{z} K(z_\mathrm{o},\pm\Omega)\,\mathrm{e}^{-(\tau_{z'}/\mu)}\,\mathrm{d}\tau \end{aligned} \tag{7.18}$$

Equation (7.18) states that the upward (downward) radiance at depth z in the canopy is a result of the attenuation of the upward (downward) direct radiance between z_o and z, plus that scattered into the beam along the path at z. Since the optical thickness, $\tau = \beta z$ we can restate Equation (7.18) in terms of optical thickness. If we consider firstly the diffuse beam components of Equation (7.18) then the diffuse flux at z in the specified direction = Diffuse flux at z_o in the specified direction + attenuated component of direct beam in the specified direction, between z_o and z, or:

$$L_1(z,\pm\Omega) = \delta(\mu - \mu_\mathrm{s})\delta(\phi - \phi_\mathrm{s})\pi F - \mathrm{e}^{-\tau/\mu} + L_\mathrm{d}(z,\pm\Omega) \tag{7.19}$$

where $\partial(\mu - \mu_\mathrm{s})$ and $\partial(\phi - \phi_\mathrm{s})$ are the differences in the cosines of the zenith distances of the scattered beam and the incident solar beam, respectively, and of the azimuths of thee same two beams. F_- is the incident solar beam directed directly downwards. If we substitute Equation (7.19) into Equation (7.15) we get:

$$\begin{aligned} \mu\frac{\partial L_\mathrm{d}(z,\pm\Omega)}{\partial z} &= L_\mathrm{d}(z,\pm\Omega) - \frac{a}{4}\sigma_\mathrm{s}(z,\Omega' \rightarrow \Omega)\cdot F_-\mathrm{e}^{-\tau/\mu} \\ &\quad - \frac{a}{4\pi}\int_0^{2\pi}\mathrm{d}\phi' \int_{-1}^{+1} \sigma_\mathrm{s}(z,\Omega' \rightarrow \Omega) L_\mathrm{d}(z,\Omega')\,\mathrm{d}\mu' \end{aligned} \tag{7.20}$$

Equation (7.20) is still a very complex equation to solve. If we apply the K–M approximations to Equation (7.20) with two parameters for the diffuse fluxes of α and γ for the absorption and scattering, respectively, and three parameters for the direct fluxes of k, s_1 and s_2, for absorption, scattering in the same (transmission) and opposite (reflection) directions, respectively, then we can form equations for each of the four fluxes:

$$\begin{aligned}
\frac{dE(-d,i,z)}{d(-\tau)} &= -(\alpha+\gamma)E(-d,i,z) + \gamma E(+d,i,z) + s_1F(-s,i,z) + s_2F(+s,i,z) \\
\frac{dE(+d,i,z)}{d(\tau)} &= -(\alpha+\gamma)E(+d,i,z) + \gamma E(-d,i,z) + s_1F(+s,i,z) + s_2F(-s,i,z) \\
\frac{dF(-s,i,z)}{d(-\tau)} &= -(k+s_1+s_2)F(-s,i,z) \\
\frac{dF(+s,i,z)}{d(\tau)} &= -(k+s_1+s_2)F(+s,i,z)
\end{aligned} \tag{7.21}$$

where $E(+d,I,z)$ is the upward diffuse radiance in the ith layer of the canopy at height z in that layer, $E(-d,I,z)$ is the downward diffuse radiance in the ith layer of the canopy at height z in that layer, $F(+s,I,z)$ is the upward direct radiance in the ith layer of the canopy at height z in that layer and $F(-s,I,z)$ is the downward direct radiance in the ith layer of the canopy at height z in that layer.

The first two terms on the right-hand side of the first two equations in Equation (7.21) state that the downward (upward) diffuse flux decreases due to its absorption and scattering and increases due to the scattering of the upward (downward) flux. The last two terms state that the downward (upward) flux is increased by the scattering of the direct fluxes. The last two equations state that the direct fluxes are reduced by their scattering and absorption.

The diffuse fluxes are related to the specific intensity by the equations:

$$\begin{aligned}
E(+d,I,z) &= \int_0^{2\pi} d\phi_s \int_0^{\pi/2} I(\tau,+\mu,\phi_s)\mu \sin\theta_s \, d\theta_s \\
E(-d,I,z) &= \int_0^{2\pi} d\phi_s \int_0^{\pi/2} I(\tau,-\mu,\phi_s)\mu \sin\theta_s \, d\theta_s
\end{aligned} \tag{7.22}$$

where the diffuse fluxes at the top of the plane parallel medium are known as the irradiance and radiant exitance, respectively. Thus, in K–M theory one integrates the fluxes over the azimuthal and zenith angles using Equation (7.22) and then solves for the integrated fluxes using Equation (7.21).

If one assumes that there is only a downward direct flux, then the resulting three-flux theory is known as the Duntley theory (Duntley 1972), involving the same five parameters as for the four-flux theory in three equations. If the direct beam is eliminated altogether, as is applicable under cloudy conditions, or when working within the canopy, then the resulting two-flux theory only contains two parameters in two equations.

The K–M theory approach to the use of the turbid medium theory to explain canopy reflectance is only one of several methods of approximation. Another approach that has been tried is by the use of Beer's Law. Beer's law can be placed in the form:

$$A = \varepsilon bc \tag{7.23}$$

where A is the absorptance of radiation as it passes through a medium for a distance b containing molecular elements of individual absorptance, ε, of concentration c in the medium. The law shows that the absorptance is linearly proportional to the distance travelled and the density of absorbing elements in the medium. If single scattering is assumed to be the dominant cause of canopy reflectance then Beer's Law can be used to describe radiant intensity at distance z into a canopy, and from this be used to build a model. The major assumptions behind Beer's Law are that the number of scattering objects in a volume of canopy is proportional to that volume and that all scattering is single scattering. In the visible part of the spectrum more than 90% of the radiation is absorbed at the first scattering, so that this assumption is reasonable. In the NIR, however, where about 90% of the radiation is either

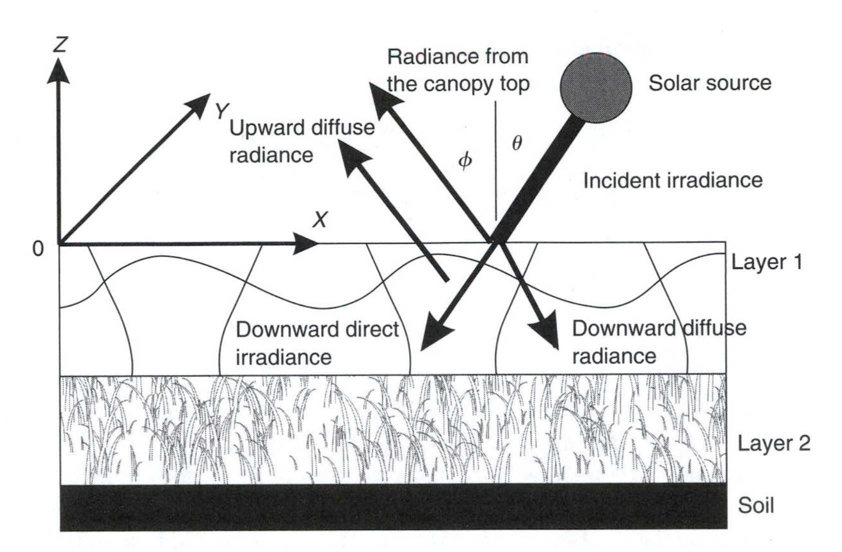

FIGURE 7.16 The co-ordinate systems, layers and fluxes that are used in the Suits canopy reflectance model (from Suits, G.H., 1972. *Remote Sensing of Environment*, 2, 117–125. With permission).

reflected or transmitted, so that up to 47% of the emerging radiation from the canopy can be due to multiple scattering, this assumption is significantly violated and thus a major source of error in any estimations.

A major limitation of the turbid medium approach is the assumption of homogeneous, horizontal, infinitely extensive layers containing a randomly distributed uniform distribution of small Lambertian scatterers. The theory thus ignores canopy structure and the finite size of the scattering and absorbing elements. Certain properties of the canopy are directly related to these properties, including the Hotspot effect, and as a consequence models based solely on the K–M theory will not capture such features in the derived BRDF data from the model.

7.4.5.2 The Suits Model

The Suits model is a multiple layer radiative transfer model that uses the Kulbelka–Munk approximation with three fluxes and five parameters, or the Duntley simplification of the K–M theory. The model components are illustrated in Figure 7.16. Each layer is a plane parallel layer of infinite extent, so that location (x, y) in the layer is not important. Within each layer the canopy elements are small and randomly scattered throughout the layer. They have set reflectance, transmittance and absorptance properties. There can be more than one type of canopy element, such as leaves or stems, each with their own reflectance, absorptance and transmittance properties. The density and orientations of the canopy elements follow a pattern that is set for each type of canopy element within each layer. In this model the phase function, or angular distribution of radiation directions is simplified by assuming that the canopy consists of vertically and horizontally reflecting, absorbing and transmitting panels, so that the canopy elements are replaced by their horizontal and vertical components. Leaf canopy elements will have reflectance, absorptance and transmittance values, whereas stem elements will only have reflectance and absorptance values, or transmittance $= 0$.

The co-ordinate system has its origin at the top of the top layer of the canopy as shown in Figure 7.16. This means that depths through the canopy are increasingly negative in value. The model assumes that there are three fluxes within the canopy. The downward direct flux is the radiant energy coming directly from the sun prior to interception in the canopy. The downward diffuse flux is radiant energy that has undergone reflection or transmission within the canopy, but which is continuing in the downward direction. The upward diffuse flux is similar, but it is travelling upwards.

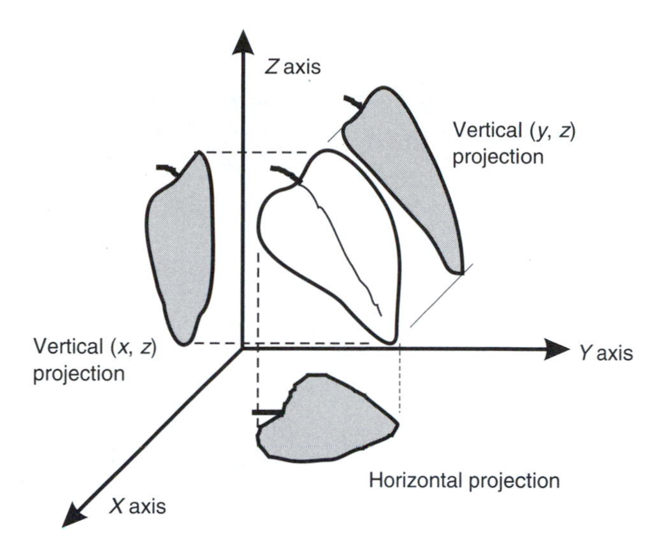

FIGURE 7.17 A leaf element and its horizontal and vertical cross-section areas

The upward diffuse flux that leaves the canopy is that which is of interest in the construction of BRDFs for the surface, and that in the direction of a sensor is of interest in modelling the response at a sensor. The boundary conditions for each layer are the direct and diffuse downward fluxes that enter the layer from the layer above and the upward diffuse flux that enters the layer from the layer below, or the soil surface for the bottom layer. The diffuse downward flux for the top layer is assumed to be zero in the original Suits model, although this can be modified.

Each layer can contain a number of different types of canopy elements, such as green and brown leaves and stems. Each type of element is oriented in space such that it can be projected onto the three planes (x, y), (x, z) and (y, z) to produce projected cross-sections on each plane as shown in Figure 7.17. These are summed to derive a horizontal and a vertical cross-section. The horizontal cross-section is simply that area projected onto the horizontal (x, y) plane. The vertical cross-section is the sum of the other two projections. These horizontal and vertical cross-section areas have to be determined from field data, for the species of interest.

For the simple geometry formulated for the Suits model, the parameters of the K–M theory are expressed in terms of the following parameters:

σ_h = average area of the projection of a vegetative component onto a horizontal plane

σ_v = average area of the projection of a vegetative component onto the two orthogonal vertical planes

n_h = number of horizontal projections per unit volume

n_v = number of vertical projections per unit volume

ρ = leaf hemispherical reflectance

τ = leaf hemispherical transmittance

ψ = relative azimuthal direction between the source and the viewing directions

θ_s = solar zenith angle

θ_o = viewing zenith angle

β = the average value of $\cos(\psi)$ and normally set to $2/\pi$

The parameters, n_h, n_v, σ_h and σ_v together with the height of the canopy, h, are related to the leaf area index (LAI) in the canopy according to Verhoef and Bunnik (1975) to a fair degree of accuracy

TABLE 7.1
Relationships between Various Parameters Occurring in the Suits Model, Canopy Parameters and the K–M Theory Parameters (Drawn from Goel, N.S. 1988. *Remote Sensing Reviews,* 4, 1–212. With permission)

Suits Model Parameters	Relationship with Canopy Parameters	K–M Theory Parameters
E_+		E_+
E_-		E_-
F_+		F_+
F_-		F_-
a	$H'(1-\tau)+V'(1-(\rho+\tau)/2)$	
b	$H'\rho+V'((\rho+\tau)/2)$	
c	$H'\rho+\beta\times V'((\rho+\tau)/2)\tan\theta_s$	s_1
c'	$H'\tau+\beta\times V'((\rho+\tau)/2)\tan\theta_s$	s_2
k	$H'+\beta\times V'\tan\theta_s$	$k+s_1+s_2$
K	$H'+\beta\times V'\tan\theta_o$	
u	$H'\tau+\beta\times V'((\rho+\tau)/2)\tan\theta_o$	
v	$H'\rho+\beta\times V'((\rho+\tau)/2)\tan\theta_o$	
w'	$H'\rho+(\beta/4)\times V'\tan\theta_o\times\tan_s[\{\sin(\psi)+(\pi-\psi)\cos(\psi)\}$ $\times\rho+\{\sin(\psi)\psi\cos(\psi)\}\tau]$	

in the relationship:

$$\text{LAI}=(h/s)((n_h\sigma_h)^2+(n_v\sigma_v)^2)^{0.5}=(h/s)(H'^2+V'^2)^{0.5} \tag{7.24}$$

where $H'=n_h\sigma_h$, $V'=n_v\sigma_v$ and s is a correction factor, which varies between 0.84 and 0.95 depending on the LAD. The vertical and horizontal projections are also related to the average leaf angle (ALA) by the relationship:

$$\text{ALA}=\arctan(V'/H') \tag{7.25}$$

The basic Kubelka–Munk Equations (7.21) have been modified in the Suits model into the form given in Equation (7.26), in which the individual variables have values as given in Table 7.1.

In the initial Suits model, ρ and τ are considered to be the same for both the upper and lower surfaces of the leaf and the only radiation incident on the canopy is the direct solar beam. Both conditions can be relaxed to allow for diffuse skylight and for different reflectances from the upper and lower surfaces of leaf elements. If there is more than one type of vegetative component in a layer, the values of the K–M theory parameters are obtained for each type separately, and then they are added together to obtain values for the canopy layer.

The K–M equations are solved for the three fluxes, using appropriate boundary conditions, needed to ensure the continuity of the downward and upward fluxes at the different layer boundaries in the canopy and to ensure that the fluxes match those that actually exist. Thus, for example, the incident direct flux should match the incident solar irradiance. These boundary conditions use one more parameter (the soil hemispherical reflectance, ρ_s) in the original Suits model, or two if the fraction of incident diffused skylight, SKYL is to be incorporated.

The simplifications made in the Suits model leads to Equation (7.26) deriving into the form:

$$\begin{aligned}\frac{dE(-d,i,z)}{dz} &= aE(-d,i,z) - bE(+d,i,z) - c'F(-s,i,z)\\ \frac{dE(+d,i,z)}{dz} &= bE(-d,i,z) - aE(+d,i,z) + cF(-s,i,z)\\ \frac{dF(-s,i,z)}{dz} &= kF(-s,i,z)\end{aligned} \tag{7.26}$$

where a, b, c, c' and k are defined in Table 7.1. These equations are used with their boundary conditions to solve for E_-, E_+ and F_- at each layer. With a single layer canopy the boundary conditions are:

$$\begin{aligned}&F(-s,i,z)(0) = E_{\text{solar}}\\ &E(-d,i,z)(0) = 0\\ &E(+d,i,z)(-h) = \rho_s[F(-s,i,z)(-h) + E(-d,i,z)(-h)]\end{aligned} \tag{7.27}$$

Using similar logic to that used in Section 2.2, Equation (7.26) can be integrated to give:

$$\begin{aligned}E(+d,i,z) &= A_i(1-f_i)\,e^{g_i z} + B_i(1+f_i)\,e^{-g_i z} + c_i\,e^{k_i z}\\ E(-d,i,z) &= A_i(1+f_i)\,e^{g_i z} + B_i(1-f_i)\,e^{-g_i z} + d_i\,e^{k_i z}\\ F(-s,i,z) &= F(-s,i-1,z_i)\,e^{k_i(z-z_{i-1})}\end{aligned} \tag{7.28}$$

in which A_i and B_i are given by the boundary conditions, that is the down welling radiance at the top of the layer and the up welling diffuse radiance at the bottom of the layer, and C_i, D_i, g_i and f_i are found by solving these equations to be:

$$\begin{aligned}C_i &= \frac{c_i(k_i - a_i) - c_i' b_i}{(k_i^2 - g_i^2)} \times E(-s,i-1,z_{i-1})\\ D_i &= \frac{c_i'(k_i + a_i) - c_i b_i}{(k_i^2 - g_i^2)} \times E(-s,i-1,z_{i-1})\\ g_i &= (a_i^2 - b_i^2)^{0.5}\\ f_i &= \left(\frac{a_i - b_i}{a_i + b_i}\right)^{0.5}\end{aligned} \tag{7.29}$$

where $F(-s, i-1, z_{i-1})$ is the direct downward radiance at the upper boundary of the ith layer. The boundary conditions at the outer boundaries are stated in Equation (7.27). These are the only boundary conditions that need to be considered if the vegetative canopy consists of just one layer. If, however, the vegetative canopy consists of two or more layers, then there are also boundary conditions at the interfaces between the layers. These boundary conditions are set so as to ensure that there is a continuous radiance flow across the boundary, involving the downward diffuse and direct fluxes at the top of the next layer and the upward diffuse flux at the bottom of the layer being considered.

Consider the simple canopies discussed in Chapter 2 illustrating some of the complexity of canopy reflectance. For these two canopies the parameters for the Suits model are given in Table 7.2.

By substitution of these parameters into Equation (7.28) we get the reflectance contributions due to the direct and diffuse fluxes and the soil as shown in Table 7.3.

TABLE 7.2
The Suits Model Parameters for the Two Simple Canopies Used in Chapter 2

Parameter	Planophile Canopy $H' = 1, V' = 0$	Erectophile Canopy $H' = 0, V' = 1$
a	$(1-\tau)$	$(1-(\rho+\tau)/2)$
b	ρ	$(\rho+\tau)/2$
c	ρ	$\beta\tan(\theta_s)(\rho+\tau)/2$
c'	τ	$\beta\tan(\theta_s)(\rho+\tau)/2$
k	1	$\beta\tan(\theta_s)$
K	1	$\beta\tan(\theta_o)$
u	τ	$\beta\tan(\theta_o)(\rho+\tau)/2$
v	ρ	$\beta\tan(\theta_o)(\rho+\tau)/2$
w'	ρ	$(\beta/4)\tan(\theta_s)\tan(\theta_o)[(\sin\psi+(\pi-\psi)\cos\psi)\rho+\sin\psi-\psi(\cos\psi)\tau]$

TABLE 7.3
The Contributions to Canopy Reflectance in Direction θ_o

Reflectance component	Planophile canopy	Erectophile canopy
ρ_i	$(1-e^{-2h}/2)$	Complex function of θ_o, θ_s and ψ
ρ_d	$(\tau E_+ + \rho E_-)e^{-h}/E_{solar}$	Complex function of θ_o, ρ and τ
ρ_{soil}	$\rho_s\{F_-(-h)+E_-(-h)\}e^{-h}/E_{solar}$	$\rho_s\{F_-(-h)+E_-(-h)\}e^{-h\times\tan\theta_s}/E_{solar}$

Table 7.2 and Table 7.3 show clearly that the planophile canopy will have negligible variations in canopy reflectance with changes in solar and sensor zenith angles, or with the relative azimuth angle, whereas these parameters will have a significant effect on the reflectance from an erectophile canopy. All canopies contain a mix of horizontal and vertical components; what varies is the proportion of the two between different types of canopies. This clearly suggests that multiple angle image data should be of use in improving classification accuracies, just as imagery taken at multiple solar elevations should also be of use in improving classification accuracy, or to invert this statement; consistencies in landcover signatures are dependent upon taking into account variations in the solar–sensor geometry.

The Suits model was the first model that gave canopy reflectance values that approximated that observed values for canopies that were relatively uniform and dense, as is the case with some agricultural crops and grasslands. The largest disagreements between the model and observations with this type of canopy occurred both very early and late in the season, when there was little green vegetative cover in the canopy and at the Hotspot. This is to be expected, as the assumptions of the model are not met under these conditions.

The limitations of the Suits model have led to the development of a large variety of other models, some of which follow the turbid medium approach, and some that follow the other approaches that are introduced below. In relation to the turbid medium approach, the SAIL model (Verhoef 1984) is one of the most used, extending the Suits model by easing some of its restrictions. The main limitation found to arise in the Suits model is the conversion of the leaf elements into horizontal and vertical components. This simplification introduced corrections that were found to be too harsh. As a result the SAIL and most subsequent Turbid Theory models replace this with a LAD model. However, more recent models have built on or extended the SAIL model and these should be considered when it is necessary to implement a canopy reflectance model.

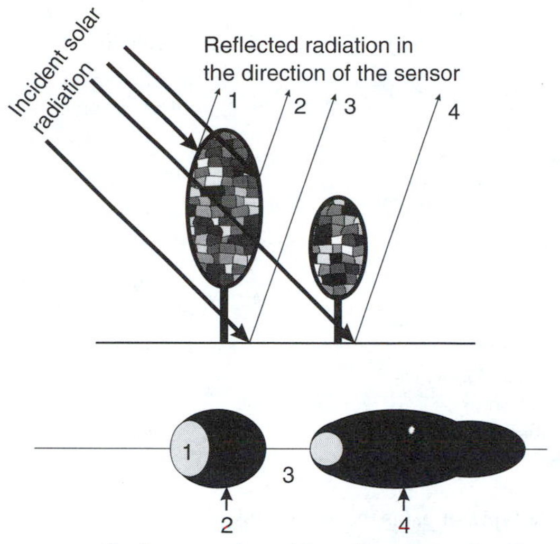

FIGURE 7.18 Geometric-optical models use a distribution of geometric shapes to represent the canopy.

7.4.6 GEOMETRIC-OPTIC (GO) MODELS

These models assume that the canopy consists of a distribution of plant objects, and that these objects have a shape that can be described geometrically (cones, spheres, cylinders), where the size and shape could vary within the canopy. Once the distribution of the geometric shapes has been set, then for a specific incidence and sensor zenith angles the areas of sunlit and shadowed canopy, sunlit and shadowed soil background can be computed. If the reflectance of each component is assumed, or measured, then the canopy reflectance can be calculated as a linear mixture sum of the four reflectance components, the proportions being the proportions of each of the four areas within the specified spatial domain being used for the calculation, such as the pixel area.

$$\rho_\lambda = p_1\rho_1 + p_2\rho_2 + p_3\rho_3 + p_4\rho_4$$

Considering Figure 7.18, where the canopy is considered to be oval shapes on stick trunks as geometric canopy objects. The objects can be of various sizes, and they can be distributed either randomly or clumped across the area to be covered. For the situation where there is no overlap between objects or their shadows, then a solution to determination of the four areas can be computed geometrically. One way to do this is to compute the canopy element geometric parameters of ellipse centre and radii $(x_c, y_c, z_c, r_x, r_y, r_z)$ and stick end points $(x_b, y_b, z_b, x_t, y_t, z_t)$ where z_b usually is set to the ground elevation or zero. For a particular solar zenith angle, compute the shadow stick and ellipse parameters. Some of the canopy shadow elements are on the canopy itself and some are on the ground surface. The projected shapes will always be ellipses where the source geometric shape is a circle or an ellipse. Use these with an orthogonal transformation to compute their projection in the direction of the sensor, and then overlay these with the sunlit elements of the canopy on top of the shadowed canopy elements and both on top of the shadowed projections onto the ground surface.

Where there is overlap then the solution becomes complex due to the mutual shadowing that occurs.

In most of the early models, the canopy shapes were considered to be opaque Lambertian reflectors. However, this has been modified in the hybrid models where the canopy elements are often considered to obey the turbid medium concepts and this theory is used to estimate the canopy element reflectance.

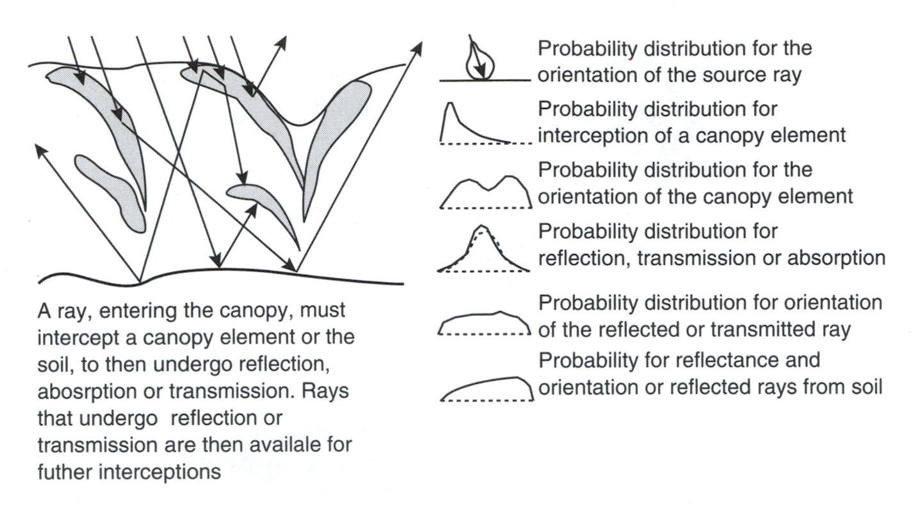

FIGURE 7.19 The types of paths that can be taken by a ray traversing through a canopy.

The approach is limited by its capacity to simulate the canopy by the closeness of the geometric shapes and their reflectance to reality. Where there is close approximation to reality then the models can provide good estimates of canopy reflectance, but not when the geometric shapes are a poor fit to the actual canopy geometry. The models can also be difficult to invert. The interested reader is directed to Li and Strahler (1985, 1986) for a detailed description of the implementation of their geometric-optical model. A good overview of geometrical-optical models is given in Chen et al. (2000).

7.4.7 Ray Tracing Models

The advent of significant computing power has created the opportunity to simulate the canopy in a more realistic way than is done with either the Radiative Transfer Theory or the geometric-optical models. There are two ways that such models have been implemented: ray tracing and radiosity models (Figure 7.19).

Ray tracing for a canopy model is similar to that used for leaf reflectance. The canopy is described either as a physical three-dimensional model of the distribution and orientation of canopy elements in space, or as a set of stochastic probability distributions in three-dimensional space. Whilst the former is favoured for leaf models, the latter is favoured for canopy models. A photon ray is then projected into the canopy, where there is a probability distribution of the chance of interception for each unit of travel through the canopy. Random numbers are used to determine whether an interception occurs in that distance element. When the ray intersects a canopy element then it is either absorbed, reflected or transmitted in accordance with stochastic probabilities, with the choice made by use of a selected random number. The absolute orientation of the canopy elements is also set by stochastic probability distributions, and random numbers allow determination of the orientation of the intercepted canopy element. Vector geometry is used with the information on the absolute orientations of the ray and the canopy element to determine the relative orientation of the ray to the element and from this a stochastic distribution is used to determine whether the ray is absorbed, reflected or transmitted. Another stochastic distribution is used to determine the relative orientation of the reflected or transmitted ray from the canopy element. Knowing the orientation of the canopy element, and the relative orientation of the emerging ray, allows the absolute orientation of the emerging ray to be calculated using vector geometry. Reflected and transmitted rays are then available for further interceptions with canopy elements, until such times as they are absorbed or they emerge from the canopy.

The fate of all rays are stored, building up a picture of the percentages that are absorbed and reflected, as well as being able to be used to determine other characteristics of the canopy such as the percentages of multiple reflections that actually occur in a canopy. From this accumulated information, a picture of the canopy BRDF can be created, once sufficient rays have been traced.

Ray tracing is computationally expensive, but they are flexible enough to be adapted to most types of canopy that may be met. However, they are not suitable for inversion.

A review of ray tracing models is given in Disney et al. (2000).

7.4.8 Radiosity or Computer Graphics Based Models

Radiosity models are based on an energy balance model similar to the radiative transfer model approach, but in which the model formulation is quite different. For radiosity models the model is formulated assuming that the canopy consists of a defined distribution of scattering and absorbing elements where subsets of this distribution illuminate each element in the canopy. A major challenge is to determine which subset of the canopy elements illuminates each canopy element. The formulation of the energy equation is of the form:

$$B_{\mathrm{i}} = E_{\mathrm{i}} + \chi_{\mathrm{i}} \sum_{j} F_{i,j} B_j \quad \text{for } j = 1, 2, 3, \ldots, 2N \text{ and } j \neq i \tag{7.30}$$

where

$$\begin{aligned} \chi_{\mathrm{i}} &= \rho_{\mathrm{i}} \quad \text{if } (\vec{n}_{\mathrm{i}} \cdot \vec{n}_j) < 0 \\ &= \tau_i \quad \text{if } (\vec{n}_{\mathrm{i}} \cdot \vec{n}_j) > 0 \end{aligned}$$

is the surface reflectance or transmittance depending on the relative orientation between the two canopy elements as determined by the dot product of their normal vectors, B_{i} is the radiant flux density (W/cm^2) or radiosity leaving a surface, E_{i} is the surface emission, usually set to zero at optical wavelengths but non-zero at thermal wavelengths, $F_{i,j}$ is the view factor specifying the fraction of radiant flux leaving surface j that illuminates surface I, and N the number of discrete two-sided surface elements to be considered.

The above description assumes that the canopy elements are Lambertian, although this constraint can be relaxed. The differences between the two approaches are illustrated in Figure 7.20.

The main challenges in the implementation of radiosity models is to compute the view factors, or the proportions of the subset of elements that illuminate each other element, and then in bringing all the equations together to solve by least squares where the number of equations are large and they are sparse in that there are many zero values in the derived least squares equations since the constants to be derived for one subset are different to those for another subset. A description of the development of radiosity models is contained in Borel et al. (1991).

7.4.9 Linear Semi-Empirical Approximations to the Physical Models

The difficulties involved in inverting the various physical models have led to a search for linear approximations to these models. The major advantage of linear models, of course, is that they are readily inverted and thus capable of being used to estimate canopy parameters from reflectance data derived from the satellite data. Linear approximations involve making assumptions that will incur acceptable errors under some conditions and not others. The models can thus only be used under a constrained range of conditions compared with the physical models. For example, the models may only work for specific landcovers, or for specific phenological stages. Clearly, such constraints need to be easily identified from an independent source for the model to have practical significance.

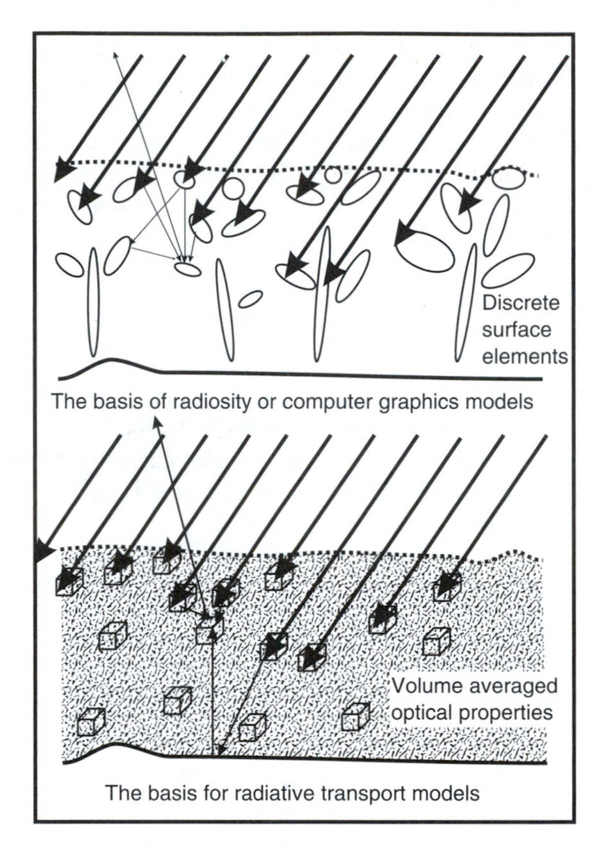

FIGURE 7.20 Comparison between the concepts that form the basis of the radiative transport and the radiosity models.

The model also needs to incorporate one or more canopy-related parameters so that the estimation of these parameters by the model enables other parameters that are of specific interest, to be estimated. Since each model will only work under a range of conditions, other models would need to be used under other conditions. The suite of models used are thus likely to be used as mosaic models, with each model being used in a part of the mosaic. The decision on which model to use has to be taken on the basis of independent information, such as landcover information. Thus, models based on different landcovers would need up to date landcover information as an input to the estimation process, so as to select the appropriate model at each pixel.

Most of these linear approximation models are formulated as a linear combination of volume and surface scattering kernels:

$$\rho(\theta_s, \theta_v, \phi, \lambda) = f_{iso}(\lambda) + f_{surf}(\lambda) \times K_{surf}(\theta_s, \theta_v, \phi) + f_{vol}(\lambda) \times K_{vol}(\theta_s, \theta_v, \phi) \tag{7.31}$$

where ρ is the estimated reflectance at illumination (θ_s) and view (θ_v) zenith angles, with a relative azimuth between them of ϕ and at wavelength, λ, K_{surf}, K_{vol} are formulae for surface and volume scattering, f_{surf}, f_{vol} are the weights assigned to the two types of kernels in the model and f_{iso} is the isometric contribution to reflectance.

By convention, most kernel models are expressed in such a way that $K_{surf} = K_{vol} = 0.0$ when $\theta_s = \theta_v = 0$, that is, when the Sun and the observer are at the zenith. Under this convention, $\rho(0, 0, 0, \lambda) = f_{iso}(\lambda)$. The kernel models are usually functions of the viewing geometry, and the goal is to estimate the values, f_{iso}, f_{surf} and f_{vol}.

Most current volume kernel models are derived from radiative transfer theory. There are two commonly used volume kernels, the $\text{Ross}_{\text{thick}}$ and the $\text{Ross}_{\text{thin}}$ kernels, where the first is designed for canopies with high levels of GLAI and the second for canopies with low GLAI values. The $\text{Ross}_{\text{thick}}$ kernel (Roujean et al. 1992) is of the form:

$$K_{\text{ross, thick}} = \frac{(\pi/2 - \xi)\cos\xi + \sin\xi}{\cos\theta_s + \cos\theta_v} - \frac{\pi}{4} \tag{7.32}$$

where ξ is the phase angle of scattering such that $\cos\xi = \cos\theta_s\cos\theta_v + \sin\theta_s\sin\theta_v\cos\phi$ and ϕ is the relative azimuth between the solar and sensor directions at the surface.

In the derivation of this formula, the LAD is assumed to be uniform in all directions. The $\text{Ross}_{\text{thin}}$ kernel (Wanner et al., 1995) is an approximation for low GLAI values:

$$K_{\text{Ross, thin}} = \frac{((\pi/2) - \xi)\cos\xi + \sin\xi}{\cos\theta_s\cos\theta_v} - \frac{\pi}{2} \tag{7.33}$$

The surface scattering kernels are usually based on geometrical-optical physical models. There are three such kernels that are commonly used. The Roujean kernel (Roujean et al. 1992) assumes scattering from a random distribution of three-dimensional bricks with isotropic scattering surfaces. Shadows are completely black and there is no mutual shadowing of objects:

$$\begin{aligned} K_{\text{brick}} = {} & \frac{1}{2\pi}\left[(\pi - \phi)\cos\phi + \sin\phi\right]\tan\theta_s\tan\theta_v \\ & - \frac{1}{\pi}[\tan\theta_s + \tan\theta_v + (\tan^2\theta_s + \tan^2\theta_v - 2\tan\theta_s\tan\theta_v\cos\phi)^{1/2}] \end{aligned} \tag{7.34}$$

The $\text{Li}_{\text{sparse}}$ and Li_{dense} kernels (Li and Strahler 1992; Strahler et al. 1994; Wanner et al. 1995) assume that the surface is covered with a random distribution of spheroidal crowns. The $\text{Li}_{\text{sparse}}$ kernel assumes no mutual shadowing between the canopy crowns:

$$K_{\text{li, sparse}} = O(\theta_s, \theta_v, \phi) - \sec\theta_s' - \sec\theta_v' + \tfrac{1}{2}(1 + \cos\xi')\sec\theta_v' \tag{7.35}$$

where

$$\theta_s' = \tan^{-1}\left(\frac{b}{r}\tan\theta_s\right) \quad \text{and} \quad \theta_v' = \tan^{-1}\left(\frac{b}{r}\tan\theta_v\right)$$

$$\cos\xi' = \cos\theta_s'\cos\theta_v' + \sin\theta_s'\sin\theta_v'\cos\phi$$

$$D' = (\tan^2\theta_s' + \tan^2\theta_v' - 2\tan\theta_s'\tan\theta_v'\cos\phi)^{1/2}$$

$$\cos t = \min\left\{1, \frac{h}{b}\frac{(D'^2 + (\tan\theta_s'\tan\theta_v'\sin\phi)^2)^{1/2}}{\sec\theta_s' + \sec\theta_v'}\right\}$$

$$O = \frac{1}{\pi}(t - \sin t\cos t)(\sec\theta_s' + \sec\theta_v')$$

The Li_{dense} kernel can accommodate mutual shadowing, and it assumes a random distribution of crown heights in the canopy. Using the same terminology as for the $\text{Li}_{\text{sparse}}$ kernel the Li_{dense} kernel is given by:

$$K_{\text{Li, dense}} = \frac{(1 + \cos\xi')\sec\theta_v'}{\sec\theta_s' + \sec\theta_v' - O} - 2 \tag{7.36}$$

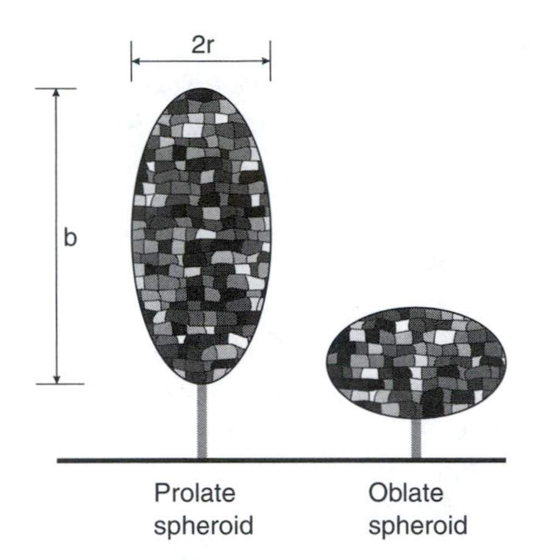

FIGURE 7.21 Ellipsoidal shapes and their dimensions.

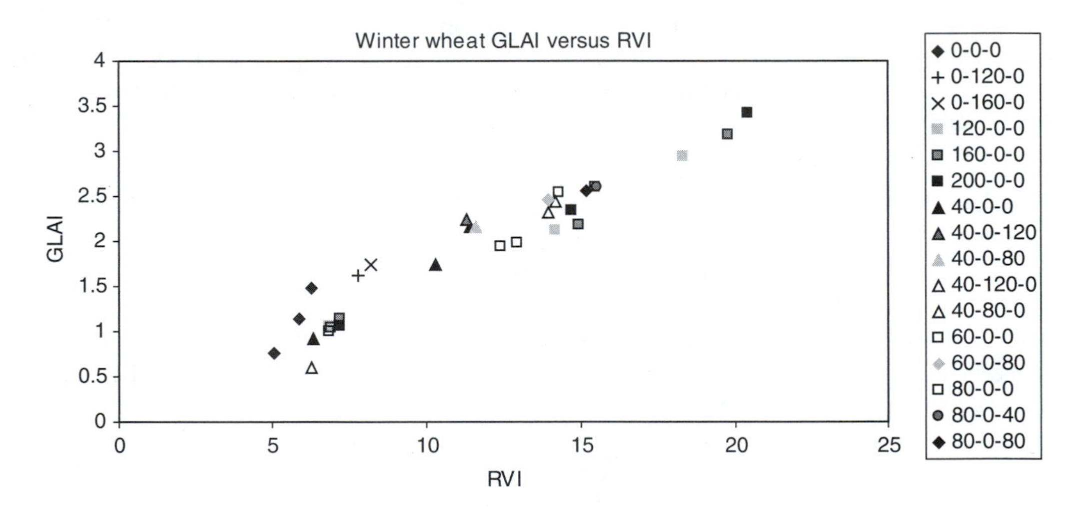

FIGURE 7.22 RVI versus GLAI plot for a winter wheat trial in Denmark, using field-based reflectance data for the derivation of the RVI values, for a variety of nutrient application trials, and for the crop prior to flowering. All the trials were on the same soil colour, and for the same variety of winter wheat.

The Li_{sparse} and Li_{dense} models contain a parameter (b/r) describing crown shape. Typical values that are adopted for this parameter are given in Table 7.4 using the figures described in Figure 7.21.

By using one surface and one volume kernels, it can be seen that there are three unknowns to be solved in Equation (7.31) at each wavelength. This set of linear equations can thus be solved with imagery containing simultaneously acquired data at three or more look angles. It would be possible to use more than two kernels, but of course, data at more look angles will be required to solve the equations. The derived equations are not always well conditioned, leading to errors in the solution due to the solutions coming to local minima. For this reason standard least squares methods are not used but rather advanced methods of solution are adopted. The interested reader is referred to Goodall (1993), Gill et al. (1991) and Pokrovsky and Roujean (2002).

At present there is only a limited understanding of the physical significance of the terms derived from the solution of kernel models. Initially the f_{surf} parameter was interpreted as being the surface reflectance due to the soil background and the f_{vol} parameter as being due to the vegetative layer. However, there are other possible interpretations, including representation of thin and thick media. Further research should clarify the physical significance of these derived parameters.

TABLE 7.4
Typical Parameter Values used for the Li Kernels

	b/r
Prolate spheroids	2.5
Oblate spheroids	0.75

At present there are two main avenues for application of the kernel based models:

1. Use the derived parameters to estimate other specific physical attributes. Thus, there is extensive work on using these models to estimate albedo and canopy structure. At present, a variety of the MODIS products are produced through the use of kernel models.
2. To remove angular effects from other variables, such as vegetation indices.

7.5 ESTIMATION OF VEGETATION PARAMETERS AND STATUS

7.5.1 INTRODUCTION

Historically, the first major effort to derive methods of estimation of canopy physical parameters from image data were empirically based methods. Most of these methods simplified the image data by converting it into a vegetation index and then building a regression relationship between the vegetation index and the canopy parameter of interest. An example of this approach is given in Section 7.5.2 below. This focus was based to a significant degree on the desire to be able to derive estimates of canopy status without the use of field data. However, limitations with the approach soon became obvious. First, the early vegetation indices were very susceptible to the soil colour, and so differences in estimates would occur on soils of different colours, particularly at lower densities of plant matter, when significant soil reflectance occurs. This led to the effort to develop indices that are not affected by soil colour. A number of indices were developed that were less sensitive to soil colour. However, the concepts behind the vegetation index approach are that the difference in reflectance between the red and the NIR is characteristic for vegetation and that the transmissivity of the NIR but not the red at higher plant canopy densities provides the capacity to continue to derive estimates of canopy variables once there is negligible red reflectance from the soil. Since soil reflectance will occur in the red and NIR at lower plant canopy densities, and NIR reflectance will occur at higher densities, it follows that soil reflectance will contribute in some way to the vegetation index. Full elimination of the effects of soil reflectance are thus going to be difficult to achieve if the index is to provide estimates across a broad range of canopy densities. A second major limitation of the approach was soon found to be that different relationships exist for different species, and indeed for different varieties within the same species. This relationship was also found to be dependent on the phenological stage of the crop or pasture since brown material in the canopy significantly degraded the ability of the vegetation indices to estimate canopy parameters and different canopy geometries and reflectances of canopy elements also interfered with the capacity of the indices. Thus, the data shown in Figure 7.22 is for one variety of Winter Wheat grown in Denmark for a number of nutrient trials, but all on the one soil type and at the one slope and gradient regime, prior to the formation and emergence of the flower heads in the crop. The emergence of the flower heads significantly changes the reflectance characteristics of many agricultural crops, and thus changes the vegetation index values that come from the crop. The last major problem with the approach was found to be that the image data needed to be accurately corrected for atmospheric effects, and converted to reflectance at the surface, since all the regression models that were developed using field data were based on reflectance at the canopy.

Whilst there were many reasons for the development of physical models of canopy reflectance, the limitations of the vegetation index approach was amongst them. A number of canopy reflectance models were developed, with the hope that they could be inverted so as to estimate canopy variables from the image data. It has been shown that this can be achieved with a number of the canopy reflectance models, but usually by making a number of simplifying assumptions that mean that the derived estimates are reasonable for only very selective conditions. Both the Suits and Sail models can be inverted, as can some of the geometrical-optical models. The more complex ray tracing and radiosity models cannot be inverted.

So this situation has led to the current effort to use the more complex physically based models to find approximations that work, under a set range of conditions. Most of the focus for this approach is in the development of kernel models as discussed in Section 7.4.9. Some of these approximations have been implemented for the routine conversion of image data into global and continental scale estimates of various physical parameters of the canopy, most notably FPAR and GLAI. That this work is only being done for these parameters that are highly correlated with variations in the red and NIR reflectances shows how difficult it will be to derive estimates of other parameters that do not enjoy such a high level of correlation.

One alternative to this approach is to reject the notion that image data can be converted into reliable estimates of physical attributes without field data. Instead of seeing images as being a replacement for field data, they could be seen as extensions of field data. It is this concept that is behind the approach of interpolation that is described below.

There are four main ways to derive estimates of canopy parameters from image data:

1. *Regression models*. The most common method of construction of regression models is by the use of field data on both the canopy reflectance and the canopy physical attributes of interest so as to derive a regression relationship between canopy reflectance and physical attributes of the canopy. If the level of correlation between the reflectance and the canopy-based parameters is satisfactory then the derived regression models can be used with image-based parameters to estimate the canopy variables, within the range of conditions that applied at the time of development of the model.
2. *Empirical interpolation*. Either image based parameters or estimates of canopy physical variables derived from regression or the inversion of canopy reflectance models can be used with a network of field data to interpolate the parameters of interest between field data values, using co-Kriging.
3. *Inverse modelling*. Canopy reflectance models are inverted so as to use them with image data to estimate canopy physical variables.
4. *Approximations of the physical models*. Derive approximations of the physical models that can be inverted and use these to derive the parameters for which the models are designed, within the range of conditions for which the model was developed.

7.5.2 Regression Models

Regression models are usually of the form:

$$\text{Estimate of physical parameter} = f(\text{reflectance-based parameters})$$

The shape of such models depends on the relationship between the reflectance and the canopy-based physical parameters. The choice of a suitable model usually comes from an analysis of a plot of the reflectance-based data against the parameters to be estimated from the reflectance-based data, typically using field collected data for this analysis, and the subsequent construction and calibration of the model. Once the form of the model has been determined, the model parameters have to be established, and this is usually done by means of least squares. Least squares finds model parameters so that the derived model has a sum of the squares of the residuals after the fitting that has the smallest possible value. For fitting of a model in this way, it is also important to test the statistical significance of the derived model parameters. If the model parameters are not statistically significant, then the model cannot be reliably used.

For example, consider the data shown in Figure 7.22. The figure shows the distribution of GLAI for the canopy of a species and observations of RVI. The data shows that there is an approximately linear relationship between the parameters. It was therefore decided to start with a simple linear regression model and see if it gave a good fit to the data. The linear regression yielded the model:

$$\text{GLAI} = 0.1622 \times \text{RVI} + 0.0596 \tag{7.37}$$

with an R^2 value of 0.925. To test the significance of the regression, we use the F-test. To do this test the hypothesis:

Null hypothesis $H_o{:}a_1 = 0$

Against the alternate hypotheses $H_1{:}a_1 \neq 0$

The null hypothesis given here states that the gain (a_1) is not significant. If this hypothesis is accepted as a result of this test, then it means that the regression is not significant. It will be recalled from Chapter 4 that MSE and MSR are two independent estimates of the variance of the residuals from the regression if H_0 is true, where MSR = SSR/(Number of independent variables) and MSE = SSE/(Number of observations – number of unknowns). For this data, MSE = 0.043 and MSR = 15.32.

If we let F = MSR/MSE, then F should be approximately one when the null hypothesis is correct since the two independent estimates of the variance should be similar in value. When F is large, then we can no longer accept that both MSR and MSE are independent estimates of the variance, and thus we need to reject the null hypothesis, and accept the alternate hypothesis.

For the example given here for simple linear regression, the numerator degrees of freedom = 1 and the denominator degrees of freedom = (31-2). From the F-statistic tables we get:

Probability	F-statistic
0.05	4.18
0.025	5.59
0.01	7.60

All of them are much lower than the computed F value of 357.7, but as the probability is decreased, so the F-value increases. As a consequence, the null hypothesis, H_0 cannot be accepted and we can confidently accept that the derived regression line is statistically significant.

Regression models of this type are dependent on the species and variety of the canopy plants, and may even be dependent on the management used. They thus attract a significant overhead in the need to develop these models under the conditions that are expected to be met in the mapping. In addition, to this, the derivation of the model is usually done using field data, so that the image data will need to be calibrated and corrected to reflectance at the surface so as to apply such models.

If the data, such as that shown in Figure 7.22, is not a good fit to a linear model, then the options are to either fit a non-linear model through the data, or to first transform the data so as to get it to fit to a linear model. For example, it can often happen that either or both axes can be converted to a logarithmic scale and the resulting data are distributed in a linear manner.

7.5.3 Empirical Interpolation

In interpolation, image-based parameters are used with field data to interpolate between the field data points to produce a map of the field parameters, by the use of co-Kriging. Imagery acquired for the purpose is used to select the field data points, so as to ensure good coverage and to cover the range of conditions that are likely to be met. The image-based parameters can then be used in co-Kriging (Chapter 6) to produce the map of the physical attributes.

Since co-Kriging assumes that there is a linear relationship between the image-based parameters used to influence the co-Kriging and the physical parameters to be mapped, non-linearities will introduce errors. One way to reduce these errors is to convert the image-based parameters into an estimate of the physical parameters by the use of a regression or by inversion of a canopy reflectance model, and then use the output of the model as the input to the co-Kriging.

The higher the level of correlation between the image-based and the canopy-based parameters, the lower the density of field data that are required for the co-Kriging. As the level of correlation decreases, this can be partially compensated by the use of a denser network of field data. In addition, since the method is anchored in the use of field data, small errors in the atmospheric correction and calibration of the image data are not likely to introduce significant errors in the derived estimates. For this reason, the method is more robust than the regression model approach, when that method is used on its own. It has the other advantage that it can still be used to estimate parameters when there is a lower level of correlation between the parameters, but at the cost of incorporating more field data. Regression models cannot be used in this way.

7.5.4 Inversion of Canopy Reflectance Models

In Section 7.4 we discussed the various forms of canopy reflectance (CR) model that have been developed, and we have seen that some of these CR models can be inverted. However, the limited amount of input spectral data that is available means that only very simple versions of these models can be inverted, and experience has shown that this usually means that the models can give reasonable estimates of canopy parameters under only a limited range of conditions. The source reflectance data can occupy three different dimensions, as well as taking the spatial dimension into account:

1. *The spectral/radiometric dimension*. Unlike some minerals that exhibit narrow absorption features in image data, vegetation exhibits two broad absorption features in the visible part of the spectrum, and a number of narrower absorption features in the NIR due to water absorption. The only distinct feature due to vegetation is the sharp red edge between the red and the NIR wavebands. As a consequence, narrow wavebands of data of vegetated surfaces are highly correlated with adjacent wavebands in the visible and NIR; only at the red edge can adjacent narrow bands be sufficiently distinct as to provide distinct information. As a consequence, the spectrally available wavebands are likely to be relatively broad bands in the blue, green, red and NIR parts of the spectrum, with a set of three or four narrow wavebands straddling the red edge. In the microwave region, shorter wavelength radar signals tend to be reflected from the top of the canopy, whilst the longer wavelength signals tend to be reflected either from within the canopy or from the soil background. As a consequence, there is independent information on the canopy in the different radar wavebands. However, it has been shown that the variance associated with a species or community is usually such that it significantly overlaps with the distribution of adjacent species or communities, making reliable discrimination difficult. This problem can be reduced by the use of imagery of the right resolution, as discussed below.

2. *The angular dimension*. It has been seen that vegetation reflectance is not Lambertian, but exhibits distinct patterns that differ between the species and communities, as well as differing with incidence angle and with the angle to the sensor, giving vegetation its BRDF characteristics. Multiple angular image data could be used for the estimation of canopy parameters, as long as the BRDF variations are taken into account either prior to or in the inversion. The maximum variability in

BRDF is in the plane containing the orientation of the Sun, so that, all other things being equal, the minimum BRDF effects are seen at right angles to this orientation, so that this is the optimum orientation to use for the analysis of multi-angle image data for model inversion.

3. *The spatial dimension.* It has also been shown that the local variance in an image is affected by the object interval relative to the image resolution, with the local variance increasing as the image resolution approaches half the size of objects in the terrain, and then decreasing, as they increase further. This variability is not due to the objects themselves, but rather their relationship to their surroundings, and so this variance should be minimised in the analysis of image data. Selection of the appropriate pixel size will thus help reduce the average local variance arising from this cause, so that a greater proportion of the variability in the data will be due to variations in conditions that may have some relationship to the factors being mapped. Selection of the right pixel size should increase the accuracy of the estimates.

4. *Polarisation dimension.* In radar imagery, volume scattering, as occurs within a canopy, leads to a strong response in cross-polarised image data, whilst single scattering leads to higher response in like polarised image data. As a consequence, the different polarisations of radar data potentially carry significant information on the surface being sensed.

We saw in Section 7.4 that the forward problem can be placed in the form:

$$\rho(\lambda, \theta_s, \phi_s) = F(a_i, v_i, b_i \mid s_i, t_i) \tag{7.2}$$

And that the inversion problem can be placed in the form:

$$\bar{v}_i = F'(s_i, a_i, \hat{v}_i, b_i, t_i, \rho) \tag{7.3}$$

In forming an inversion, there are at least four issues that have to be addressed:

1. *The number of parameters versus the number of observations.* We have seen that the solution of a set of equations requires that the number of equations, m is the same as or greater than the number of unknowns, n. We have also seen that the number of image-based variables that may be used is strictly limited, thus limiting the number of physical variables, or combinations of variables, that can be estimated. This means that models that use only a few canopy parameters are the most likely to be able to be inverted successfully.

2. *The mathematical invertibility of the model.* To invert a numerical model it is essential that the inversion be well posed. A well-posed inversion is one that has a solution; the solution is unique and depends continuously on the data. The first condition is obvious. Some non-linear inversions can have a number of local minima in the criteria used to derive a solution. Thus with least squares, there may be a number of solutions each with their local minima in the sums of squares. If the initial values are in the catchment of a local minimum, the solution will converge to that local minimum even though that local minimum may not be as low as another local minimum. Under these conditions the global minima will not be found. The third condition requires continuous data between the possible solutions; continuous functions are one aspect of this and errors in the data or the model are another aspect that needs to be dealt with. Proving that a model can be inverted is very difficult. One way suggested by Goel and Stebel (1983) is to use the forward model with a set of data to determine canopy reflectances. Once this has been done, then use these reflectances in the inverse model, to see if the source canopy physical attributes are accurately estimated. If they are, then this test should be conducted for a range of physical attributes, and if they give reliable results, then it can probably be assumed that the model can be inverted.

3. *Local versus global minimisation.* Many non-linear models contain a number of local minima in their sums of squares, where one of these local minima is the correct value and the others represent errors as discussed above. This problem arises when the starting values are in the domain of the wrong local minima, sending the solution to that local minimum, rather than to the correct local minimum.

4. *Stability of the solution*. Many models are not stable, which means that small differences, or errors, can introduce large responses. If a model is not stable then small differences in the input values will yield large differences in the output solutions. When this occurs, then the model is not suitable as an inversion model.

The simplest way to invert a model of the form $y = b_0 + b_1x$ is to convert it to the form $x = (y - b_0)/b_1$. However, such a form of inversion can only be practised with the simplest of models. Thus, for most model inversions an alternate approach must be taken.

7.5.4.1 Linearisation for Least Squares

The simplest alternative approach is to linearise the model and then use least squares to find the parameters of this linear model. This is done by assuming that the model parameters have starting values that are close to the actual values. The least squares adjustment finds adjustments to these starting values to give new values that are closer to the correct values. These corrections are then used to update the starting values in a new iteration, which continues until the sums of squares resulting from the least squares adjustment tends to oscillate around so value. Consider a non-linear model of the form:

$$y = a_0 + a_1 \sin(a_2x) \tag{7.38}$$

To linearise this model assume that the starting values for a_0, a_1 and a_2 have been set, so that the model gives an estimated value for the dependent variable, y, given input values for the independent variable, x. Now the correct or observed values of y are given by:

$$y + \delta y = y_{\text{observed}} = (a_0 + \delta a_0) + (a_1 + \delta a_1) \sin\{(a_2 + \delta a_2)x\} \tag{7.39}$$

Expand this equation to give:

$$y + \delta y = a_0 + \delta a_0 + (a_1 + \delta a_1)\{\sin a_2x \cos \delta a_2x + \cos a_2x \sin \delta a_2x\} \tag{7.40}$$

In this equation there are known starting values for a_0, a_1 and a_2 and the unknown corrections $\delta a_0, \delta a_1$ and δa_2 are small so that $\cos \delta a_2x \approx 1$ and $\sin \delta a_2x \approx \delta a_2x$ to give:

$$y + \delta y = a_0 + \delta a_0 + a_1 \sin a_2x + a_1x \cos a_2x \times \delta a_2 + \sin a_2x\delta a_1 + x \cos a_2x \times \delta a_1\delta a_2$$

Since the changes are assumed to be small, their product is assumed to be negligible, that is, $\delta a_1\delta a_2 = 0$. This gives the final linear form of the model:

$$y + \delta y = a_0 + a_1 \sin a_2x + \delta a_0 + a_1x \cos a_2x \times \delta a_2 + \sin a_2x \times \delta a_1 \tag{7.41}$$

Now remove Equation (7.38) from Equation (7.41) to give:

$$\delta y = \delta a_0 + a_1x \cos a_2x \times \delta a_2 + \sin a_2x \times \delta a_1 \tag{7.42}$$

which is the linear version of Equation (7.38). In this equation, $y_{\text{observed}} = y_{\text{computed}} + \delta y$ so that: $\delta y = y_{\text{observed}} - y_{\text{computed}} = y_{\text{observed}} - a_0 - a_1 \sin a_2x$, which can be substituted into Equation (7.41) to give the observation equations needed to solve for the model parameters.

$$\delta a_0 + a_1x \cos a_2x \times \delta a_2 + \sin a_2x \times \delta a_1 = y_{\text{observed}} - a_0 - a_1 \sin a_2x \tag{7.43}$$

In Equation (7.43) the delta terms are the unknowns. Equation (7.43) can be solved by means of least squares, adopting initial estimates of the canopy parameter values and the model variables, and iteratively adjusting these variables until the sums of squares derived by the least squares stops decreasing, but tends to oscillate about some value. It is possible, of course, to simply do this with image data of sufficient dimensionality to solve the equations. However, this will lead to a very computer-intensive solution. If possible, some of the model variables should also be found, possibly by the use of field data collected for the purpose, thereby reducing the number of unknowns to be found by the least squares solution.

In conducting such a least squares adjustment, it may be desired to weight the observations, that is to minimise the weighted sums of squares:

$$S = \sum w_i \varepsilon_i^2$$

To implement this, the individual observation equations are multiplied by their weights, during the derivation of the normal equations. Thus, a weight of two, for example, is equivalent to having two of that equation in the adjustment relative to an equation with a weight of 1. Weights can be used for a variety of purposes. Thus, if observations in some wavebands have smaller variances, then the variances may be used as an inverse of the weight, giving more emphasis to the wavebands with lower variances. Alternatively, the variables in the model may have different sensitivity levels, and so the variables may be weighted as a function of their sensitivity. The application of weights means that those observations with the higher weight will suffer smaller changes compared to the other variables.

In implementing such a solution, the major problems that are likely to arise are the number of equations to the number of unknowns and the rate of convergence of the adjustment. If the number of equations is the same as the number of unknowns, then in theory it should be possible to solve the equations, but in practice it may not be possible, because of the approximations inherent in the model and because of errors in the data. Therefore, it is essential that the number of equations exceeds the number of unknowns, and preferably be double the number of unknowns. The other problem is the rate of convergence. Some formulations converge very slowly or indeed may not converge at all. When this occurs, then this approach is not suitable and a more rigorous method of solution has to be found.

7.5.4.2 The Levenberg–Marquandt (L–M) Method of Solution

The Levenburg–Marquandt method of solution exploits the fact that least squares minimises the sums of squares and so the surface of the sums of squares function has a trough at its optimum value. The algorithm seeks to find this trough by seeking for the location of zero gradient by progressing down the steepest descent from the current position, in steps, towards zero gradient.

If our data is of the form:

$$y = f(t; a) \tag{7.44}$$

where a are the M parameters of the function. This equation thus has the merit function from least squares of:

$$\chi(a) = \sum_{i=1}^{N} \left[\frac{y_i - f(t_i; a)}{\sigma_i} \right]^2 \tag{7.45}$$

Containing N observations. This merit function represents the sums of squares that least squares will minimise. The gradient of the merit function will be zero when the merit function is a minimum, that is:

$(\partial\chi/\partial a_k) = 0$ is the condition that has to be met. Now:

$$\frac{\partial\chi}{\partial a_k} = -2\sum_{i=1}^{N} \frac{[y_i - f(t_i;a)]}{\sigma_i^2}\frac{\partial f(t;a)}{\partial a_k} \tag{7.46}$$

which is an M vector in the as. The second partial derivative is:

$$\frac{\partial^2\chi}{\partial a_k \partial a_l} = 2\sum_{i=1}^{N}\frac{1}{\sigma_i^2}\left[\frac{\partial f(t;a)}{\partial a_k} \times \frac{\partial f(t;a)}{\partial a_l} - \{y_i - f(t_i;a)\}\frac{\partial^2 y}{\partial a_k \partial a_l}\right] \tag{7.47}$$

which is an $M \times M$ array in the as. It is normal to ignore the second derivative as it gives very small to zero contributions that get lost in the iterative sequence. It has been found that the second derivatives can in fact introduce instabilities when there are large outliers in the data. As a consequence, it is usually used in the form:

$$\frac{\partial^2\chi}{\partial a_k \partial a_l} = 2\sum_{i=1}^{N}\frac{1}{\sigma_i^2}\left[\frac{\partial f(t;a)}{\partial a_k} \times \frac{\partial f(t;a)}{\partial a_l}\right] \tag{7.48}$$

It is conventional to remove the factor of 2 and make:

$$\beta_k = -0.5 \times \frac{\partial\chi}{\partial a_k} \quad \text{and} \quad \alpha_{kl} = 0.5 \times \frac{\partial^2\chi}{\partial a_k \partial a_l} \tag{7.49}$$

If $\chi(a)$ is differentiable and sufficiently smooth so that the Taylor's expansion is valid, then $\chi(a)$ can be approximated by a quadratic function when the parameters give a solution close to the minimum, that is:

$$\chi(a + \delta a) \approx \chi(a) - d \times \delta a^{\mathrm{T}} + 0.5\delta a^{\mathrm{T}} \times D \times \delta a^{\mathrm{T}} + \varepsilon \tag{7.50}$$

In which d is an M-vector of the partial derivatives of $f(t;a)$ in terms of the parameters a and D is the $M \times M$ Hessian matrix of second order partial derivatives, connected to Equation (7.49) by:

$$\alpha = D/2 \tag{7.51}$$

The α matrix, equal to half the Hessian matrix, is often called the curvature matrix. If the approximation is close, then we can find the best estimate in one step by using:

$$a_{\text{best}} = a_{\text{current}} + D^{-1} \times [-\nabla\chi(a_{\text{current}})] \tag{7.52}$$

In practice a_{current} will only be an approximation to a_{best}, and so the process has to start from initial values and proceed iteratively to better values:

$$a_{\text{next}} = a_{\text{current}} - k \times \nabla\chi(a_{\text{current}}) \tag{7.53}$$

where the constant, k is set so that it does not overshoot the trough. It may thus start at larger values and become smaller as the gradient decreases.

Differentiation of Equation (7.50) gives $(\partial\chi/\partial a) = -d + aD = 0$ so that we get $D \times a = d$ which can be solved to find a. This can be done by the use of the set of M linear equations:

$$\sum_{l=1}^{M} \alpha_{kl} \times \delta a_l = \beta_k \tag{7.54}$$

This set is solved for the increments of a_l that are added to the current values of the a_l to give the next values for a_l to be used in the next iteration. Equation (7.54) is the inverse Hessian method of finding the minima. The steepest descent method uses the formulae:

$$\delta a_l = \text{constant} \times \beta_l \tag{7.55}$$

The Levenberg–Marquandt method varies smoothly between the extremes of these two methods where Equation (7.54) works satisfactorily when there is poor correspondence and Equation (7.55) works best when there is close correspondence. The L–M method is implemented in the steps:

1. Compute $\chi(a)$.
2. Pick a modest value for λ say let $\lambda = 0.001$.
3. Solve Equation (7.54) for δa and evaluate $\chi(a + \delta a)$.
4. If $\chi(a + \delta a) \geq \chi(a)$ then increase λ by a factor of ten or larger and repeat step 3.
5. If $\chi(a + \delta a) < \chi(a)$ then decrease λ by a factor of ten or larger and repeat step 3.
6. Repeat until the change in $\chi(a)$ is less than one at which point the iterations should be stopped, but not after a step in which $\chi(a)$ increases.
7. Once this best solution has been found, then set $\lambda = 0$ and find the matrix $C = \alpha^{-1}$ as the estimated covariance matrix of the standard errors in the fitted parameters.

7.5.5 Estimation Based on Linear Approximations to the CR Models

It was shown in Section 7.4.9 that these models are of the form:

$$R = f_{\text{iso}} + f_{\text{surf}} \times K_{\text{surf}} + f_{\text{vol}} \times K_{\text{vol}} \tag{7.56}$$

This equation is made as a function of θ_s, θ_v, ϕ, GLAI, s and ρ_s. The values for θ_s, θ_v and ϕ are known or can be calculated and so these values can be used in the equations, leaving the equations with three variables, $f_{\text{iso}}, f_{\text{surf}}$ and f_{vol}. The functional relationship depends on the kernels used, so the analyst should choose between these kernels on the basis of the cover characteristics in relation to the kernel assumptions. Since different pixels are likely to contain different cover types, different kernels may need to be used for the analysis of different pixels, necessitating the use of suitable landcover data in the analysis.

7.6 CLASSIFICATION OF VEGETATION

It can be seen that the distribution of response from vegetation can be affected by a great variety of factors:

1. *Within community differences.* Differences in canopy structure and canopy element reflectance due to differences in plant or leaf age, the effects of different levels of spatially and temporally distributed stress on the vegetation, the influences of microclimate and soil variation, the effects of disease or predation and the variation between species within the community, where the different

species will occupy different ecological niches, with the attendant variation in reflectance that this implies, either over time or at the same time.

2. *Between community differences.* In addition to the variation between individual plants of the same species or community, there are differences between communities as a result of differences in canopy structure, leaf reflectance and phenology.

3. *Variation as a function of data acquisition.* The average local variance (ALV) Function (Chapter 4) shows that the local variance is a function of the pixel size relative to the object sizes in the scene. If the pixel size is very small, or very large, relative to the typical object size, then the local variance will be small. As the pixel size approaches the size of typical objects in a scene, the local variance increases. Thus, the variance within a community is a function of image resolution.

4. *Variation in solar and sensor azimuth and zenith angles.*

The goal of classification is to assign the pixels to one of a finite set of classes that have physical meaning to the end user. This definition involves consideration of not only how to maximise the discrimination between the classes, but also the possibility of changes occurring in the landcover or landuse classes during the period of image acquisition for the classification since this will lead to inconsistent temporal reflectance characteristics relative to either the starting or the ending landcovers and hence introducing errors in the classification. Most vegetation grows within a defined seasonal period. In general, the vegetation type remains constant within this period. Natural communities will also tend to remain the same between seasons whereas agricultural landcovers can change dramatically between seasons. With most vegetation this period is a regular seasonal period controlled by either radiation or available moisture, whilst with some communities the season is an opportunity event, usually controlled by rainfall, as is the case with much desert vegetation. However, the general rule still applies that within this season the community remains constant. As a consequence, imagery can be used within the whole of this period for the discrimination between communities. Indeed, imagery should be used within this whole period, as the differences in phenology that occur between communities may be more spectrally obvious at some stages during the season than at other stages. However, the season for one species or community may not match the season for other species or communities, and going outside of the season for either community may introduce errors in the classification due to changes in the community that occur at the end of the season.

Discrimination between the classes is maximised if the within class variance is small, but the between class mean differences are large. The between class mean differences are made as large as possible by the selection of optimum wavebands at the optimum times during the common seasonal period, as discussed above. The within class variances are made as small as possible by the selection of the most suitable resolution for the classification.

Currently, most classification uses the strategy of selection of a classifier algorithm, typically from those discussed in Chapter 4, trains the algorithm, either with the use of field data collected for the purpose, or from analysis of the spectral data itself, and then conducts the classification. If field data are used, then the final classes, which may in fact be groups of the classes derived during the classification, are related to the physical classes identified at the selection of sites in the field. If the classification is based on clustering of the image data, then the allocation of the derived statistical classes to physical classes is achieved by the use of field data, or the analyst makes the assignment based on experience. This approach is based on the assumption that the prior experience or the sample of field data collected for the purpose provides a valid basis for assignment of pixels at other locations to those classes, based on the spectral values in those pixels. We have seen that there are many factors that are affecting the response of a pixel, and so this assumption is a weak assumption. Further, since there is no way of checking the validity of the assumption, classification is a process of extrapolation. Extrapolation is dangerous because there are no bounds to the errors that can occur, there is usually no way of constraining the errors within specific limits and the process itself contains no self-checking mechanisms which can reveal errors that have occurred. It would be much safer if image classification, were conducted within the paradigm of interpolation rather than extrapolation. The main avenue for the development of interpolation-based classification methods is by the incorporation of knowledge or information on the physical environment into the classification process.

7.6.1 Incorporation of Environmental Knowledge into the Classification Process

Simple examples of where environmental knowledge can improve the classification process includes inclusion of elevation data when it is known that one or more communities occupy specific elevation ranges or slope conditions. More sophisticated knowledge can be incorporated through the use of decision rules incorporated into the classification process. Such an approach may take the main decision process out of the domain of simple image processing, and place it in the domain of the GIS environment where the decision process calls for the use of data from images as well as from the GIS environment.

An illustration of the approach is that reported by McCloy et al. (1989) for the monitoring of irrigated rice crops in South Eastern Australia. In temperate Australia rice is grown under irrigation in summer, in areas that have negligible summer rain, but long hot and dry periods that are suitable for the growth of the crop. The irrigation areas are located on the inland floodplains of rivers draining from the Great Dividing Range that runs down the eastern side of Australia. Temperatures range from day time peak average values of 35°C in summer down to day time average peak values of 10°C in winter. Other annual crops, particularly wheat and vegetables, are grown in the area under irrigation. There are extensive horticultural plantings in the area.

The monitoring procedure starts with an initial mask in which pixels are assigned either 0 (cannot be rice) or 1 (could grow rice). All horticulture areas, non-arable areas including urban development, and swamps and rivers, as well as areas outside of the study area, are assigned 0. Areas of permanent water within the irrigation area, such as dams, are assigned 0 in this mask.

The procedure involves two steps for each image acquired throughout the growing season:

1. The first step is classification of the individual image. This classification assigns each pixel to either be within or without a flooded crop response domain, depending upon the pixel response characteristics. All of the flood-irrigated crops have a response distribution that is a linear combination of water and green canopy response, with the combination depending on the stage of the crop. The mixed pixel algorithm is used to discriminate pixels that occupy this domain. Pixels that are classified as being within this domain are assigned a value of 1 in the classification, and all other areas are assigned a value of 0. Essentially, the image classification is discriminating water from other landcovers, particularly early in the season. Since water is spectrally very distinct from all other landcovers, this stage achieves high levels of accuracy.

2. The second step is to discriminate rice from the other flood irrigated crops. Only rice stays permanently within this domain of water to green canopy for the whole of the growing season. The other irrigated crops move in and out as they are flood irrigated. So the second step was to assign pixels to rice only if they stayed within the water to green canopy domain on two sequential images during the growing season, where typically five or six images were acquired during this period. Those areas that were assigned a value 1 in the image classification have their mask values incremented 1 higher in the mask, achieving values of 2 or more. Those areas that are assigned 0 in the image classification, but have values higher than 1 in the mask are re-assigned 1 in the mask. They essentially go back to the starting position. Areas with values of 4 or higher are considered to be rice, that is, they are areas that were classified as being in the water to canopy response domain on three or more sequential images in the image series used in the analysis. The process is illustrated in Figure 7.23.

This procedure clearly assumes that non-rice areas of flood irrigation are very unlikely to be in the water to canopy response domain in a sequential pair of images during the establishment phase. Rice

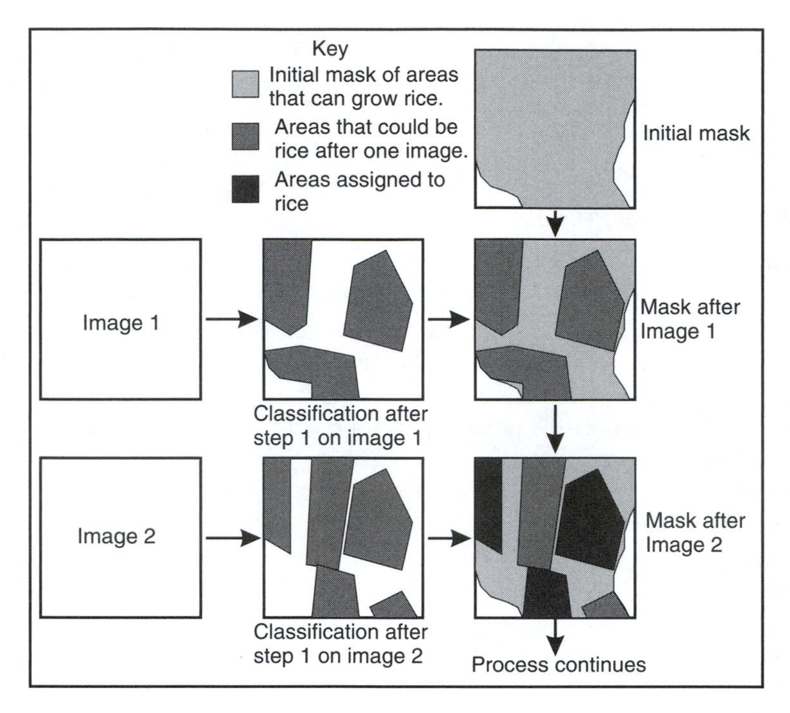

FIGURE 7.23 The steps in the classification of rice by incorporation of agronomic knowledge into the classification process.

will only diverge from this domain when physiological changes in the plants start to effect canopy reflectance.

This set of rules sets a very limited demand on the image data and used agronomic criteria to discriminate the areas of interest from other similar areas. Such rules suited this particular situation, but clearly would not suit other situations. However, other sets of rules may be applicable to other situations, and by their use the demands on the image data may be made less onerous and hence more reliable and accurate. By setting such rules for the classification, you are also setting constraints, and so the process is moving away from the unchecked characteristics of extrapolation and towards the constrained characteristics of interpolation.

7.6.2 Incorporation of Environmental Data into the Classification Process

Another way to move away from extrapolation is to incorporate environmental data into the classification process. One source of such information on agricultural areas is farmer data. In Denmark farmers provide information on crop areas each year, providing information on the Blocks in which the fields are located. The Blocks are physical units in the landscape with semi-permanent boundaries, such as creeks, rivers, roads and hedges, unlike fields that can change from year to year. As a consequence, there is information on the crops grown to Block resolution, but not to the fields within the Block. Such information will contain some errors, since farmer decisions are sometimes changed, and sometimes mistakes are made (Figure 7.24).

Within each Block there are a number of fields; by image segmentation the locations of the fields are known within the block, but the crop type on each parcel is not known. Two criteria can be used to allocate a crop type: coming from the farmer data, to the segments in the Block; the areas and the spectral signatures.

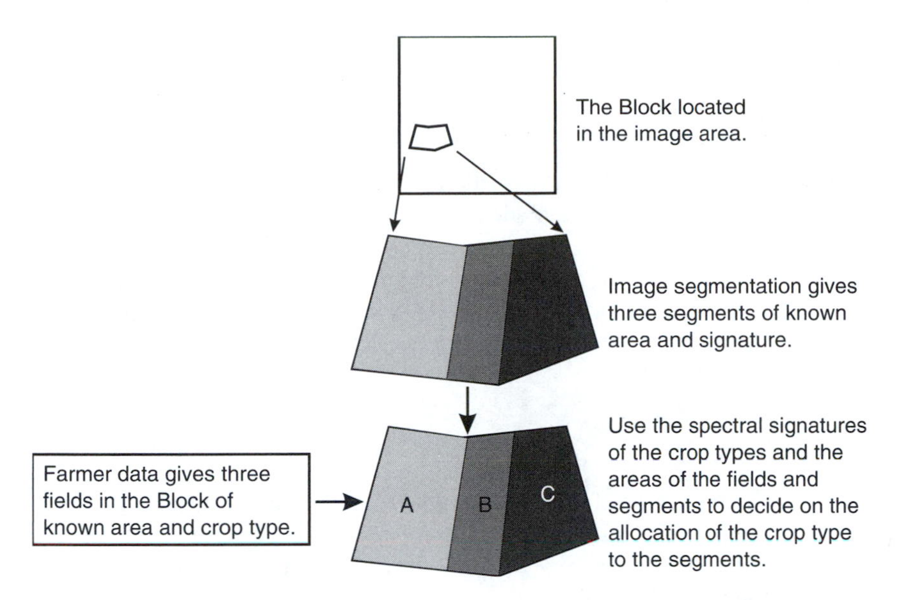

FIGURE 7.24 Incorporation of farmer data into the classification process.

The process moves away from extrapolation because the farmer data tells us the number of fields in the Block, the area of the fields and their crop types. The image segmentation tells us the number of segments that have been found, their areas and signatures. The assignment of cover types to segments is then a process of using the areas of the fields and the segments as one basis for a decision, and using the signatures of the segments in comparison with the signatures for the crop types. The signatures of the major crop types can be derived either from a set of training areas for each crop type, or by the use of a set of Blocks for which there is only one field within the Block, and having sufficient of these to provide signatures for each crop type. Those Blocks that contain confusions can be left until last as the classification of the Blocks can be used to improve the signatures for each crop type.

7.7 THE ANALYSIS OF VEGETATION PHENOLOGY

Phenology is defined as the timing and rate of change in plants and animals as a result of the seasonal forces acting on them. The major seasonal forces operating on plants include radiation, temperature and rainfall. The effects of these major forces are often further modified by other factors that usually have a more localised impact, including slope and aspect, soil conditions and altitude. The variations in the major forces is a major factor for the evolution of plant communities as we know them, and the variations that we see in the phenological characteristics of the different types of communities. The major phenological stages of wheat are typical of many annual grasses and are indicative of the major phenological stages in most vegetation species:

1. Germination
2. Seedling
3. Tillering
4. Stem elongation or jointing
5. Booting
6. Heading
7. Flowering or anthesis

8. Milk
9. Dough
10. Ripening

Several systems have been developed to provide numerical designations for growth and developmental stages. Among these, the Feekes, Zadoks, and Haun scales are used the most frequently. The Zadoks scale (Table 7.5) provides the most complete description of wheat plant growth stages. It uses code based on ten major stages that can be sub-divided, making it particularly suited for computerisation. When using the Zadoks scale, the main growth stages, for example, seedling versus tillering, should be identified before proceeding to a description of the secondary stages, for example, seedling leaf number or tiller number. The major stages are:

7.7.1 Germination

Germination starts with the uptake of water (imbibition) by a wheat kernel that has lost its post-harvest dormancy. Plant development is resumed once the embryo is fully imbibed. With the resumption of growth, the radicle and coleoptile emerge from the seed. The first three seminal roots are produced and then the coleoptile elongates pushing the growing point toward the soil surface.

7.7.2 Seedling Stage

The seedling stage begins with the appearance of the first leaf and ends with the emergence of the first tiller (Figure 7.25 and Figure 7.26). Up to six seminal roots and three leaves support the plant at this stage. The crown of the plant usually becomes noticeably distinct after the third leaf has emerged.

7.7.3 Tillering

The roots, leaves, tillers and spikelets on the head of the wheat plant develop from primodia at nodes. While the first tiller is not produced until the third leaf has fully emerged, the appearance of later tillers is usually synchronised with the emergence of each subsequent new leaf that develops on the main shoot.

7.7.4 Stem Elongation

The nodes from which leaves develop are telescoped at the crown during the tillering stage. Once jointing starts, the internode region elongates, moving the nodes and the growing point upward from the crown to produce a long stiff stem that will carry the head. Appearance of the first node (Zadoks stage 31) can usually be detected without dissecting the plant by pressing the base of the main (largest) stem between your fingers.

The stem elongation or jointing stage comes to an end with the appearance of the last (flag) leaf.

7.7.5 Booting Stage

The developing head within the sheath of the flag leaf becomes visibly enlarged during the booting stage. The booting stage ends when the first awns emerge from the flag leaf sheath and the head starts to force the sheath open.

7.7.6 Heading Stage

The heading stage extends from the time of emergence of the tip of the head from the flag leaf sheath to when the head has completely emerged but has not yet started to flower.

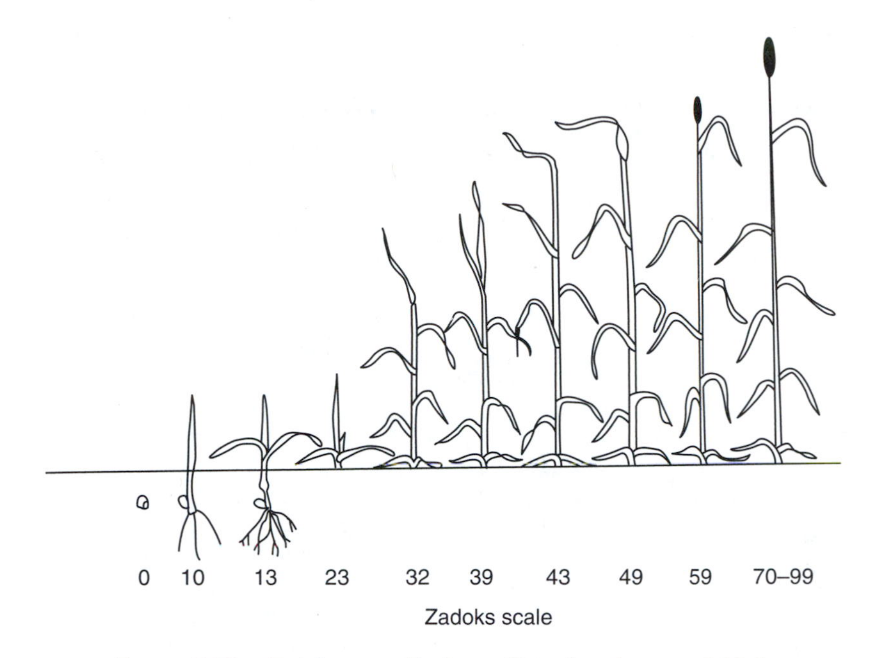

Figure 7.25 Zadoks stage 13 of a seedling, three leaves unfolded.

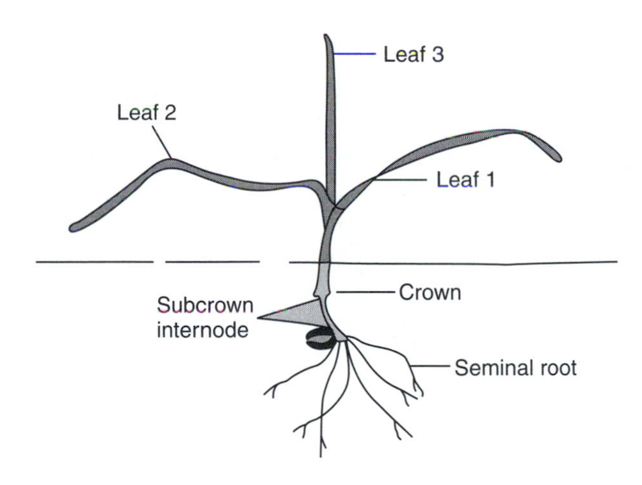

Figure 7.26 The Zadoks stages in the growth of winter wheat.

7.7.7 Flowering or Anthesis Stage

The flowering or anthesis stage lasts from the beginning to the end of the flowering period. Pollination and fertilisation occur during this period. All heads of a properly synchronised wheat plant flower within a few days and the embryo and endosperm begin to form immediately after fertilisation.

7.7.8 Milk Stage

Early kernel formation occurs during the milk stage. The developing endosperm starts as a milky fluid that increases in solids as the milk stage progresses. Kernel size increases rapidly during this stage.

TABLE 7.5
Zadoks Growth Stages for winter wheat, Depicted Diagrammatically in Figure 7.25

Germination		Seedling		Tillering		Stem elongation		Booting	
Dry seed	00	1st leaf emerged	10	Main shoot only	20	Pseudo-stem erection	30		
Water uptake started	01	1st leaf unfolded	11	Plus 1 tiller	21	1st node detectable	31	Flag leaf extending	41–44
Imbibition complete	02	2 leaves unfolded	12	Plus 1 tiller	22	2nd node detectable	32	Boot just swollen	45
Radicle emerged	05	3 leaves unfolded	13	Plus 1 tiller	23	3rd node detectable	33	Flag leaf sheath opening	47
Coleoptile emerged	07	4 leaves unfolded	14	Plus 1 tiller	24	4th node detectable	34	First awns visible	49
Leaf at coleoptile tip	09	5 leaves unfolded	15	Plus 1 tiller	25	5th node detectable	35		
		6 leaves unfolded	16	Plus 1 tiller	26	6th node detectable	36		
		7 leaves unfolded	17	Plus 1 tiller	27	Flag leaf emergence	37		
		8 leaves unfolded	18	Plus 1 tiller	28	Flag leaf collar	39		
		9 or more leaves unfolded	19		29				

Heading		Flowering		Milk		Dough		Ripening	
First head spikelet	50	Beginning	60	Kernel watery	71	Early dough	83	Hard kernel	91
1/4 head emerged	53	Flowering 1/2 complete	65	Early milk	73	Soft dough	85	Kernel loosening in daytime	93
1/2 head emerged	57	Flowering complete	69	Medium milk	75	Hard dough	87	Seed dormant	95
Emergence complete	59			Late milk	77			Seed not dormant	97
								Secondary dormancy lost	99

7.7.9 Dough Development Stage

Kernel formation is completed during the dough development stage. The kernel accumulates most of its dry weight during dough development. The transport of nutrients from the leaves, stems and spike to the developing seed is completed by the end of the hard dough stage. The developing kernel is physiologically mature at the hard dough stage even though it still contains approximately 30% water.

7.7.10 Ripening Stage

The seed loses moisture, and any dormancy it may have had, during the ripening stage.

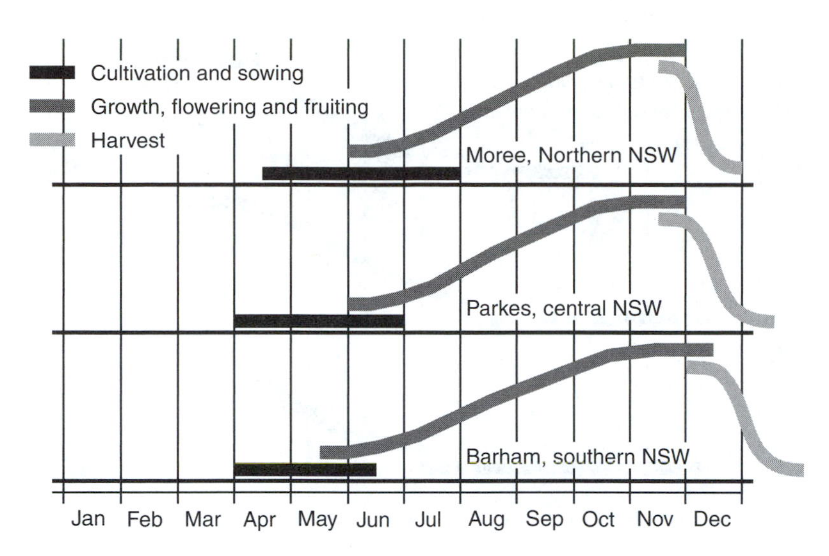

FIGURE 7.27 Crop calendars for Wheat in the vicinity of three towns in New South Wales — Moree (26°S), Parkes (29°S) and Barham (31°S) indicating the influence of temperature on crop phenology.

Different species can exhibit quite different characteristics in their phenological stages than those described above for Winter Wheat. However, most species go through the five main stages of:

1. *Vegetative stage*, starting with the emergence of new leaves and progressing through their maturation and the period of maximum growth in the plant.
2. *Flowering*, with the emergence of the flowers through to its completion with their disappearance.
3. *Seed set*, where the seed or fruit progress through immature to green stages and then ripening or maturation of the seed.
4. *Seed drop*, when the mature seed or fruit are dispersed in some way.
5. *Dormancy*, The plant returns to a period of relatively little growth, usually due to low temperatures or a lack of moisture.

The characteristics of these stages varies considerably depending on the influence of the main drivers on the vegetation in an area, as shown in Table 7.5.

Another way to depict the phenology of agricultural crops is by means of graphs as shown in Figures 7.27 to Figure 7.31, also called crop calendars. Figures 7.27 to 7.31 have been constructed from field observations, whilst Figure 7.32 has been constructed from the output of a crop growth model. What can be deduced from studies of the phenology of different species in the same ecosystems is that no two species can occupy the same ecological niche. One will survive and one will fail. Species need to occupy different niches. The more energy there is in the system, the smaller the niche required to survive, and indeed competition forces species to tailor their competitive capacity to a niche that best suits them. Thus, you find a greater range of species and greater specialisation in high energy systems like tropical rainforests, and fewer species and less specialisation in lower energy systems. From a remote sensing point of view, this means that a community will exhibit a range of conditions, growth, flowering and fruiting, due to the different species within that community, all of them ultimately constrained by the limits of the weather and the other factors discussed earlier. However, these individual plants are not discrete from each other in the canopy. Some will be natural occupiers of lower levels in the canopy, whilst others struggle in this part of the canopy until they get an opportunity to reach the full light of day. This creates a very complex response signal that is probably impossible to separate into its component parts from image data. What image data can

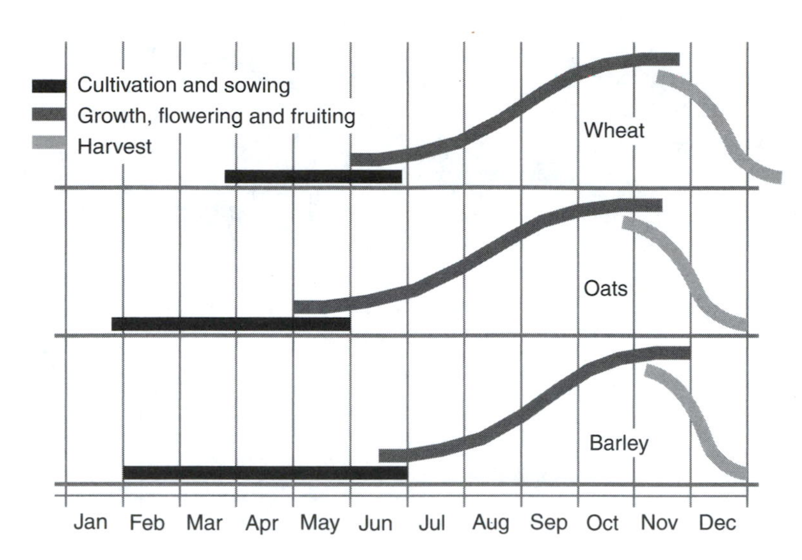

FIGURE 7.28 Crop calendars from three different cereal crops at the same location near Parkes (NSW) (29°S).

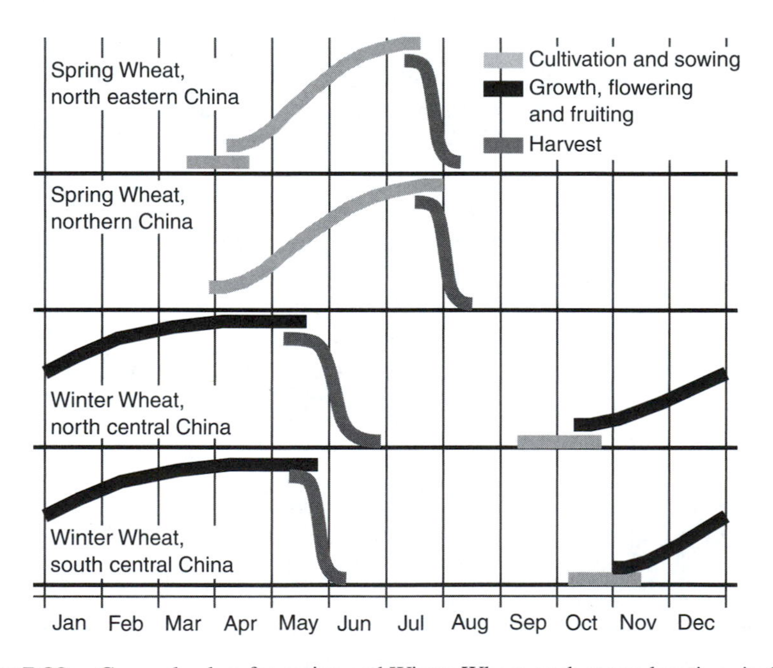

FIGURE 7.29 Crop calendars for spring and Winter Wheat, each at two locations in China.

give is a very accurate, detailed and spatially extensive picture of the phenological conditions in a community.

The growth of vegetation throughout its growth cycle is governed by predictable and unpredictable factors. The predictable factors are those that do not change significantly throughout the growth cycle, such as the slope, aspect, the soil and its inherent soil fertility. The unpredictable factors are those that do change, and in ways that are cannot be predicted, such as disease, predation, rainfall and temperature. In any given region, there is a typical distribution associated with all these factors, and the net effect of them on crops is what is reflected in the crop calendars shown in the figures above. In this context, in relation to agriculture, the application of fertiliser will be deemed to be

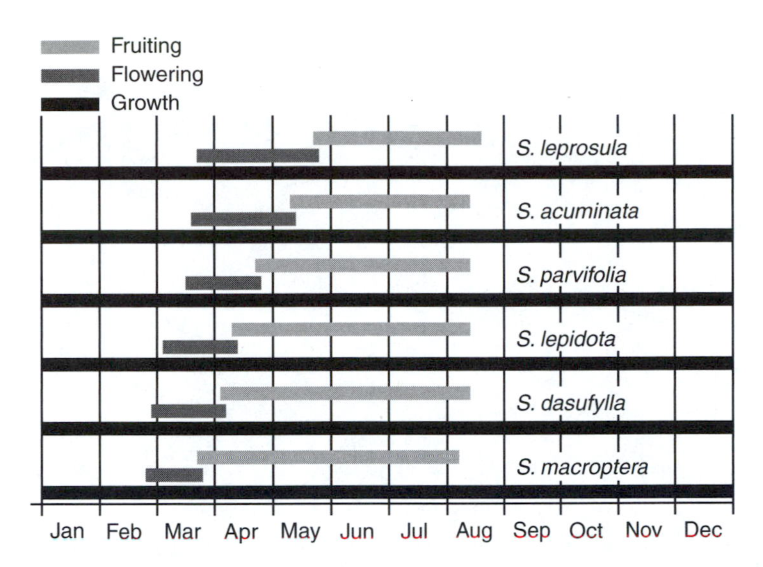

FIGURE 7.30 Sequential flowering and simultaneous fruiting in related species of *Shorea* during a year with gregarious flowering (Longman 1985).

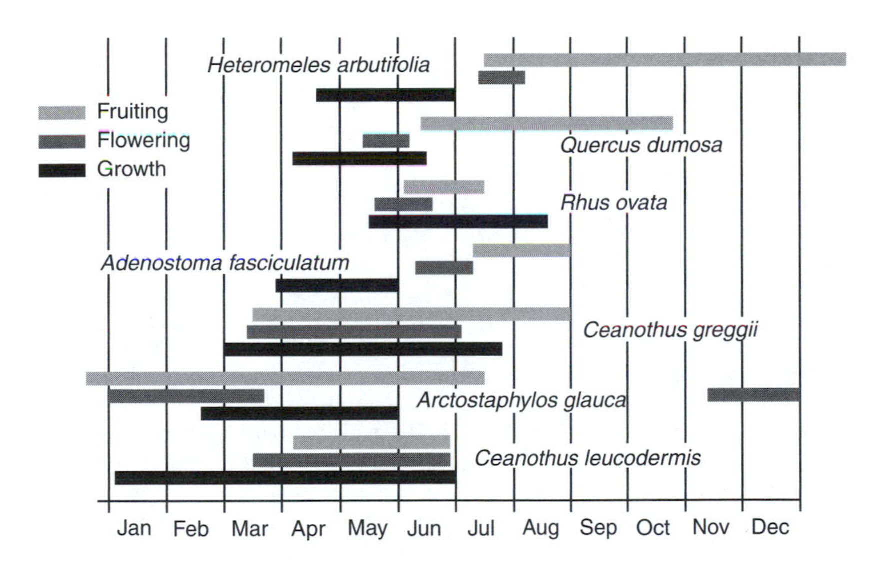

FIGURE 7.31 Canopy development, flowering and fruiting of the dominant chaparral shrubs in California (Mooney et al. 1977).

predictable factor, since it is under the control of the manager, whatever the level of application that is used.

If the unpredictable factors are not a constraint to growth, then the crop will grow up to its optimum growth potential for the predictable factors. Since these do not change, this optimum growth will also be the same from year to year, although it clearly will change from place to place. Under these conditions, the temporal growth profile will be the same from year to year at the one location, but of course the shape of this curve will change from location to location. Figure 7.32 shows the temporal growth profile form winter wheat grown in Denmark under conditions of negligible unpredictable constraints to growth that is irrigated, with adequate fertiliser applied and treated against disease. This curve marks the maximum growth that can be expected for this crop in this location.

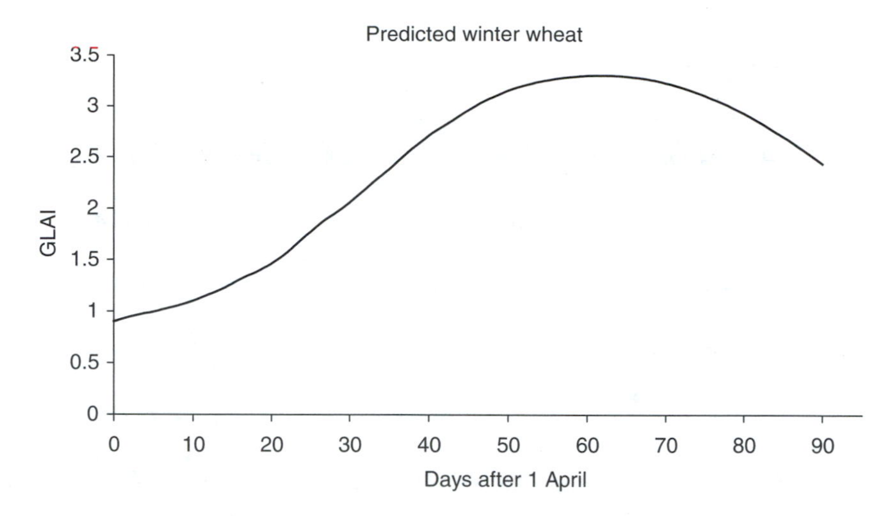

Figure 7.32 The GLAI temporal profile for winter wheat grown in Germany under conditions of adequate fertiliser and irrigation, indicating the optimum growth that can be expected for winter wheat, as estimated using the Daisy crop growth model (from Svendsen et al. 1995. *Ecological Modeling*, 81, 197–212. With permission.)

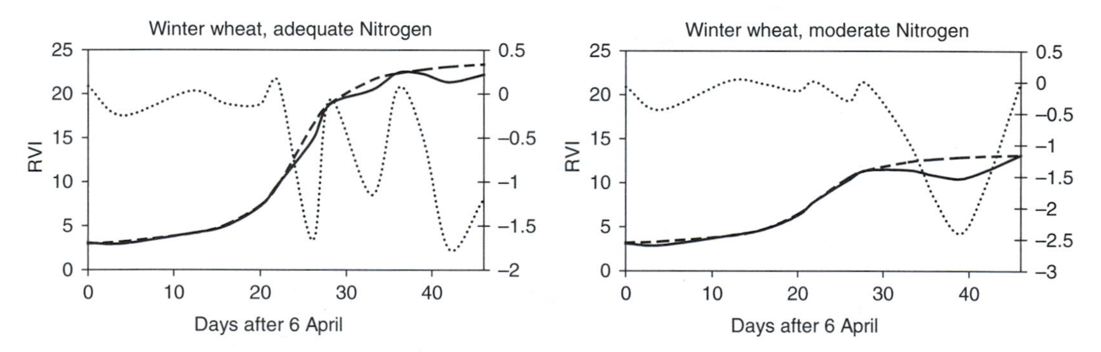

Figure 7.33 Temporal RVI profiles for winter wheat in Denmark, with an envelope curve indicating optimum growth and the differences between them, derived from field observed data. The vertical axis indicates values in ratio vegetation index. The data has been acquired in the field using field-based instruments. The discrepancy between the two curves at about days 26 and 32 are due to water on the leaves of the plants, and at day 42 is due to moisture stress.

The unpredictable factors impose constraints on growth that reduce the growth below this optimum curve, where the degree of reduction is a function of the magnitude and duration of the constraint. Figure 7.33 shows winter wheat crops in Denmark, with the optimum growth profile superimposed on the actual crop temporal profile. It can be seen that there are periods of dips away from the optimum profile sat stages in the growth of the crop. These dips are due to small deficiencies in soil moisture, that have been sufficient to affect the crop growth curve, but not sufficient to do permanent damage to the crop.

The shape of the optimum growth curve does change with soil fertility, as controlled by the use of fertiliser, and Figure 7.33 shows that the difference between the optimum and the actual growth curves can be measured with adequate data, opening the possibility for estimating both the predictable status of the crop and the impact of the unpredictable factors on the crop.

7.8 CONCLUDING REMARKS

The early euphoria associated with remote sensing in the 1970s was initially replaced with considerable scepticism, which in turn has now been replaced with a more realistic recognition of the limitations of images on their own in recording information about a very complex and spatially and temporally dynamic process. At this point, after about 40 yr of research and development, we can extract accurate and reliable information from image data about relatively simple components of the environment and relatively inaccurate or unreliable information about the more complex issues in the environment that many resource managers are interested in. The main response to this situation at present has been to either not use the data at all, or to use it as an indicator of conditions, with the real information being derived by other means, but benefiting from the reduced search requirements indicated by the information derived from the image data. For example, remote sensing images are currently used in Europe for the verification of farmer claims for environmental subsidy by classifying the agricultural landcovers and comparing these with farmer supplied information as part of their applications. Discrepancies are then visited in the field and the field visit becomes the legally valid evidence of the actual landuse in the specific field.

This current situation is unsatisfactory. Only when information derived from image data at a resolution, accuracy and reliability that allows that information to stand on its own two feet, will remote sensing come of age. Whilst the information derived from imagery is only used to reduce the laborious field checking work, then this information might be nice to have but it is not essential. However, once the information can stand on its own, then the field verification work, as described above, becomes unnecessary and the image-based information becomes an essential component of resource management.

There are many reasons for this situation. Imagery records but a small subset of the total characteristics of the imaged environment, yet we expect this subset to tell us all that we want to know about that environment. Logically, the chances of quite different conditions providing similar values in this subset are quite high, as in fact often happens. Yet despite this, in remote sensing we have tended to try and use images on their own, with limited external or auxiliary data or information. It is probable that images recording but a small subset of the characteristics of an environment will never be able to provide the information that we require, and so it becomes necessary to see how the image data should be embedded with other data or information and knowledge for the extraction of the information that we require. It is probable that we are expecting too much from images on their own; if we place lower demands on them and higher demands on other sources of data or information and knowledge, then we may find that satisfactory solutions are readily available.

Another issue is the availability of consistent data. Whilst the problem of atmospheric contamination has always been recognised and software to correct for the stable atmospheric constituents has been around for some time, it is only in the last 10 yr that a solution have been found to the routine correction of image data for the effects of atmospheric water vapour and there is still not a satisfactory solution to the aerosol problem. Both atmospheric water vapour and aerosols are highly variable, both spatially and temporally, and so satellite-based solutions, using suitable imagery acquired at the same time as the images to be used for surface information extraction, is the only truly suitable solution. So far no satellite-based systems can be said to be truly satisfactory, although the experimental satellite systems Terra and Aqua may contain adequate data if suitable systems for the corrections for aerosols can be developed and accepted.

A third major issue concerns the way that we manage resources. Most managers focus on the information that they require to manage their segment of the environment. Yet, the costs of information derived from satellite-based imagery are such that these costs may only be realistic if different managers share in the information extraction process. There are major advantages to this, particularly in terms of achieving both production and environmentally oriented goals are concerned. Historically, production-oriented managers have seen environmentally oriented managers as a threat to their capacity to be competitive. However, the community will not allow environmental concerns to disappear, and indeed constraints on production may become more stringent; under these conditions production managers will adapt to this reality and one form of adaptation is to share in the information extraction process so that all sides are using consistent data in their desire to be efficient and competitive on the one hand and to provide for the sustainable use of resources on the other hand.

These complexities and challenges should not mask the importance of image data in providing critical information for the management of resources, since the characteristics of image data make it the only way that certain types of information can be realistically collected. There is thus still some way to go to realise the real potential of image data. Towards this end there are stil many research questions that need to be adequately resolved. For example, the current NASA research focus is on:

1. How are regional and global ecosystems changing?
2. What trends in atmospheric constituents and solar radiation are driving global climate?
3. What changes are occurring in global landcover and landuse, and what are driving these changes?
4. How do ecosystems, landcover and the BGC cycles respond to and affect global environmental change?
5. What are the consequences of landcover and landuse change for human societies and the sustainability of ecosystems?

7.9 ADDITIONAL READING

There is a wide range of books that deal with terrestrial ecology and the associated BGC processes, at vary levels of depth and detail. The reader is encouraged to seek the latest of these to find those that best suit their interests and knowledge levels. For those who want a thorough treatment, but who have little knowledge of these disciplines, the books by Chapin et al. (2002) and Butcher et al. (2000) are worth considering.

For those who wish to better understand the interaction of radiation with matter, there is a detailed treatment of the topic in Myneni et al. (1989), focussing on the process of photon transportation through a leaf with the general goal of developing physical models of the process, and their subsequent inversion. A good, but now out of date, treatment of the modelling of canopy transport and reflectance is given in Goel (1988). For more up to date material the reader is referred to the extensive literature on the topic.

A good summary of kernel models as in 1998 is given in Jupp (1998).

REFERENCES

Allan, W.A., Gausman, H.W., Richardson, A.J., and Thomas, J.R., 1969. "Interaction of isotropic light with a compact plant leaf," *Journal of the Optical Society of America*, 59, 1376–1379.

Allan, W.A., Gale, T.V. and Richardon, A.J., 1970. "Plant canopy irradiance specified by the Duntley Equations," *Journal of Optics Society of America*, 60, 372–376.

Butcher, S.S., Charlson, R.J., Orians, G. and Wolfe, G.V., 2000. "Global biogeochemical cycles," *International Geophysics Series*, Vol. 50, Academic Press, London.

Borel, C.C., Gerstl, S.A.W. and Powers, W.J., 1991. "The radiosity method in optical remote sensing of structured 3-D surfaces," *Remote Sensing of Environment*, 36, 13–44.

Chandrasekhar, S., 1950. *Radiative Transfer*, Clarendon Press, Oxford.

Chapin, F.S., Matson, P. and Mooney, H.A., 2002. *Principles of Terrestrial Ecosystem Ecology*, Springer-Verlag, New York.

Chen, J.M., Li, X., Nilson, T. and Strahler, A.H., 2000. "Recent advances in geometrical-optical modeling and its applications," *Remote Sensing Reviews*, 18, 227–262.

Disney, M.I., Lewis, P. and North, P.R.J., 2000. "Monte Carlo ray tracing in optical canopy reflectance modeling," *Remote Sensing Reviews*, 18, 163–196.

Duntley, S.Q., 1972. "The optical properties of difuse materials," *Journal of Optical Society of America*, 32, 61–70.

Egbert, D.D., 1977. "A practical method for correcting bi-directional reflectance variation," *Proceedings of Machine Processing Remotely Sensed Data Symposium*, 178–189.

Gabry's-Mizera, H., 1976. "Model considerations of the light conditions in non-cylindrical plant cells (*Funaria hygrometrica* and *Lemna trisulca* leaves)," *Photochemistry and Photobiology*, 24, 453–461.

Gill, P.E., Murray, W. and Wright, M.H., 1991. *Numerical Linear Algebra and Optimization*, Vol. 1, Addison Wesley, Redwood City, CA.

Goel, N.S., 1988. "Models of Vegetative canopy reflectance and their use in estimation of biophysical parameters from reflectance data," *Remote Sensing Reviews*, 4, 1–212.

Goel, N.S. and Stebel, D.E., 1983. "Inversion of vegetation canopy reflectance models for estimating agronomic variables. I. Problem definition and initial results using the Suits model," *Remote Sensing of Environment*, 13, 487–507.

Goel, N.S. and Thompson, R.L., 2000. "A snapshot of canopy reflectance models and a universal model for the radiation regime," *Remote Sensing Reviews*, 18, 197–225.

Goodal, C.R., 1993. "Computation using the QR decomposition," *Handbook in Statistics, Statistical Computing*, C.R. Rao (Ed), Vol 9, Elsevier/North Holland, Amsterdam.

Govaerts, Y. and Verstraete, M.M., 1998. "Raytran: a Monte Carlo ray tracing model to compute light scattering in three dimensional heterogeneous media," *IEEE Transactions on Geoscience and Remote Sensing*, 36, 493–505.

Jacquemoud, S. and Baret, F., 1990. "Prospect: a model of leaf optical properties spectra," *Remote Sensing of Environment*, 34, 75–91.

Jacquemoud, S., Ustin, S.L., Verdebout, J., Schmuck, G., Andreoli, G. and Hosgood, B., 1996. "Estimating leaf biochemistry using the PROSPECT leaf optical properties model," *Remote Sensing of Environment*, 56, 194–202.

Jacquemoud S., Bacour, C., Poilve, H. and Frangi, J.P., 2000. "Comparison of four radiative transfer models to simulate plant canopy reflectance — direct and inverse mode," *Remote Sensing of Environment*.

Jupp, D.L.B., 1998. "A compendium of kernel and other (semi-empirical BRDF models," *CSIRO, Office of Space Science Applications*, Earth Observation Centre, Canberra, Australia.

Li, X. and Strahler, A.H., 1985. "Geometrical-optical modeling of a coniferous forest canopy," *IEEE Transactions of Geoscience and Remote Sensing*, GE-23, 207–221.

Li, X. and Strahler, A.H., 1986. "Geometrical-optical bi-directional reflectance modeling of a coniferous forest canopy," *IEEE Transactions of Geoscience and Remote Sensing*, GE-24, 281–293.

Li, X. and Strahler, A.H., 1992. "Geometrical-optical bi-directional reflectance modeling of the discrete crown vegetation canopy: effect of crown shape and mutual shadowing," *IEEE Transactions on Geoscience and Remote Sensing*, 30, 276–292.

Longman, 1985.

McCloy, K.R., Smith F.R. and Robinson, M.R., 1989. "Monitoring rice areas using Landsat MSS data," *International Journal of Remote Sensing*, 8, 741–749.

Myneni, R.B., Asrar, G. and Kanemasu, E.T., 1989. "The theory of photon transport in leaf canopies," *Theory and Applications of Optical Remote Sensing*, Ghassem Asrar (Ed), chap. 5, John Wiley Series in Remote Sensing, New York.

Pokrovsky, O and Roujean, J.L., 2002. "Land surface albedo retrieval via kernel based BRDF modeling. I. Statistical inversion method and model comparison," *Remote Sensing of Environment*, 84, 100–119.

Ross, J., 1975. "Radiative transfer in plant communities," *Vegetation and the Atmosphere*, Academic Press, New York.

Roujean, J.L., Leroy, M. and Deschamps, P.V., 1992. "A bi-directional model of the Earth's surface for the correction of remotely sensed data," *Journal of Geophysical Research*, 97, 20455–20468.

Sauer, R.H. and Unest, D.W., 1976. "Phenology of steppe plants in wet and dry years," *Northwest Scientist*, 50, 133–139.

Strahler, A.H., Li, X., Liang, S., Muller, J.P., Barnsley, M.J. and Lewis, P., 1994. "MODIS BRDF/Albedo product: algorithm Theoretical Basis Document," *NASA EOS_MODIS Document*, vol. 2.1, p. 55.

Suits, G.H., 1972. "The calculation of the directional reflectance of a vegetative canopy," *Remote Sensing of Environment*, 2, 117–125.

Svendsen, H., Hansen, S. and Jensen, H.E., 1995. "Simulation of crop production, water and nitrogen balances in two German agro-ecosystems using the Daisy model," *Ecological Modeling*, 81, 197–212.

Verhoef, W., 1984. "Light scattering by leaf layers with application to canopy reflectance modeling: the SAIL model," *Remote Sensing of Environment*, 16, 125–141.

Verhoef, W. and Bunnik, N.J.J., 1975. "A model study on the relations between crop characteristics and canopy reflectance. 1. Deduction of crop parametersfrom crop multispectral data. 2. Detectability of variations in crop parameters by multispectral scanning," NIWARS Publication 13, Delft, The Netherlands, 1–89.

Walthal, C.L., Norman, J.M., Welles, J.M., Campbell, G. and Blad, B.L., 1985. "Simple equation to approximate the bi-directional reflectance from vegetative canopies and bare soil surfaces," *Applied Optics*, 24, 383–387.

Wanner, W., Li, X. and Strahler, A.H., 1995. "On the derivation of kernels for kernel driven models of bi-directional reflectance," *Journal Geophysical Research*, 100, 21077–21089.

8 The Management of Spatial Resources and Decision Support

8.1 INTRODUCTION

So far this book has dealt with methods of acquiring and analysing spatial data and information. The methods of acquisition include the digitisation of maps, collection of field data and the analysis of remotely sensed images. Other sources of information that have been mentioned include statistics, records and other types of observations. All these sources of information have a role to play in information systems for resource management.

The acquired information needs to be properly managed, analysed, displayed and maintained to provide management information, with many of these tasks being conducted within geographic information systems (GIS). GIS should also be capable of supporting statistical analyses, facilitating the use of predictive models and conveying to the potential user the quality of the information that is being displayed by the system.

The human element is the third component in resource management information systems (RMIS). Who makes decisions and how are those decisions made? What type of information do managers require and how must it be presented? How are the decisions best conveyed to those who must implement them, and what role does the information system play in conveying the decisions, and relevant supporting information, to the working staff? In short, how should an institution's information base be structured and operated so as to assist resource managers to make decisions that best enable the institution achieve its objectives.

> A Resource Management Information System is defined as a functioning *information system*, consisting of facilities, staff and data, designed to support the *proper management* of physical *resources* by the provision of appropriate information to decision makers, to facilitate their understanding of the conditions relevant to making decisions.

The definition does not restrict an RMIS to spatially extensive data such as images and maps, but it does envisage that these data types will be an important component of such systems because of the importance of spatial information in the management of physical, environmental resources.

The *information system* contained within this definition consists of the technical facilities, data and staff necessary to implement and operate the information system. As a system it will consist of an integrated set of components that must include:

1. The acquisition of data, it's conversion into information and the acquisition of information.
2. The storage, maintenance and ongoing management of the information base, including its auditing, editing and updating, verification and assessment, provision of backup and documentation on the system and its functions. Documentation will usually include both technical and non-technical documentation.
3. The analysis and presentation of derived information, as statistics and maps, to resource managers.
4. The facilities necessary to conduct these tasks. Information may be used to make decisions and then used to convey background information on those decisions to staff who must implement the decisions. It is essential that the facilities necessary to convey the information to where it is required, when it is required, be included in the system.

5. Training and user support as is necessary to ensure that the information is properly and efficiently used. The level of training and user support required will clearly depend on (a) the role of the recipients in either the conduct of the resource information systems or the use of the information in the decision making process, and (b) their skills levels prior to the training.
6. Integration of the information system into the resource management structure by adaptation of the information systems to that structure. The implementation and use of the information system will also influence the way management decisions are made, that is, the management style will adapt to this new information source.
7. System management to ensure the effective and efficient management of the whole information system: staff, facilities, information flow and utilisation.

The information system is an integral component of the management structure of the organisation. It should be tailored to the needs of the organisation. To execute this design requires a good appreciation of the role of information systems in organisations, how to design and implement information systems and in the role of information in the management of resources.

Fundamental to the design of an RMIS is an understanding of how managers make decisions. It is a crucial component in ensuring that RMIS provide the optimum information set, in the right form, at the right time, to the appropriate resource manager concerned with the particular resource management decision. The management of environmental resources can be very complex. There are always a number of different perspectives from which a problem can be viewed. At the minimum these involve the perspectives of the resources themselves, society and the resource manager. But there may be quite different perspectives on an issue by different groups with each of these three broad categories.

Recognising these complex perspectives has meant that society has often defined constraints or set specifications for areas that can be used for one purpose but not for another purpose. Thus, National Parks are a form of constraint for some purposes, simply because they cannot be used for those purposes. At another level, landuse zoning is another form of constraint. Within an allowed use of land there are still many perspectives on how the land can and should be managed. This leads to the establishment of criteria, which define the suitability of an area for a task. There are many types of criteria that can be met. In terms of biodiversity, certain birds species may require stands of trees of a minimum size or area, and at a maximum separation from the next adjacent stands, for the area to be suitable for them. This becomes a criterion to be evaluated. Other types of criteria can be slope, distance, surface roughness and so forth. Finally, there may need to be resolution of conflict between competing uses of the land, or multiple conflicting objectives.

To assist a resource manager deal with these complex tasks, that have a spatial component, decision support tools are being developed within the GIS environment. In this chapter we will learn about how these tools can be used to assist resources managers address complex multi-constraint, multi-criteria, multi-objective management decisions.

The objective of this chapter is thus to impart an appreciation of the characteristics of information itself, its role in resource management and the use of decision support systems (DSS) in resource management, so that the student can better design and implement operational resource information systems. To do this the chapter is structured so as to:

1. Review the characteristics of current management practices
2. Analyse the nature of information
3. Analyse the nature of resource management decision making
4. Consider the characteristics of resource management
5. Understand the nature of spatial DSSs
6. Develop principles governing the design of RMIS.

8.2 THE NATURE OF MANAGEMENT OF RURAL PHYSICAL RESOURCES

8.2.1 Introduction

Pressures on natural resources continue to increase. This pressure comes from two main sources: from increased demand and from increased capacity to use and exploit these resources. Increased demand comes from another set of sources: from increasing populations and from increasing expectations. Whilst populations are currently stabilising in many western countries, this may not remain the case, and nor is it the situation in most developing countries where populations continue to increase. Expectations of rising standards of living also continue to increase, most significantly not only in the developing countries, but also in the developed countries. Rising expectations may prove to be a much bigger threat to natural resources than the direct threat of population growth.

These rising expectations are matched by an increasing capacity to exploit the land, primarily through better technology but also through better knowledge and understanding of how systems work. Whereas, land used to be allowed to lie fallow, this is no longer possible in many countries, so that the land becomes utilised up to and possibly beyond its long-term capacity.

These pressures on natural resources are also being reported back to society as a whole through the media and through the various information services, including the Web. Thus, issues of degradation and damage to the land are increasingly being brought into the living rooms of suburban houses. Suburban homes are often a key source of political power in democracies. As the suburban population sees what is happening to the land, they mobilise to counter what they see as degradation. Political survival will demand that action be taken to deal with issues in the country, driven by the concerns of suburban society. It is this concern that has given rise to the current demands for sustainable land using practices in Europe and much of the remainder of the developed world. This demand is being translated into rules, regulations and practices designed to protect what are seen as key elements of the environment.

It can be seen that the original pressures to produce economically have not been removed, but have been overlaid with additional pressures to manage land in a sustainable manner. There are thus many competing pressures that combine to make many land using decisions very complex and fraught with political, ecological and economic risk, where many of them will involve many criteria on which to base the decision which must meet the needs of multiple objectives. Through this chapter runs the theme of the need to understand the management process and the use tools that will assist in the proper management of resources.

8.2.2 Levels of Management

Management functions can be grouped into three major levels:

1. Strategic
2. Regional
3. Operational

8.2.2.1 Strategic Management

> Strategic management is defined as planning the future directions for the corporation through establishing the corporations long-term objectives and mobilising its resources so as to achieve these objectives.

Strategic management is normally implemented by developing corporate objectives and goals, establishing the structure appropriate to the achievement of these goals, and then setting the goals of the units within the corporation. This type of activity is the primary concern of the Board of Directors and senior managers in an organisation, including national and regional governments.

Strategic management activities are concerned with the environments (physical, economic, social and political) within which the corporation operates. It involves assessing how these may change in the future, the potential impact of these changes on the corporation, and how the corporation needs to respond and operate under evolving conditions. *Strategic management requires extensive information about environments that are external to the corporation*. Strategic management, in considering these external environments, will need to consider many factors over which the corporation has no control, and many of which are quite unpredictable since the corporation will usually have incomplete information on the factors and their environments. *Strategic management requires the use of models that can accommodate the impact of unexpected and unpredictable factors. It generally requires the use of unstructured models*.

Generally, the information that is used in strategic management will be of low spatial resolution, generalised and qualitative because of the way that the information is to be used, and the extensive range of information that has to be considered.

Typical information used in strategic management includes:

1. *Statistical information* as collected by central Statistical Bureaus providing national and regional level information on different aspects of the community. Remote sensing and GIS can improve the collection and accuracy of this information by providing sampling strata appropriate to the information being collected, and in some cases by the actual collection of the sampled information itself.
2. *Broad-scale monitoring information* as collected by satellite at the global and regional level. Remote sensing with field data are the basis of most of these monitoring programs, such as with weather prediction.
3. *Institutional information* as collected from documents and publications including economic, social and political information about the institution and its relationships with other institutions, including governments.

8.2.2.2 Regional

Regional management translates strategic objectives and plans into implementable programs within the units of the organisation, whilst matching current operations to both current and future resource needs.

Regional management involves structuring the unit to achieve the set goals, marshalling the staff resources to the achievement of these objectives, and ensuring that the unit continues to operate at peak levels of productivity. This strategic and tactical level of activity is the concern of middle management.

Regional management is concerned with both optimising productivity within the whole of the corporation, and with maintaining the resources to ensure the long-term viability of the corporation and its environment. It is in the position to influence all components of the corporation that impact on the resources of an area, and thus it is in the position to balance resource maintenance with productivity, within the general policy framework established by the corporation's strategic plan. Regional management is not concerned with the over arching focus of strategic planning, but with a more direct focus on the corporation as a whole, and its interactions with its immediate environment. *Regional management is concerned with more localised external and internal information on all aspects of the environment within which the corporation operates.* Regional management concern with both maintenance and productivity means that it needs to integrate the needs of the corporation with those of others groups or corporations in the environment; it is concerned with very complex yet rigorous decision processes. It will require the necessary research and development support to evaluate these decision processes.

Regional management is concerned with predicting future effects of corporation activities so as to assess decision options for the corporation. This prediction must come from the use of quantitative models of processes, requiring quantitative information to drive these models. The resolution of the information must be compatible with being able to assess the impact of decisions on units of management. This means that regional management requires information that is sufficiently accurate and with a resolution that they can get summarised information at the level of the unit of production, where the unit of production on a farm is the paddock, and the stand within a forest area. The regional manager thus needs information on the nature of the landcover in the field or the stand, and on the condition of the cover within that field or that stand.

> Regional management requires that the quantitative and qualitative information that it uses has a resolution sufficient to provide information about, or within, the smallest units of management within the corporation.

Regional management is the level of management above individual units of production; it is the first level of management that can address resource degradation issues since these generally flow across these units of production, with actions in one unit having impacts on other units, often at some other time. Regional management thus requires rigorous databases and predictive models to assess potential impacts and revise corporate tactical objectives so as to adjust operational management to take these impacts into account. Since this process will involve resolution of conflict between operational units, it will often require accurate and detailed quantitative information.

In a very important sense, regional management is the main beneficiary of these new technologies of remote sensing and GISs. For the first time, regional management has the tools necessary to assess how the individual units of production are operating in a rigorous way, and respond to that information by the formulation of better tactical plans for those units of production, so as to better utilise and maintain the resources of the institution or region.

Regional management is the level of management that will have to deal with resolving conflicts of interest between the demands of the sustainable use of land on the one hand, and the needs for maintaining productivity on the other. Whilst strategic management may use indicators as the basis of the formulation and verification of the effectiveness of policies, such indicators are not adequate to the management of resources. For this task regional managers require much more detailed and extensive information.

8.2.2.3 Operational

> The purpose of the operational management of resources is to optimise productivity within the unit of production whilst meeting the goals and objectives set for it by regional management.

Operational management involves the direction of staff in activities directly involved in managing the resources of the unit of production, whether it be a farm, a factory or a shop. Operational management is concerned with day-to-day decisions. It requires quick decisions, often on site, using primarily on site or internal information. It requires models that can be implemented by the manager on site, and that can respond to events as they occur. Information requirements at this level are for high-resolution yet simple information since the decisions are usually made quickly using empirical models.

The relationship between these categories of management and their information needs are set down in Table 8.1 and shown diagrammatically in Figure 8.1. As the level of management gets higher up the organisational structure, the information needs change as indicated in Table 8.1. The information needs become more summarised and more oriented around strategic planning issues

TABLE 8.1
Relationship between Management Category and the Attributes of the Information Required for Management Purposes

Information Characteristic	Management Activity: Strategic	Regional	Operational
Coverage	Global	Regional	Local
Accuracy	High average accuracy, low local accuracy	High accuracy to the field level	High local accuracy
Level of detail (resolution)	Generalised	5–100 m	0.01–2 m
Time horizon	Future oriented, 2–20 yr	Immediate future oriented; the current and next seasons	Solve current problems; present
Sources	Primarily external	External and internal	Internal
Data type	Qualitative	Quantitative	Qualitative and quantitative
Model type	Unstructured	Structured	Unstructured and structured

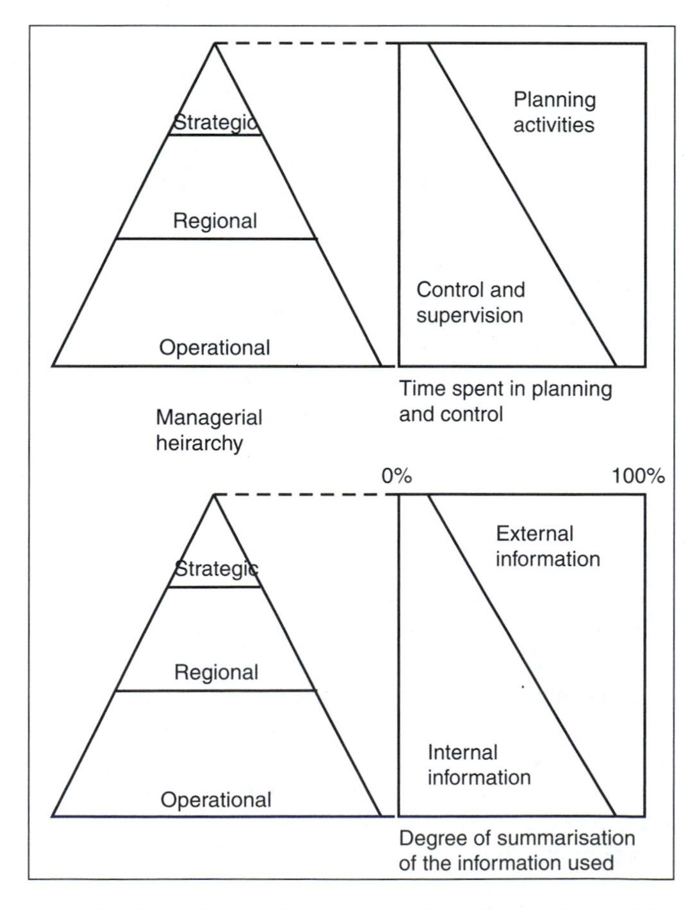

FIGURE 8.1 Management levels relating to planning, control and information in (a) relation of managerial position to the time spent in planning and control, and (b) relation of managerial position to the summarisation of information.

and less around detailed management tasks. The mix of information also changes the higher up the management structure the information is required; it becomes more dependent on external sources of information, more predictive, subjective and qualitative, simply because the number and range of factors and how they change over time both increase and become harder to predict.

8.2.3 The Role of Remote Sensing and GIS in Resource Management

8.2.3.1 Strategic Management

We have seen that the information bases for strategic management are often sampled data. Typical sampled data include that collected by national statistical offices and indeed this information is primarily designed for use in strategic management. Another form of sampled data are the point monitoring observations, that can be very detailed and accurate summaries of the status of an area. Water quality as measured at water monitoring sites provides integrated information on water quality upstream of the site. If the water quality is inadequate, then the sampling site itself may tell very little about the number of sources and their contributions. Other information must be sought so as to deal with the problem.

Remote sensing and GIS can assist in the collection and analysis of statistical and point data in a number of ways:

1. *Defining strata for sampling*. If the information to be collected is unevenly distributed across the area of interest then stratification of the area and random sampling within each strata will provide a better estimate of the measured parameters than will a simple random sampling. If these strata are expressed in remotely sensed images in a way that allows discrimination between them, then images are an appropriate medium for defining the strata. This is typically the case if the strata are related to landuse, landcover or landcover condition, as applies to many forms of rural census. Typical strata could be:
   (a) Landuse intensity for agriculture and other census data collection where these strata would be identified by the visual interpretation of suitable imagery.
   (b) Vegetative vigour as a basis for estimating yield.
   Once the strata have been identified then they would be incorporated into a GIS so that they can be used with attribute data collected in some way to derive estimates of the attributes for each strata and for the whole area of interest.
2. Sampling within the remotely sensed data itself so as to provide strategic planning level information.

8.2.3.2 Regional Management

Whilst strategic and local management levels have been well established for a long time, regional management is a new and evolving level of management in the form that is discussed in this book. There has been a level of regional management for some time, but it has historically focussed on human oriented issues, like education, policing, medical services, treatment of sewage and removal of rubbish. With increased societal concern for maintenance of the environment, and the sustainable management of resources, society is laying down additional demands and responsibilities on regional management. As a consequence of this new and evolving responsibility, regional management is both deficient of an independent, objective source of information and yet it is the key management level in the maintenance of resources. It is deficient of suitable information sources since most of its information, in relation to resource management, comes from the operational control units that report to it. This information will tend to be inconsistent, since it is collected by many different people, with widely varying interests in the information being collected, and often collected at different times. Operational management perspectives are closely focussed on the productivity of the unit. The impacts of operational decisions on other operational units are of marginal interest to the operational manager, particularly if responses to those impacts affect productivity. However, maintenance of resources requires a careful consideration and balancing of these impacts. Regional management is therefore in the position that it requires rigorous quantitative information if it is to balance the demands of operational managers against the potential cost of their actions either in terms of degradation or impact on other units of production.

Regional management is the level of management that can benefit most effectively from the information that is derived from airborne and satellite data, and incorporated in a GIS since:

1. The derived information is at a finer resolution than the units of operational control so that it provides information down to the operational control level. This enables the regional manager to relate to the operational manager at a summarisation of information at the unit of production level.
2. The derived information is extensive enough to cover all units of operational control that are of concern to regional management.

The definition of regional management should be interpreted so as to include:

1. District Agricultural Extension Staff and commercial agricultural consultants where they are concerned with farms as operational units.
2. Banks, valuers, stock and station agents and local government who deal with farms as operational units.
3. Forestry departments and timber millers who deal with logging concession operators as operational units.
4. Local and regional government concerned with the maintenance of regional and environmental resources.
5. Community and government-based integrated resource management cooperatives that deal with individual landowners or lessees as operational units.
6. Staff of commercial rural industries (herbicides, fertilisers, farm machinery) who deal with farmers and other land users as individual units of operational control.

The information support required by regional management does involve some inventory, but it will more importantly involve monitoring and prediction, to assess performance and environmental impact, and for planning and design purposes. Further, it will involve the integration of this information with many other types of information in dealing with multiple criteria and multiple objective issues. It is for these reasons that decision support tools are essential components of the facilities required by the regional resource manager.

8.2.3.3 Operational Management

Operational management can use remote sensing and GIS when the resolution in the data is appropriate to the task, and the derived information are provided on time. This requirement will often see satellite data being used for meteorological, marine and fisheries purposes since these activities cover extensive areas at the operational control level. Agriculture, forestry and other land-based activities are more likely to use large-scale photography or video image data for operational management purposes.

The demand of operational management is for current information, so as to address current management issues. This requirement places a significant demand on the system to provide the information, to an agreed level of accuracy and reliability, within a set time frame. This time constraint creates an advantage for video data relative to photography because it does not loose valuable time in development and processing.

8.3 THE PROCESS OF DECISION MAKING IN RESOURCE MANAGEMENT

The process of decision making is still the subject of considerable scientific investigation. However, making a decision in resource management is generally considered to involve the following sequence of events:

1. Selecting an appropriate model of events
2. Collecting the necessary information to parameterise the model
3. Use the parameterised model to identify decision choices
4. Make the decision

In each of these steps may lie uncertainty and conflict and hence risk. Risk should be seen as an inherent component of any decision process.

8.3.1 Selecting an Appropriate Model of Events

Prior to any decision making, the decision maker has to be aware of the need to make a decision. Generally, either of two types of reasons may trigger recognition of this need: *problem detection* or *opportunity seeking*. Problem detection arises when a collection of facts or evidence suggests to someone that some conditions are deviant from those expected, that is, the facts or evidence indicate conditions different to the norm. Opportunity seeking arises when circumstances suggest that there may be opportunities for the decision maker.

The decision maker will observe the facts or evidence as brought to him by an employee in the process or by an outsider. The resource manager will then, from the description of conditions, decide which model is appropriate for the conditions under consideration. The resource manager may need further information before selecting from amongst alternate models to select a model for use in making the decision. The model at this stage will be a *generalised model*, that is, the model will not be parameterised so as to meet specific conditions.

8.3.2 Parameterise the Model

The *generalised model* needs to be parameterised to create a *localised model*. This *localised model* needs to be tested to ensure that it is appropriate for the making of the necessary decision. Data has to be used for this purpose. This process may involve the collection of more data and it may reveal inadequacies in the model, requiring either modification of the model, or its rejection and the acceptance of some other model. The process of model selection and parameterisation is thus an iterative one, in which the experience of the resource manager is a critical factor in the selection and effective parameterisation of the model.

Consider the case where the manager who wants to travel from A to B. A topographic map is an appropriate generalised model. The specific map of the area is parameterisation of this model, allowing the manager to then choose his route. However, if the route is through a town and is to be taken at rush hour, then parameterisation may involve a time component so that a different route may be chosen during rush hour to that chosen at other times.

8.3.3 Identify Decision Choices

The parameterised model leads to a better understanding of the conditions existing when the need to make a decision was identified. Often this understanding will require the use of other models, since the decisions that are made will have impacts that can only be assessed by the use of other models. These other models may also need to be parameterised before they can be applied to the specific conditions that have been met. Thus, the resource manager uses the understanding generated by the model to assess whether he has adequate knowledge and understanding to make the required decision. Once this condition has been met then the alternate decision choices need to be identified. Each alternative choice must be examined under the following criteria:

1. Does the choice solve the problem?
2. Is it technically, economically and practically feasible?
3. Does it conform to common practices and meet community legal requirements?
4. Does it comply with community time constraints?
5. Who will be affected and what will these effects be?
6. What will be the community reaction to the solution?
7. What are the environmental and social costs and benefits?

From this process of considering the advantages and disadvantages of each solution, the decision maker then selects one choice from the alternatives.

8.3.4 Making the Choice

Making the choice is often a complex task because the decision may involve choosing between:

1. *Multiple criteria*. In many cases the outcome is not measured by one, but by several variables. Often some of these variables cannot be easily quantified, so that the choice becomes one between trade-offs that may not be easily measurable.
2. *Uncertainty of effects*. Often there is a level of uncertainty as to the effects of the different choices, and again it may not be possible to quantify this level of uncertainty.
3. *Conflicts of interest*. The different objectives that need to be met may involve conflicts of interest for the organisation, the manager or for other staff. A choice may benefit some parts of an organisation, but disbenefit other parts, and the implications of these conflicts needs to be taken into account when selecting between choices.
4. *Control*. It is not sufficient to make a decision unless that decision can be adequately supported. Is there sufficient information to make the follow up decisions that will need to be made? Is there enough capital to allocate to the decision, and to cope with contingencies that might arise? Can the decision be reversed if necessary, and at what cost?

8.3.5 Types of Models and Decisions

Models, and the decisions that flow from them, can be identified as being either *structured* or *unstructured* in form. *Unstructured models and decisions* involve intuition, common sense and heuristics; they often depend on trial and error. The relevant data will often be qualitative rather than quantitative, and it may be vague. The decisions rarely replicate previous decisions, and so they have a high degree of uniqueness and are difficult to program. Their characteristics are that they:

1. Cannot be subjected to quantitative scientific analyses.
2. Are usually very generalised with low resolution and accuracy; they can be vague.
3. Are not predictive in a quantitative sense.

4. Can respond to the unexpected and outside influences; they can be flexible and adaptable.
5. Can incorporate factors that are hard to quantify.

Structured models and decisions are based on clear logic, they are often quantitative, the factors involved and the outcomes are usually well defined, they can be routine, repetitive and are often amenable to programming. Their characteristics are that they:

1. Can be subjected to rigorous quantitative scientific analyses.
2. They have or require high levels of resolution, accuracy and precision to be most useful.
3. Can be predictive.
4. Cannot work outside the premises of the model; they can be inflexible.
5. Cannot respond easily to outside influences.

Structured models can be either qualitative or quantitative in form. Qualitative models that have been made more rigorous through being verbalised and thus subject to analysis by others can be structured models. Quantitative models, parameterised by quantitative data, will usually be structured models. Indeed the more rigorous the data, from nominal to ratio, the more rigorous or structured the resulting model is likely to be.

8.4 DECISION SUPPORT SYSTEMS AND THEIR ROLE IN DECISION MAKING

Decision support systems are defined as systems that provide help in making decisions in complex situations by using data and models to evaluate one or alternative modes of solution of a management issue so as to reduce the alternatives to a simple comparison for the manager to then evaluate and to use as one of the bases of his final decision. DSSs are based on computer technology so that they will, by their nature, use quantitative data and models. From Table 8.1 it can be seen that the level of management that will have the greatest need for quantitative analysis and decision support tools is regional management and as a consequence, regional managers will be the greatest users of decision support technology. DSS can be used at the other two management levels. Generally, it will not be used at the strategic level because of the nature of the information being used and the types of models used in developing strategic decisions. Generally, it will not be used at the local management level, because of the nature of the data used in making fast and local decisions. DSSs contain a number of key components:

1. *They do not make decisions*, but provide relative information on the alternative solutions in a way that facilitates comparison by the resource manager. This means that they provide sufficient information on the alternatives to enable the manager to evaluate these alternatives. Since most decisions must come back at some stage to economic costs and benefits, most DSSs provide information in these terms. However, this does not have to be the case as other forms of cost and benefit can be assessed and may be more useful under some circumstances, for example, energy flows or time usage.
2. *It must use data and models to mimic the aspects of the environment that are of interest.* This means that DSS will usually be based on computer systems and use numerical models.
3. *It is interactive*, so that the user, in assessing the results, can pose other queries to the DSS and reformulate some of the queries that have been posed and find out how the changes affect the evaluation.
4. *Used to confront complex issues*. Most straightforward issues can be readily solved, without the need to resort to complex support systems, and so DSS are designed to assist with

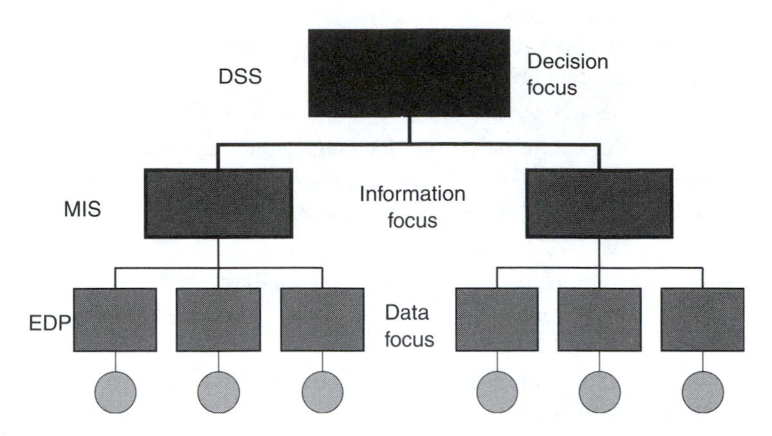

FIGURE 8.2 A view of the relationship between electronic data processing, management information systems and decision support systems.

complex situations where the choices may not be clear and the potential effects or responses to those decisions will often be far from clear.

A reasonable way to view DSS is to see it as a natural evolution out of the way that computer-based use of data has evolved from better management of data through the reformatting of data into useful information, to a focus on the use of information in decision making, as shown in Figure 8.2. Electronic Data Processing (EDP) had as its focus the structuring of digital data in a consistent and coherent way for subsequent use. It was focussed on the data, its storage, retrieval and use. EDP naturally had a strong focus on databases, their structure, functionality and usability. The basic characteristics of EDP included:

1. A focus on data, its storage, processing and flows
2. Efficient transaction processing
3. Scheduled and optimised computer runs
4. Integrated files for related functions or tasks
5. Summary reports to management

Management Information Systems (MIS) were not so much interested in the data, which were assumed to be in place, but rather in the information required by management to understand the state of the organisation and to make decisions. It was concerned with structuring the data into information that was more useful to management, by means of summarising data, deriving trends through analysis of data and so forth. Just as the database was a key component of EDB, the Spreadsheet and advanced database analysis functions are key components of MIS. The elements of MIS include:

1. A focus on information for middle management
2. Structured information flow
3. An integration of EDP jobs by function
4. Generation of analyses, inquiry results and reports

The decision support system uses the structured data with advanced analytical tools, in particular with complex models, to provide integrated information on an option or among options for a resource manager. It needs to be flexible and interactive, it needs to be able to interrogate models to see what the likely outcomes are of specific proposed actions and to assess the costs and benefits of those proposed actions. DSSs are not designed to support the decision making of senior managers, so much as to support regional managers dealing with complex but very specific issues, where this complexity may be in the physical environment or the societal and economic environments. They are therefore most useful when dealing with situations that are too complex to solve without supporting systems that can analyse components of the system and provide information on the costs/benefits of alternative modes of action.

There is now an extensive literature on decision theory in the Management Science and Operations Research areas, but little work has been done in this field in the GIS area, with the exception of that implemented within the IDRISI GIS and Image Processing package (henceforward called IDRISI). Accordingly, much of the material used in this section is heavily influenced by the pioneering work done by IDRISI in the development of decision support tools in a GIS environment (Eastman 1999).

8.4.1 Definitions

- *Decision.* A Decision is a choice between alternatives. The alternatives may include different courses of action, different hypotheses about the character of a feature, different classifications and so forth.
- *Decision frame.* The set of alternatives are called the *decision frame*. Thus, the decision frame for a zoning decision may be between the alternative land uses (for example, commercial, residential, recreational, industrial).
- *Candidate set.* The decision frame must be distinguished from the individuals to whom the decisions will be applied, called the *candidate set*. The candidate set for the above zoning decision will be all of those land parcels to which the decision will apply.
- *Decision set.* Finally, the *decision set* is that set of all individuals that are assigned a specific alternative from those available by the decision. Thus, for example, all the land parcels that are assigned to residential area form the one decision set and all those parcels that are assigned to the recreational area are another decision set.
- *Criterion.* A *criterion* is some basis on which a decision is to be measured and evaluated. It is usually the evidence used to assign an individual to a particular decision set. Criterion can be of two kinds; factors and constraints, and they can be based upon the attributes of the individuals, or on the attributes of the decision set.
- *Factor.* A *factor* is a criterion that enhances or detracts from the suitability of a specific alternative for the activity under consideration. It is most commonly measured on a continuous scale, for example, it may be more costly to remove trees by logging on steeper slopes and so slope is a factor in considering the suitability of an area for logging.
- *Constraint.* A *constraint* serves to limit the alternatives under consideration. For example, in the above zoning it may be illegal to build industrial buildings in a residential area, so this becomes a constraint. In the case of the forestry company looking for area of lower slope, if there are National park or recreational areas to which logging is not allowed, then these become a constraint on the options that are available to the company. In many cases constraints can be expressed as Boolean maps, or maps with only (0, 1) values, where 0 may represent the areas of constraint and 1 all other areas. Another form of constraint may be put in terms of some characteristic that must be met. For example, if we refer again to the logging company, who may specify that flat areas of less than 5 ha and more than 5 km from adjacent flat areas may not be realistic to log. This condition would be constructed as a constraint.

  Although factors and constraints may be viewed as being very different types of criteria, IDRISI (Eastman 1999) take the view that they are part of a continuum of variation in the criterion used to influence decisions. For example, the hard or sharp boundaries that may come from constraints, as discussed above, may be seen as the *crisp* limit defining decisions being made within a decision frame. Softer, or fuzzy, criteria can be used instead of crisp criteria when they are more suitable, by incorporation of the concepts of factors into the constraints criterion.

- *Decision rule.* A *decision rule* is the procedure by which criteria are selected and combined so as to arrive at a particular evaluation, from which a decision is made. A decision rule may be as simple as setting a threshold on a single criterion, such as, "all slopes greater than 35° will not be logged, used for agriculture, or developed for residential or commercial purposes." On the other hand, decision rules can depend on the comparison of several multiple criteria. A decision rule will usually be structured so as to combine the criteria in such a way as to end up with a single composite value, and a statement of how this value or index is to be used to make assignments, that is to make decisions. For example, the suitability of an area for the different types of agricultural crops is dependent on physical and economic criteria. The physical criteria include soil, slope, aspect and climate, combined in a model to give the physical suitability of a location for that crop. Each crop has a different model as they have different dependencies on the physical resources. The economic criteria include costs of preparation, management, harvest and delivery to market, where the first three are models that are dependent on the area of crop and the last is a model that is dependent on the distance to market. The physical models will give a relative suitability for each crop. This information then needs to be combined so as to get areas for each crop and use these with the economic models to determine the suitability of each area for each crop.
- *Choice function and choice heuristic.* Decision rules usually contain either a choice function, as described above, or a choice heuristic. A *choice function* provides a mathematical means of comparing alternatives. Since they involve some form of optimisation, such as maximising or minimising some criterion, they theoretically require that each alternative be evaluated in turn. However, techniques exist that enable the analyst to just evaluate the most likely alternatives. A *choice heuristic* specifies a procedure to be followed rather than a function to be evaluated. Since they are simpler to implement, they are often the method of choice.
- *Evaluation.* The process of applying a decision rule is called an *evaluation*. Decisions can be of four forms as shown in Figure 8.3. Whilst one may sometimes be concerned with single criterion, single objective decisions, this will be the exception rather than the rule, particularly as the simpler forms of these decisions can be readily conducted using a map without the need to utilise a GIS system as such. Most simpler forms of decision rule will involve multi-criteria, single objective evaluation, where *multi-criteria* evaluation occurs when there are multiple criteria that have to be met in the evaluation of the decision rule, but a single objective has to be met. Most forms of decision criteria that involve a single objective arise when there is only one candidate set that is concerned with the same resources. When there is more than one candidate set concerned with the same resources, then often *multi-objective* evaluation has to be conducted. It is possible that situations can arise that involve multi-objective with single criteria, but most multi-objective evaluations will involve multi-criteria analysis as well.

	Single criterion	Multi criterion
Single objective		
Multiple objective		

FIGURE 8.3 Types of decisions.

8.4.2 Multi-Criteria Evaluation

Multi-criteria evaluation (MCE) is concerned with integrating multiple criteria so as to form a single index of evaluation. The simplest form of multi-criteria evaluation involves the reduction of the individual criteria to a Boolean result that can then be combined with the other criteria using logical decision rules. However, for continuous factors, a weighted combination is often used. With a Weighted Linear Combination (WLC), factors are combined by applying a weight to the individual factors and then summing the results to yield a suitability map.

$$S = \sum w_i x_i \tag{8.1}$$

where w_i are the weights applied to the criterion score of factor x_i. In cases where Boolean constraints, c_i, also apply, multiplying the suitability can modify the procedure, S, by the product of the constraints, as long as the value of 1 is used for acceptance in the Boolean data sets.

$$S = \sum w_i x_i \prod c_i \tag{8.2}$$

where $\prod c_i$ represents the product of the different c_i values. If one of the c_i values is zero, then the product will be zero.

There are four major steps that have to be completed in deriving the factor x_i values for use in Equation (8.1):

1. Identify the factors and describe the model that will be used to relate that factor values as held to derive suitability values. Consider the case of a firm seeking land for residential development. The factors that they may wish to consider include landuse (different costs of purchase and preparation), soil type (costs of constructive services and roads), distance to shops (effect on sale price), slope (costs of construction) and access to transportation (costs of construction). For this problem it would be most logical to convert all the factor values into money terms. The landuse classes will be assigned a cost value based on experience, as will each of the soil classes from the soil map. The slope data will be derived from a digital elevation model and the slopes converted into a cost using a non-linear model built for this purpose. The impact of distance to shops will be based on valuations and their distance to shops being converted into a model for this purpose. The distance from transport will be based on the costs of construction derived from experience.
2. Create the numerical relationship between the factor and its suitability. The factor variables are converted into cost values that are then converted into suitability, with increasing values linearly proportional to suitability. Some cost variables are proportional to suitability whilst others are inversely proportional. In the same way, some variables are linearly related to the cost and hence suitability, whilst others are non-linearly related and have to be converted using a non-linear model.
3. Scale the suitability data so that all the suitability maps for the different factors cover the same range. If all of the suitability maps are based on costs, then they are likely to be scaled automatically. However, if they are based on different criteria, then they will need to be scaled to cover the same range.
4. Determine the weights to be applied to each factor. These weights are often chosen from experience or from discussion with users. Thus, the firm seeking land for development may wish to place a higher weight on ongoing costs rather than on once-off costs, since their potential buyers will do so in considering the purchase of a block of land in the

development. The weights should sum to one, so that the derived suitability scores cover the same range as the source factor suitability maps.

Once the individual criteria have been normalised in this way, then they need to be combined so as to produce a final value that is related to the suitability of the pixel in terms of the objective. There are many different ways of combining criteria to derive a single index, of which three will be considered here:

1. *Boolean*. When a set of criteria have been reduced to Boolean valued maps, and a Boolean criteria is suitable then the product of the individual maps gives the result as the Boolean map showing as 1 for the areas that meet the criteria. Boolean methods of deciding on suitability are extreme deciders. If one sets the Boolean criteria by use of the .AND. function, then all factors must be suitable for the results to be suitable. This approach yields very risk averse or conservative solutions. For example, the *if* test: IF(Landuse == 1)AND(Soiltype == sandyloam)AND(slope $< 5°$), will only be passed for areas that meet all three criteria. If, on the other hand the Boolean decisions are made using the .OR. logical operator, then only one of the criteria needs to be suitable to give a suitable result, and so this is a risk taking solution. Using the previous example, the *if* test: IF(Landuse == 1)OR(Soiltype == sandyloam)OR(slope $< 5°$), means that only one of the three criteria have to be met for the area to be deemed acceptable. If one does not want to be forced into either of these extreme situations, then Boolean criteria have to be replaced by one or other of the following approaches.
2. *Weighted linear combinations*. The WLC aggregation method multiplies each normalised factor by its factor weight and then sums the result as introduced above. The factor weights indicate the relative degree of importance of each factor in determining the suitability of an objective. Factor weights can provide a capacity to trade-off between the factors in a decision. For example, factors with high factor weights can trade-off or compensate for the low factor values in other factors, whilst factors with low factor weights will only have a small influence on the decision, even if they have high suitability values. If the factor weights sum to one, then the derived value will be in the same range as the normalised data. This result is then multiplied by each constraint in turn so as to mask out unsuitable areas.
3. *Ordered weighted average*. The ordered weighted average (OWA) method is similar to the WLC option, with the exception that a second set of order weights are used, which control the manner in which the weighted factors are aggregated. Order weights determine the overall level of trade-off that is to be allowed in the development of suitability.

First, we will discuss the implementation of WLC-MCE and then OWA-MCE. To implement WLC-MCE, it is necessary to:

1. *Identify the constraints and factors*.
2. *Set the constraints as Boolean images*. Convert each constraint image into a Boolean image in which areas that cannot be used are assigned the value of 0 and areas that can be used are assigned the value of 1.
3. *Convert the factor images into factor values*. All the factors have to have values that cover the same range, for example, 0 to 255, and with increasing values indicating increasing suitability. For example, a decision on selecting a route for a road may have constraints that the road is to be at least 500 m from towns and 200 m from residential buildings and the other land uses may be assigned weights as a function of the costs of purchase of the land. In this case, these costs have to be determined, scaled to the set range and then applied to the landuse map to create the landuse-based weights for use in the WLC-MCE process.

Other factors may have a continuous function that is related in some way to the function. Thus, in the above example, slopes may be proportional to the costs of construction, and thus inversely proportional to the required factor values.

4. *Select the factor weights*. For economic factors, the weights may be set proportional to the costs of construction. Environmental factors have weights assigned that are based on other criteria. Such weights could be based on a questionnaire on the issue in the community, on an assessment of the scarcity of the environmental component or on its perceived importance to other parts of the environment. The weights must sum to 1.0.
5. *Combine the factors* and their weights in accordance with Equation (8.1).

WLC-MCE only allows full trade-off between the factors. There are many instances where it is not desirable to allow full trade-off between the factors, but only trade-off when some factors are clearly dominant. To address this need, the technique of OWA-MCE has been developed. Figure 8.4 shows the options that one can select in conducting multi-criteria evaluation in IDRISI. The user needs to specify any constraints, where these have been previously set up as Boolean files with values of 0 representing areas not accessible, and 1 for areas that can be utilised if they are suitable. The factors are listed and their weights given. In this example, there are two constraints and six factors that need to be considered.

With OWA-MCE, a second set of weights is used. Instead of this second set of weights being assigned to factors, they are assigned to the ordered and weighted derived factor values at a pixel. With OWA-MCE, the Equation (8.1) is used to compute each $w_i x_i$ value. These are then ordered starting from the highest value to the lowest value. The order weights are then applied to this ordered list to derive the final S′ value, so that the highest value gets the first weight, the second highest value the second weight and so forth. Let us say that we have three factors, *A*, *B* and *C*, with weights for WLC-MCE of 0.35, 0.45 and 0.2, respectively. If we had values of (25, 32 and 75) in *A*, *B* and *C* at a pixel, then WLC-MCE would give:

$$S = \sum_{i=1}^{3} w_i x_i = (0.35 \times 25 + 0.45 \times 32 + 0.2 \times 75) = (8.75 + 14.4 + 15.0) = 38.15$$

Now this is not a high score if the factor cost functions cover the range 0 to 255; it should be noted that the *S* values will also cover the same range as the cost functions if the weights sum to

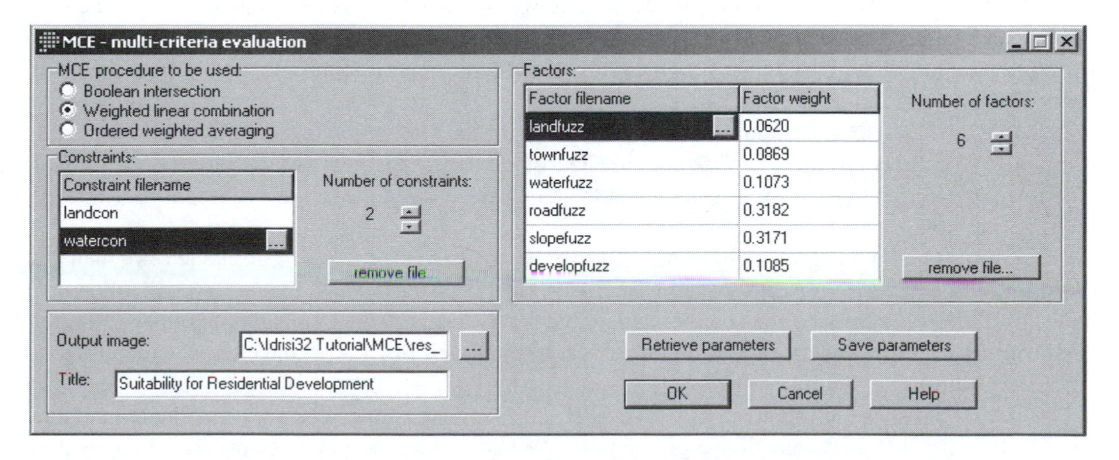

FIGURE 8.4 The IDRISI interface for multi-criteria evaluation.

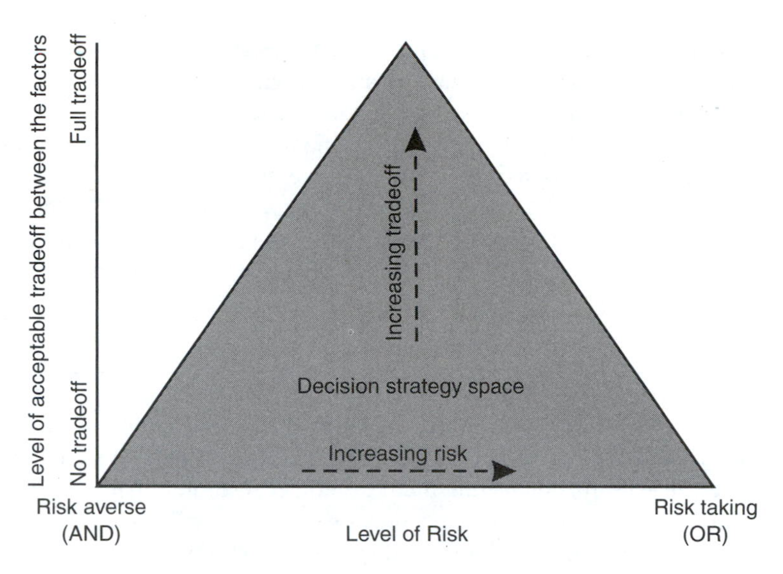

FIGURE 8.5 The decision space in multi-criteria evaluation.

unity. Further it may be noted that the highest factor-based score value (15.0) was from the lowest weighted factor, and it may be decided that the users do not want to give full trade-off when this occurs, so OWA-MCE may be implemented to deal with this situation. In doing so, order weights of (0.6, 0.3 and 0.1) may be chosen. When the components of WLC-MCE are ordered in sequence from the highest to the lowest, we get (15.0, 14.4, 8.75), and so these are then combined with the order weights using the same form of equation as in Equation (8.1) to give:

$$S' = (0.6 \times 15.0 + 0.3 \times 14.4 + 0.1 \times 8.75) = 14.195$$

which has effectively reduced the suitability of the pixel from 38 to 14. Again, if the order weights sum to unity, then the derived suitability will cover the same range as the source suitability maps. If the ordered weights were all made to be the same, 0.3333 in this case with three factors, then the result will be the same as for WLC-MCE, so equal ordered weights gives the simple weighted linear combination. This is the apex of the triangular decision space (Figure 8.5). As the weights are changed from this, the location moves down into the triangular decision space. Weights of (1, 0, 0) for the example of three factors considered here would thus give full assignment to only the one factor, and so this represents no trade-off, or it represents a location on the base of the triangle. If the ordered weights were given as (1, 0, 0), then this would represent the suitability only taking the most suitable factor into account. It represents the Boolean "OR" situation. If the ordered weights were set as (0, 0, 1), then this would indicate the suitability for the least suitable factor, and represents the suitability for the Boolean "AND" situation. Clearly, other arrangements of the ordered weights of the form (0,1,0), will give positions that lie along the base between the left and the right base apexes of the triangle.

As the ordered weights change from the form (1, 0, 0, . . .) to ($1/n$, $1/n$, $1/n$, for n factors), then the position in the triangle moves from the base line towards the apex. Thus, the analyst can adjust the level of trade-off to that the client thinks is suitable for the problem at hand, and of course, different levels of trade-off can be tried to ensure that the solution found fits in with the clients perceptions of realistic solutions.

Figure 8.6 shows the user interface in IDRISI for OWL-MCE. It can be seen that the constraints and factors are as for weighted linear combination. What are new are the set of ordered weights that are to be used and which the user needs to insert prior to implementation. In this case there are six

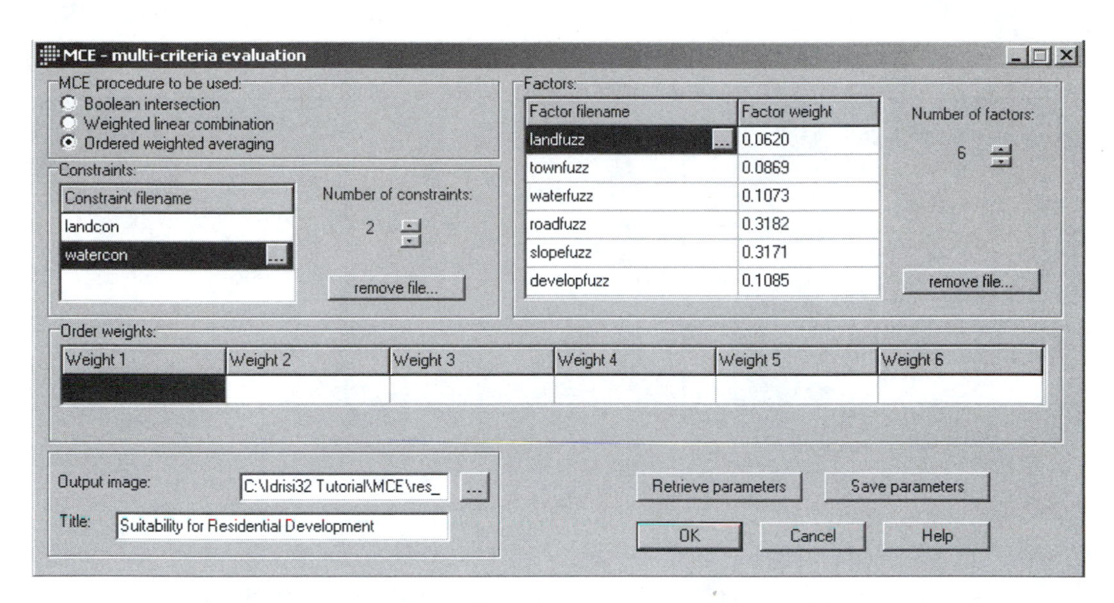

FIGURE 8.6 The IDRISI user interface for ordered weighted average MCE.

factors and so there will be six ordered weights. If each ordered weight is given as (1/6), then the result will be the same as WLC-MCE.

8.4.3 MULTI-OBJECTIVE EVALUATION

Multi-objective evaluation will have to deal with situations of complimentary or conflicting objectives. When situations of complimentary objectives arise, then the evaluation can be conducted for each objective in turn, and then the derived suitability for each objective are weighted and summed to derive a final suitability. This final suitability is then multiplied by the various constraints to give areas that are suitable for the combination of multiple complimentary objectives.

However, it is normal that the objectives are not complimentary at all, but rather are conflicting. Again, if they are conflicting, but they can be prioritised, then the suitability can be multiplied by the priority of each objective to find the areas best suited for each objective. However, it is usual that conflicting objectives cannot be prioritised in this way, but that some form of compromise solution needs to be found. The approach used to solve this problem that will be described in this text is taken from the IDRISI reference (Eastman 1999).

Consider each suitability map as forming one axis in n-dimensional space, where there are n conflicting objectives to be dealt with. Here we will consider just two conflicting objectives, with each of the three diagrams in Figure 8.7 representing this space from different perspectives. Each cell can be located within this decision space according to its suitability level in terms of each objective. The right-hand column in the left-hand diagram represents suitability values that give high suitability for Objective 1, and the top row represents suitability for Objective 2. Areas that are suitable for one objective but not the other can be readily assigned to that objective to which they are suitable. Conflict thus only arises in those areas that are suitable for both objectives, or the Region of Conflict in the diagram.

To resolve the area of conflict, a partitioning of the space is conducted. The central diagram in Figure 8.7 shows that the suitability space can also be partitioned into areas that are closer to the ideal point for Objective 1 and Objective 2, respectively. The simplest form of partition is to allocate pixels in the area of conflict to that Ideal point to which they are the closest. This may be represented by the *first solution decision surface* in the right-hand diagram in Figure 8.7. If this approach yields

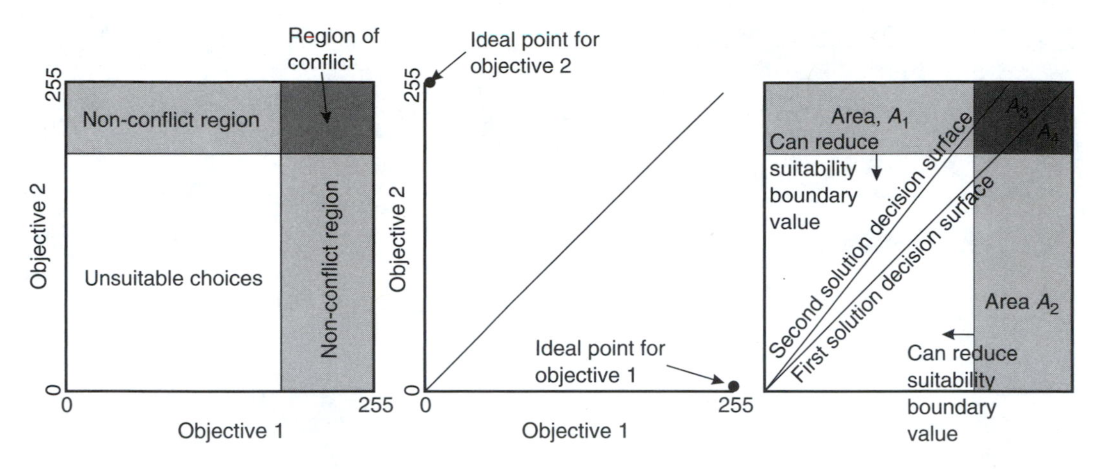

FIGURE 8.7 Decision surfaces in situations of conflict.

sufficient satisfactory areas to meet objectives 1 and 2, then it yields a satisfactory solution. However, in general it will still result in insufficient areas to meet one or more of the objectives and as such it represents an unsatisfactory solution. When this occurs then a number of options are available:

1. If the solution gives one objective more area than was necessary to meet the requirements of the objective, but less to the other, then the distances can be weighted, giving a slightly higher weight to the disadvantaged objective relative to the other objective. The effect of this is to rotate the decision line so that a larger portion of the area of conflict is allocated to the disadvantaged objective. The two decision lines in the right-handed diagram in Figure 8.7 would represent this approach. If, however, both objectives were dissatisfied with the solution, then weighting the decision surface will not give satisfaction to either, although it may change the level of dissatisfaction.
2. The best solution when this happens is to enlarge the areas that are suitable by lowering the standards of what are considered suitable to both objectives. This will increase the area given to each objective, and it will also increase the area of conflict. This area of conflict contains better areas, that are areas of higher suitability, and so it needs to then be partitioned in the same manner as before. It should be noted that partitioning the conflict area by means of the use of weights should only be done once one objective has been satisfied and the other still needs to be satisfied.

8.4.3.1 Legal Issues in the Use of Data and Information

In remote sensing and GIS, software and data in various forms are integrated together to create information. This information, in conjunction with other information is then used as the basis of resource management decisions. If the decisions being made are related to the resources that are owned by those who are also making the decisions, then conflict will not usually arise, although it can do so when the decisions of one manager are seen by others to affect their operations. However, we have seen that remote sensing and GIS have a natural domain in the management of regional resources. In this context we have seen that regional resource managers are attempting to balance the desires and aspirations of the individual resource manager to maximise productivity or production against community desires to maintain and to protect the environment. These community aspirations are written into legally binding laws and statutes that form the basis of the regional resource manager's decision-making process. By definition, these community-based aspirations will sometimes restrict the courses of action that are available to the individual resource manager. It is clear that there is considerable scope for conflict to arise in such situations and as a consequence, there is considerable scope for individuals or individual resource managers to challenge the decisions that are made by regional resource managers.

The sequence involved in making resource management decisions is thus to:

1. Define the types of information that are required to support the decision making process.
2. Implement a process to derive this information using a combination of software and data in various forms.
3. Use the information in making the decision.
4. Convey this decision to the resource managers that are affected.
5. Record these decisions in a publicly available manner.

Each of these steps involves legal implications. Most of these implications are very similar to those that apply in similar areas of human endeavour, and so they will not be dealt with in detail here. However, the way spatial information is used in making regional resource management decisions contains new elements that need to be considered in some more detail. Section 8.4.4 will introduce some of the concepts that have evolved in society to deal with the protection of intellectual property, including software and data. However, images may also contain information that third parties consider to be confidential, and this issue will be considered in Section 8.4.5. Finally, the spatial information derived from imagery and GIS is used as the basis of decisions that will affect resource managers, and some of the legal implications of this are considered in Section 8.4.6.

8.4.4 Use of Data Supplied by a Third Party for Use by the Analyst

Image processing and GIS analysis transform data in an infinite variety of ways so that sometimes it is very difficult to determine what was the source data for the transformed data or information. When this occurs then serious questions can arise as to the rights of the supplier of the data. Consider three very typical situations:

1. An image is enhanced and used to extract information by visual interpretation.
2. What rights does the vendor supplying the source image data have over the derived image data, and subsequently over the derived information? An image is classified and then rectified. The classified information is combined with other GIS data to provide farm level information.
3. An image is rectified, calibrated, corrected for atmospheric effects before being converted into vegetation index values that are in turn used to derive an estimate of leaf area index (LAI). This image is then used to select field sites at which the LAI is determined, and the image data is used to interpolate between the field sites by means of co-Kriging.

It can be extremely difficult to know where is an appropriate place to change the level of rights that the supplier of data has over the derived products. At present the various suppliers have established their own contractual arrangements to protect their investment. There are a number of means by which a vendor can protect data and information of which the most common are (Longhorn et al. 2002):

1. Copyright
2. Legal protection of databases
3. Patents
4. Trade secrets
5. Trademarks

8.4.4.1 Copyright

The term "copyright" originally referred to the right to copy the literary works of authors. Copyright protects any works in the literary, scientific and artistic domain from being copied without the express permission of the authors of the works. The definition of works has been extensively enlarged over the last decades, so that it includes written, drawn, painted, architectural, sculptural, engraved, printed, computer program as works that may be in the form of illustrations, maps, plans, sketches and three-dimensional works relative to geography, topography, architecture and science. Unlike patents, copyright does not protect ideas, but it protects products, or the form of expression of those ideas. The copyright means that works cannot be either reproduced or translated without the express agreement of the author of the works. A translation will require its own copyright, but this does not release the translator from the need to abide by the copyright rights of the originating author.

In most countries the copyright owner does not need to formally register the works, nor to display copyright notices, for his rights to be recognised in law. The duration of copyright also varies between countries, whilst the Berne convention set protection till the life of the author plus 50 yr.

In the spatial disciplines, some of the data will be in the public domain, and thus does not attract copyright protection, but some may belong to a government agency that is acting under a strong cost recovery regime, whilst others may be privately owned. The suppliers of data in all three categories may all have quite different perceptions of how there data may be accessed and used by third parties, so that it is essential that the users of this data be aware of all the data sets that are being used, and the contractual arrangements that the owners of the data wish to make with data users.

8.4.4.2 Legal Protection of Databases

There is much debate about appropriate means to protect databases, recognising that some databases represent a huge investment in the compilation and accumulation of facts, whilst others are simply compilations of data that are neither copyrightable nor patentable. In 1996 the European Union adopted a directive that provided protection for databases in the form:

1. Protection is not based on an extension of copyright or patent.
2. It protects "databases in an form," where a database is defined as "a collection of independent works, data or other materials arranged in a systematic or methodical way and individually accessible by electronic or other means."
3. The protection is, "... for the maker of a database which shows that there has been qualitatively and/or quantitatively a substantial investment in either the obtaining, verification or presentation of the contents to prevent extraction and/or re-utilisation of the whole or a substantial part, evaluated qualitatively and/or quantitatively, of the contents of the database."
4. Extraction is defined as "... the permanent or temporary transfer of all or a substantial part of the contents of a database to another medium by any means or in any form." Re-utilisation means, "... any form of making available to the public all or a substantial part of the contents of a database by the distribution of copies, by renting, by on-line or other forms of transmission."
5. In place of a fair use concept, the directive prohibits, "... repeated and systematic extraction and/or re-utilisation of substantial parts of ... the database ... which conflict with normal exploitation of the database or which unreasonably prejudice the legitimate interests of the maker of the database.
6. Users may extract or re-utilise substantial parts of the contents of a database for private, teaching or scientific purposes as long as the source is acknowledged and the extent of the copying is justified by the non-commercial purposes for which the data is to be used.
7. The period of protection is 15 yr, that can be extended if the database has undergone extensive modification or updating.

8. Databases that, "by reason of the selection or arrangement of their contents, constitute the authors own intellectual creation," are still protected by copyright.

8.4.4.3 Patents

Contrasting with the focus of copyright on the form of expression, patents focus on the intellectual effort as represented by inventions. A patent must be lodged with the national patent office in a way that will ensure that it will provide protection to the inventor. There is a cost in the registration of patents, just as there is a cost in searching to find out if another inventor has not previously patented the idea or invention, and then in the construction of the patent application. Patents are thus expensive to establish, with the cost varying from country to country, just as the period of protection can also vary between countries. Patents need to be established in each country for which protection is sought.

If a patent is infringed, the only remedy is usually a civil action by the party that created the patent, or their representatives. This contrasts with copyright that also carries criminal penalties.

The patent system is designed to protect the invention of both products and processes. In many countries software can also be patented, even though the product itself can be copyrighted. However, patents cannot be given for a product or process that has been described in the public domain, such as in trade and professional journals.

The reason why some software houses patent software has to do with the difference between copyright and patents. If software is only covered by copyright, then the copyright protects the owner from direct copying of the software without permission. However, a third party can then construct their own software using the ideas that are embedded in the copyrighted software. Patents would stop this activity, or enable the originating inventor to receive benefit from his invention by giving approval to use the patent under conditions that are acceptable to the original inventor.

8.4.5 Protection of Confidentiality of Data about Third Parties

The land surfaces that have been imaged are of interest to other people or groups in their society. This interest can be financial, as in land ownership, cultural, as applies to most communal land or societal, where a society defines the land that is owned in some way by members of that society. These other people will often have various forms of real or imagined vested interests in some of the land that has been covered by images. Images are, by definition, of parts of the globe over which other people have various forms of real or imagined vested interests. There are thus issues related to the invasion of privacy that occurs when the data is acquired and subsequently analysed. For many projects, the data is related to actual field data that is acquired for this or for some other purpose. There may thus also exist issues of the rights to privacy of the individuals who have provided data either directly and knowingly or indirectly and unknowingly.

In many countries, information on the actual management of local resources, such as farm expenditures, farming practices and property boundaries, are legally viewed to be personal data. In many countries such personal data can only be collected, used, held and distributed under tightly controlled conditions, where these conditions may well vary depending on the nature of the data itself. Failure to abide by these rules can lead to criminal prosecution.

The U.S. Federal Geographic Data Committee recommends that the following actions be taken in relation to personal information held in spatial or geographic databases (http://www.fgdc.gov/fgdc/policies/privacypolicy.pdf):

1. When personal information is collected directly, at the time of collection the agency should inform individuals of:
   (a) Why the information is being collected and the legal authority to do so.

   (b) How the information will be used and protected as to confidentiality, integrity and quality.
   (c) The consequences of providing or withholding the requested information.
   (d) How to correct personal information if it lacks sufficient quality to ensure fairness in its use.
   (e) The opportunity to remain anonymous when appropriate and any rights of redress.
   (f) The records retention schedule of the agency.
2. Personal information is to be acquired and used only in ways that respect an individual's privacy.
3. Personal information should be collected only as needed to support current or planned activities.
4. Agency staff should be aware of the privacy implications of GIS technology.
5. Technical and managerial controls will be used to protect the confidentiality and integrity of personal information.
6. Agencies should ensure the integrity of personal information, including prevention of alteration or destruction of such information held in or linked to geospatial databases.
7. Such information should be as accurate, timely, complete and relevant as possible for the purposes for which it is acquired and used.

8.4.6 The Use of Information Derived from the Analysis of Data

Resource management occurs at three broad levels, strategic, regional and local. Of these three, regional management is quite unique in that regional managers are attempting to reach a balance between the wishes of the individual local resource manager to maximise productivity or production on the one hand, and the wishes of the community. This situation can naturally lead to conflict of interest at the regional level, not only between the wishes of different local managers, but also between the desires of individual managers and the community.

Regional managers are responsible for implementing community desires as expressed through legally binding laws and statutes. Any challenge to them will thus need to show that either the law or statute is faulty in some way, or that its implementation has been at fault. One possible way in which its implementation may be at fault is through the use of information in the decision process that is in error. A challenge made on this basis would need to show that the information used was in error and that the information was critical to the decision. The goal of the prosecutor of such an action would be to prove that these elements were correct, that is, the information was in error and that it was critical to the decision, with the intention of having the decision over turned by the court.

There are a number of examples of where remotely sensed data are used to support this type of resource management right now. Remotely sensed data are used to identify areas of cropping and clearing in the rangelands of New South Wales (Australia) as there are clear legal restrictions on both activities in these fragile environments and agencies have been established to ensure that these legal requirements are obeyed. The European Union uses remotely sensed data to check on the claims of farmers seeking agricultural subsidies, so as to reduce the opportunities to defraud the state by the few that may consider doing so.

In both cases landuse information is derived from remotely sensed image data. However, in both cases the information derived from remotely sensed imagery is considered to be of insufficient accuracy to stand a challenge in court. The solution is to use the information derived from the image data as a filter, to identify fields that may be in use in ways that are different to that indicated in other information, where this other information is the approvals given for cropping or clearing in the Western Division of New South Wales or the farmer applications in the case of the European Union agricultural subsidy. When fields are identified in this way as being outside the approved usage, then field visits must be taken to those fields to check on the actual conditions in the fields, and the field

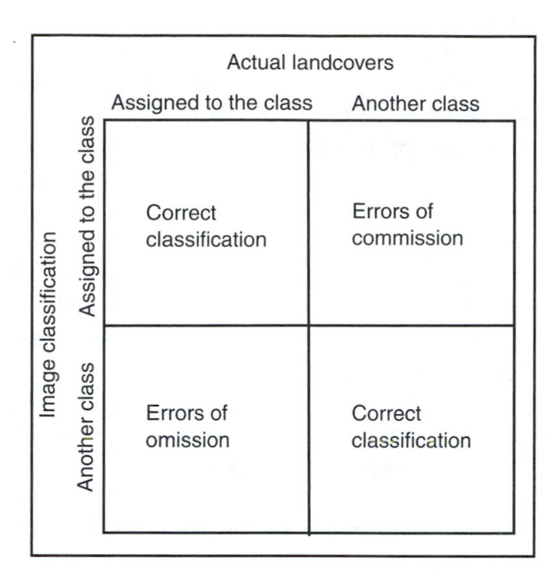

FIGURE 8.8 Classification versus actual conditions — errors of omission and commission.

visit is taken as the legally binding observation. Such an approach means that the image data reduces the field visits from, say a regular sample of the properties in the area, to only those who appear to be outside their legal entitlement.

Such an approach is a valid way to use image data, but it contains a number of characteristics that are less than satisfactory for some situations:

1. *Errors of omission and commission*. We saw in Chapter 4 that classification of image data involves errors of omission and commission, as shown in Figure 8.8. When a field is used illegally, but an error of omission occurs, that is, the classification is into another class, then the illegality will not be detected from the image classification if the other class is legally acceptable. Thus, errors of omission may not be detected. Errors of commission will show up as illegal activities, but the inspection will show that this was an error in the classification. We saw in Chapter 4 that the total level of errors in a classification cannot be eliminated, but there is some capacity to move the decision surface so that the mix of errors of omission and commission are changed. Thus, in the above case, the decision surface could be moved so as to increase the errors of commission and reduce the errors of omission. Such a process, however, may suggest to those not familiar with the manipulation being conducted that the total errors are larger than are in fact the case, as they will assume that the errors of omission and commission are of similar magnitude.
2. *Constraints on the imagery that can be used and the time available*. The requirement to visit the fields while the observer can observe the infringement usually means that the classification has to be conducted relatively early in the season. This means that imagery at harvest time is not available for use in the classification. Harvest time is usually a period of dry weather, giving better chances of acquiring images, it covers a shorter period than cultivation and sowing, and the changes are just as dramatic as at sowing, so that more of the changes can usually be detected on fewer images. It is thus a good time to acquire image data for the classification of agricultural crops, and will yield higher accuracy classifications than would otherwise be achieved. However, this option is not available if the classification has to be completed in time to allow field visits prior to harvest. In addition to this constraint, the time constraint on the field visits may also incur costs.

For both reasons, it would be better to develop a better way to incorporate information derived from image data into the decision-making process. One of the main ways to achieve this is to derive the information at an accuracy that is acceptable in a court of law. Courts of law generally use the criteria that information can be accepted if it "is beyond reasonable doubt." At present the meaning of this in relation to image data has not been tested in many countries and so it is impossible to say what this means in terms of accuracy and reliability. Indeed, what is deemed to be "beyond reasonable doubt" may vary somewhat from country to country. The only criteria that can be used as a measure of what may be acceptable to a court of law is the situation with regards to topographic maps. Topographic maps normally obey the 95/95 accuracy criterion, which means 95% accurate, 95% of the time. If topographic maps are accepted as legal documents in a country, then it is possible that similar levels of accuracy would be accepted for information derived from other remotely sensed data sources. If this is the case, then information derived from remotely sensed data would need to show that it meets this standard through a statistically valid accuracy assessment, supplied as an integral part of the information derived from the data.

There are a number of advantages of being able to derive legally acceptable information from image data, including:

1. Reduction in the cost and complexity of field verification and validation, as discussed above.
2. Creates the possibility of historical mapping for legally acceptable purposes.
3. Creates the opportunity for the development of new applications for which the constraints of in-season field verification and validation are too severe.

Such a development will truly mean that remote sensing has, "come of age."

8.5 OTHER PROJECT MANAGEMENT TOOLS

There are many other tools available to the resource manager. Most of these tools have been developed quite independently of the use of spatial data and so they are dealt with in suitable texts on Project Management. They will not be dealt with in this book, the reader being referred to the literature should they be interested in this topic.

8.6 CONCLUDING REMARKS

This chapter covers one of the most exciting, critical yet newly evolving aspects of resource information systems. It is by no means an established component of existing software systems. Only is it explicitly integrated into the software of one system, the IDRISI system. Whilst much of this can be done in some other systems, in general, the user will have to know a lot about what he wants to do as he will get very little help from the software. This is not the case with IDRISI. I hope that this situation changes in the near future with other vendors.

You, as a resource manager, have now been given tools to derive spatial information, integrate it with other information so as to derive resource management information. You have been shown how models get used in such information systems and how to integrate such models. Finally, you have been introduced to the way resources are managed, the information needs of the different levels of management, how these spatial tools fit into this management structure and how you can use these tools to assist managers deal with situations of multiple criteria and multiple objectives so as to find the best solutions to specific situations. I wish you a productive use of these tools in your career.

FURTHER READING

The reader is referred to the extensive literature on project management tools for information on Gant Charts, PERT Charts, and other project management tools. For those who wish to better understand the use of decision support tools in a GIS environment, the reader is referred to Eastman (1999).

REFERENCES

Eastman, J.R., 1999. *Guide to GIS and Image Processing*, Vols 1, 2, Clark Labs, Clark University, Worcester.

Longhorn, R.A., Henson-Apollonio, V. and White, J.W., 2002. *Legal Issues in the Use of Geospatial Data and Tools for Agriculture and Natural Resource Management: A Primer*, International Maize and Wheat Improvement Centre (CIMMYT), Mexico, DF.

http://www.fgdc.gov/fgdc/policies/privacypolicy.pdf

Index

B

C

E

F

G

H

I

J

K

L

M

P

Q

R

S

T

U

X

Y

Z